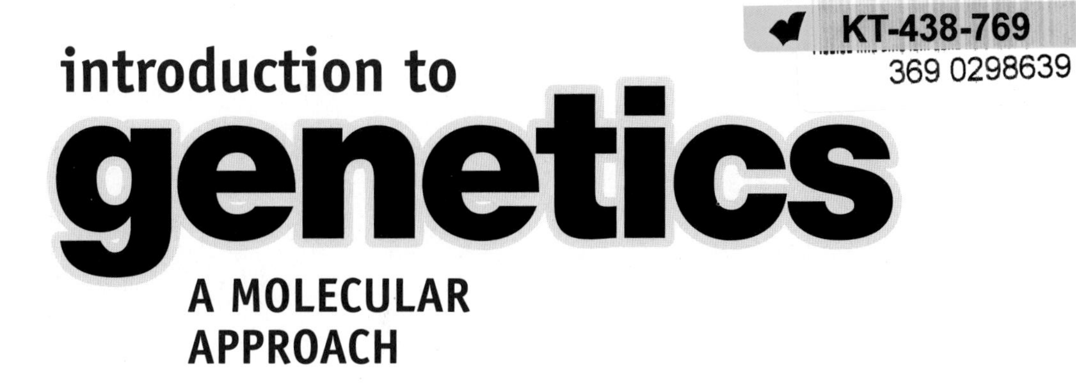

introduction to genetics
A MOLECULAR APPROACH

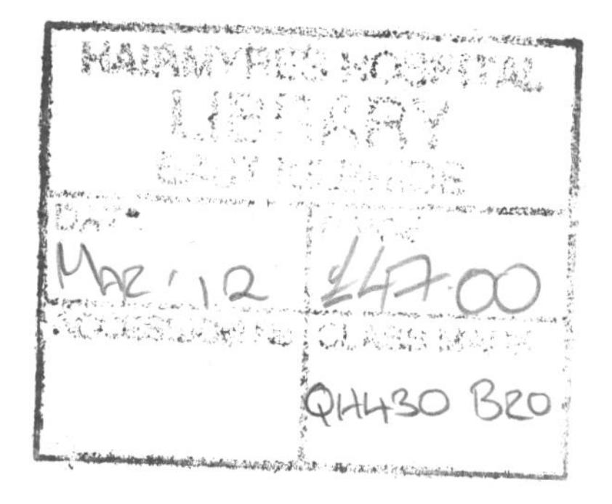

A MOLECULAR APPROACH

Terry Brown

Garland Science
Vice President: Denise Schanck
Editor: Elizabeth Owen
Editorial Assistants: David Borrowdale and Louise Dawnay
Production Editor and Layout: EJ Publishing Services
Illustrator and Cover Design: Matthew McClements, Blink Studio, Ltd.
Copyeditors: Richard K. Mickey and Bruce Goatly
Proofreader: Jo Clayton

Cover image courtesy of Richard Wheeler

ISBN 978-0-8153-6509-9

Library of Congress Cataloging-in-Publication Data

Brown, Terry.
Introduction to genetics : a molecular approach / Terry Brown.
p. cm.
Includes index.
ISBN 978-0-8153-6509-9 (alk. paper)
1. Molecular genetics. I. Title.
QH442.B77 2012
576.5--dc23
2011024255

Published by Garland Science, Taylor & Francis Group, LLC, an informa business, 711 Third Avenue, New York, NY 10017, USA, and 2 Park Square, Milton Park, Abingdon, OX14 4RN, UK.

Printed in the United States of America

15 14 13 12 11 10 9 8 7 6 5 4 3 2 1

Visit our website at http://www.garlandscience.com

About the Author
I became fascinated with the natural world when I was very young. I began my research career studying the effects of metal pollution on microorganisms and the tolerance that some plants display to high concentrations of toxic metals. I then became excited by DNA and worked on mitochondrial genes in fungi in order to learn the new (in those days) techniques for gene cloning and DNA sequencing. I contributed to the discovery of mitochondrial introns and to work that described the base-paired structure of these introns. I then became interested in ancient DNA and was one of the first people to carry out DNA extractions with bones and preserved plant remains. This work has required close collaboration with archaeologists, and has led to my current interests in the origins of agriculture, genetic profiling of archaeological skeletons, and the evolution of disease.

I obtained my PhD from University College London in 1977 and then worked in New York, Oxford, Colchester, and Manchester before beginning in 1984 as a Lecturer in Biotechnology at the University of Manchester Institute of Science and Technology (UMIST). I was appointed Professor of Biomolecular Archaeology in 2000 and was Head of Biomolecular Sciences at UMIST from 2002–2004. I was then Associate Dean in the Faculty of Life Sciences of the University of Manchester until 2006, before taking a break from administration in order to have more time to do research.

My other undergraduate textbooks include *Genomes*, Third Edition (Garland Science), *Gene Cloning and DNA Analysis: An Introduction*, Sixth Edition (Wiley-Blackwell) and, with Keri Brown, *Biomolecular Archaeology: An Introduction* (Wiley-Blackwell).

PREFACE

There are so many genetics texts available in the bookshops that the author of an entirely new one has a duty to explain why his own contribution should be necessary. In my case the decision to write a genetics text was prompted by my strong feeling that genetics is today inexorably centered on DNA, and that the teaching of genetics should reflect this fact. The theme of this book is therefore the progression from *molecules* (DNA and genes) to *processes* (gene expression and DNA replication) to *systems* (cells, organisms, and populations). This progression reflects both the basic logic of life and the way in which modern biological research is structured. My experience in teaching (and of once being taught) an introductory course in genetics has led me to believe that this "molecular approach" enables students who might otherwise be daunted by the intricacies of genetics to gradually build up their confidence in the subject. The molecular approach is particularly suitable for the large number of students for whom genetics is a part of a broader degree course in, for example, biology, biochemistry, biomedical sciences, or biotechnology.

The difficulty in attempting to write an introductory textbook, in any subject, lies in presenting the material in an understandable fashion without falling into the trap of over-simplification. To be of value the book should ensure that the basic facts and concepts are grasped by the reader, and yet should provide a sufficient depth of knowledge to stimulate the student's interest and to engender the desire to progress on to more advanced aspects of the subject. With an introductory text in genetics these objectives are perhaps relatively easy to attain, as even the most fundamental facts are fascinating and, in my experience at least, most undergraduates arrive already primed with a curiosity about genes. I hope that this book will help to turn that curiosity into a lifelong pursuit.

ACKNOWLEDGMENTS

I would like to thank the reviewers who provided helpful comments on the original proposal for *Introduction to Genetics: A Molecular Approach*, and who gave detailed feedback on chapters from the various iterations that the book went through before evolving into its final form.

Shivanthi Anandan (Drexel University, USA); Thierry Backeljau (University of Antwerp, Belgium); Edward L. Bolt (University of Nottingham, UK); Laura C. Bridgewater (Brigham Young University, USA); John Bright (Sheffield Hallam University, UK); Kuttalaprakash Chudalayandi (Birla Institute of Technology and Science, Pilani, India); H. Neval Erturk (Converse College, USA); Bill Field (Aston University, UK); Paula L. Fischhaber (California State University, Northridge, USA); Adrian J. Hall (Sheffield Hallam University, UK); Ralph Hillman (Temple University, USA); Eric A. Hoffman (University of Central Florida, USA); David T. Kirkpatrick (University of Minnesota, USA); Sarah Lewis (University of Bristol, UK); Cindy S. Malone (California State University, Northridge, USA); Patrick H. Masson (University of Wisconsin-Madison, USA); Mike J. McPherson (University of Leeds, UK); Melissa Michael (University of Illinois at Urbana-Champaign, USA); Roger L. Miesfeld (University of Arizona, USA); Marcus Munafò (University of Bristol, UK); Philip Oliver (University of Cambridge, UK); Christine Rushlow (New York University, USA); Inder Saxena (The University of Texas at Austin, USA); Stephanie C. Schroeder (Webster University, USA);

Robert J. Slater (University of Hertfordshire, UK); Tahar Taybi (Newcastle University, UK); John J. Taylor (Newcastle University, UK); Jeffrey Townsend (Yale University, USA); David Veal (University of the West of England, UK); Jemima Whyte (Stanford University California, UK).

Thanks also go to the many Garland staff who have contributed to the creation of *Introduction to Genetics: A Molecular Approach* and have helped convert my own contribution—words not necessarily in the right order and scribbled diagrams—into an actual textbook.

Finally, but not least, I would like to thank my wife Keri for putting up with "not another book."

Terry Brown
Manchester, UK

NOTE TO THE READER

I have tried to make *Introduction to Genetics: A Molecular Approach* as user friendly as possible. The book therefore includes a number of devices intended to help the reader and to make the book an effective teaching aid.

Organization of *Introduction to Genetics: A Molecular Approach*

The book is divided into three parts. In Part I we examine the function of the gene as a unit of biological information. First, we must become familiar with the structure of DNA (Chapter 2) and with the way in which genes are organized within DNA molecules (Chapter 3). Then we will be ready to follow the process, called gene expression, which results in the information contained in a gene being utilized by the cell. We will study the way in which DNA is copied into RNA (Chapter 4), and we will look in detail at the roles in the cell of the different types of RNA molecule that are made (Chapters 5 and 6). We will then discover how the genetic code is used to direct the synthesis of protein molecules whose structures and functions are specified by the information contained in the genes (Chapters 7 and 8). Finally, we will examine how all of these events are controlled so that only those genes whose information is needed are active at a particular time (Chapter 9).

In Part II we will study the role of the gene as a unit of inheritance. To do this, we will ask three questions. The first is how a complete set of the genes is passed to the daughter cells when the parent cell divides. We will therefore study the mechanism by which DNA molecules replicate so that new, identical copies are made (Chapter 10), and we will examine how DNA molecules, and the genes that they contain, are passed on to the progeny when an animal or plant cell divides (Chapter 11), when a bacterium divides (Chapter 12), and during the infection cycle of a virus (Chapter 13). The second question concerns the inheritance of genes during sexual reproduction. We will study how DNA molecules are passed from parents to their offspring (Chapter 14), and we will investigate how the genes on these DNA molecules specify the biological characteristics of the offspring in such a way that these offspring resemble, but are not identical to, their parents (Chapter 15). The third question concerns the link between the inheritance of genes and the evolution of species. To answer this question we will study how the information contained in a gene can change due to errors that are sometimes made during DNA replication and as a result of mutation (Chapter 16), and we will examine how the new gene variants that are created in this way can spread through a population (Chapter 17).

In Part III, we will explore some of the areas of research that are responsible for the high profile that genetics has in our modern world. The first topic that we will study is the role of genes in development. How do genes control the pathway that begins with a fertilized egg cell and ends with an adult organism? Finding the answer to this question is one of the biggest challenges in all of genetics (Chapter 18). Then we will devote three chapters to our own species. In Chapter 19 we will look closely at the human genome, and in particular we will ask what it is about our genome that makes us special. Then we will look at the ways in which defects in genes can give rise to inherited diseases such as cystic fibrosis and to cancer (Chapter 20). In Chapter 21 we will examine how genetic profiles are obtained and why these have become so important in forensic biology. We will also learn how genetics can be used to study human evolution and to trace the routes taken by early humans as they migrated out of Africa and colonized the globe. We will then look at the important applications of genetics in industry and agriculture, in the production of important pharmaceuticals and in the design of genetically modified crops (Chapter 22). This will lead us into some of the controversial aspects of modern genetics. We must not ignore these controversies and so in the final chapter we will examine some of the ethical issues raised by genetics, and we will ask how these issues should be debated so that the controversies can be resolved.

Organization of chapters

The chapters include Research Briefings, Questions and Problems, and Further Reading lists, all designed to help you in your exploration of genetics.

Research Briefings

Each Research Briefing is a self-contained illustration of the importance of research in genetics. Some of the Research Briefings describe classic projects of the past or present, such as the discovery of the structure of DNA by Watson and Crick in 1953 (Research Briefing 2.1), and the research currently being carried out on the genetics of Neanderthals (Research Briefing 21.1). A few describe a single important method or group of techniques, such as the polymerase chain reaction (Research Briefing 10.2), and the methods used to map the positions of genes on chromosomes (Research Briefing 15.2). Others describe research strategies, such as the design of a project to work out in which cells a particular gene is expressed (Research Briefing 3.1), and the strategies used to identify a gene that causes an inherited disease (Research Briefing 20.1). The overall aim of the Research Briefings is to show you how research in genetics is conducted and how past research has established what we look on as the "facts" about genetics. Each Research Briefing relates to the information contained in the chapter in which it is placed, and can be read as part of that chapter, or the Research Briefings can be studied separately in order to gain a comprehensive overview of research methods.

Questions and Problems

Each set of Questions and Problems is divided into three sections, designed to help you in different ways in individual and group study programs. The first section asks you to define the key terms encountered in a chapter. All the terms are defined in the Glossary and you should check that you can remember the definitions yourself; they are provided as flashcards in the online Student Resources. If you can remember the definitions, then you have an excellent grasp of the main facts for that particular chapter. You should then move on to the Self-study Questions, which are aimed to test not just your recall of the facts, but also your understanding of the concepts behind those facts. Each Self-study Question can be answered in 50–100 words, or possibly by a table or annotated diagram. The questions cover the entire content of each chapter in a fairly straightforward manner. You can check your answers by comparing them with the relevant parts of the text, or by going to the online Student Resources site where you'll find either full written answers or hints on how to answer some questions. You can use the Self-study Questions to work systematically through a chapter, or you can select individual ones in order to confirm that you have the correct understanding of a specific topic.

Finally, there are Discussion Topics. These vary in nature and in difficulty. The simplest can be answered by a well-directed search of the genetics literature, the intention being that you advance your learning a few stages from where the book leaves off. In some cases the questions point you forward to issues that will be discussed later in the book, to help you see how the basic information that we deal with in the early chapters is relevant to the more complex topics that we study later on. Other problems require you to evaluate a statement or a hypothesis, based on your understanding of the material in the book, possibly supplemented by reading around the subject. These problems are intended to make you think carefully about the subject, and perhaps to realize for yourself that often there are hidden complexities that are not immediately apparent. A few problems are very difficult, in some cases to the extent that there is no solid answer to the question posed. These are designed to stimulate debate and speculation, stretching your knowledge and that of your colleagues with whom you discuss the problems. If you find these discussions stimulating, then you will know that you have become a real geneticist. Tips to help you with the Discussion Topics can be found on the online Student Resources site.

Further Reading

The reading lists at the end of each chapter are intended to help you obtain further information, for example when writing extended essays or dissertations on particular topics. In some cases, I have appended a few words summarizing the particular value of each entry, to help you decide which ones you wish to seek out. The lists are not all-inclusive and I encourage you to spend some time searching your library and the internet for other books and articles. Browsing is an excellent way to discover interests that you never realized you had!

Glossary

I am very much in favor of glossaries as learning aids and I have provided an extensive one for *Introduction to Genetics: A Molecular Approach*. Every term that is highlighted in bold in the text is defined in the Glossary, along with a number of additional terms that you might come across when referring to books or articles in the Further Reading sections. An online version of the glossary can also be found on the Student Resources site.

STUDENT AND INSTRUCTOR RESOURCES WEBSITES

Accessible from www.garlandscience.com, these websites provide learning and teaching tools created for *Introduction to Genetics: A Molecular Approach*. The Student Resources site is open to everyone, and users have the option to register in order to use book-marking and note-taking tools. The Instructor Resources site requires registration and access is available only to qualified instructors. To access the Instructor Resources site, please contact your local sales representative or email science@garland.com.

Below is an overview of the resources available for this book. On the website, the resources may be browsed by individual chapters and there is a search engine. You can also access the resources available for other Garland Science titles.

Student Resources

The following resources are available on the Student Resources site:

Animations and Videos

Animations and videos have been carefully selected and created to dynamically illustrate important concepts from the book, and make many of the more difficult topics accessible.

Quizzes

Each chapter contains a multiple-choice quiz, written by Sheryl L. Fuller-Espie, Cabrini College (USA), to test comprehension.

Flashcards

Each chapter contains a set of flashcards that allow students to review key terms from the text. The flashcards form the answers to the definition questions.

Self-study Questions and Discussion Topics

Answers to, or hints on how to answer, the Self-study Questions and Discussion Topics at the end of each chapter are given.

Glossary

The complete glossary from the book is available on the website and can be searched and browsed as a whole or sorted by chapter.

INSTRUCTOR RESOURCES

The following resources are available on the Instructor Resources site:

The Art of Introduction to Genetics

The images from the book are available in two convenient formats: MS PowerPoint® and JPEG. They have been optimized for display on a computer. Figures are searchable by figure number, figure name, or by keywords used in the figure legend from the book.

Animations and Videos

The animations and videos that are available to students are also available on the Instructor Resources website in two formats. The WMV formatted movies are created for instructors who wish to use the movies in PowerPoint presentations on Windows® computers; the QuickTime formatted movies are for use in PowerPoint for Apple computers or Keynote® presentations. The movies can easily be downloaded to your PC using the "download" button on the movie preview page.

Instructor's Lecture Outlines

The section headings, concept headings, and figures from the text have been integrated into PowerPoint presentations.

Question Bank

Written by Sheryl L. Fuller-Espie, Cabrini College (USA), the question bank contains over 400 multiple-choice, fill-in-the-blank, matching, true-false, and written-answer questions that can be used for creating tests, quizzes, or questions for personal response systems (that is, "clickers"). The question bank is available in MS Word® format and pre-loaded into Diploma® generator software, which is fully compatible with the major course management systems. Questions are organized by chapter and type, and can be additionally categorized by the instructor according to difficulty or subject. Diploma software allows questions to be scrambled (to create multiple tests) and edited, and new questions to be added.

CONTENTS

DETAILED CONTENTS

The Scope of Modern Genetics CHAPTER 1

Genetics plays a central role in modern society. Googling "genetics" on a typical day in the early twenty-first century reveals that during the previous week:

- Scientists discover new genetic link to Alzheimer's disease
- Genetic map points way to a better soybean
- Genetic evidence snares man over unsolved rape

These news stories illustrate the scope of genetics in today's world. Genetics is enabling diseases to be cured or treated so the patient's quality of life is improved. Genetics is the key to improving the nutritional value and productivity of the world's crops. And genetics can even catch criminals. It is no surprise that genetics has been described as the most living of the life sciences.

1.1 WHAT IS GENETICS?

Genetics is the study of **heredity**, the process by which characteristics are passed from parents to offspring so that all organisms, human beings included, resemble their ancestors. The central concept of genetics is that heredity is controlled by a vast number of factors, called **genes**, which are physical particles present inside living cells.

Genes are units of **biological information**. The entire complement of genes in an organism contains the total amount of information needed to construct a living, functioning example of that organism. Some genes are responsible for the visible characteristics of the organism, such as eye color in humans. Other genes are responsible for biochemical activities that are discernible only when the physiology of the organism is examined. Often, we become aware of the function of such genes only when they fail to work properly. An example is the human gene *CFTR*, which contains part of the biological information for transport of chloride ions into and out of cells. If the *CFTR* gene is defective, then chloride transport breaks down, leading to inflammation and mucus accumulation in the lungs. These are the primary symptoms of the disease called cystic fibrosis, and it was through study of the underlying cause of cystic fibrosis that the *CFTR* gene was first discovered. CFTR in fact stands for "cystic fibrosis transmembrane regulator," a term describing the biological information contained in this gene.

Because genes are units of biological information they are also units of **inheritance**. During **sexual reproduction**, genes of the parents are incorporated into the fertilized egg cell. These genes provide the fertilized egg with the full complement of biological information that it needs in order to develop into a new living organism. This new organism therefore inherits physical and biochemical characteristics from both its parents.

An understanding of genetics therefore requires an understanding of genes. The purpose of this book is to help you to acquire that understanding.

Genes are units of biological information

The first topic we will study is the most fundamental of all. How is biological information stored in genes? This problem perplexed early geneticists during the first part of the twentieth century, and it was not until the 1950s that the first step in solving the puzzle was made when it was discovered

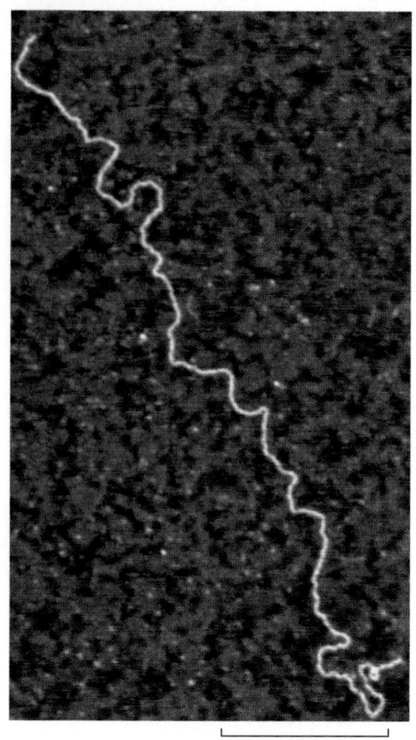

FIGURE 1.1 A molecule of DNA from the yeast *Saccharomyces cerevisiae*, visualized by atomic-force microscopy. (Courtesy of Jim de Yoreo, Lawrence Berkeley National Laboratory. This work was published in R. W. Friddle, J. E. Klare, S. S. Martin et al., *Biophys. J.* 86: 1632–1639, 2004.)

that genes are made of **DNA** (Figure 1.1). Our exploration of genetics must therefore start with DNA. In Chapter 2 we will study the structure of DNA and investigate how this structure enables DNA molecules to contain, in the form of a four-letter language made up of A, C, G, and T, the instructions for making a living organism (Figure 1.2).

Understanding the structure of DNA is, however, only the first step in understanding how biological information is contained in genes. When we start to look more closely at the structures of individual genes we realize that DNA contains different types of biological information, with genes specifying characteristics such as eye color and ability to transport chloride ions into and out of cells falling into just one category. There are also regulatory genes, genes that control the activities of other genes. One particular type of regulatory gene was not discovered until the 1990s. There are also genes that contain no biological information. These are thought to be evolutionary relics, genes that were once active but are no longer needed by the organism. Another complication in our study of genes is that in some genes the biological information is split into segments and the segments can be combined in different ways to alter the message being sent to the cell. We will examine all of these aspects of gene structure in Chapter 3.

The biological information in genes is read by the process called gene expression

By studying DNA and genes we will understand what is currently known about the way in which biological information is stored. The next question we must address is how this information is made available to the cell. How, for example, is the biological information in the *CFTR* gene read so that the cell can actually move chloride ions across cell membranes?

The transfer of biological information from gene to cell is called **gene expression** (Figure 1.3). For all genes, the process begins with the transfer of information from a DNA molecule into an **RNA** molecule. In Chapter 4 we will discover that DNA and RNA are very similar types of molecule and that this step of gene expression, called **transcription**, is quite straightforward in chemical terms. We will also learn that the RNA molecules that are made by transcription fall into different groups based on their function. Some RNA molecules are short-lived messengers that direct the second stage of gene expression, in which the gene's biological information, now contained in its RNA molecule, is used to direct the synthesis of a protein. This type of RNA is called **messenger RNA** (**mRNA**), and we will study its structure, and how it is prepared for its role in protein synthesis, in Chapter 5. Other RNA molecules, over 90% of all those that are made by

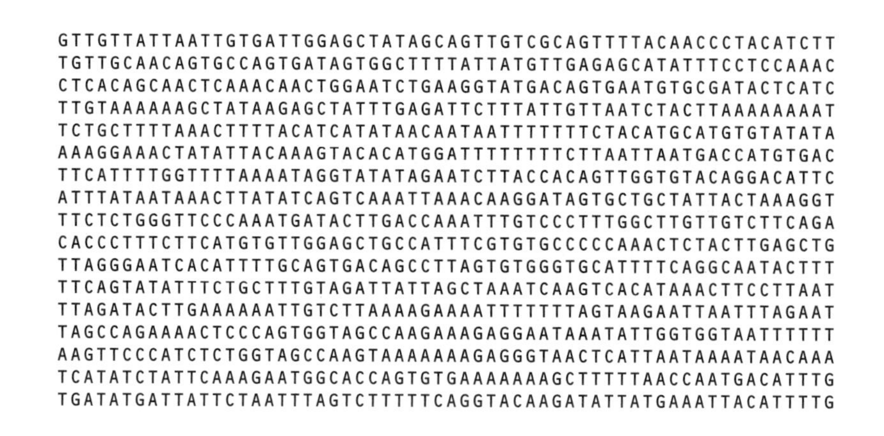

FIGURE 1.2 Part of the DNA sequence of the human *CFTR* gene. The biological information contained in DNA is in the form of a language made up of the letters A, T, G, and C. The entire *CFTR* gene contains 4443 letters, of which 1020 are shown here.

transcription, are the end products of gene expression and perform their own functions in the cell. Two important types, **ribosomal RNA (rRNA)** and **transfer RNA (tRNA)**, are involved in converting the information carried by the mRNAs into protein. We will look at the features of rRNA and tRNA in Chapter 6, and then in Chapters 7 and 8 we will examine exactly how these two types of RNA work together to synthesize proteins by **translation** of the mRNAs.

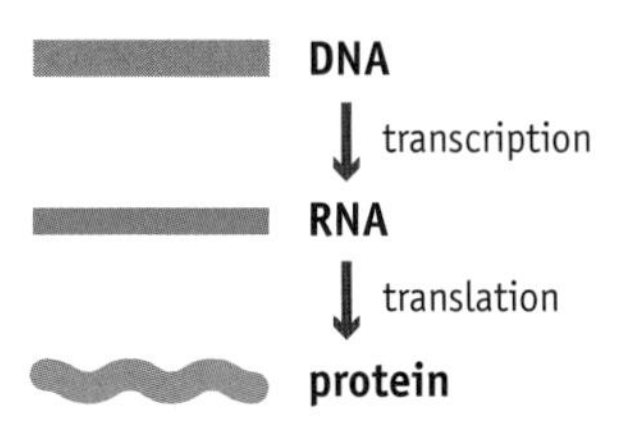

FIGURE 1.3 Gene expression. In its simplest form, gene expression can be looked on as a two-step process. The first step is the transcription of DNA into RNA, and the second step is the translation of some of the RNA molecules into proteins.

The expression of individual genes can be switched on and off

The entire complement of genes in a single cell represents a staggering amount of biological information. The information carried in some genes is needed by the cell at all times, but other genes have more specialized roles. Gene expression can therefore be controlled so that only those genes whose units of biological information are needed are active at any particular time.

Even the simplest organisms are able to control the expression of their genes in order to respond to changes in the environment. Many bacteria, for example, deal with sudden increases in temperature by switching on a set of genes whose protein products help to protect the cell from damage (Figure 1.4). In multicellular organisms, individual cells change their gene expression patterns in response to hormones, growth factors, and other regulatory molecules. Chemical signals between cells are therefore able to coordinate the activities of groups of cells in a manner that is beneficial to the organism as a whole.

In Chapter 9, we will study the way in which gene expression is controlled. We will learn that gene regulation involves more than simply switching genes on and off. It is also necessary to modulate the rates of expression of those genes that are switched on. We will discover that, in response to these complex requirements, a myriad of different ways of controlling gene expression have evolved, but that the most important of these acts at the very beginning of the gene expression pathway, by controlling whether or not a gene is transcribed into RNA.

Genes are also units of inheritance

In Part II we will study how genes act as units of inheritance. We might think that this simply involves sexual reproduction, but that is only part of the story. The much more frequent type of reproduction is the division of a parent cell into two daughters (Figure 1.5). This is the type of reproduction that gives rise to the vast majority of cells in the human body, with

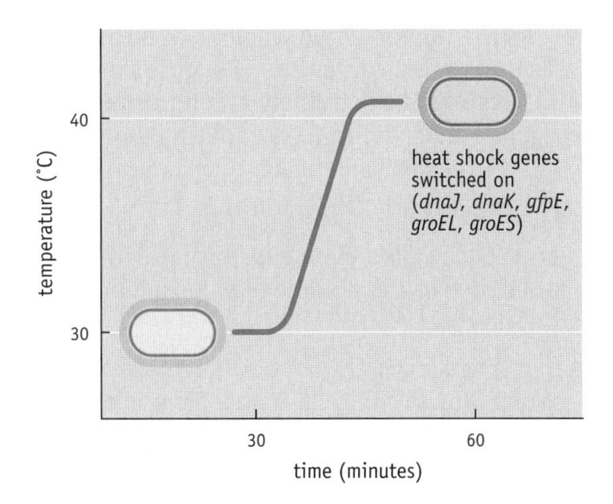

FIGURE 1.4 The heat shock response of a bacterium such as *Escherichia coli*. An increase in temperature from 30°C to 42°C results in the switching on of genes such as *dnaJ* and *dnaK*. The proteins coded by these genes help to protect the cell from damage.

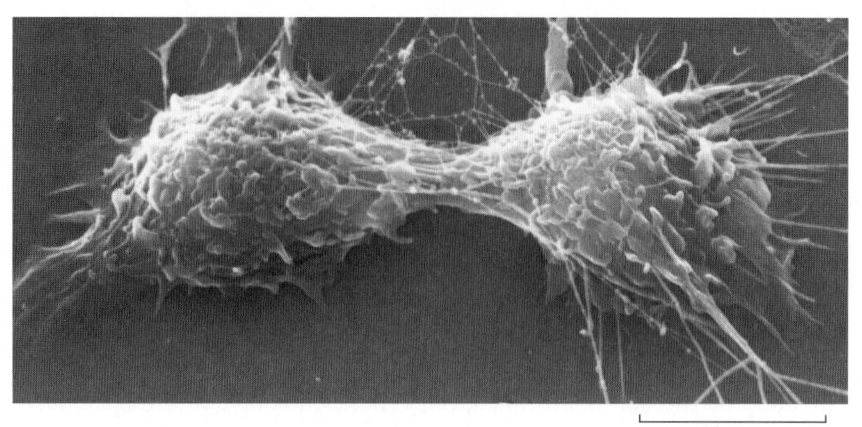

FIGURE 1.5 Cell division. This picture was taken with a scanning electron microscope and shows an animal cell in the final stages of cell division. (Courtesy of Guenter Albrecht-Buehler, Northwestern University.)

10^{17} divisions needed to produce all the cells needed by a human being during its lifetime. It is also the way that, for example, bacteria reproduce. During each of these cell divisions, a complete set of genes must be passed from the parent to both of the two daughter cells. Any error would be disastrous, as it would give rise to a lineage of cells that possessed an incomplete or altered complement of the organism's biological information. We must therefore begin our study of genes as units of inheritance by investigating how genes are inherited during cell division.

Our first task will be to understand how genes are replicated. If one cell divides into two, then clearly a copy has to be made of every gene. Because genes are made of DNA, the key to making copies of genes lies with the replication of DNA molecules. When we examine DNA in Chapter 2 we will see that the structure of a DNA molecule provides an obvious means for its replication, and that recognition of this fact was the "eureka" moment that convinced geneticists that genes are made of DNA. In Chapter 10, we will study the replication process in more detail. First, we will look at the overall pattern of replication and ask how a single DNA molecule can give rise to two identical daughter molecules. Then we will study the biochemistry and enzymology of DNA replication. What proteins are involved and how are the new DNA molecules synthesized?

By studying replication of DNA we do not, however, fully explain how genes are inherited by daughter cells during cell division. We must also understand how a full set of the replicated DNA molecules is passed to the daughter cells. This requires that we study how DNA molecules are organized inside the cells of different types of organism, and how replication of the DNA is coordinated with division of the cell. In Chapter 11 we will do this for the inheritance of genes during cell division in humans and other higher life forms, and in Chapter 12 we will look at the equivalent, but very different, processes that underlie inheritance of genes by bacteria. We must also, in Chapter 13, explore the extremely different mechanisms by which viruses pass their genes on to their progeny during their infection cycles.

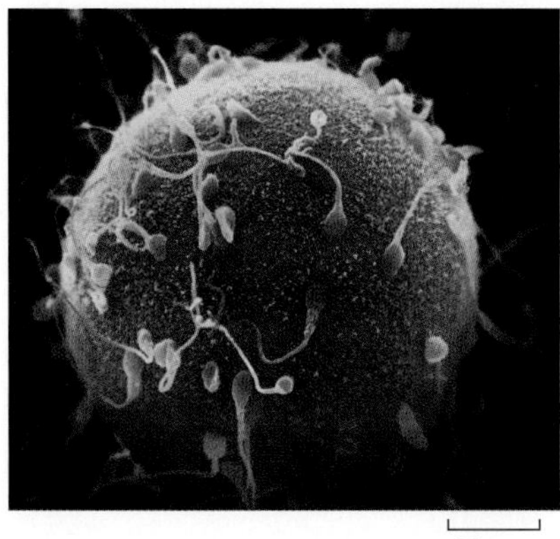

FIGURE 1.6 Fertilization of a female human egg cell. This scanning electron micrograph shows that many sperm cells attach to the outer surface of an egg, but only one will actually fertilize it. (Courtesy of D. Phillips/Science Photo Library.)

Children inherit genes from their parents

Once we have a firm grasp of the events occurring during cell division, we can move with confidence to the complexities posed by the inheritance of genes during sexual reproduction.

Sexual reproduction is preceded by a specialized type of cell division that produces the male and female sex cells. One male and one female cell then fuse to give a fertilized egg, which then develops into a new version of the organism (Figure 1.6). To understand this process, we must answer two questions. First, how are DNA molecules inherited during sexual reproduction? In Chapter 14 we will follow through the steps in sexual reproduction and understand what happens to the DNA molecules at each stage. We will discover that, unlike the type of cell division that occurs during asexual reproduction, the formation of sex cells does not involve simply the replication of DNA molecules and the passage of the daughter molecules into the new cells that are formed when the parent cell divides. Sexual reproduction is more complicated because it also provides an opportunity for DNA molecules to exchange segments by **recombination** (Figure 1.7). We will study how recombination occurs and investigate the effects that it has on the structures of the DNA molecules inherited by the offspring that result from sexual reproduction.

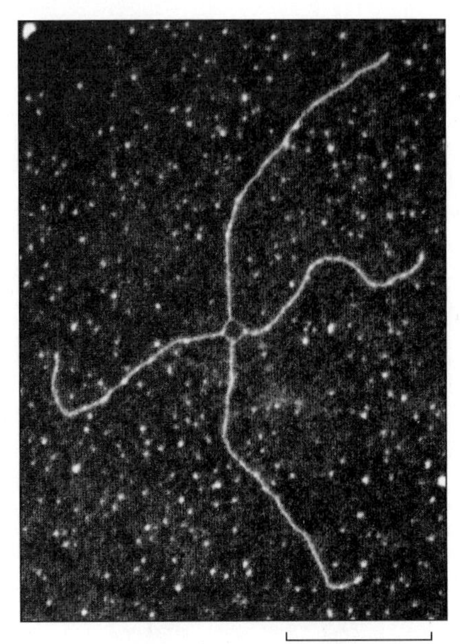

FIGURE 1.7 Two DNA molecules that have attached to one another in order to exchange segments by recombination. (Courtesy of Huntington Potter, University of South Florida, and David Dressler.)

The second question we must answer concerns the inheritance of genes rather than DNA molecules. A child is not an exact image of its mother or father, but instead a composite of the two parents—"she has her mother's eyes". This means that brothers and sisters are not identical, but they share a family resemblance (Figure 1.8). The same is true for all organisms that reproduce sexually. In all of these species the offspring inherit features from both of the two parents. In Chapter 15 we will investigate the ways in which the genes inherited from the two parents interact in the offspring, and how these interactions give rise to the particular set of characteristics displayed by an individual that results from sexual reproduction.

The inheritance of genes underlies evolution

There remains one final topic that we must investigate to complete our study of genes as units of inheritance. How do genes change over time? We know that DNA molecules can undergo structural alterations, called **mutations**. These may be caused by various chemicals present in the environment, or may be brought about, for example, by physical agents such as ultraviolet radiation from the sun. Many mutations are repaired by the cell soon after they occur, but a few slip through, which means that the DNA molecules that are passed to the offspring are not always precise copies of the parental molecules. Sometimes a mutation occurs within a gene, possibly causing a subtle change to the biological information contained in that gene (Figure 1.9).

In Chapter 16 we will investigate the various ways in which mutations can occur, and the mechanisms that cells have for correcting them. We will also look at the effects that a mutation can have on a gene, and in Chapter 17

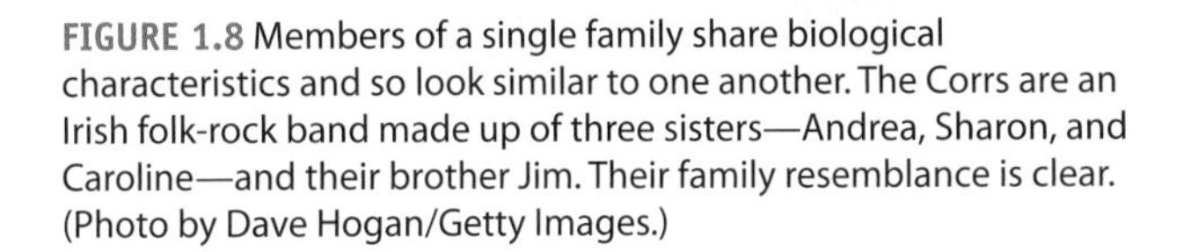

FIGURE 1.8 Members of a single family share biological characteristics and so look similar to one another. The Corrs are an Irish folk-rock band made up of three sisters—Andrea, Sharon, and Caroline—and their brother Jim. Their family resemblance is clear. (Photo by Dave Hogan/Getty Images.)

FIGURE 1.9 A mutation can change the biological information contained in a gene. On the left is a wild ear of wheat and on the right is an ear of domesticated wheat. The difference between the two is that the wild ear shatters when it becomes mature, so the seeds break away and fall to the ground. In domestic wheat, the ear does not shatter, so the seeds stay attached to the plant, making it easier to collect them. The nonshattering ear is caused by a single mutation in a gene whose biological information specifies the rigidity of the structure that attaches the seed to the ear. (Courtesy of George Willcox, Université de Lyon II. From K. Tanno and G. Willcox *Science* 311: 1886, 2006. With permission from AAAS.)

we will examine the possible fate of the new versions of genes that arise in this way. To do this we will have to investigate how genes are inherited not by individuals but by populations. We will discover that the genetic features of a population can change over time, and that the cumulative effect of these changes, called **microevolution**, underlies the processes by which new species arise.

1.2 GENETICS IN OUR MODERN WORLD

We began this chapter with three headlines illustrating the importance of genetics in today's society. In Part III we will explore some of the areas of research that are responsible for the high profile that genetics has in our modern world.

The first topic that we will study is thc role of genes in development. How do genes control the pathway that begins with a fertilized egg cell and ends with an adult organism? Finding the answer to this question is one of the biggest challenges in all of genetics. In Chapter 18 we will learn that advances in understanding developmental processes in humans have been made by studying **model organisms** such as the fruit fly, *Drosophila melanogaster* (Figure 1.10). Although the developmental pathway for a fruit fly is much simpler than that of a human, the important genes that control development are very similar in both types of organism. Through use of model organisms, and the combined endeavors of geneticists, cell biologists, physiologists, and biochemists, our knowledge of development has made great leaps forward in recent years.

FIGURE 1.10 The fruit fly *Drosophila melanogaster*. The fruit fly is an important model organism for research in many areas of biology, including genetics. (Courtesy of Nicolas Gompel, Institut de Biologie du Développement de Marseille-Luminy.)

Most of us share the opinion that our own species is the most important on the planet. Chapters 19 through 21 are therefore devoted to the genetics of *Homo sapiens*. In Chapter 19 we will study the human **genome** (Figure 1.11). This is the entire complement of DNA molecules in a human cell, containing all our genes and therefore all the biological information needed to make a human being. What genes do we possess, how are they arranged, and how do they make us special when compared, for example, with a chimpanzee?

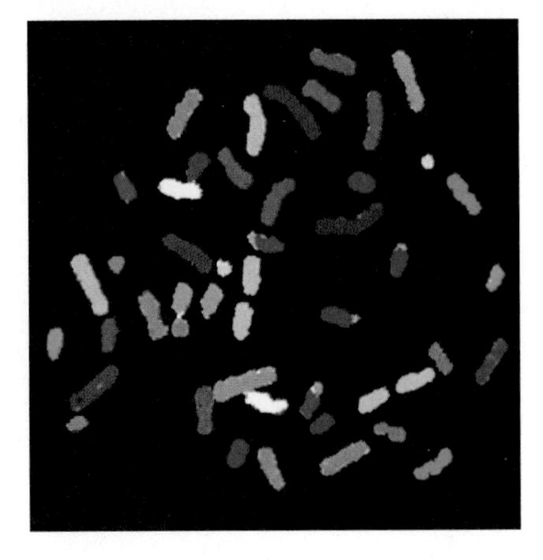

FIGURE 1.11 The human genome. The DNA molecules that make up our genome are contained in a set of chromosomes, which are shown here. Each chromosome has been "painted" a different color so the individual ones can be identified when observed with the light microscope. The painting is carried out with fluorescent dyes designed so that a different dye binds to each chromosome. (From E. Schröck, S. du Manoir, T. Veldman et al., *Science* 273: 494–497, 1996. With permission from AAAS.)

Probably the most important way that genetics is used to benefit humankind is in medical research. Over 6000 **inherited diseases** are known, diseases that are caused by defects in the genome and which, like other genetic features, can be passed from parents to offspring. Approximately 1 in every 200 children who are born suffer from one or another of these disorders. Other diseases, including many cancers, also result from malfunctioning of one or more genes. In Chapter 20 we will look at the progress that geneticists are making in understanding these diseases and devising ways of treating them.

How is genetics used to catch criminals? In Chapter 21 we will examine how genetic profiles are obtained and why these have become so important in forensic biology. We will also learn how similar techniques enable geneticists to study human evolution and to trace the routes taken by early humans as they migrated out of Africa and colonized the globe.

Genetics also has applications in industry and agriculture. Important proteins that are needed to treat diseases, or which have other industrial applications, can be synthesized in large amounts by transferring the gene that codes for the protein into a bacterium or other type of microorganism. Insulin, the protein that is used to treat diabetes, is a good example. The human insulin gene has been transferred to *Escherichia coli*, and these genetically engineered bacteria are now used as a cheap means of producing insulin for use by diabetics (Figure 1.12). Genetic engineering is also being used to create improved crops, ones that give higher yields or have higher nutritional values (Figure 1.13). These industrial and agricultural applications of genetics are covered in Chapter 22.

FIGURE 1.12 Part of an industrial plant for the production of human insulin from genetically engineered *Escherichia coli* bacteria. (Photo by Felix Denis. With permission from Sanofi-Aventis.)

FIGURE 1.13 Golden rice, on the right, compared with white rice. Golden rice has been genetically modified to synthesize increased amounts of β-carotene, a precursor of vitamin A. Golden rice is designed for consumption in parts of the world where people suffer from vitamin A deficiency. (Courtesy of Golden Rice Humanitarian Board [www.goldenrice.org].)

Each of these applications of genetics raises ethical issues that we must not ignore. It might be possible one day to eradicate inherited diseases by **gene therapy**, in which the defective gene is taken out of the patient's genome and replaced by the correct, functioning gene. But the same technology might be used to replace genes that are not defective with "better" ones, to produce "designer" babies. Genetic profiles are a powerful means of catching criminals, but does that mean that everybody's genetic profile should be stored on an international database? Genetically modified crops might have higher yields and better nutritional qualities, but could they be harmful to the environment? We will attempt to tackle these ethical issues in Chapter 23.

KEY CONCEPTS

- Genetics is the study of heredity. Heredity is the process by which characteristics are passed from parents to offspring so that all organisms resemble their ancestors.
- Heredity is controlled by genes. Genes are physical particles present inside living cells.
- Genes are units of biological information. The entire complement of genes in an organism contains the total amount of information needed to construct a living, functioning example of that organism.
- Genes are also units of inheritance. The transmission of genes from parents to offspring ensures that those offspring inherit the biological characteristics of their parents.
- Genetics is having a huge impact in our modern world. It has important applications in medicine, industry, forensics, and agriculture.
- As students of genetics, we must be aware of the ethical issues raised by some of the ways in which genetics is being applied in today's world.

PART I

GENES AS UNITS OF BIOLOGICAL INFORMATION

We begin our study of genetics with DNA. We start with DNA because genes are made of DNA, or, to be more precise, DNA is the **genetic material**. By studying the structure of DNA we can immediately understand how genes are able to fulfill their two related functions in living organisms, as units of biological information and as units of inheritance. This is the **molecular approach** to genetics, which we will be following in this book. The molecular approach enables us to make a logical, step-by-step progression from the structure of DNA to the process by which the biological information contained in DNA is released to the cell when it is needed. An understanding of DNA structure is also the best starting point for investigating how DNA molecules are copied and passed to offspring during reproduction, and for understanding how DNA molecules can change over time, enabling evolution to take place. The most accessible route into the more complex intricacies of genetics therefore starts with the structure of DNA.

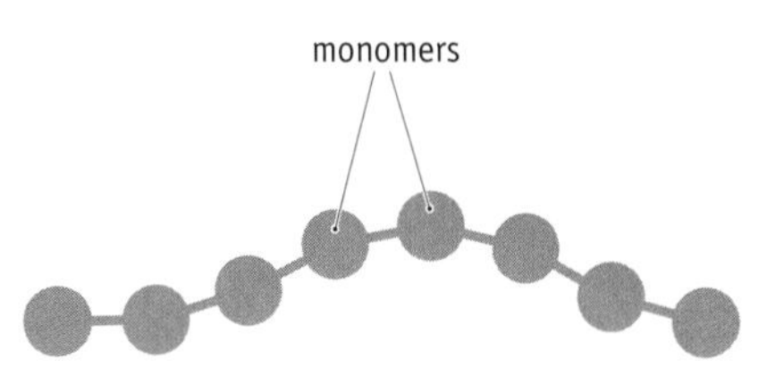

FIGURE 2.1 DNA is a linear polymer. In this depiction, each bead in the chain is an individual monomer.

In this chapter, then, we will study the structure of DNA and ask how this structure enables DNA to carry out the two functions of genes. We will learn how biological information is encoded in DNA and how copies can be made of DNA molecules. We will also examine how scientists can read the biological information contained in an individual DNA molecule by DNA sequencing, one of the most important techniques in modern biological research.

2.1 THE STRUCTURE OF DNA

DNA (**deoxyribonucleic acid**) is a **polymer**, a long, chainlike molecule made up of subunits called **monomers** (Figure 2.1). Many important biological molecules, including not only DNA but also proteins, polysaccharides, and lipids, are polymers of one type or another. In DNA, the monomers are called **nucleotides**, and these are linked together to form a **polynucleotide** chain that can be hundreds, thousands, or even millions of nucleotides in length. First we will study the structure of a nucleotide, and then we will examine how nucleotides are joined together to form a polynucleotide.

Nucleotides are the basic units of a DNA molecule

The basic unit of the DNA molecule is the nucleotide. Nucleotides are found in the cell either as components of nucleic acids or as individual molecules. Nucleotides have several different roles and are not just used to make DNA. For example, some nucleotides are important in the cell as carriers of energy used to power enzymatic reactions.

The nucleotide is itself quite a complex molecule, being made up of three distinct components. These are a sugar, a nitrogenous base, and a phosphate group (Figure 2.2). We will look at each of these in turn.

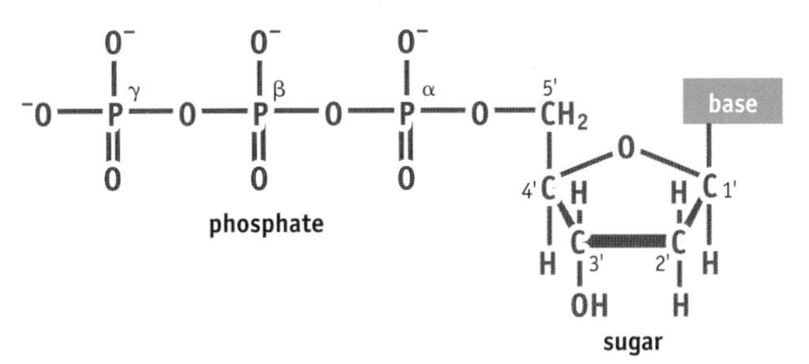

FIGURE 2.2 The components of a deoxyribonucleotide.

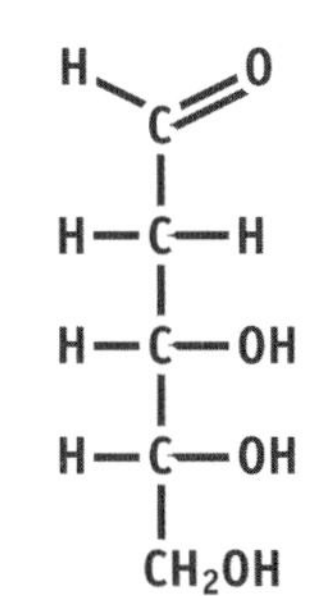

Fischer structure

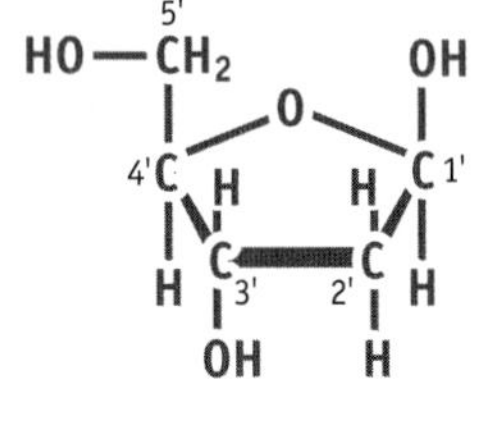

Haworth structure

FIGURE 2.3 The two structural forms of 2′-deoxyribose.

The sugar component of the nucleotide is a **pentose**. A pentose is a sugar that contains five carbon atoms. The particular type of pentose present in the nucleotides found in DNA is called 2′-deoxyribose. Pentose sugars can exist in two forms: the straight chain, or Fischer structure; and the ring, or Haworth structure (Figure 2.3). It is the ring form of 2′-deoxyribose that occurs in the nucleotide.

The name 2′-deoxyribose indicates that the standard ribose structure has been altered by replacement of the hydroxyl group (–OH) attached to carbon atom number 2′ with a hydrogen group (–H). The carbon atoms are always numbered in the same way, with the carbon of the carbonyl group (–C=O), which occurs at one end of the chain structure, numbered 1′. It is important to remember the numbering of the carbons because it is used to indicate the positions at which other components of the nucleotide are attached to the sugar. It is also important to realize that the numbers are not just 1, 2, 3, 4, and 5, but 1′, 2′, 3′, 4′, and 5′. The upper-right stroke is called a "prime," and the numbers are called "one-prime," "two-prime," and so on. The prime is used to distinguish the carbon atoms in the sugar from the carbon and nitrogen atoms in the nitrogenous base, which are numbered 1, 2, 3, and so on.

The nitrogenous bases are single- or double-ring structures that are attached to the 1′ carbon of the sugar. In DNA any one of four different nitrogenous bases can be attached at this position. These are called **adenine** and **guanine**, which are double-ring **purines**, and **cytosine** and **thymine**, which are single-ring **pyrimidines**. Their structures are shown in Figure 2.4. The base is attached to the sugar by a **β-*N*-glycosidic bond** attached to nitrogen number 1 of the pyrimidine or number 9 of the purine.

A molecule comprising the sugar joined to a base is called a **nucleoside**. This is converted into a nucleotide by attachment of a phosphate group to the 5′ carbon of the sugar. Up to three individual phosphates can be attached in series. The individual phosphate groups are designated α, β, and γ, with the α phosphate being the one attached directly to the sugar.

The full names of the four different nucleotides that polymerize to form DNA are:

2′-deoxyadenosine 5′-triphosphate

2′-deoxyguanosine 5′-triphosphate

2′-deoxycytidine 5′-triphosphate

2′-deoxythymidine 5′-triphosphate

Normally, however, we abbreviate these to dATP, dGTP, dCTP, and dTTP, or even just to A, G, C, and T, especially when writing out the sequence of nucleotides found in a particular DNA molecule.

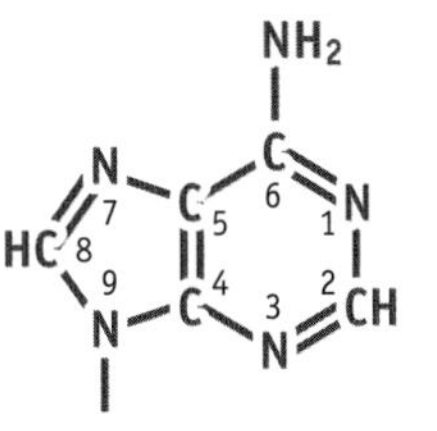

adenine (**A**)

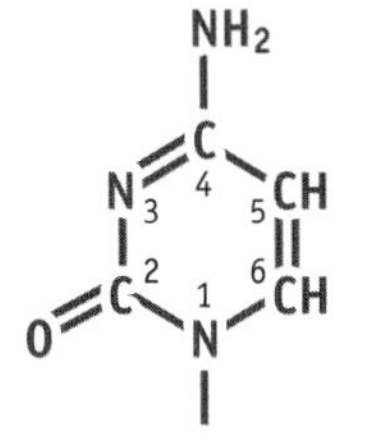

cytosine (**C**)

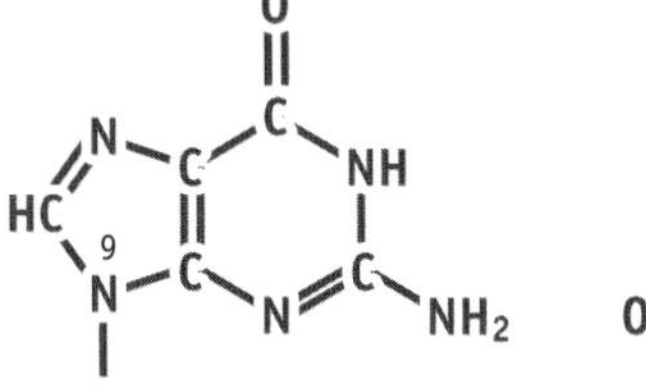

guanine (**G**)

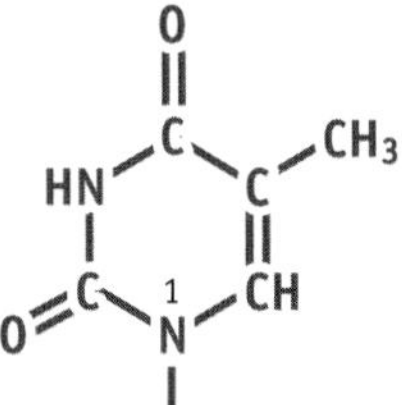

thymine (**T**)

FIGURE 2.4 The structures of the four bases that occur in deoxyribonucleotides.

Nucleotides join together to make a polynucleotide

The next stage in building up the structure of a DNA molecule is to link the individual nucleotides together to form a polymer. This polymer is called a polynucleotide and is formed by attaching one nucleotide to another through the phosphate groups.

The structure of a trinucleotide, a short DNA molecule comprising three individual nucleotides, is shown in Figure 2.5. The nucleotide monomers are linked together by joining the α phosphate group, attached to the 5′ carbon of one nucleotide, to the 3′ carbon of the next nucleotide in the chain. Normally a polynucleotide is built up from nucleoside triphosphate subunits, so during polymerization the β and γ phosphates are cleaved off. The hydroxyl group attached to the 3′ carbon of the second nucleotide is also lost. The linkage between the nucleotides in a polynucleotide is called a **phosphodiester bond**, "phospho-" indicating the presence of a phosphorus atom and "diester" referring to the two ester (C–O–P) bonds in each linkage. To be precise we should call this a 3′-5′ phosphodiester bond so that there is no confusion about which carbon atoms in the sugar participate in the bond.

An important feature of the polynucleotide is that the two ends of the molecule are not the same. This is clear from an examination of Figure 2.5. The top of this polynucleotide ends with a nucleotide in which the triphosphate group attached to the 5′ carbon has not participated in a phosphodiester bond and the β and γ phosphates are still in place. This end is called the **5′-P terminus** or simply the **5′ terminus**. At the other end of the molecule the unreacted group is not the phosphate but the 3′ hydroxyl. This end is called the **3′-OH** or **3′ terminus**. The chemical distinction between the two ends means that polynucleotides have a direction, which can be looked on as 5′ → 3′ (down in Figure 2.5) or 3′ → 5′ (up in Figure 2.5). An important consequence of the polarity of the phosphodiester bond is that the chemical reaction needed to extend a DNA polymer in the 5′ → 3′ direction is different from that needed to make a 3′ → 5′ extension. All of the enzymes that make new DNA polynucleotides in living cells carry out 5′ → 3′ synthesis. No enzymes that are capable of catalyzing the chemical reaction needed to make DNA in the opposite direction, 3′ → 5′, have ever been discovered.

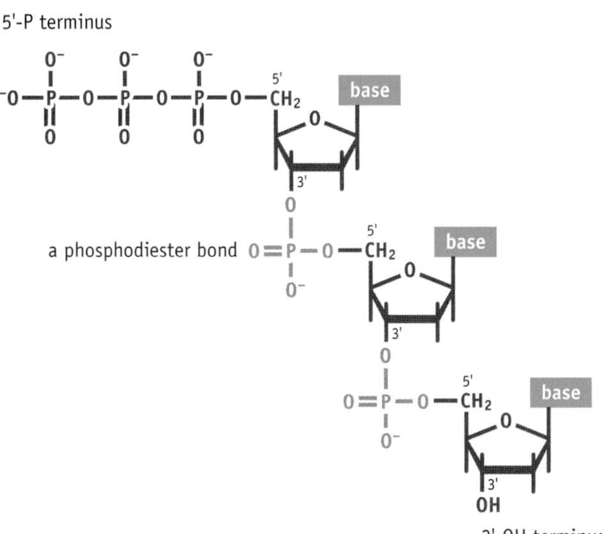

FIGURE 2.5 The structure of a trinucleotide, a DNA molecule comprising three individual nucleotides.

There is apparently no limitation to the number of nucleotides that can be joined together to form an individual DNA polynucleotide. Molecules containing several thousand nucleotides are frequently handled in the laboratory, and the DNA molecules in chromosomes are much longer, sometimes several million nucleotides in length. In addition, there are no chemical restrictions on the order in which the nucleotides can join together. At any point in the chain the nucleotide could be A, G, C, or T.

In living cells, DNA is a double helix

In living cells, DNA molecules almost always contain two polynucleotides, wrapped around one another to form the famous double-helix structure discovered by James Watson and Francis Crick in 1953 (Research Briefing 2.1). The double helix is a complicated structure (Figure 2.6), but the key facts about it are not too difficult to understand.

First we will look at the basic features of the double helix. The two polynucleotides in the helix are arranged in such a way that their sugar–phosphate "backbones" are on the outside of the helix and their bases are on the inside. The bases are stacked on top of each other rather like a pile of plates. The two polynucleotides are **antiparallel**, meaning that they run in different directions, one oriented in the 5′ → 3′ direction and the other in the 3′ → 5′ direction. The polynucleotides must be antiparallel in order to form a stable helix. Molecules in which the two polynucleotides run in the same direction are unknown in nature. The double helix is right-handed, so if it were a spiral staircase then the banister (that is, the sugar–phosphate backbone) would be on your right-hand side as you were climbing upward. The final point is that the helix is not absolutely regular. Instead, on the outside of the molecule we can distinguish a **major** and a **minor groove**. These two grooves are clearly visible in Figure 2.6.

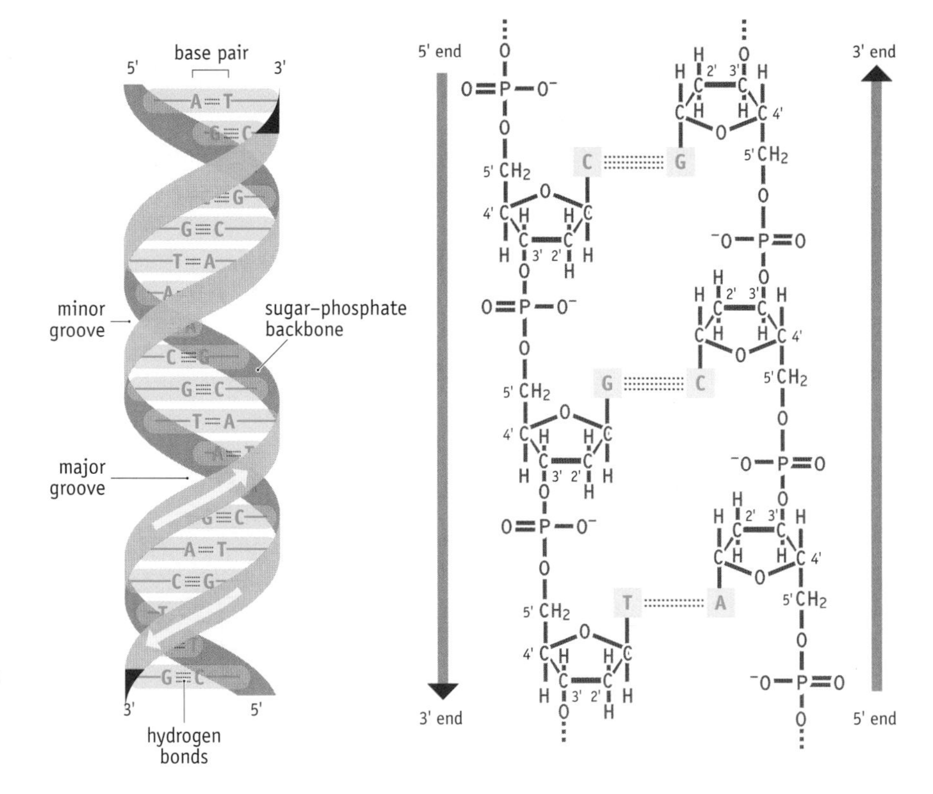

FIGURE 2.6 The double-helix structure of DNA. On the left the double helix is drawn with the sugar–phosphate backbone of each polynucleotide shown as a gray ribbon with the base pairs in green. On the right the chemical structure for three base pairs is given.

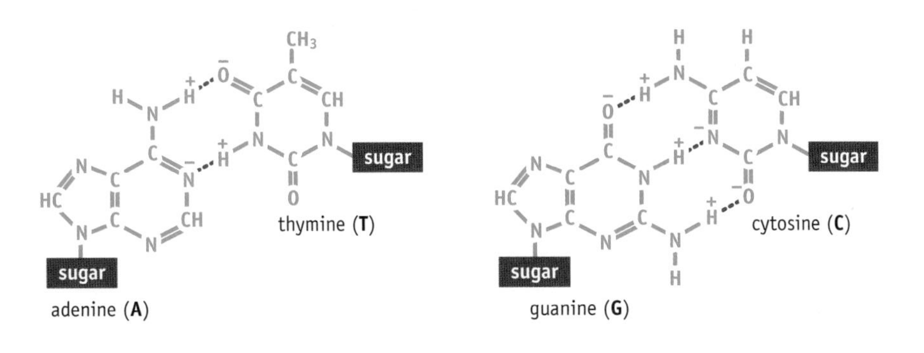

FIGURE 2.7 Base pairing. A base-pairs with T, and G base-pairs with C. The bases are drawn in outline, with the hydrogen bonds indicated by dotted lines. Note that a G–C base pair has three hydrogen bonds whereas an A–T base pair has just two.

Those are the key facts about the double helix. There remains just one additional feature to consider, but this one is the most important of all. Within the helix, an adenine in one polynucleotide is always adjacent to a thymine in the other strand, and similarly guanine is always adjacent to cytosine. This is called **base pairing** and involves the formation of **hydrogen bonds** between an adenine and a thymine, or between a cytosine and a guanine. A hydrogen bond is a weak electrostatic attraction between an electronegative atom (such as oxygen or nitrogen) and a hydrogen atom attached to a second electronegative atom. The base pairing between adenine and thymine involves two hydrogen bonds, and that between guanine and cytosine involves three hydrogen bonds (Figure 2.7).

The two base-pair combinations—A base-paired with T, and G base-paired with C—are the only ones that are permissible. This is partly because of the geometries of the nucleotide bases and the relative positions of the atoms that are able to participate in hydrogen bonds, and partly because the pair must be between a purine and a pyrimidine. A purine–purine pair would be too big to fit within the helix, and a pyrimidine–pyrimidine pair would be too small. Because of the base pairing, the sequences of the two polynucleotides in the helix are **complementary**—the sequence of one polynucleotide determines the sequence of the other.

The double helix exists in several different forms

The double helix shown in Figure 2.6 is referred to as the **B-form** of DNA. This is the structure that was studied by Watson and Crick, and it represents by far the most common type of DNA in living cells, but it is not the only version that is known. Other versions are possible because the three-dimensional structure of a nucleotide is not entirely rigid. For example, the orientation of the base relative to the sugar can be changed by rotation around the β-*N*-glycosidic bond (Figure 2.8). This has a significant effect on the double helix, as it alters the relative positioning of the two polynucleotides. The relative positions of the carbons in the sugar can also be changed slightly, affecting the conformation of the sugar–phosphate backbone.

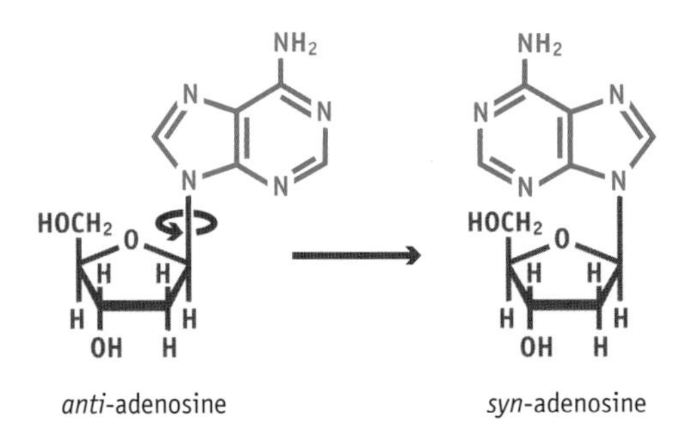

FIGURE 2.8 The structures of *anti*- and *syn*-adenosine, which differ in the orientation of the base relative to the sugar component of the nucleotide. Rotation around the β-*N*-glycosidic bond can convert one form into the other. The three other nucleotides also have *anti* and *syn* conformations. In A- and B-DNA, all the bases are in the *anti* conformation, but in Z-DNA there is a mixture of the two types.

RESEARCH BRIEFING 2.1

The discovery of the double helix

The discovery of the double helix, by James Watson and Francis Crick of Cambridge University, UK, in 1953, was the most important breakthrough in twentieth-century biology. The nature of the double helix revealed how genes can replicate, a puzzle that had seemed almost impossible to solve just a few years earlier. According to Watson in his book *The Double Helix*, the work was a desperate race against the famous American biochemist Linus Pauling, who initially thought that DNA was a triple helix. This mistake gave Watson and Crick the time they needed to complete their description of the double-helix structure.

When Watson and Crick began their work, the structures of the nucleotides, and the way these are linked together to form a polynucleotide, were already known. What was not known was the actual structure of DNA in a living cell. Was it a single polynucleotide, perhaps folded up in some way? Or were there two or more polynucleotides in a DNA molecule?

To solve the structure of DNA, Watson and Crick used **model building**—they built a scale model of what they thought a DNA molecule must look like. The model had to obey the laws of chemistry, which meant that if a polynucleotide was coiled in any way then its atoms must not be placed too close together. It was equally vital that the model take account of the results of other investigations into DNA structure. One of these studies was carried out by Erwin Chargaff at Columbia University in New York, the other, by Rosalind Franklin at King's College, London.

Chargaff's base ratios paved the way for the correct structure

Erwin Chargaff became interested in DNA in the 1940s, when scientists first realized that DNA *might* be the genetic material. He decided to use a new technique, called **paper chromatography**, to measure the amounts of each of the four nucleotides in DNA from different tissues and organisms. In paper chromatography, a mixture of compounds is placed at one end of a paper strip, and an organic solvent, such as *n*-butanol, is then allowed to soak along the strip. As the solvent moves it carries the compounds with it, but at different rates depending on how strongly each one absorbs into the paper matrix (Figure 1).

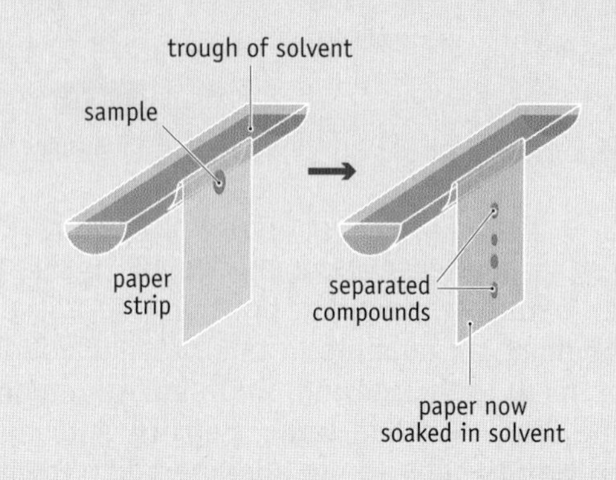

FIGURE 1 Paper chromatography.

Chargaff purified DNA from different sources and treated each sample with acid to break the molecules into their component nucleotides (Figure 2). He then used paper chromatography to separate the nucleotides in each mixture so their concentrations could be measured. The results were quite startling. They revealed a simple relationship between the proportions of the nucleotides in any one sample of DNA. The relationship is that the number of adenines equals the number of thymines, and the number of guanines equals the number of cytosines. In other words, A – T and G = C.

Chargaff did not speculate to any great extent, at least not in his publications, about the relevance of these base ratios to the structure of DNA. In fact, Watson and Crick appear to have been unaware of his results until they met Chargaff when he visited Cambridge University. But once they became aware of the A = T and G = C relationship, they knew this had to be accounted for in their model of DNA.

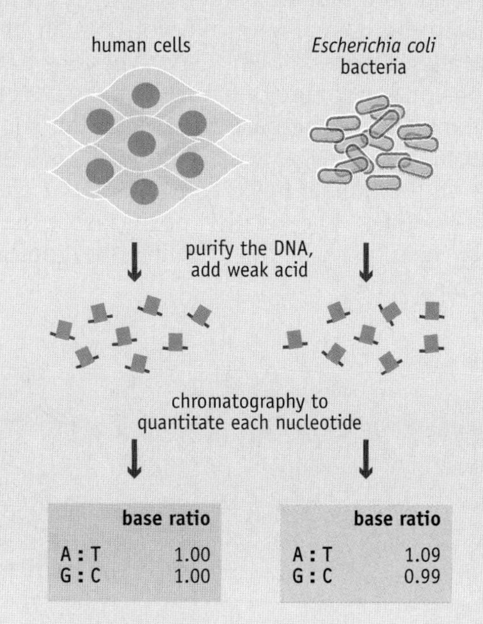

FIGURE 2 Chargaff's experiments.

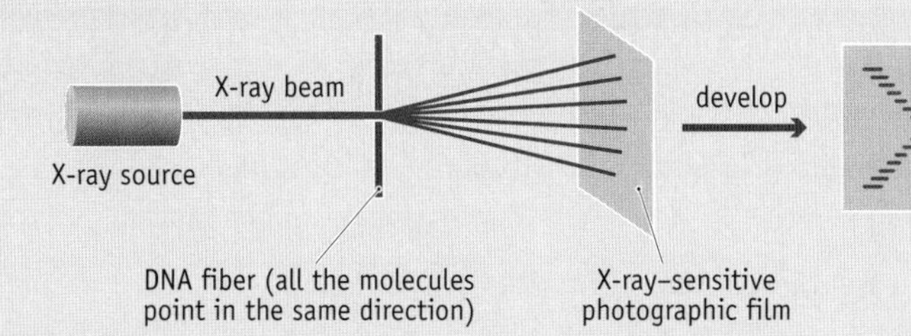

FIGURE 3 X-ray diffraction analysis.

X-ray diffraction analysis indicates that DNA is a helical molecule

The second piece of evidence available to Watson and Crick was the **X-ray diffraction pattern** obtained when a DNA fiber is bombarded with X-rays. X-rays have very short wavelengths—between 0.01 and 10 nm—comparable with the spacings between atoms in chemical structures. When a beam of X-rays is directed onto a DNA fiber some of the X-rays pass straight through, but others are diffracted and emerge at a different angle (Figure 3). As the fiber is made up of many DNA molecules, all positioned in a regular array, the individual X-rays are diffracted in similar ways, resulting in overlapping circles of diffracted waves which interfere with one another. An X-ray–sensitive photographic film placed across the beam reveals a series of spots and smears, called the X-ray diffraction pattern.

X-ray diffraction pictures of DNA fibers were made by Rosalind Franklin, during 1952, using techniques previously developed by Maurice Wilkins (Figure 4). The pictures immediately showed that DNA is a helix, and mathematical calculations based on them revealed that the helix has two regular periodicities of 0.34 nm and 3.4 nm. But how do these deductions relate to Chargaff's base ratios and to the actual structure of DNA?

Pulling together the evidence

Watson and Crick put together all the experimental data concerning DNA and decided that the only structure that fitted all the facts was the double helix shown in Figure 2.6. The main difficulty was deciding how many polynucleotides were present in a single molecule. This could be estimated from the density of the DNA in a fiber. Several measurements of DNA fiber density had been reported, and they did not agree. Some suggested that there were three polynucleotides in a single molecule, others suggested two. Pauling thought the first set of measurements were correct, and devised a triple-helix structure that was completely wrong. Watson and Crick decided it was more likely to be two.

Once two polynucleotides had been decided on, it became clear that the sugar–phosphate backbone had to be on the outside of the molecule. This was the only way the various atoms could be spaced out appropriately within the models that Watson and Crick built. The models also indicated that the two strands had to be antiparallel and the helix right-handed. The X-ray diffraction data enabled the dimensions of the helix to be set. The periodicity of 0.34 nm indicated the spacing between individual base pairs, and that of 3.4 nm gave the distance needed for a complete turn of the helix.

What about Chargaff's base ratios? These were the key to solving the structure. Watson realized, on the morning of Saturday, March 7, 1953, that the pairs formed by adenine–thymine and guanine–cytosine have almost identical shapes (see Figure 2.7). These pairs would fit neatly inside the double helix, giving a regular spiral with no bulges. And if these were the only pairs that were allowed, then the amount of A would equal the amount of T, and G would number the same as C. Everything fell into place, and the greatest mystery of biology—how genes can replicate—had been solved.

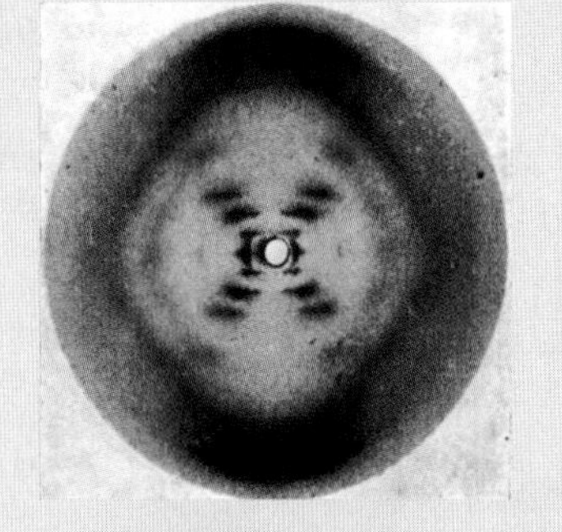

FIGURE 4 Franklin's "photo 51" showing the diffraction pattern obtained with a fiber of DNA. The cross shape indicates that DNA has a helical structure, and the relative positions of the various dots and smears enable the periodicities within the molecule to be calculated. (From R. Franklin and R. G. Gosling, *Nature* 171: 740–741, 1953. With permission from Macmillan Publishers Ltd.)

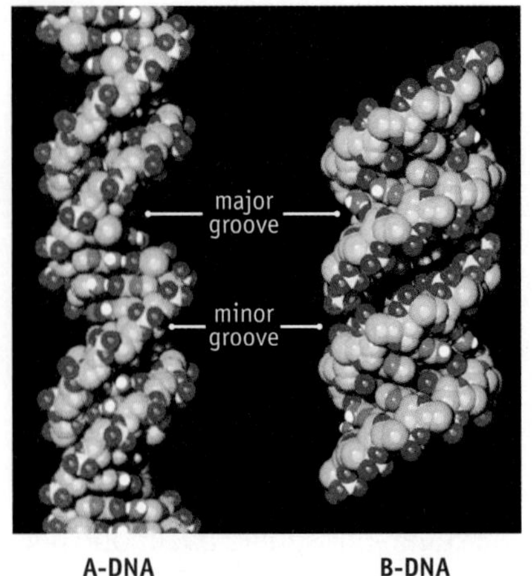

FIGURE 2.9 Comparison between the A- and B-DNA versions of the double helix. These are space-filling models in which the individual atoms are represented by spheres. The atoms of the sugar–phosphate backbone are shown in green and red, and those in the bases in blue and white. (From J. Kendrew (ed.) Encyclopedia of Molecular Biology. Oxford: Blackwell, 1994. With permission from Wiley–Blackwell.)

TABLE 2.1 FEATURES OF THE DIFFERENT CONFORMATIONS OF THE DNA DOUBLE HELIX

Feature	B-DNA	A-DNA	Z-DNA
Type of helix	Right-handed	Right-handed	Left-handed
Number of base pairs per turn	10	11	12
Distance between base pairs (nm)	0.34	0.29	0.37
Distance per complete turn (nm)	3.4	3.2	4.5
Diameter (nm)	2.37	2.55	1.84
Major groove	Wide, deep	Narrow, deep	Flat
Minor groove	Narrow, shallow	Wide, shallow	Narrow, deep

The common, B-form of DNA has 10 base pairs (abbreviated as 10 bp) per turn of the helix, with 0.34 nm between adjacent base pairs and hence a pitch (the distance needed for one complete turn) of 3.40 nm. The diameter of the helix is 2.37 nm. The second type of DNA double helix to be discovered, called the **A-form**, is more compact, with 11 bp per turn, 0.29 nm between each two base pairs, and a diameter of 2.55 nm (Table 2.1). But comparing these dimensions does not reveal what is probably the most significant difference between these two conformations of the double helix. This is the extent to which the internal regions of the DNA molecule are accessible from the surface of the structure. A-DNA, like the B-form, has two grooves, but with A-DNA the major groove is even deeper, and the minor groove shallower and broader (Figure 2.9).

In Chapter 9 we will learn that expression of the biological information contained within a DNA molecule is mediated by DNA-binding proteins that attach to the double helix and regulate the activity of the genes contained within it. To carry out their function, each DNA-binding protein must attach at a specific position, close to the gene whose activity it must influence. The protein can achieve this, with at least some degree of accuracy, by reaching down into a groove, within which it can "read" the DNA sequence without opening up the helix by breaking the base pairs (Figure 2.10). This is possible with B-DNA, but easier with A-DNA because the bases are more exposed in the minor groove. It has been suggested that some DNA-binding proteins induce formation of a short stretch of A-DNA when they attach to a DNA molecule.

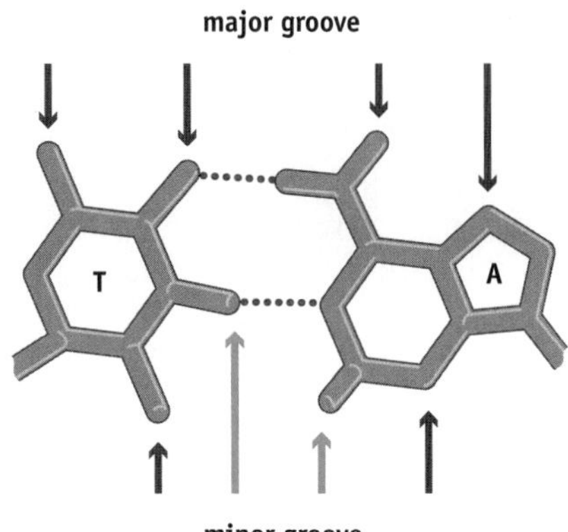

FIGURE 2.10 A DNA-binding protein can recognize individual base pairs without opening up the double helix. This drawing shows an A–T base pair in outline, with arrows indicating the chemical features that can be recognized by accessing the base pair via the major groove (above) and minor groove (below). Using these features, a binding protein can recognize that the base pair is an A–T rather than a G–C. In the major groove the distribution of these chemical features also enables the orientation of the base pair to be identified, so the strand containing the A can be distinguished from the one containing the T. For some time it was believed that this was not possible in the minor groove, because only the two features indicated by the black arrows were known to be present. The two additional features indicated by the green arrows have recently been discovered, and these are thought to enable the orientation of the pair to be discernible via the minor groove.

FIGURE 2.11 The structure of the Z-DNA version of the double helix. The atoms of the sugar–phosphate backbone are shown in green and red, and those in the bases in blue and white. (From J. Kendrew (ed.) Encyclopedia of Molecular Biology. Oxford: Blackwell, 1994. With permission from Wiley–Blackwell.)

A third type, **Z-DNA**, is more strikingly different. In this structure the helix is left-handed, not right-handed as it is with A- and B-DNA, and the sugar–phosphate backbone adopts an irregular zigzag conformation (Figure 2.11). Z-DNA is more tightly wound, with 12 bp per turn and a diameter of only 1.84 nm (Table 2.1). It is thought to form around regions of B-DNA that have become slightly unwound, as occurs when a gene is being transcribed into RNA. Unwinding results in torsional stress, which might be relieved to some extent by forming the more compact Z version of the helix (Figure 2.12).

2.2 THE MOLECULAR EXPLANATION OF THE BIOLOGICAL ROLE OF DNA

To fulfill its role as the genetic material, a DNA molecule must possess properties that enable the genes it contains to act as units of biological information and units of inheritance. We will now ask ourselves how these requirements are met by the structure of the double helix.

Biological information is contained in the nucleotide sequence of a DNA molecule

DNA is able to act as a store of biological information because of its polymeric structure and because there are four different nucleotides. The order of nucleotides in a DNA molecule—the **DNA sequence**—is, in essence, a *language* made up of the four letters A, C, G, and T. The biological information contained in genes is written in this language, which we call the **genetic code**. The language is read through the process called **gene expression**.

There are no chemical restrictions on the order in which the nucleotides can join together in a DNA molecule. At any point the nucleotide could be A, G, C, or T. This means that a polynucleotide just 10 nucleotides in length could have any one of 4^{10} = 1,048,576 different sequences (Figure 2.13). The average length of a gene is about 1000 nucleotides. This length of DNA can exist as 4^{1000} different sequences, which we will look on as simply a very big number, greater than the supposed number of atoms in the observable universe, which is a paltry 10^{80}. Bear in mind that we can also have genes 999 and 1001 nucleotides in length, as well as many other lengths, each of these lengths providing its own immense number of possible DNA sequence variations. The early geneticists at the start of the twentieth century were mystified by the ability of genes to exist in many different forms,

FIGURE 2.12 Regions of Z-DNA might form on either side of an underwound segment of B-DNA, in order to relieve the torsional stress that is created.

AGCTAAGGGT
ACCTAAGGGT
AACTAAGGGT
ATCTAAGGGT
AGGTAAGGGT
AGTTAAGGGT
ACGTAAGGGT
ACGAAAGGGT
AGCTTCGGGT
AGCTGGGGGT
AGCTAAGGGA
TCAATTTTAA

FIGURE 2.13 Twelve of the 1,048,576 different possible sequences for a DNA molecule 10 nucleotides in length. The orange nucleotides are ones that are different from the equivalent nucleotide in the topmost sequence.

in order to account for all those present in the myriad of species alive today and which lived in the past. An answer was so difficult to imagine that some biologists wondered whether genes really were physical structures inside cells. Perhaps they were just abstract entities whose invention by geneticists made it possible to explain how biological characteristics are passed from parents to offspring. Now that we understand the structure of DNA the immense variability required by the genetic material is no puzzle at all.

Complementary base pairing enables DNA molecules to replicate

In order to act as units of inheritance, genes must be able to replicate. Copies of the parent's genes must be placed in the fertilized egg cell so that this cell receives the information it needs to develop into a new living organism that displays a mixture of the biological characteristics of its parents. Genes must also be replicated every time a cell divides, so that each of the two daughter cells can be given a complete copy of the biological information possessed by the parent cell.

Before the structure of the double helix was known, the ability of genes to make copies of themselves was an even greater mystery than their ability to exist in an almost infinite number of different forms. No chemical capable of replication had ever been found in the natural world. The discovery of the double helix provided a clear and obvious solution to the problem. The key lies with the complementary base pairing that links nucleotides that are adjacent to one another in the two strands of the double helix. The rules are that A can base-pair only with T, and G can base-pair only with C. This means that if the two polynucleotides in a DNA molecule are separated, then two perfect copies of the parent double helix can be made simply by using the sequences of these preexisting strands to dictate the sequences of the new strands (Figure 2.14).

In living cells, DNA molecules are copied by enzymes called **DNA polymerases**. A DNA polymerase builds up a new polynucleotide by adding nucleotides one by one to the 3′ end of the growing strand, using the base-pairing rules to identify which of the four nucleotides should be added at each position (Figure 2.15). This is called **template-dependent DNA synthesis**. It ensures that the double helix that is made is a precise copy of

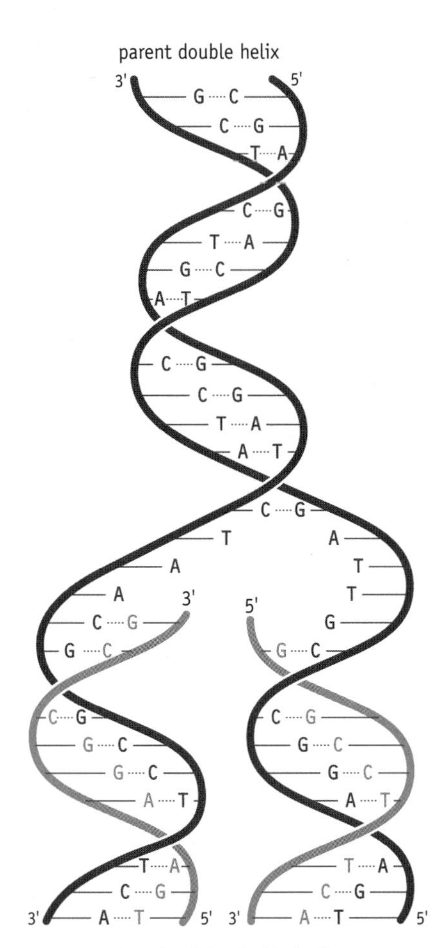

FIGURE 2.14 Complementary base pairing provides a means for making two copies of a double helix. The polynucleotides of the parent double helix are shown in blue, and the newly synthesized strands are in orange.

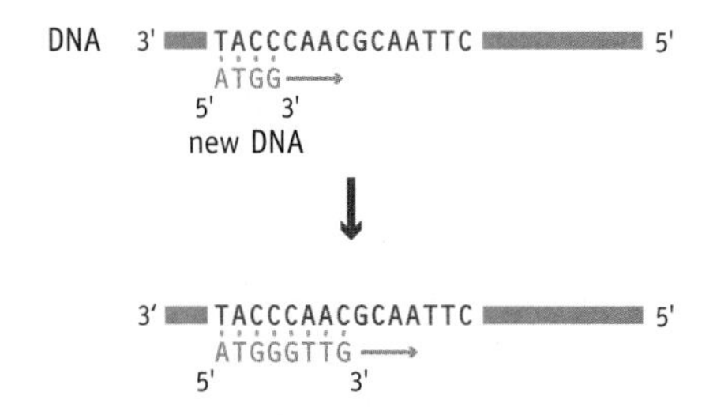

FIGURE 2.15 Template-dependent DNA synthesis. The new DNA, shown in orange, is extended by adding nucleotides one by one to its 3′ end.

the double helix from which the original polynucleotide was obtained. The structure of DNA therefore explains how genes are able to replicate and hence to act as units of inheritance.

2.3 HOW DNA IS SEQUENCED

Now that we understand the structure of DNA we can begin to appreciate why **DNA sequencing** is so important in modern biology. By working out the sequence of nucleotides we can gain access to the biological information contained in a DNA molecule. From the sequence it might be possible to identify the genes present in the DNA molecule, and possibly to deduce the functions of those genes. We should therefore complete our study of DNA by examining the methods used to obtain nucleotide sequences.

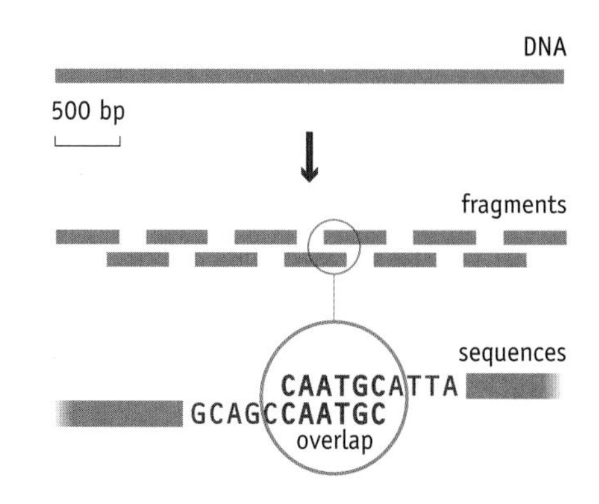

FIGURE 2.16 Assembling long DNA sequences from the sequences of short fragments. The original DNA molecule is broken into smaller fragments. These are sequenced individually, and overlaps between these sequences identified in order to assemble the complete sequence of the starting molecule.

The DNA molecules present in living cells are extremely long, possibly many millions of base pairs in length. It is not yet possible to work out the sequence of a molecule of this length in a single experiment. It might be possible one day, as DNA sequencing technology is advancing by leaps and bounds, but at present this is far beyond our capabilities. In fact, the maximum amount of sequence that can be read in a single experiment is just 750 bp. The strategy in a sequencing project is therefore to break these long DNA molecules down into many short fragments, each of which is individually sequenced. The sequence of the original molecule is then assembled by searching for overlaps between the sequences of the short fragments (Figure 2.16). Immediately we can understand why big sequencing projects make use of automated techniques, with the sequencing experiments carried out by robots, and computer programs used to find the overlaps and build up the contiguous blocks of sequence, or **contigs**.

Clone libraries contain DNA fragments ready to be sequenced

The first step in a DNA sequencing project is to prepare a collection of short DNA fragments ready for sequencing. This is done by **DNA cloning**, and the resulting collection is called a **clone library**.

DNA cloning is a sophisticated technology, but we do not need to delve too deeply into it in order to understand how it is used to prepare a clone library. We will learn much more about it when we study the industrial applications of genetics in Chapter 22. First, we must break down the DNA molecules that we wish to sequence into the shorter fragments from which the library will be prepared. Often the DNA is fragmented by **sonication**, a technique that uses high-frequency sound waves to make random cuts in the molecules (Figure 2.17).

The next task is to separate the fragments resulting from sonication so that each one can be sequenced individually. This is the actual cloning part of the procedure, and it requires a **vector**, a DNA molecule that is able to replicate inside a cell of the bacterium *Escherichia coli*. The vector is usually a modified version of a naturally occurring, self-replicating DNA molecule called a **plasmid**. In its replicative state the plasmid is a circular molecule, but at the start of the cloning experiment it is obtained in a linear form. The linear vector molecules are mixed with the DNA fragments, and a **DNA ligase** enzyme is added, which joins DNA molecules together end-to-end. Various combinations are produced, including some in which a vector molecule has become linked to one of the DNA fragments and the two ends have then joined together to produce a circular **recombinant** vector (Figure 2.18A).

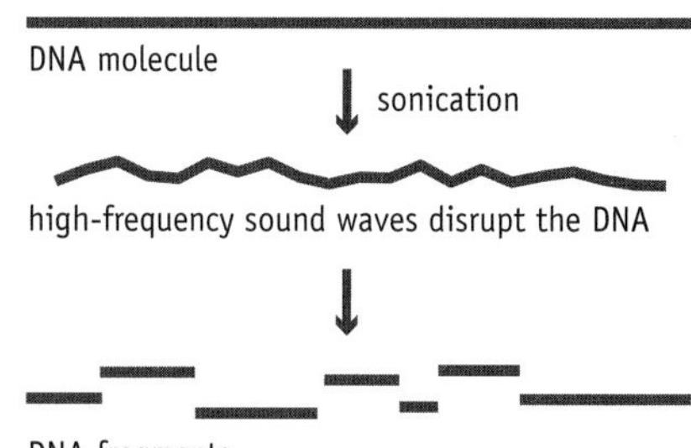

FIGURE 2.17 Using sonication to break a DNA molecule into fragments.

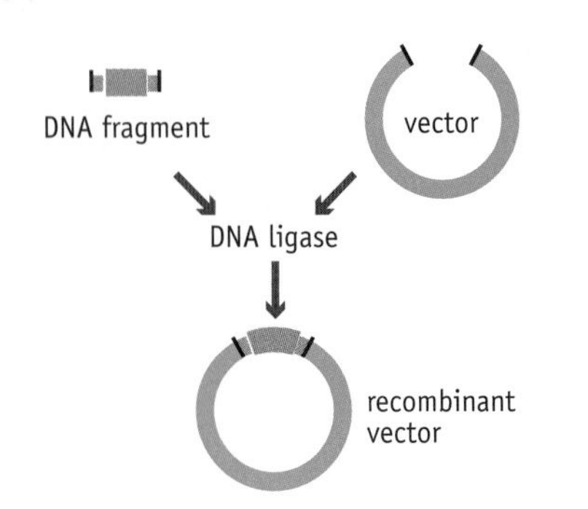

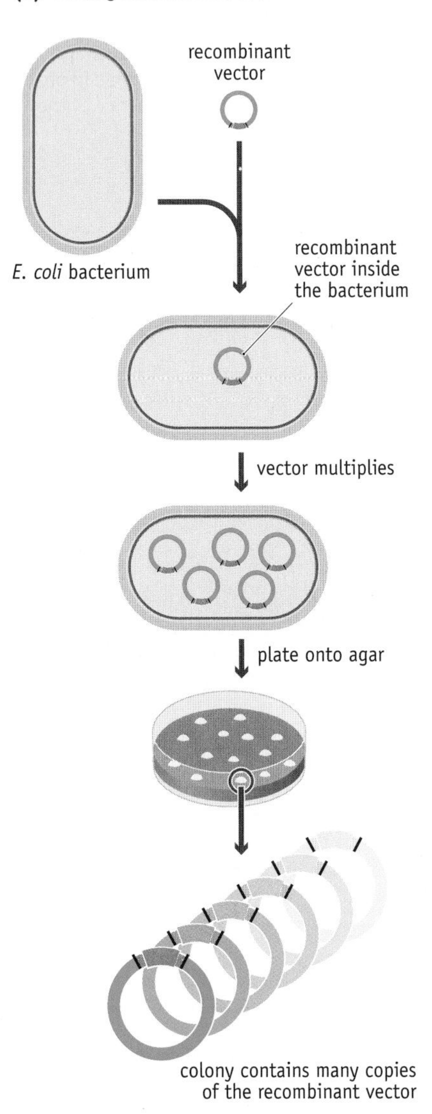

FIGURE 2.18 The steps in DNA cloning. (A) A recombinant DNA molecule is constructed by ligation of a DNA fragment into a cloning vector. (B) In the actual cloning stage, the recombinant vector is taken up by a bacterium. The recombinant vector is then replicated until there are several hundred copies per cell. On a solid agar medium, the bacterium will divide many times to produce a colony. The colony will contain many copies of the recombinant vector, these molecules being the direct descendents of the molecule taken up by the original bacterium.

In the next stage of the cloning procedure, the molecules resulting from ligation are mixed with *E. coli* cells, which have been chemically treated so that they are **competent**, a term used to describe cells that are able to take up DNA molecules from their environment (Figure 2.18B). Some of the molecules will be taken up by the bacteria—but usually no more than one molecule by any individual cell. Linear molecules, and circular molecules that lack the vector sequences, cannot be propagated inside a bacterium. Circular vectors, on the other hand, including the recombinant ones, are able to replicate, so that multiple copies are produced, exactly how many depending on the identity of the cloning vector, but usually several hundred per cell.

The bacteria are now spread onto a solid agar medium so that individual cells can divide and give rise to colonies. Each colony will contain many copies of the vector, these molecules being the direct descendents of the original molecule taken up by the bacterium that gave rise to the colony. If that original molecule was a recombinant vector, then the colony will contain multiple copies of a single DNA fragment, separate from all the other fragments that were present in the starting mixture. Other colonies will contain other cloned fragments. This collection of colonies is the clone library.

Each fragment in a clone library is individually sequenced

There are several methods for DNA sequencing, but most experiments are carried out using the **chain termination** method, which was first developed in the 1970s and has been used, with modifications and improvements, ever since. In the chain termination method, the sequence is obtained by making copies of a cloned DNA fragment in a way that enables the position of each of the four nucleotides to be identified. Chain termination sequencing therefore makes use of a purified DNA polymerase, the type of enzyme that makes copies of DNA molecules.

First the recombinant vector molecules are purified from one of the clones in the clone library. A short piece of DNA called an **oligonucleotide**, usually about 20 nucleotides in length, is then attached to the vector part of the recombinant DNA molecule, just adjacent to the site into which the cloned fragment has been inserted (Figure 2.19A). This oligonucleotide acts as a **primer** for synthesis of a new DNA strand. The new strand is made by the DNA polymerase and extends over the cloned fragment, so that a copy is made of this fragment.

DNA polymerases can make polynucleotides that are several thousand nucleotides in length, but in a sequencing experiment synthesis of the new strand is terminated prematurely. This occurs because as well as the four standard deoxynucleotides, a small amount of each of four **dideoxynucleotides** (ddNTPs—ddATP, ddCTP, ddGTP, and ddTTP) is added to the reaction. A dideoxynucleotide differs from a deoxynucleotide in that the –OH group

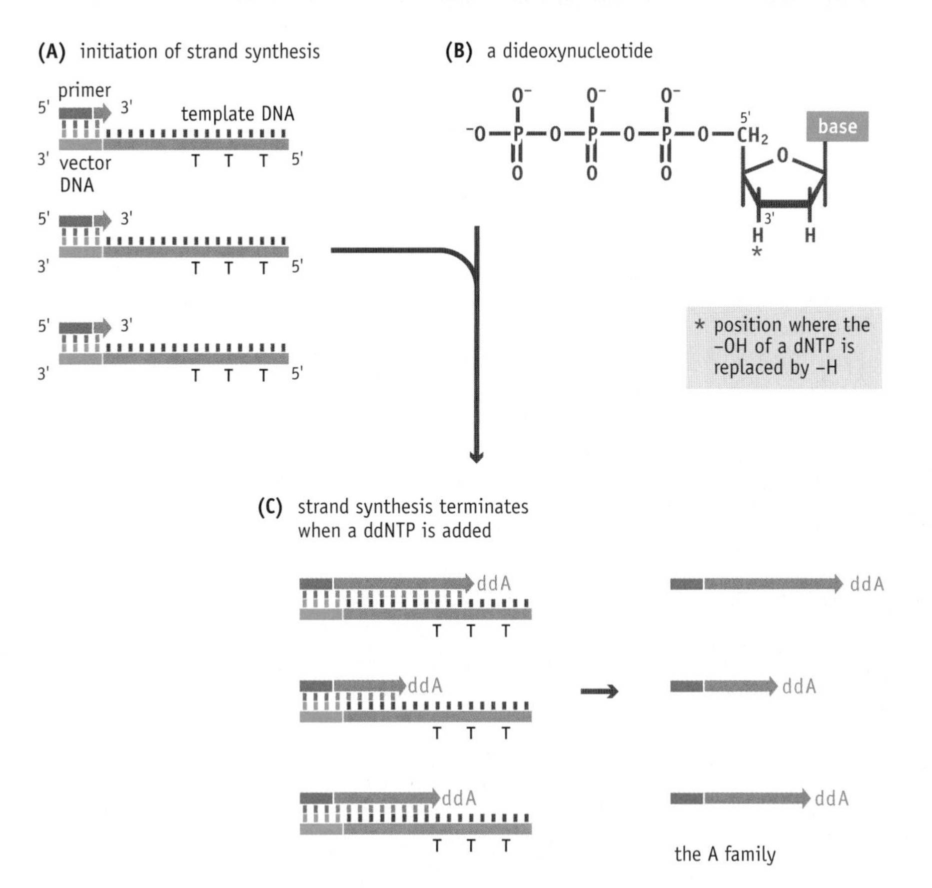

FIGURE 2.19 The steps in chain termination DNA sequencing. (A) Chain termination sequencing involves the synthesis of new strands of DNA that are complementary to a single-stranded template. Strand synthesis begins at the primer, and proceeds in the direction indicated by the orange arrow. (B) Strand synthesis does not proceed indefinitely because the reaction mixture contains small amounts of each of the four dideoxynucleotides. These block further elongation because they have a hydrogen atom rather than a hydroxyl group attached to the 3′ carbon. (C) Incorporation of ddATP results in chains that are terminated opposite Ts in the template. This generates the A family of terminated molecules. Incorporation of the other dideoxynucleotides generates the C, G, and T families.

at the 3′ position is replaced by an H. The polymerase enzyme does not discriminate between deoxy- and dideoxynucleotides, but once incorporated a dideoxynucleotide blocks further elongation because it lacks the 3′-OH group needed to form a connection with the next nucleotide (Figure 2.19B).

Because the normal deoxynucleotides are also present, in larger amounts than the dideoxynucleotides, the strand synthesis does not always terminate close to the primer. In fact, several hundred nucleotides might be polymerized before a dideoxynucleotide is eventually incorporated. The result is a set of "chain-terminated" molecules, all of different lengths, and each ending in a dideoxynucleotide whose identity indicates the nucleotide—A, C, G, or T—that is present at the equivalent position in the DNA that has been copied (Figure 2.19C).

Capillary gel electrophoresis is used to read the sequence

To read the DNA sequence, all that we have to do is identify the dideoxynucleotide at the end of each chain-terminated molecule. This might sound difficult, but in fact it is easy. The technique involved is called **capillary gel electrophoresis**.

A capillary gel is a long, thin tube of **polyacrylamide**, typically 80 cm in length with a diameter of 0.1 mm, contained in a plastic sleeve (Figure 2.20A). The mixture of chain-terminated molecules is placed at one end of the capillary and an electric current is applied. The migration rate of each molecule through the capillary gel depends on its electric charge. Each phosphodiester bond in a DNA molecule carries a single negative electric charge (see Figure 2.5), so the overall negative charge of a polynucleotide is directly proportional to its length. The chain-terminated molecules therefore migrate through the capillary gel at different rates, the shortest ones

RESEARCH BRIEFING 2.2

Genes are made of DNA

Today we are so familiar with the fact that DNA is the genetic material that it comes as quite a surprise to learn that this idea was considered ridiculous by most biologists until the 1940s, and that experimental proof that human genes are made of DNA was not obtained until the 1970s. Why did it take so long to establish this fundamental fact of genetics?

At first it was thought that genes might be made of protein

The first speculations about the chemical nature of genes were prompted by the discovery in the very early years of the twentieth century that genes are contained in chromosomes. **Cytochemistry**, in which cells are examined under the microscope after staining with dyes that bind specifically to just one type of biochemical, showed that chromosomes are made of DNA and protein, in roughly equal amounts. One or the other must therefore be the genetic material.

In deciding whether it was protein or DNA, biologists considered the properties of genes and how these properties might be provided for by the two types of compound. The most fundamental requirement of the genetic material is that it be able to exist in an almost infinite variety of forms. Each cell contains a large number of genes, several thousand in the simplest bacteria, and tens of thousands in higher organisms. Each gene specifies a different biological characteristic, and each presumably has a different structure. The genetic material must therefore have a great deal of chemical variability.

This requirement appeared not to be satisfied by DNA, because in the early part of the twentieth century it was thought that all DNA molecules were the same. On the other hand, it was known, correctly, that proteins are highly variable polymeric molecules, each one made up of a different combination of 20 chemically distinct amino acids. There are many different proteins, distinct from one another by virtue of their different amino acid sequences. Proteins therefore possess the variability that would be required by the genetic material. Not surprisingly, biologists during the first half of the twentieth century concluded that genes were made of protein and looked on the DNA component of chromosomes as much less important—perhaps a structural material, needed to hold the protein "genes" together.

Two experiments suggested that genes might be made of DNA

The errors regarding DNA structure lingered on, but by the late 1930s it had become accepted that DNA, like protein, has immense variability. The notion that protein was the genetic material initially remained strong, but was eventually overturned by the results of two experiments.

The first of these was carried out by Oswald Avery, Colin MacLeod, and Maclyn McCarty, of Columbia University, New York. They studied what researchers of the time were calling the **transforming principle**. They prepared an extract from dead cells of *Streptococcus pneumoniae*, a bacterium that causes pneumonia. Something in the extract was known to transform a harmless strain of *S. pneumoniae* into one capable of causing the disease. This transforming principle must therefore contain genes that provide the bacteria with the biological characteristics they need to cause pneumonia. Avery and his colleagues showed in 1944 that the active component of the extract, the transforming principle, is DNA (Figure 1).

The second experiment was performed in 1952 by Alfred Hershey and Martha Chase, at Cold Spring Harbor, New York. They showed that when a bacterium is infected with a virus, the DNA of the virus enters the cell, but most of the virus protein stays outside (Figure 2). This was a vital observation because, during the infection cycle, the genes of the infecting viruses are used to direct synthesis of new viruses, and this synthesis occurs within the bacterial cells. If it is the DNA of the infecting viruses that enters the

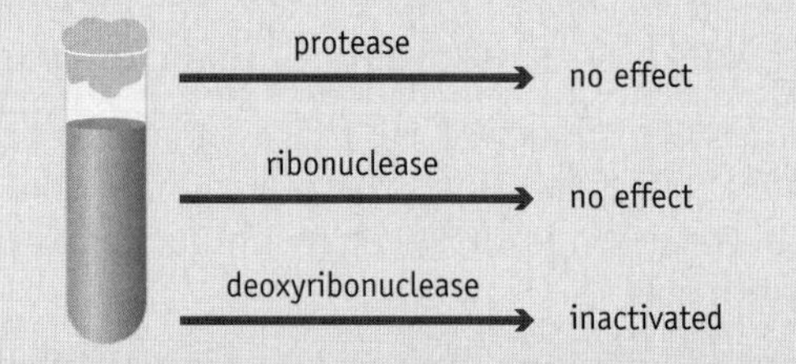

FIGURE 1 Avery and his colleagues treated extracts containing the transforming principle with a protease enzyme, which specifically degrades protein, with a ribonuclease, which breaks down RNA, and with a deoxyribonuclease, which degrades DNA. The protease and the ribonuclease had no effect on the ability of the extract to transform harmless *S. pneumoniae* bacteria. The deoxyribonuclease, on the other hand, inactivated the transforming principle. The active component of the transforming principle must therefore be DNA.

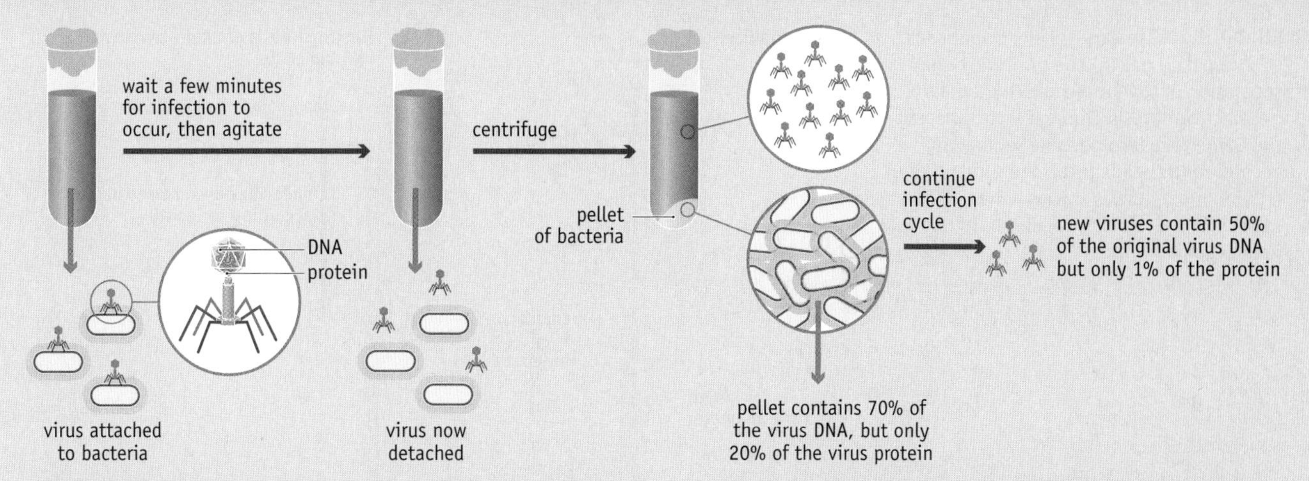

FIGURE 2 The Hershey–Chase experiment showed that the genes of T2 bacteriophage, a type of virus that infects *E. coli*, are made of DNA. Hershey and Chase knew that T2 bacteriophage was made of DNA and protein. We now know that the DNA contained a protein "head," which is attached to a "body" and "legs," also made of protein. Hershey and Chase added some T2 bacteriophages to a culture of *E. coli* bacteria, and then waited a few minutes to allow the viruses to inject their genes into the cells. They then agitated the culture in a blender in order to detach the empty virus particles from the surfaces of the bacteria. The culture was then centrifuged, which collects the bacteria plus virus genes as a pellet at the bottom of the tube, but leaves the empty virus particles in suspension. Hershey and Chase found that the bacterial pellet contained most of the viral DNA, but only 20% of the viral protein. Hershey and Chase also allowed some of the bacteria to complete their infection cycles. The new viruses that were produced inherited 50% of the DNA from the original T2 bacteriophages, but only 1% of the protein.

cells, then it follows that the virus genes must be made of DNA. Hershey and Chase also showed that the new viruses that are produced contain DNA, but only small amounts of protein, from the original infecting viruses. In other words, the new viruses inherit DNA but not protein from their parents.

The double helix convinced biologists that genes are made of DNA

Although from our perspective the Avery and the Hershey–Chase experiments provided the key results to tell us that genes are made of DNA, biologists at the time were not so easily convinced. Both experiments had limitations that enabled skeptics to argue that protein could still be the genetic material. For example, there were worries about the specificity of the **deoxyribonuclease** enzyme that Avery used to inactivate the transforming principle. This result, a central part of the evidence that the transforming principle is DNA, would be invalid if the enzyme contained trace amounts of a contaminating **protease** and so was also able to degrade protein.

The Hershey–Chase experiment is less open to criticism, but Hershey and Chase did not believe that it proved that genes are made of DNA. In the paper describing their results, they state, "The chemical identification of the genetic part [of the virus] must wait, however, until some questions ... have been answered."

Whether these two experiments actually proved that genes are made of DNA is not really important. They made biologists much more receptive to the notion that DNA *might* be the genetic material. This meant that when the double-helix structure of DNA was discovered by Watson and Crick in 1953, revealing how genes can replicate, the scientific world immediately accepted that genes really are made of DNA.

It was not proved that human genes are made of DNA until cloning was invented

If we wish to apply the most rigorous scientific principles, then we might argue that the Avery experiment shows that bacterial genes are made of DNA, and the Hershey–Chase experiment does the same for virus genes, but neither tells us anything about the genes of higher organisms such as humans.

It was not until 1979 that an experiment was carried out that showed that human genes are made of DNA. David Goeddel, working for the biotechnology company Genentech, used DNA cloning to transfer the gene for somatotrophin, one of the human growth hormones, into *Escherichia coli*. The bacteria acquired the ability to make somatotrophin, showing that the DNA that had been transferred did indeed contain the human gene. The purpose of this experiment was to develop a means of using bacteria to produce the large amounts of somatotrophin needed for its clinical use in the treatment of growth defects, and it is rightly looked on as one of the major breakthroughs in the development of the biotechnology industry. The fact that it also finally laid to rest the question about the chemical nature of the genetic material, first asked some 75 years earlier, went unnoticed.

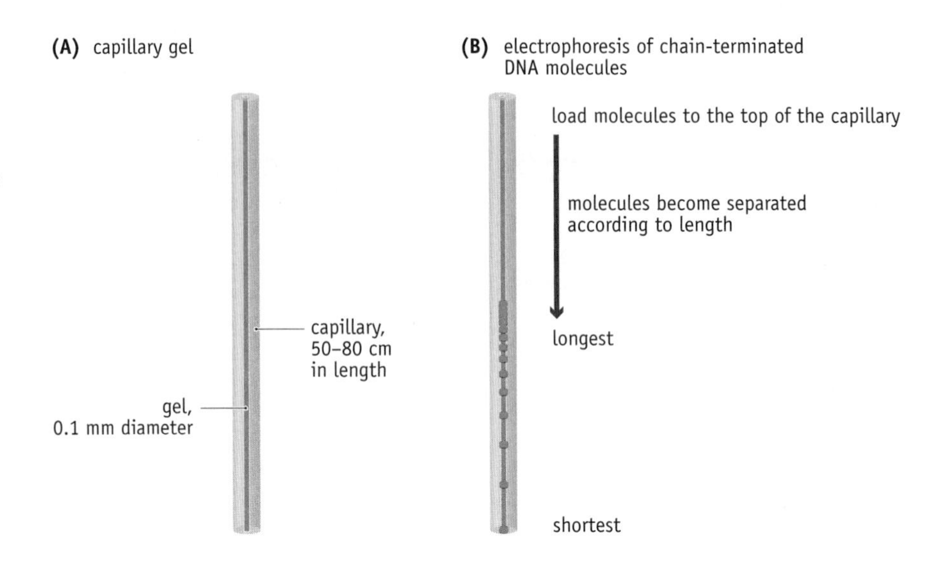

FIGURE 2.20 Capillary gel electrophoresis. (A) A capillary gel. (B) The principle behind separation of DNA molecules by capillary gel electrophoresis. During electrophoresis, the DNA molecules become separated according to their lengths, the shortest ones moving most quickly through the gel. The molecules therefore form a series of bands in the capillary tube.

most quickly and the longer ones the slowest. By the time they get to the end of the capillary they have formed bands, each band containing the molecules that are one nucleotide longer than those in the preceding band (Figure 2.20B).

Each of the four dideoxynucleotides emits a fluorescent signal of a different wavelength. A fluorescence detector is therefore used to identify which of the four dideoxynucleotides is present in each band. By identifying whether the dideoxynucleotide is an A, a C, a G, or a T, the detector is able to build up the DNA sequence of the cloned fragment nucleotide by nucleotide (Figure 2.21). The sequence can be printed out for examination by the operator, or entered directly into a storage device for future analysis.

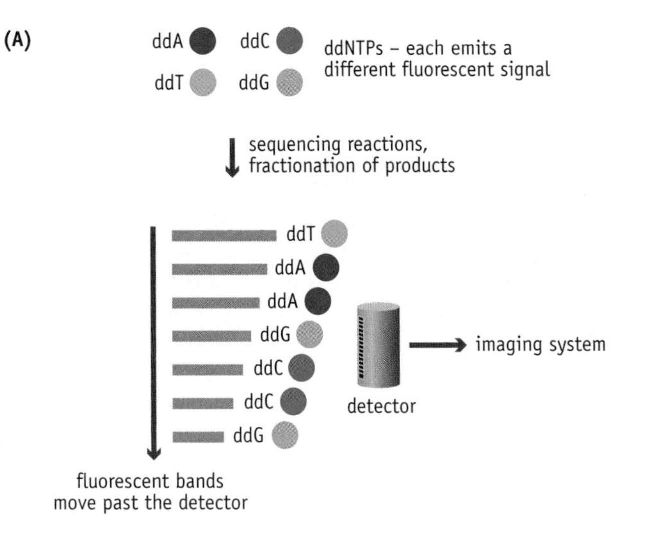

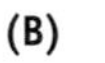

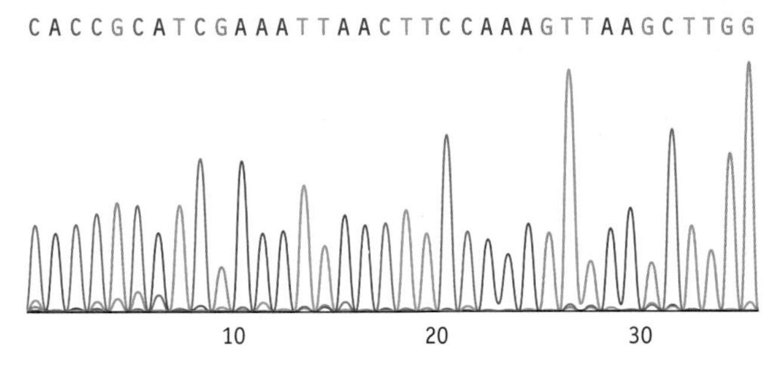

FIGURE 2.21 Reading the sequence generated by a chain termination experiment. (A) Each dideoxynucleotide emits a different fluorescent signal. During electrophoresis, the chain-terminated molecules move past a fluorescence detector, which identifies which dideoxynucleotide is present in each band. The information is passed to the imaging system. (B) A DNA sequencing printout. The sequence is represented by a series of peaks, one for each nucleotide position. In this example, a red peak is an A, blue is a C, orange is a G, and green is a T.

As mentioned above, a maximum of about 750 bp of sequence can be obtained in a single experiment. Automated sequencers with multiple capillaries working in parallel can read up to 96 different sequences in a 2-hour period, which means that 864,000 bp of information can be generated per machine per day. This, of course, requires round-the-clock technical support, with robotic devices used to prepare the sequencing reactions and to load the chain-terminated molecules into the sequencers. If such a factory approach can be established and maintained, then the data needed to sequence DNA molecules that are millions of base pairs in length can be generated in just a few weeks.

KEY CONCEPTS

- The structure of DNA reveals how genes are able to play their dual roles as units of biological information and as units of inheritance.
- DNA is a polymer in which the individual units are called nucleotides. There are four different nucleotides, usually referred to as A, C, G, and T, the abbreviations of their full chemical names. Nucleotides can be linked together in any order.
- The sequence of nucleotides in a DNA molecule is a language made up of the four letters A, C, G, and T. The biological information contained in a gene is written in this language.
- The double-helix structure reveals how DNA molecules are able to replicate. The two strands of the double helix are held together by hydrogen bonds, in such a way that A can base-pair only with T and G can pair only with C. This means that if the two polynucleotides in a DNA molecule are separated, then two perfect copies of the parent double helix can be made as the sequences of these preexisting strands dictate the sequences of new complementary strands.
- There are variations of the double-helix structure, called A-, B-, and Z-DNA. B-DNA is the most common type in the cell.
- Modern techniques of DNA sequencing are able to read the order of nucleotides in DNA molecules. Only a short sequence, of up to 750 bp, can be obtained in a single experiment, but automated systems enable many experiments to be carried out at once.

QUESTIONS AND PROBLEMS (Answers can be found at www.garlandscience.com/introgenetics)

Key Terms

Write short definitions of the following terms:

3′-OH terminus
5′-P terminus
β-*N*-glycosidic bond
adenine
A-form
antiparallel
base pairing
B-form
capillary gel electrophoresis
chain termination
clone library
competent
complementary
contigs
cytochemistry
cytosine
deoxyribonuclease
deoxyribonucleic acid
dideoxynucleotide
DNA cloning
DNA ligase
DNA polymerase
DNA sequence
DNA sequencing
gene expression
genetic code
genetic material

guanine
hydrogen bond
major groove
minor groove
model building
molecular approach
monomer
nucleoside
nucleotide
oligonucleotide
paper chromatography
pentose
phosphodiester bond
plasmid
polyacrylamide
polymer
polynucleotide
primer
protease
purine
pyrimidine
recombinant plasmid
sonication
template-dependent DNA synthesis
thymine
transforming principle
vector
X-ray diffraction pattern
Z-DNA

Self-study Questions

2.1 Draw the structure of a nucleotide.

2.2 What are the complete chemical names of the four nucleotides found in DNA molecules?

2.3 Draw a fully annotated diagram of the structure of a short DNA polynucleotide containing each of the four nucleotides.

2.4 Explain why the two ends of a polynucleotide are chemically distinct.

2.5 Outline the two major types of experimental analysis that laid the foundations for the deduction of the structure of DNA by Watson and Crick.

2.6 What are the important features of the double-helix structure?

2.7 What is meant by complementary base pairing, and why is it important?

2.8 If the sequence of one polynucleotide of a DNA double helix is 5′-ATAGCAATGCAA-3′, what is the sequence of the complementary polynucleotide?

2.9 Thirty percent of the nucleotides in the DNA of the locust are As. What are the percentage values for: (a) T, (b) G + C, (c) G, (d) C?

2.10 DNA from the fungus *Neurospora crassa* has an AT content of 46%. What is the GC content?

2.11 What are the main differences between the A- and B-forms of DNA?

2.12 Describe how Z-DNA differs from the A- and B-forms, and outline one possible function of Z-DNA in living cells.

2.13 Explain how DNA provides the variability needed by the genetic material.

2.14 Draw a diagram to illustrate the process called template-dependent DNA synthesis.

2.15 Explain why in the laboratory a long DNA molecule is initially sequenced as a set of shorter contigs.

2.16 Outline how a clone library is prepared.

2.17 Describe how a chain termination DNA sequencing experiment is carried out.

2.18 Explain how a DNA sequence is read by capillary gel electrophoresis.

2.19 Why did biologists originally think that protein is the genetic material?

2.20 Outline the two experiments carried out in the 1940s and 1950s that indicated that genes are made of DNA.

2.21 To what extent is the statement "Genes are made of DNA" consistent with the results of the Avery and Hershey–Chase experiments?

Discussion Topics

2.22 Is the statement "Genes are made of DNA" universally correct?

2.23 An A–T base pair is held together by two hydrogen bonds and a G–C base pair by three hydrogen bonds. In which parts of the genome might you expect to find AT-rich sequences?

2.24 Discuss why the double helix gained immediate universal acceptance as the correct structure for DNA.

2.25 The human genome has a GC content of 40.3%. In other words, 40.3% of the nucleotides in the genome are either G or C. The GC contents for different organisms vary over a wide range. The genome of the malaria parasite, *Plasmodium falciparum*, has a GC content of just 19.0%, whereas that of the bacterium *Streptomyces griseolus* is 72.4%. Speculate on the reasons why the GC contents for different species should be so different.

2.26 Explore the reasons why, in the early twentieth century, some biologists thought that genes were abstract entities invented by geneticists to explain how biological characteristics are passed from parents to offspring.

2.27 In Research Briefing 2.2, we noted that a formal proof that human genes are made of DNA was not obtained until 1979 when a human DNA molecule was first transferred into *E. coli* and shown to direct synthesis of a human protein. What other experiments could be carried out to demonstrate that DNA is the genetic material in humans?

2.28 The scheme for DNA replication shown in Figure 2.14 is the same as that proposed by Watson and Crick immediately after their discovery of the double-helix structure. Many biologists thought that this process would be impossible in a living cell, especially for the circular DNA molecules present in many bacteria. Why was this?

2.29 A DNA polymerase builds up a new polynucleotide by adding nucleotides one by one to the 3′ end of the growing strand. Enzymes that make DNA in the opposite direction, by adding nucleotides to the 5′ end, are unknown. This fact complicates the process by which a double-stranded DNA molecule is replicated. Explain.

FURTHER READING

Brown TA (2010) Gene Cloning and DNA Analysis: An Introduction. Oxford: Wiley-Blackwell. *Includes details of DNA cloning and sequencing.*

Hershey AD & Chase M (1952) Independent functions of viral protein and nucleic acid in growth of bacteriophage. *J. Gen. Physiol.* 36, 39–56. *One of the original papers that showed that genes are made of DNA.*

Maddox B (2002) Rosalind Franklin: The Dark Lady of DNA. London: HarperCollins. *A biography of one of the key people involved in discovery of the double-helix structure, who sadly died just a few years later.*

McCarty M (1985) The Transforming Principle: Discovering That Genes Are Made of DNA. London: Norton. *Personal account by one of the scientists who worked with Avery.*

Olby R (1974) The Path to the Double Helix. London: Macmillan. *A scholarly account of the research that led to the discovery of the double helix.*

Prober JM, Trainor GL, Dam RJ et al. (1987) A system for rapid DNA sequencing with fluorescent chain-terminating dideoxynucleotides. *Science* 238, 336–341. *The chain termination method for DNA sequencing, as it is used today.*

Rich A & Zhang S (2003) Z-DNA: The long road to biological function. *Nat. Rev. Genet.* 4, 566–572.

Sanger F, Nicklen S & Coulson AR (1977) DNA sequencing with chain terminating inhibitors. *Proc. Natl. Acad. Sci. USA* 74, 5463–5467. *The first description of chain termination sequencing.*

Watson JD (1968) The Double Helix. London: Atheneum. *The most important discovery of twentieth-century biology, written as a soap opera.*

Watson JD & Crick FHC (1953) Molecular structure of nucleic acids: A structure for deoxyribose nucleic acid. *Nature* 171, 737–738. *The scientific report of the discovery of the double-helix structure.*

Genes CHAPTER 3

In Chapter 2 we studied the structure of DNA and asked how this structure enables genes to play their two roles in living cells, as units of biological information and as units of inheritance. We learned that biological information is encoded in the nucleotide sequence of a DNA molecule, and that complementary base pairing enables DNA molecules to be replicated so that copies can be passed from parents to offspring. Now we must focus more closely on the genes themselves. Knowing that biological information can be encoded in a nucleotide sequence is only the first step in understanding how that information is used by the cell. To take the next step forward we must examine the units of biological information that are contained in individual genes.

The first question that we will address in this chapter is the most fundamental of all. What is the nature of the information contained in an individual gene? We know that information is encoded in the form of a nucleotide sequence, but what is this information for? Answering this question will introduce us to **gene expression**, the process by which the information contained in a gene is utilized by the cell.

The second question we address is one that puzzled the early geneticists over a century ago. Why are the members of a species subtly different from one another even though they all possess the same set of genes? Why, for example, do some people have blue eyes when others have brown? In answering this question we will discover that small alterations in the nucleotide sequence can inactivate a gene or alter the precise nature of the information that it contains.

Finally, we will ask why some genes are present in multiple copies. In answering this question we will begin to understand how genes evolve.

3.1 THE NATURE OF THE INFORMATION CONTAINED IN GENES

We know that the information contained in a gene is encoded in its nucleotide sequence. But what is the information for? Before we can begin to answer this question we must take a closer look at the molecular structures of individual genes.

Genes are segments of DNA molecules

There are many more genes in a cell than there are DNA molecules. Humans, for example, have over 20,000 genes but only 25 different DNA molecules. A single DNA molecule therefore carries a large number of genes. In humans the number ranges from 37 for the shortest of the 25 DNA molecules to over 3000 for the longest. A gene is therefore a segment of a DNA molecule.

The shortest genes are less than 100 base pairs (bp) in length, and the longest over 2,400,000 bp. To simplify these larger numbers we use the terms **kilobase pair** (kb) for 1000 bp and **megabase pair** (Mb) for 1 million bp. According to this notation the longest genes are over 2400 kb, or 2.4 Mb, in length. But such huge genes are quite uncommon—in the human genome the average gene length is 10 to 15 kb. The genes are separated from one another within the DNA molecule by **intergenic DNA** (Figure 3.1). Again, there is enormous variation in the actual distances between genes, but in the human genome the average is 25 to 30 kb.

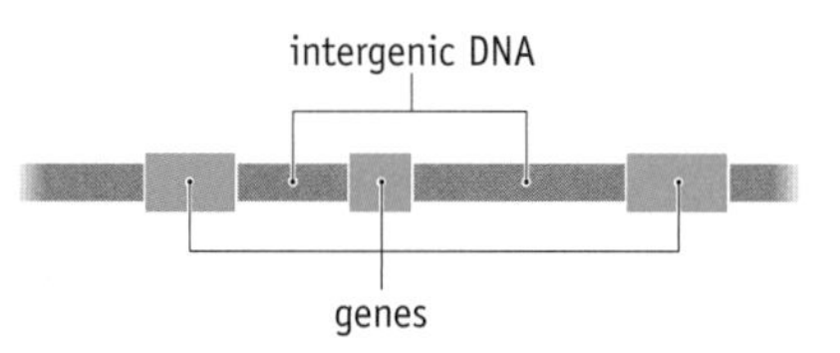

FIGURE 3.1 Genes are separated from one another by intergenic DNA.

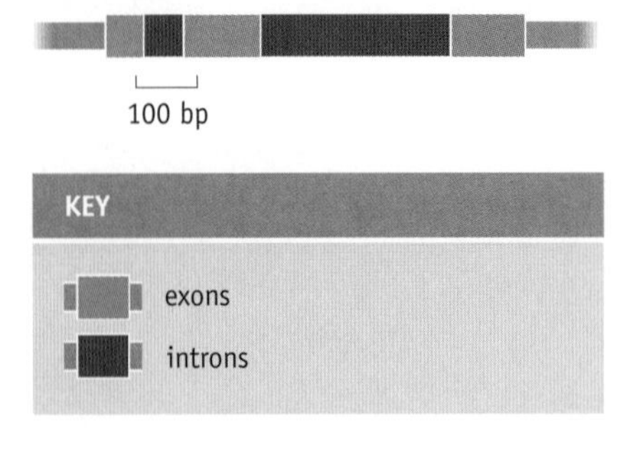

FIGURE 3.2 Exons and introns in a discontinuous gene.

Describing the range of gene lengths as between 100 bp and 2.4 Mb is rather misleading when our focus is on the information content of individual genes. Merely stating the length difference might imply that the longest genes contain 24,000 times as much information as the shortest, which is not the case. This is because the lengths of many genes are inflated by the presence within them of long tracts of DNA that do not contain any biological information (Figure 3.2). In a **discontinuous gene** (also called a split or mosaic gene) the sections containing biological information are called **exons** and the intervening sequences are referred to as **introns**. Discontinuous genes are common in higher organisms. More than 95% of all human genes contain at least one intron, and the average number is nine.

In most discontinuous genes, the introns are much longer than the exons. In the longest genes, the introns added together make up over 90% of the length of the gene. These genes therefore contain much less biological information than suggested by their overall lengths. When just the exons are considered, the longest "gene" that has so far been discovered is 103 kb, and the average length of "genes" in all organisms is about 1.2 kb. The average unit of biological information can therefore be encoded in a nucleotide sequence just 1200 letters in length.

Genes contain instructions for making RNA and protein molecules

Genes contain biological information, but on their own they are unable to release that information to the cell. Its utilization requires the coordinated activity of **enzymes** and other kinds of proteins, which participate in the series of events that make up gene expression.

Gene expression is conventionally looked on as a two-stage process (Figure 3.3). All genes undergo the first stage of gene expression, which is called **transcription** and which results in synthesis of an RNA molecule. Transcription is a simple copying reaction. RNA, like DNA, is a polynucleotide, the only chemical differences being that in RNA the sugar is ribose rather than 2′-deoxyribose, and that the base thymine is replaced by the base uracil (U), which, like thymine, base-pairs with adenine (Figure 3.4). During transcription of a gene, one strand of the DNA double helix acts as

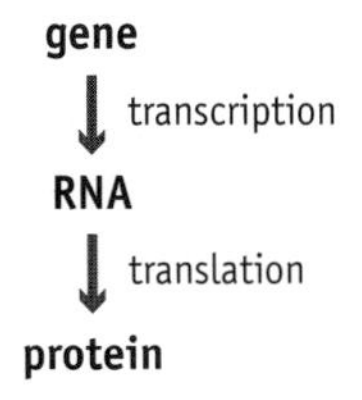

FIGURE 3.3 The conventional representation of gene expression as a two-stage process.

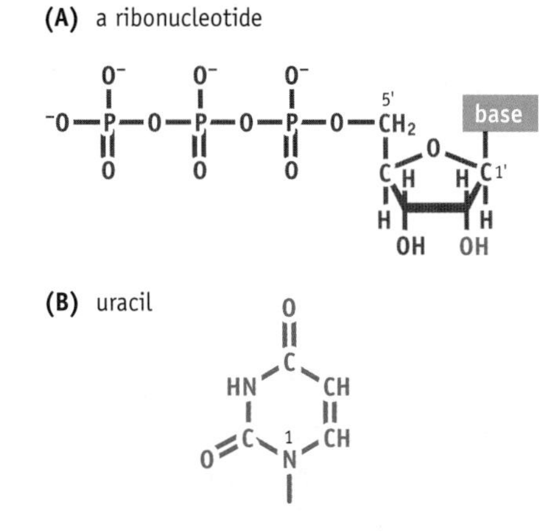

FIGURE 3.4 The chemical differences between DNA and RNA. (A) RNA contains ribonucleotides, in which the sugar is ribose rather than 2′-deoxyribose. The difference is that a hydroxyl group rather than a hydrogen atom is attached to the 2′ carbon. (B) RNA contains the pyrimidine called uracil instead of thymine.

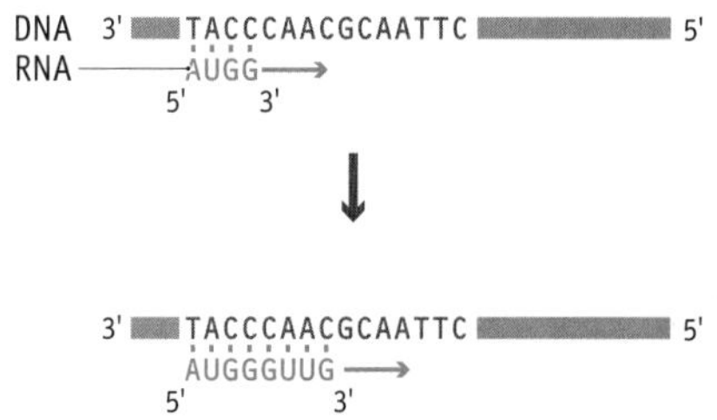

DNA 3' TACCCAACGCAATTC 5'
RNA AUGG
5' 3'

3' TACCCAACGCAATTC 5'
AUGGGUUG
5' 3'

FIGURE 3.5 Template-dependent RNA synthesis. The RNA molecule, shown in blue, is extended by adding nucleotides one by one to its 3′ end. The process is very similar to template-dependent DNA synthesis, which was illustrated in Figure 2.15.

a template for synthesis of an RNA molecule whose nucleotide sequence is determined, by the base-pairing rules, by the DNA sequence (Figure 3.5).

For some genes the RNA transcript is itself the end product of gene expression. For others the transcript is a short-lived message that directs a second stage of gene expression, called **translation**. During translation the RNA molecule (called a **messenger RNA** or an **mRNA**) directs synthesis of a protein. A protein is another type of polymeric molecule but quite different from DNA and RNA. In a protein the monomers are called amino acids, and there are 20 different ones, each with its own specific chemical properties. When a protein is made by translation, its amino acid sequence is determined by the nucleotide sequence of the mRNA. Each triplet of adjacent ribonucleotides specifies a single amino acid of the protein, the identity of the amino acid corresponding to each triplet being set by the **genetic code** (Figure 3.6).

Protein synthesis is the key to expression of biological information

For all genes the endpoint of gene expression is synthesis of either an RNA molecule or a protein molecule. How can this simple process enable the information contained in the genes to specify the biological characteristics of a living organism?

The answer lies with proteins. Proteins with different amino acid sequences can have quite different chemical properties that enable them to play a variety of roles (Table 3.1). Some amino acid sequences, for example, result in fibrous proteins that give rigidity to the framework of an organism. Collagen is an example of one of these **structural proteins** (Figure 3.7). Collagen is found in the bones, tendons, and cartilage of vertebrates, making up about 20% of the dry weight of the skeleton. Without their collagen component, bones would be fragile and unable to support the body mass. A second example is keratin, present in hair and feathers, and also a major component of the exoskeleton of arthropods such as crabs. In contrast, **motor proteins** have amino acid sequences that give them flexibility. They

TABLE 3.1 THE FUNCTIONAL DIVERSITY OF PROTEINS

Type of protein	Examples
Structural proteins	Collagen, keratin
Motor proteins	Myosin, dynein
Catalytic proteins (enzymes)	Hexokinase, DNA polymerase
Transport proteins	Hemoglobin, serum albumin
Storage proteins	Ovalbumin, ferritin
Protective proteins	Immunoglobulins, thrombin
Regulatory proteins	Insulin, somatostatin, somatotrophin

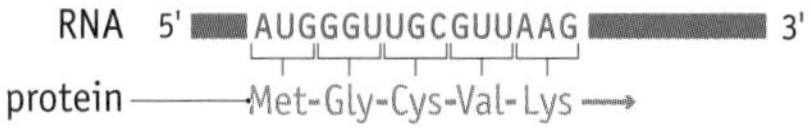

RNA 5' AUGGGUUGCGUUAAG 3'
protein Met-Gly-Cys-Val-Lys

FIGURE 3.6 During translation, each triplet of nucleotides in the RNA specifies a particular amino acid, the identity of this amino acid being set by the genetic code. The amino acids indicated here by their three-letter abbreviations are methionine, glycine, cysteine, valine, and lysine.

FIGURE 3.7 Part of a molecule of collagen, a structural protein. A collagen molecule is made up of three protein polymers wrapped around each other in a triple helix. This helical arrangement gives the protein a high tensile strength. (After D. S. Goodsell, Our Molecular Nature. New York: Springer-Verlag, 1996. With permission from Springer Science and Business Media.)

are able to change their shape, enabling organisms to move around. The muscle protein myosin is a motor protein, as is dynein in cilia and flagella (Figure 3.8).

Other types of proteins have quite different functions. Enzymes are proteins whose amino acid sequences enable them to catalyze the multitude of cellular reactions that bring about the release and storage of energy and the synthesis of new compounds (Figure 3.9). Other proteins have transport functions and carry compounds around the body, the most important in mammals being hemoglobin, which carries oxygen in the bloodstream (Figure 3.10), and serum albumin, which transports fatty acids. Some proteins help to store molecules for future use by the organism. These include ovalbumin, which stores amino acids in egg white, and ferritin, which stores iron in the liver. A vast range of proteins have protective functions and guard against infectious agents and injury. Examples in mammals are immunoglobulins and other antibodies, which form complexes with foreign molecules (such as proteins on the surface of an invading virus), and thrombin and other components of the blood clotting mechanism (Figure 3.11).

There are also **regulatory proteins** that control cellular activities. These include well-known hormones such as insulin, which regulates glucose metabolism in vertebrates, and the two growth hormones somatostatin and somatotrophin. Hormones are extracellular proteins. Although made inside a cell they are released so they can travel around the body and convey their regulatory messages to other cells (Figure 3.12). Sometimes the message is that certain genes must be switched on and others switched off. Proteins are therefore not just the products of gene expression. Some of them also control the way in which biological information is released to the cell.

Proteins can therefore provide organisms with their structure, their ability to move, and their ability to carry out biochemical reactions. Proteins carry compounds around the body, store those that are not immediately needed, and protect the organism from disease. Proteins also control cellular activities including the gene expression processes that determine which proteins are synthesized. The units of biological information contained in

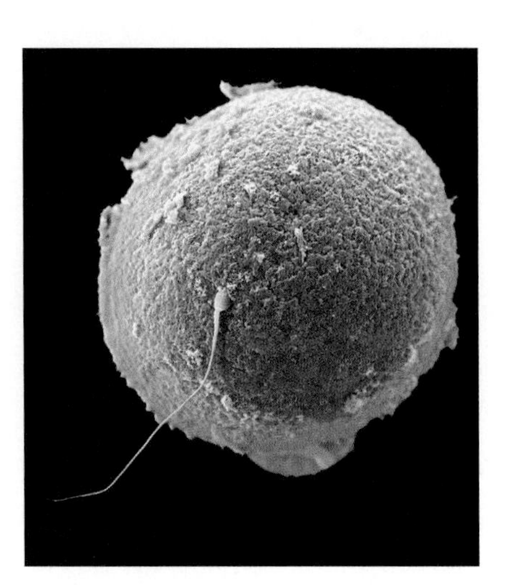

FIGURE 3.8 A single human sperm attempting to fertilize a female egg cell. The sperm cell, shown in blue, has a long, thin flagellum, which contains multiple copies of the motor protein called dynein. By changing their shape, these dynein proteins generate a series of waves that move along the flagellum from its base to its tip. This wavelike motion propels the sperm forward, enabling it to swim in search of the female cell. (Courtesy of Eye of Science/ Science Photo Library.)

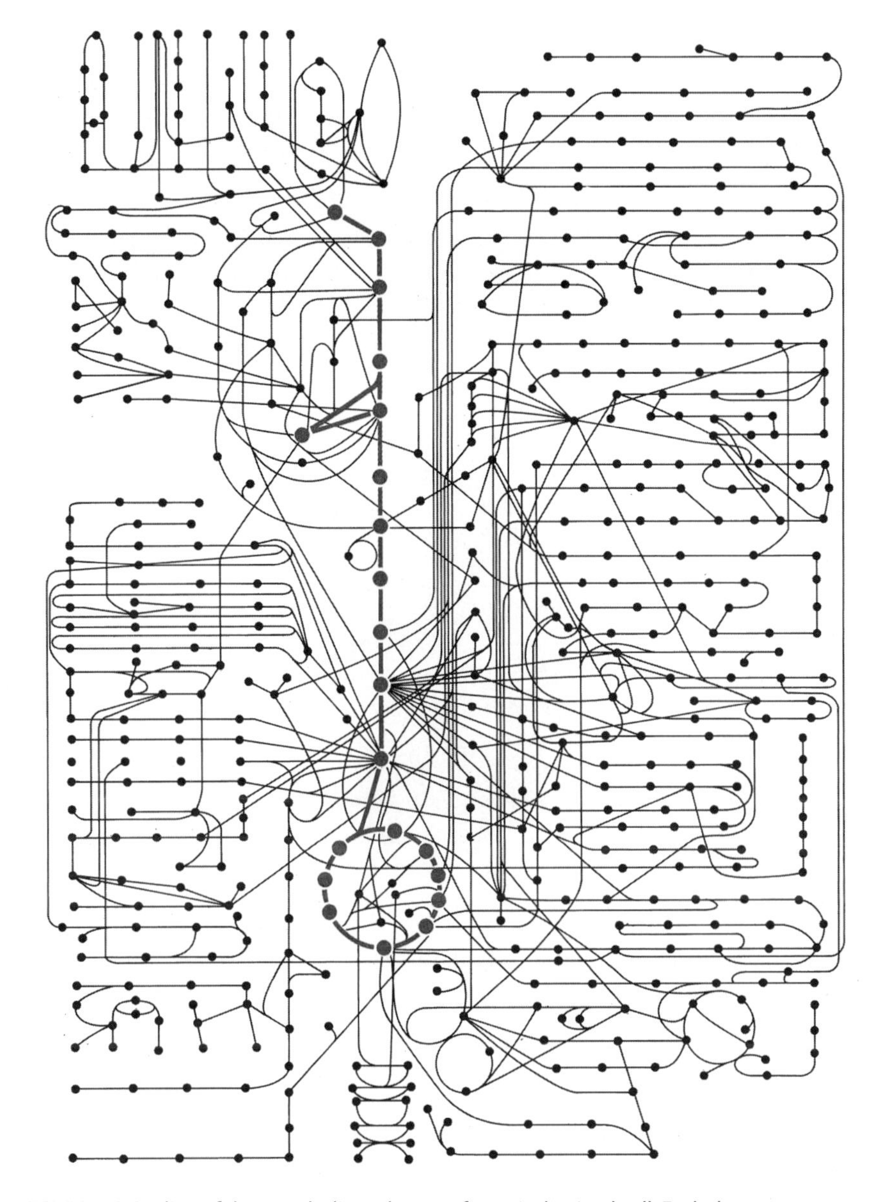

FIGURE 3.9 Outline of the metabolic pathways of a typical animal cell. Each dot represents a different biochemical compound. The lines indicate the steps in the network, each of these steps resulting in conversion of one compound into another. There are approximately 500 steps, each catalyzed by a different enzyme. The glycolysis pathway and the citric acid cycle, which provide energy for metabolism, are shown in red. The pathways shown in black provide substrates for energy generation or build up molecules such as amino acids and nucleotides from smaller precursors. (Adapted from B. Alberts et al., Essential Cell Biology, 3rd ed. New York: Garland Science, 2010.)

genes give rise to this multiplicity of function by encoding the amino acid sequences of all the proteins that a cell is able to make. In this way, the genes are able to specify the biological characteristics of a living organism.

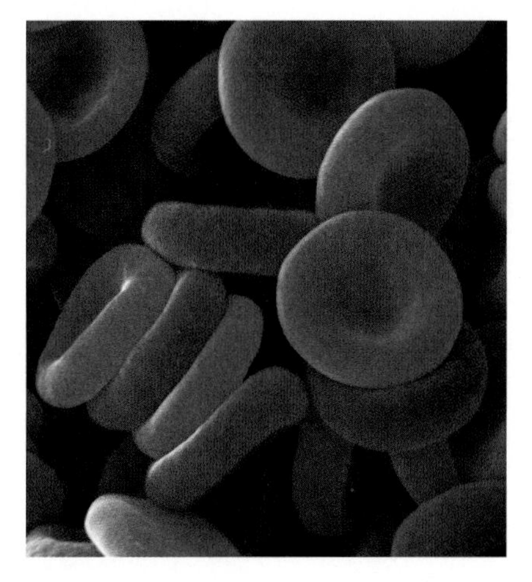

FIGURE 3.10 Red blood cells, which contain large amounts of hemoglobin and hence are able to carry oxygen from one part of the body to another, via the bloodstream. (Courtesy of Power and Syred/Science Photo Library.)

The RNA molecules that are not translated into protein are also important

Only a small fraction of a cell's RNA, usually no more than 4% of the total, is messenger RNA. The vast bulk is **noncoding RNA**, molecules that are not translated into protein but instead play direct roles in the cell as RNA.

FIGURE 3.11 Representation of an immunoglobulin protein, a type of antibody, which forms part of the human body's defense against invading viruses, bacteria, and other infectious agents. (© http://visualscience.ru/en/, 2010.)

Clearly we must know what noncoding RNA does in order to fully understand the nature of the biological information contained in genes.

There are relatively few genes that specify noncoding RNA, even though noncoding RNA is much more abundant than mRNA. This is because noncoding RNA comprises just a few types of molecule (Figure 3.13). The most abundant of all is **ribosomal RNA**, or **rRNA**. There are only four different rRNA molecules in humans, but together they can make up over 80% of the total RNA in an actively dividing cell. These molecules are components of ribosomes, the structures on which protein synthesis takes place. **Transfer RNAs**, or **tRNAs**, are small molecules that are also involved in protein synthesis, carrying amino acids to the ribosome and ensuring these are linked together in the order specified by the nucleotide sequence of the mRNA that

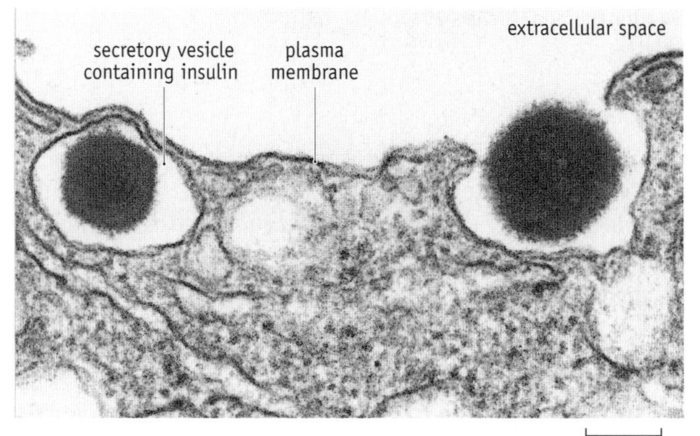

FIGURE 3.12 Electron micrograph showing insulin being secreted by β-cells in the human pancreas. (Courtesy of Lelio Orci, University of Geneva.)

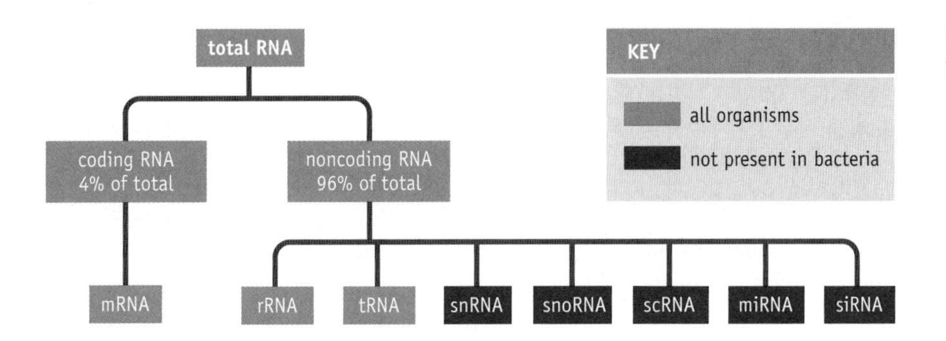

FIGURE 3.13 The various types of RNA that are found in living cells.

is being translated. Most organisms have genes for between 30 and 50 different tRNAs.

Ribosomal and transfer RNAs are present in all organisms. The other noncoding RNA types are more limited in their distribution, and in particular are absent in bacteria (see Figure 3.13). Some of these types of RNA are named according to their locations in the cell. So, for example, there is a category called **small nuclear RNA** (**snRNA**), which is found in the nucleus. There are 15 to 20 different snRNAs, and most are involved in **RNA splicing**, the process by which copies of introns are removed from mRNA molecules and the exons, the parts containing biological information, are joined together. Splicing is an essential part of the expression pathway for a discontinuous gene because the exons have to be linked before the mRNA can be translated into a protein (Figure 3.14).

There are also **small nucleolar RNAs** (**snoRNAs**) found in the nucleoli, the parts of the nucleus in which rRNA is transcribed. Extra chemical groups, such as methyl groups, must be added to rRNA molecules before they can be assembled into ribosomes, and snoRNAs aid this process. Finally, **small cytoplasmic RNAs** (**scRNAs**) are a diverse group including molecules involved in the transport of secreted proteins out of the cell.

This list was thought to be complete until quite recently. Now we know that there are two other categories of noncoding RNA whose existence was not suspected until the 1990s. These are the **microRNAs** (**miRNAs**) and **small interfering RNAs** (**siRNAs**). Both are involved in the control of gene expression, and the more we find out about them the more important they become. They cause certain mRNAs to be "silenced" so they cannot be translated, possibly by attaching to these mRNAs through base pairing and causing them to be degraded. We will learn more about this process, called **RNA interference**, in Chapter 5.

3.2 VARIATIONS IN THE INFORMATION CONTENT OF INDIVIDUAL GENES

It is a fundamental fact of life that not all the members of a species are identical to one another. It is obvious to us that this is the case with humans, and it is equally true for all species.

How is it possible for individuals to have their own distinctive biological characteristics when everybody has the same set of genes? Part of the distinctiveness might be due to environmental factors, dietary differences, and the like, but this is only a small part of the answer. Biological characteristics are variable because the genes that code for these characteristics are themselves variable. Our next task is to understand the molecular basis of this variability.

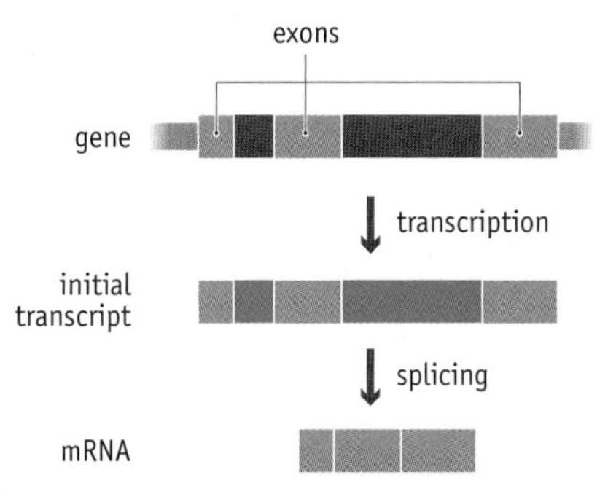

FIGURE 3.14 The mRNA of a discontinuous gene must be spliced before it can be translated into protein.

Genetic variants are called alleles

Genetics began with systematic studies of the inheritance of variable biological characteristics, carried out by Gregor Mendel and others from the mid-nineteenth century onward. These early geneticists coined the word **allele** to describe the variants of a biological characteristic. A good example is the round versus wrinkled peas that Mendel studied. Here, the biological characteristic is "pea shape," and there are two alleles, "round" and "wrinkled" (Figure 3.15A).

We now know that the round and wrinkled alleles are specified by two variants of a gene involved in starch synthesis. The main biochemical difference between round and wrinkled peas is that the la tter have a higher sucrose content, and as a consequence absorb more water while developing in the pod. When the pea reaches maturity some of this water is lost, causing the pea to collapse and become wrinkled. Round peas, which have less sucrose but more starch, do not take up so much water when they are developing and so become less dehydrated when they mature. As a result they stay round (Figure 3.15B).

The key difference between wrinkled and round peas therefore lies in the relative amounts of sucrose and starch that they contain. This in turn depends on the activities of a series of enzymes that in the "normal" (round) pea convert the bulk of the sucrose into starch (Figure 3.15C). Each of these enzymes is, of course, coded by a gene. One of these genes is called *SBE1* and codes for starch branching enzyme type 1 (hence the name). In wrinkled peas this enzyme is absent. This means that less sucrose is converted into starch and the pea collapses in the pod. Why is the enzyme absent? Because in wrinkled peas an extra piece of DNA has become inserted into the *SBE1* gene (Figure 3.15D). This extra piece of DNA interrupts the nucleotide sequence of the gene and prevents it from being expressed correctly. As a result no starch branching enzyme type 1 is made.

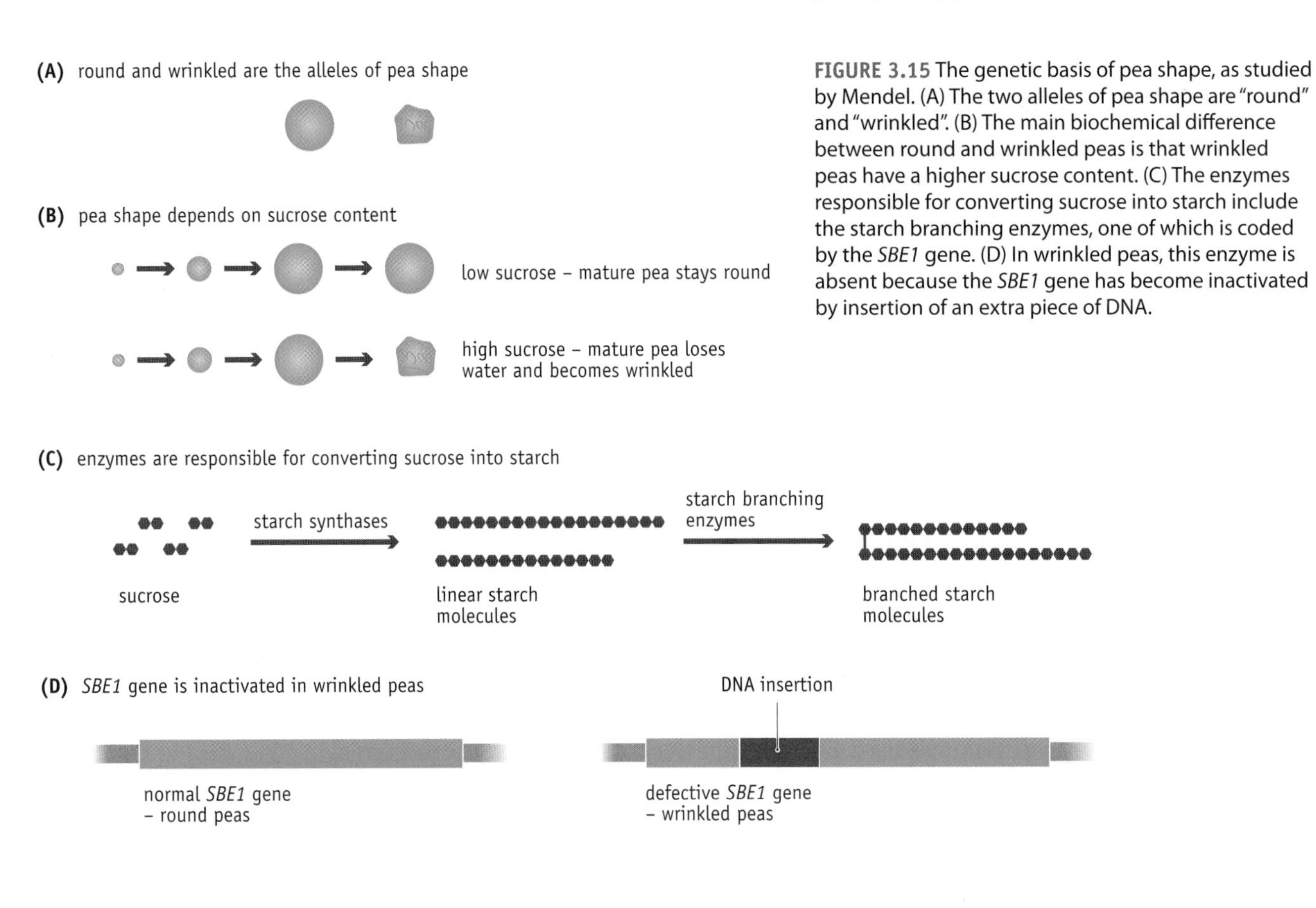

FIGURE 3.15 The genetic basis of pea shape, as studied by Mendel. (A) The two alleles of pea shape are "round" and "wrinkled". (B) The main biochemical difference between round and wrinkled peas is that wrinkled peas have a higher sucrose content. (C) The enzymes responsible for converting sucrose into starch include the starch branching enzymes, one of which is coded by the *SBE1* gene. (D) In wrinkled peas, this enzyme is absent because the *SBE1* gene has become inactivated by insertion of an extra piece of DNA.

We therefore understand the molecular basis for the round and wrinkled alleles of the pea shape characteristic. One allele, for round peas, is specified by the correct, functional copy of the gene, and the second allele, for wrinkled peas, arises when that gene is nonfunctional. Strictly speaking, the term "alleles" refers to the alternative forms of the biological characteristic, not the gene variants. Scientists, like all others who use a specialized nomenclature, sometimes tend to be lazy with commonly used terms, and nowadays we also refer to the two versions of the gene as alleles.

There may be many variants of the same gene

Some genes come in a whole range of variant forms. To explore this added complexity we will look at the human *CFTR* gene, which codes for the cystic fibrosis transmembrane regulator protein. This is a membrane-bound protein involved in transport of chloride ions into and out of human cells. As its name indicates, if this protein does not function correctly then the disease called cystic fibrosis might develop, characterized by inflammation and mucus accumulation in the lungs.

The *CFTR* gene is 250 kb in length, but most of that size is due to 26 introns (Figure 3.16A). If these are ignored, then the coding exons on their own make up 4440 bp, specifying a protein that is 1480 amino acids in length—remember that each amino acid in a protein is coded by a triplet of three adjacent nucleotides. Three common variants of *CFTR* result in defective proteins that may give rise to cystic fibrosis (Figure 3.16B). In the most common variant three nucleotides are deleted from positions 1514, 1515, and 1516 in the coding sequence. As a result the protein that is synthesized is missing an amino acid. This small change prevents the protein from attaching to the cell membrane, so that the protein cannot carry out its function. Cystic fibrosis might result.

Now things become a little more complicated. There is another variant of the *CFTR* gene in which the nucleotide at position 1624 is changed from a G to a T. This is called a **single nucleotide polymorphism**, or **SNP**. The SNP causes the mRNA to be degraded so no cystic fibrosis transmembrane regulator protein is made. Once again the protein function is absent and cystic fibrosis might result.

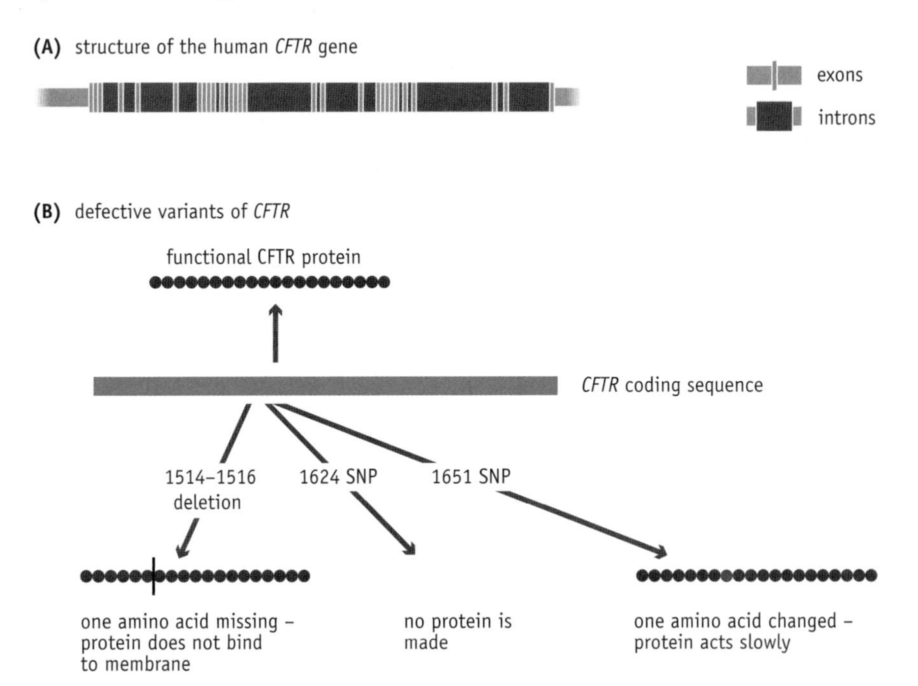

FIGURE 3.16 Variants of the human *CFTR* gene. (A) The structure of the *CFTR* gene, showing the positions of the exons and introns. There are 27 exons and 26 introns, one intron between each pair of adjacent exons. Some of the introns are very short and are not clearly visible when the gene is drawn at this scale. (B) The coding sequence of the *CFTR* gene—the gene minus its introns—is indicated by the blue line. The normal gene gives rise to the functional CFTR protein, as shown above the coding sequence. The proteins produced by the three most common defective variants of the gene are illustrated below the coding sequence.

Another *CFTR* variant has a SNP at position 1651, which does not affect the length of the protein that is made, but changes one of the amino acids. This alters the kinetic properties of the protein so that it now transports chloride ions at only 4% of the rate displayed by the normal protein. Again, this can lead to cystic fibrosis. Although the protein is still able to transport chloride ions it does so at such a low rate that the physiological function of the protein is lost. In total, over 1400 variants of the *CFTR* gene that give rise to a dysfunctional cystic fibrosis transmembrane regulator protein are known.

There are also other variants that do not affect the ability of the gene to give rise to a functional protein. The nucleotide sequence changes present in some of these variants result in no change in the amino acid sequence of the protein, and in others they change it in a way that has no impact on the protein's function.

With a gene–protein system as complex as this, the use of the word "alleles" to describe the alternative forms of the biological characteristic *and* the underlying gene variants can make things seem more complicated than they are. Does the *CFTR* gene have two alleles ("healthy" and "cystic fibrosis"), or does it have 1400, the number of gene variants that can give rise to the "cystic fibrosis" biological characteristic? Do we include the gene variants that do not affect the function of the protein, in which case there are many more than 1400 "alleles"? To avoid these contradictions, geneticists increasingly use the word **haplotype** to describe a sequence variant of a gene. According to this nomenclature, the *CFTR* gene has more than 1400 haplotypes, some giving rise to the healthy allele and some to the cystic fibrosis allele.

3.3 FAMILIES OF GENES

Since the earliest days of DNA sequencing it has been known that **multigene families**—groups of genes of identical or similar sequence—are common in many organisms. The members of these gene families therefore encode identical or similar units of biological information. Why do these families exist?

Multiple gene copies enable large amounts of RNA to be synthesized rapidly

Families of identical genes are generally assumed to specify RNA molecules or proteins that are needed in large quantities by the cell, continually or at certain times, such as when the cell is dividing. Gene expression is not an instantaneous process, and there is a limit to the rate at which the product of an individual gene can be synthesized.

The identities of **simple multigene families**, ones in which all the genes are the same, support this idea. In most organisms, there are multiple copies of the rRNA genes. The bacterium *Escherichia coli*, for example, has seven copies of each of its three rRNA genes. Experiments have suggested that five copies of each rRNA gene are needed to maintain the number of ribosomes needed when the bacterium is dividing at its maximum rate in an unchanging environment. Under these conditions, there may be as many as 20,000 ribosomes in the cell. All seven gene copies are needed when the bacterium is adapting to a new environment, presumably because adaptation requires the synthesis of many new proteins that were not needed under the preexisting conditions.

identical genes

FIGURE 3.17 A tandem array.

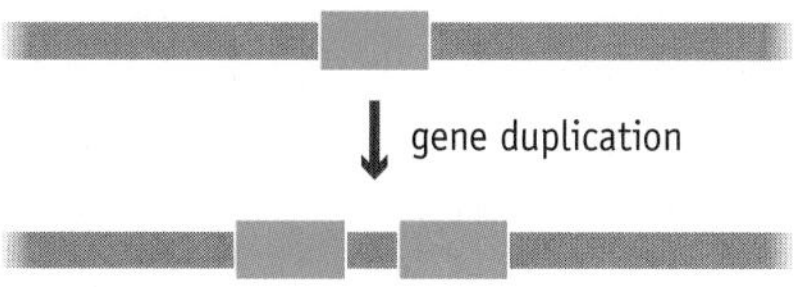

gene duplication

FIGURE 3.18 Gene duplication.

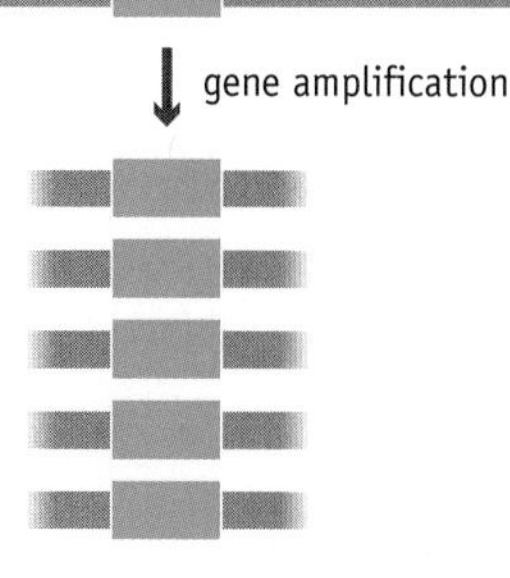

FIGURE 3.19 Gene amplification. The new gene copies resulting from amplification exist as independent molecules, rather than being attached to one of the chromosomes.

Higher organisms also have multiple copies of their rRNA genes. In humans, there are about 280 copies of three of the rRNA genes, and over 2000 copies of the fourth. These copies are grouped into large clusters where many genes are arranged head-to-tail in a **tandem array** (Figure 3.17). This arrangement probably does not aid the rate at which the genes can be transcribed, but is more a reflection of the process that led to the multiple copies. Each cluster is thought to have evolved by multiple rounds of gene duplication, individual genes being duplicated into two copies placed side by side (Figure 3.18).

In some organisms, even these large numbers of gene copies are apparently unable to meet the demand for ribosomes at certain times. In some amphibians, the 450 or so copies of the rRNA genes in the genome are increased to up to 16,000 copies by **gene amplification**. This involves replication of the rRNA genes into multiple DNA copies which subsequently exist as independent molecules not attached to the chromosomes (Figure 3.19). Gene amplification typically occurs in oocytes that are actively developing into mature egg cells.

In some multigene families the genes are active at different stages in development

In **complex multigene families**, the individual gene members, although having similar nucleotide sequences, are sufficiently different to code for proteins with distinctive properties. These proteins are therefore able to play slightly different roles. The flexibility provided by these families of genes is sometimes exploited during development of the organism.

The mammalian globin genes are one of the best examples of this type of multigene family. The globins are the blood proteins that combine to make hemoglobin, each molecule of hemoglobin being made up of two α-type and two β-type globin polypeptides. Both of these globin types, α and β, exist as a family of related molecules differing from one another at just a few amino acid positions.

In humans there are two α-like globins, α and ζ, and five β-like globins, β, δ, A_γ, G_γ, and ϵ. These proteins are coded by two multigene families (Figure 3.20). The genes were among the first to be sequenced, back in the late 1970s. The sequence data showed that the genes in each family are similar

FIGURE 3.20 The human globin gene families. The genes in each family are active at different stages in human development. The α_1 and α_2 genes are the only ones that give rise to identical proteins.

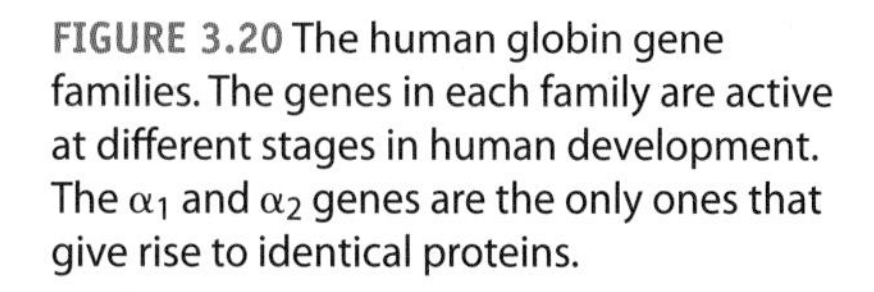

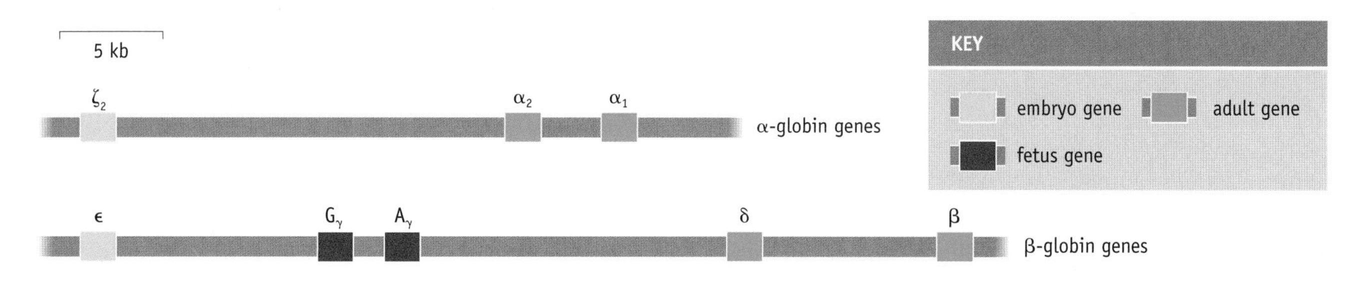

RESEARCH BRIEFING 3.1

Studying the expression profile of a gene

The human β-globin genes provide an excellent example of **differential gene expression**.

A differentially expressed gene is switched on only when the biological information that it contains is needed. At all other times the gene is inactive. With the β-globin genes, the differential expression profile is linked to development, but other genes display tissue specificity, being active in some cell types but inactive in others. Identifying the expression profile of a gene is important in many areas of genetics. In particular, knowing in which tissues a newly discovered gene is active often provides the first clue that leads eventually to an understanding of the function of that gene.

How can the expression profile of a gene be worked out? The answer is to search for the presence of its mRNA in different tissues or at different developmental stages. If the gene is inactive, then it is not transcribed into mRNA. The presence of a gene's mRNA in a tissue sample therefore indicates that the gene is being expressed in that tissue. We will explore how this type of research is carried out.

A gene's mRNA can be detected by hybridization probing

Tissue extracts can be tested for the presence of a particular mRNA by **hybridization probing**. The basis of this technique is the ability of any two complementary polynucleotides to form base pairs with one another. This can occur not only between single-stranded DNA molecules to form the DNA double helix, but also between single-stranded RNA molecules and between combinations of one DNA and one RNA strand (Figure 1). The formation of base pairs between any two polynucleotides is called **nucleic acid hybridization**.

The first step in hybridization probing is to prepare samples of RNA from the tissues being studied, and to spot these samples onto a nylon membrane (Figure 2). The RNA in the samples becomes bound to the membrane. To determine which samples contain a particular mRNA, the membrane is probed with DNA whose sequence matches that of the gene we are interested in. This DNA could be the gene itself, or a segment of it, purified by DNA cloning. Alternatively it could be an oligonucleotide that has been synthesized in the laboratory and whose sequence is the same as a part of the gene. The probe will hybridize with any tissue extracts that contain a complementary RNA sequence, revealing the tissues that contain mRNA copies of the gene.

The probe must be labeled so it can be detected after hybridization

We need a way of being able to recognize those samples on the membrane with which the probe hybridizes. Before use, the probe DNA must therefore be modified by attachment of a marker that emits a signal that enables the bound probe to be located. This is called **labeling**.

The traditional method of labeling a DNA probe is to use radioactive phosphorus. Atoms of radioactive phosphorus can be incorporated into the sugar–phosphate backbone of a DNA molecule, so that these molecules in turn become radioactive. After being applied to the membrane, the nonhybridized probe molecules are washed away, leaving only those molecules that are bound to RNA contained in the samples. The location of the probe is then determined by **autoradiography**, in which a sheet of X-ray-sensitive film is placed over the membrane. The samples that contain hybridized RNA are revealed as dark spots on the film after it is developed (Figure 3).

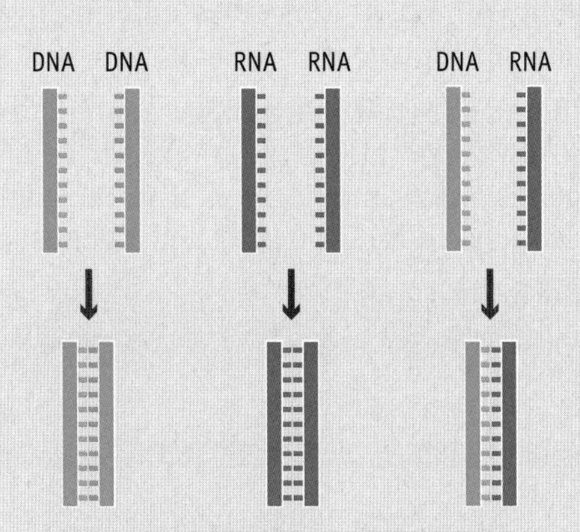

FIGURE 1 Nucleic acid hybridization. Hybridization can occur between any pair of complementary polynucleotides, such as two complementary DNA molecules (left), two complementary RNAs (center), or a gene and its mRNA (right).

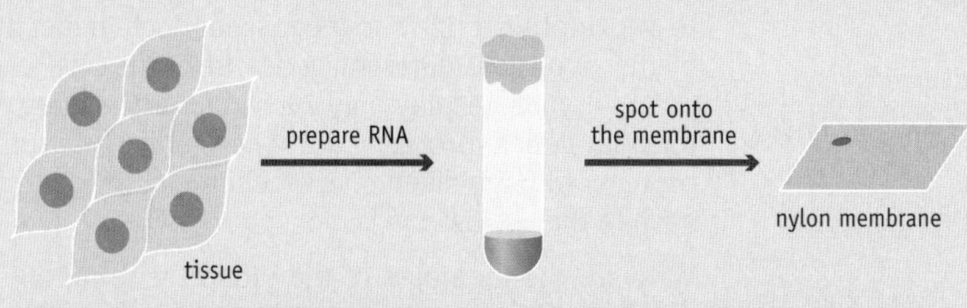

FIGURE 2 Preparing an RNA dot blot. This method is called "dot blotting" because a spot or "dot" of RNA is "blotted" onto the membrane. After blotting, the RNA is bound tightly to the membrane by exposure to ultraviolet radiation, which causes some of the nucleotide bases to form covalent bonds with the membrane surface. The presence of these covalent bonds does not affect the ability of the RNA molecules to hybridize to the probe.

The problem with radioactive labeling is that it presents a health hazard to the experimenter, and an environmental hazard when the waste material is disposed. A number of methods for labeling DNA with nonradioactive markers have therefore been developed. In one of these, molecules of the enzyme called alkaline phosphatase are attached to the probe DNA. After hybridization, the location of the enzyme on the membrane is detected through its ability to degrade a special substrate called luminol. Degradation of luminol results in the emission of light, which can be recorded on normal photographic film in a manner similar to autoradiography. Fluorescent labels are also popular and are used in methods such as DNA sequencing (see Figure 2.21), but are less easy to use with membrane-bound DNA.

Expression profiling was used to identify the *CFTR* gene

In order to illustrate the importance of expression profiling in research, we will examine how the technique was used to help identify the human *CFTR* gene. This is the gene which, when defective, can give rise to cystic fibrosis. The work was carried out in 1989 by geneticists at the Howard Hughes Medical Institute in Maryland, at the University of Michigan, and at The Hospital for Sick Children in Toronto. They had previously identified a **candidate gene**, one they thought might be the one they were looking for. Expression profiling was one of the tests that were carried out to confirm that this was indeed the *CFTR* gene.

RNA extracts were prepared from various tissues from autopsy specimens. These tissues were those of the pancreas, lung, colon, brain, placenta, liver, adrenal gland, testis, parotid gland, and kidney, and of nasal polyps from the mucous membranes of the nose and sinuses. A final RNA preparation was made from cultured epithelial cells, which have a similar gene expression profile to the sweat glands in the skin. The extracts were probed with a clone of the candidate gene, which had been radioactively labeled. When the autoradiographs were developed, the radioactive signals showed that the probe had hybridized mainly with the RNA extracts from the pancreas and nasal polyps, with weaker signals detected from the lung, colon, cultured epithelial cells, placenta, liver, and parotid gland. This expression profile matches the one that we would expect for the gene that is responsible for cystic fibrosis. Cystic fibrosis is usually looked on as a lung disease because the buildup of mucus in the lungs is the major symptom displayed by patients. Most of this mucus is synthesized by the basal polyps, and so we would expect the *CFTR* mRNA to be present in nasal polyp extracts. Cystic fibrosis is also associated with an increased secretion of mucus by the pancreas, so again we expect the mRNA for the *CFTR* gene to be present in this tissue. The expression profile therefore provided strong evidence that the candidate gene was indeed the one responsible for cystic fibrosis, enabling the research to begin to focus on the possible defects that might occur in the gene and give rise to the disease.

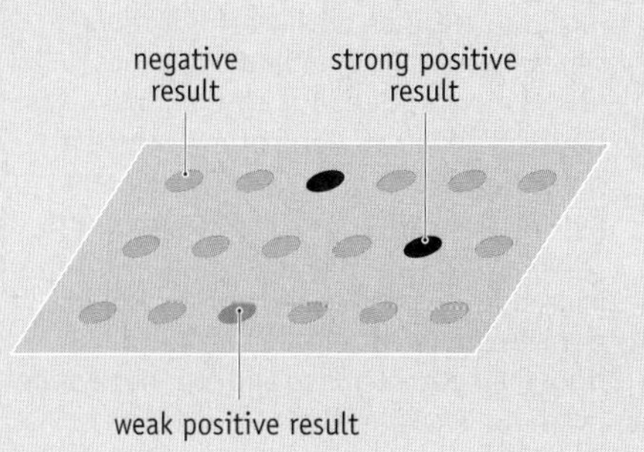

FIGURE 3 Typical example of an autoradiograph obtained when an RNA dot blot is probed with radioactively labeled DNA. Dark spots on the autoradiograph indicate the positions of those RNA samples that hybridize with the probe. These samples come from tissues that are expressing the gene contained in the probe and which therefore contain mRNA that has been transcribed from that gene. If the spot is very dark, then the sample contains a relatively large quantity of the mRNA and we can conclude that the gene is very active in that tissue. If the spot is less dark, giving only a weak positive signal, then the conclusion is that smaller amounts of mRNA are present and that the gene, although active, is not being transcribed very rapidly. A negative result is obtained when the spot contains no mRNA, showing that the gene is not switched on in that particular tissue.

to one another, but by no means identical. In fact the nucleotide sequences of the two most different genes in the β-type cluster, coding for the β- and ϵ-globins, display only 79.1% identity. Although this is similar enough for both genes to specify a β-type globin, it is sufficiently different for the proteins to have distinctive biochemical properties. Similar variations are seen in the α-cluster.

Why are the members of the globin gene families so different from one another? The answer was revealed when the expression patterns of the individual genes were studied. It was discovered that the genes are active at different stages in human development. For example, in the β-type cluster, gene ϵ is expressed in the early embryo, G_γ and A_γ (whose protein products differ by just one amino acid) in the fetus, and δ and β in the adult (see Figure 3.20). The different biochemical properties of the resulting globin proteins are thought to reflect slight changes in the physiological role that hemoglobin plays during the course of human development.

The globin families reveal how genes evolve

Multigene families are thought to arise by gene duplication (see Figure 3.18). In a simple multigene family, the sequences of the gene copies stay the same after duplication, so all the members are identical. Exactly how this occurs is not completely understood, because our expectation is that sequences of the individual members of a family should gradually change over time. This is because DNA replication is not an entirely error-free process, and because some chemicals and physical agents, such as ultraviolet radiation, can cause DNA sequence alterations by the process called **mutation**. A complex multigene family, in which the members have different but related sequences, is therefore the normal outcome of a series of gene duplications.

Mutations and replication errors accumulate over time, so two members of a gene family that have very similar sequences are likely to have arisen from a duplication that occurred more recently than one giving rise to two genes with less similar sequences (Figure 3.21). The pattern of gene duplications that gave rise to the genes we see today can therefore be deduced. When we apply this logic to the β-globin family, we infer that G_γ and A_γ are the result of the most recent duplication, as these are the two most similar genes in the cluster, and that β and δ arose from the preceding duplication.

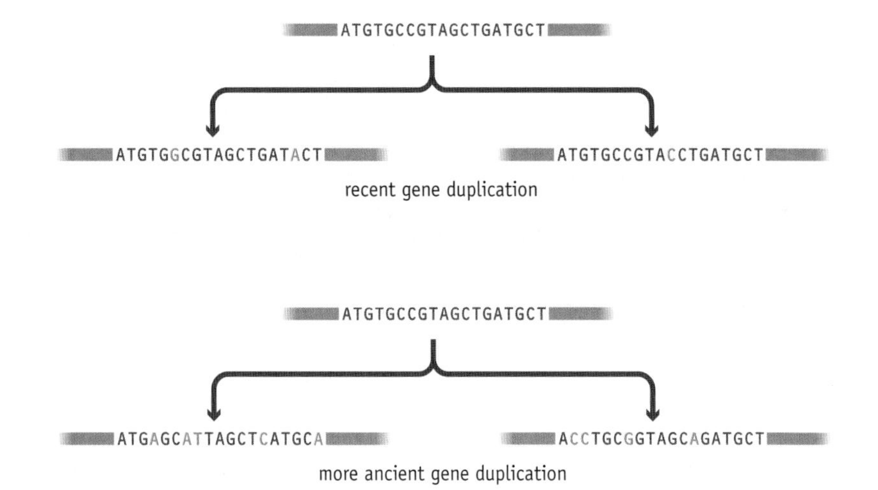

FIGURE 3.21 Two genes resulting from a recent duplication are likely to be more similar to one another than two genes from a more ancient duplication. The two new genes resulting from the recent gene duplication differ from their parent gene at two and one nucleotide positions, respectively. These nucleotides are shown in orange. In contrast, the genes resulting from the more ancient duplication differ from their parent at five and four nucleotide positions, respectively.

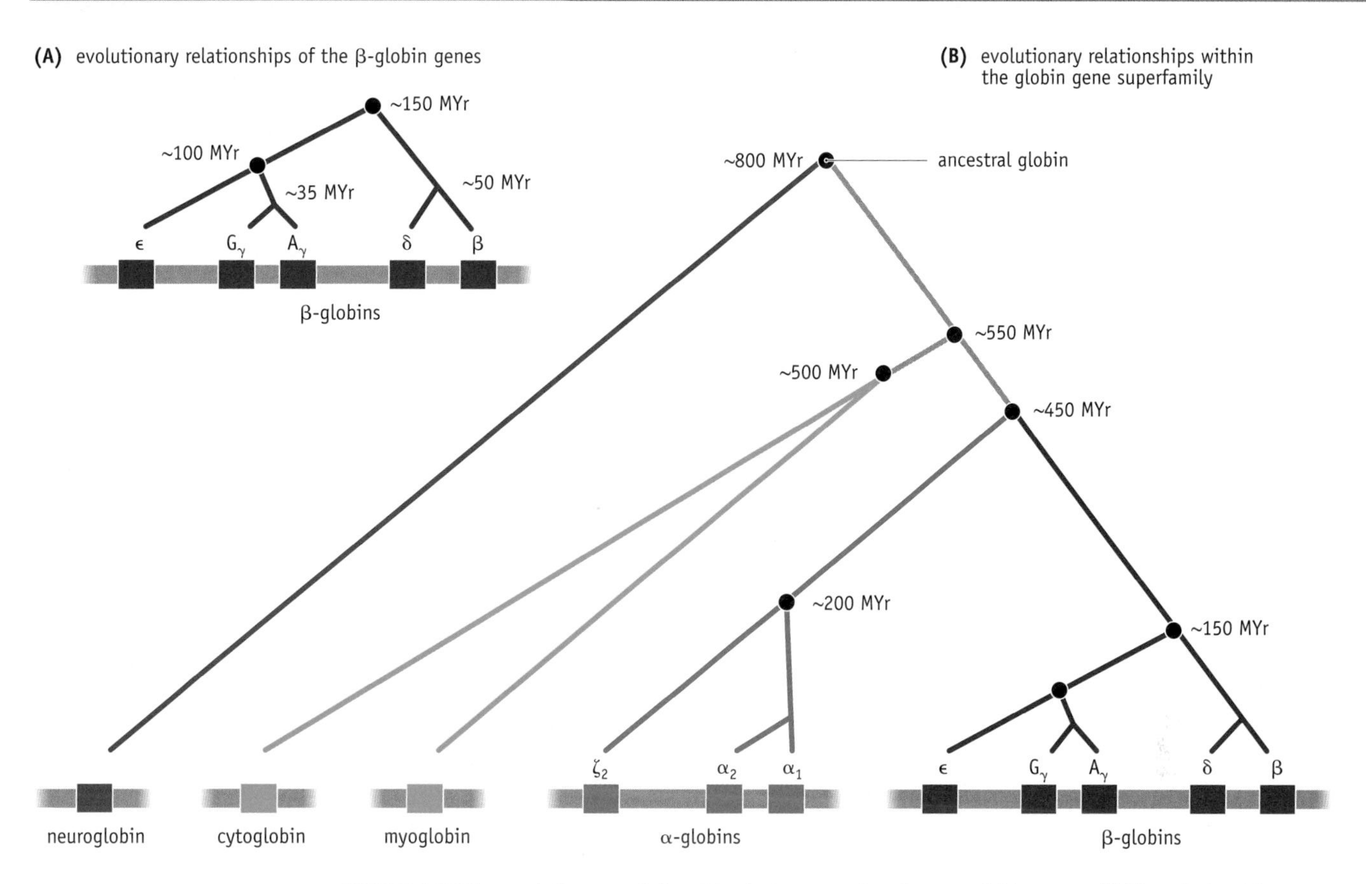

FIGURE 3.22 The evolutionary relationships between various human globin genes. (A) The relationships within the β-globin family. (B) The relationships within the globin gene superfamily. MYr, million years ago.

We can take the analysis a little bit further. In Chapter 21 we will learn how it is possible to work out the rate at which nucleotide sequence changes accumulate in genes. This rate can be converted into a **molecular clock** that enables dates to be assigned to gene duplication events. When we apply the molecular clock to the β-globin genes we discover that the initial gene duplication that gave rise to the first two members of this family occurred approximately 150 million years ago. The times of the subsequent duplications can also be dated (**Figure 3.22A**).

If we compare not just the β-globins but the members of the α family as well, then we can make further deductions. The first duplication giving rise to the α-globins took place about 200 million years ago (**Figure 3.22B**). The very first α- and β-globin genes appeared as a result of a duplication 450 million years ago. Even further back there were duplications that gave rise to genes whose protein products are now synthesized in tissues other than blood. These include neuroglobin, from the brain, myoglobin from muscles, and cytoglobin, which is present in a number of tissues. Each of these proteins is involved in oxygen binding in one way or another, similar to the function of the α- and β-globins in blood cells.

The globin gene clusters also tell us something else about the evolution of genes. When we examine the α and β families, we see several genes whose nucleotide sequences have changed to such an extent that they have lost their function altogether. There are four such genes in the human α family, called $\psi_{\chi 1}$, $\psi_{\alpha 1}$, $\psi_{\alpha 2}$, and θ; and one, ψ_{β}, in the β cluster (**Figure 3.23**). These are **pseudogenes**, a type of evolutionary relic. They are related to

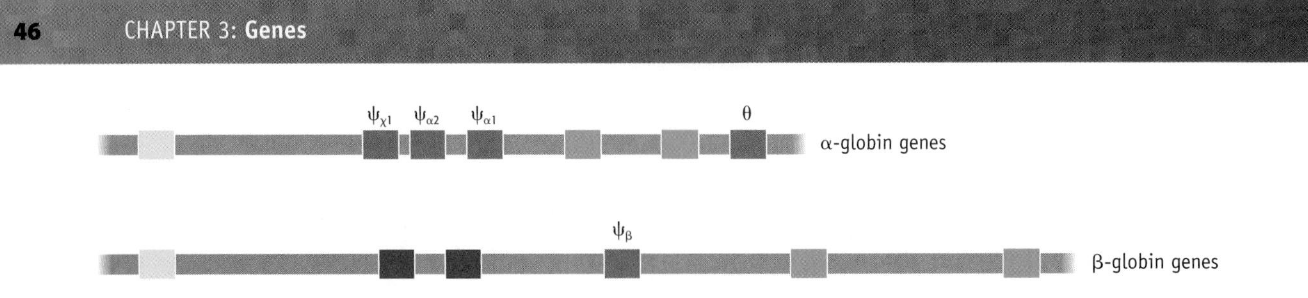

FIGURE 3.23 Pseudogenes in the human globin gene clusters.

the other genes in their clusters and arose from them by duplication, but since then they have acquired sequence changes that have resulted in their loss of function. Interestingly, the δ gene of the β cluster, which is an active gene in humans, is a pseudogene in some other mammals, including mice and rabbits. The gene clusters have therefore followed different evolutionary pathways in the lineages leading to these other species.

KEY CONCEPTS

- A gene is a segment of a DNA molecule. The shortest genes are less than 100 bp in length, and the longest ones are over 2,400,000 bp.
- Some genes are discontinuous. The biological information that they contain is split into sections called exons, separated by intervening sequences called introns.
- Utilization of the biological information contained in a gene requires the coordinated activity of enzymes and other proteins, in a series of events referred to as gene expression.
- All genes undergo the first stage of gene expression, which is called transcription and which results in synthesis of an RNA molecule.
- For some genes the RNA transcript is the end product of gene expression. For others the transcript is a short-lived message that directs the second stage of gene expression, called translation, during which the RNA molecule directs synthesis of a protein.
- Protein synthesis is the key to expression of biological information. Proteins of different types have many different functions. By coding for different proteins, genes are able to specify the biological characteristics of a living organism.
- Many biological characteristics are variable because the genes that code for these characteristics are themselves variable. The variants are called alleles or haplotypes.
- Groups of genes of identical or similar sequence are common in many organisms. These are called multigene families.
- In a simple multigene family all the genes are identical. Many of these families specify RNA molecules or proteins that are needed in large quantities by the cell. In a complex multigene family the genes have similar sequences and code for related but slightly different proteins. Often these genes are active at different developmental stages.
- Multigene families evolve by gene duplication. The series of gene duplications giving rise to a complex multigene family can be deduced by making comparisons between the nucleotide sequences of the individual genes.

QUESTIONS AND PROBLEMS (Answers can be found at www.garlandscience.com/introgenetics)

Key Terms

Write short definitions of the following terms:

allele
autoradiography
candidate gene
complex multigene family
differential gene expression
discontinuous gene
enzyme
exon
gene amplification
gene expression
genetic code
haplotype
hybridization probing
intergenic DNA
intron
kilobase pair
labeling
megabase pair
messenger RNA
microRNA
molecular clock
motor protein
multigene family
mutation
noncoding RNA
nucleic acid hybridization
pseudogene
regulatory protein
ribosomal RNA
RNA interference
RNA splicing
simple multigene family
single nucleotide polymorphism
small cytoplasmic RNA
small interfering RNA
small nuclear RNA
small nucleolar RNA
structural protein
tandem array
transcription
transfer RNA
translation

Self-study Questions

3.1 What is a gene?

3.2 Distinguish between the terms base pair, kilobase pair, and megabase pair. How many base pairs would there be in a gigabase pair?

3.3 Why does a human gene that is 2.4 Mb in length not contain 24,000 times the amount of information present in a gene 100 bp in length?

3.4 What is the difference between an exon and an intron? If a gene has six introns, then how many exons would it have?

3.5 Outline the process by which the information contained in a gene is used to direct synthesis of a protein.

3.6 Write down the nucleotide sequences of the messenger RNA molecules that would be transcribed from these two genes:

5′-TACCCATGATTTCCCGCGATATATATGCCTGCAATC-3′
5′-TACCCTTGGTTTCGCGCGATGTATATACCAGGAATC-3′

3.7 Give examples of the various roles that proteins play in living organisms.

3.8 What is the difference between coding and noncoding RNA?

3.9 List the types of noncoding RNA found in human cells and briefly describe the functions of each one.

3.10 Write a description of the molecular basis for the pea shape characteristic studied by Mendel.

3.11 Give three examples of variants of the *CFTR* gene that give rise to cystic fibrosis.

3.12 Distinguish between the terms "allele" and "haplotype."

3.13 What is the difference between a simple and a complex multigene family?

3.14 Give one example of a simple multigene family in which the individual genes are arranged in a tandem array. What is the biological purpose of this arrangement?

3.15 Give an example of gene amplification.

3.16 Draw a diagram showing the organization of the human α- and β-globin gene families.

3.17 Why are the members of the globin gene families so different from one another?

3.18 What do comparisons between the DNA sequences of the human globin genes tell us about the process by which these gene families evolved?

3.19 What is the special feature of the gene called ψ_β in the β-globin cluster?

3.20 Explain how hybridization probing is used to determine the expression profile of a gene.

3.21 Describe the various types of marker that can be used to label DNA molecules.

3.22 How was expression profiling used in identification of the *CFTR* gene?

Discussion Topics

3.23 Discontinuous genes are common in higher organisms but virtually absent in bacteria. Discuss the possible reasons for this.

3.24 Eukaryotic cells expend a lot of energy in removing introns from the transcripts of discontinuous genes. Yet introns seem to play no role of their own. Discuss.

3.25 What would be the implications for theories of evolution if it were discovered that information can flow from proteins to RNA and thence to DNA?

3.26 There appears to be no biological reason why a DNA polynucleotide could not be directly translated into protein, without the intermediary role played by mRNA. What advantages do cells gain from the existence of mRNA?

3.27 Scientists studying the origins of life have proposed that RNA was the first type of nucleic acid to evolve in the primordial soup, with DNA appearing later. How does the existence of mRNA fit in with this theory?

3.28 Human cells are able to make at least 1000 different types of microRNA, and these molecules play a central role in controlling the expression of human genes. As there are so many microRNAs and they are so important, why was their existence not suspected until the 1990s?

3.29 Expression of different versions of the α- and β-globin genes at different developmental stages is thought to reflect changes in the physiological role played by hemoglobin during human development. What are these different physiological roles? Do you think this theory can be defended?

3.30 In addition to expression profiling, what methods might have been used to confirm that the candidate *CFTR* gene was the correct one?

FURTHER READING

Balakirev ES & Ayala FJ (2003) Pseudogenes: are they "junk" or functional DNA? *Annu. Rev. Biochem.* 37, 123–151.

Brown DD & Dawid IB (1968) Specific gene amplification in oocytes: oocyte nuclei contain extrachromosomal replicas of genes for ribosomal RNA. *Science* 160, 272–280.

Catania F & Lynch M (2008) Where do introns come from? *PLoS Biol.* 6, e283.

Crick FHC (1970) Central Dogma of molecular biology. *Nature* 227, 561–563. *Summarizes the process of gene expression from DNA to protein.*

Efstratiadis A, Posakony JW, Maniatis T et al. (1980) The structure and evolution of the human beta-globin gene family. *Cell* 21, 653–668.

Fincham JRS (1990) Mendel—now down to the molecular level. *Nature* 343, 208–209. *Describes the molecular basis for round and wrinkled peas.*

Fritsch EF, Lawn RM & Maniatis T (1980) Molecular cloning and characterization of the human α-like globin gene cluster. *Cell* 19, 959–972.

Kumar S (2005) Molecular clocks: four decades of evolution. *Nat. Rev. Genet.* 6, 654–662. *Describes in detail the basis of the use of the molecular clock to date events in the past.*

Mattick JS & Makunin IV (2006) Non-coding RNA. *Hum. Mol. Genet.* 15, R17–R29.

Riordan JR, Rommens JM, Kerem B et al. (1989) Identification of the cystic fibrosis gene: cloning and characterization of complementary DNA. *Science* 8, 1066–1073. *Expression profiling of the CFTR gene.*

Tsui L-C (1992) The spectrum of cystic fibrosis mutations. *Trends Genet.* 8, 392–398.

Transcription of DNA to RNA CHAPTER 4

In Chapter 3 we learned that the biological information contained in a gene is encoded in its nucleotide sequence, and that this information is utilized by the cell by the process called gene expression. The end product of gene expression is either a noncoding RNA or a protein that has a specific function in the cell. Through synthesis of these end products the genes are able to specify the biological characteristics of the organism.

The essential link between the biological information contained in the genes and the characteristics of the living organism is therefore made by the gene expression pathway. We must now turn our attention to this pathway and examine exactly how DNA is transcribed into RNA and how mRNA is translated into protein. As we are taking the modern, molecular approach to genetics, we must acquire much more than a simple overview of this process, and instead must look closely at the events that occur at each step in gene expression. An important component of genetics research in the twenty-first century is describing those events in greater detail—and understanding in particular how the expression of individual genes is controlled in healthy cells and how errors in the control processes might lead to diseases such as cancer.

The details of the gene expression pathway and how it is controlled will take up the next six chapters. The present chapter deals with the first step, the transcription of DNA to RNA. First, however, we must look at gene expression in outline and identify the key features of the pathway in different types of organism.

4.1 GENE EXPRESSION IN DIFFERENT TYPES OF ORGANISM

A major challenge for a student of genetics is learning the variations in the gene expression pathways found in different types of organism. In some types, the pathway is much less complicated than in others. Those studying genetics for the first time may feel a natural temptation to concentrate on the simplest example of gene expression, as this is, obviously, the easiest to remember. But then we would be ignoring substantial parts of the pathway operating in humans, for example—and all other "higher" life forms, for that matter. Genetics is not just about humans, but humans are such important organisms, at least to us, that any serious course in genetics must have a strong focus on humans. In this book we will not shy away from the human complexities. To put these in context, we begin by comparing the gene expression pathways for different types of organism.

There are three major groups of organisms

Traditionally, biologists have divided organisms into two groups, called **prokaryotes** and **eukaryotes**. Prokaryotes and eukaryotes are distinguished by their fundamentally different cellular organization (Figure 4.1). Prokaryotes, which include bacteria, lack an extensive cellular architecture, and their DNA is not enclosed in a distinct structure. In contrast, the typical eukaryotic cell is usually larger and more complex, having a membrane-bound nucleus containing the chromosomes, and having other distinctive membranous organelles such as mitochondria, vesicles, and Golgi bodies.

Most prokaryotes are unicellular, though in some species individual cells can associate together to form larger structures, such as the chains of cells

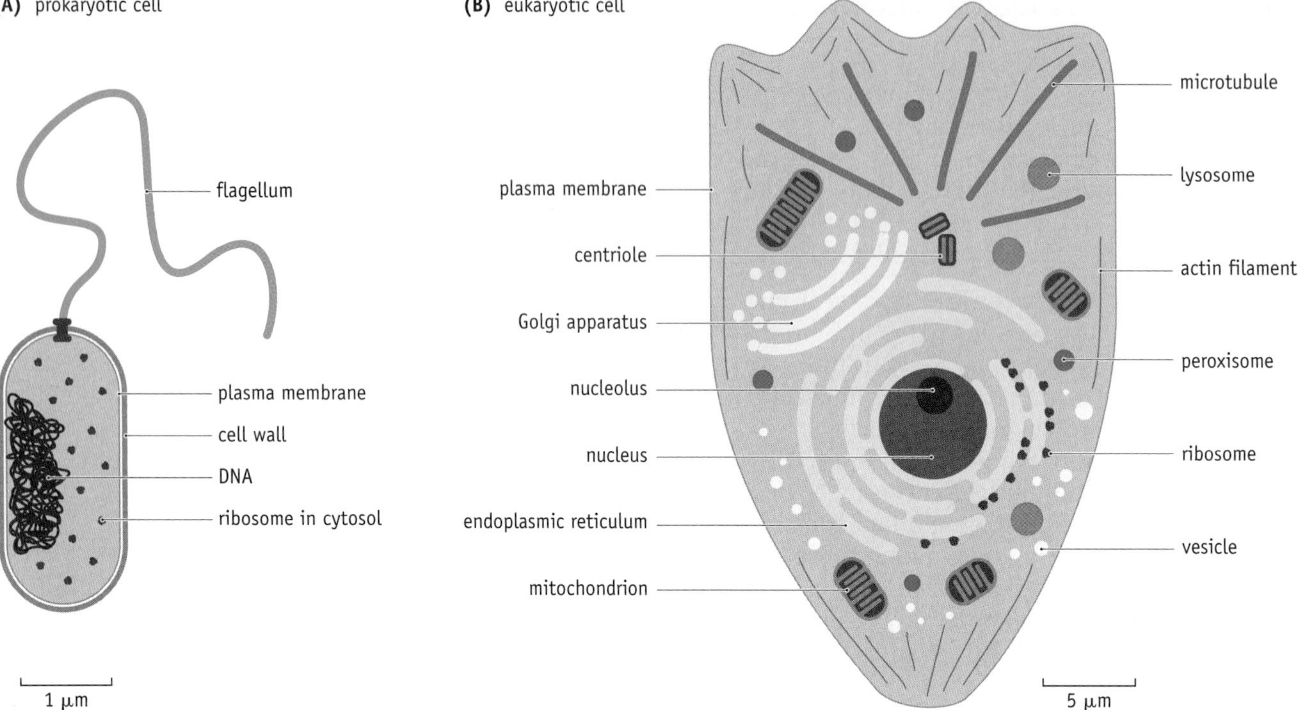

FIGURE 4.1 The different cellular organizations of prokaryotes and eukaryotes. (A) A typical bacterial cell. (B) An example of an animal cell. (Adapted from B. Alberts et al., Essential Cell Biology, 3rd ed. New York: Garland Science, 2010.)

formed by *Anabaena* (Figure 4.2). Eukaryotes can be unicellular or multicellular and comprise all the macroscopic forms of life such as plants, animals, and fungi (Figure 4.3). Humans are therefore eukaryotes.

Until 1977 it was thought that all prokaryotes were more or less the same. It was recognized that the group included a great diversity of organisms, but the differences were thought to be variations on a theme. This assumption has now been overturned, and it is accepted that the prokaryotes include two distinct groups of organisms, the **bacteria** and the **archaea**. The bacteria comprise most of the prokaryotes that are familiar to us as people and as scientists, such as the pathogens *Mycobacterium tuberculosis*, which causes tuberculosis in humans, *Vibrio cholerae*, responsible for cholera, and *Escherichia coli*, some strains of which can cause food poisoning but which is more familiar to geneticists as a bacterium that is frequently studied in the laboratory (Figure 4.4).

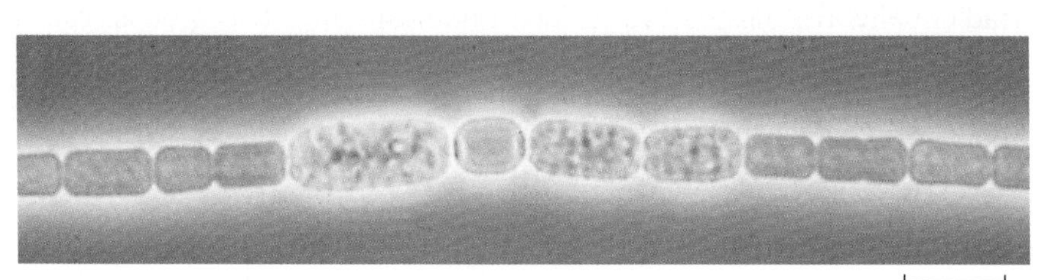

FIGURE 4.2 Part of a chain of cells formed by the photosynthetic bacterium *Anabaena cylindrica*. (Courtesy of David Adams, University of Leeds.)

FIGURE 4.3 Examples of multicellular eukaryotic species. (A) A gray wolf, *Canis lupus*. (B) The Colorado potato beetle, *Leptinotarsa decemlineata*. (C) Tulips. (D) *Strophariaceae*, a type of fungus. (A, courtesy of U.S. Fish & Wildlife Service, John & Karen Hollingsworth; B, courtesy of U.S. Department of Agriculture, Scott Bauer.)

The archaea are less well known, partly because they are not widespread in nature. They include methanogens, which live at the bottom of lakes and other bodies of water and release methane, which they synthesize as a metabolic by-product. Other types of archaea are confined to extreme environments, such as hot springs where the temperature can be 60°C or higher, brine pools and high-salt lakes such as the Dead Sea, and acidic

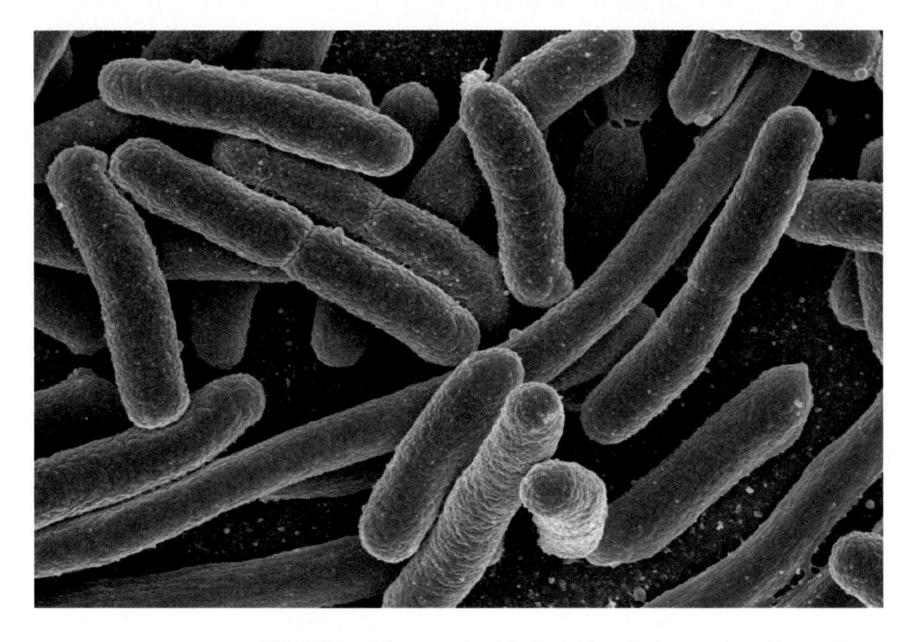

FIGURE 4.4 Micrograph of *Escherichia coli*, the species of bacteria that is frequently studied in the laboratory. (Courtesy of Rocky Mountain Laboratories, NIAID, NIH.)

FIGURE 4.5 An example of an extreme environment, likely to harbor members of the archaea. The picture shows an acidic mine drainage in Spain. (Courtesy of Carol Stoker, NASA Ames Research Center.)

streams emerging from old mining works (Figure 4.5). All of these environments are hostile to most other forms of life.

We now look on the eukaryotes, bacteria, and archaea as the three main types of living organism, with each group displaying its own distinctive genetic features. Of the three groups, the archaea are by far the least well studied, and at present the gene expression pathway operating in these organisms is not understood in detail. So we must focus our attention on the differences in gene expression in bacteria and eukaryotes.

Gene expression is more than simply "DNA makes RNA makes protein"

In its simplest form, gene expression can be looked on as a two-step process, conveniently described as "DNA makes RNA makes protein." The first step, transcription, results in synthesis of mRNAs and noncoding RNAs, and the second step, translation, uses the mRNAs to direct synthesis of proteins (Figure 4.6).

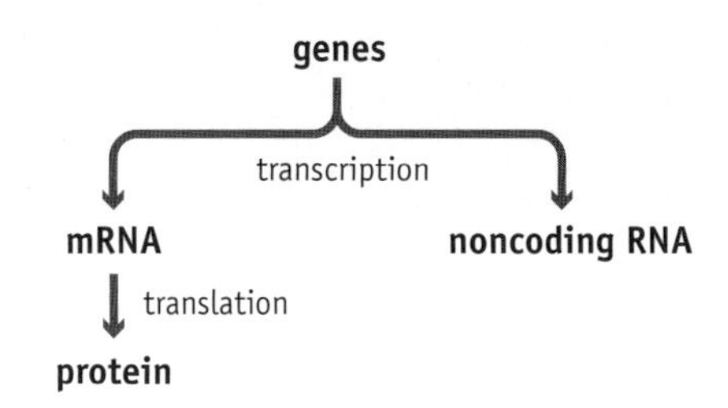

FIGURE 4.6 Gene expression described as a two-step process. The first step is synthesis of mRNA and noncoding RNA by transcription, and the second step is translation of the mRNA into protein.

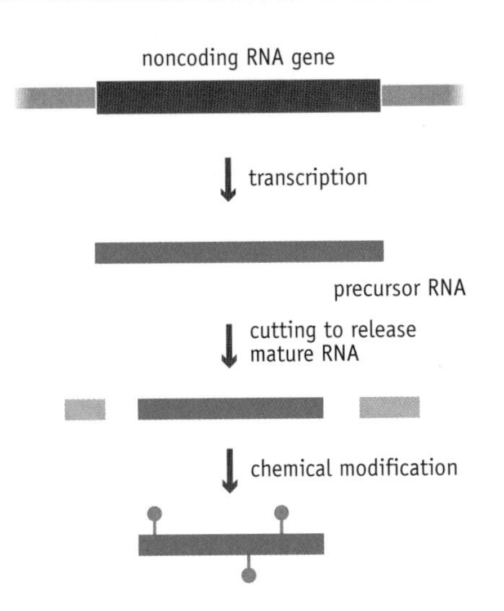

FIGURE 4.7 Processing of the primary transcripts of noncoding RNA genes. Following transcription, the primary transcript, or precursor RNA, is cut into segments to release the mature, functional RNAs. Some of these RNAs may also be modified by addition of extra chemical groups.

For bacteria, "DNA makes RNA makes protein" is a reasonably accurate description of the process as a whole. Its one weakness is that it implies that the noncoding RNAs made during the first step are immediately able to take up their functions in the cell. In reality, the primary transcripts of these noncoding RNAs have to be cut into segments to release the functional molecules, and additional chemical groups also have to be added to some of them (Figure 4.7). These **processing** events are an important part of the gene expression pathway for bacterial noncoding RNAs.

In eukaryotes, gene expression is much more than "DNA makes RNA makes protein" (Figure 4.8). In eukaryotes, we place much greater emphasis on the start of the process, when a set of proteins, including the enzyme that carries out transcription, is assembled on the DNA adjacent to the gene that is going to be expressed. This preliminary step is more complicated in eukaryotes compared with bacteria, largely because eukaryotes have more sophisticated mechanisms for controlling the expression of individual genes and because many of these control processes operate by regulating the assembly of this **transcription initiation complex**. The same is true for the initiation of translation, which is a second important control point in eukaryotic gene expression, and hence is more complex than the equivalent event in bacteria. This is not to suggest that gene expression is not regulated in bacteria. It certainly is, and both transcription and translation are controlled, but the regulatory processes in bacteria have not evolved

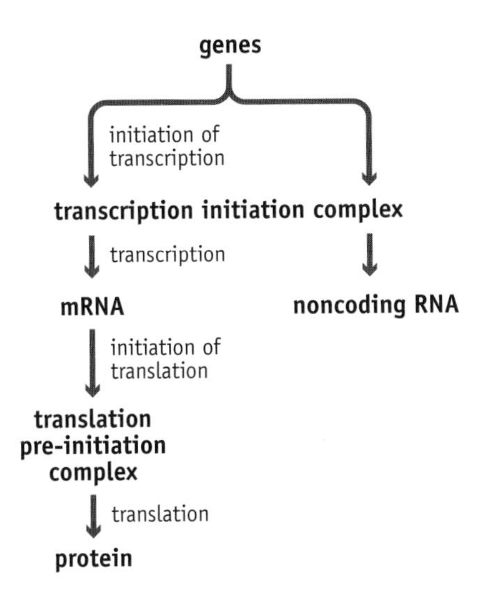

FIGURE 4.8 The more complex series of events involved in gene expression in eukaryotes.

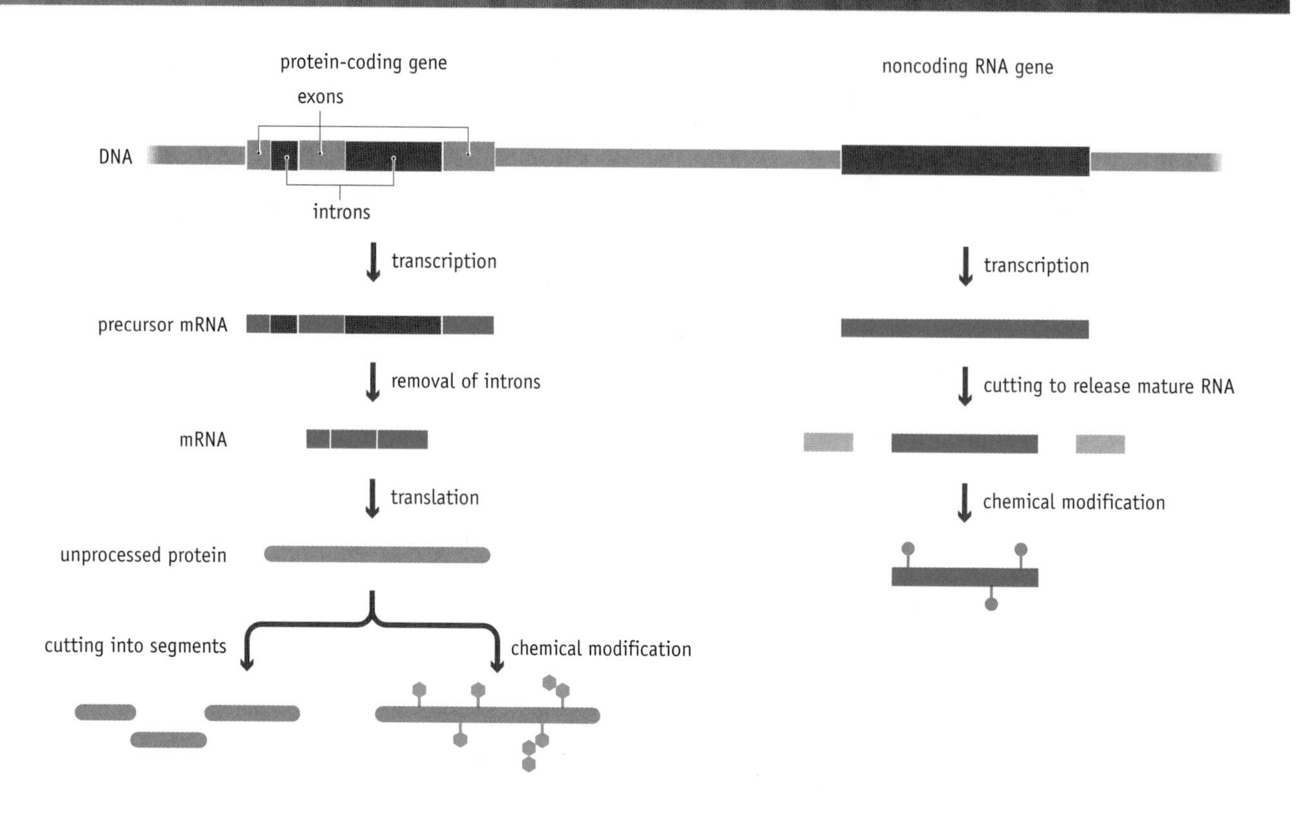

FIGURE 4.9 RNA and protein processing in eukaryotes. On the right, we see the same processing events for noncoding RNA as occur in bacteria. On the left, the processing of eukaryotic mRNAs and proteins is illustrated. If the gene is discontinuous, then the initial transcript is a precursor mRNA whose introns must be removed to give the mature mRNA. After translation of this mRNA, the initial protein product may be cut into segments, and may undergo chemical modification. Processing of mRNA is common in eukaryotes, but virtually unknown in bacteria because very few bacterial genes contain introns. Protein processing is also common in eukaryotes but much less frequent in bacteria.

the degrees of complexity and sophistication that we see in eukaryotes. This will become clear when we examine these regulatory mechanisms, in both types of organism, in Chapter 9.

A second difference is that processing events are more extensive in eukaryotes (Figure 4.9). In addition to the processing of the noncoding RNAs, eukaryotic mRNAs must be processed to remove introns from the transcripts of discontinuous genes. This is essential because the exons—the coding segments—must be linked together before the mRNA can be translated. In eukaryotes, many proteins are also processed, some by cutting into segments and some by **post-translational chemical modification**. These modifications include addition of chemical groups ranging in size from methyl residues ($-CH_3$) to large side chains made up of 5 to 10 sugar units. As with noncoding RNA processing, these protein processing events are an integral part of gene expression, because the end product, in this case the protein, is not functional until after it has been processed.

Besides the synthesis of RNA and protein, we must also consider the degradation, or **turnover**, of these molecules. Turnover is not simply a way of getting rid of unwanted molecules. It plays an active role in controlling gene expression by influencing the amount of an RNA or protein that is present in the cell at any particular time.

4.2 ENZYMES FOR MAKING RNA

An enzyme that transcribes DNA into RNA is called a **DNA-dependent RNA polymerase**. We can usually shorten this term to **RNA polymerase** without worrying about ambiguity. There are also **RNA-dependent RNA polymerases**, involved in replication of some viral RNAs, but it is usually clear from the context which RNA polymerase is meant. RNA polymerases are the central component of the transcription process, and before going any further we will therefore familiarize ourselves with the properties of these enzymes in both prokaryotes and eukaryotes.

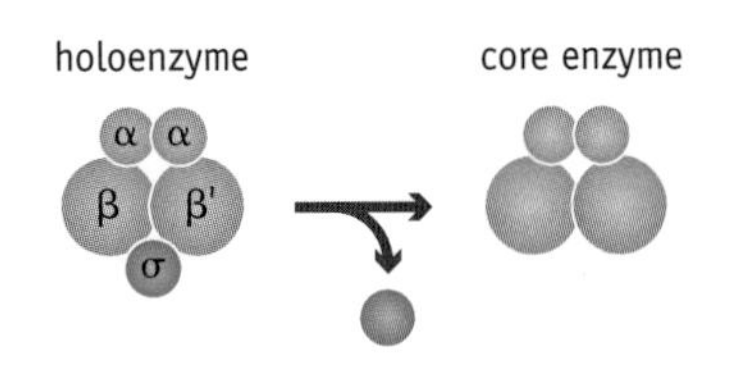

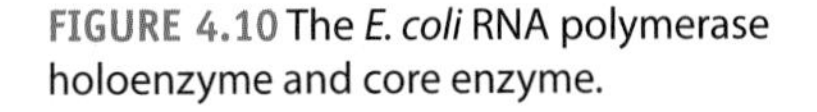
FIGURE 4.10 The *E. coli* RNA polymerase holoenzyme and core enzyme.

The RNA polymerase of *Escherichia coli* comprises five subunits

Like many of the proteins involved in gene expression, an RNA polymerase has to perform several tasks, as we will see when we study the events occurring during transcription, later in this chapter. It is not surprising, therefore, that the *E. coli* RNA polymerase is a large protein made up of several different polypeptide subunits.

Each *E. coli* cell contains about 7000 RNA polymerase molecules, between 2000 and 5000 of which may be actively involved in transcription at any one time. The structure of the enzyme is described as $\alpha_2\beta\beta'\sigma$, meaning that each molecule is made up of two α subunits plus one each of β, the related β′, and σ (Figure 4.10). This version of the enzyme is called the **holoenzyme** and has a molecular mass of 480 kD. It is distinct from a second form, the **core enzyme** (molecular mass 395 kD), which lacks the σ subunit and is just $\alpha_2\beta\beta'$. The two versions of the enzyme have different roles during transcription, as will be described below.

Eukaryotes possess more complex RNA polymerases

All *E. coli* genes are transcribed by the same type of RNA polymerase enzyme. In contrast, transcription of eukaryotic nuclear genes requires three different RNA polymerases. These are called **RNA polymerase I**, **RNA polymerase II**, and **RNA polymerase III**. The eukaryotic enzymes are larger than the *E. coli* version, each being made up of 8 to 12 subunits and having a total molecular mass in excess of 500 kD. Structurally, these polymerases are related to one another, the three largest subunits being similar in each enzyme and some of the smaller subunits being shared by more than one enzyme. The three largest subunits appear to be equivalent to the α, β, and β′ subunits of the *E. coli* enzyme. This has been deduced by comparing the amino acid sequences of the subunits from *E. coli* and yeast (which is a unicellular eukaryote) and by examining their exact roles during transcription.

Although the three eukaryotic RNA polymerases have related structures, their functions are quite distinct. Each works on a different set of genes, with no interchangeability (Table 4.1). Most research attention has been directed at RNA polymerase II, as this is the one that transcribes genes that code for proteins and hence synthesizes mRNA. But this is not its only role because RNA polymerase II also transcribes genes specifying the snRNAs and miRNAs, two types of noncoding RNA.

The other two types of eukaryotic RNA polymerase synthesize only noncoding RNA. RNA polymerase I transcribes a set of adjacent genes specifying three of the four rRNA molecules present in eukaryotic ribosomes. The fourth of these rRNAs is synthesized by RNA polymerase III, which also transcribes genes for tRNAs, snoRNAs, and a single type of snRNA.

TABLE 4.1 FUNCTIONS OF THE THREE EUKARYOTIC RNA POLYMERASES

Polymerase	Genes transcribed
RNA polymerase I	28S, 5.8S, and 18S rRNA genes
RNA polymerase II	Protein-coding genes, most snRNA genes, miRNA genes
RNA polymerase III	Genes for 5S rRNA, tRNAs, snoRNAs, U6-snRNA

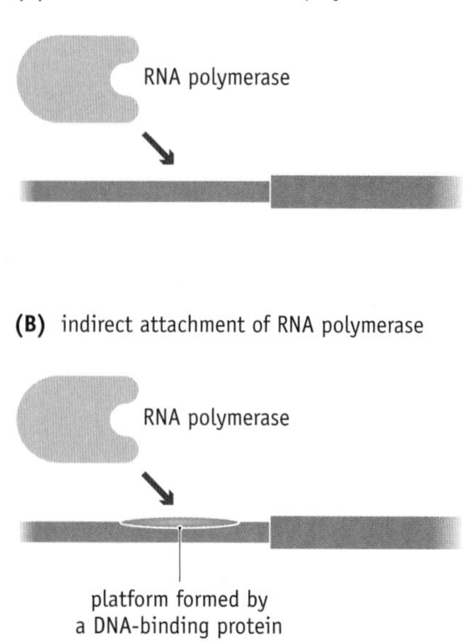

FIGURE 4.11 Two ways in which RNA polymerases bind to their promoters. (A) Direct attachment of the RNA polymerase to the DNA molecule, as occurs in bacteria. (B) Indirect attachment of the RNA polymerase via a platform formed by a DNA-binding protein, as occurs in eukaryotes.

4.3 RECOGNITION SEQUENCES FOR TRANSCRIPTION INITIATION

It is essential that transcription occurs at the correct positions on a DNA molecule. RNA polymerase enzymes must transcribe genes rather than random pieces of DNA, and must begin transcription near the start of a gene rather than in the middle. This means that the initial binding of an RNA polymerase to a DNA molecule must occur at a specific position, just in front (**upstream**) of the gene to be transcribed. Positions where transcription should begin are therefore marked by target sequences that are recognized either by the RNA polymerase itself or by a DNA-binding protein which, once attached to the DNA, forms a platform to which the RNA polymerase binds (Figure 4.11).

Bacterial RNA polymerases bind to promoter sequences

In bacteria, the target sequence for RNA polymerase attachment is called the **promoter**. A promoter is a short nucleotide sequence that is recognized by the bacterial RNA polymerase as a point at which it should bind to DNA in order to begin transcription. Promoters occur just upstream of genes, and nowhere else.

The sequences that make up the *E. coli* promoter were first identified by comparing the regions upstream of over 100 genes. It was assumed that promoter sequences would be very similar for all genes and so should be recognizable when the upstream regions were compared. These analyses showed that the *E. coli* promoter is made up of two distinct components, the –35 box and the –10 box (Figure 4.12). The latter is also called the **Pribnow box**, after the scientist who first discovered it. The names refer to the locations of the boxes on the DNA molecule relative to the position at which transcription starts. The nucleotide at this point is labeled +1 and is anywhere between 20 and 600 nucleotides upstream of the start of the coding region of the gene. The spacing between the two boxes is important, because it places the two motifs on the same face of the double helix, facilitating their interaction with the DNA-binding component of the RNA polymerase.

The sequence of the –35 box is described as 5′-TTGACA-3′, and that of the –10 box as 5′-TATAAT-3′. These are **consensus sequences** and so describe the "average" of all promoter sequences in *E. coli*. When we say that the consensus of the –35 box is 5′-TTGACA-3′, we mean that when all known promoter sequences are examined, the first nucleotide is most often a "T", as is the second nucleotide, the third is most often a "G", and so on. This means that the actual sequence upstream of any particular gene might be slightly different from the consensus (Table 4.2). These variations,

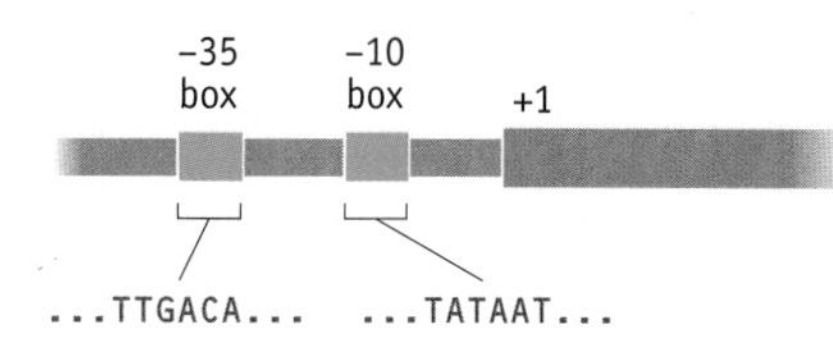

FIGURE 4.12 The two components of an *E. coli* promoter. The position at which transcription starts is labeled "+1". The two sequences are the consensus sequences of the –35 box and –10 box.

TABLE 4.2 SEQUENCES OF *E. COLI* PROMOTERS

	Sequence	
Promoter	–35 box	–10 box
Consensus	5′-TTGACA-3′	5′-TATAAT-3′
Lactose promoter*	5′-TTTACA-3′	5′-TATGTT-3′
Tryptophan promoter*	5′-TTGACA-3′	5′-TTAACT-3′

*These are the promoters for the lactose and tryptophan operons, which we will study in Chapter 9.

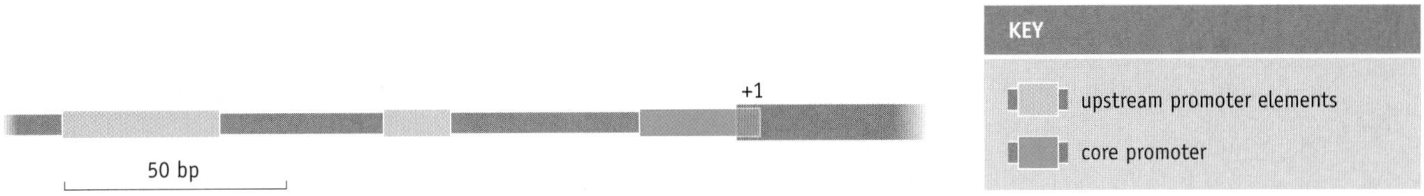

FIGURE 4.13 The components of a typical promoter for a eukaryotic RNA polymerase. The core promoter is the attachment point for the complex of proteins containing the RNA polymerase. Most core promoters are preceded by a series of upstream promoter elements that play an ancillary role in initiation of transcription.

together with less well-defined sequence features around the transcription start site and the following 50 or so nucleotides, affect the efficiency of the promoter. Efficiency is defined as the number of productive initiations that are promoted per second, a productive initiation being one that results in the RNA polymerase leaving the promoter and beginning synthesis of a full-length transcript. The exact way in which the sequence of the promoter affects initiation is not known, but it is clear that different promoters vary 1000-fold in their efficiencies. The most efficient promoters (called **strong promoters**) direct 1000 times as many productive initiations as the weakest promoters. We refer to these as differences in the **basal rate** of transcript initiation.

Eukaryotic promoters are more complex

In eukaryotes, the term "promoter" is used to describe all the sequences that are important in initiation of transcription of a gene. For some genes these sequences can be numerous and diverse in their functions. They include the **core promoter**, which is the site at which the initiation complex containing the RNA polymerase is assembled, as well as one or more short sequences found at positions upstream of the core promoter (Figure 4.13). Assembly of the initiation complex on the core promoter can usually occur in the absence of the upstream elements, but only in an inefficient way.

Each of the three eukaryotic RNA polymerases recognizes a different type of promoter sequence. Indeed, it is the difference between the promoters that defines which genes are transcribed by which polymerase.

The promoters for RNA polymerase II, the enzyme that transcribes protein-coding genes, are the most complicated, some stretching for several kb upstream of the transcription start site (Figure 4.14). The RNA polymerase II core promoter consists of two main segments. The first of these is the –25 or **TATA box**, which has the consensus sequence 5′-TATAWAAR-3′. In

+1

50 bp

...TATAWAAR...
TATA box

...YCANTYY...
Inr sequence

FIGURE 4.14 A typical promoter for a gene transcribed by RNA polymerase II. The core promoter is made up of two segments, the TATA box and Inr sequence. The consensus sequences of these two segments are shown.

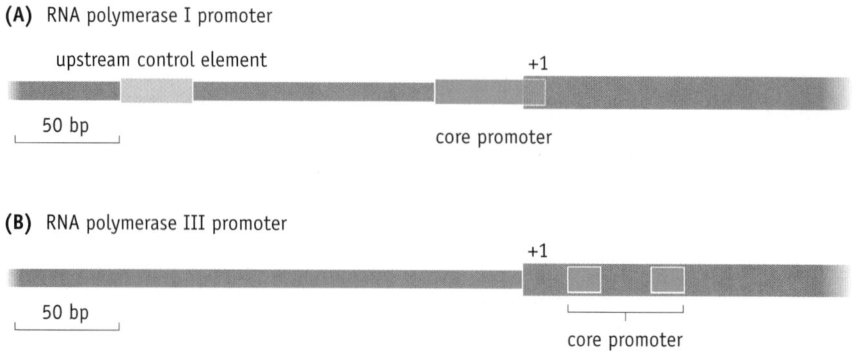

(A) RNA polymerase I promoter

upstream control element
+1
50 bp
core promoter

(B) RNA polymerase III promoter

+1
50 bp
core promoter

FIGURE 4.15 Promoters for genes transcribed by RNA polymerases I and III. (A) A typical promoter for RNA polymerase I consists of a core promoter that spans the point where transcription will begin, and a single upstream control element. (B) RNA polymerase III promoters are variable. In this example, the core promoter is made up of two segments located within the gene.

this sequence W indicates either A or T, either of which is equally likely to occur at this position, and R is a purine, A or G. The second part of the core promoter is the **initiator (Inr) sequence**, which is located around nucleotide +1. In mammals the consensus of the Inr sequence is 5′-YCANTYY-3′, where Y is a pyrimidine, C or T, and where N is any of the four nucleotides. Some genes transcribed by RNA polymerase II have only one of these two components of the core promoter, and some, surprisingly, have neither. The latter are called "null" genes. They are still transcribed, although the start position for transcription is more variable than for a gene with a TATA and/or an Inr sequence. Besides the components of the core promoter, genes transcribed by RNA polymerase II also have various upstream sequence elements to which regulatory proteins bind. These will be described when we study the control of gene expression in Chapter 9.

The promoters for RNA polymerases I and III are not so complicated (Figure 4.15). RNA polymerase I promoters consist of a core promoter region spanning the transcription start point, between nucleotides –45 and +20, and a sequence called an **upstream control element** about 100 bp further upstream. RNA polymerase III promoters are variable, falling into at least three categories. Two of these categories are unusual in that the important sequences are located within the genes whose transcription they promote. Usually these sequences span 50 to 100 bp and comprise two segments whose sequences are similar in all promoters of a particular type, separated by a variable region.

Some eukaryotic genes have more than one promoter

In eukaryotes, an additional level of complexity is seen with some protein-coding genes that have **alternative promoters**. This means that transcription of the gene can begin at two or more different sites, giving rise to mRNAs of different lengths.

An example is provided by the human dystrophin gene, which has been extensively studied because defects in this gene result in the genetic disease called Duchenne muscular dystrophy. The dystrophin gene is one of the largest known in the human genome, stretching over 2.4 Mb and containing 78 introns. It has at least seven alternative promoters (Figure 4.16). The different promoters are active in different parts of the body, such as the

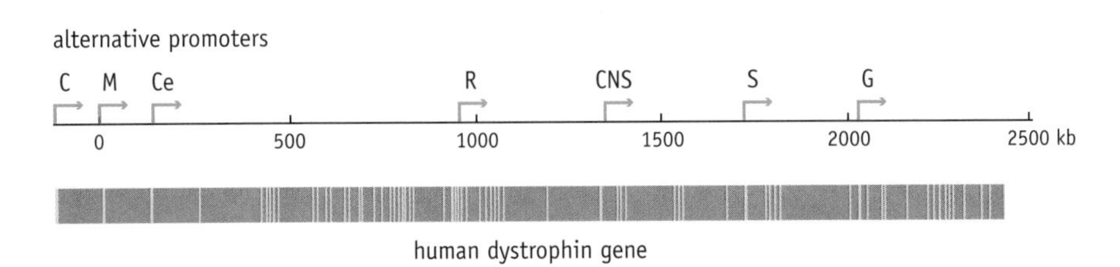

FIGURE 4.16 Alternative promoters. The positions of seven alternative promoters for the human dystrophin gene are shown. Abbreviations indicate the tissue within which each promoter is active: C, cortical tissue; M, muscle; Ce, cerebellum; R, retinal tissue (and also brain and cardiac tissue); CNS, central nervous system (and also kidney); S, Schwann cells; G, general (most other tissues). The exons of the gene are shown in yellow, and the introns in gray.

brain, muscle, and the retina, enabling different versions of the dystrophin protein to be made in these various tissues. Presumably the biochemical properties of these variants are matched to the needs of the cells in which they are synthesized.

In addition to tissue-specific patterns of expression like that directed by the alternative promoters of the dystrophin gene, alternative promoters can also generate related versions of a protein at different stages in development, and also can enable a single gene to direct synthesis of two or more proteins at the same time in a single tissue. This last point indicates that although they are usually referred to as *alternative* promoters, these are, more correctly, *multiple* promoters, as more than one may be active at a particular time. Indeed, this may be the normal situation for many genes. For example, over 10,500 promoters have been shown to be active in human fibroblast cells, but these promoters drive expression of fewer than 8000 genes. A substantial number of genes in these cells are therefore being expressed from two or more promoters.

4.4 INITIATION OF TRANSCRIPTION IN BACTERIA AND EUKARYOTES

It is useful to divide the transcription process into three phases, referred to as initiation, elongation, and termination. Of these, initiation is looked on as the most important phase because it is a key control point in the gene expression pathway. Regulation of transcript initiation often determines whether or not a gene is active in a particular cell at a particular time.

In a general sense, initiation of transcription operates along the same lines with any of the four types of RNA polymerase (Figure 4.17). The bacterial polymerase and the three eukaryotic enzymes all begin by attaching, directly or via accessory proteins, to their promoter or core promoter sequences. Next, the resulting **closed promoter complex** is converted into an **open promoter complex** by breakage of a limited number of base pairs adjacent to the transcription initiation site. Finally, the RNA polymerase moves away from the promoter. This last step is more complicated than it might appear, because some attempts by the polymerase to leave

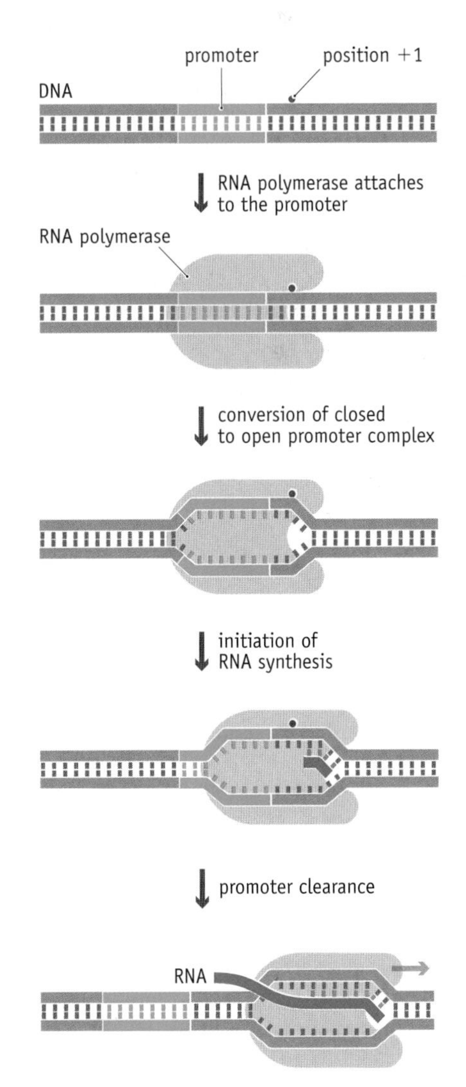

FIGURE 4.17 Generalized scheme for the events occurring during initiation of transcription. The diagram shows initiation of transcription by a bacterial RNA polymerase, but equivalent events take place with each of the three eukaryotic enzymes. Attachment of the RNA polymerase to the promoter (shown in green) is followed by the breakage of base pairs adjacent to the transcription initiation site (indicated by the red dot) to give the open promoter complex. RNA synthesis then begins. Initiation of transcription is completed by 'promoter clearance', which occurs when the RNA polymerase has made a stable attachment to the DNA and has begun to synthesize a full-length transcript.

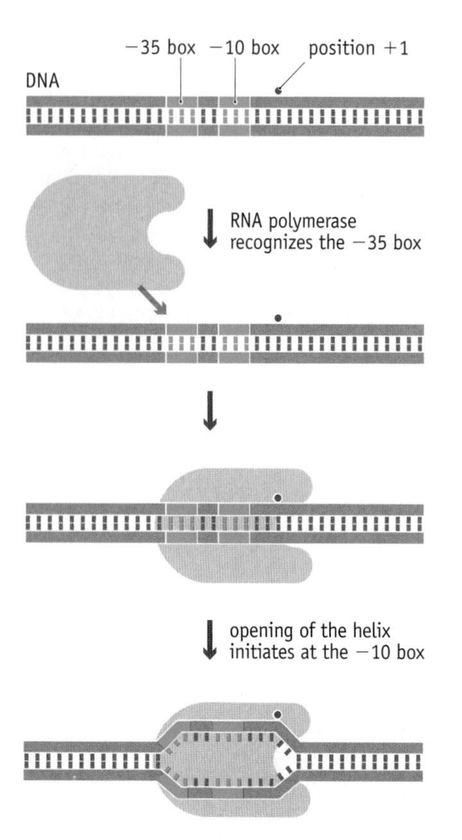

FIGURE 4.18 Details of the initiation of transcription in *E. coli*. The RNA polymerase recognizes the –35 box. The breakage of base pairs that converts the closed promoter complex into the open complex is thought to occur initially within the –10 box.

the promoter region are unsuccessful and lead to truncated transcripts that are degraded soon after they are synthesized. True completion of the initiation stage of transcription therefore occurs when the RNA polymerase has made a stable attachment to the DNA and has begun to synthesize a full-length transcript.

Although the scheme shown in Figure 4.17 is correct in outline for all four polymerases, the details are different for each one. We will begin with the more straightforward events occurring in *E. coli* and other bacteria, and then move on to the ramifications of initiation in eukaryotes.

The σ subunit recognizes the bacterial promoter

In *E. coli*, a direct contact is formed between the promoter and the RNA polymerase. The sequence specificity of the polymerase resides in its σ subunit. The "core enzyme," which lacks this component, can make only loose and nonspecific attachments to DNA.

Studies of *E. coli* promoters have shown that the ability of the RNA polymerase to bind can be affected if changes are made to the –35 box, whereas changes to the –10 box affect the conversion of the closed promoter complex into the open form. These results have led to the model for *E. coli* initiation shown in Figure 4.18, where recognition of the promoter occurs through an interaction between the σ subunit and the –35 box. This interaction forms a closed promoter complex in which the RNA polymerase spans some 80 bp of DNA, from upstream of the –35 box to downstream of the –10 box. The closed promoter complex is converted to the open form by the combined action of the β′ and σ subunits, which break the base pairs within the –10 box. The model is consistent with the fact that the –10 boxes of different promoters are comprised mainly or entirely of A–T base pairs, which are linked by just two hydrogen bonds and are therefore weaker than G–C pairs, which have three hydrogen bonds (see Figure 2.7).

Opening up of the helix requires that contacts be made between the polymerase and the nontranscribed strand of the gene, again with the σ subunit playing a central role. However, the σ subunit is not all-important, because it usually (but not always) dissociates soon after initiation is complete, converting the holoenzyme to the core enzyme, which carries out the elongation phase of transcription. Initially the core enzyme covers some 60 bp of the DNA, but soon after the start of elongation the polymerase undergoes a second conformational change, reducing its footprint to just 30 to 40 bp.

Formation of the RNA polymerase II initiation complex

How does initiation of transcription in *E. coli* compare with the equivalent processes in eukaryotes? Examining transcription by RNA polymerase II will show us that eukaryotic initiation involves more proteins and has added complexities.

As illustrated in Figure 4.11, an important difference between initiation of transcription in *E. coli* and in eukaryotes is that eukaryotic polymerases do not directly recognize their core promoter sequences. For genes transcribed by RNA polymerase II, the initial contact is made by the protein called **transcription factor IID**, or **TFIID**. This is another multisubunit complex, made up of the **TATA-binding protein** (**TBP**) and at least twelve **TBP-associated factors**, or **TAFs**. Structural studies of TBP show that it has a saddlelike shape that wraps partially around the double helix, forming a platform onto which the remainder of the initiation complex can be

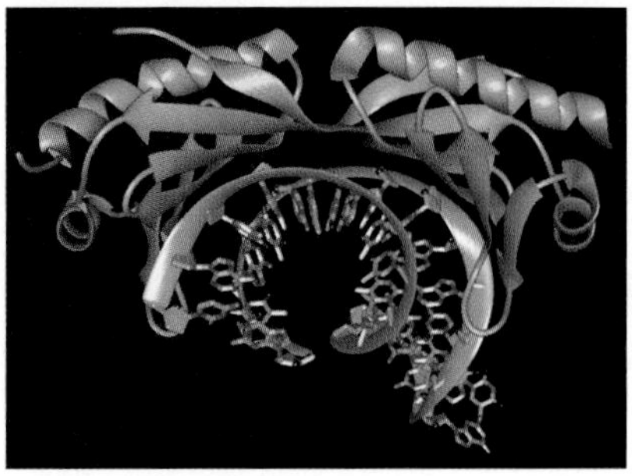

FIGURE 4.19 The role of the TATA-binding protein (TBP) in initiation of transcription in eukaryotes. TBP is a dimer of two identical subunits, shown here in brown. The protein attaches to the core promoter, forming a platform onto which the initiation complex can be assembled. The DNA is shown in silver—in this view we are looking directly along the double helix. (Courtesy of Song Tan, Pennsylvania State University.)

assembled (Figure 4.19). The TAFs assist in attachment of TBP to the TATA box and, in conjunction with other proteins called **TAF- and initiator-dependent cofactors (TICs)**, possibly also participate in recognition of the Inr sequence, especially at those promoters that lack a TATA box.

The next step results in attachment of RNA polymerase II to the complex (Figure 4.20). This involves two more transcription factors. These are TFIIB, which attaches to the DNA-bound TBP and forms the structure that RNA polymerase II recognizes, and TFIIF, which binds to the RNA polymerase and prevents it from attaching to the DNA at the wrong place. Finally, two more transcription factors, TFIIE and TFIIH, are recruited to complete the initiation complex. TFIIH is particularly important, as it is a **helicase**, a type of enzyme that is able to break the base pairs that hold the two strands of the double helix together. This activity is needed to convert the closed promoter complex into the open form.

The initiation complex must be activated before it will begin to transcribe the DNA to which it is attached. Activation involves addition of phosphate groups to the largest subunit of RNA polymerase II, specifically to a series of amino acids within the part of this protein referred to as the **C-terminal domain (CTD)**. Once these phosphates have been attached, the polymerase is able to leave the initiation complex and begin synthesizing RNA.

After departure of the polymerase, at least some of the transcription factors detach from the core promoter, but TFIID and TFIIH remain, enabling re-initiation to occur without the need to rebuild the entire assembly from the beginning. Re-initiation is therefore a more rapid process than primary initiation, which means that once a gene is switched on, transcripts can be initiated from its promoter with relative ease until such a time as a new set of signals switches the gene off.

Initiation of transcription by RNA polymerases I and III

We know rather less about the way in which transcription is initiated at RNA polymerase I and III promoters, but it is clear that in some respects the events are similar to those seen with RNA polymerase II. One of the most striking similarities is that TBP, first identified as the key promoter-locating

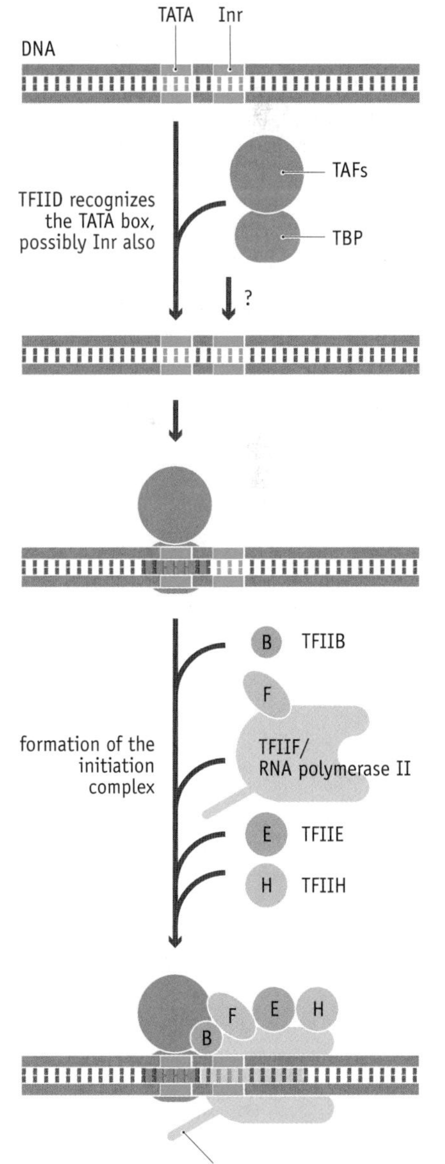

FIGURE 4.20 Initiation of transcription by RNA polymerase II. The first step is attachment of transcription factor IID (TFIID) to the core promoter. TFIID is made up of the TATA-binding protein (TBP) and the TBP-associated factors (TAFs). TFIID recognizes the TATA box and attaches to the DNA at that point. Additional proteins might also enable the Inr sequence to be recognized as an attachment site. In the second step, the initiation complex is completed by attachment of the RNA polymerase. This step involves at least four additional transcription factors TFIIB, E, F, and H. Transcription begins when the initiation complex is activated by attachment of phosphate groups to the C-terminal domain (CTD) of the RNA polymerase.

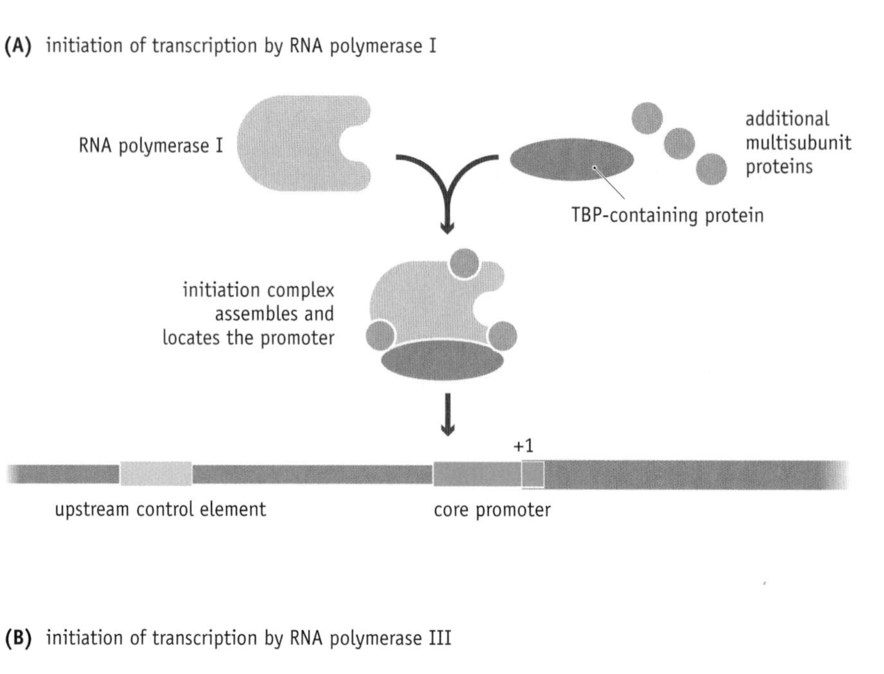

FIGURE 4.21 Initiation of transcription by RNA polymerase I and RNA polymerase III. (A) RNA polymerase I forms an initiation complex with four multisubunit proteins, one of which contains TBP. This initiation complex then attaches to the RNA polymerase I core promoter. (B) Initiation by RNA polymerase III begins with attachment of the TBP component of TFIIIB to the core promoter. The RNA polymerase then attaches to the bound TBP.

component of the RNA polymerase II initiation complex, is also involved in initiation of transcription by the two other eukaryotic RNA polymerases.

The RNA polymerase I initiation complex is made up of the polymerase and four additional multisubunit proteins, one of which contains TBP. These proteins locate the core promoter and the upstream control element and hence direct RNA polymerase I to its correct attachment point (**Figure 4.21A**). Originally it was thought that the initiation complex was built up in a stepwise fashion, but recent research suggests that RNA polymerase I binds to its four protein partners before locating the promoter, the entire assembly attaching to the DNA in a single step.

TBP is also a subunit of the transcription factor called TFIIIB, which is involved in assembly of RNA polymerase III initiation complexes. The variability displayed by the promoters for this polymerase is reflected in the processes used for assembly of the initiation complexes, a different set of transcription factors being required for each category of promoter. TFIIIB provides the common link by making the initial contact with the core component of each promoter and, via its TBP subunit, providing the platform onto which RNA polymerase III is attached (**Figure 4.21B**).

4.5 THE ELONGATION PHASE OF TRANSCRIPTION

Once successful initiation has been achieved, the RNA polymerase begins to synthesize the transcript. Synthesis of an RNA molecule involves polymerization of ribonucleotide subunits (NTPs) and can be summarized as:

$$n(\text{NTP}) \rightarrow \text{RNA of length } n \text{ nucleotides} + (n-1)(\text{PP}_i)$$

The reaction is shown in detail in Figure 4.22. During each nucleotide addition, the β and γ phosphates are removed from the incoming nucleotide, and the hydroxyl group is removed from the 3′ carbon of the nucleotide at

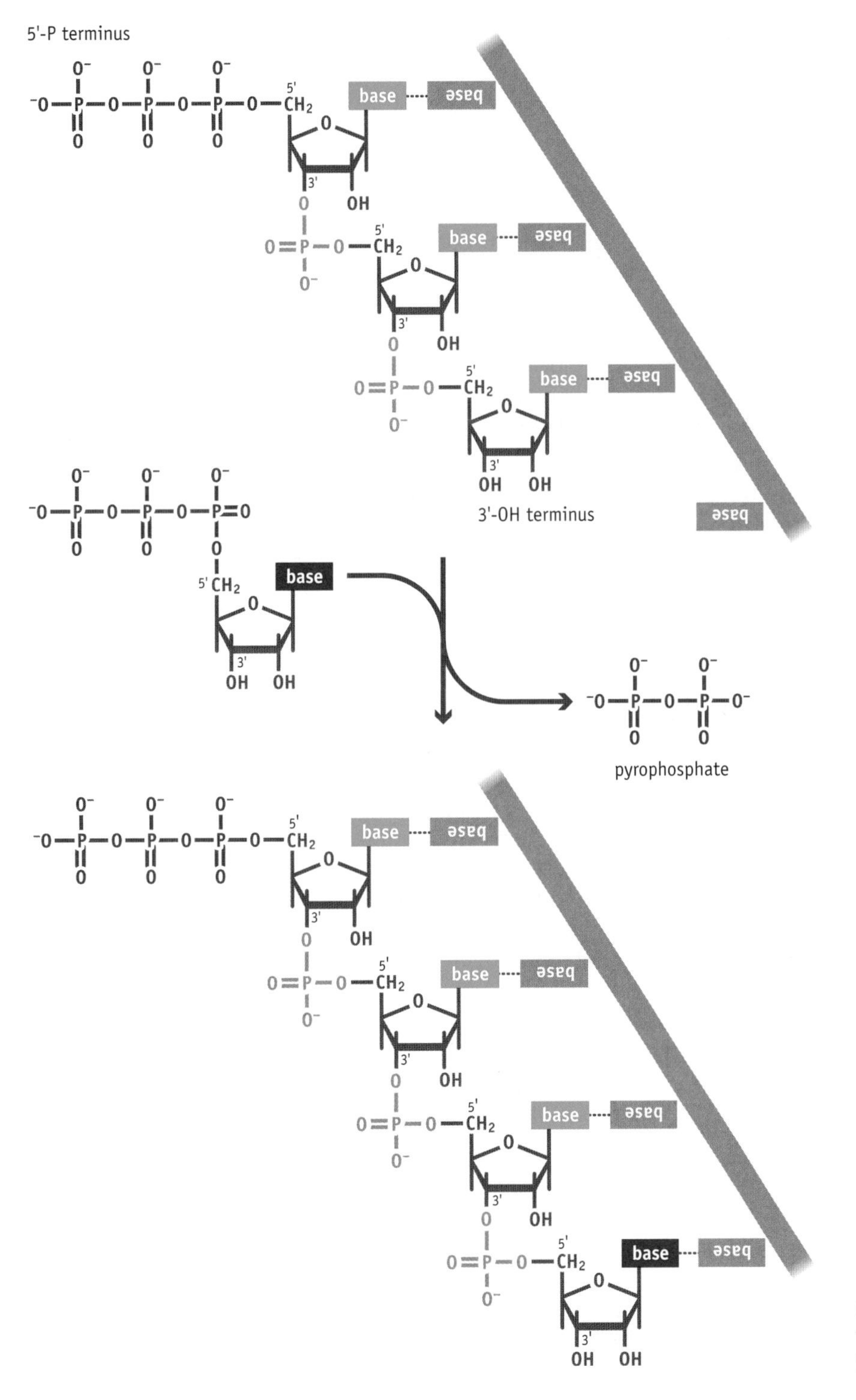

FIGURE 4.22 The polymerization reaction that results in synthesis of an RNA polynucleotide.

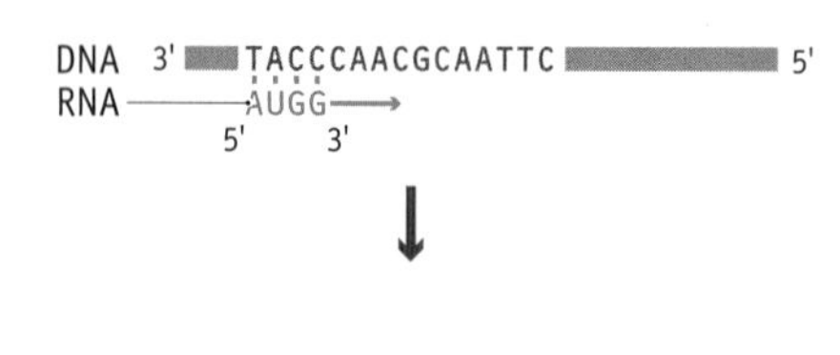

FIGURE 4.23 Template-dependent RNA synthesis. The RNA molecule, shown in blue, is extended by adding nucleotides one by one to its 3′ end.

the end of the chain. This results in loss of a pyrophosphate molecule (PP_i) for each bond formed. In transcription, the chemical reaction is modulated by the presence of the DNA template, which directs the order in which the individual ribonucleotides are polymerized into RNA, A base-pairing with T or U, and G base-pairing with C (Figure 4.23). The RNA transcript is therefore built up in a step-by-step fashion in the 5′ → 3′ direction, new ribonucleotides being added to the free 3′ end of the existing polymer. Remember that in order to base-pair, complementary polynucleotides must be antiparallel. This means that the transcribed strand of the gene must be read in the 3′ → 5′ direction.

Bacterial transcripts are synthesized by the RNA polymerase core enzyme

During the elongation stage of transcription, the bacterial RNA polymerase is in its core enzyme form, denoted as $\alpha_2\beta\beta'$. The σ subunit that has the key role in initiation has usually left the complex at this stage. The RNA polymerase covers about 30 to 40 bp of the template DNA, including the **transcription bubble** of 12 to 14 bp, within which the growing transcript is held to the transcribed strand of the DNA by approximately eight RNA–DNA base pairs (Figure 4.24).

The RNA polymerase has to keep a tight grip on both the DNA template and the RNA that it is making in order to prevent the transcription complex from falling apart before the end of the gene is reached. However, this grip must not be so tight as to prevent the polymerase from moving along the DNA. Structural studies of actively transcribing RNA polymerase enzymes have shown that the DNA molecule lies between the β and β′ subunits, within a trough on the enclosed surface of β′ (Figure 4.25). The active site for RNA synthesis also lies between these two subunits, with the nontranscribed strand of DNA looping away at this point. The RNA transcript extrudes from the polymerase via a channel formed partly by the β and partly by the β′ subunit.

The polymerase does not synthesize its transcript at a constant rate. Instead, synthesis is discontinuous, with periods of rapid elongation interspersed with brief pauses during which the active site of the polymerase undergoes a slight structural rearrangement. A pause rarely lasts longer than 6 milliseconds, and might be accompanied by reverse movement of the polymerase (**backtracking**) along the template. Pauses occur randomly rather than being caused by any particular feature of the template DNA. Pausing plays an important role in transcript termination, as described below, but whether or not this is its only function is currently unknown.

All eukaryotic mRNAs have a modified 5′ end

The fundamental aspects of transcript elongation are the same in bacteria and eukaryotes. The one major distinction concerns the precise structure

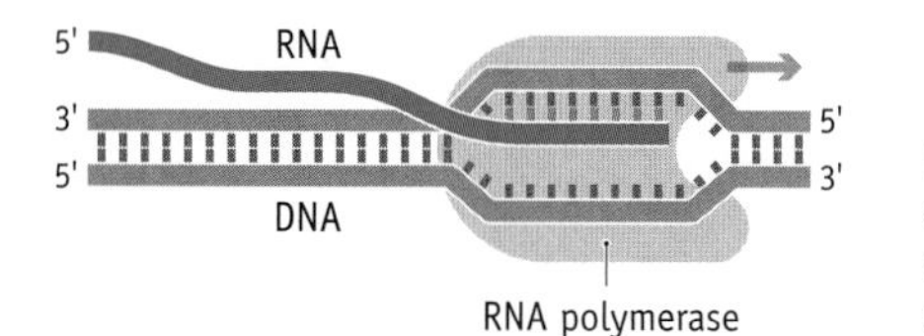

FIGURE 4.24 The *E. coli* transcription elongation complex. The arrow indicates the direction in which the polymerase moves along the DNA. In reality, the polymerase is larger than drawn here, and covers 30–40 bp of the template DNA.

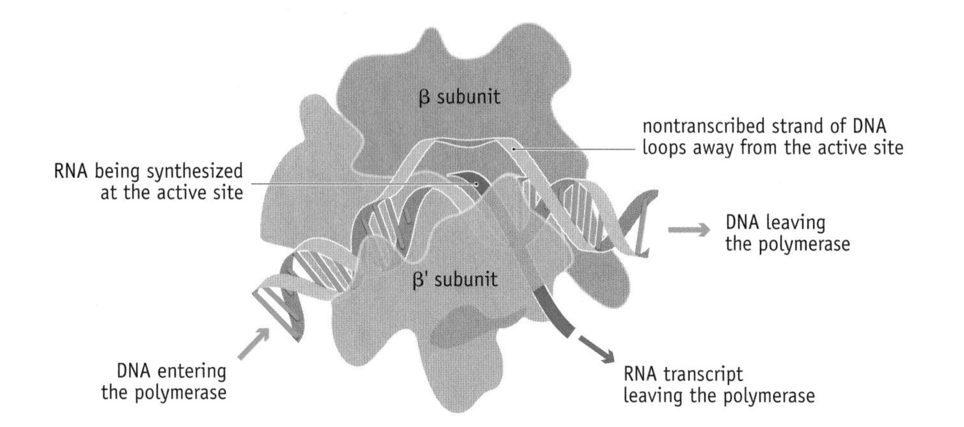

FIGURE 4.25 Synthesis of RNA within a bacterial RNA polymerase. The β and β' subunits of the RNA polymerase are depicted in gray, the double helix is colored green, and the RNA transcript is blue. The positions at which the DNA enters and leaves the complex are marked, as is the exit point for the RNA transcript. Note how the nontranscribed strand of the DNA molecule loops away from the active site at which RNA is being synthesized. Compare this depiction of the polymerase with the schematic drawing shown in Figure 4.24.

of the 5′ end of a eukaryotic mRNA. There is usually a triphosphate group at the 5′ end of a polynucleotide (see Figure 2.5). This applies both to DNA polynucleotides and to the RNA molecules that are synthesized by transcription. As a simple notation, we can describe the 5′ end as pppNpN ..., where N is the sugar–base component of the nucleotide and p represents a phosphate group.

The RNAs made by RNA polymerase II are exceptions to this rule. With these molecules, the 5′ terminus has a more complex chemical structure described as 7-MeGpppNpN ..., where 7-MeG is the nucleotide carrying the modified base 7-methylguanosine. This is referred to as the **cap structure**, and it is put in place soon after the RNA polymerase leaves the promoter, before the mRNA reaches 30 nucleotides in length.

The first step in "capping" is addition of the extra guanosine to the extreme 5′ end of the RNA. Rather than occurring by normal RNA polymerization, capping involves a reaction between the 5′ triphosphate of the terminal nucleotide and the triphosphate of a GTP nucleotide. The γ phosphate of the terminal nucleotide (the outermost phosphate) is removed, as are the β and γ phosphates of the GTP, resulting in a 5′–5′ bond (Figure 4.26A). The reaction is carried out by the enzyme **guanylyl transferase**.

The second step of the capping reaction converts the new terminal guanosine into 7-methylguanosine by attachment of a methyl group to nitrogen number 7 of the purine ring, this modification catalyzed by **guanine methyltransferase**. The two capping enzymes make attachments with the RNA polymerase and it is possible that they are intrinsic components of the transcription complex during the early stages of RNA synthesis.

The 7-methylguanosine structure is called a **type 0 cap** and is the most common form in unicellular eukaryotes such as yeast. In higher eukaryotes, including humans, additional modifications occur (Figure 4.26B).

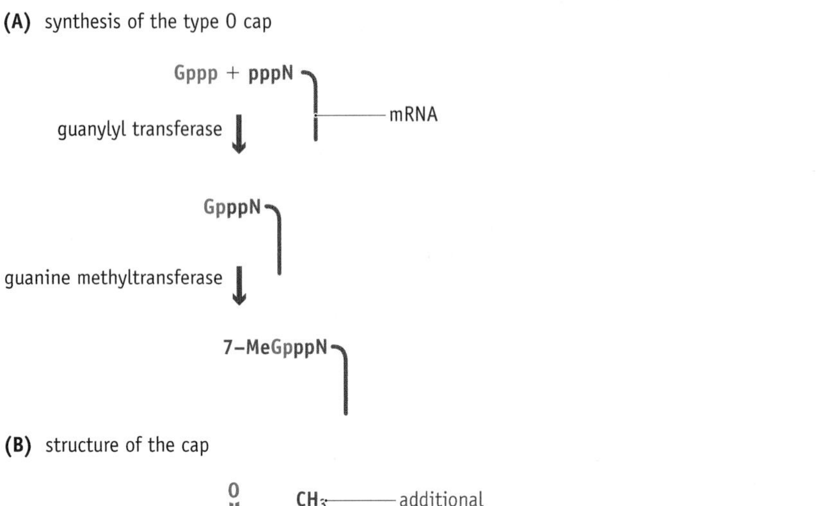

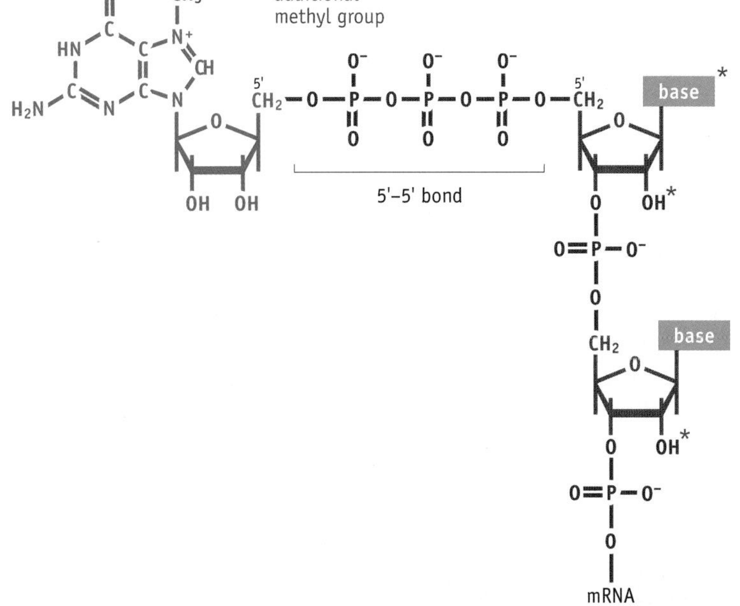

FIGURE 4.26 The cap structure at the 5′ end of a eukaryotic mRNA. (A) Synthesis of the type 0 cap. The nucleotide at the 5′ end of the mRNA is depicted as "pppN" to indicate that it has a terminal triphosphate. The GTP molecule, in blue, is depicted as "Gppp", again indicating that it has a triphosphate group. The type 0 cap is synthesized in two steps. First the GTP is attached to the terminal nucleotide, forming a 5′–5′ triphosphate linkage comprising one phosphate from the GTP and two from the terminal nucleotide of the mRNA. In the second step a methyl group is added to position 7 of the guanine base. (B) The detailed structure of the type 0 cap. The red stars indicate the positions where additional methyl groups might be added to produce the type 1 and type 2 cap structures.

The most common of these affects what is now the second nucleotide in the transcript, where the hydrogen of the 2′–OH group is replaced with a methyl group. This results in a **type 1 cap**. If this second nucleotide is an adenosine, then a methyl group might also be added to the NH_2 attached to carbon number 6 of the purine. Finally, another 2′–OH methylation might occur at the third nucleotide position, resulting in a **type 2 cap**.

All RNAs synthesized by RNA polymerase II are capped in one way or another. This means that in addition to mRNAs, the snRNAs that are transcribed by this enzyme are also capped (see Table 4.1). The cap may be important for export of mRNAs and snRNAs from the nucleus, but its best-defined role is in translation of mRNAs (Chapter 8).

Some eukaryotic mRNAs take hours to synthesize

The longest bacterial genes are only a few kb in length and can be transcribed in a matter of minutes by the bacterial RNA polymerase, which has a polymerization rate of several hundred nucleotides per minute. In contrast, RNA polymerase II can take hours to transcribe a single gene, even though it can work at up to 2000 nucleotides per minute. This is because the presence of multiple introns in many eukaryotic genes means that considerable lengths of DNA must be copied. For example, the transcript of the human dystrophin gene is 2400 kb in length and takes about 20 hours to synthesize.

The extreme length of eukaryotic genes places demands on the stability of the transcription complex. RNA polymerase II on its own is not able to meet these demands. When the purified enzyme is studied, its polymerization rate is found to be less than 300 nucleotides per minute and the transcripts that it makes are relatively short. In the nucleus, the transcription complex is stabilized by **elongation factors**, proteins that associate with the polymerase after it has moved away from the promoter. Thirteen elongation factors are currently known in mammalian cells. Their importance is shown by the effects of mutations that disrupt the activity of one or another of the factors. Inactivation of the elongation factor called CSB, for example, results in Cockayne syndrome, a disease characterized by developmental defects such as mental retardation. Disruption of a second factor, ELL, causes acute myeloid leukemia.

4.6 TERMINATION OF TRANSCRIPTION

Current thinking views transcription as a discontinuous process, with the polymerase pausing regularly and "making a choice" between continuing elongation by adding more ribonucleotides to the transcript, or terminating by dissociating from the template. Which choice is selected depends on which alternative is more favorable in thermodynamic terms. This means that, in order for termination to occur, the polymerase has to reach a position on the template where dissociation is more favorable than continued RNA synthesis.

Hairpin loops in the RNA are involved in termination of transcription in bacteria

Bacteria use two distinct strategies for transcription termination, but a common feature of both is that the termination site is marked by an **inverted repeat** in the DNA sequence. An inverted repeat is a segment of DNA or RNA in which a sequence is followed by its reverse complement (Figure 4.27). If the two halves of the inverted repeat are separated by a few intervening nucleotides, as is the case at a termination site, then

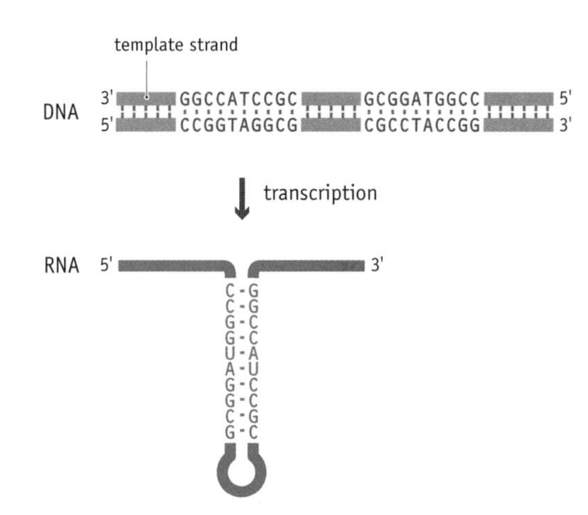

FIGURE 4.27 An inverted repeat in a double-stranded DNA molecule can give rise to a hairpin loop in a single-stranded RNA molecule.

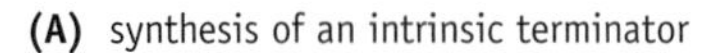

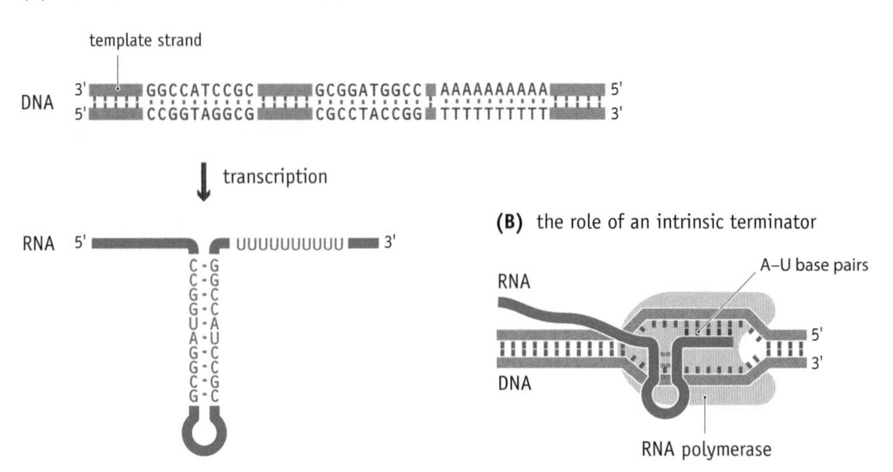

FIGURE 4.28 Termination at an intrinsic terminator. (A) Synthesis of an intrinsic terminator involves transcription of an inverted repeat sequence followed by a series of A nucleotides. (B) An intrinsic terminator is thought to weaken the DNA–RNA interaction because the presence of the hairpin loop reduces the number of contacts between the template DNA and the transcript, and those contacts that are made are via relatively weak A–U base pairs.

intrastrand base pairing between the two sequence components can form a hairpin loop in a single-stranded polynucleotide, in this case the RNA transcript.

At about half the positions in *E. coli* at which transcription terminates, the inverted repeat is followed by a run of deoxyadenosine nucleotides in the nontranscribed strand (Figure 4.28). These are called **intrinsic terminators**, and the hairpin loops that they specify are relatively stable, more so than the DNA–RNA pairing that normally occurs within the transcription bubble. This means that formation of the hairpin loop is favored, reducing the number of contacts between the template DNA and transcript, and weakening the overall DNA–RNA interaction. The interaction is further weakened when the run of As in the DNA is transcribed, because the resulting A–U base pairs have only two hydrogen bonds each, compared with three for each G–C pair. The net result is that detachment of the transcript is favored over continued elongation.

Structural studies suggest that it is also possible that the RNA hairpin makes contact with a flap structure on the outer surface of the RNA polymerase β subunit, adjacent to the exit point of the channel through which the RNA emerges from the complex (Figure 4.29). Movement of the flap could affect the positioning of amino acids within the active site, possibly leading to breakage of the DNA–RNA base pairs and termination of transcription.

The second type of bacterial termination signal is **Rho-dependent**. Signals of this type usually include an inverted repeat as seen at intrinsic terminators, although the hairpin that is formed is less stable and there is no run of

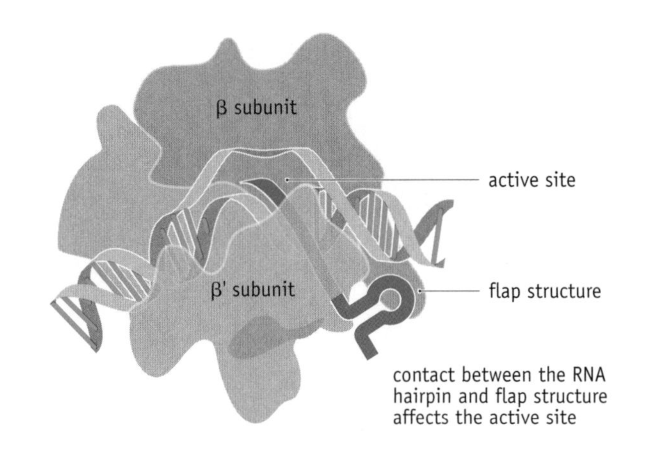

FIGURE 4.29 A model for termination of transcription in bacteria. A flap structure on the surface of the RNA polymerase could mediate termination by responding to the presence of the hairpin loop in the RNA that is leaving the polymerase. Movement of the flap might affect the positioning of amino acids at the active site of the polymerase.

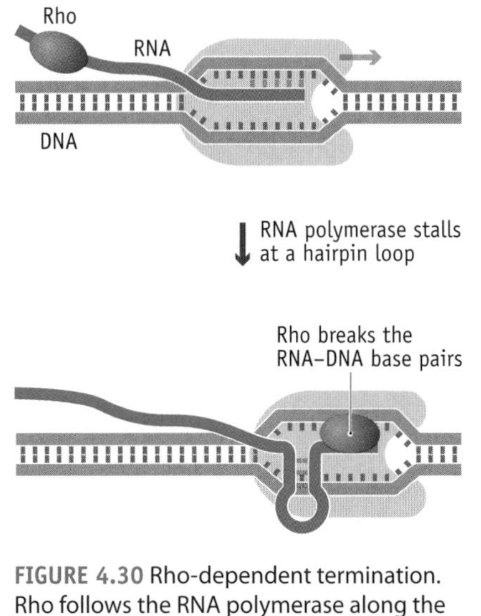

FIGURE 4.30 Rho-dependent termination. Rho follows the RNA polymerase along the transcript. When the polymerase stalls at a hairpin, Rho catches up and breaks the RNA–DNA base pairs, releasing the transcript.

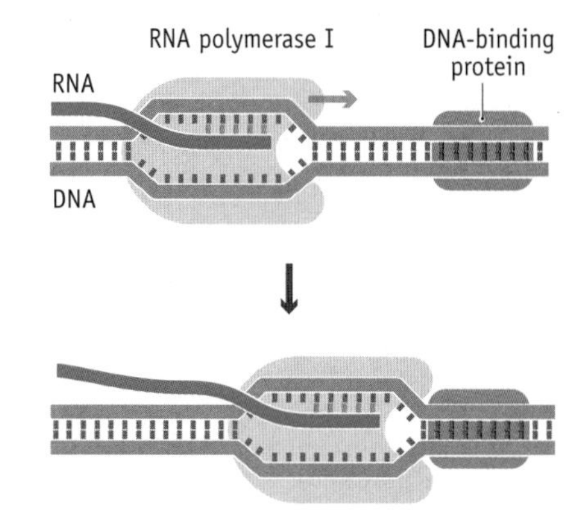

FIGURE 4.31 A possible scheme for termination of transcription by RNA polymerase I. In this model, the polymerase becomes stalled because its progress is blocked by the presence of the DNA-binding protein.

As in the transcribed strand of the DNA. Termination requires the activity of a protein called Rho, which attaches to the transcript and moves along the RNA toward the polymerase. If the polymerase continues to synthesize RNA, then it keeps ahead of the pursuing Rho, but at the termination signal the polymerase stalls (Figure 4.30). Exactly why has not been explained, but presumably the hairpin loop that forms in the RNA is responsible in some way. The result is that Rho is able to catch up. Rho is a helicase, which means that it actively breaks base pairs, in this case between the DNA and the transcript, resulting in termination of transcription.

Eukaryotes use diverse mechanisms for termination of transcription

The three eukaryotic RNA polymerases each use different mechanisms for transcript termination. With RNA polymerase I, termination involves a DNA-binding protein that attaches to the DNA at a recognition sequence located 12 to 20 bp downstream of the point at which transcription terminates (Figure 4.31). Exactly how the bound protein causes termination is not known, but a model in which the polymerase becomes stalled because of its blocking effect has been proposed. Even less is known about RNA polymerase III termination. A run of adenosines in the template is implicated, but the process does not involve a hairpin loop and so is not analogous to the intrinsic termination system in bacteria.

With RNA polymerase II, termination is accompanied by addition of a series of adenosine nucleotides, called the **poly(A) tail**, to the 3′ end of the mRNA that is being synthesized (Figure 4.32). These adenosines are not transcribed from the DNA—there is no equivalent run of deoxythymidines in the transcribed DNA strand—but instead are added by a DNA-*in*dependent RNA polymerase called **poly(A) polymerase**. This polymerase does not act at the extreme 3′ end of the transcript, but at an internal site which is cleaved to create a new 3′ end to which the poly(A) tail is added. In mammals, this **polyadenylation site** is located between 10 and 30 nucleotides

mRNA poly(A) tail
5' AAAAAAAAAAAAA 3'

FIGURE 4.32 A eukaryotic mRNA has a poly(A) tail at its 3′ end.

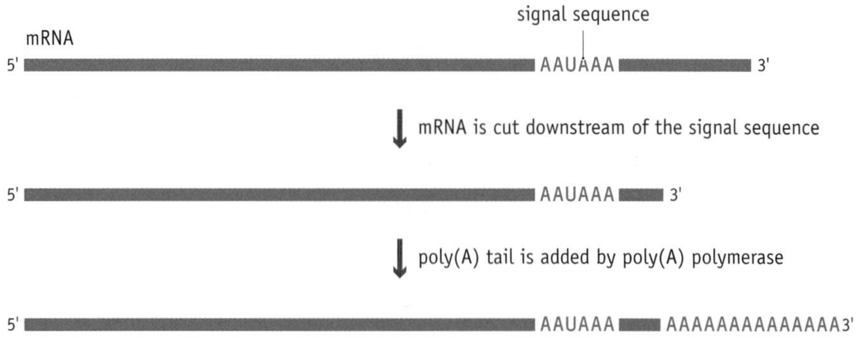

FIGURE 4.33 Polyadenylation of a eukaryotic mRNA. The mRNA is cut at a position between 10 and 30 nucleotides downstream of a signal sequence, and the poly(A) tail added at this polyadenylation site.

downstream of a signal sequence, almost always 5′-AAUAAA-3′ (Figure 4.33). Most genes have more than one polyadenylation site, which means that termination can occur at different positions, resulting in mRNAs with identical coding properties but with distinctive 3′ ends. The different polyadenylation sites appear to be used in different tissues, suggesting that **alternative polyadenylation** could be an important mechanism for establishing tissue-specific patterns of gene expression.

The signal sequence is the binding site for a multisubunit protein called the **cleavage and polyadenylation specificity factor** (**CPSF**). CPSF attaches to the polymerase complex during initiation of transcription, and by riding along the template with RNA polymerase II it is able to bind to the signal sequence as soon as it is transcribed, initiating the polyadenylation reaction. It has been suggested that the nature of the contact between CPSF and the polymerase changes when the poly(A) signal sequence is located, and that this change alters the properties of the elongation complex so that termination becomes favored over continued RNA synthesis. As a result, transcription stops soon after the poly(A) signal sequence has been transcribed.

KEY CONCEPTS

- There are three main types of living organism. These are eukaryotes, bacteria, and archaea.
- Gene expression is sometimes described as "DNA makes RNA makes protein," but in reality is more complicated than this, especially in eukaryotes. Additional steps include the processing of the initial RNAs and proteins into their functional forms, and the turnover of these molecules.
- The enzymes that carry out transcription are called DNA-dependent RNA polymerases. Bacteria have just one type of RNA polymerase, but eukaryotes have three, each transcribing a different set of genes. Protein-coding genes in eukaryotes are transcribed by RNA polymerase II.
- Transcription must initiate at the correct positions on a DNA molecule, so that genes are transcribed rather than random pieces of DNA.

Positions where transcription should begin are marked by sequences called promoters, which are recognized either by the RNA polymerase itself or by a DNA-binding protein that forms a platform to which the RNA polymerase binds.

- The promoters for RNA polymerase II comprise a core element to which the RNA polymerase attaches, and various upstream sequences that are involved in regulation of transcription initiation.
- Initiation of transcription involves opening up a short stretch of the DNA to form an open promoter complex, followed by the start of RNA synthesis and movement of the RNA polymerase away from the promoter region.
- An RNA polymerase does not synthesize its transcript at a constant rate. Instead, synthesis is discontinuous, with periods of rapid elongation interspersed with brief pauses of 6 milliseconds or less.
- All transcripts that are made by RNA polymerase II have a chemical modification called the cap structure at the 5′ end.
- In order for termination to occur, the RNA polymerase has to reach a position on the template where dissociation is more favorable than continued RNA synthesis.
- Termination of transcription by RNA polymerase II is accompanied by attachment of a poly(A) tail to the 3′ end of the transcript.

QUESTIONS AND PROBLEMS (Answers can be found at www.garlandscience.com/introgenetics)

Key Terms

Write short definitions of the following terms:

alternative polyadenylation
alternative promoter
archaea
backtracking
bacteria
basal rate of transcript initiation
C-terminal domain
cap structure
cleavage and polyadenylation specificity factor
closed promoter complex
consensus sequence
core enzyme
core promoter
DNA-dependent RNA polymerase
elongation factor
eukaryote
guanine methyltransferase
guanylyl transferase
helicase
holoenzyme
initiator sequence
intrinsic terminator
inverted repeat
open promoter complex
poly(A) polymerase
poly(A) tail
polyadenylation site
post-translational chemical modification
Pribnow box
processing
prokaryote
promoter
Rho-dependent terminator
RNA-dependent RNA polymerase
RNA polymerase
RNA polymerase I
RNA polymerase II
RNA polymerase III
strong promoter
TAF- and initiator-dependent cofactors
TATA-binding protein
TATA box
TBP-associated factor
transcription bubble
transcription factor IID
transcription initiation complex
turnover
type 0 cap
type 1 cap
type 2 cap
upstream
upstream control element

Self-study Questions

4.1 Describe the distinguishing features of prokaryotic and eukaryotic cells. What are the names given to the two distinct groups of prokaryotes, and what are the differences between them?

4.2 Why is "DNA makes RNA makes protein" not a good description of gene expression in eukaryotes?

4.3 What is the difference between a DNA- and an RNA-dependent RNA polymerase? Which type is involved in transcription?

4.4 Distinguish between the core and holoenzyme versions of the *E. coli* RNA polymerase. What are the roles of the two versions in transcription?

4.5 What are the roles of the three eukaryotic RNA polymerases?

4.6 Define the term "promoter." Draw annotated diagrams to illustrate the structures of the promoters for the three eukaryotic RNA polymerases and for the *E. coli* enzyme.

4.7 Propose a suitable consensus for the following set of sequences:

```
ATAGACATA
ATAGCCATT
ATACACTTA
AGAGAGAAT
ATAGACATA
TTTGACATT
ATAAATATA
```

4.8 What is the link between the sequence of an *E. coli* promoter and the basal rate of transcription for the gene transcribed from that promoter?

4.9 Give an example of a human gene that is under the control of alternative promoters.

4.10 Explain the roles of the two components of the *E. coli* promoter during initiation of transcription. Be sure to make clear the difference between the closed and open versions of the promoter–RNA polymerase complex.

4.11 What is the role of the σ subunit during promoter recognition in *E. coli*?

4.12 Write an essay on assembly of the RNA polymerase II initiation complex. As part of your essay, compile a table giving the names of the main proteins or groups of proteins involved in assembly of this complex, along with a summary of the role of each one.

4.13 What is the importance of the C-terminal domain of RNA polymerase II?

4.14 How does TATA-binding protein provide a link between the initiation processes of all three eukaryotic RNA polymerases?

4.15 Draw a detailed diagram of the chemical reaction involved in the synthesis of RNA from individual nucleotides.

4.16 Outline the important features of the elongation phase of transcription in *E. coli*.

4.17 Describe the series of events that result in capping of a eukaryotic mRNA.

4.18 What are the roles of the elongation factors during eukaryotic transcription?

4.19 Explain what is meant by an "inverted repeat."

4.20 Which of the following RNA molecules would be able to form a hairpin structure?

```
5'-ACGUUUGGCAGUCCAAACU-3'
5'-ACCUACCUACACCUAGCUUUGUGAGCUAGCUUGUG-3'
5'-GGCUAGGUCGAAACGACCUAGCC-3'
```

4.21 Distinguish between the events that are thought to occur during transcription termination at intrinsic and Rho-dependent terminators.

4.22 Outline our current knowledge regarding termination of transcription by RNA polymerases I and III.

4.23 Draw a series of diagrams to illustrate how a eukaryotic mRNA becomes polyadenylated.

Discussion Topics

4.24 Post-translational chemical modification can alter the function of a protein. Does this contravene the dogma that biological information is encoded in genes?

4.25 Some species of bacteria possess more than one σ subunit. For instance, *Bacillus subtilis* (a bacterium able to produce spores) has at least four. What might be the role of these additional σ subunits?

4.26 With some types of virus, transcription of the host's genes ceases shortly after infection. All the cell's RNA polymerase enzymes start transcribing the virus genes instead. Suggest events that might underlie this phenomenon.

4.27 Construct a hypothesis to explain why eukaryotes have three RNA polymerases. Can your hypothesis be tested?

4.28 Speculate on the reasons why the TATA-binding protein is involved in initiation of transcription by each of the three eukaryotic RNA polymerases.

4.29 An operon is a group of bacterial genes that are adjacent to one another in the genome and are transcribed together as a single mRNA, from a promoter located upstream of the first gene in the cluster. Often the genes in an operon code for proteins that work together in a single biochemical pathway, such as synthesis of an amino acid. What might be the advantage of having genes grouped into operons?

FURTHER READING

Banerjee S, Chalissery J, Bandey I & Sen R (2006) Rho-dependent transcription termination: more questions than answers. *J. Microbiol.* 44, 11–22.

Bujord H (1980) The interaction of *E. coli* RNA polymerase with promoters. *Trends Biochem. Sci.* 5, 274–278.

Buratowski S (2009) Progression through the RNA polymerase II CTD cycle. *Mol. Cell* 36, 541–546. *Describes recent discoveries regarding the role of the C-terminal domain of RNA polymerase II during transcription.*

Cougot N, van Dijk E, Babajko S & Seeraphin B (2004) "Cap-tabolism." *Trends Biochem. Sci.* 29, 436–444. *mRNA capping.*

Geiduschek EP & Kassavetis GA (2001) The RNA polymerase III transcription apparatus. *J. Mol. Biol.* 310, 1–26.

Green MR (2000) TBP-associated factors (TAF_{II}s): multiple, selective transcriptional mediators in common complexes. *Trends Biochem. Sci.* 25, 59–63.

Klug A (2001) A marvellous machine for making messages. *Science* 292, 1844–1846. *Description of the bacterial RNA polymerase.*

Manley JL & Takagaki Y (1996) The end of the message—another link between yeast and mammals. *Science* 274, 1481–1482. *Polyadenylation.*

Russell J & Zomerdijk JCBM (2005) RNA-polymerase-I-directed rDNA transcription, life and works. *Trends Biochem. Sci.* 30, 87–96.

Sandelin A, Carninci P, Lenhard B et al. (2007) Mammalian RNA polymerase II core promoters: insights from genome-wide studies. *Nat. Rev. Genet.* 8, 424–436. *Alternative promoters.*

Tora L & Timmers HT (2010) The TATA box regulates TATA-binding protein (TBP) dynamics in vivo. *Trends Biochem. Sci.* 35, 309–314. *Details of the recognition of the TATA box by TBP.*

Toulokhonov I, Artsimovitch I & Landick R (2001) Allosteric control of RNA polymerase by a site that contacts nascent RNA hairpins. *Science* 292, 730–733. *A model for termination of transcription in bacteria involving the flap structure on the outer surface of the RNA polymerase.*

Travers AA & Burgess RR (1969) Cyclic re-use of the RNA polymerase sigma factor. *Nature* 222, 537–540. *The first demonstration of the role of the σ subunit.*

Types of RNA Molecule: Messenger RNA CHAPTER 5

Transcription, as described in Chapter 4, results in synthesis of RNA molecules whose nucleotide sequences are set, according to the base-pairing rules, by the sequences of the genes from which they are copied. These RNA molecules can be grouped into two categories, messenger RNA (mRNA) and noncoding RNA. Messenger RNA is transcribed from protein-coding genes and so undergoes the second stage of gene expression, translation. It therefore acts as the intermediate between a gene and its protein product. Noncoding RNA includes a variety of transcripts that do not code for proteins but instead play roles in the cell as RNA molecules. Now we must look at these two types of RNA in more detail, mRNA in this chapter, and noncoding RNA in the next.

Geneticists now use the word **transcriptome** to refer to the mRNA content of a cell. This is a useful term that helps us to move away from thinking about the expression of individual genes and instead to consider the coordinated activity of groups of genes. In particular, by studying transcriptomes rather than individual mRNAs, researchers have been able to make important discoveries about the way in which gene expression patterns change when a cell becomes cancerous. Our initial objective in this chapter will therefore be to understand how transcriptomes are studied.

We must also understand the processes that determine the composition of a transcriptome. These involve more than simply the synthesis of new mRNAs by transcription. The degradation of existing mRNAs is equally important. This is an area of research that has moved rapidly forward over the last few years thanks to the discovery in eukaryotes of new types of noncoding RNA that participate in the breakdown of mRNAs that are no longer needed.

The processes that influence the composition of a eukaryotic transcriptome also include **splicing**, the mechanism by which introns are removed from the transcripts of discontinuous genes and the exons "spliced" together. Splicing was initially looked on as a necessary but not particularly exciting process, but that view changed with the discovery that many genes can follow **alternative splicing** pathways. These different pathways enable a single primary transcript to give rise to different mRNAs containing different combinations of exons and, of course, specifying different proteins (Figure 5.1). The particular splicing pathways that are operational in a cell greatly influence the composition of that cell's transcriptome. We must therefore look at splicing in some detail.

Finally, the composition of a transcriptome can be influenced by processes that change the nucleotide sequences of existing mRNAs. We will examine these processes, which are called **RNA editing**, at the end of this chapter.

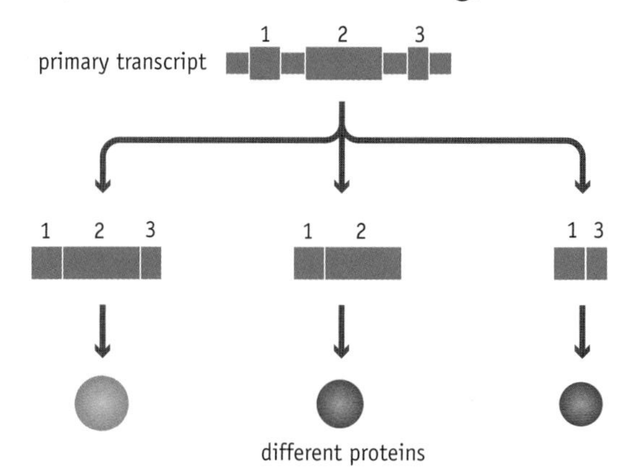

FIGURE 5.1 Alternative splicing enables a gene to specify more than one protein.

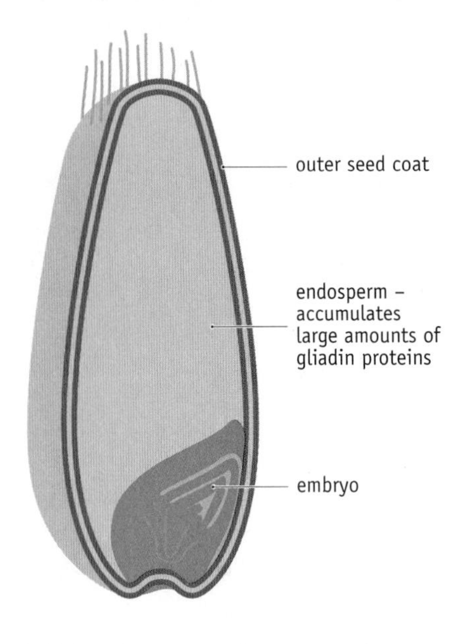

FIGURE 5.2 Gliadin proteins accumulate within the endosperm of a wheat seed.

5.1 TRANSCRIPTOMES

Although some geneticists like to think of the transcriptome as all of the RNA in a cell, including the noncoding component, the term was originally used in 1997 to describe just the mRNA, and this is the definition that we will adopt. It is the most useful definition because the mRNA, although making up less than 4% of the total RNA in most cells, is the most significant component because it specifies the proteins that the cell is able to synthesize. A transcriptome therefore provides the link between the genes present in a cell and the proteins that the cell makes, which in turn determines the cell's biochemical capabilities.

Even in the simplest organisms such as bacteria and yeast, many genes are active at any one time. Transcriptomes can therefore contain copies of hundreds, if not thousands, of different mRNAs. Usually, each mRNA makes up only a small fraction of the transcriptome as a whole, with the most common mRNA in the transcriptome rarely contributing more than 1% of the total. Exceptions are those cells that have highly specialized biochemistries, which are reflected by transcriptomes in which one or a few mRNAs predominate. Wheat seeds are an example. The cells in the endosperm of wheat seeds synthesize large amounts of the gliadin proteins, which accumulate in the dormant grain and provide a source of amino acids for the germinating seedling (Figure 5.2). Within the developing endosperm, the gliadin mRNAs can make up as much as 30% of the transcriptomes of certain cells.

The important features of a transcriptome are therefore the identities of the mRNAs that it contains, and their relative amounts. We will therefore begin by examining how the mRNAs in a transcriptome can be identified and quantified.

Transcriptomes are studied by microarray analysis

There are several ways of characterizing a transcriptome, but the most frequently used method is called **microarray analysis**. A microarray is a small slip of glass or silicon, typically 2 cm by 2 cm, containing a large number of different DNA molecules absorbed on to the surface in an ordered array (Figure 5.3). These molecules are usually short oligonucleotides up to about 125 nucleotides in length, each of which has been chemically synthesized at its appropriate position in the microarray. With the latest technology it is possible to squeeze half a million sites onto a single microarray, each site containing multiple copies of a particular oligonucleotide.

The sequences of the oligonucleotides in an array match the sequences of the genes in the organism whose transcriptome is being studied. As most genes are much longer than 125 bp, the oligonucleotides are clearly not complete gene copies, but their sequences are designed so that each one is unique to an individual gene. The sequence of the oligonucleotide is present only in that one gene, and not in any others.

How does a microarray carrying gene-specific sequences help us to identify and quantify the mRNAs in a transcriptome? The answer is quite simple. Any two polynucleotides whose sequences are complementary will attach

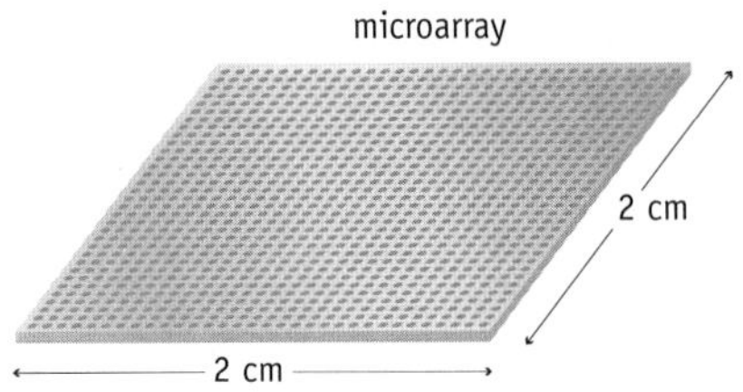

microarray
2 cm
2 cm

FIGURE 5.3 A microarray. Each dot indicates a position at which a different DNA molecule has been absorbed into the surface of this glass or silicon chip.

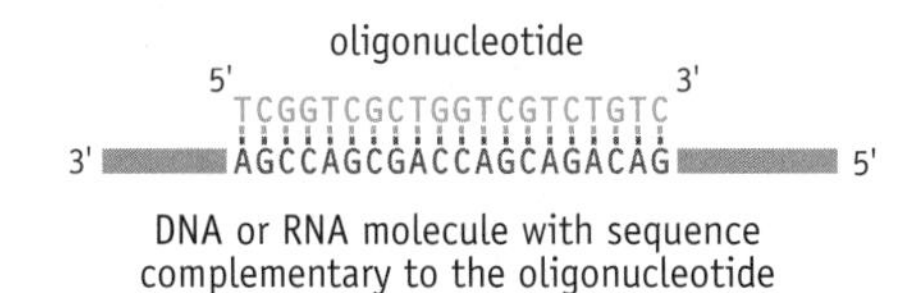

FIGURE 5.4 Nucleic acid hybridization. A short oligonucleotide is shown hybridized to a longer DNA or RNA molecule.

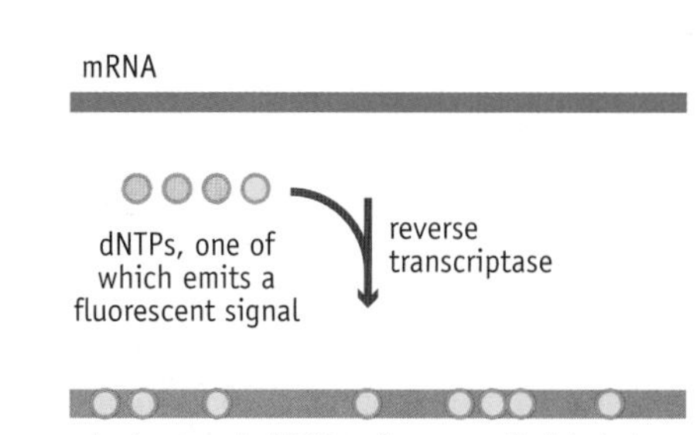

FIGURE 5.5 Conversion of an mRNA into a labeled DNA molecule. Reverse transcriptase copies an RNA molecule into a single-stranded DNA molecule. If one of the dNTPs provided as substrates for DNA synthesis is labeled, then the resulting DNA molecule will also be labeled.

to one another by base pairing (Figure 5.4). This means that an mRNA can hybridize to a copy of the gene from which it was transcribed, or to a short oligonucleotide whose sequence matches a part of that gene. An mRNA could therefore hybridize to the surface of a microarray, but only at a position containing oligonucleotides whose sequences match the gene from which it was transcribed.

Now that we understand the principle we can move on to some of the technical details. Messenger RNAs are easily degraded and so can be difficult to handle in the laboratory. To avoid this problem, mRNA is rarely applied directly to a microarray. Instead, DNA copies are made, using an RNA-dependent DNA polymerase called **reverse transcriptase** (Figure 5.5). This copying reaction is carried out in the test tube. The mRNA is extracted from the cells and mixed with the enzyme and each of the four deoxynucleotides. One of these nucleotides is chemically modified so that it emits a fluorescent signal. This modification does not affect the ability of the nucleotide to be incorporated into the DNA molecules that are being made, but it means that these molecules are labeled—they give off fluorescent signals.

The DNA copy of the transcriptome is now applied to the microarray. The positions at which hybridization occurs can easily be located by scanning with a fluorescence detector (Figure 5.6). These positions indicate which genes hybridize to components of the transcriptome, and hence identify the mRNAs in the transcriptome. The relative amounts of the different mRNAs in the transcriptome can also be worked out, because abundant mRNAs give rise to more intense fluorescent signals.

Transcriptome analysis has been important in cancer studies

Microarray analysis and other methods for studying transcriptomes have been responsible for many advances in genetics over the last 10 years. Perhaps the most important of these have been the new insights it has given into the changes in gene expression that occur during cancer.

Cancer is a group of diseases characterized by uncontrolled cell division. Cancers are often described according to the part of the body they affect, and we are all familiar with terms such as "lung cancer" or "colon cancer." This classification hides the fact that a single tissue can be affected by different types of cancer, and understanding which one is present in a patient is

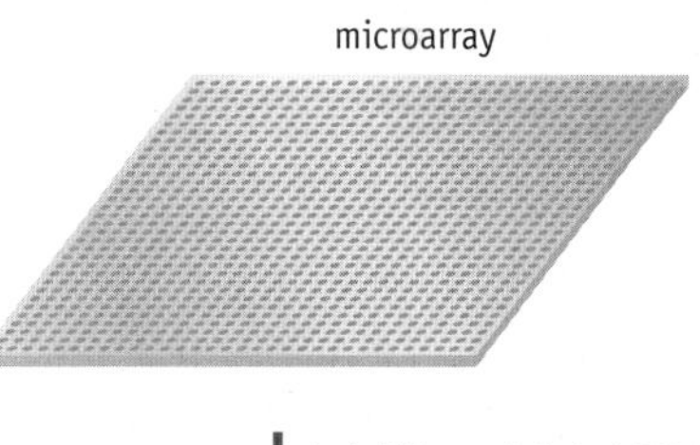

FIGURE 5.6 Microarray analysis. The labels have been detected on the microarray by confocal laser scanning and the intensity converted into a pseudocolor spectrum, with red indicating the greatest hybridization, followed by orange, yellow, green, blue, indigo, and violet, the last representing the background level of hybridization. (From T. Strachan, M. Abitbol, D. Davidson and J. S. Beckman, *Nat. Genet.* 16: 126–132, 1997. With permission from Macmillan Publishers Ltd.)

RESEARCH BRIEFING 5.1

'Omes

Since the 1990s, geneticists have increasingly focused on the genome as a whole, rather than the activity of individual genes. The genome is the entire genetic complement of an organism, made up of one or more lengthy DNA molecules, and comprising all that organism's genes and all of the DNA between those genes. The shift of attention to genomes was prompted by the need to understand complex physiological processes such as the functioning of the brain and the conversion of normal tissues into cancerous ones. These processes involve the coordinated activities of many hundreds or thousands of genes, and could not be addressed just by studying individual or small groups of genes.

The shift from individual genes to the entire genome has been accompanied by a similar advance in the way we think about gene expression. The expression pathway for a single gene involves transcription of that gene into an mRNA, followed by translation of the mRNA into a protein (Figure 1A). Expression of the genome as a whole is more complex. Now we have a set of genes being transcribed into a collection of mRNAs, possibly many thousand different ones, which are translated into an equally large number of different proteins. Two new terms, **transcriptome** and **proteome**, were introduced by geneticists in the 1990s to describe these products of genome expression (Figure 1B).

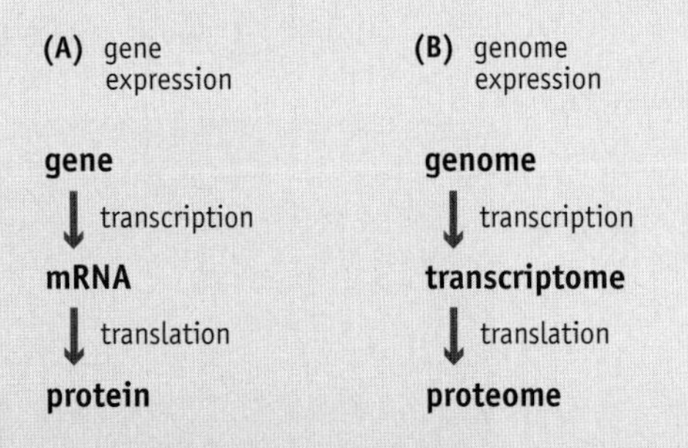

FIGURE 1 (A) Gene and (B) genome expression. The gene expression pathway outlines how the biological information contained in a single gene is used to direct synthesis of a single protein molecule. The genome expression pathway describes the process by which the entire complement of biological information contained within the genome specifies the synthesis of the set of proteins present in a particular cell at a particular time.

Geneticists and other biologists have become very enthusiastic about what are loosely called **'omes**, and several new types of 'ome, some of rather dubious usefulness, have been invented in recent years. Understanding these jargon words can be quite difficult. We will examine exactly what we mean by genome, transcriptome, and proteome, and then look briefly at some of the other 'omes that are becoming important in modern genetics.

The genome is the entire complement of DNA molecules in a cell

We have established that the genome is the entire genetic complement of an organism, comprising all that organism's genes and all of the DNA between those genes. In some organisms, the entire genome is contained within a single DNA molecule. This is the case with the bacterium *Escherichia coli*, whose 4377 genes are all located in a circular DNA molecule of 4639 kb.

In other organisms, the genome is divided into two or more DNA molecules. The human genome, for example, is made up of 24 linear DNA molecules with a total length of approximately 3200 Mb and containing about 20,000 genes, and a single circular molecule of 16.5 kb that carries another 37 genes. The 24 linear molecules are contained in the chromosomes present in the nucleus of a human cell (Figure 2). The 24 chromosomes consist of the two sex chromosomes, X and Y, and 22 others called **autosomes**. The vast majority of human cells have a nucleus containing two copies of each autosome plus a pair of sex chromosomes, XX for females or XY for males.

The circular DNA molecule is not found in the nucleus, but instead is located in the mitochondria, the energy-generating organelles of human cells. This **mitochondrial genome** is present in multiple copies, about 8000 per cell, 10 or so in each mitochondrion.

Transcriptomes and proteomes

Now we will look at the RNA and protein equivalents of the genome. The transcriptome, as we learned earlier in this chapter, is sometimes defined as all of the RNA in a cell, including the noncoding component, but is more usefully looked on as just the mRNA. The transcriptome is therefore the product of genome expression, and is made up of the individual transcripts of many different protein-coding genes.

The proteome is the collection of proteins in a cell. These are synthesized by translation of the mRNAs present in the transcriptome and determine the cell's biochemical and physiological properties. We will return to the proteome in Chapter 8 when we study protein synthesis. In Research Briefing 8.1 we will examine the methods used to identify the proteins in a proteome and to make comparisons between the proteomes of two different types of cell.

The genomic version of "DNA makes RNA makes protein" is "the genome makes the transcriptome makes the proteome," as summarized in Figure 1B. The genome expression pathway therefore describes the process by which the entire complement of biological information contained within the genome specifies the synthesis of the set of proteins present in a particular cell at a particular time.

Other 'omes

What of the other 'omes that have proliferated in biology since the start of the new millennium? There are three of these that we should become familiar with.

The first is the **metabolome**, which is defined as the complete collection of metabolites present in a cell under a particular set of conditions. The metabolome therefore reflects the biochemical activities occurring in a cell. These activities are specified by the proteome, so there is a clear link between genome expression and the metabolome. Some geneticists go so far as to look on the metabolome as the final product of genome expression.

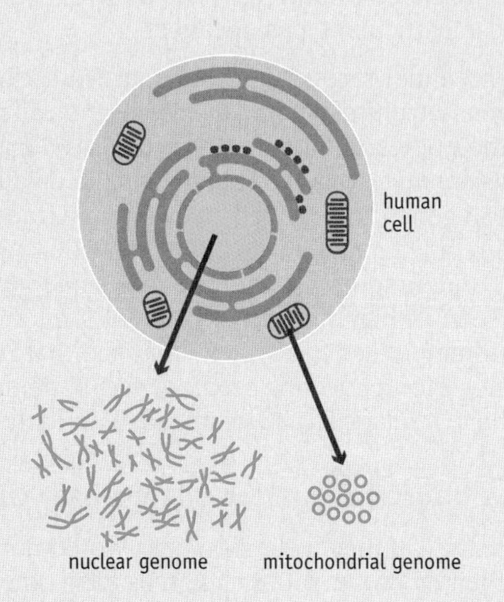

FIGURE 2 The nuclear and mitochondrial components of the human genome. The mitochondria are the energy-generating organelles present in eukaryotic cells.

A **metagenome** is rather different. This is the collection of genomes present in an environmental sample that contains many different organisms. The term was first used by microbiologists, who realized that there are many bacterial species in nature that have never been identified because they cannot be grown in laboratory cultures. Instead, these species have been detected by bulk sequencing of DNA samples prepared from different environments. One of the first environments to be studied in this way was the Sargasso Sea. Bacterial DNA was prepared from 1500 liters of surface water. The metagenome obtained by sequencing this DNA contained segments of the genomes of over 1800 species, of which 148 were totally new. Metagenomics therefore enables us to study the genomes of organisms that nobody has ever seen.

Metagenomics leads us to the final 'ome that we will consider. The **microbiome** is slightly more complicated because the "-biome" part of the term comes from ecology and refers to a community of organisms occupying a single environment. A microbiome is therefore any microbial community. It includes the community of bacteria and other microbes that live on the surface of and inside an animal such as a human being. It has been estimated that the human microbiome comprises some 10^{14} bacteria and other microbes, which is 10 times the number of the individual's own cells in an adult body. Hundreds or thousands of species make up the human microbiome, and many of these species have not yet been identified. The composition of the microbiome is thought to reflect the health status of a person. This is not just because the presence of a pathogenic species might lead to a gut disorder or other disease. The microbiome also influences our ability to withstand infections and to absorb certain nutrients from the food we eat. Geneticists are attempting to use metagenomics to characterize the human microbiome, in order to understand how changes in the microbiome influence our health and may be an indicator of disease.

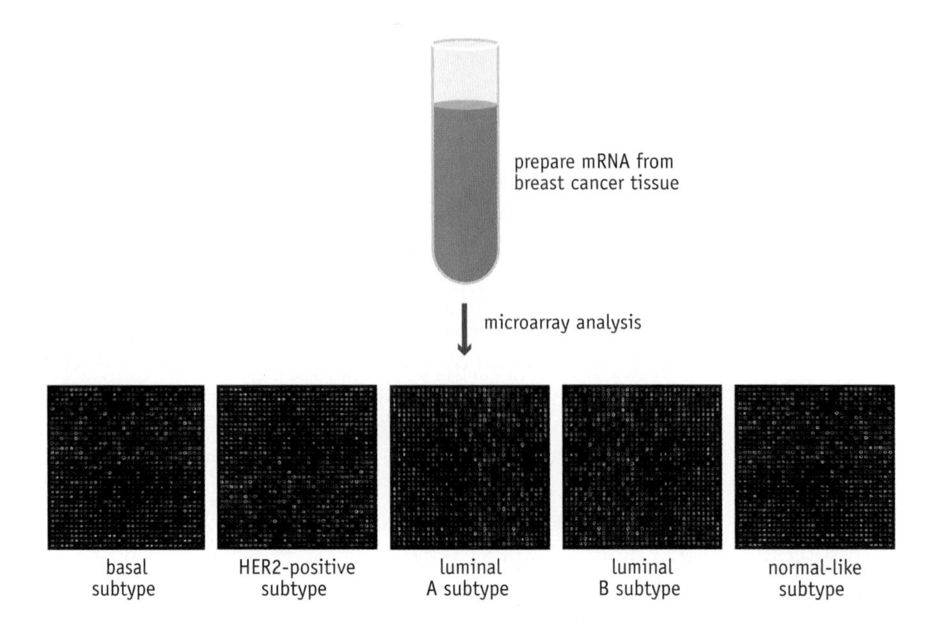

FIGURE 5.7 The use of microarray analysis to identify different subtypes of breast cancer. Close examination of these 5 microarrays shows that the hybridization patterns are different, indicating that the transcriptomes for the 5 subtypes have different mRNA compositions. The breast cancer tissue that is being studied, whose mRNA has been prepared and is contained in the test tube at the top, gives the hybridization pattern associated with the luminal A subtype. Identification of the subtype enables the appropriate treatment regime to be designed for the patient.

one of the keys to treatment and possible recovery. Transcriptome analyses have been particularly important in characterizing the different versions of a cancer. Breast cancer, for example, can be divided into five subtypes, each associated with different patterns of gene expression as revealed by analysis of their transcriptomes (Figure 5.7).

Detailed examination of the breast cancer transcriptomes has also identified mRNAs that might throw light on the underlying causes of some of the clinical features of the disease. An interesting example is the discovery that the transcriptome of the most difficult subtype to treat contains mRNAs more normally found in tissues that are recovering from trauma such as wounding. These mRNAs have also been found in the more fatal types of stomach and lung cancer. Researchers are now trying to understand how expression of the genes for these wound response mRNAs is linked to the aggressive nature of these particular types of cancer.

A critical factor in the potential recovery of a patient from cancer is the avoidance of **metastasis**, which results in cells from one cancer spreading to other places in the body and initiating new tumors (Figure 5.8). In breast cancers, a high risk of metastasis has been associated with a specific expression pattern for 186 genes, as revealed by examining transcriptomes for the presence of and relative abundances of their mRNAs. Again, these mRNA patterns are not unique to breast cancers. The same gene expression patterns are also seen in highly invasive types of lung and prostate cancer. Other transcriptome profiles have been identified that help to predict the long-term prognosis for recovering breast cancer patients, enabling women more likely to have a relapse to be identified at an early stage.

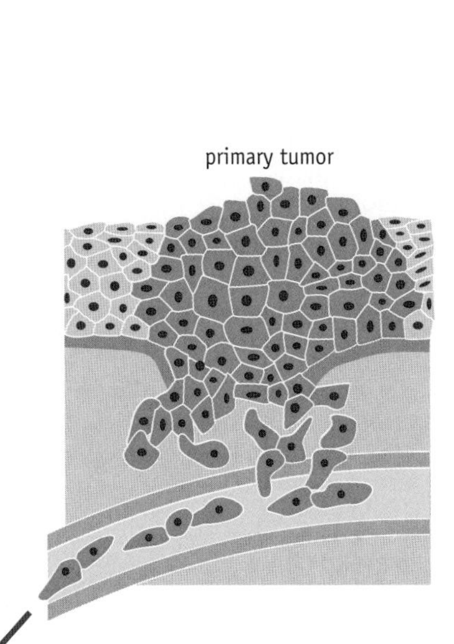

FIGURE 5.8 Metastasis. Cancerous cells from a primary tumor are shown entering the bloodstream. They can therefore move to other parts of the body where they may initiate secondary tumors. (Adapted from B. Alberts et al., Essential Cell Biology, 3rd ed. New York: Garland Science, 2010.)

The composition of a transcriptome can change over time

Transcriptomes are not static. If they were then cells would not be able to respond to changes in their environment or to signals from other cells, nor would they be able to differentiate into specialized types during the development of an organism. We must therefore examine how the composition of a transcriptome can change.

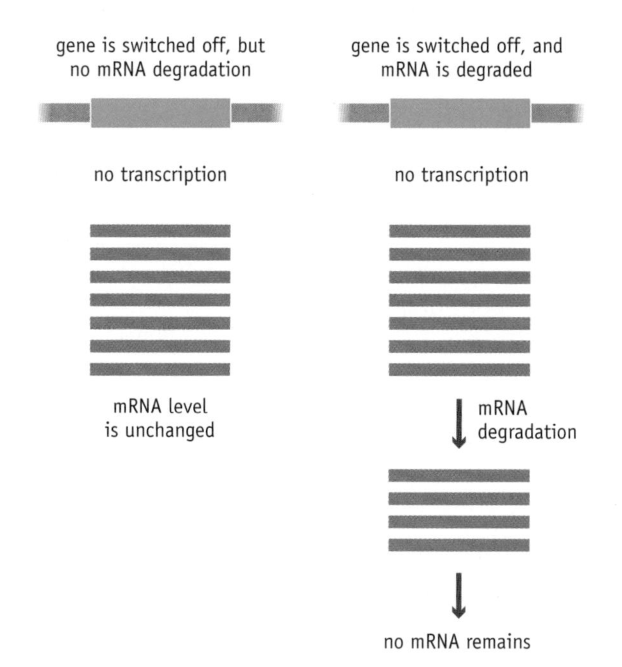

FIGURE 5.9 The influence of mRNA degradation on the composition of a transcriptome. On the left, a gene has been switched off but its mRNA has not been degraded. The mRNA therefore remains in the transcriptome, and the protein coded by the gene will still be synthesized. On the right, we see the importance of mRNA degradation. Now the mRNAs of the inactive gene are removed from the transcriptome, so synthesis of the protein stops, reflecting the fact that the gene has been switched off.

The most obvious way to change the composition of a transcriptome is to switch on a new gene. Transcription of a gene that previously was silent will add a new type of mRNA to the transcriptome. This is such an obvious way of changing the mRNA content of a cell that we might imagine that it is the key to the dynamism of transcriptomes. In fact, the ability to remove mRNAs from the transcriptome is equally important. If mRNAs were stable, never being degraded or being turned over only very slowly, then switching off the transcription of a gene would have no immediate effect on the synthesis of the protein coded by that gene (Figure 5.9). The gene's mRNA would persist in the transcriptome even though the gene was no longer being transcribed, and repeated rounds of translation of that mRNA would maintain the levels of the protein coded by the gene. Turnover therefore enables control processes that act on transcription to achieve both up- and down-regulation of individual mRNAs.

The rate of degradation of an mRNA can be estimated by determining its **half-life** in the cell. These estimates show that there are considerable variations between and within organisms. Most bacterial mRNAs have a half-life of only a few minutes and so are turned over very quickly. The short half-life reflects the rapid changes in protein synthesis patterns that are needed by an actively growing bacterium with a generation time of 20 minutes or so. Eukaryotic mRNAs are longer-lived, with half-lives of, on average, 10 to 20 minutes for yeast and several hours for mammals.

Within individual cells the variations are almost equally striking. Some yeast mRNAs have half-lives of only 1 minute whereas for others the figure is more like 35 minutes. Examples of exceptionally long-lived mRNAs are also known. For instance, globin mRNA in human reticulocytes is almost fully stable. In these cells regulation of globin gene expression is not important because the maximum rate of globin synthesis is required virtually all the time. Other long-lived mRNAs are being discovered as different types of specialized cell are studied.

There are various pathways for nonspecific mRNA turnover

Over the years, biochemists have identified a number of processes that degrade mRNAs in bacteria or eukaryotes, but until recently all of these

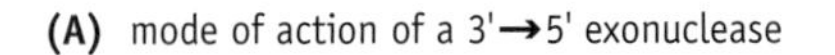

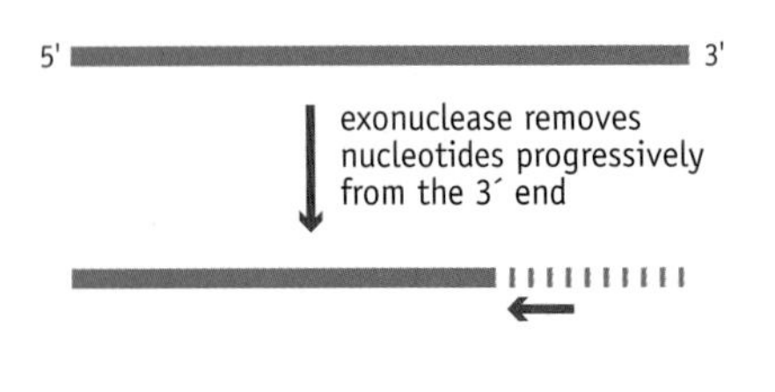

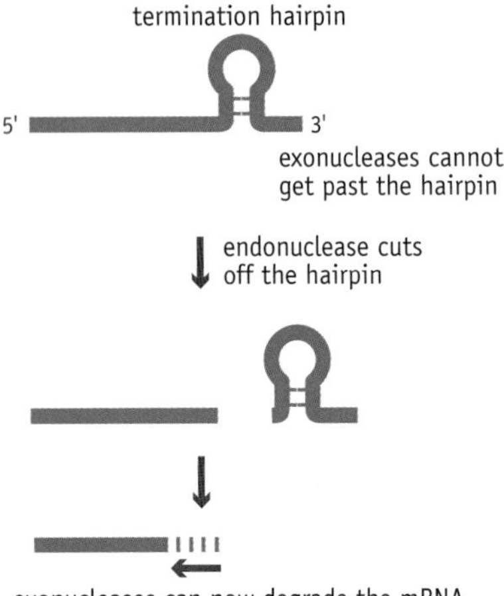

FIGURE 5.10 Degradation of mRNAs in bacteria. (A) The enzymes responsible for mRNA degradation are exonucleases that remove nucleotides one by one from the 3′ end of the molecule.
(B) Most bacterial mRNAs have a hairpin structure near their 3′ end, which prevents progress of the exonuclease. The first step in mRNA degradation is therefore removal of this hairpin by an endonuclease.

appeared to be nonspecific with regard to which mRNAs they targeted. One of the most important developments in genetics over the last 10 years has been the identification in eukaryotes of pathways that target individual mRNAs. Through their ability to remove particular transcripts from a transcriptome these pathways form a previously unrecognized mechanism for control of gene expression. First we will look at the apparently nonspecific pathways, and then we will focus on the newly discovered ones that target individual mRNAs.

Studies of mutant bacteria whose mRNAs have extended half-lives have identified a range of **ribonucleases** that are thought to be involved in mRNA degradation. Most of these enzymes are **exonucleases** that degrade mRNAs by removing nucleotides one by one from the 3′ end of the molecule (Figure 5.10A). These exonucleases are unable to progress past the hairpin structure, involved in termination of transcription, that is present near the 3′ end of most bacterial mRNAs (see Figure 4.27), and hence cannot gain access to the coding part of the transcript. The first step in bacterial mRNA degradation is therefore thought to be removal of the 3′ terminal region, including the hairpin, by an **endonuclease**, an enzyme able to cut internal phosphodiester bonds (Figure 5.10B). This internal cut exposes a new 3′ end from which the exonucleases can enter the coding region, destroying the functional activity of the mRNA.

In eukaryotes, the situation is more complicated, as several different mRNA degradation pathways have been identified and their individual roles are not yet fully understood. Eukaryotic mRNAs that have been incorrectly synthesized, possibly because of an error during splicing, appear to be degraded by a process called **mRNA surveillance**. The "surveillance" mechanism involves a complex of proteins that scans the mRNA for the nucleotide sequences that mark the position where translation of the mRNA should stop. These are called **termination codons**, and we will look at them more closely when we study the genetic code in Chapter 7. The termination codon should normally be near the 3′ end of the mRNA, within the final exon if the transcript is from a discontinuous gene (Figure 5.11). If the surveillance proteins find a termination codon in the wrong place, then the cap structure is removed from the 5′ end of the mRNA and the molecule degraded from this end by an exonuclease.

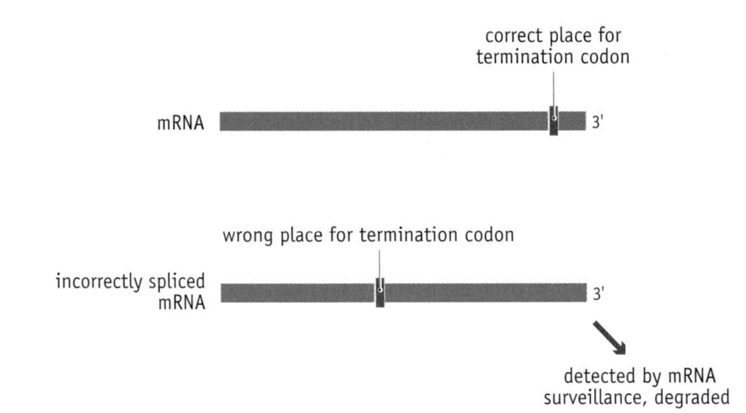

FIGURE 5.11 The basis to mRNA surveillance. If a transcript has been correctly synthesized then its termination codon will be near the 3′ end of the mRNA. The surveillance proteins are able to identify if the termination codon is in the wrong place, as might happen if the mRNA has been incorrectly spliced. These mRNAs are degraded.

FIGURE 5.12 The deadenylation-dependent decapping pathway for degradation of an mRNA. The poly(A) tail and the cap structure are removed and the mRNA is degraded from its 5′ end.

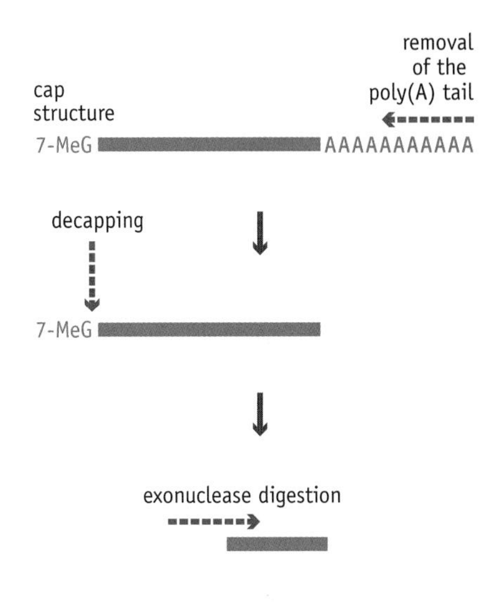

Eukaryotes are also able to degrade mRNAs that are no longer being translated into protein. This system is called **deadenylation-dependent decapping**. The first step is removal of the poly(A) tail, but the cap structure is also removed and the mRNA is degraded from its 5′ end (Figure 5.12). Whether or not a particular mRNA is degraded is probably determined by the ability of the decapping enzyme to gain access to the cap structure, which in turn depends on whether or not other proteins are bound to it. Proteins are bound when an mRNA is being translated, so these "active" mRNAs are protected from decapping and degradation. When an mRNA becomes inactive the binding proteins detach from the cap and the decapping enzyme can gain access. This pathway is therefore able to remove from the transcriptome those mRNAs whose translation products are no longer needed by the cell.

Individual mRNAs in eukaryotes are degraded by the Dicer protein

Now we turn our attention to the processes that result in the specific degradation of individual mRNAs. These pathways were discovered by a roundabout route. Some viruses have genomes that are made of double-stranded RNA, or which replicate via a double-stranded RNA intermediate. For several years it has been known that many eukaryotes possess RNA degradation mechanisms that protect their cells from attack by these viruses. The process is called **RNA silencing** or **RNA interference**, and involves a ribonuclease called **Dicer**, which cuts the viral genomes into **short interfering RNAs** (**siRNAs**) of 21 to 28 nucleotides in length (Figure 5.13A).

The action of Dicer inactivates the virus genome, but what if the virus genes have already been transcribed? If this has occurred then the harmful effects of the virus will already have been initiated and RNA silencing would appear to have failed in its attempt to protect the cell from damage. Remarkably, there is a second stage of the silencing process that is directed specifically at the viral mRNAs. The siRNAs produced by cleavage of the viral genome are separated into individual strands. One strand of each siRNA then base-pairs to any viral mRNAs that are present in the cell. The double-stranded regions that are formed are target sites for assembly of the **RNA-induced silencing complex** (**RISC**), which cleaves and hence silences the mRNAs (Figure 5.13B).

A link between silencing of viral RNAs and specific degradation of endogenous mRNAs was made when it was discovered that many eukaryotes have more than one type of Dicer protein. The fruit fly, *Drosophila melanogaster*, for example, has two Dicer enzymes. It turns out that the second type of Dicer in *Drosophila* works not with viral RNAs, but with endogenous molecules called **foldback RNAs**, which are coded by the fruit fly DNA and

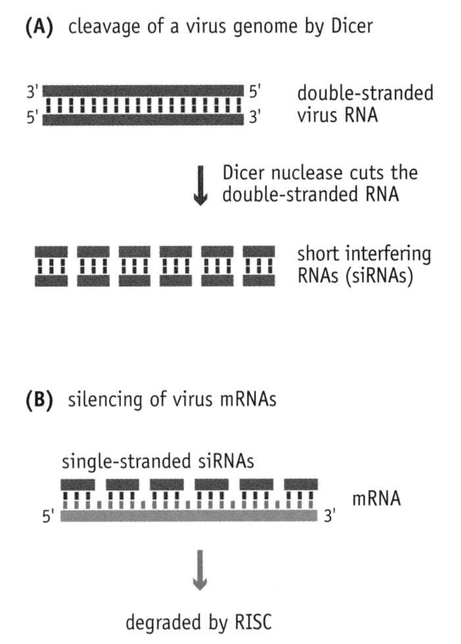

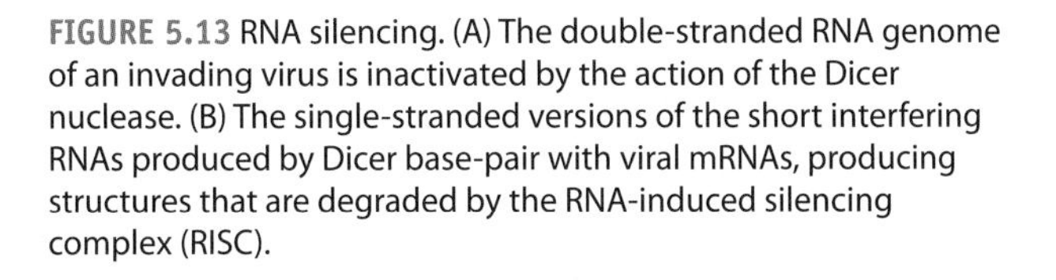
FIGURE 5.13 RNA silencing. (A) The double-stranded RNA genome of an invading virus is inactivated by the action of the Dicer nuclease. (B) The single-stranded versions of the short interfering RNAs produced by Dicer base-pair with viral mRNAs, producing structures that are degraded by the RNA-induced silencing complex (RISC).

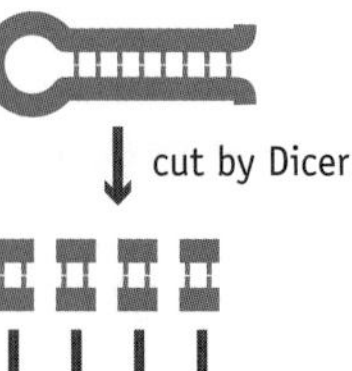

FIGURE 5.14 Synthesis of microRNAs in *Drosophila*.

synthesized by RNA polymerase II. The name "foldback" is given to these RNAs because they can form intrastrand base pairs giving rise to a hairpin structure (Figure 5.14). This hairpin can be cut by Dicer, releasing short double-stranded molecules, each about 21 bp, called **microRNAs**. Each microRNA is complementary to part of a cellular mRNA and hence base-pairs with this target, stimulating assembly of an RISC. Attachment of the microRNA therefore leads to cleavage and silencing of the mRNA.

Human cells are able to make up to 1000 different microRNAs. Many of these are able to silence more than one type of mRNA, possibly because those mRNAs all have identical target sequences, or possibly because a precise match between microRNA and mRNA is not needed for induction of the RISC. If a few **mismatches** are allowed—positions within the hybrid where base pairs do not form (Figure 5.15)—then the range of sequences that can be recognized by a single microRNA would be greatly increased.

Current thinking is that microRNAs silence sets of mRNAs that are no longer needed when a cell's environment changes or when the cell enters a new phase of a differentiation pathway. The specific degradation of these unwanted mRNAs, combined with the addition of new mRNAs by switching on of new genes, enables the composition of the transcriptome to be altered to meet the new demands being placed on the cell.

5.2 REMOVAL OF INTRONS FROM EUKARYOTIC mRNAS

In bacteria the mRNA molecules that are translated are direct copies of the genes from which they are transcribed. In eukaryotes the situation is different, as many mRNAs are coded by discontinuous genes, ones that contain introns (see Figure 3.2). Transcription produces a faithful copy of the gene, so if the gene contains introns then the initial transcript includes copies of these. These introns must be excised from the transcript and the exons joined together in the correct order before the transcript can function as a mature mRNA. This process is called splicing.

Splicing occurs in the nucleus. This is evident when the RNA fractions present in the nucleus and cytoplasm are compared. The nucleus can be divided into two regions, the **nucleolus**, in which ribosomal RNA genes are transcribed, and the **nucleoplasm**, where other genes, including those for mRNA, are transcribed. The nucleoplasmic RNA fraction is called **heterogeneous nuclear RNA** or **hnRNA**, the name indicating that it is made up of a complex mixture of RNA molecules, many over 20 kb in length. The mRNA in the cytoplasm is also heterogeneous, but its average length is only 2 kb. If the mRNA in the cytoplasm is derived from hnRNA, then evidently the primary transcripts are being shortened before the mRNA leaves the nucleus (Figure 5.16).

The existence of introns as noncoding segments within genes was not suspected until 1977, when DNA sequencing was first used on a large scale. Initially it was thought that hnRNA was converted to mRNA simply by trimming the ends of the molecules. The discovery of introns prompted a great

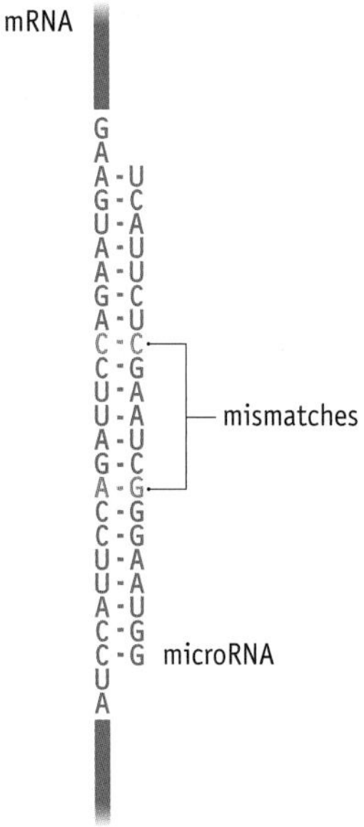

mRNA

mismatches

microRNA

FIGURE 5.15 Mismatches in the hybrid formed between a microRNA and an mRNA.

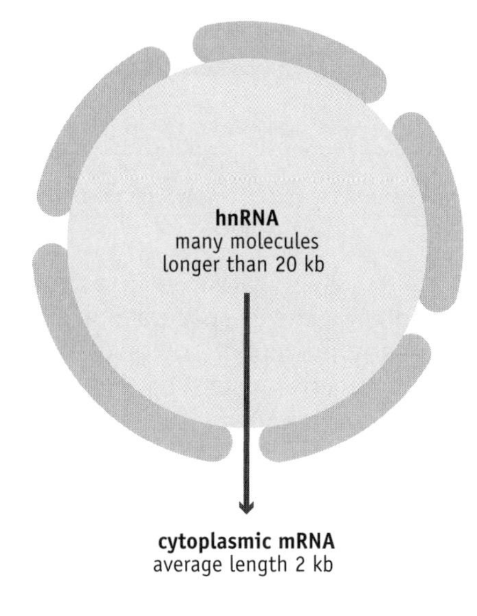

FIGURE 5.16 Size comparisons between hnRNA and mRNA indicate that mRNA molecules are shortened before leaving the nucleus.

deal of research aimed at describing the splicing process, and although this research continues, a fairly complete picture was available by 2000. Another layer of complexity was then added with the realization that many genes can follow alternative splicing pathways, enabling a single gene to code for more than one protein product. Research since 2000 has focused on mapping the alternative pathways for individual genes, and attempting to understand how the choice is made between them.

We begin by examining how introns are removed from transcripts, and we will then look at what is currently known about alternative splicing.

Intron–exon boundaries are marked by special sequences

When introns were first discovered, it was immediately realized that there must be a very precise mechanism for recognizing the boundary between an exon and an intron and cutting the RNA at this position. An error of one nucleotide in either direction would make it impossible for the mRNA to be translated into its correct protein product (Figure 5.17). Research was therefore directed at understanding how splice sites are recognized.

As more and more genes were sequenced it became clear that in the vast majority of cases, the first two nucleotides of the intron sequence are 5′-GU-3′ and the last two 5′-AG-3′ (Figure 5.18). As intron sequences started to accumulate in the databases it was further realized that the GU–AG motifs are merely parts of longer consensus sequences that span the 5′ and 3′ splice sites. These consensus sequences vary in different types of eukaryote. In vertebrates they can be described as

5′ splice site: 5′-AG↓GUAAGU-3′

3′ splice site: 5′-PyPyPyPyPyPyNCAG↓-3′

In these designations, Py is one of the two pyrimidine nucleotides (U or C), N is any nucleotide, and the arrow indicates the exon–intron boundary.

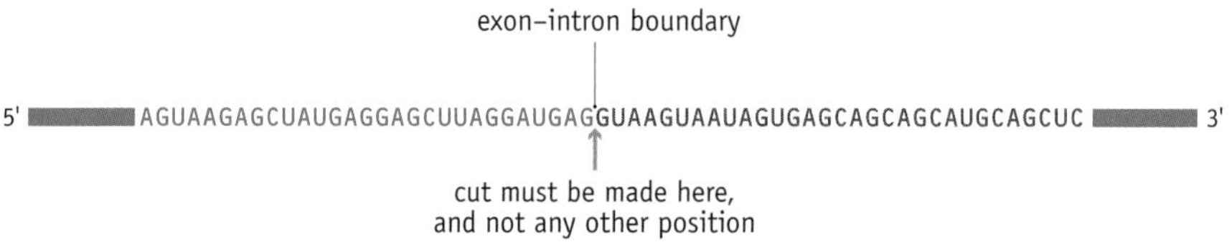

exon–intron boundary

5′ AGUAAGAGCUAUGAGGAGCUUAGGAUGAGGUAAGUAAUAGUGAGCAGCAGCAUGCAGCUC 3′

cut must be made here,
and not any other position

FIGURE 5.17 The importance of precision when making a cut at an exon–intron boundary. The cut must be made at precisely the position shown. An error in either direction would result in a change in the sequence of the mature mRNA. An mRNA that has been incorrectly spliced in this way would no longer be able to direct synthesis of its protein product.

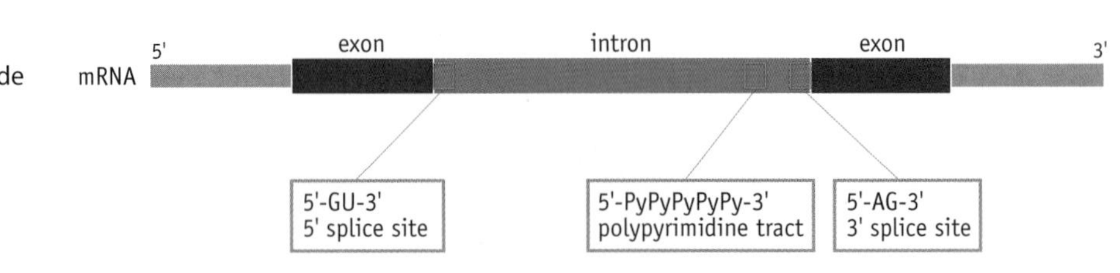

FIGURE 5.18 Conserved sequences in vertebrate introns. Py, pyrimidine nucleotide (U or C).

The consensus sequence for the 3′ splice site is therefore six pyrimidines in a row, followed by any nucleotide, followed by a C and then the AG. Remember that these are consensus sequences. The actual sequences might be slightly different, although the GU and AG components are almost always present.

The sequences around these two splice sites were originally thought to be the only regions of nucleotide similarity. It was then discovered that in the yeast *Saccharomyces cerevisiae* there is another conserved sequence, consensus 5′-UACUAAC-3′, located between 18 and 140 nucleotides upstream of the 3′ splice site. Other lower eukaryotes, such as fungi, have a similar sequence, but the motif is not present in vertebrate introns. Vertebrates do, however, have a **polypyrimidine tract** in the same general region, this tract having no clear consensus sequence but, as its name suggests, usually made up mainly of pyrimidine nucleotides (see Figure 5.18). Although located at similar positions, the polypyrimidine tract and the 5′-UACUAAC-3′ sequence are not functionally equivalent.

The splicing pathway

Once it was understood how the splice sites could be recognized, attention became focused on the series of events that result in excision of an intron and linking together of the two adjacent exons.

In outline, the splicing pathway involves three steps (Figure 5.19). The first step is cleavage at the 5′ splice site, which, because it is the first to be cut, is sometimes called the **donor site**. The resulting free 5′ end is then attached to an internal branch site within the intron to form a lariat structure. In yeast the branch site is the last A in the UACUAAC sequence. In vertebrates the branch site is also an A, but as vertebrate introns do not have UACUAAC boxes it is not clear how this A is selected. The lariat is formed by creating a phosphodiester bond between the 5′ carbon of the first nucleotide of the intron (the G of the GU motif) and the 2′ carbon of the internal

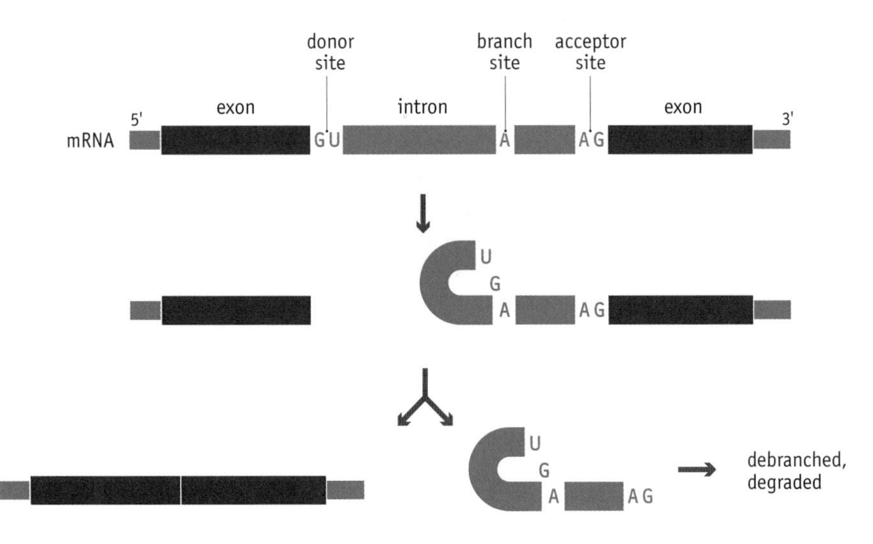

FIGURE 5.19 Splicing in outline. The first step is cleavage at the donor site. The resulting free 5′ end is then attached to an internal branch site within the intron to form a lariat structure. The acceptor site is then cleaved and the two exons joined together.

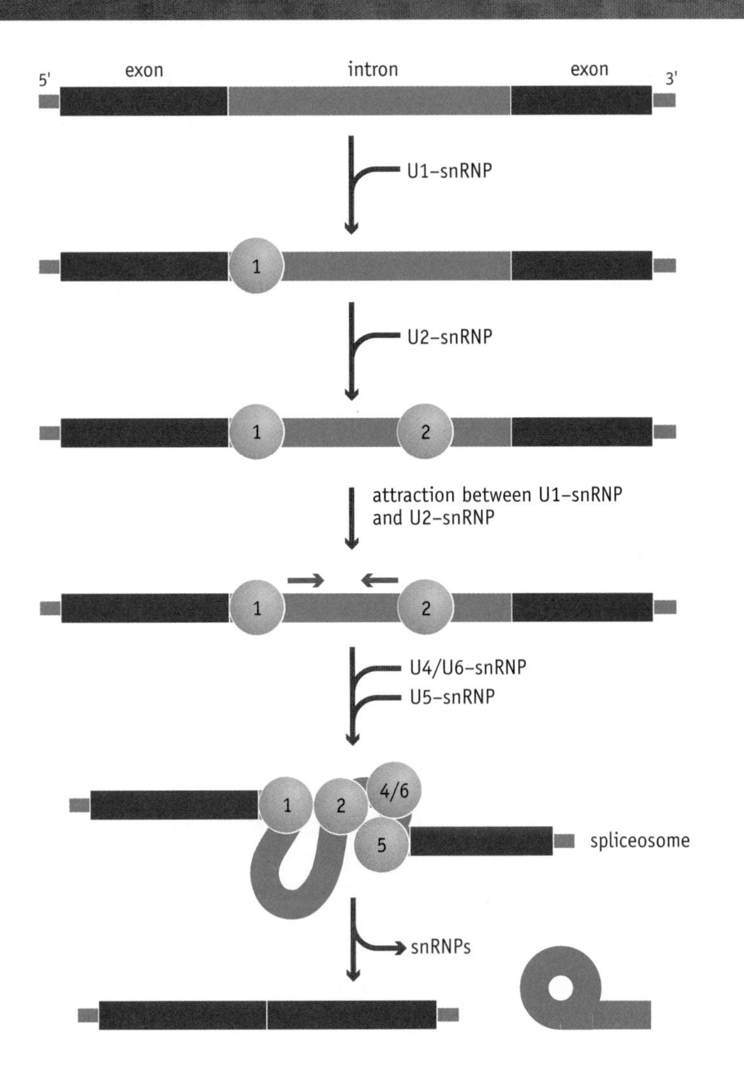

FIGURE 5.20 The roles of the snRNPs during splicing. To begin the splicing process, U1–snRNP binds to a donor site. U2–snRNP then attaches to the branch site. U1– and U2–snRNPs move toward one another, bringing the donor site close to the branch point. U5–snRNP and U4/U6–snRNP then attach, forming the spliceosome. The donor and acceptor sites are now cleaved and the two exons joined together. Other proteins are also involved in the splicing process, but their roles are less clear. These include the $U2AF^{35}$ protein, which attaches to the polypyrimidine tract and helps direct U2-snRNP to the branch site.

adenosine. The 3′ splice site—the **acceptor site**—is then cleaved and the two exons joined together. The intron is released as the lariat structure, which is subsequently converted back to linear RNA and degraded.

The splicing reaction is carried out by a set of RNA–protein complexes called **small nuclear ribonucleoproteins** (**snRNPs**). Each of these contains several proteins and one or two of the noncoding snRNAs. There are a number of different snRNAs in vertebrate nuclei, not all of them involved in splicing, the most abundant being the ones called U1, U2, U3, U4, U5, and U6 . The U stands for "uracil-rich." They are quite short, varying in size from 106 nucleotides for U6 to 217 nucleotides for U3.

A splicing reaction initiates when U1–snRNP binds to a donor site (Figure 5.20). This attachment is made, at least in part, by RNA–RNA base pairing, as U1–snRNA contains a sequence that is complementary to the consensus for a donor site. U2–snRNP then attaches to the branch site, probably not by base pairing but instead by an interaction between the branch site and one of the proteins associated with the U2–snRNP. The U1– and U2–snRNPs have an affinity for each other, and this draws the donor site toward the branch point. The other snRNPs involved in splicing (U5–snRNP and U4/U6–snRNP) also attach themselves to the intron at this stage, resulting in a large complex called the **spliceosome**. The cutting and joining reactions that excise the intron and ligate the exons take place within the spliceosome. Once completed the spliceosome dissociates into its component snRNPs and the spliced mRNA is released.

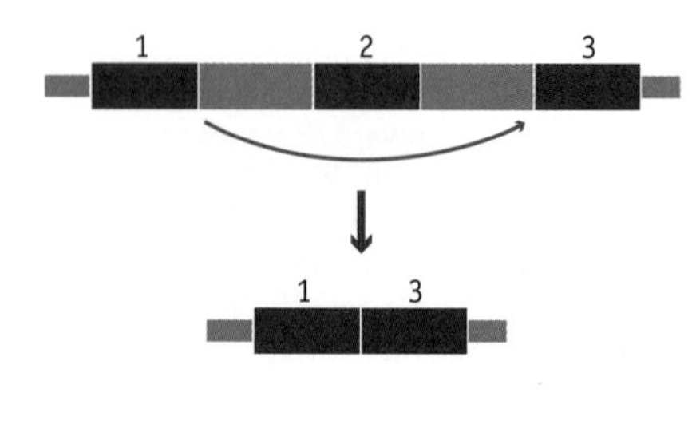

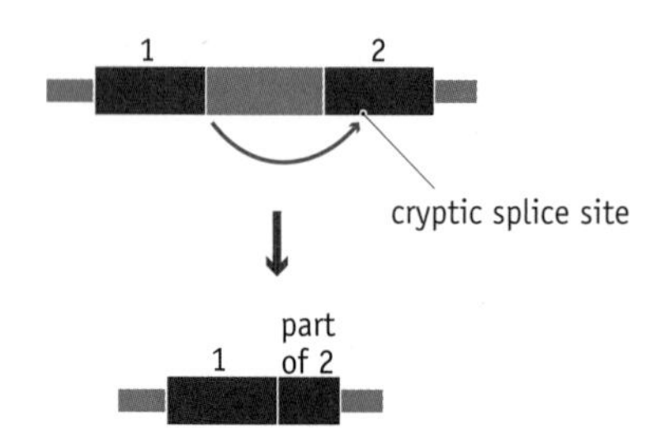

FIGURE 5.21 Two aberrant forms of splicing. (A) In exon skipping the aberrant splicing results in an exon being lost from the mRNA. (B) When a cryptic splice site is selected, part of an exon might be lost from the mRNA, as shown here, or if the cryptic site lies within an intron then a segment of that intron will be retained in the mRNA.

It is still not clear how errors are avoided during splicing

In a chemical sense, RNA splicing is not a great challenge for the cell. It simply involves two internal phosphodiester bonds being cut, and two new ones being made. The difficulty lies with the topological problems. The first of these is the substantial distance that might lie between splice sites, possibly a few tens of kb, representing 100 nm or more if the mRNA is in the form of a linear chain. A means is therefore needed of bringing the splice sites into proximity.

The second topological problem concerns selection of the correct splice site. All splice sites are similar, so if a transcript contains two or more introns then there is the possibility that the wrong splice sites could be joined, resulting in **exon skipping**—the loss of an exon from the processed mRNA (Figure 5.21A). Equally unfortunate would be selection of a **cryptic splice site**, a site within an intron or exon that has sequence similarity with the consensus motifs of real splice sites (Figure 5.21B). Cryptic sites are present in most mRNAs and must be ignored by the splicing apparatus.

The discovery of the GU and AG sequences at the two ends of an intron, and the gradual piecing together of the role of snRNPs in the splicing pathway, tell us how splicing occurs, but do not help us understand how the necessary degree of precision is achieved so that cuts are made at exactly the right places and nonadjacent exons are not joined together by mistake. These aspects of splicing are still poorly understood, but it has become clear that **SR proteins** are important in splice site selection. SR proteins were first implicated in splicing when it was discovered that they are components of the spliceosome. They appear to carry out several functions during the earliest steps in formation of the spliceosome, the period when the correct exon–intron junctions are being identified. SR proteins also interact with **exonic splicing enhancers** (**ESEs**), which are purine-rich sequences located in the exon regions of a transcript. We are still at an early stage in our understanding of ESEs and their counterparts, the **exonic splicing silencers** (ESSs), but their importance in controlling splicing is clear from the discovery that several human diseases, including one type of muscular dystrophy, are caused by mutations in ESE sequences.

Alternative splicing is common in many eukaryotes

When introns were first discovered it was imagined that each gene would always give rise to the same mRNA. This assumption was found to be incorrect when it was shown that some genes can follow two or more alternative splicing pathways, enabling the initial transcript to be processed into related but different mRNAs, these directing synthesis of related but different proteins. In some organisms alternative splicing is uncommon, only three examples being known in yeast, but in higher eukaryotes it is much more prevalent. At least 80% of all the human genes undergo alternative splicing.

For some genes there are just two or three alternative splicing pathways, but others can give rise to many mRNAs. An example is provided by the human *slo* gene. This gene codes for a membrane protein that regulates the entry and exit of potassium ions into and out of cells. The gene has 35 exons, eight of which are optional and are involved in alternative splicing (Figure 5.22). The alternative splicing pathways involve different combinations of these eight optional exons, leading to over 500 distinct mRNAs, each specifying a membrane protein with slightly different functional properties. The human *slo* genes are active in the inner ear and determine the auditory properties of the hair cells on the basilar membrane of the cochlea. Different hair cells respond to different sound frequencies between 20 and 20,000 Hz, their individual capabilities determined in part by the properties of their Slo proteins. Alternative splicing of *slo* genes in cochlear hair cells therefore determines the auditory range of humans.

FIGURE 5.22 The human *slo* gene. The gene comprises 35 exons, shown as boxes. The 27 exons shown in orange are present in all *slo* mRNAs. The eight green exons are optional, and appear in different combinations in different *slo* mRNAs. There are 8! = 40,320 possible splicing pathways, and hence 40,320 possible mRNAs, but only some 500 of these are thought to be synthesized in the human cochlea.

Alternative splicing is also important in developmental processes in some organisms. We will work through an example to illustrate this and to understand some of the strategies used to select alternative splice sites. The example is important because it is responsible for determining whether *Drosophila* eggs develop into male or female fruit flies. The key protein that determines sex in *Drosophila* is called DSX, which has a slightly different structure in male and female fruit flies. DSX is an activator of transcription initiation, and the male and female versions activate different sets of genes (**Figure 5.23**). Those activated by the male DSX protein cause the egg to differentiate into a male fly, and those activated by the female DSX protein result in a female fly.

Whether the male or the female version of DSX is synthesized is determined by a three-step pathway (**Figure 5.24A**). The first step involves the gene called *sxl*, whose transcript contains an optional exon (exon 3, in the figure). This exon contains a stop signal for translation, so when the exon is included in the mRNA a truncated protein is produced. In females, but not males, the splicing pathway is such that this exon is skipped, so the mRNA codes for a full-length, functional SXL protein.

The SXL protein, which is present and functional only in the female, affects splicing of a second transcript, from the *tra* gene of the fruit fly, by blocking the acceptor site of that gene's first intron (**Figure 5.24B**). The spliceosome is unable to locate this site and instead moves to a cryptic site in exon 2. Part of this exon is therefore omitted from the female mRNA. In this instance, cryptic site selection is not an error, but instead codes for a functional TRA protein, which will be present only in the female.

In the final step in the pathway, the TRA protein—which is present in functional form only in the female—affects splicing of the transcript of the *dsx* gene (**Figure 5.24C**). In males, without the influence of TRA, exon 4 of this transcript is skipped. In females, TRA stabilizes the attachment of SR proteins to an ESE located within exon 4 so that this exon is retained in the mRNA. Exon 5 is lost in the female because the intron between exons 4 and 5 has no donor site, meaning that exon 5 cannot be ligated to the end of exon 4. The male and female versions of the *dsx* mRNA therefore have

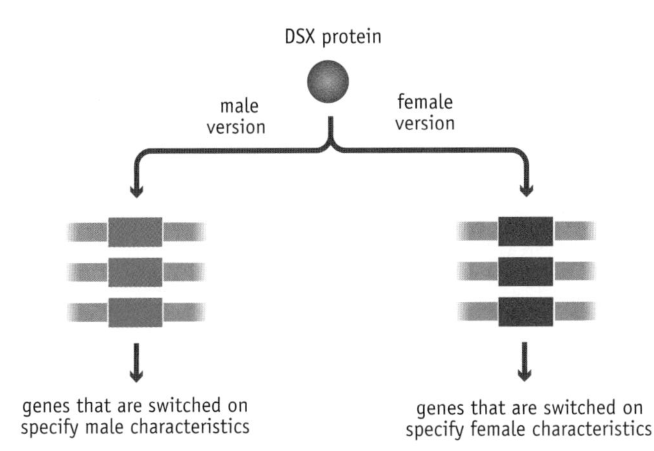

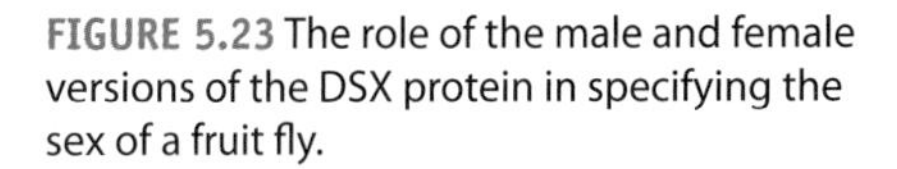

FIGURE 5.23 The role of the male and female versions of the DSX protein in specifying the sex of a fruit fly.

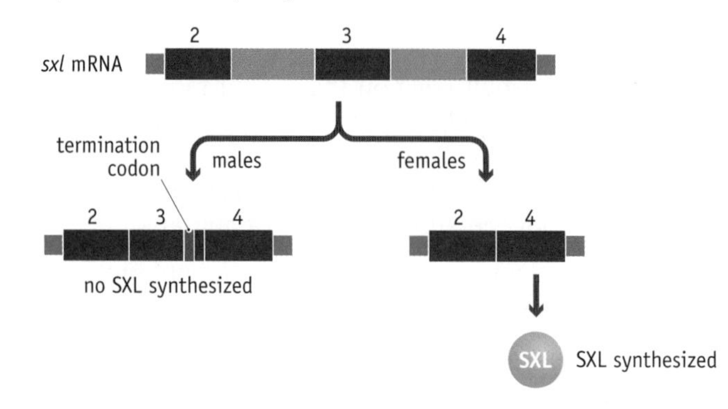

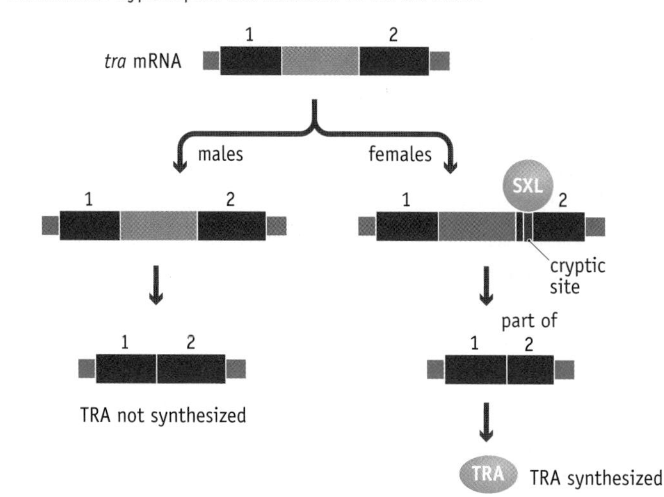

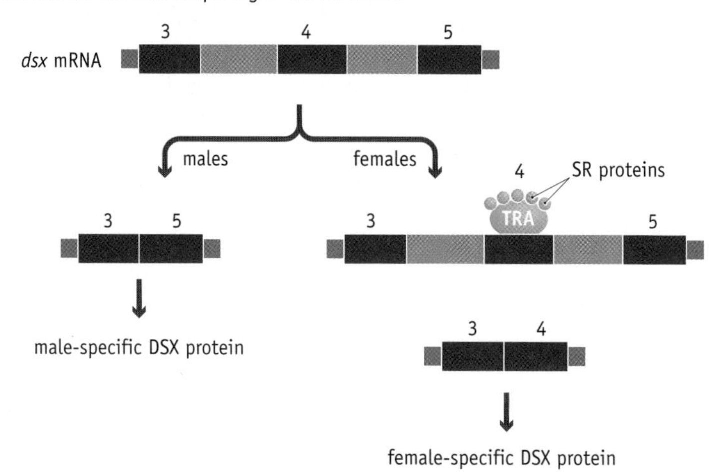

FIGURE 5.24 The series of alternative splicing events involved in sex determination in fruit flies. (A) The process begins with sex-specific alternative splicing of the *sxl* mRNA. In males all exons are present in the mRNA, but this means that a truncated protein is produced because exon 3 contains a termination codon. In females, exon 3 is skipped, leading to a full-length, functional SXL protein. (B) In females, SXL blocks the acceptor site in the first intron of the *tra* mRNA and directs splicing to a cryptic splice site in exon 2. This results in an mRNA that codes for a functional TRA protein. In males, there is no SXL so the acceptor site is not blocked and a dysfunctional mRNA is produced. (C) In males, exon 4 of the *dsx* mRNA is skipped. The resulting mRNA codes for a male-specific DSX protein. In females, TRA stabilizes the attachment of SR proteins to an exonic splicing enhancer located within exon 4, so this exon is not skipped, resulting in the mRNA that codes for the female-specific DSX protein. The female *dsx* mRNA ends with exon 4 because the intron between exons 4 and 5 has no donor site, meaning that exon 5 cannot be ligated to the end of exon 4.

different compositions, the male version comprising exons 1, 2, 3, and 5, and the female version being made up of exons 1, 2, 3, and 4. These mRNAs specify the male and female versions of the DSX protein.

Trans-splicing links exons from different transcripts

In the examples of splicing that we have considered so far the two exons that are joined together are both located within the same transcript. In a

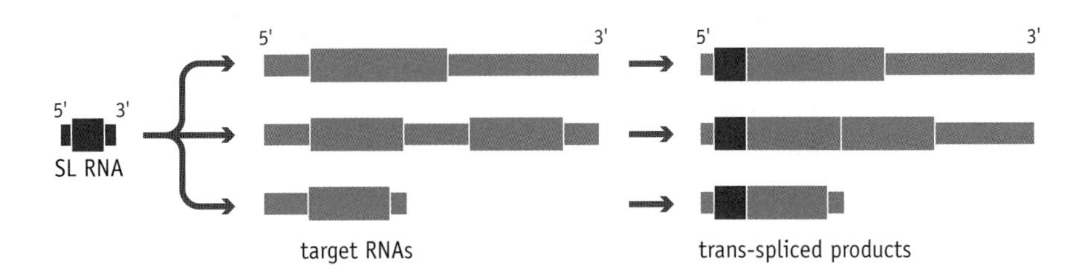

FIGURE 5.25 Trans-splicing. The diagram shows the SL RNA being trans-spliced onto the 5′ ends of three different target RNAs.

few organisms, splicing also takes place between exons that are contained within different RNA molecules. This is called **trans-splicing**, and it occurs with some genes in the nematode worm *Caenorhabditis elegans*, and in trypanosomes, the protozoan parasites of vertebrates that cause sleeping sickness in humans.

In trans-splicing, the same short segment becomes attached to the 5′ end of each member of a set of mRNAs (Figure 5.25). The transcript that donates this leader segment is called the **spliced leader RNA** (**SL RNA**). In *C. elegans* there are two SL RNAs, SL1 and SL2, both approximately 100 nucleotides in length and each containing a 22-nucleotide sequence that is attached to the 5′ end of the target mRNA. The splicing reaction proceeds in a manner very similar to the standard scheme shown in Figure 5.19, although as the splicing partners are different molecules a forked structure is formed instead of the lariat.

The reasons for trans-splicing are mysterious. About 70% of all *C. elegans* mRNAs become trans-spliced, most by the SL1 RNA, but in all cases trans-splicing simply adds an extra segment to the leader region of the mRNA, in front of the part that actually codes for the protein. So the protein specified by the mRNA is unaffected by the splicing event. One possibility is that addition of the new segment is necessary before the mRNA can be translated, in which case trans-splicing could be a means of controlling which mRNAs in the *C. elegans* transcriptome can be translated at any particular time.

5.3 EDITING OF mRNAS

The final type of event that can affect the composition of a transcriptome is **RNA editing**. This results in alteration of the nucleotide sequence of an mRNA by insertion of new nucleotides, or by deletion or changing of existing ones.

There are various types of RNA editing

RNA editing was discovered in 1986 during studies of genes located in the mitochondria of the parasitic protozoan *Trypanosoma brucei*. It was realized that many of these mRNAs are specified by **cryptogenes**—sequences lacking some of the nucleotides present in the mature mRNAs. The initial transcripts from these cryptogenes are processed by insertions of U nucleotides. The editing is quite extensive, with one or more Us being added at a number of different positions to give rise to the functional mRNA (Figure 5.26). Specificity is controlled by **guide RNAs**, short RNA molecules that

5′ GGGGAAAGAUAUUGAUGAAAAGGA 3′

insertion of Us

5′ GUGGUGAAUUUAUUGUUUAUAUUGAUUGAUUAUAAGGUUA 3′

FIGURE 5.26 Editing of a *Trypanosoma* cryptogene.

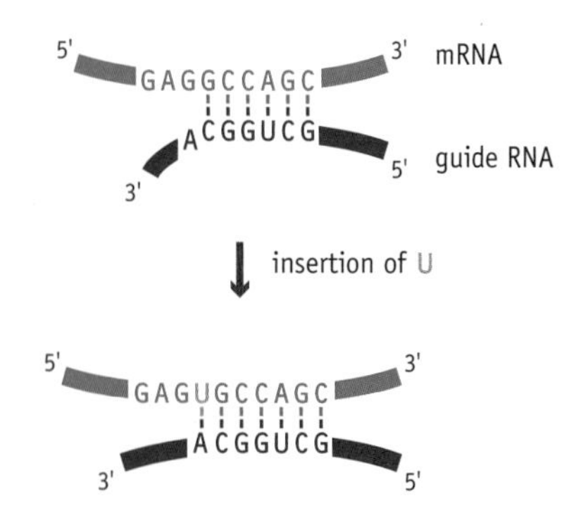

FIGURE 5.27 The role of the guide RNAs in cryptogene editing. A guide RNA base-pairs to the mRNA in the region to be edited. The A at the 3′ end of the guide sequence indicates where a U must be inserted into the mRNA.

base-pair to the mRNA in the region to be edited (Figure 5.27). The guide RNAs contain As at the positions where Us must be inserted into the mRNA and so indicate the exact positions at which editing must take place. This process, called **pan-editing**, appears to be unique to trypanosomes and related protozoans.

Since its initial discovery, examples of mRNA editing have been found in a variety of organisms, including humans and other mammals. The best studied of these involves the human mRNA for apolipoprotein-B. In liver cells this mRNA directs synthesis of a 4563–amino acid protein, called apolipoprotein-B100, which is involved in lipid transport in the bloodstream. Intestinal cells also make apolipoprotein-B, but in this case the protein is the B48 version, which is shorter (only 2153 amino acids) and has slightly different biochemical properties. At first it was thought that the B100 and B48 versions were coded by different genes, but now we know that the same gene specifies both types, the intestinal B48 protein arising by RNA editing. What happens is that in intestinal cells a C within the mRNA is deaminated, by removal of the $-NH_2$ group attached to carbon number 4, converting it into uracil (Figure 5.28). This results in the triplet CAA changing into UAA, which is a signal for the termination of mRNA translation. The consequence is that the protein coded by the edited intestinal mRNA is shorter than that coded by the unedited mRNA present in the liver.

The cytidine deaminase enzyme that carries out the C $\rightarrow$ U transformation locates the correct position on the mRNA by recognizing the nucleotide sequence on either side of the editing site, specificity being ensured because this 21-bp recognition sequence does not occur elsewhere in the apolipoprotein-B mRNA. Equally important, the sequence appears to be absent in other human mRNAs, so the deaminase does not edit the wrong mRNA by mistake.

KEY CONCEPTS

- The mRNA content of the cell is referred to as the transcriptome. The important features of a transcriptome are the identities of the mRNAs that it contains, and their relative amounts.
- Transcriptomes contain copies of hundreds or thousands of different mRNAs. Usually, each mRNA makes up only a small fraction of the transcriptome. Exceptions are those cells that have highly specialized biochemistries, which are reflected by transcriptomes in which one or a few mRNAs predominate.
- The composition of a transcriptome can change over time. Transcription of a gene that previously was silent will add a new type of mRNA to the transcriptome. Turnover of mRNAs means that when a gene is switched off its mRNA will disappear from the transcriptome.
- Most of the processes for mRNA degradation appear to be nonspecific with regard to which mRNAs they target. An exception is the silencing of specific eukaryotic mRNAs by a degradation pathway mediated by microRNAs.
- Transcription produces a faithful copy of a gene, so if the gene contains introns then the initial transcript includes copies of these. These introns must be excised from the transcript and the exons joined together in the correct order before the transcript can function as a mature mRNA.
- Introns are removed from mRNAs by a three-step pathway carried out by a set of RNA–protein complexes called small nuclear ribonucleoproteins.

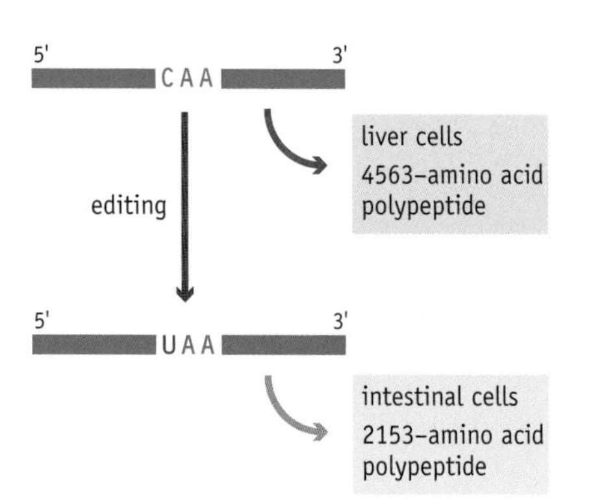

FIGURE 5.28 Editing of the human apolipoprotein-B mRNA.

- Many genes can follow two or more alternative splicing pathways, enabling the initial transcript to be processed into related but different mRNAs, these directing synthesis of related but different proteins.
- The nucleotide sequence of an mRNA can be altered by RNA editing, which involves the insertion of new nucleotides, or the deletion or alteration of existing ones.

QUESTIONS AND PROBLEMS (Answers can be found at www.garlandscience.com/introgenetics)

Key Terms

Write short definitions of the following terms:

acceptor site
alternative splicing
autosome
cryptic splice site
cryptogene
deadenylation-dependent decapping
Dicer
donor site
endonuclease
exon skipping
exonic splicing enhancer
exonic splicing silencer
exonuclease
foldback RNA
guide RNA
half-life
heterogeneous nuclear RNA
metabolome
metagenome
metastasis
microRNA
microarray analysis
microbiome
mismatch
mitochondrial genome
mRNA surveillance
nucleic acid hybridization
nucleolus
nucleoplasm
'ome
pan-editing
polypyrimidine tract
proteome
reverse transcriptase
ribonuclease
RNA editing
RNA-induced silencing complex
RNA interference
RNA silencing
short interfering RNA
small nuclear ribonucleoprotein
spliced leader RNA
spliceosome
splicing
SR protein
termination codon
transcriptome
trans-splicing

Self-study Questions

5.1 Define the term "transcriptome." How are a cell's proteome and metabolome related to its transcriptome?

5.2 Describe how microarray analysis is used to characterize a transcriptome. Why is mRNA not directly applied to the microarray?

5.3 Write a short essay on the applications of microarray analysis in studies of cancer.

5.4 Why is mRNA degradation an essential part of the processes that enable transcriptomes to change over time?

5.5 Describe the differences in half-life that are seen when bacterial and eukaryotic mRNAs are compared. What variations in half-life are seen in individual cells and organisms?

5.6 What is the difference between an exonuclease and an endonuclease?

5.7 Why is an exonuclease that degrades RNA in the $3' \rightarrow 5'$ direction unable to degrade a bacterial mRNA?

5.8 Explain how termination codons are involved in the eukaryotic RNA degradation process called mRNA surveillance.

5.9 Describe the deadenylation-dependent decapping pathway for degradation of eukaryotic RNA.

5.10 What is Dicer, and what does it do?

5.11 What is the relationship between a foldback RNA and a microRNA?

5.12 Why is it necessary to remove introns from an mRNA before translation occurs?

5.13 Explain why geneticists were able to conclude that splicing occurs in the nucleus.

5.14 Describe the key sequence features of intron–exon boundaries. What additional sequences within an intron are important in the splicing pathway?

5.15 Give a detailed description of the series of events involved in the RNA splicing pathway.

Your answer should explain why the donor and acceptor sites are given those names, and should also make a clear distinction between the roles of different snRNPs.

5.16 What processes are thought to ensure that the correct splice sites are selected during splicing?

5.17 Give an example of alternative splicing in human cells.

5.18 Draw a detailed series of diagrams that illustrate the role of alternative splicing in sex determination in *Drosophila*.

5.19 Explain, with an example, what is meant by "trans-splicing."

5.20 Describe the role of the guide RNAs in pan-editing.

5.21 Explain how RNA editing leads to synthesis of two different versions of human apolipoprotein-B.

Discussion Topics

5.22 The transcriptome and proteome are looked on as, respectively, an intermediate and the end-product of genome expression. Evaluate the strengths and limitations of these terms for our understanding of genome expression.

5.23 To what extent can transcriptome studies identify the functions of genes?

5.24 How might the length of the poly(A) tail influence the half-life of a eukaryotic mRNA?

5.25 In addition to the GU–AG introns described in this chapter, some eukaryotic genes include AU–AC introns. Obtain some information on these AU–AC introns and discuss the similarities they display with the GU–AG versions and how they have been used in studies of the splicing pathway.

5.26 Discuss the questions raised by the discovery of RNA editing.

FURTHER READING

Blencowe BJ (2000) Exonic splicing enhancers: mechanism of action, diversity and role in human genetic diseases. *Trends Biochem. Sci.* 25, 106–110.

Blencowe BJ (2006) Alternative splicing: new insights from global analyses. *Cell* 126, 37–47.

Bushati N & Cohen SM (2007) MicroRNA functions. *Annu. Rev. Cell Dev. Biol.* 23, 175–205.

Cheang MCU, van de Rijn M & Nielsen TO (2008) Gene expression profiling of breast cancer. *Annu. Rev. Pathol. Mech. Disease* 3, 67–97. *Applications of transcriptome analysis in cancer studies.*

Coller J & Parker R (2004) Eukaryotic mRNA decapping. *Annu. Rev. Biochem.* 73, 861–890. *Decapping and RNA degradation.*

Kaddurah-Daouk R, Kristal BS & Weinshilboum RM (2008) Metabolomics: a global biochemical approach to drug response and disease. *Annu. Rev. Pharmacol. Toxicol.* 48, 653–683.

Mello CC & Conte D (2004) Revealing the world of RNA interference. *Nature* 431, 338–342.

Padgett RA, Grabowski PJ, Konarska MM & Sharp PA (1985) Splicing messenger RNA precursors: branch sites and lariat RNAs. *Trends Biochem. Sci.* 10, 154–157. *A good summary of the basic details of RNA splicing.*

Smith HC & Sowden MP (1996) Base-modification mRNA editing through deamination: the good, the bad and the unregulated. *Trends Genet.* 12, 418–424. *Apolipoprotein-B and similar editing systems.*

Stoughton RB (2005) Applications of DNA microarrays in biology. *Annu. Rev. Biochem.* 74, 53–82.

Stuart KD, Schnaufer A, Ernst NL & Panigrahi AK (2005) Complex management: RNA editing in trypanosomes. *Trends Biochem. Sci.* 30, 97–105.

Venter JC, Remington K, Heidelberg JF et al. (2004) Environmental genome shotgun sequencing of the Sargasso Sea. *Science* 304, 66–74. *An example of metagenomics.*

Wahl MC, Will CL & Lührmann R (2009) The spliceosome: design principles of a dynamic RNP machine. *Cell* 136, 701–718. *Review of RNA splicing with focus on the role of the spliceosome.*

Types of RNA Molecule: Noncoding RNA CHAPTER 6

Only a small fraction of a cell's RNA, usually no more than 4% of the total, comprises messenger RNA. The vast bulk is made up of **noncoding RNA**, molecules that are not translated into protein but instead play their roles in the cell as RNA. A typical bacterium contains 0.05 to 0.10 pg of noncoding RNA, making up about 6% of its total weight. A mammalian cell, being much larger, contains more noncoding RNA, 20 to 30 pg in all, but this represents only 1% of the cell as a whole.

We have already met three types of noncoding RNA—siRNA, microRNA, and snRNA—in Chapter 5, where we looked at the roles of these molecules in mRNA degradation and RNA splicing. In this chapter our attention will be on two additional types of noncoding RNA, called ribosomal and transfer RNA, both of which play central roles in translation of mRNAs into protein. We will study those roles in detail in Chapters 7 and 8. First, we must become familiar with the structures of rRNA and tRNA molecules, and we must understand how the initial transcripts of rRNA and tRNA genes are converted into the functional versions of these RNAs.

6.1 RIBOSOMAL RNA

Ribosomal RNA molecules are components of **ribosomes**, which are large, multimolecular structures that act as factories for protein synthesis. During this second stage in gene expression, ribosomes attach to mRNA molecules and migrate along them, synthesizing proteins as they go (Figure 6.1).

Ribosomes are made up of rRNA molecules and proteins, and are extremely numerous in most cells. An actively growing bacterium may contain over 20,000 ribosomes, comprising about 80% of the total cell RNA and 10% of the total protein. Originally, ribosomes were looked on as passive partners in protein synthesis, merely the structures on which translation occurs. This view has changed over the years, and ribosomes are now considered to play an active role, with one of the rRNAs catalyzing the synthesis of bonds between amino acids, the central chemical reaction occurring during translation.

Our understanding of ribosome structure has gradually developed over the last 50 years as more and more powerful techniques have been applied to the problem. We will follow through this research to see what it has revealed.

Ribosomes and their components were first studied by density gradient centrifugation

Originally called "microsomes," ribosomes were first observed in the early decades of the twentieth century as tiny particles almost beyond the resolving power of light microscopy. In the 1940s and 1950s, the first electron micrographs showed that bacterial ribosomes are oval-shaped, with dimensions of 29 nm × 21 nm. Eukaryotic ribosomes are slightly larger, varying a little in size depending on species but averaging about

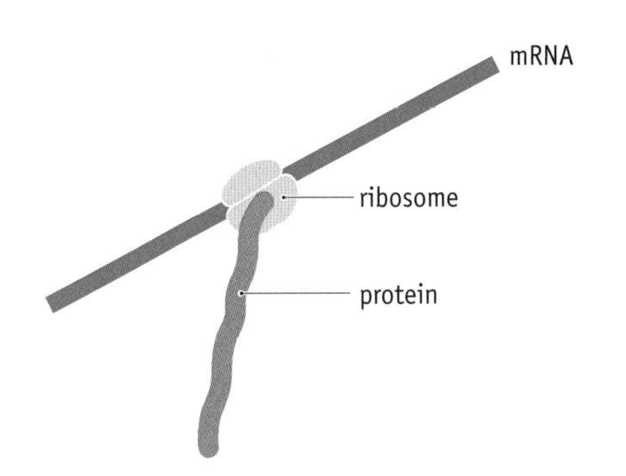

FIGURE 6.1 A ribosome attaches to an mRNA molecule and translates it into a protein.

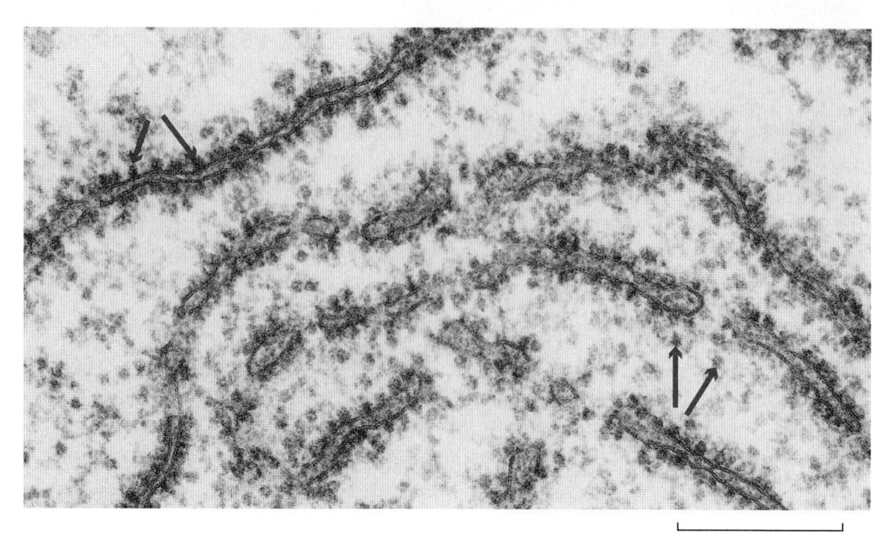

FIGURE 6.2 Electron micrograph showing ribosomes inside an animal cell. The ribosomes appear as black dots, indicated by the red arrows. Some occur free in the cytoplasm, and others are attached to membranes. (Courtesy of the late Professor George Palade.)

32 nm × 22 nm (Figure 6.2). In the mid-1950s, the discovery that ribosomes are the sites of protein synthesis stimulated attempts to define the structures of these particles in greater detail.

The initial progress in understanding the detailed structure of the ribosome was made by analyzing the particles by **density gradient centrifugation** (Figure 6.3). In this procedure, a cell extract is not centrifuged in a normal aqueous solution, but instead a sucrose solution is layered into the tube in such a way that a density gradient is formed, the solution being more concentrated and hence denser toward the bottom of the tube. The cell extract is placed on the top of the gradient and the tube centrifuged at a very high speed, at least 500,000 × *g* for several hours. Under these conditions, the rate of migration of a cell component through the gradient depends on its **sedimentation coefficient**. The sedimentation coefficient is expressed as a svedberg (S) value, after the Swedish chemist The Svedberg, who built the first ultracentrifuge in the early 1920s. The S value is dependent on several factors, notably molecular mass and shape.

The centrifugation studies showed that eukaryotic ribosomes have a sedimentation coefficient of 80S, and bacterial ribosomes, reflecting their smaller size, are 70S (Figure 6.4). Each type of ribosome is made up of two subunits. In eukaryotes these subunits are 60S and 40S, and in bacteria they are 50S and 30S. There are no mistakes here! Sedimentation coefficients are not additive because they depend on shape as well as mass, so it is perfectly acceptable for the intact ribosome to have an S value less than the sum of its two subunits.

In eukaryotes, the large subunit contains three rRNA molecules, called the 28S, 5.8S, and 5S rRNAs. The bacterial large subunit has just two rRNAs, of 23S and 5S. This is because in bacteria the equivalent of the eukaryotic

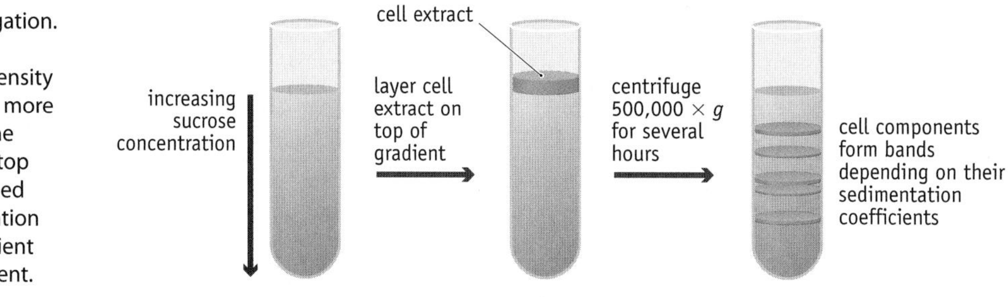

FIGURE 6.3 Density gradient centrifugation. A sucrose solution is layered into a centrifuge tube in such a way that a density gradient is formed, the solution being more concentrated toward the bottom of the tube. The cell extract is placed on the top of the gradient and the tube centrifuged at a very high speed. The rate of migration of a cell component through the gradient depends on its sedimentation coefficient.

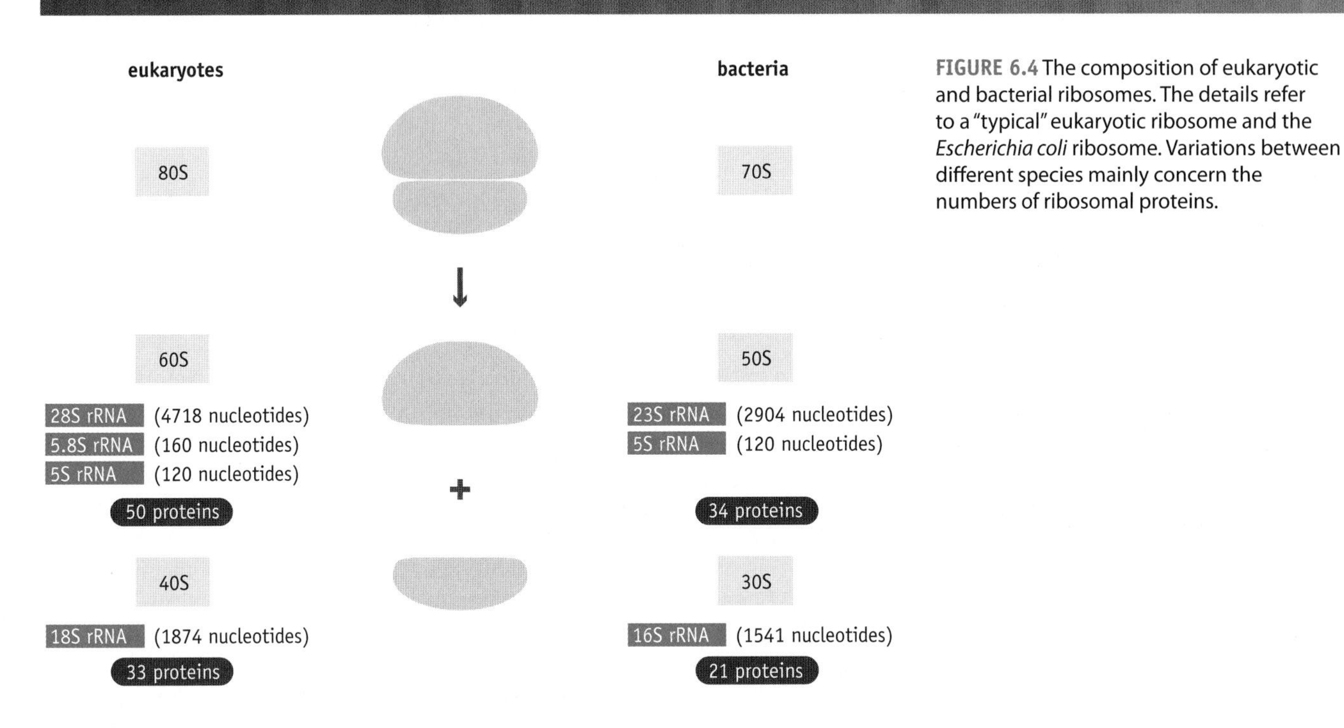

FIGURE 6.4 The composition of eukaryotic and bacterial ribosomes. The details refer to a "typical" eukaryotic ribosome and the *Escherichia coli* ribosome. Variations between different species mainly concern the numbers of ribosomal proteins.

5.8S rRNA is contained within the 23S molecule. The small subunit contains just a single rRNA in both eukaryotes and bacteria, 18S in eukaryotes and 16S in bacteria. Both subunits contain a variety of **ribosomal proteins**, the numbers detailed in Figure 6.4. The ribosomal proteins of the small subunit are called S1, S2, etc., and those of the large subunit are L1, L2, etc. There is just one of each protein per ribosome, except for L7 and L12, which are present as dimers.

Understanding the fine structure of the ribosome

The traditional view of ribosome structure is that the rRNA molecules act as a scaffolding onto which the proteins, which provide the functional activity of the ribosome, are attached. To fulfill this role the rRNA molecules must be able to take up a stable three-dimensional structure. This is achieved by inter- and intramolecular base pairs, with different rRNAs of a subunit base-pairing in an ordered fashion with each other, and also, more important, with different parts of themselves.

A two-dimensional representation of the base-paired structure of the *Escherichia coli* 16S rRNA, the component of the small subunit of the bacterial ribosome, is shown in Figure 6.5. This model has been developed by a combination of computational and experimental approaches. First, the nucleotide sequence of the *E. coli* 16S rRNA was examined using computer programs designed to identify the regions that could form base pairs with one another. Comparisons were also made between the *E. coli* sequence and the slightly different sequences of the equivalent rRNAs of other bacteria, as it would be expected that the regions of base pairing in the *E. coli* molecule would also be present in other bacteria. The computer predictions were then compared with the results of experiments in

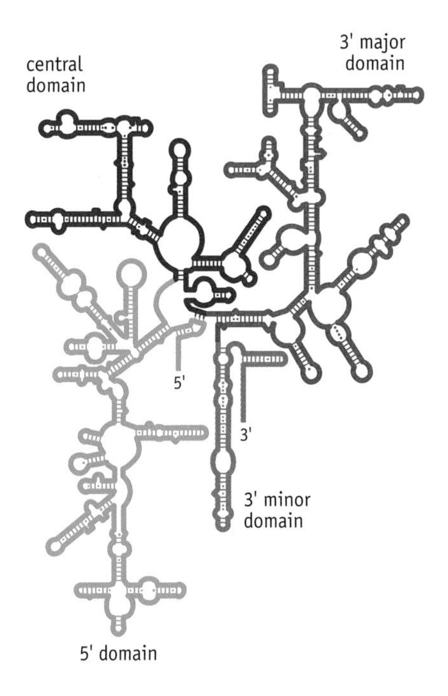

FIGURE 6.5 The base-paired structure of the *Escherichia coli* 16S rRNA. The rRNA is a single continuous molecule, but the base-paired structure is looked on as comprising four domains, as shown by the different colors. Besides the standard base pairs (G–C, A–U), nonstandard base pairs (e.g., G–U) can also form in double-stranded regions of an RNA molecule, usually at the edge of a non-base-paired region. These nonstandard pairs are shown as dots.

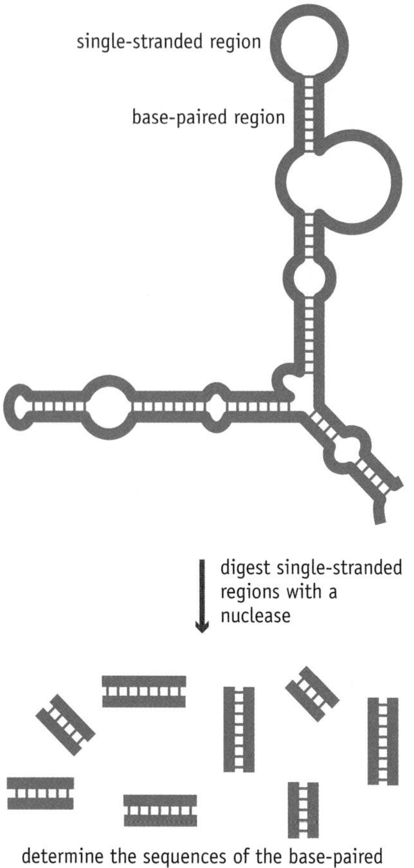

FIGURE 6.6 Using a single-strand–specific ribonuclease to identify the base-paired regions in an rRNA molecule.

which the base-paired rRNA was treated with ribonucleases that act on single-stranded RNA but have no effect on base-paired regions (Figure 6.6). The double-stranded regions remaining after treatment with these ribonucleases were examined to see whether they corresponded with those predicted by the computer model.

Additional experiments have located those regions of the 16S rRNA to which the protein components of the *E. coli* small subunit attach. One important technique here is **nuclease protection**, which identifies regions of an rRNA protected from attack by a ribonuclease enzyme. A complex between an rRNA and an individual ribosomal protein is treated with a nonspecific ribonuclease, one that digests both single- and double-stranded RNA. If no protein is present then the rRNA is completely degraded into mononucleotides, because every phosphodiester bond in the molecule is open to attack (Figure 6.7A). When the protein is bound this does not happen. The ribonuclease cannot gain access to all the phosphodiester bonds, because some are shielded ("protected") by the bound protein. After digestion the ribonuclease is inactivated and the bound protein removed, revealing two or more intact segments of rRNA whose sequences indicate the parts of the rRNA to which the protein was attached (Figure 6.7B).

Electron microscopy and X-ray crystallography have revealed the three-dimensional structure of the ribosome

The base-paired structures illustrated by Figure 6.5 are two-dimensional representations that tell us little about the real three-dimensional structure of the ribosome. For many years progress in moving from two- to

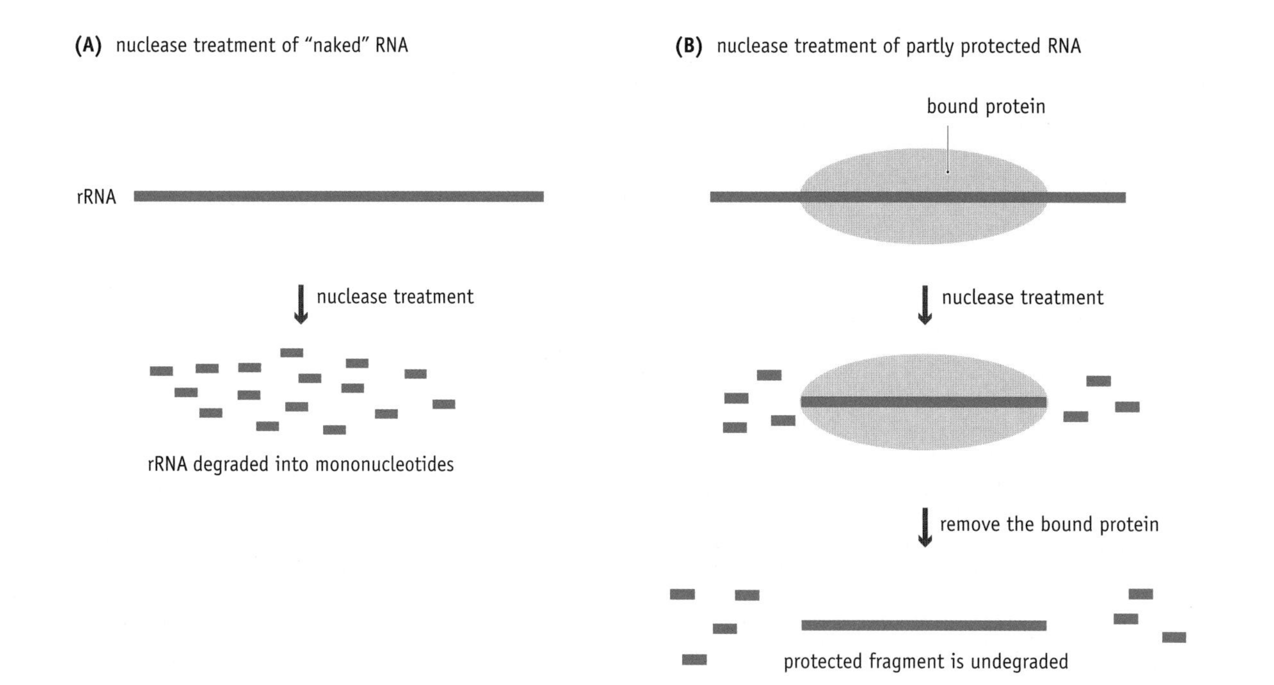

FIGURE 6.7 The principle behind the nuclease protection technique for identifying protein binding sites on an RNA molecule. (A) If no protein is present, the nuclease can attack every phosphodiester bond, and the rRNA is completely degraded into mononucleotides. (B) When a protein is bound to the RNA, the ribonuclease cannot gain access to all the phosphodiester bonds, because some are protected by the protein.

three-dimensional representation of ribosomes was slow, but over the last 10 years there have been remarkable advances due to the application of modern electron microscopy and X-ray diffraction techniques.

Ribosomes are so small that they are close to the resolution limit of the electron microscope, and in the early days of this technique the best that could be achieved were approximate three-dimensional reconstructions built up by analyzing the fuzzy ribosome images that were obtained. As electron microscopy gradually became more sophisticated, the overall structure of the ribosome could be resolved in greater detail, and the development of innovations such as **immunoelectron microscopy** provided important new information. Before examination by immunoelectron microscopy, ribosomes are first mixed with antibodies that bind specifically to individual ribosomal proteins (Figure 6.8). The antibodies, which are themselves proteins, are labeled with an electron-dense material such as gold particles so that they stand out strongly in the resulting images, enabling the positions of the ribosomal proteins to be located on the ribosome surface.

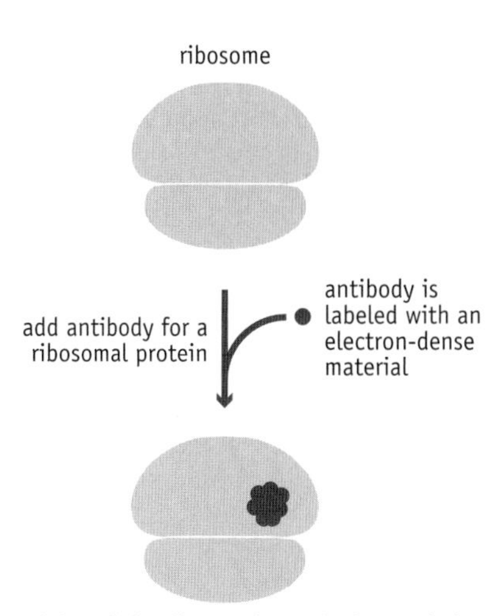

FIGURE 6.8 Identifying the position of a protein on a ribosome by immunoelectron microscopy.

The most exciting insights into ribosome structure have resulted from X-ray diffraction analysis. With an object as large as a ribosome, the diffraction patterns that are obtained are very complex and the data analysis is a huge task, but modern techniques have proved to be up to the challenge. Structures have now been deduced for entire ribosomes, including ones attached to mRNA and tRNAs (Figure 6.9), greatly increasing our understanding of how protein synthesis works, as we will see in Chapter 8.

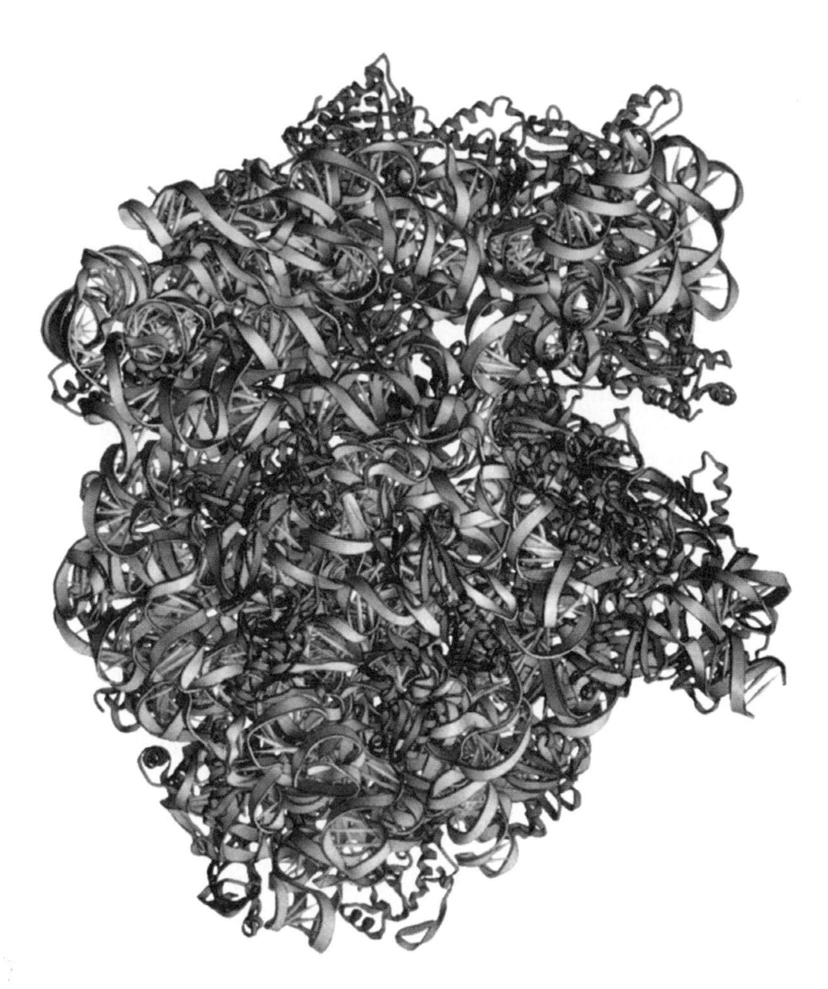

FIGURE 6.9 The bacterial ribosome. The picture shows the ribosome of the bacterium *Thermus thermophilus*. The small subunit is at the top, with the 16S rRNA in light blue and the small subunit ribosomal proteins in dark blue. The large subunit rRNAs are in gray and the proteins in purple. The gold area is the A site—the point at which tRNAs enter the ribosome during protein synthesis. (From M. B. Mathews and T. Pe'ery, *Trends Biochem. Sci.* 26: 585–587, 2001. With permission from Elsevier.)

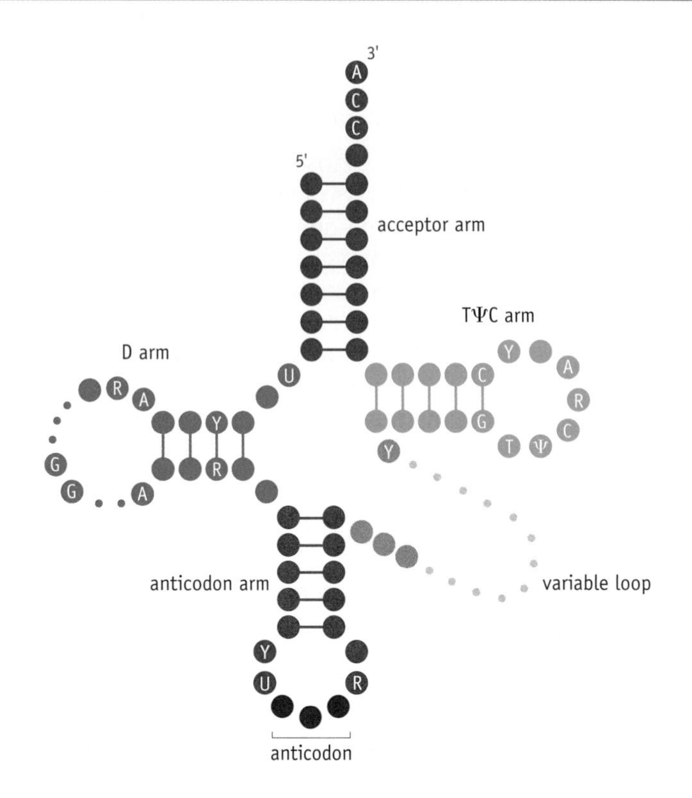

FIGURE 6.10 The cloverleaf structure of a tRNA. Invariant nucleotides (A, C, G, T, U, and Ψ, where Ψ = pseudouridine) and semi-invariant nucleotides (R, purine; Y, pyrimidine) are indicated. Optional nucleotides not present in all tRNAs are shown as smaller dots.

6.2 TRANSFER RNA

Transfer RNA molecules are also involved in protein synthesis, but the part they play is completely different from that of rRNA. They form the link between the mRNA that is being translated and the protein that is being synthesized. We will learn exactly how tRNA carries out its function in Chapters 7 and 8 when we deal with the genetic code and the mechanism of protein synthesis. First, in this chapter, we will look at the structures of tRNA molecules.

All tRNAs have a similar structure

Transfer RNA molecules are relatively small, most between 74 and 95 nucleotides in length. Each organism synthesizes a number of different tRNAs, each in multiple copies. However, virtually every tRNA molecule in every organism can be folded into a similar base-paired structure referred to as the **cloverleaf** (Figure 6.10).

The cloverleaf structure has five components. The first is the **acceptor arm**, which is formed by seven base pairs between the 5′ and 3′ ends of the molecule. During protein synthesis an amino acid is attached to the 3′ end of the acceptor arm of the tRNA, to the adenosine of the invariant CCA terminal sequence.

The second part of the cloverleaf structure is the **D arm**, named after the modified nucleotide dihydrouridine (Figure 6.11), which is always present in this structure. Next is the **anticodon arm**. "Anticodon" is an annoying jargon term, but its meaning will become clear in Chapter 7 when we look at the role of this part of the tRNA in protein synthesis.

Following the anticodon arm is the **extra** (or **optional**, or **variable**) **loop**, which may be a loop of just three to five nucleotides or a much larger hairpin structure of 13 to 21 nucleotides with up to five base pairs in the stem. Types of tRNA with the smaller loop are called class I, and make up 75%

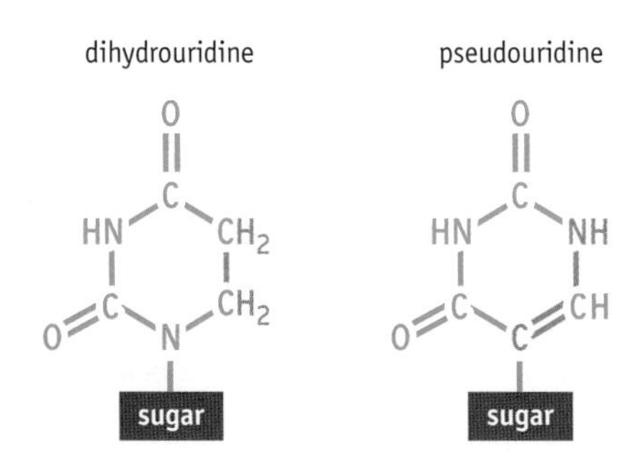

FIGURE 6.11 The structures of two modified nucleotides found in tRNAs. Both are modified versions of uridine, the standard U nucleotide found in RNA. The differences from the structure of uridine are indicated in brown.

of all tRNAs. Those with the larger loop are class II. The final part of the tRNA is the **TΨC arm**, named after the sequence thymidine–pseudouridine–cytosine, which is always present. Pseudouridine is another modified nucleotide (Figure 6.11).

The cloverleaf structure can be formed by virtually all tRNAs. In addition to having this common secondary structure, different tRNAs also display a certain amount of nucleotide sequence conservation. Some positions are invariant and in all tRNAs they are occupied by the same nucleotide. Others are semi-invariant and always contain the same type of nucleotide, either one of the pyrimidine nucleotides or one of the purine nucleotides. The invariant and semi-invariant positions are shown in Figure 6.10.

You should keep in mind that although the cloverleaf is a convenient way to draw the structure of a tRNA, it is only a representation. In the cell, tRNAs have a different three-dimensional structure. This structure has been determined by X-ray diffraction analysis and is shown in Figure 6.12. The base pairs in the arms of the cloverleaf are still present in the three-dimensional structure, but several additional base pairs form between nucleotides in the D and TΨC loops, which appear widely separated in the cloverleaf. These additional base pairs cause the molecule to be folded into a compact L-shaped conformation. Many of the nucleotides involved in these extra base pairs are the invariant or semi-invariant ones that are the same in different tRNAs. Note that the three-dimensional conformation places the acceptor arm and the anticodon loop at opposite ends of the molecule.

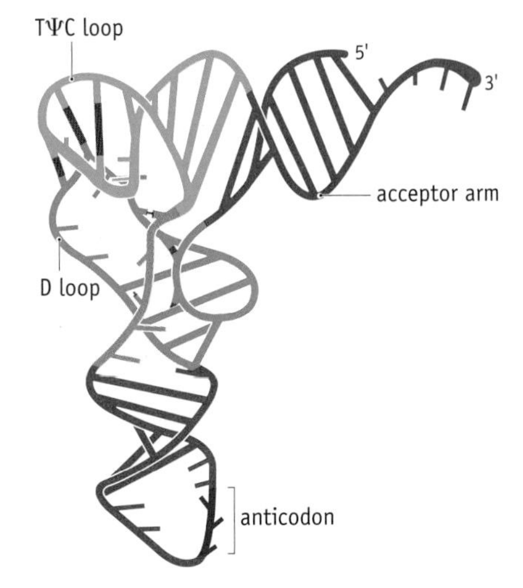

FIGURE 6.12 The three-dimensional structure of a tRNA. Additional base pairs, shown in black and mainly between the D and TΨC loops, fold the cloverleaf structure into this L-shaped configuration. The color scheme is the same as in Figure 6.10.

6.3 PROCESSING OF PRECURSOR rRNA AND tRNA MOLECULES

The rRNAs and tRNAs that play active roles in protein synthesis are quite different from the RNA molecules that are initially transcribed from the rRNA and tRNA genes. These initial transcripts are precursor molecules, longer than the mature RNAs and often containing more than one rRNA and/or tRNA sequence (Figure 6.13). These precursors must therefore be cut into smaller pieces to release the rRNAs and tRNAs. Most rRNAs and tRNAs also contain unusual nucleotides that are not present in the initial transcripts. We have already discovered that two unusual nucleotides—dihydrouridine and pseudouridine—are present in important regions of the tRNA cloverleaf structure (see Figure 6.11), and there are others in tRNA molecules and some in rRNAs also. These unusual nucleotides are produced by chemical modification of the initial transcripts.

All of these processing reactions must be carried out with precision. The cuts must be made at exactly the right positions in the precursor molecules, because if they are not then the rRNA and tRNA molecules that are released will have extra sequences, or may lack segments, and in either event might not be able to carry out their functions in the cell. The chemical modifications must be equally precise. The reasons for many of these modifications

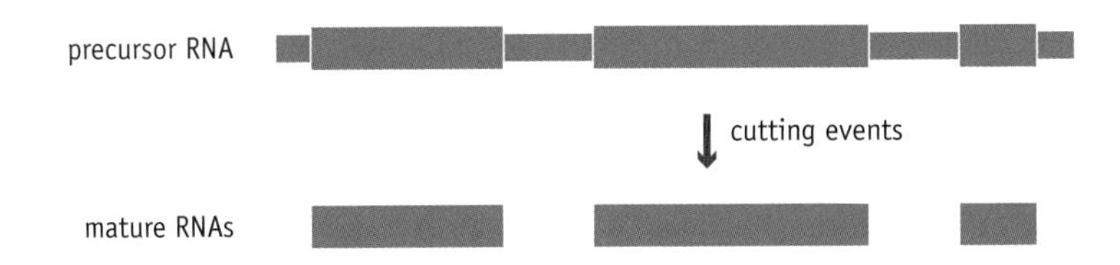

FIGURE 6.13 The initial transcripts of rRNA and tRNA genes are precursor molecules that must be cut into smaller pieces to release the mature RNAs.

are unknown, but we assume they have important roles because each copy of a particular type of tRNA or rRNA is modified at exactly the same nucleotide positions.

We must therefore examine the processing of precursor rRNA and tRNA molecules, not just the events themselves but also the mechanisms that enable the necessary precision to be achieved.

Ribosomal RNAs are transcribed as long precursor molecules

Each ribosome contains one copy of each of the different rRNA molecules, three rRNAs for the bacterial ribosome or four for the eukaryotic version. The most efficient system would be for the cell to synthesize equal numbers of each of these molecules. Of course the cell could make different amounts of each one, but this would be wasteful because some copies would be left over when the least abundant rRNA was all used up.

Synthesis of an equal number of each rRNA molecule is assured by having an entire complement of rRNA molecules transcribed together as a single unit. The product of transcription, the primary transcript, is therefore a long RNA precursor, the **pre-rRNA**, containing each rRNA separated from the next by a short spacer. In bacteria all three rRNAs are transcribed into a single pre-rRNA with a sedimentation coefficient of 30S, containing the mature molecules in the order 16S–23S–5S (Figure 6.14A). Cutting events are therefore needed to release the mature rRNAs. The same is true for eukaryotic rRNA, with the exception that only the 28S, 18S, and 5.8S genes are transcribed together (Figure 6.14B). The 5S genes occur elsewhere on the eukaryotic chromosomes and are transcribed independently of the main unit.

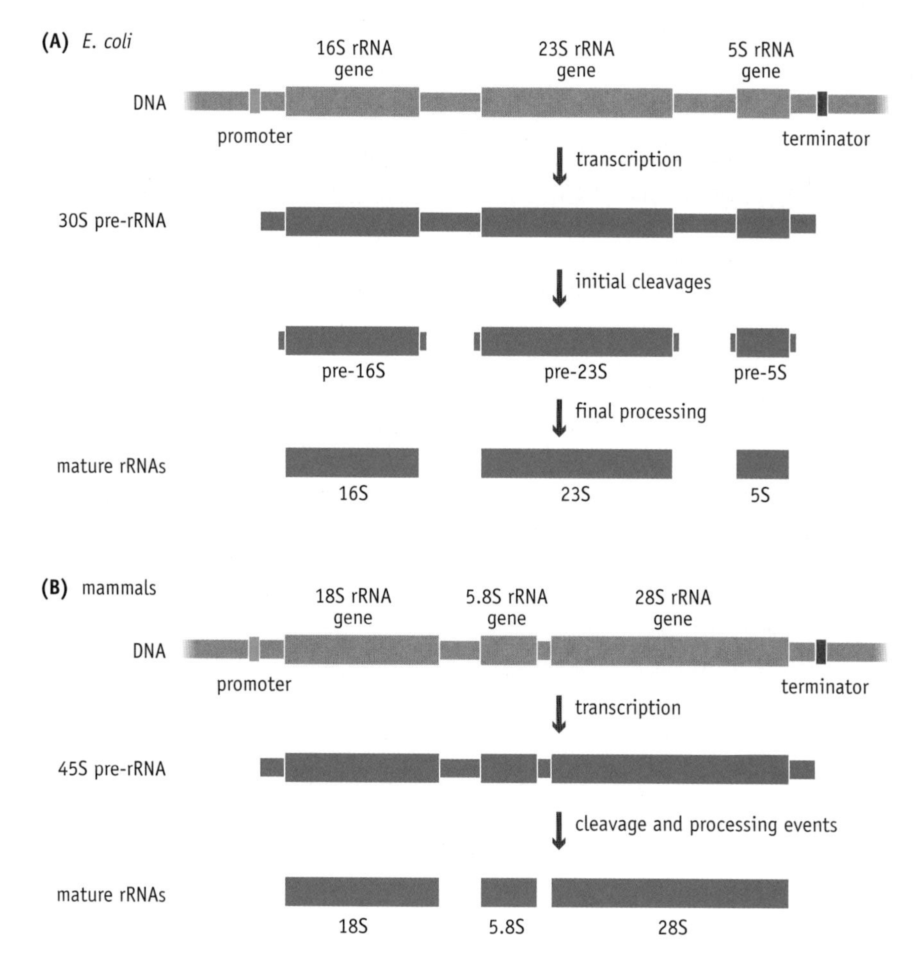

FIGURE 6.14 Processing of pre-rRNA in (A) *E. coli* and (B) mammals.

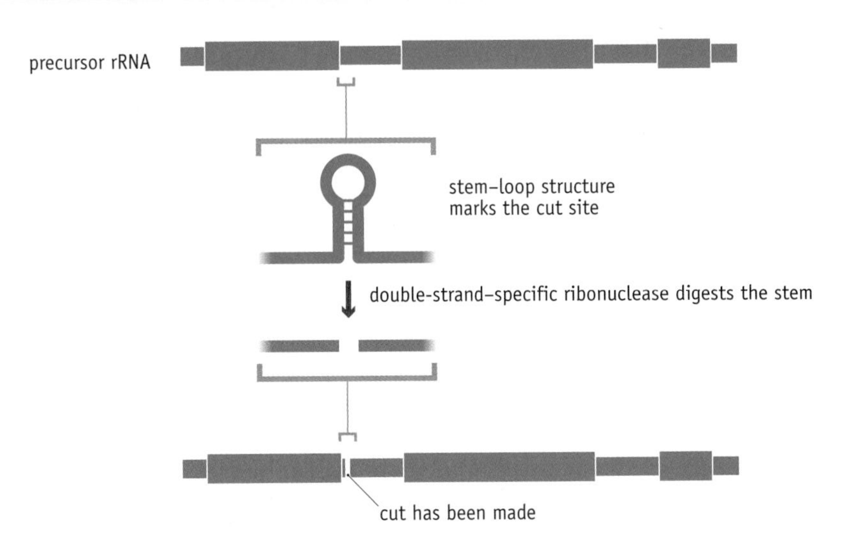

FIGURE 6.15 The role of a double-strand–specific ribonuclease in cutting a precursor rRNA molecule.

A variety of ribonucleases are involved in cutting the pre-rRNA molecules so that the mature rRNAs are released. Most of these are double-strand–specific. They cut the pre-rRNA by digesting short segments of double-stranded RNA formed by base pairing between different parts of the precursor (Figure 6.15). The base pairing, which of course is determined by the sequence of the pre-rRNA, therefore ensures that these cuts are made at the correct positions.

Cells not only need equal numbers of each rRNA, most cells need lots of them. An actively growing *E. coli* cell contains 20,000 ribosomes and divides once every 20 minutes or so. Therefore, every 20 minutes it needs to synthesize 20,000 new ribosomes, an entire complement for one of the two daughter cells. This necessitates a considerable amount of rRNA transcription, to such an extent that a single transcription unit would not be able to meet the demand. In fact, the *E. coli* chromosome contains seven copies of the rRNA transcription unit. In eukaryotes there can be an even greater demand for rRNA synthesis, and 50 to 5000 identical copies of the rRNA transcription unit are present depending on species. In eukaryotes these units are usually arranged into multigene families, with large numbers of copies following one after the other, separated by nontranscribed spacers (Figure 6.16).

Under certain circumstances, the demand for rRNA synthesis is so great that the transcriptional abilities of the existing genes in a genome are outstripped. In some eukaryotic cells (for example, amphibian oocytes) an additional strategy known as **gene amplification** may be called upon. This involves replication of rRNA genes into multiple DNA copies which subsequently exist as independent molecules not attached to the chromosomes (Figure 6.17). Transcription of the amplified copies then produces additional rRNA molecules. Gene amplification is not restricted to rRNA genes but also occurs with a few other genes whose transcription is required at a greatly enhanced rate in certain situations.

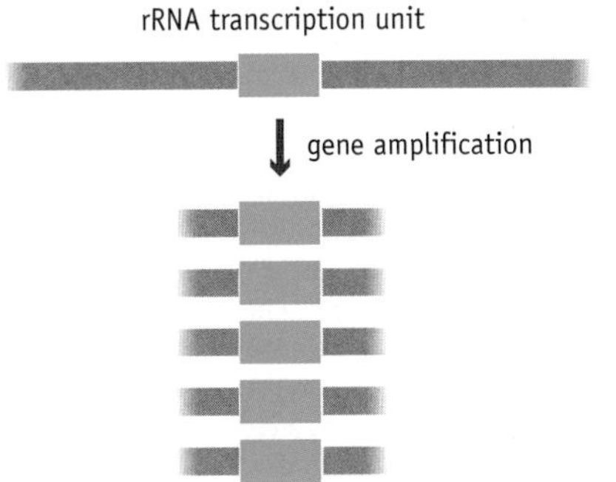

FIGURE 6.17 Production of multiple copies of rRNA genes by gene amplification. The amplified DNA copies exist as independent molecules and are not attached to any of the chromosomes.

multiple rRNA transcription units

nontranscribed spacers

FIGURE 6.16 A tandem array of rRNA transcription units. Each green box is a transcription unit comprising one set of 18S, 5.8S, and 28S rRNA genes.

Transfer RNAs are also cut out of longer transcription units

With tRNAs, unlike rRNA, there is no obvious reason why the cell would need equal amounts of the different types that it can make. Nonetheless, in both bacteria and eukaryotes tRNAs are transcribed initially as precursor tRNA, which is subsequently processed to release the mature molecules. In *E. coli* there are several separate tRNA transcription units, some containing just one tRNA gene and some with as many seven different tRNA genes in a cluster. In some bacteria, tRNA genes also occur as infiltrators in the rRNA transcription units. This is the case with *E. coli*, which has either one or two tRNA genes between the 16S and 23S genes in each of its seven rRNA transcription units (see Figure 6.14A).

All pre-tRNAs are processed in a similar though not identical way (**Figure 6.18**). The tRNA sequence within the precursor molecule adopts its base-paired cloverleaf structure, and two additional hairpin structures form, one on either side of the tRNA. Processing begins with a cut by ribonuclease E or F, forming a new 3′ end just upstream of one of the hairpins. Ribonuclease D, which is an exonuclease, trims seven nucleotides from this new 3′ end and then pauses while ribonuclease P makes a cut at the start of the cloverleaf, forming the 5′ end of the mature tRNA. Ribonuclease D then removes two more nucleotides, creating the 3′ end of the mature molecule.

All mature tRNAs must end with the trinucleotide 5′-CCA-3′. With some tRNAs the terminal CCA is present in the pre-RNA and is not removed by ribonuclease D, but with some other pre-tRNAs this sequence is absent, or is removed by the processing ribonucleases. If the CCA is absent or removed during processing, it has to be added by one or more template-independent RNA polymerases such as **tRNA nucleotidyltransferase**.

Transfer RNAs display a diverse range of chemical modifications

Besides being cut out of their pre-RNAs, tRNAs are also processed by chemical modification. This involves conversion of certain nucleotides into unusual forms by alteration of their chemical structures. The reasons for many of these modifications are unknown, although roles have been assigned to some specific cases. Some modifications in the D and TΨC

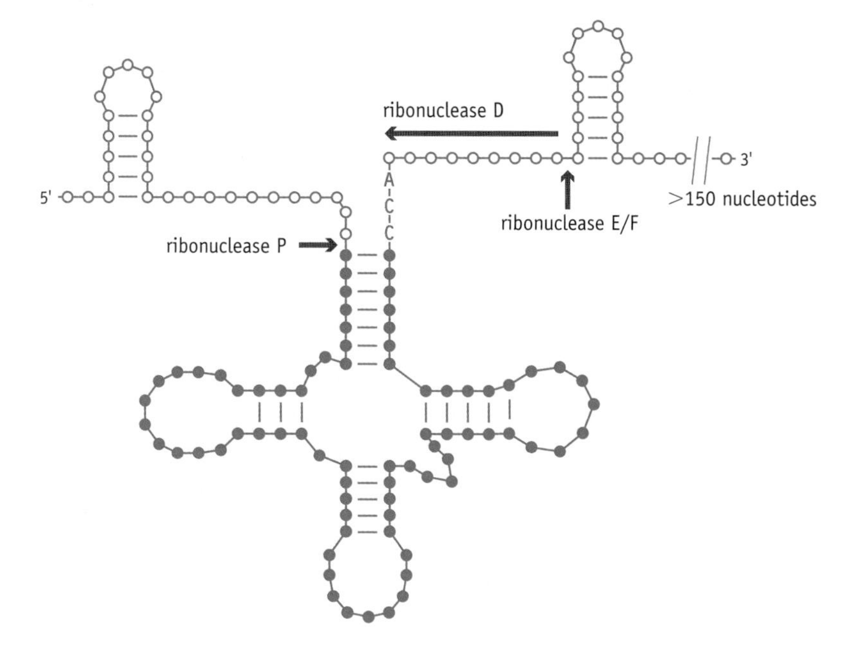

FIGURE 6.18 Processing of a pre-tRNA. The nucleotides of the mature tRNA are shown as closed circles, and those of the remainder of the precursor molecule as open circles. The tRNA has adopted its base-paired structure, and two additional hairpins have formed in the pre-tRNA, one on either side of the cloverleaf. Ribonuclease E or F makes a cut adjacent to one of these hairpins, forming a new 3′ end. Ribonuclease D then removes seven nucleotides from this new 3′ end. Ribonuclease P makes a cut at the start of the cloverleaf, forming the 5′ end of the mature tRNA. Ribonuclease D then removes two more nucleotides, creating the 3′ end of the mature molecule. In this example, the trinucleotide 5′-CCA-3′ is present in the pre-tRNA, and is not removed from the 3′ end of the mature tRNA. If this sequence is absent or removed during processing, then it is added by tRNA nucleotidyltransferase.

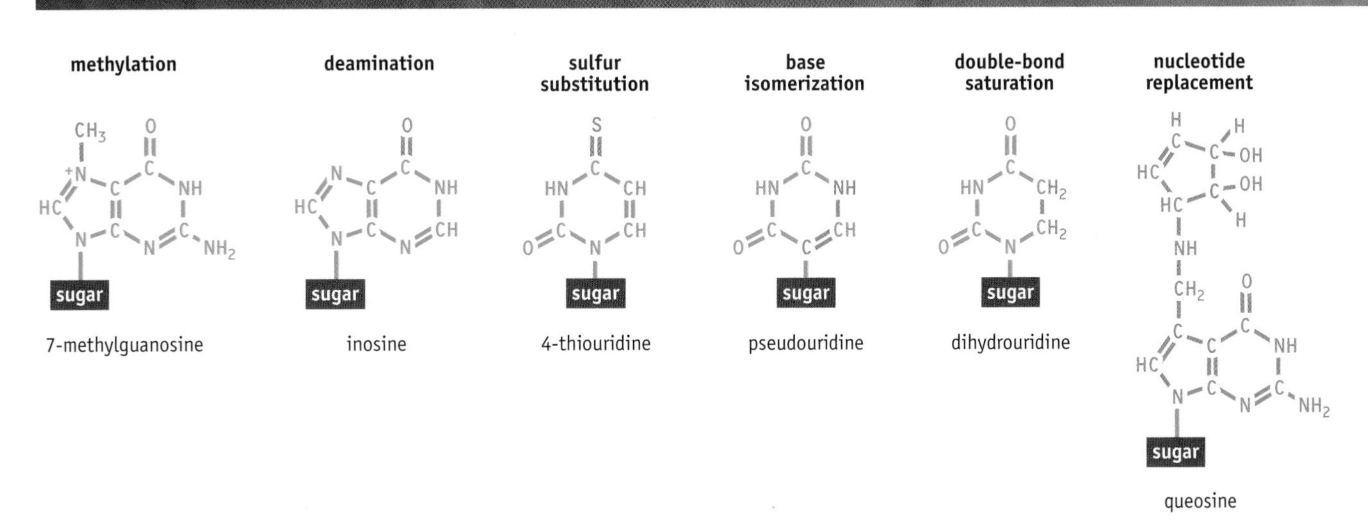

FIGURE 6.19 Examples of chemical modifications occurring with nucleotides in tRNAs. The differences between these modified nucleotides and the standard ones from which they are derived are shown in brown.

arms are recognized by the enzymes that attach an amino acid to the tRNA. These modifications therefore provide some of the specificity that ensures that the correct amino acid is attached to the correct tRNA. Some modifications in the anticodon arm are critical for the correct reading of the genetic code, as we will see in Chapter 7.

A number of different types of chemical modification are known in tRNAs (Figure 6.19). One of the most common is methylation, which involves the addition of one or more methyl groups ($-CH_3$) to the base or sugar component of the nucleotide. Examples include the conversion of guanosine to 7-methylguanosine. Some modifications involve removal of groups from the original nucleotide, such as deamination, which is the removal of an amino group ($-NH_2$), as in the conversion of adenosine into inosine. Replacement of an oxygen atom with a sulfur gives a thio-substituted nucleotide, an example being 4-thiouridine, obtained by sulfur substitution of uridine.

More complex modifications include base rearrangements, which result in the positions of atoms in the purine or pyrimidine ring becoming changed, as in the conversion of uridine to pseudouridine. Another type of rearrangement is double-bond saturation, which involves conversion of a double bond in the base to a single bond. An example is the modification of uridine to give dihydrouridine. A few modifications involve replacement of an entire base with a different, more complex one. The nucleotide called queosine is produced in this way from guanosine.

Over 50 types of chemical modification have been discovered so far in tRNAs. The enzymes that carry out these modifications are thought to recognize particular nucleotide sequences or base-paired structures in the tRNA, or possibly a combination of both, and so modify only the appropriate nucleotides.

Ribosomal RNAs are also modified, but less extensively

Chemical modification is not simply a feature of tRNAs. Ribosomal RNAs are also modified, though not to such a great extent.

In eukaryotic rRNAs the two most common types of modification are 2′-*O*-methylation, in which the hydrogen of the –OH group attached to the 2′ carbon is replaced by a methyl group (Figure 6.20), and the conversion of uridine to pseudouridine. Modification by 2′-*O*-methylation occurs at just over 100 positions in each set of mature human rRNAs—in other words, at about one in every 70 nucleotides. These modified positions are,

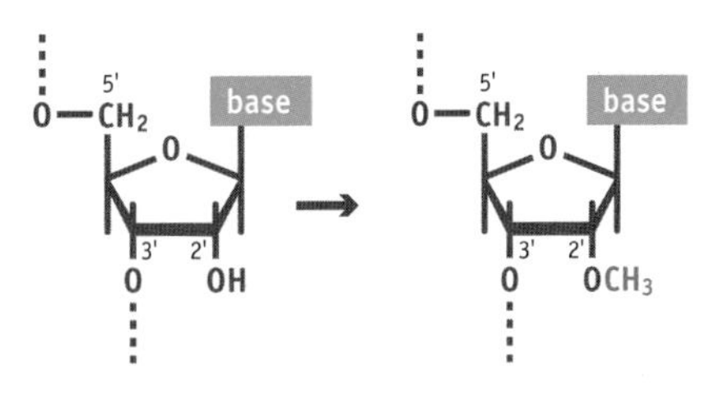

FIGURE 6.20 2′-*O*-methylation of a nucleotide.

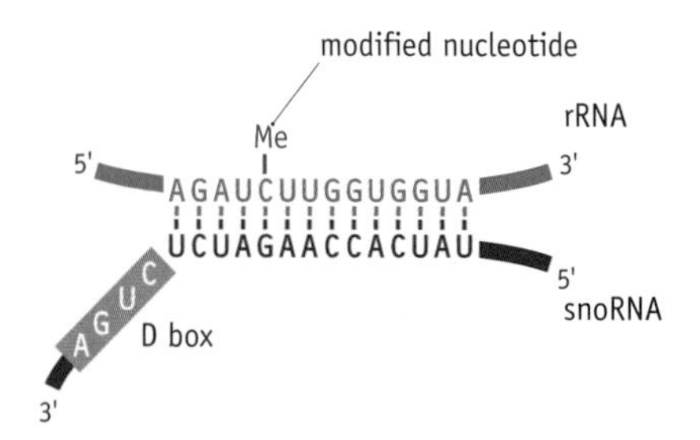

FIGURE 6.21 The role of a snoRNA in methylation of a specific nucleotide in an rRNA molecule. The D box of the snoRNA is highlighted. Modification always occurs at the base pair five positions away from the D box. Note that the interaction between the rRNA and the snoRNA involves an unusual G–U base pair, which is permissible between RNA polynucleotides.

to a certain extent, the same in different species, and some similarities in modification patterns are even seen when bacteria and eukaryotes are compared, although bacterial rRNAs are less heavily modified than eukaryotic ones. The functions of such modifications have not been identified, although most occur within those parts of rRNAs thought to be most critical to the activity of these molecules in ribosomes. Modified nucleotides might, for example, be involved in rRNA-catalyzed reactions such as synthesis of peptide bonds.

In bacteria, rRNAs are modified by enzymes that directly recognize the sequence and/or structure of the regions of RNA that contain the nucleotides to be modified. Often two or more nucleotides in the same region are modified at once. In eukaryotes a more complex machinery exists for ensuring that the modifications are made at the correct positions. The nucleolus (the part of the eukaryotic nucleus where rRNA transcription and processing occur) contains short RNAs, between 70 and 100 nucleotides in length, called small nucleolar RNAs (snoRNAs), that are involved in the modification process. By base pairing to the relevant region, snoRNAs pinpoint positions at which the pre-rRNA must be methylated. The base pairing involves only a few nucleotides, not the entire length of the snoRNA, but these nucleotides are always located immediately upstream of a sequence called the D box, which is present in all snoRNAs (Figure 6.21). The base pair involving the nucleotide that will be modified is five positions away from the D box.

The first snoRNAs to be discovered were all involved in methylation. It was then realized that a different family of snoRNAs carries out the same guiding role in conversion of uridines to pseudouridines. These snoRNAs do not have D boxes but still have conserved motifs that can be recognized by the modifying enzyme, and each is able to form a specific base-paired interaction with its target site, indicating the nucleotide to be modified.

There is a different snoRNA for each modified position in a pre-rRNA, except possibly for a few sites that are close enough together to be dealt with by a single snoRNA. This means that there must be a few hundred snoRNAs per cell. At one time this seemed unlikely because very few snoRNA genes could be located. Now it appears that only a fraction of all the snoRNAs are transcribed from standard genes, most being specified by sequences within the introns of other genes and released by cutting up the intron after splicing (Figure 6.22).

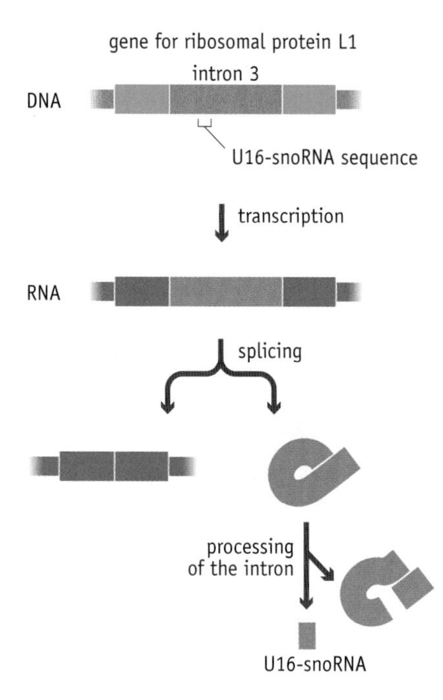

6.4 REMOVAL OF INTRONS FROM PRE-rRNAS AND PRE-tRNAS

Introns are found not only in genes that are transcribed into mRNA, but also in some eukaryotic rRNA and tRNA genes. These introns are not the same as those found in mRNA. They lack the characteristic consensus sequences, such as GU and AG at the splice sites, and they are not associated with spliceosomes. The mechanisms by which they are spliced are therefore different from the process that we studied in Chapter 5.

FIGURE 6.22 Most snoRNAs are specified by sequences within the introns of other genes. This example shows synthesis of the human U16-snoRNA by processing of the intron spliced from the mRNA from ribosomal protein L1.

Splicing of pre-rRNAs and pre-tRNAs appears simply to be a part of the series of processing events needed to produce the mature molecules. There is no equivalent of alternative splicing with rRNA and tRNA transcripts. But these introns are still interesting, especially the ones in rRNAs, as their discovery forced scientists to reappraise one of the fundamental notions of biology.

Some rRNA introns are enzymes

The fundamental notion that was overturned by rRNA introns was the assumption that only proteins can have enzymatic activity and hence catalyze biochemical reactions. This is because the introns present in some pre-rRNAs are enzymes. The biochemical reaction that they catalyze is their own splicing.

Introns are not common in rRNA genes but are found in certain protozoa, notably the ciliate *Tetrahymena*. The introns of these organisms are able to fold up by intramolecular base pairing into a complex structure in which the two splice sites are brought close together (Figure 6.23). The intron is then cut out and the two exons joined together in the complete absence of any protein molecules. The intron catalyzes the reaction by acting as an RNA enzyme, or **ribozyme**. The *Tetrahymena* self-splicing intron was the first known example of a ribozyme and caused quite a stir when it was discovered in 1982, as many biochemists at that time were unwilling to believe that RNA could have enzymatic activity.

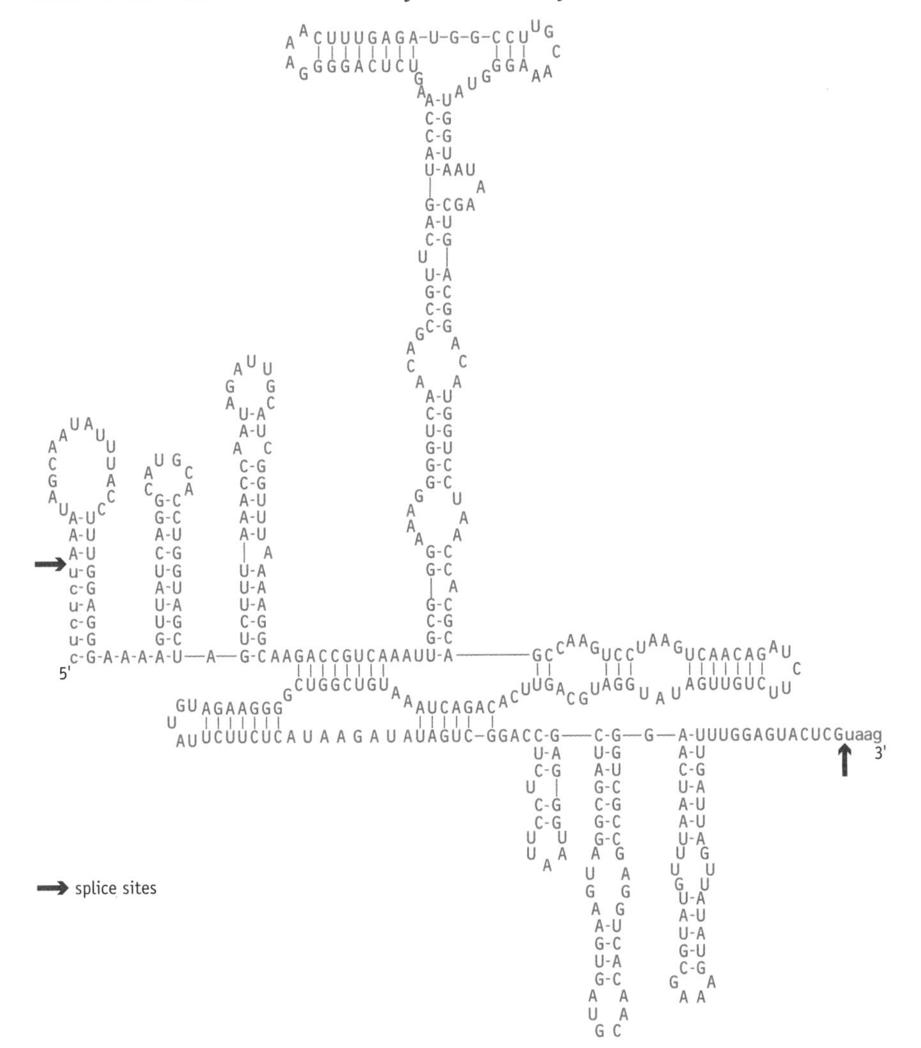

FIGURE 6.23 The base-paired structure of the *Tetrahymena* rRNA intron. The sequence of the intron is shown in capital letters, with the exons in lowercase. Additional interactions fold the intron into a three-dimensional structure that brings the two splice sites (indicated by the arrows) close together.

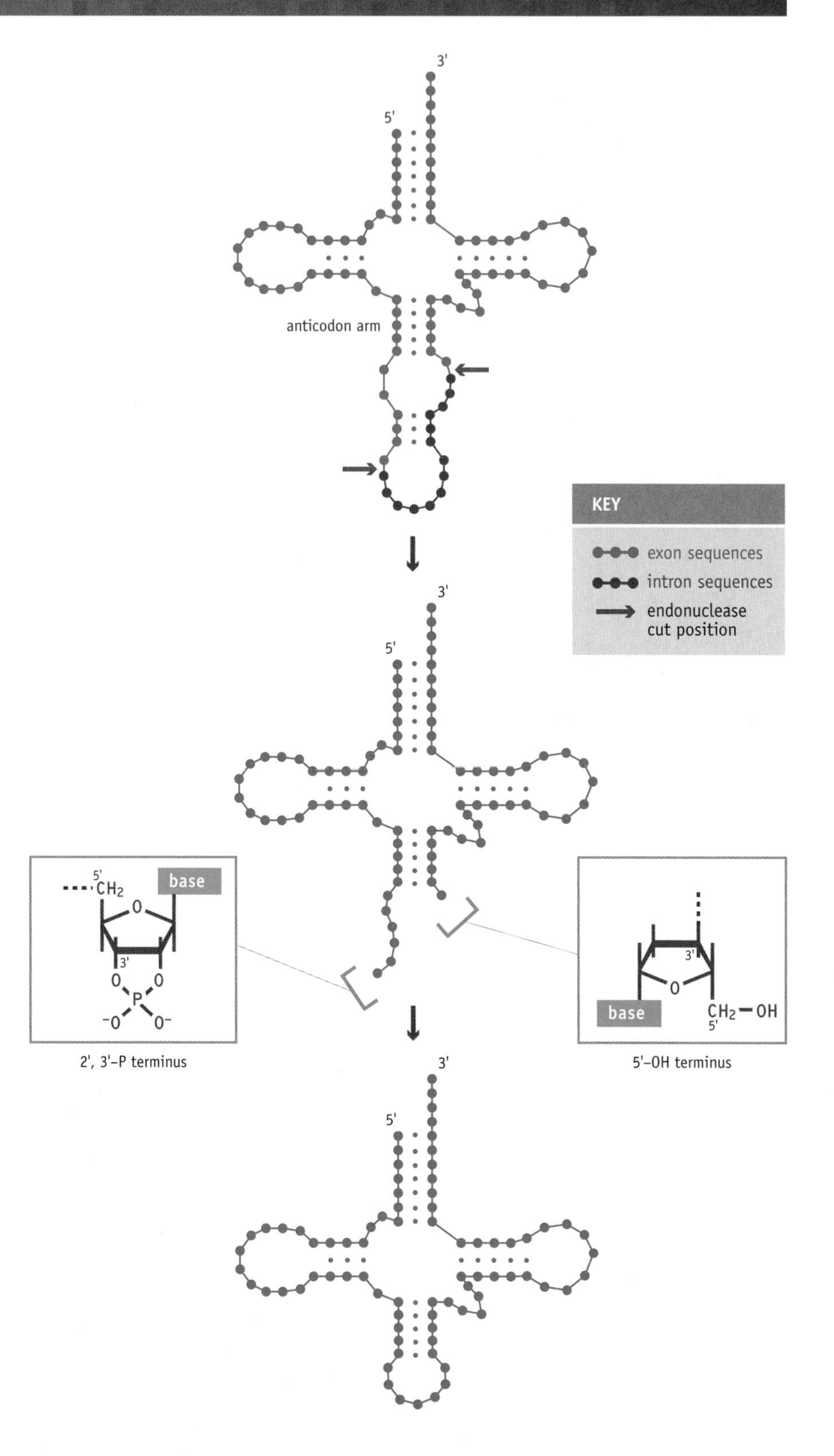

FIGURE 6.24 Removal of an intron from a pre-tRNA. In the top drawing, the tRNA has folded into its cloverleaf structure, but with an extra stem-loop, formed by the intron, in the anticodon arm. The nucleotides of the mature tRNA are shown in blue, and those of the intron in black. Cuts are made by endonucleases at the exon–intron boundaries, indicated by red arrows. The unusual structures that are left at the cut sites are shown in the boxes. These are converted to standard 3′–OH and 5′–P termini, which are ligated to one another to complete the splicing process.

The self-splicing intron is a member of the "Group I" class of introns. Group I introns are also found in the rRNA genes of other protozoa and in genes in the mitochondrial DNA of fungi and yeast. They are characterized by their ability to take up the base-paired structure displayed by the *Tetrahymena* intron, but only the *Tetrahymena* intron has been shown to splice efficiently in the absence of proteins. This does not mean that the other Group I introns are not ribozymes. The splicing reaction might still be catalyzed by the intron itself, with the proteins that are required in the cell acting only to stabilize the base-paired structure.

Some eukaryotic pre-tRNAs contain introns

Transfer RNA introns are relatively common in lower eukaryotes but less frequent in vertebrates—introns are present in only 6% of all human tRNA genes. Introns in eukaryotic pre-tRNAs are 14 to 60 nucleotides in length and are usually found at the same position in the transcript, within the anticodon arm. The intron sequence is variable, but includes a short region complementary to part of the anticodon arm. Base pairing between the complementary sequences forms a short stem between two loops in the unspliced pre-tRNA (Figure 6.24).

Unlike the mRNA and Group I introns, removal of pre-tRNA introns involves an endonuclease. This enzyme contains four nonidentical subunits, one of which uses the structure of the base-paired intron as a guide to identify the correct positions at which the RNA should be cut. The upstream and downstream cuts are then made by two of the other enzyme subunits. Cleavage leaves an unusual cyclic phosphate structure attached to the 3′ end of the upstream exon, and a hydroxyl group at the 5′ end of the downstream exon. Before the ends can be joined, the cyclic phosphate must be converted to a 3′–OH end, and the 5′–OH terminus to 5′–P. The two ends are held in proximity by the natural base pairing adopted by the tRNA sequence, and are ligated together. The enzymatic activities needed to convert the ends and join them together are all provided by a single protein.

KEY CONCEPTS

- Most of the RNA in a cell is noncoding. Noncoding RNAs are molecules that are not translated into protein but instead play their roles in the cell as RNA.
- Ribosomal RNA is the most abundant type of noncoding RNA. Ribosomal RNA molecules are components of ribosomes, which are large, multimolecular structures that act as factories for protein synthesis.
- Transfer RNA molecules are also involved in protein synthesis, but the part they play is completely different from that of rRNA. Transfer RNAs form the link between the mRNA that is being translated and the protein that is being synthesized.
- Almost all tRNAs can adopt a base-paired two-dimensional structure called the cloverleaf. In the cell, the cloverleaf folds into a compact L-shaped conformation.
- The rRNAs and tRNAs that play active roles in protein synthesis are initially transcribed as precursor molecules. The precursors are processed by cutting events, which might include RNA splicing, and chemical modification of certain nucleotides.
- The precursor rRNA molecules contain copies of three or more types of rRNA. This ensures that equal amounts of each rRNA are synthesized.

- Chemical modification of tRNA and bacterial rRNA is carried out by enzymes that recognize the sequence and/or structure of the regions of RNA that contain the nucleotides to be modified. Modification of eukaryotic rRNA is aided by snoRNAs, which bind to the rRNA and guide the modifying enzyme to the correct nucleotide.
- Some rRNAs contain introns that are self-splicing. These were the first types of ribozyme to be discovered and forced a reappraisal of the notion that enzymes are always made of protein.

QUESTIONS AND PROBLEMS (Answers can be found at www.garlandscience.com/introgenetics)

Key Terms

Write short definitions of the following terms:

acceptor arm
anticodon arm
cloverleaf
D arm
density gradient centrifugation
extra (or optional, or variable) loop
gene amplification
immunoelectron microscopy
noncoding RNA
nuclease protection
pre-rRNA
ribosomal protein
ribosome
ribozyme
sedimentation coefficient
tRNA nucleotidyltransferase
TΨC arm

Self-study Questions

6.1 What is the definition of noncoding RNA?

6.2 Explain how density gradient centrifugation is used to measure the sizes of cell components. How does the structure of a cell component influence its sedimentation coefficient?

6.3 What are the sizes of typical bacterial and eukaryotic ribosomes?

6.4 Construct two tables giving details of the RNA and protein components of bacterial and eukaryotic ribosomes.

6.5 Describe how methods that make use of nucleases have been used to study the structure of the ribosome.

6.6 An RNA molecule has the sequence:

```
5'-AUUAGCAUGUAAAUGCUAUCUGGCAAC-3'
```

After digestion with a single-strand–specific ribonuclease, the following base-paired molecule is obtained:

```
5'-AGCAU-3'
   |||||
3'-UCGUA-5'
```

What was the base-paired structure of the original molecule?

6.7 A second RNA molecule has the following sequence

```
5'-CUCCUCAAAUGAUUCAGUGCAUCUAACCC
   UAUUUAAACGCACUAGCUCAUGAGAGGAG-3'
```

After digestion with the same enzyme, the following are obtained:

```
5'-UAA-3'  5'-AUGA-3'  5'-AGUGC-3'
   |||        ||||        |||||
3'-AUU-5'  3'-UACU-5'  3'-UCACG-5'

5'-CUCCUC-3'
   ||||||
3'-GAGGAG-5'
```

Draw the structure of the original molecule.

6.8 Describe the methods that have been used and are being used to study the three-dimensional structure of the ribosome. What are the applications and limitations of each technique?

6.9 Draw and annotate the cloverleaf structure of tRNA.

6.10 To what extent is the cloverleaf a true representation of the actual structure of tRNA?

6.11 Here is the sequence of the yeast alanine tRNA:

5′-GGGCGUGUGGCGUAGUCGGUAGCGC
GCUCCCUUIGCIUGGGAGAGGUCUC
CGGUUCGAUUCCGGACUCGUCCACCA-3′

Make a fully annotated drawing of a possible cloverleaf structure for this tRNA.

6.12 Give one possible reason why the pre-rRNAs contain copies of more than one gene.

6.13 Describe the cutting events involved in processing of pre-rRNA.

6.14 Why do cells require large amounts of rRNA? How are these demands met?

6.15 Describe how a tRNA molecule is cut out of its primary transcript.

6.16 List six types of chemical modification that occur with nucleotides in tRNA. In each case, draw the structure of an example of a nucleotide resulting from the modification.

6.17 What are the possible roles of the chemical modifications found in tRNAs?

6.18 What types of chemical modification are found in eukaryotic rRNA?

6.19 Describe how rRNAs are chemically modified in bacteria.

6.20 How does chemical modification of rRNA in eukaryotes differ from the process occurring in bacteria?

6.21 Why is the *Tetrahymena* rRNA intron referred to as "self-splicing"?

6.22 Describe how introns are removed from pre-tRNA molecules.

Discussion Topics

6.23 To what extent have studies of ribosome structure been of value in understanding the detailed process by which proteins are synthesized?

6.24 Ribosomal and transfer RNA molecules are relatively long-lived in the cell. In contrast, mRNAs are subject to quite rapid turnover rates. Discuss the possible reasons and consequences.

6.25 In eukaryotes, the 28S, 18S, and 5.8S genes are transcribed together, but the 5S genes are transcribed separately. What implications does this have for the hypothesis that the most efficient system is for the cell to synthesize equal numbers of each of its rRNAs?

6.26 What are the possible reasons for the observation that virtually all tRNA molecules in all organisms adopt a similar base-paired structure?

6.27 Discuss the reasons why tRNA and rRNA molecules are chemically modified.

6.28 The existence of ribozymes is looked on as evidence that RNA evolved before proteins and therefore at one time, during the earliest stages of evolution, all enzymes were made of RNA. Assuming that this hypothesis is correct, explain why some ribozymes persist to the present day.

FURTHER READING

Björk GR, Ericson JU, Gustafsson C et al. (1987) Transfer RNA modification. *Annu. Rev. Biochem.* 56, 263–287. *Information on modified nucleotides in tRNAs.*

Cech TR (1990) Self-splicing of group I introns. *Annu. Rev. Biochem.* |59, 543–568. *Review of self-splicing introns.*

Chndramouli P, Topf M, Ménétret J-F et al. (2008) Structure of the mammalian ribosome at 8.7Å resolution. *Structure* 16, 535–548. *A recent detailed study of ribosome structure.*

Clark BFC (2001) The crystallization and structural determination of tRNA. *Trends Biochem. Sci.* 26, 511–514. *Determination of the three-dimensional structure of a tRNA.*

Holley RW, Apgar J, Everett GA et al. (1965) Structure of a ribonucleic acid. *Science* 147, 1462–1465. *The first complete sequence of a tRNA and discovery of the cloverleaf structure.*

Hopper AK, Pai DA & Engelke DR (2010) Cellular dynamics of tRNAs and their genes. *FEBS Lett.* 584, 310–317. *Review of tRNA processing including splicing.*

Kruger K, Grabowski PJ, Zaug AJ et al. (1982) Self-splicing RNA: autoexcision and autocyclization of the ribosomal RNA intervening sequence of Tetrahymena. *Cell* 31, 147–157. *The first indication that the Tetrahymena intron is a ribozyme.*

Tollervey D (1996) Small nucleolar RNAs guide ribosomal RNA methylation. *Science* 273, 1056–1057.

Venema J & Tollervey D (1999) Ribosome synthesis in *Saccharomyces cerevisiae*. *Annu. Rev. Genet.* 33, 261–311. *Extensive details on rRNA processing.*

Yusupov MM, Yusupova GZ, Baucom A et al. (2001) Crystal structure of the ribosome at 5.5Å resolution. *Science* 292, 883–896. *An early and important X-ray crystallography study of ribosome structure.*

The Genetic Code CHAPTER 7

The three major types of RNA molecule that are produced by transcription—messenger, ribosomal, and transfer RNA—work together to synthesize proteins by the process called translation. The key feature of translation is that the sequence of amino acids in the protein being synthesized is specified by the sequence of nucleotides in the mRNA molecule that is being translated (Figure 7.1).

mRNA
5' AUGGGUUGCGUUUUG 3'
Met-Gly-Cys-Val-Leu →
protein

FIGURE 7.1 During translation, each triplet of nucleotides in the RNA specifies a particular amino acid, the identity of this amino acid being set by the genetic code. The amino acids indicated here by their three-letter abbreviations are methionine, glycine, cysteine, valine, and leucine.

We will study the mechanism by which proteins are synthesized in Chapter 8. Before doing that we must cover two underlying aspects of translation, both of which are vital to our understanding of the molecular basis of genetics. First, we must look more closely at proteins. We established in Chapter 3 that proteins with different amino acid sequences can have quite different chemical properties and that this enables them to play a variety of roles in the cell. The link between a protein's amino acid sequence and its function is therefore a vital part of the process by which the biological information contained in a gene is expressed and made use of by the cell. To understand how a protein functions, we must understand the nature of this link. The first part of this chapter is therefore devoted to a short overview of protein structure.

The second underlying aspect of translation that we must examine is the genetic code. The genetic code is the set of rules that determines which sequence of nucleotides specifies which sequence of amino acids. It is therefore central to the translation of mRNA into protein and demands our detailed attention. Not only must we be aware of the features of the genetic code, we must also understand how the rules of the genetic code are enforced during the translation of an individual mRNA. That enforcement is the role of tRNA, whose function in protein synthesis we will look at in the last part of this chapter.

7.1 PROTEIN STRUCTURE

A protein, like a DNA molecule, is a linear unbranched polymer. In proteins the monomeric subunits are called **amino acids**, and the resulting polymers, or **polypeptides**, are rarely more than 2000 units in length.

As with DNA, the key features of protein structure were determined in the first half of the twentieth century, this phase of protein biochemistry culminating in the 1940s and early 1950s with the elucidation of the major conformations or **secondary structures** taken up by polypeptides. In recent years, interest has focused on how these secondary structures combine to produce the complex three-dimensional shapes of proteins.

Amino acids are linked by peptide bonds

Twenty different amino acids are found in protein molecules (Table 7.1). Each has the general structure shown in Figure 7.2, comprising a central α carbon atom to which four groups are attached. These are a hydrogen atom, a carboxyl group ($-COO^-$), an amino group ($-NH_3^+$), and an **R group**, which is different for each amino acid (Figure 7.3).

The R groups vary considerably in chemical complexity. For glycine the R group is simply a hydrogen atom, whereas for tyrosine, phenylalanine, and tryptophan the R groups are complex aromatic side chains. The majority of R groups are uncharged, though two amino acids have negatively

TABLE 7.1 THE 20 AMINO ACIDS FOUND IN PROTEINS

Amino acid	Abbreviation	
	Three-letter	One-letter
Alanine	Ala	A
Arginine	Arg	R
Asparagine	Asn	N
Aspartic acid	Asp	D
Cysteine	Cys	C
Glutamic acid	Glu	E
Glutamine	Gln	Q
Glycine	Gly	G
Histidine	His	H
Isoleucine	Ile	I
Leucine	Leu	L
Lysine	Lys	K
Methionine	Met	M
Phenylalanine	Phe	F
Proline	Pro	P
Serine	Ser	S
Threonine	Thr	T
Tryptophan	Trp	W
Tyrosine	Tyr	Y
Valine	Val	V

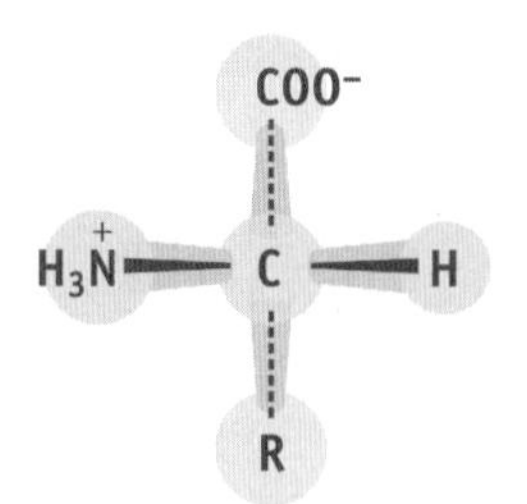

FIGURE 7.2 The general structure of an amino acid. The four bonds have a tetrahedral arrangement around the central carbon, so the amino group and hydrogen atom are, in effect, above the plane of the paper, and the carboxyl and R groups are below it.

charged R groups (aspartic acid and glutamic acid) and three have positively charged R groups (lysine, arginine, and histidine). Some R groups are **polar** (such as those of serine and threonine) and are attracted to an aqueous environment, because of the polarity of the water molecule. Other R groups are **nonpolar** (such as those of alanine, valine, and leucine) and lack an affinity for water. These differences mean that although all amino acids are closely related, each has its own specific chemical properties.

The 20 amino acids whose R groups are shown in Figure 7.3 are the ones that are conventionally looked on as being specified by the genetic code. They are therefore the amino acids that are linked together when mRNA molecules are translated into proteins. However, these 20 amino acids do not on their own represent the limit of the chemical diversity of proteins. At least two additional amino acids—selenocysteine and pyrrolysine (Figure 7.4)—can be inserted into a polypeptide chain during protein synthesis, their insertion directed by a modified reading of the genetic code. Also, after a protein has been synthesized, some amino acids might be modified by the addition of new chemical groups, for example, by phosphorylation, or by attachment of large side chains made up of sugar units.

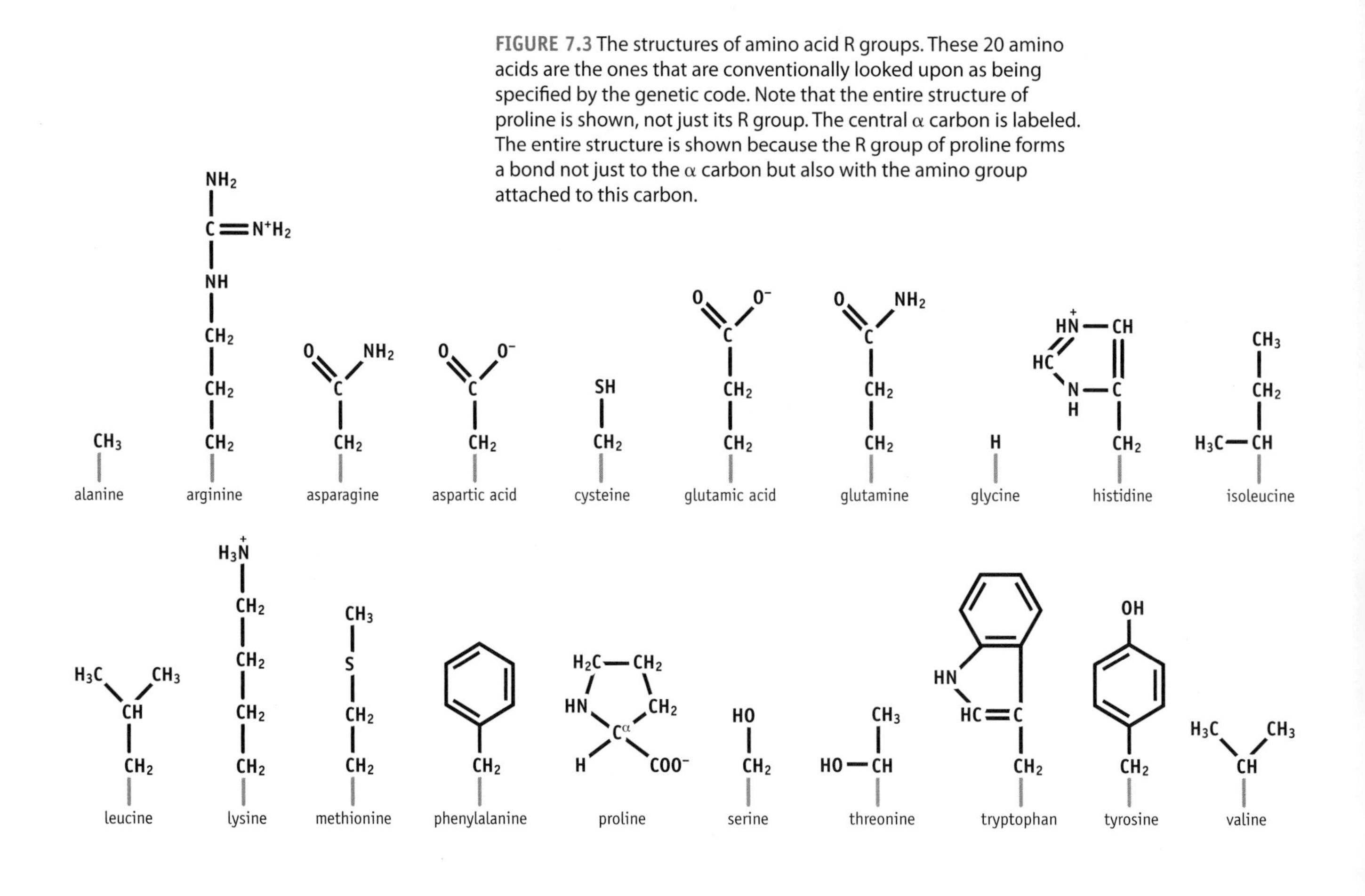

FIGURE 7.3 The structures of amino acid R groups. These 20 amino acids are the ones that are conventionally looked upon as being specified by the genetic code. Note that the entire structure of proline is shown, not just its R group. The central α carbon is labeled. The entire structure is shown because the R group of proline forms a bond not just to the α carbon but also with the amino group attached to this carbon.

FIGURE 7.4 The structures of the R groups of selenocysteine and pyrrolysine. The parts shown in brown indicate the differences between these amino acids and cysteine and lysine, respectively.

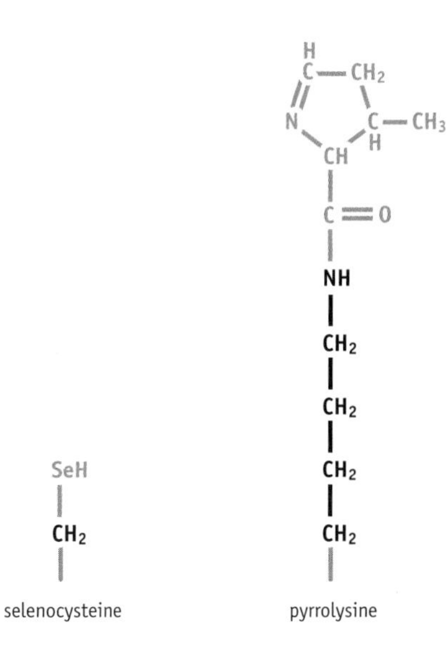

The polymeric structure of a polypeptide is built by linking a series of amino acids by **peptide bonds**, formed by condensation between the carboxyl group of one amino acid and the amino group of a second amino acid (Figure 7.5A). The structure of a tripeptide, comprising three amino acids, is shown in Figure 7.5B. Note that, as with a polynucleotide, the two ends of the polypeptide are chemically distinct. One has a free amino group and is called the **amino, NH_2**, or **N terminus**, and the other has a free carboxyl group and is called the **carboxyl, COOH**, or **C terminus**. The direction of the polypeptide can therefore be expressed as either N → C (left to right in Figure 7.5B) or C → N (right to left in Figure 7.5B).

There are four levels of protein structure

Proteins are traditionally looked on as having four distinct levels of structure. These levels are hierarchical, the protein being built up stage by stage, with each level of structure depending on the one below it.

The first of these structural levels, called the **primary structure**, is the linear sequence of amino acids. The next level, the **secondary structure**, refers to the different conformations that can be taken up by the polypeptide. The two main types of secondary structure are the **α helix** and the **β sheet** (Figure 7.6). These are stabilized mainly by hydrogen bonds that form between different amino acids in the polypeptide. Usually different regions of a polypeptide take up different secondary structures, so that the whole is made up of a number of α helices and β sheets, together with less organized regions.

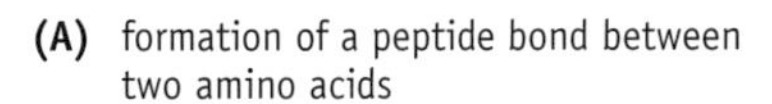

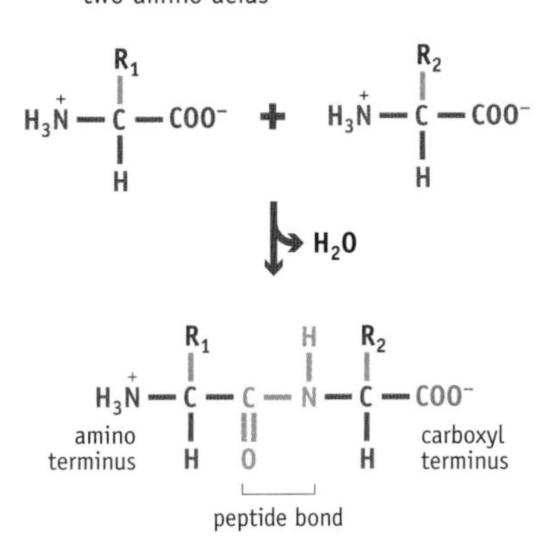

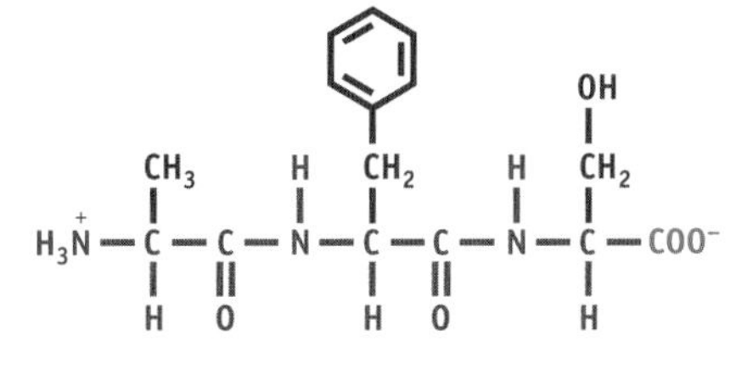

FIGURE 7.5 Amino acids are linked together by peptide bonds. (A) The chemical reaction that results in two amino acids becoming linked together by a peptide bond. (B) The structure of a tripeptide with the sequence alanine-phenylalanine-serine.

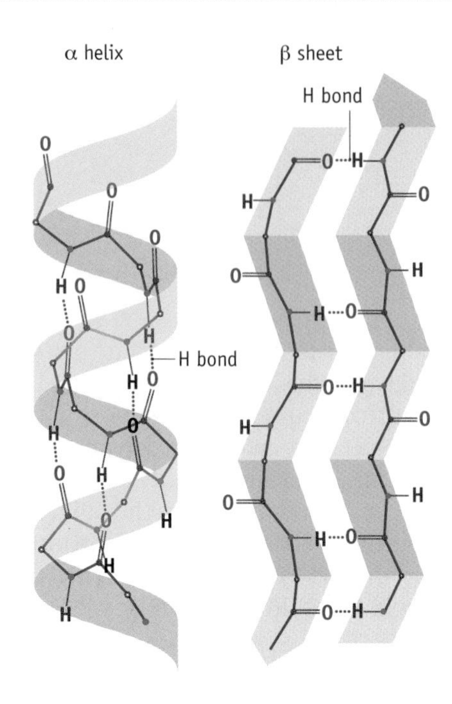

FIGURE 7.6 The two main secondary structural units found in proteins: the α helix, and the β sheet. The polypeptide chains are shown in outline. Hydrogen bonds are shown by dotted lines. The R groups have been omitted for clarity. The β sheet conformation that is shown is antiparallel, the two chains running in opposite directions. Parallel β sheets also occur.

The **tertiary structure** results from folding of the secondary structural components of the polypeptide into a three-dimensional configuration (Figure 7.7). The tertiary structure is stabilized by various chemical forces, notably hydrogen bonding between individual amino acids, electrostatic interactions between the R groups of charged amino acids, and hydrophobic forces. The hydrophobic forces dictate that amino acids with nonpolar side groups must be shielded from water by embedding within the internal regions of the protein. There may also be covalent linkages called **disulfide bridges** between cysteine amino acids at various places in the polypeptide (Figure 7.8).

Finally, the **quaternary structure** involves the association of two or more polypeptides, each folded into its tertiary structure, into a multisubunit protein. Not all proteins form quaternary structures, but it is a feature of many proteins with complex functions, including many involved in gene expression. Some quaternary structures are held together by disulfide bridges between the different polypeptides, resulting in stable multisubunit proteins that cannot easily be broken down into their component parts. Other quaternary structures comprise looser associations of subunits stabilized by hydrogen bonding and hydrophobic effects, which means that these proteins can revert to their component polypeptides, or change their subunit composition, according to the functional requirements. The

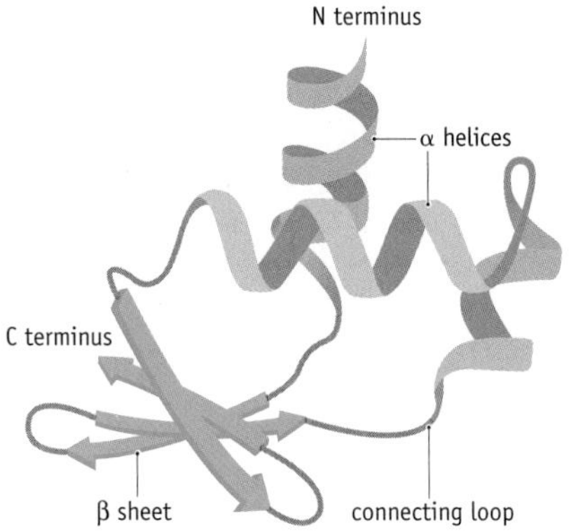

N terminus
α helices
C terminus
β sheet
connecting loop

FIGURE 7.7 The tertiary structure of a protein. This imaginary protein structure comprises three α helices, shown as coils, and a four-stranded β sheet, indicated by the arrows.

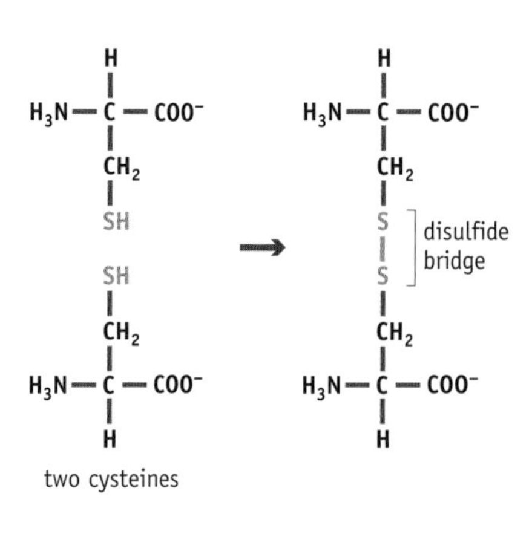

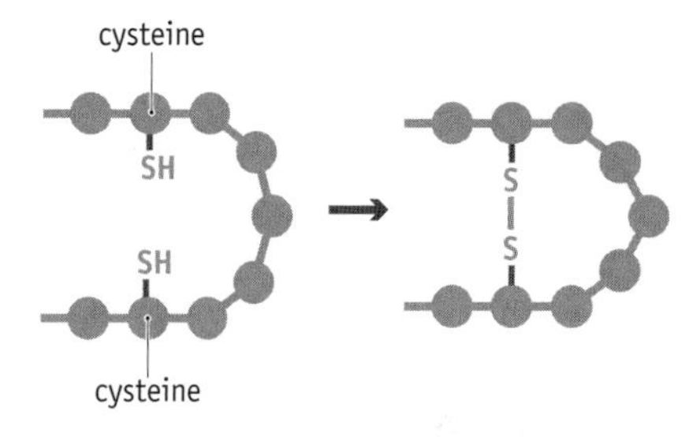

FIGURE 7.8 Disulfide bridges. The upper drawing shows the chemical structure of a disulfide bridge. Below is the effect that formation of a disulfide bridge can have on the structure of a polypeptide.

quaternary structure may involve several molecules of the same polypeptide or may comprise different polypeptides. An example of the latter is the bacterial RNA polymerase, whose subunit composition is described as $\alpha_2\beta\beta'\sigma$ (see Figure 4.10). In some cases the quaternary structure is built up from a very large number of polypeptide subunits, to give a complex array. The best examples are the protein coats of viruses, such as that of tobacco mosaic virus, which is made up of 2130 identical protein subunits (Figure 7.9).

The amino acid sequence is the key to protein structure

Each of the higher levels of protein structure—secondary, tertiary, and quaternary—is specified by the primary structure, the amino acid sequence itself. This is most clearly understood at the secondary level, where it is recognized that certain amino acids, because of the chemical and physical properties of their R groups, stimulate the formation of an α helix, whereas others promote formation of a β sheet. Conversely, certain amino acids more frequently occur outside regular structures and may act to determine the endpoint of a helix or sheet. These factors are now so well understood that rules to predict the secondary structures taken up by amino acid sequences have been developed.

Although less well characterized, it is nonetheless clear that the tertiary and quaternary structures of a protein also depend on the amino acid sequence.

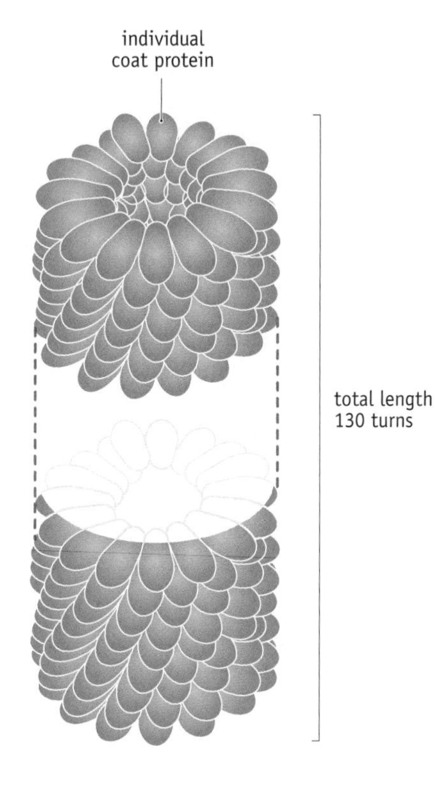

FIGURE 7.9 The tobacco mosaic virus coat, which is made up of 2130 identical protein subunits.

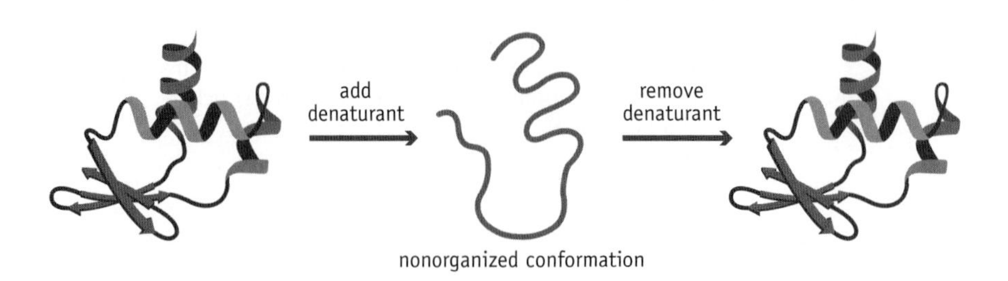

FIGURE 7.10 A denatured protein can regain its tertiary structure after removal of the denaturant.

The interactions between individual amino acids at these levels are so complex that predictive rules, although attempted, are still unreliable. However, it has been established for some years that if a protein is **denatured**, for instance, by mild heat treatment or adding a chemical denaturant such as urea, so that it loses its higher levels of structure and takes up a nonorganized conformation, it still retains the innate ability upon **renaturation** (by cooling down again, for example) to re-form spontaneously the correct tertiary structure (Figure 7.10). Once the tertiary structure has formed, subunit assembly into a multimeric protein again occurs spontaneously. This shows that the instructions for the tertiary and quaternary structures must reside in the amino acid sequence.

Amino acid sequence also determines protein function

The amino acid sequence does not only determine the secondary, tertiary, and quaternary structures of a protein. Through these it also specifies the protein's function. This is because a protein, in order to perform its function, must interact with other molecules, and the precise nature of these interactions is set by the overall shape of the protein and the distribution of chemical groups on its surface.

Understanding the interactions of proteins with other molecules (including other proteins) is a complex area of modern biochemistry, but the basic principle that amino acid sequence determines function is easily illustrated. Let us consider one of the many proteins that must attach themselves to a DNA molecule in order to perform their function. These DNA-binding proteins form a large and diverse group that includes, for instance, RNA polymerase and the repressors and activators that regulate initiation of transcription. An example is Cro, a regulatory protein that controls expression of a number of genes of the **bacteriophage** called λ. Bacteriophages are viruses that infect bacteria, *E. coli* in the case of λ.

Cro is an example of the **helix–turn–helix** family of DNA-binding proteins, the name indicating that the binding motif is made up of two α helices separated by a turn (Figure 7.11). The latter is not a random conformation but a specific structure, referred to as a **β turn**, made up of four amino acids, the second of which is usually glycine. This turn, in conjunction with the first α helix, positions the second α helix on the surface of the protein in an orientation that enables it to fit inside the major groove of a DNA molecule. This second α helix is therefore called the **recognition helix**, because it makes the vital contacts with the DNA.

The active form of the Cro protein is a dimer. The two recognition helices, one from each polypeptide in the dimer, are exactly 3.4 nm apart and

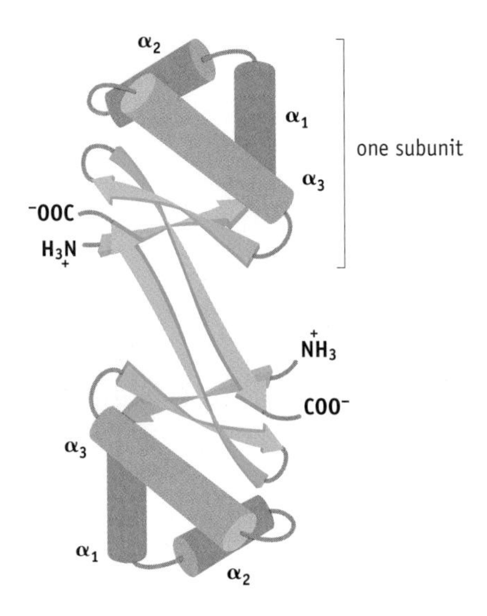

FIGURE 7.11 Structure of the Cro protein. The protein is a dimer. Each subunit has three α helices. The helix–turn–helix structure comprises the helices labeled α_2 and α_3. The recognition helix, which makes the important contacts with the DNA, is α_3.

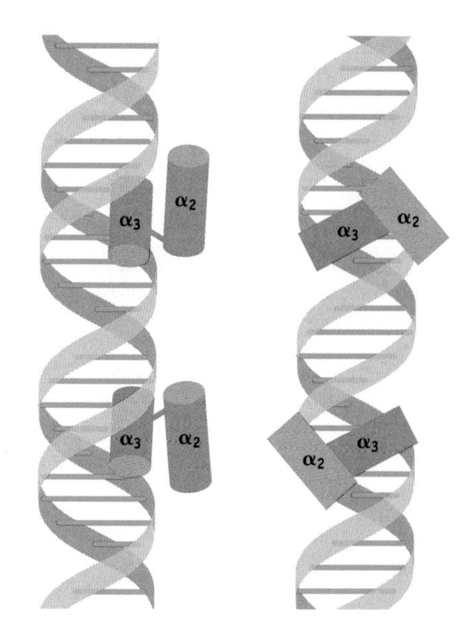

FIGURE 7.12 The two recognition helices of a Cro dimer fit perfectly into two adjacent sections of the major groove of a DNA double helix. Only the recognition helices are shown, the rest of the dimeric protein being left out for clarity. The two drawings show two different views of the helix, rotated 90°.

therefore fit into two adjacent sections of the major groove of a DNA molecule (Figure 7.12). Hence the shape of the protein is critical to its function. If either recognition helix were absent or if the helices were oriented incorrectly on the surface of the protein, then Cro would not be able to bind to DNA. But this is only part of the story. Cro does not attach randomly to the λ DNA molecule. Like the vast majority of regulatory proteins it binds to specific sequences adjacent to the genes whose expression it controls. This requires that precise contacts be made between chemical groups in the recognition helices of the Cro protein and the bases in the DNA sequence to which Cro binds. If the helices have the wrong chemical groups, or if the groups are positioned incorrectly, then the binding sequence will not be recognized.

The function of Cro therefore depends on its amino acid sequence in three ways. First, the recognition structure is present because the amino acid sequence of that particular part of the Cro polypeptide promotes formation of an α helix. Second, the amino acid sequence of the protein as a whole specifies a quaternary structure in which the two recognition helices of the dimeric protein are oriented in precisely the correct way. And finally, the amino acid sequence of each recognition helix provides the particular combination of R groups that enables the protein to recognize its specific binding sequence. There is therefore a precise link between the amino acid sequence of the Cro polypeptide and the function of the dimeric Cro protein.

7.2 THE GENETIC CODE

Now that we have a clear grasp of the link between a protein's amino acid sequence and its function, we can turn our attention to the rules that govern the order in which amino acids are joined together during translation of an mRNA. These rules are called the genetic code.

Understanding the genetic code was the main preoccupation of geneticists in the years immediately after it was realized that genes are made of DNA and that biological information is contained in the nucleotide sequences of DNA molecules. The culmination of this work came in 1966 when the genetic code operating in the bacterium *Escherichia coli* was completely solved. Since then the code itself has been studied less intensively, but important new discoveries have still been made, including variations of the code that operate in some species, including humans. We will first review the key features of the genetic code and then examine the coding variations that are known.

There is a colinear relationship between a gene and its protein

When research into the genetic code began in the 1950s, it was more or less assumed that there is a colinear relationship between a gene and its protein. By "colinear" we mean that the order of nucleotides in the gene correlates directly with the order of amino acids in the corresponding protein (Figure 7.13). This is clearly the most straightforward way in which genes could code for proteins.

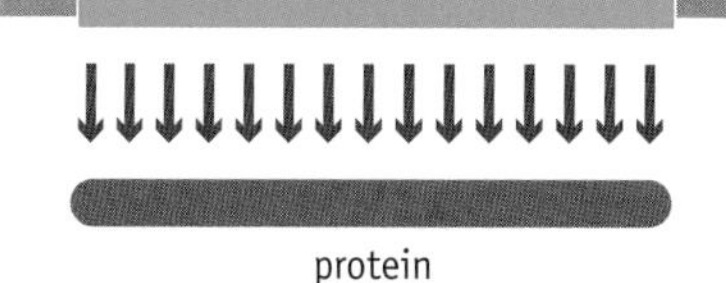

FIGURE 7.13 The relationship between a gene and its protein is colinear. The order of nucleotides in the gene correlates directly with the order of amino acids in the protein.

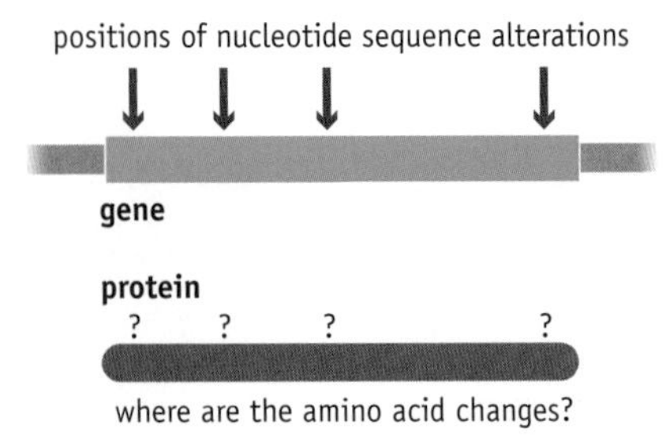

FIGURE 7.14 An experimental strategy for testing whether a gene is colinear with its protein.

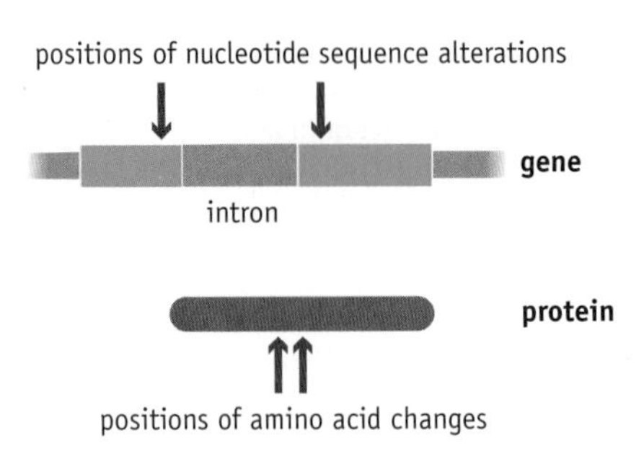

FIGURE 7.15 A gene that contains an intron is not, strictly speaking, colinear with its protein.

Assumptions are dangerous in science and must always be tested by experiments. The easiest way to test the hypothesis that genes and proteins are colinear would be to alter the nucleotide sequence of a gene at a specific point and see whether the resulting change in the amino acid sequence of the corresponding protein occurs at the same relative position or elsewhere (Figure 7.14). This type of experiment was first carried out with the *E. coli* gene coding for one of the two subunits—subunit A—of the enzyme called tryptophan synthetase. As its name implies, this enzyme catalyzes the final step in the biochemical pathway that results in synthesis of tryptophan. The experiments confirmed that a change in the subunit A nucleotide sequence gives rise to an amino acid alteration at the equivalent position in the subunit A protein. The results showed that the subunit A gene is colinear with the subunit A protein, with the amino terminus of the protein corresponding to the 5′ end of the gene.

Establishing that genes and proteins are colinear was an important step forward in understanding the genetic code. It is fortunate that these experiments were done with bacteria and not with a eukaryote. The presence of introns means that a discontinuous eukaryotic gene is not, strictly speaking, colinear with its protein. The relationship is only linear, because two nucleotide changes on either side of an intron, which could be several kb apart, will result in amino acid alterations that are much closer in the corresponding protein (Figure 7.15). The relationship is even more complicated with genes that have alternative splicing pathways. Assumptions are indeed dangerous in science.

Each codeword is a triplet of nucleotides

A second fundamental question about the genetic code is the size of the codeword, or **codon**, the group of nucleotides that code for a single amino acid. Codons cannot be just single nucleotides (A, T, G, or C) because that would allow only four different codewords when 20 are required, one for each of the 20 amino acids found in proteins. Similarly, a doublet code (codons such as AT, TA, TT, GC, etc.) seems unlikely as this would contain only $4^2 = 16$ different codons. However, the next stage up, a triplet code (codons AAA, AAT, TAT, GCA, etc.), would be feasible as this would yield $4^3 = 64$ codewords, which would be more than enough.

As with colinearity, the hypothesis that codons are triplets of nucleotides was first tested by experiments with *E. coli*. These experiments made use of **proflavin**, one of the **acridine dyes**, a group of chemicals that cause base-pair deletions or additions in double-stranded DNA molecules

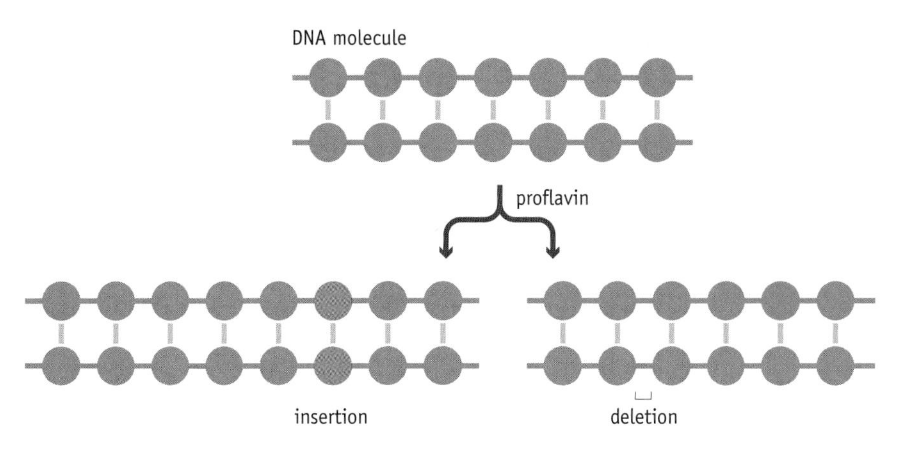

FIGURE 7.16 An acridine dye such as proflavin can cause an insertion or a deletion in a double-stranded DNA molecule.

(Figure 7.16). The rationale was as follows. Some proteins contain segments where the amino acid sequence can be changed without altering the function of the protein. What if a series of insertions and/or deletions are introduced into the region of a gene coding for one of these tolerant segments of a protein? If each codeword is a triplet of nucleotides, then a single insertion or deletion would give rise to a nonfunctional protein, because all the codewords downstream of the insertion or deletion would be altered, including those in the nontolerant segment following the tolerant region (Figure 7.17). Two insertions or deletions (although not one of each) would have the same effect. But three insertions or deletions in the tolerant region would maintain the correct **reading frame** in the nontolerant region and would be predicted to have no effect on the function of the protein.

An elegant experiment of this kind was first carried out successfully with a gene from the *E. coli* bacteriophage called T4. This work established the triplet nature of the code, and although assumptions are still dangerous, this particular one holds for all organisms.

The genetic code is degenerate and includes punctuation codons

Now that we have established the principles of (co)linearity and triplet codons, we can focus on the central question of which codons specify which amino acids. With 64 possible codons and just 20 amino acids one expectation might be that the genetic code is **degenerate**, which in this context means that some amino acids are specified by more than one codon. We

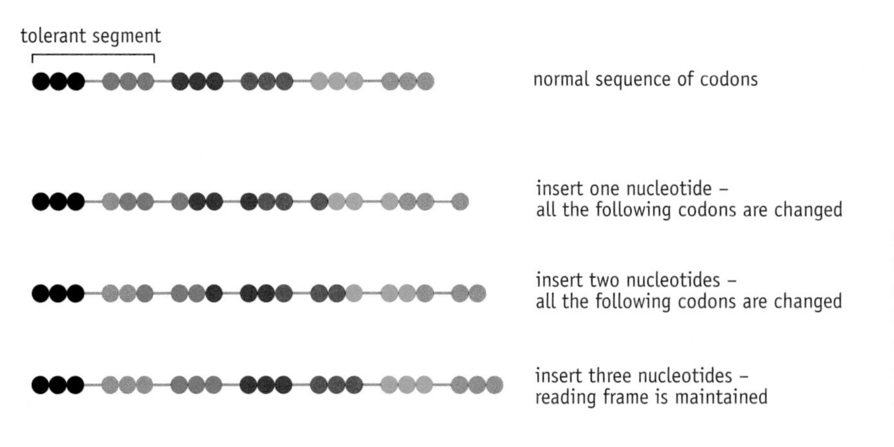

FIGURE 7.17 The rationale behind an experiment to test whether codons are triplets of nucleotides. A series of codons are shown, each in a different color. Insertion of a single nucleotide changes all the following codons, as does an insertion of two nucleotides. Insertion of three nucleotides leaves the reading frame intact.

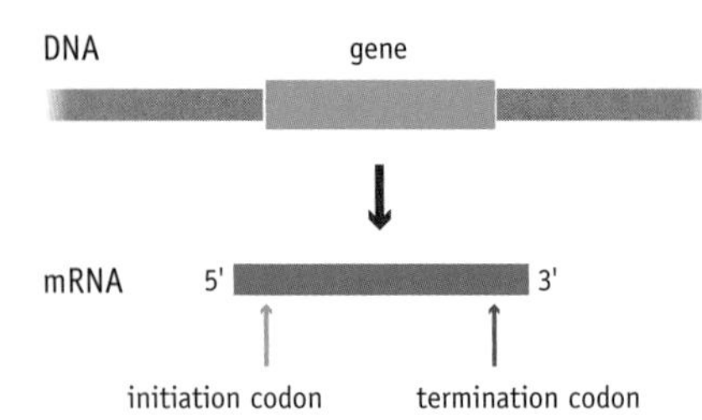

FIGURE 7.18 The positions of the punctuation codons in an mRNA.

might also anticipate that—as turned out to be the case—the code will include **punctuation codons**, special ones that indicate the start and end of the nucleotide sequence that must be translated into protein (Figure 7.18).

The work that led to elucidation of the genetic code was carried out during 1961–1966, and at that time was the most exciting and talked-about research in all of biology (Research Briefing 7.1). The first codon to be assigned a meaning was 5′-UUU-3′, which specifies phenylalanine. This was worked out by showing that poly(U), an RNA molecule that contains just U nucleotides, directs synthesis of polyphenylalanine. Equivalent experiments enabled 5′-AAA-3′ to be assigned to lysine and 5′-CCC-3′ to proline. For unexplained reasons poly(G) gave no protein product. Gradually, more sophisticated approaches were developed, and eventually the code shown in Figure 7.19 was built up.

As predicted, the code is degenerate. All amino acids except methionine and tryptophan are specified by more than one codon, so that all the possible triplets have a meaning, despite there being 64 triplets and only 20 amino acids. Most synonymous codons are grouped into families. For example, 5′-GGA-3′, 5′-GGU-3′, 5′-GGG-3′, and 5′-GGC-3′ all code for glycine. This similarity between synonymous codons is relevant to the way the code is deciphered during protein synthesis, as we will see when we examine the role of tRNA later in this chapter.

The genetic code also contains punctuation codons. The triplet 5′-AUG-3′ occurs at the start of most genes and marks the position where translation should begin. This triplet is therefore the **initiation codon**, and because it codes for methionine, most newly synthesized polypeptides have this amino acid at the amino terminus, though the methionine may subsequently be removed after the protein has been made. Note that 5′-AUG-3′ is the only codon for methionine, so AUGs that are not initiation codons may be found in the internal region of a gene. With a few genes a different triplet such as 5′-GUG-3′ or 5′-UUG-3′ is used as the initiation codon.

Codon		Codon		Codon		Codon	
UUU	Phe	UCU	Ser	UAU	Tyr	UGU	Cys
UUC		UCC		UAC		UGC	
UUA	Leu	UCA		UAA	stop	UGA	stop
UUG		UCG		UAG		UGG	Trp
CUU	Leu	CCU	Pro	CAU	His	CGU	Arg
CUC		CCC		CAC		CGC	
CUA		CCA		CAA	Gln	CGA	
CUG		CCG		CAG		CGG	
AUU	Ile	ACU	Thr	AAU	Asn	AGU	Ser
AUC		ACC		AAC		AGC	
AUA		ACA		AAA	Lys	AGA	Arg
AUG	Met	ACG		AAG		AGG	
GUU	Val	GCU	Ala	GAU	Asp	GGU	Gly
GUC		GCC		GAC		GGC	
GUA		GCA		GAA	Glu	GGA	
GUG		GCG		GAG		GGG	

FIGURE 7.19 The genetic code.

TABLE 7.2 NONSTANDARD CODONS IN HUMAN MITOCHONDRIAL DNA

Codon	Should code for	Actually codes for
UGA	Stop	Tryptophan
AGA, AGG	Arginine	Stop
AUA	Isoleucine	Methionine

Three triplets, 5′-UAA-3′, 5′-UAG-3′, and 5′-UGA-3′, do not code for amino acids and instead act as **termination codons**. One of these three always occurs at the end of a gene at the point where translation must stop.

The genetic code is not universal

It was originally thought that the genetic code must be the same in all organisms. The argument was that, once established, it would be impossible for the code to change because giving a new meaning to any single codon would result in widespread disruption of the amino acid sequences of an organism's proteins.

This reasoning seems sound, so it is surprising that, in reality, the genetic code is not universal. The code shown in Figure 7.19 holds for the vast majority of genes in the vast majority of organisms, but deviations are widespread. In particular, genes present in mitochondrial DNA often use a nonstandard code (Table 7.2). This was discovered when DNA sequencing was first applied to human mitochondrial DNA. It was shown that several human mitochondrial genes contain the sequence 5′-UGA-3′, which normally codes for termination, at internal positions where protein synthesis is not expected to stop. Comparison with the amino acid sequences of the proteins coded by these genes showed that 5′-UGA-3′ is a tryptophan codon in human mitochondria, and that this is just one of four code deviations in this particular genetic system. Mitochondrial genes in other organisms also display code deviations, although at least one of these—the use of 5′-CGG-3′ as a tryptophan codon in plant mitochondria—is probably corrected by RNA editing before translation occurs.

Nonstandard codes are also known in the nuclear genes of lower eukaryotes. Often a modification is restricted to just a small group of organisms and frequently it involves reassignment of the termination codons (Table 7.3). Modifications are less common among bacteria.

A second type of code variation is **context-dependent codon reassignment**, which occurs when the protein to be synthesized contains either selenocysteine or pyrrolysine. Proteins containing pyrrolysine are rare, and are probably present only in some archaea and a very small number of bacteria, but proteins containing selenocysteine are widespread in many

TABLE 7.3 EXAMPLES OF NONSTANDARD CODONS IN NUCLEAR GENOMES

Organism	Codon	Should code for	Actually codes for
Several protozoa	UAA, UAG	Stop	Glutamine
Candida rugosa (yeast)	CUG	Leucine	Serine
Euplotes spp. (ciliated protozoan)	UGA	Stop	Cysteine

RESEARCH BRIEFING 7.1

The genetic code

The elucidation of the genetic code—working out which codons specify which amino acids—was the greatest challenge for biological research in the early 1960s. Although identification of the meaning of individual codons was initially looked on as an almost intractable problem, the entire code operating in *Escherichia coli* was worked out in the years 1961–1966. This rapid advance was made possible by two new techniques that were developed in response to the need to find ways of studying the genetic code.

The first of these innovations was the development of methods for chemical synthesis of RNA molecules. The rationale was that if protein synthesis could be directed by an RNA molecule of known nucleotide sequence, then it would be possible to assign individual codons by looking at the amino acid sequence of the protein that was synthesized. To do this it would be necessary to make artificial RNA molecules of known or predictable sequence. This became possible with the discovery by Severo Ochoa of **polynucleotide phosphorylase**, an enzyme that degrades RNA in the cell, but which, in the test tube, will catalyze the reverse reaction and synthesize RNA. This reaction does not require a DNA template and is unrelated to transcription.

The second innovation was the development of **cell-free protein synthesis** systems. To use an artificial RNA molecule as a message for translation, a cell extract able to synthesize proteins is needed. Such a cell-free, or *in vitro*, system must contain all the cellular components necessary for protein synthesis, including ribosomes, tRNAs, amino acids, and the like. It must, however, lack endogenous mRNA so that protein synthesis occurs only when the artificial message is added.

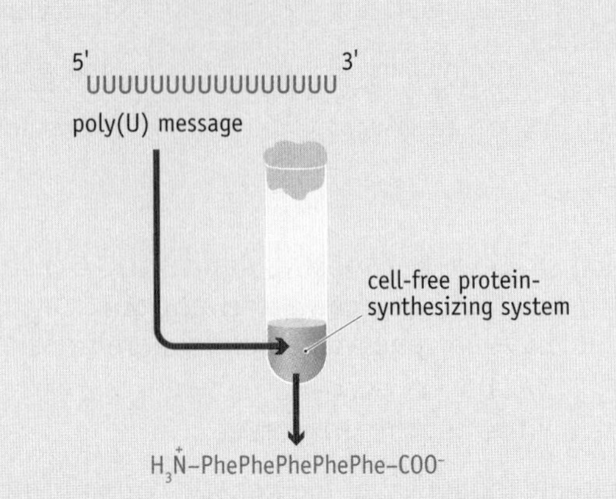

FIGURE 1 Cell-free translation of poly(U) gives polyphenylalanine.

Cell-free protein synthesis with homopolymers identified the first codons

Marshall Nirenberg and Heinrich Matthaei at the National Institutes of Health Laboratories in the United States eventually perfected a cell-free system, prepared from *E. coli*, able to synthesize proteins specific to an added artificial RNA message. According to scientific legend, Matthaei worked through the night of Friday, May 26, 1961, and around 6:00 the following morning discovered that when a homopolymer made up entirely of uridine nucleotides—poly(U)—was added to the cell-free system, polyphenylalanine was synthesized. The first codon could be assigned: UUU for phenylalanine (Figure 1).

Poly(U) was used in this first experiment because homopolymer RNAs are relatively simple to synthesize. If the reaction mixture from which polynucleotide phosphorylase builds up an RNA molecule contains only U nucleotides, then the only possible RNA that can be made is poly(U). The three other RNA homopolymers were also synthesized, and added individually to samples of the cell-free system. From these experiments, Matthaei discovered that AAA codes for lysine and CCC for proline. For reasons that have never been entirely clear, poly(G) would not work and so the amino acid coded by GGG could not at this stage be identified.

Random heteropolymers enabled more codons to be identified

The next step was to construct heteropolymers, artificial RNA molecules containing more than just one nucleotide. This can be achieved by polymerizing mixtures of nucleotides with polynucleotide phosphorylase. The problem is that the random nature of the polymerization means that the actual sequence of the resulting RNA molecule is not known. For example, a random heteropolymer of A and C contains eight different codons—AAA, AAC, ACA, CAA, ACC, CCA, CAC, CCC—but these can occur in any order in the RNA molecule. When used in the cell-free system this poly(A,C) heteropolymer directed synthesis of a protein that contained six amino acids—proline, histidine, threonine, asparagine, glutamine, and lysine. Clearly these six amino acids are coded by the eight possible codons, but which is coded by which?

An answer can be obtained, at least partially, by using different amounts of each nucleotide in the reaction mixture for RNA synthesis. If the ratio of C to A is 5:1, then the probability of a CCC codon occurring is much higher than the probability of an AAA. In fact, the frequencies of each of the eight possible codons can be worked out and compared with the amounts of each amino acid in the resulting polypeptide (Figure 2). The codon allocations provided by this method are not definite but are statistically probable and can be cross-checked by changing the composition of the heteropolymer. These techniques allowed Nirenberg, Matthaei, and Ochoa to propose meanings for many of the 64 codons of the genetic code.

Completion of the code required additional innovations

Although homopolymers and random heteropolymers allowed much of the genetic code to be worked out, unambiguous identification of each codon required three additional types of experiment.

The first of these made use of ordered heteropolymers, which were first synthesized by H. Gobind Khorana. These are made by polymerizing not mononucleotides, but known dinucleotides such as AC. This would give poly(AC), whose sequence is ACACACAC, and which contains two codons, ACA and CAC. Trinucleotides such as UGU could also be polymerized, giving the sequence UGUUGUUGU, with the codons UGU, GUU, and UUG. The proteins produced by these messages allowed the meaning of several of the more difficult codons to be determined.

The next breakthrough was the development, by Nirenberg and Philip Leder in 1964, of a modification of the cell-free protein-synthesizing system called the **triplet binding assay**. It was discovered that purified ribosomes can attach to an mRNA molecule of only three nucleotides, and having done so will then bind the amino acid–tRNA molecule that is specified by that particular codon (Figure 3). As triplets of known sequence could be synthesized in the laboratory, it was possible with this binding assay to check the previously assigned codons and to allocate virtually all of the remaining ones.

The final remaining ambiguity concerned the termination codons. It was suspected that these were UAA, UAG, and UGA, because no aminoacyl-tRNA would attach to these codons when they were used in the triplet binding assay. Confirmation that these are indeed the termination codons was provided by Sydney Brenner and coworkers at Cambridge, UK, through the study of "suppression", which results in one or another of the termination codons being recognized as a codon for an amino acid. We will look more closely at suppression in Chapter 16.

random heteropolymer comprising 5C:1A

probability of **C** being included in a triplet = **5/6**
probability of **A** being included in a triplet = **1/6**

possible triplets	*probability*	
CCC	$(5/6)^3$	= 57.9%
CCA, CAC, ACC	$(1/6)(5/6)^2$	= 11.6%
CAA, ACA, AAC	$(1/6)^2(5/6)$	= 2.3%
AAA	$(1/6)^3$	= 0.4%

results of cell-free protein synthesis

amino acid	*amount in polypeptide*	*interpretation*
proline	69%	**CCC** + one of **CCA, CAC, ACC**
threonine	14%	one of **CCA, CAC, ACC** + one of **CAA, ACA, AAC**
histidine	12%	one of **CCA, CAC, ACC**
asparagine	2%	one of **CAA, ACA, AAC** for each
glutamine	2%	
lysine	1%	**AAA**

FIGURE 2 Analysis of the results of a typical experiment with a random heteropolymer.

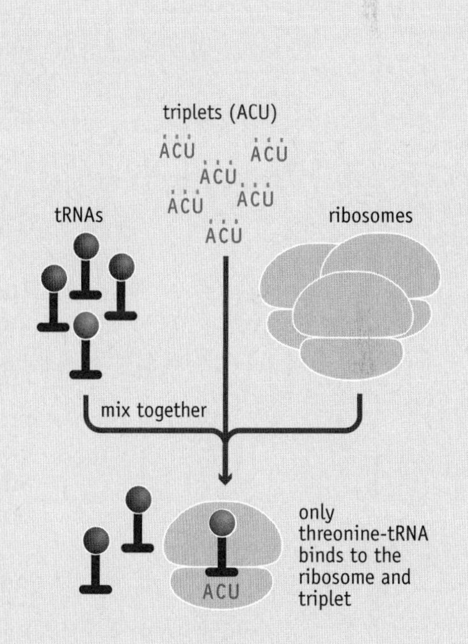

FIGURE 3 The triplet binding assay. The assay was based on the discovery that a purified ribosome will attach to a triplet RNA molecule and then bind a tRNA that carries the amino acid specified by that triplet. In this example, the triplet is ACU and the tRNA that is bound carries threonine, the amino acid coded by ACU. The assay is performed with a series of amino acid-tRNA mixtures, tested one after the other, each mixture containing one amino acid that is radioactively labeled. A positive result is signaled by the ribosomes becoming radioactive. This indicates that the labeled amino acid in that particular mixture is the one specified by the triplet being tested.

TABLE 7.4 EXAMPLES OF PROTEINS THAT CONTAIN SELENOCYSTEINE

Organisms	Protein
Mammals	Glutathione peroxidase
	Thioredoxin reductase
	Iodothyronine deiodinase
Bacteria	Formate dehydrogenase
	Glycine reductase
	Proline reductase
Archaea	Hydrogenase
	Formyl-methanofuran dehydrogenase

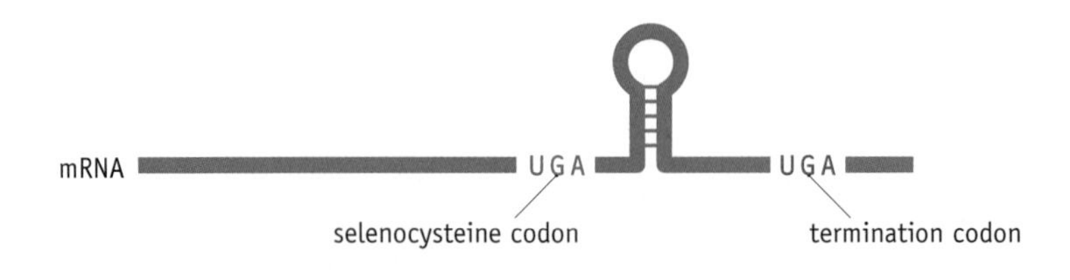

FIGURE 7.20 A 5′-UGA-3′ codon that specifies selenocysteine is distinguished from a termination codon by the presence of a hairpin loop in the mRNA. This drawing shows a bacterial mRNA, in which the hairpin is adjacent to the selenocysteine codon.

organisms (Table 7.4). One example is the enzyme glutathione peroxidase, which helps protect the cells of humans and other mammals against oxidative damage. Selenocysteine is coded by 5′-UGA-3′, which therefore has a dual meaning because it is still used as a termination codon in the organisms concerned. A 5′-UGA-3′ codon that specifies selenocysteine is distinguished from true termination codons by the presence of a hairpin loop structure in the mRNA, positioned just downstream of the selenocysteine codon in prokaryotes and in the 3′ untranslated region (i.e., the part of the mRNA after the termination codon) in eukaryotes (Figure 7.20). Recognition of the selenocysteine codon requires interaction between the hairpin and a special protein that is involved in translation of these mRNAs. A similar system probably operates with pyrrolysine, which is specified by a second termination codon, 5′-UAG-3′.

7.3 THE ROLE OF tRNAS IN PROTEIN SYNTHESIS

Transfer RNAs perform the key role of ensuring that the rules laid down by the genetic code are followed when an mRNA is translated into a protein. To do this, tRNAs form a physical link between the mRNA and the protein that is being synthesized, binding to both the mRNA and the growing protein (Figure 7.21). To understand the role of tRNAs we must therefore examine **aminoacylation**, the process by which the correct amino acid is attached to each tRNA, and **codon–anticodon recognition**, the interaction between tRNA and mRNA.

Aminoacyl-tRNA synthetases attach amino acids to tRNAs

Bacteria contain 30 to 45 different tRNAs, and eukaryotes have up to 50. As only 20 amino acids are designated by the genetic code, this means that all organisms have at least some **isoaccepting tRNAs**, different tRNAs that are specific for the same amino acid. The terminology used when describing tRNAs is to indicate the amino acid specificity by a superscript suffix, with the numbers 1, 2, etc., distinguishing different isoacceptors. According to this notation, two tRNAs specific for glycine would be written as $tRNA^{Gly1}$ and $tRNA^{Gly2}$.

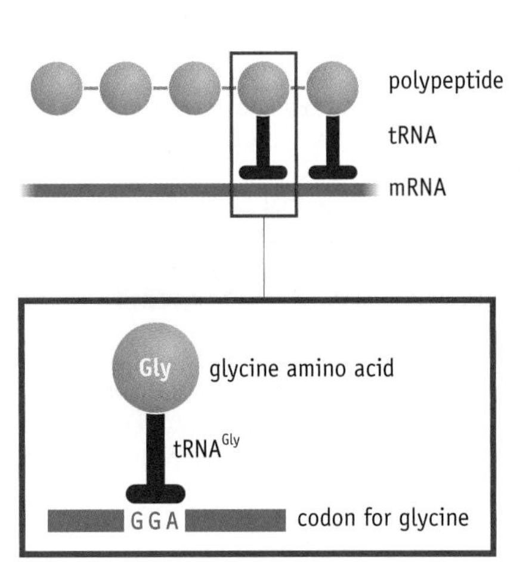

FIGURE 7.21 The role of tRNA in translation. The top drawing shows two tRNAs forming an attachment between the polypeptide and the mRNA. The lower drawing shows that the amino acid that is attached to the tRNA is the one specified by the codon that the tRNA recognizes.

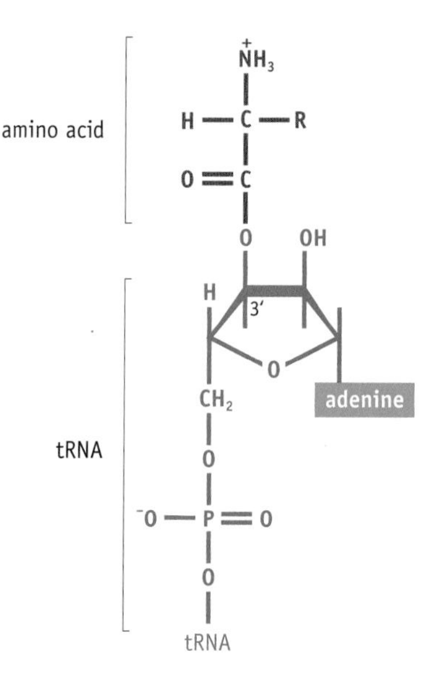

FIGURE 7.22 Aminoacylation, or charging, of a tRNA. The result of aminoacylation by a class II aminoacyl-tRNA synthetase is shown, the amino acid being attached via its –COOH group to the 3′–OH of the terminal nucleotide of the tRNA. A class I aminoacyl-tRNA synthetase attaches the amino acid to the 2′–OH group.

Each tRNA molecule forms a covalent linkage with its specific amino acid by a process called aminoacylation or **charging**. The amino acid becomes attached to the end of the acceptor arm of the tRNA cloverleaf. The linkage forms between the carboxyl group of the amino acid and the 2′-OH or 3′-OH group of the terminal nucleotide of the tRNA (Figure 7.22). Remember that this terminal nucleotide is always A, because all tRNAs have the sequence 5′-CCA-3′ at their 3′ ends (see Figure 6.10).

Aminoacylation is catalyzed by a group of enzymes called the **aminoacyl-tRNA synthetases**. In most cells there is a single aminoacyl-tRNA synthetase for each amino acid, meaning that one enzyme can charge each member of a series of isoaccepting tRNAs. Although aminoacyl-tRNA synthetases are a fairly heterogeneous group of enzymes, each catalyzes the same reaction (Figure 7.23). The energy required to attach the amino acid to the tRNA is provided by cleavage of ATP to adenosine monophosphate (AMP) and pyrophosphate. The reaction takes place in two distinct steps, the first resulting in an activated amino acid intermediate in which the carboxyl group has formed a link with AMP. This intermediate remains bound to the enzyme until the AMP is replaced in the second stage of the reaction by the tRNA molecule, producing aminoacyl-tRNA and free AMP.

Aminoacylation must be carried out accurately. The correct amino acid must be attached to the correct tRNA if the rules of the genetic code are to be followed during protein synthesis. Each aminoacyl-tRNA synthetase forms an extensive interaction with its tRNA, with contacts made to the acceptor arm and anticodon loop as well as to individual nucleotides in the D and TψC arms. These interactions enable the enzyme to distinguish the specific sequence features of different tRNAs and hence to recognize the correct one. The interaction between enzyme and amino acid is, of necessity, less extensive, amino acids being much smaller than tRNAs, and presents a greater problem with regard to specificity because several pairs of amino acids are structurally similar. Errors do therefore occur, at a very low rate for most amino acids but possibly as frequently as one aminoacylation in 80 for difficult pairs such as isoleucine and valine. Most of these errors are corrected by the aminoacyl-tRNA synthetase before the charged tRNA is released.

Unusual types of aminoacylation

In most organisms, aminoacylation is carried out by the process just described, but a few unusual events have been documented. These include a number of instances where the aminoacyl-tRNA synthetase attaches the incorrect amino acid to a tRNA, this amino acid subsequently being transformed into the correct one by a second, separate chemical reaction.

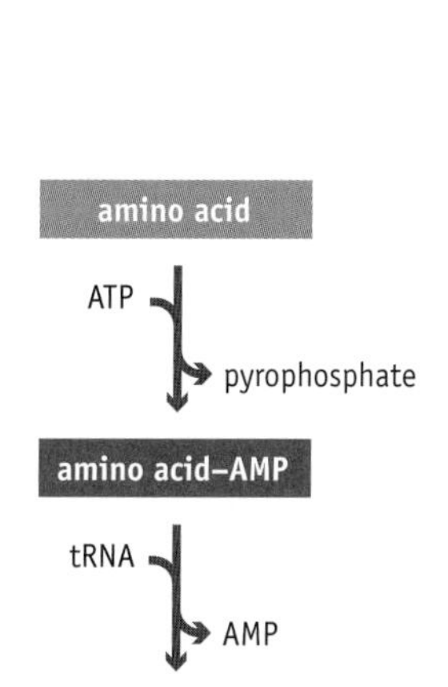

FIGURE 7.23 The two-step reaction catalyzed by an aminoacyl-tRNA synthetase. The first step results in an activated amino acid intermediate whose carboxyl group has formed a link with AMP. In the second stage of the reaction, the AMP is replaced by the tRNA molecule, producing aminoacyl-tRNA and free AMP.

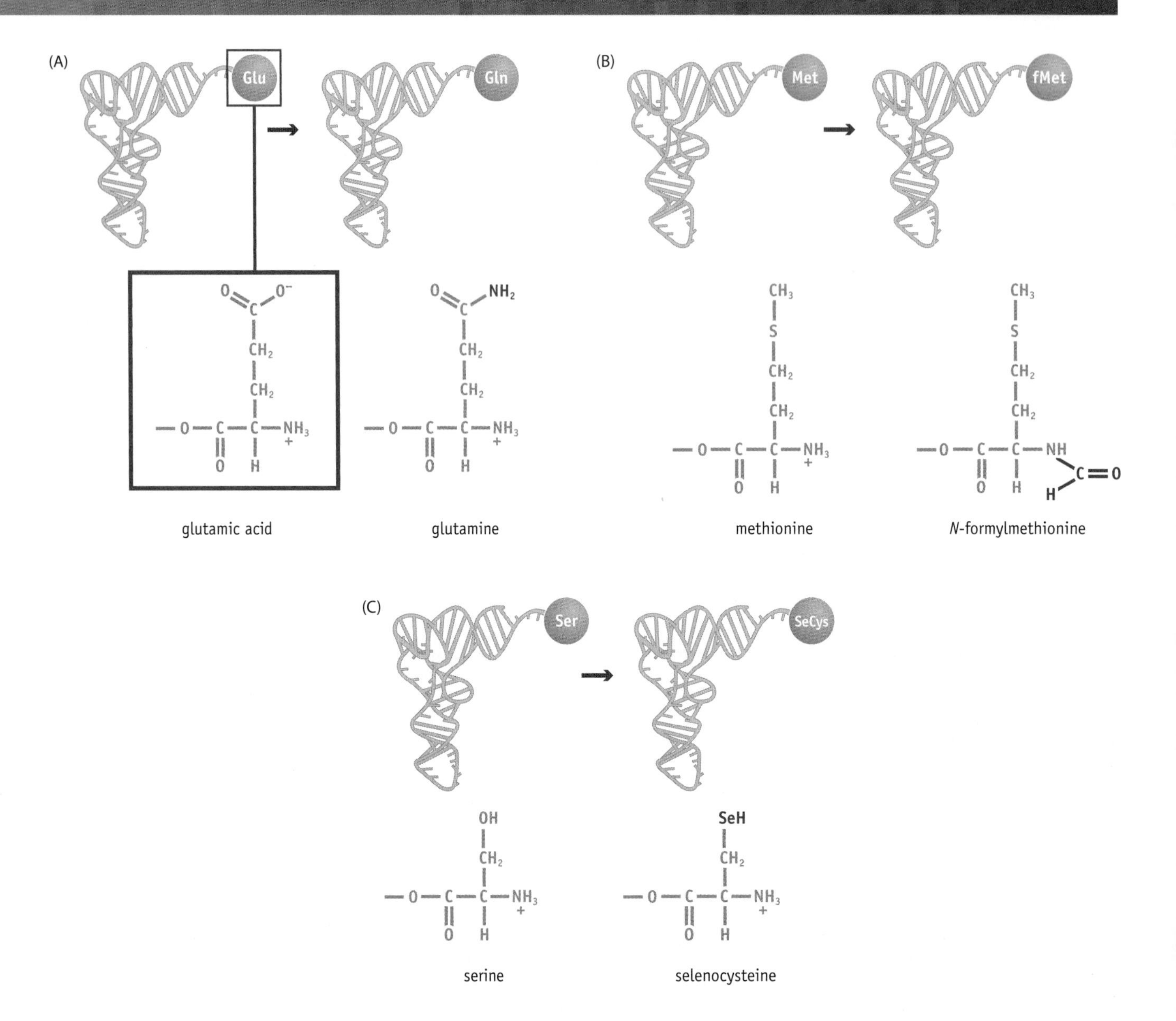

FIGURE 7.24 Unusual types of aminoacylation. (A) In some bacteria, $tRNA^{Gln}$ is aminoacylated with glutamic acid, which is then converted to glutamine by an aminotransferase enzyme. (B) The special tRNA used in initiation of translation in bacteria is aminoacylated with methionine, which is then converted to *N*-formylmethionine. (C) $tRNA^{SeCys}$ in various organisms is initially aminoacylated with serine.

This unusual type of aminoacylation was first discovered in the bacterium *Bacillus megaterium*, which lacks an aminoacyl-tRNA synthetase capable of attaching glutamine to $tRNA^{Gln}$. Instead, the glutamic acid tRNA synthetase attaches a glutamic acid to $tRNA^{Gln}$, this "mistake" being corrected by a second enzyme (an aminotransferase) that replaces the $-O^-$ group of the glutamic acid with an $-NH_2$ group, converting it into glutamine (**Figure 7.24A**). The same process takes place in various other bacteria, although not *E. coli*.

In the example given above, the amino acid resulting from the correction process is one of the 20 specified by the genetic code. There are also cases where chemical modification results in an unusual amino acid. One such chemical modification results in the conversion of methionine to *N*-formylmethionine (**Figure 7.24B**). This produces a special aminoacyl-tRNA which, as we will see in Chapter 8, is used during initiation of bacterial protein synthesis.

Chemical modification is also responsible for synthesis of tRNAs aminoacylated with selenocysteine. Context-dependent reading of 5′-UGA-3′ triplets as selenocysteine codons involves a special $tRNA^{SeCys}$, but there is no aminoacyl-tRNA synthetase that is able to attach selenocysteine to this tRNA. Instead, the tRNA is aminoacylated with a serine by the seryl-tRNA synthetase, and then modified by replacement of the –OH group of the serine with an –SeH, to give selenocysteine (Figure 7.24C).

The mRNA sequence is read by base pairing between the codon and anticodon

Aminoacylation represents the first level of specificity displayed by a tRNA. Once the correct amino acid has been attached to the acceptor arm of the tRNA, the aminoacylated molecule must complete the link between mRNA and protein by recognizing and attaching to the correct codon, one coding for the amino acid that it carries.

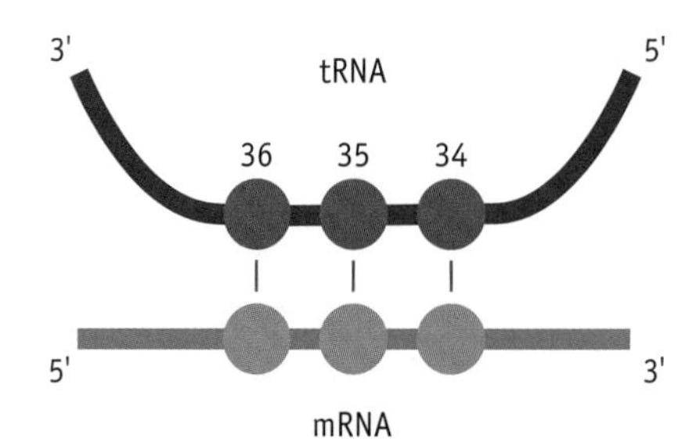

FIGURE 7.25 The interaction between a codon and its anticodon. The anticodon is complementary to the codon and can therefore attach to it by base pairing. The numbers refer to the positions of the nucleotides within the tRNA sequence, position 1 being the nucleotide at the extreme 5′ end.

Codon recognition is a function of the anticodon loop of the tRNA, specifically, of the triplet of nucleotides called the anticodon (see Figure 6.10). This triplet is complementary to the codon and can therefore attach to it by base pairing (Figure 7.25). The specificity of the genetic code is ensured because the anticodon present on a particular tRNA is one that is complementary to a codon for the amino acid with which the tRNA is charged.

From what has been said so far it might be imagined that there are 61 different types of tRNA molecule in each cell, one for each of the codons that specify an amino acid. In fact it has been known since the early 1960s that there are substantially fewer than 61 different tRNA molecules, usually between 30 and 50 depending on the organism. The explanation for this is provided by the **wobble hypothesis**. This hypothesis is based on the fact that the anticodon loop is just that, a loop, and the anticodon itself is not a perfectly linear trinucleotide. The short double helix formed by base pairing between the codon and anticodon does not therefore have the precise configuration of a standard RNA helix. Instead, its dimensions are slightly altered. As a result, nonstandard base pairs can form at the "wobble position," between the third nucleotide of the codon and the first nucleotide of the anticodon.

Because of the wobble pairing a single anticodon may be able to base-pair with more than one codon. This means that a single tRNA might decode more than one member of a codon family. However, the base-pairing rules do not become totally flexible at the wobble position, and only a few types of unusual base pairs are allowed. G–U base pairs are a common example. By allowing G to pair with U as well as C, an anticodon with the sequence 3′-▼▼G-5′ can base-pair with both 5′-▲▲C-3′ and 5′-▲▲U-3′ (Figure 7.26). Similarly, the anticodon 3′-▼▼U-5′ can base-pair with both 5′-▲▲A-3′ and 5′-▲▲G-3′. The consequence is that the four members of a codon family (e.g., 5′-GCN-3′, all coding for alanine) can be decoded by just two tRNAs.

A second type of wobble involves inosine, one of the modified nucleotides present in tRNA. Inosine can base-pair with A, C, and U. The triplet 3′-UAI-5′ is sometimes used as the anticodon in a tRNA^{Ile} molecule because

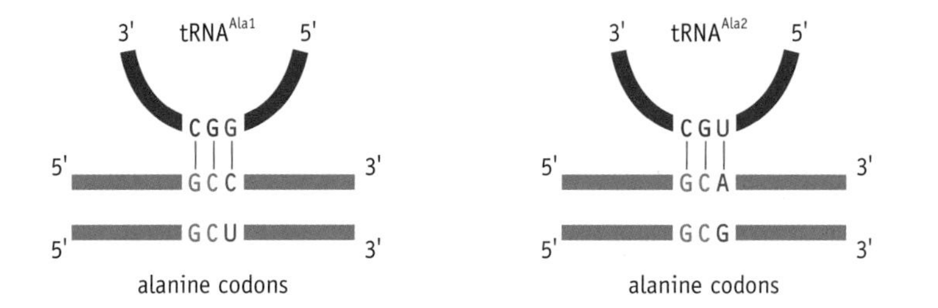

FIGURE 7.26 Wobble involving a G–U base pair. In this example, wobble enables the four-codon family for alanine to be decoded by just two tRNAs. The wobble position is highlighted in magenta.

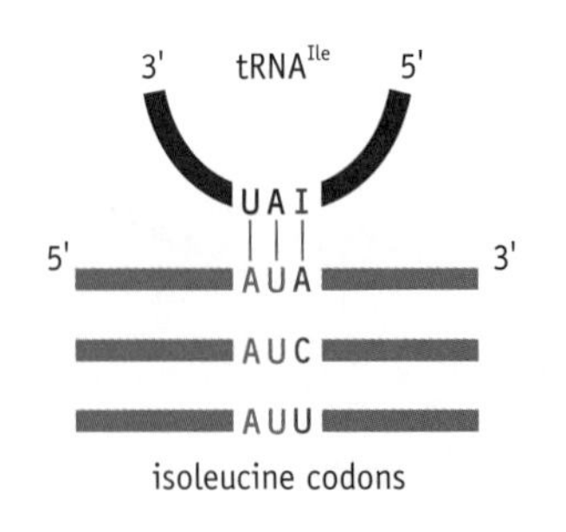

FIGURE 7.27 Wobble involving inosine. The wobble position is highlighted in magenta.

it pairs with 5′-AUA-3′, 5′-AUC-3′, and 5′-AUU-3′ (**Figure 7.27**). These triplets form the three-codon family for isoleucine in the standard genetic code.

Wobble reduces the number of tRNAs needed in a cell by enabling one tRNA to read two or possibly three codons. Hence bacteria can decode their mRNAs with as few as 30 tRNAs. Eukaryotes also make use of wobble but in a restricted way. The human genome, which in this regard is fairly typical of higher eukaryotes, has 48 tRNAs. Of these, 16 are predicted to use wobble to decode two codons each, with the remaining 32 being specific for just a single triplet. Although reducing the number of tRNAs that are needed, wobble does not violate the rules of the genetic code, and the protein that is made during translation is always synthesized strictly in accordance with the nucleotide sequence of the relevant mRNA.

KEY CONCEPTS

- A protein is a linear unbranched polymer. The monomeric subunits are called amino acids, and the resulting polymers, or polypeptides, are rarely more than 2000 units in length.
- Twenty different amino acids are found in protein molecules. Their R groups vary considerably in chemical complexity. These differences mean that although all amino acids are closely related, each has its own specific chemical properties.
- Proteins have four distinct levels of structure. These levels are hierarchical, the protein being built up stage by stage, with each level of structure depending on the one below it. Each of the higher levels of protein structure is specified by the amino acid sequence.
- Through the higher levels of structure, the amino acid sequence also specifies the function of a protein. This is because a protein, in order to perform its function, must interact with other molecules, and the precise nature of these interactions is set by the overall shape of the protein and the distribution of chemical groups on its surface.
- The genetic code contains the rules that govern the order in which amino acids are joined together during translation of an mRNA.
- The codewords of the genetic code are three-letter triplets called codons. Most amino acids are specified by more than one codon. There are also initiation and termination codons.
- The genetic code is not universal.
- Transfer RNAs ensure that the rules laid down by the genetic code are followed when an mRNA is translated into a protein.
- The correct amino acid must be attached to each tRNA. This is ensured by the specificity of the aminoacyl-tRNA synthetase enzymes.
- The aminoacylated tRNA must recognize and attach to the correct codon, one coding for the amino acid that it carries. Codon recognition is a function of the anticodon of the tRNA. The anticodon is complementary to the codon and can therefore attach to it by base pairing.
- Wobble reduces the number of tRNAs needed in a cell by enabling one tRNA to read two or possibly three codons. Wobble does not violate the rules of the genetic code, and the protein that is made during translation is always synthesized strictly in accordance with the nucleotide sequence of the relevant mRNA.

QUESTIONS AND PROBLEMS (Answers can be found at www.garlandscience.com/introgenetics)

Key Terms

Write short definitions of the following terms:

α helix
β sheet
β turn
acridine dye
amino acid
amino terminus (or NH_2 terminus, or N terminus)
aminoacyl-tRNA synthetase
aminoacylation
bacteriophage
carboxyl terminus (or COOH terminus, or C terminus)
cell-free protein synthesis
charging
codon
codon–anticodon recognition
context-dependent codon reassignment
degenerate
denaturation
disulfide bridge
helix–turn–helix
initiation codon
isoaccepting tRNAs
nonpolar
peptide bond
polar
polynucleotide phosphorylase
polypeptide
primary structure
proflavin
punctuation codon
quaternary structure
R group
reading frame
recognition helix
renaturation
secondary structure
termination codon
tertiary structure
triplet binding assay
wobble hypothesis

Self-study Questions

7.1 Draw the structure of an amino acid and indicate which parts of the molecule are the same for each amino acid and which parts are variable.

7.2 How are amino acids linked together to form a polypeptide?

7.3 Describe the four levels of protein structure. Your answer should include an indication of the types of chemical interaction that are important at each structural level.

7.4 What experimental evidence supports the contention that the instructions for the higher levels of protein structure reside in the amino acid sequence? Using Cro as an example, explain how the function of a protein is specified by its amino acid sequence.

7.5 Explain what is meant by colinearity between gene and protein. How was colinearity first demonstrated for genes and proteins in *E. coli*?

7.6 What considerations suggested that each codon would comprise three nucleotides? How was this fact proven experimentally?

7.7 Calculate the number of codons that would be possible if each codon contained four nucleotides.

7.8 Explain what is meant by degeneracy with respect to the genetic code.

7.9 List the key features of the genetic code.

7.10 A messenger RNA contains the following sequence:

```
5'-AUGUUAGCUGAUCCGGAAAUGAUG
   UUAUAUAUAAUAUAUGCCCAAUAG-3'
```

What would be the amino acid sequence of the protein specified by this mRNA?

7.11 Distinguish between the way homopolymers and heteropolymers were used in the experiments that elucidated the genetic code.

7.12 List the codons that would be contained in a random heteropolymer comprising A and G nucleotides. What amino acids would a polypeptide synthesized from this heteropolymer contain?

7.13 Explain why the triplet binding assay was important in studies of the genetic code.

7.14 Give examples of nonstandard genetic codes found in eukaryotes.

7.15 Using examples, explain what is meant by context-dependent codon reassignment.

7.16 Explain how tRNA molecules ensure that protein synthesis follows the rules laid down by the genetic code.

7.17 Outline the terminology used to distinguish between isoaccepting tRNAs.

7.18 Describe how an amino acid is attached to a tRNA molecule.

7.19 Give three examples where the incorrect amino acid is attached to a tRNA, this amino acid subsequently being transformed into the correct one.

7.20 Explain why fewer than 61 tRNAs are sufficient to decode the entire genetic code.

7.21 Draw a series of diagrams to illustrate the codon–anticodon interactions that occur during wobble involving G–U base pairs and inosine.

Discussion Topics

7.22 The 20 amino acids shown in Figure 7.3 are not the only amino acids found in living cells. Devise a hypothesis to explain why these 20 amino acids are the only ones that are specified by the genetic code. Can your hypothesis be tested?

7.23 Discuss the reasons why polypeptides can take up a variety of structures whereas polynucleotides cannot.

7.24 During the 1950s it was suggested that adjacent codons may overlap, so that ACAUG might contain two codons: ACA and AUG. Design an experiment to test this proposition.

7.25 How can the genetic code change if an alteration in a codon assignment is likely to cause an alteration to every protein in the cell?

7.26 Most organisms display a distinct codon bias in their genes. For instance, of the four codons for proline, only two (CCU and CCA) appear at all frequently in genes of *Saccharomyces cerevisiae*. CCC and CCG are less common. It has been suggested that a gene that contains a relatively high number of unfavored codons might be expressed at a relatively slow rate. Explain the thinking behind this hypothesis and discuss its ramifications.

7.27 In human mitochondria, protein synthesis requires only 22 different tRNAs. What implications does this have for the rules governing codon–anticodon interactions in this system?

7.28 Discuss the connection between the wobble hypothesis and the degeneracy of the genetic code.

FURTHER READING

Agris PF, Vendeix FAP & Graham WD (2007) tRNA's wobble decoding of the genome: 40 years of modification. *J. Mol. Biol.* 366, 1–13. *Review of the development of the wobble hypothesis.*

Branden CI & Tooze J (1998) Introduction to Protein Structure, 2nd ed. Abingdon, UK: Garland Science.

Brennan RG & Matthews BW (1989) Structural basis of DNA-protein recognition. *Trends Biochem. Sci.* 14, 286–290. *The binding of the Cro repressor to DNA.*

Crick FHC, Barrett FRSL, Brenner S & Watts-Tobin RJ (1961) General nature of the genetic code for proteins. *Nature* 192, 1227–1232. *The experimental proof of the triplet nature of the code.*

Hall BD (1979) Mitochondria spring surprises. *Nature* 282, 129–130. *Reviews the first reports of unusual genetic codes in mitochondrial genes.*

Nirenberg MW & Leder P (1964) RNA codewords and protein synthesis. *Science* 145, 1399–1407. *Results of the triplet binding assay.*

Nirenberg MW & Matthaei JH (1961) The dependence of cell-free protein synthesis in *E. coli* upon naturally occurring or synthetic polyribonucleic acids. *Proc. Natl. Acad. Sci. USA* 47, 1588–1602. *An early publication on elucidation of the genetic code.*

Ling J, Reynolds N & Ibba M (2009) Aminoacyl-tRNA synthesis and translational quality control. *Annu. Rev. Microbiol.* 63, 61–76. *The role of aminoacyl-tRNA synthetases with focus on how accuracy of aminoacylation is ensured.*

Low SC & Berry MJ (1996) Knowing when not to stop: selenocysteine incorporation in eukaryotes. *Trends Biochem. Sci.* 21, 203–208. *The questions raised by the discovery that UGA can code for both stop and selenocysteine.*

Yanofsky C, Carlton BC, Guest JR et al. (1964) On the colinearity of gene structure and protein structure. *Proc. Natl. Acad. Sci. USA* 51, 266–272. *The first demonstration of colinearity.*

We now reach the stage of the gene expression pathway where the information contained in the nucleotide sequence of an mRNA is translated into a protein. We have already learned that this process follows the rules laid down by the genetic code, with tRNAs enforcing these rules. Now we must examine the mechanism by which all this happens. The first part of this chapter is therefore about the role of the ribosome in protein synthesis.

It is often forgotten that translation is only the first stage in protein synthesis. Before becoming functional in the cell, the linear amino acid sequence that results from translation must be folded into its correct secondary, tertiary, and, possibly, quaternary structures. In some cases it must also undergo processing, possibly by chemical modification or, less frequently, by removal of some segments of the polypeptide chain (Figure 8.1). These post-translational events are inherent steps in the final stage of gene expression, because until they have been carried out the biological information contained in the gene has not been made fully available to the cell. We will therefore spend some time becoming familiar with the most important of these post-translational processes.

The collection of proteins in a cell is called the **proteome**. Within the proteome, the identity and relative abundance of individual proteins represent a balance between the synthesis of new proteins and the degradation of existing ones (Figure 8.2). Exactly as with mRNAs in the transcriptome, degradation is needed so that individual proteins can be down- as well as up-regulated by changing their rate of synthesis. The terminal step in the expression pathway for a gene is therefore the process by which its protein, when no longer needed, is removed from the proteome. We will conclude this chapter with a brief overview of protein degradation.

8.1 THE ROLE OF THE RIBOSOME IN PROTEIN SYNTHESIS

Messenger RNAs are translated within the structures called ribosomes. We will follow through the series of events involved in initiation of translation, elongation of the polypeptide chain, and termination of translation. These events are similar in bacteria and eukaryotes, though the details are different, most strikingly during the initiation phase.

Initiation in bacteria requires an internal ribosome binding site

The main difference between initiation of translation in bacteria and eukaryotes is that in bacteria the translation initiation complex is built up

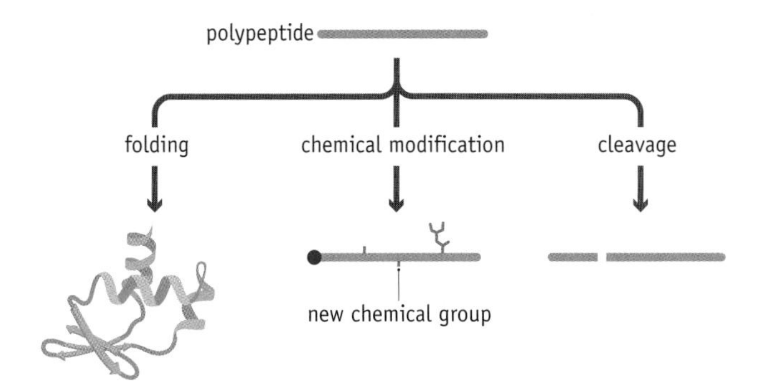

FIGURE 8.1 A summary of protein processing events. All proteins must be folded in order to become functional, and some are also processed by chemical modification and/or cleavage of the polypeptide into segments.

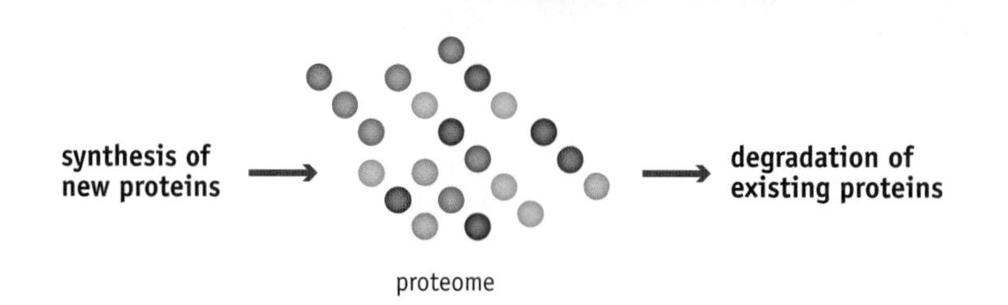

FIGURE 8.2 The composition of a proteome reflects the balance between synthesis of new proteins and degradation of existing ones. In this drawing, dots of different colors represent different proteins.

directly at the initiation codon, the point at which protein synthesis will begin, whereas eukaryotes use a more indirect process for locating the initiation point.

When not actively participating in protein synthesis, ribosomes separate into their subunits, which remain in the cytoplasm waiting to be used for a new round of translation. In bacteria, the process initiates when a small subunit attaches to the **ribosome binding site** (also called the **Shine–Dalgarno sequence**). This is a short sequence, consensus 5′-AGGAGGU-3′ in *E. coli* (Table 8.1), located about 3 to 10 nucleotides upstream of the initiation codon, the point at which translation will begin (Figure 8.3). The ribosome binding site is complementary to a region at the 3′ end of the 16S rRNA, the one present in the small subunit, and it is thought that base pairing between the two is involved in the attachment of the small subunit to the mRNA.

Attachment to the ribosome binding site positions the small subunit of the ribosome over the initiation codon (Figure 8.4). As described in Chapter 7, this codon is usually 5′-AUG-3′, which codes for methionine, although 5′-GUG-3′ and 5′-UUG-3′ are sometimes used. All three codons can be recognized by the same initiator tRNA, the last two by wobble. The initiator tRNA ($tRNA_i$) joins the small subunit of the ribosome along with a molecule of GTP, which will be used as an energy source for the final step of initiation. This initiator tRNA is the one that was aminoacylated with methionine and subsequently modified by conversion of the methionine to *N*-formylmethionine (see Figure 7.24). The modification attaches a formyl group (–COH) to the amino group, which means that only the carboxyl of the initiator methionine is free to participate in peptide bond formation. This ensures that polypeptide synthesis can take place only in the N $\rightarrow$ C direction.

The formyl group remains attached until translation has proceeded into the elongation phase, but it is then removed from the growing polypeptide, either on its own or along with the rest of the initial methionine. Note that the $tRNA_i^{Met}$ is able to decode only the initiation codon. It cannot enter

TABLE 8.1 EXAMPLES OF RIBOSOME BINDING SITE SEQUENCES IN *E. COLI*

Gene	Codes for	Ribosome binding sequence	Nucleotides from the initiation codon
E. coli consensus	—	5′-AGGAGGU-3′	3–10
Lactose operon	Lactose utilization enzymes	5′-AGGA-3′	7
galE	Hexose-1-phosphate uridyltransferase	5′-GGAG-3′	6
rplJ	Ribosomal protein L10	5′-AGGAG-3′	8

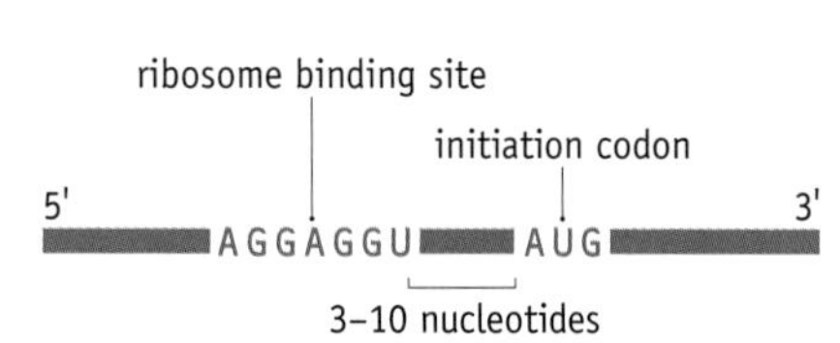

FIGURE 8.3 The ribosome binding site for bacterial translation. In *Escherichia coli*, the ribosome binding site is located between 3 and 10 nucleotides upstream of the initiation codon.

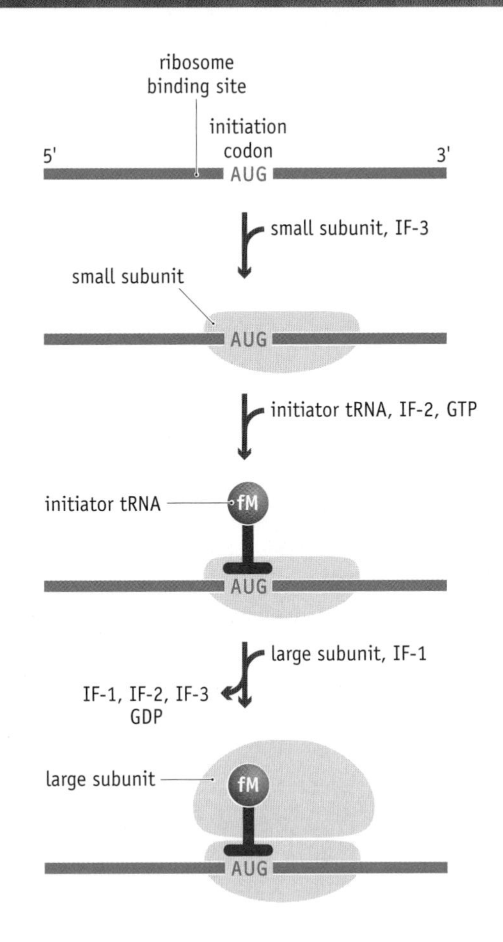

FIGURE 8.4 Initiation of translation in *E. coli*. The process begins when the small subunit of the ribosome attaches to its binding site on the mRNA. The small subunit is accompanied by IF-3, which prevents the large subunit from attaching until the appropriate time. The initiator tRNA is now brought to the complex by IF-2, along with a molecule of GTP. The initiation process is completed by attachment of the large subunit. IF-1 is also involved at this stage but its role is unclear. Attachment of the large subunit requires energy, obtained by conversion of the bound GTP to GDP, and results in release of the initiation factors. fM, *N*-formylmethionine.

the complete ribosome during the elongation phase of translation. During elongation, the internal 5′-AUG-3′ codons are recognized by a different $tRNA^{Met}$ carrying an unmodified methionine.

In addition to the small subunit of the ribosome and the *N*-formylmethionine tRNA, initiation also requires three proteins, called **initiation factors**, which are not permanent components of the ribosome, but which attach at the appropriate times in order to perform their functions (Table 8.2). The roles of two of these initiation factors are fairly well understood. The one called IF-2 directs the initiator tRNA to its correct position on the small subunit of the ribosome, and IF-3 prevents the large subunit from joining the complex until it is needed, which is important as the complete ribosome is unable to initiate translation. The function of IF-1 is less clear. It attaches to the complex toward the end of the initiation phase and appears to coordinate the transition from initiation to elongation, but exactly how it does

TABLE 8.2 FUNCTIONS OF INITIATION FACTORS IN BACTERIA

Initiation factor	Function
IF-1	Unclear. Might cause conformational changes that prepare the small subunit for attachment to the large subunit, or might prevent premature entry of the second aminoacyl-tRNA.
IF-2	Directs the initiator tRNA to its correct position in the initiation complex.
IF-3	Prevents premature reassociation of the large and small subunits of the ribosome.

RESEARCH BRIEFING 8.1

Studying the proteome

Proteome studies are important because of the central role that the proteome plays as the link between the genome and the biochemical capability of the cell. Characterization of the proteomes of different cells is therefore the key to understanding how the genome operates and how dysfunctional genome activity can lead to diseases. Transcriptome studies can only partly address these issues. Examination of the transcriptome tells us which genes are active in a particular cell, but gives a less accurate reflection of the proteins that are present. This is because the factors that influence the proteome include not only the amount of mRNA that is available, but also the rate at which the mRNAs are translated into protein and the rate at which the proteins are degraded. Additionally, the protein that is the initial product of translation might not be active, because some proteins must undergo post-translational modification before becoming functional. Determining the amount of the *active* form of a protein is therefore critical to understanding the biochemistry of a cell or tissue.

The methodology used to study proteomes is called **proteomics**. Strictly speaking, proteomics is a diverse set of techniques used not only to identify the proteins that are present in a proteome, but also to study the functions of individual proteins and their localization within the cell. The particular technique that is used to study the composition of a proteome is called **protein profiling** or **expression proteomics**.

Protein profiling is based on two methods—**protein electrophoresis** and **mass spectrometry**—both of which have long pedigrees but which were rarely applied together in the pre-proteomics era.

Proteins can be separated by two-dimensional gel electrophoresis

To characterize a proteome it is first necessary to prepare pure samples of its constituent proteins. A mammalian cell may contain 10,000 to 20,000 different proteins, so a highly discriminating separation system is needed.

Polyacrylamide gel electrophoresis is the standard method for separating the proteins in a mixture. Depending on the composition of the gel and the conditions under which the electrophoresis is carried out, different chemical and physical properties of proteins can be used as the basis for their separation. The most common technique makes use of the detergent called sodium dodecyl sulfate (SDS). This detergent has two effects. It causes each protein to unfold, and it adds negative charges to each one, the number of negative charges depending on the length of the protein. This means that if electrophoresis is carried out in the presence of SDS, the proteins in a mixture separate according to their molecular masses, the smallest proteins migrating more quickly toward the positive electrode.

Another way of separating proteins is by **isoelectric focusing**. Now the gel contains chemicals that form a pH gradient when the electrical charge is applied. In this type of gel, a protein migrates to its **isoelectric point**, the position in the gradient where its net charge is zero.

In protein profiling, these methods are combined in **two-dimensional gel electrophoresis**. In the first dimension, the proteins are separated by isoelectric focusing. The gel is then soaked in SDS and rotated by 90°, and a second electrophoresis, separating the proteins according to their sizes, is carried out at right angles to the first (Figure 1). Several thousand proteins can be separated in a single gel.

Proteins are identified by peptide mass fingerprinting

After electrophoresis, staining the gel reveals a complex pattern of spots, each one containing a different protein. The second stage of protein profiling is to identify the proteins. This used to be a difficult

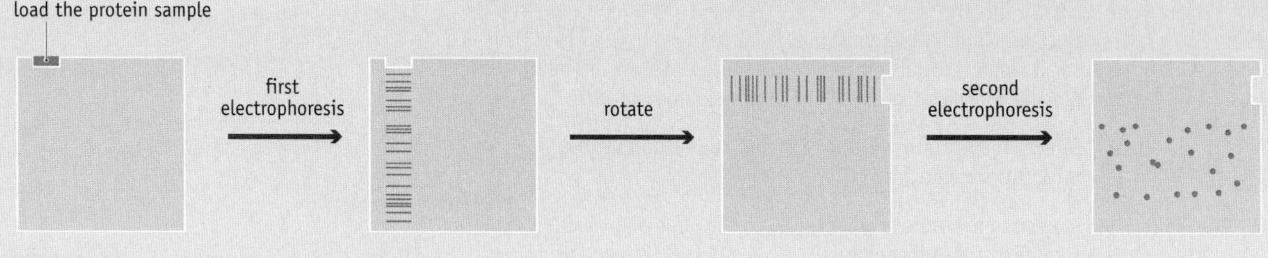

FIGURE 1 Two-dimensional gel electrophoresis.

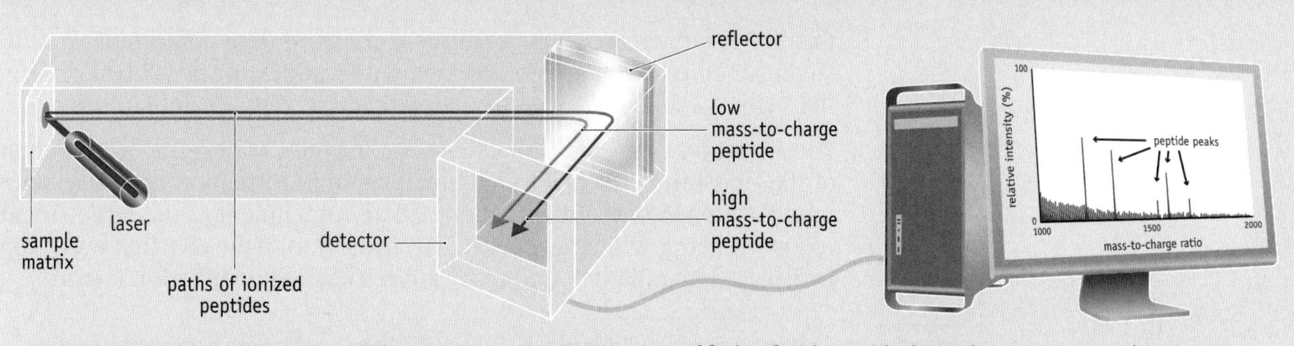

FIGURE 2 Separation of peptides of different mass by MALDI-TOF. The peptides are ionized by energy from a laser and accelerated down the column to the reflector and onto the detector. The time of flight of each peptide depends on its mass-to-charge ratio. The data are visualized as a spectrum.

proposition but has been made much easier by the development of **peptide mass fingerprinting**.

Peptide mass fingerprinting was made possible by advances in mass spectrometry. In mass spectrometry, the compound under study is ionized by exposure to a high-energy field. The ratio between the mass and electric charge of the resulting molecules is then used to identify the compound. The standard technique cannot be used with proteins because they are too large to be ionized effectively, but **matrix-assisted laser desorption ionization time of flight** (**MALDI-TOF**) gets around this problem, at least with proteins of up to 50 amino acids in length. Of course, most proteins are much longer than 50 amino acids, and it is therefore necessary to break them into fragments before examining them by MALDI-TOF. The usual approach is to digest the protein with a protease such as trypsin, which cleaves proteins immediately after arginine or lysine amino acids. With most proteins, this results in a series of peptides 5 to 75 amino acids in length.

Once a peptide is ionized, its mass-to-charge ratio is determined from its "time of flight" within the mass spectrometer as it passes from the ionization source to the detector (Figure 2). The mass-to-charge ratio enables the molecular mass of the peptide to be worked out, which in turn allows its amino acid composition to be deduced. This compositional information is compared with the amino acid sequences of all the proteins that the organism can make, which will be known if the complete genome sequence is available. The protein that has been isolated from the proteome can therefore be identified.

Two proteomes can be compared by isotope-coded affinity tag labeling

If two proteomes are being compared, then a key requirement is to identify proteins that are present in different amounts. These differences may reveal important features of the biochemistry of one or another tissue. If the differences are relatively large, then they will be apparent simply by comparing the pair of stained gels obtained for the two proteomes. But important changes in the biochemical properties of a proteome can result from relatively minor changes in the amounts of individual proteins. Methods for detecting small-scale changes are therefore essential.

One possibility is to label the constituents of the two proteomes with different fluorescent markers, and then run them together in a single two-dimensional gel. Visualization of the two-dimensional gel at different wavelengths enables the intensities of equivalent spots to be judged more accurately than is possible when two separate gels are obtained. A more accurate alternative is to label each proteome with an **isotope-coded affinity tag** (**ICAT**). This is a type of marker that can be obtained in two different forms, one containing normal hydrogen atoms and the other containing deuterium, the heavy isotope of hydrogen. The normal and heavy versions can be distinguished by mass spectrometry, enabling the relative amounts of a protein in two proteomes that have been mixed together to be determined during the MALDI-TOF stage of the profiling procedure (Figure 3).

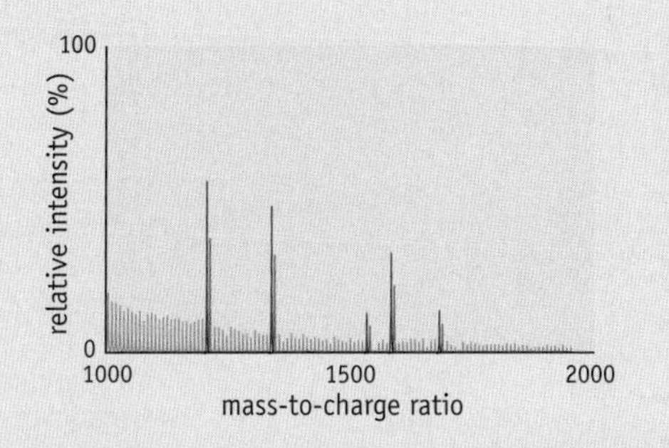

FIGURE 3 Analyzing two proteins by ICAT. Peaks resulting from peptides containing normal hydrogen are shown in red, and those from peptides containing deuterium are shown in blue.

this is not known. Possibly it causes a change in the conformation of the small subunit, enabling the large subunit to attach, or possibly it prevents the second aminoacyl-tRNA from entering the complex until it is needed.

The initiation phase of translation is completed when the large subunit of the ribosome attaches to the small subunit, forming a complete ribosome positioned over the initiation codon. Attachment of the large subunit requires energy, which is generated by hydrolysis of the GTP that was bound earlier during initiation, and results in release of the initiation factors.

Initiation in eukaryotes is mediated by the cap structure and poly(A) tail

Only a small number of eukaryotic mRNAs have internal ribosome binding sites. Instead, with most mRNAs the small subunit of the ribosome makes its initial attachment at the 5′ end of the molecule and then **scans** along the sequence until it locates the initiation codon.

The process works as follows. The first step involves assembly of the **pre-initiation complex**, the principal components of which are the small subunit of the ribosome and the initiator tRNA (Figure 8.5A). As in bacteria, the initiator tRNA is distinct from the normal $tRNA^{Met}$ that recognizes internal 5′-AUG-3′ codons, but, unlike the initiator tRNA of bacteria, it is aminoacylated with normal methionine, not the formylated version.

After assembly, the pre-initiation complex attaches to the cap structure at the extreme 5′ end of the mRNA. The **initiation complex**, as it is now called, then scans along the molecule to find the initiation codon (Figure 8.5B). The leader regions of eukaryotic mRNAs can be several tens, or even hundreds, of nucleotides in length and often contain several 5′-AUG-3′ triplets that the complex must ignore before it reaches the true initiation codon. This discrimination is possible because the initiation codon, but not the upstream 5′-AUG-3′ triplets, is contained in a short consensus sequence, 5′-ACCAUGG-3′, referred to as the **Kozak consensus**.

There are two further aspects of eukaryotic translation initiation that we must consider. The first is the involvement of the poly(A) tail of the mRNA, which somehow promotes the binding of the pre-initiation complex to the cap structure. We know that the poly(A) tail is involved at this stage because the length of the tail influences the extent of translation initiation that occurs with a particular mRNA, and removal of the tail is one of the steps that lead to inactivation of an mRNA whose translation product is no longer needed. But exactly how the poly(A) tail associates with the pre-initiation complex is not yet known.

The second issue is the role of the plethora of initiation factors possessed by eukaryotes, at least 13 at the last count (Table 8.3). Five of these—eIF-1, eIF-1A, eIF-2, eIF-2B, and eIF-3—are components of the pre-initiation complex. A further three—eIF-4A, eIF-4E, and eIF-4G—form a structure called

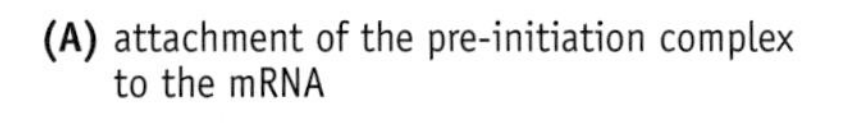

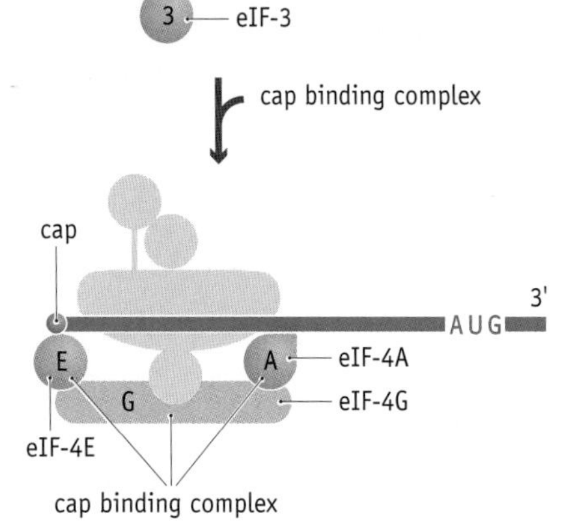

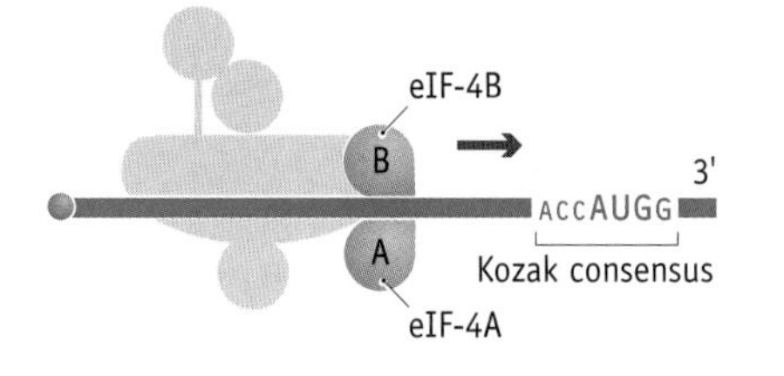

FIGURE 8.5 Initiation of translation in eukaryotes. (A) Assembly of the pre-initiation complex and its attachment to the mRNA. The pre-initiation complex comprises the small subunit of the ribosome, the initiator tRNA, and various initiation factors including eIF-2 and eIF-3. The pre-initiation complex attaches to the cap structure at the 5′ end of the tRNA, assisted by the cap binding complex, which is made up of eIF-4A, eIF-4E, and eIF-4G. (B) The initiation complex scans along the mRNA until it reaches the initiation codon, which is located within the Kozak consensus sequence. Two of the initiation factors, eIF-4A and eIF-4B, are involved in the scanning process. M, methionine.

TABLE 8.3 FUNCTIONS OF INITIATION FACTORS IN EUKARYOTES

Initiation factor	Function
eIF-1	Component of the pre-initiation complex
eIF-1A	Component of the pre-initiation complex
eIF-2	Binds to the initiator tRNA within the pre-initiation complex; phosphorylation of eIF-2 results in a global repression of translation
eIF-2B	Component of the pre-initiation complex
eIF-3	Component of the pre-initiation complex; makes direct contact with eIF-4G and so forms the link with the cap binding complex
eIF-4A	Component of the cap binding complex; a helicase that aids scanning by breaking intramolecular base pairs in the mRNA
eIF-4B	Aids scanning, possibly by acting as a helicase that breaks intramolecular base pairs in the mRNA
eIF-4E	Component of the cap binding complex, possibly the component that makes direct contact with the cap structure at the 5′ end of the mRNA
eIF-4F	The cap binding complex, comprising eIF-4A, eIF-4E, and eIF-4G, which makes the primary contact with the cap structure at the 5′ end of the mRNA
eIF-4G	Component of the cap binding complex; forms a bridge between the cap binding complex and eIF-3 in the pre-initiation complex; in at least some organisms, eIF-4G also forms an association with the poly(A) tail
eIF-4H	In mammals, aids scanning in a manner similar to eIF-4B
eIF-5	Aids release of the other initiation factors at the completion of initiation
eIF-6	Associated with the large subunit of the ribosome; prevents large subunits from attaching to small subunits in the cytoplasm

the **cap binding complex** (unhelpfully also called eIF-4F), which makes the initial contact with the 5′ end of the mRNA and mediates subsequent attachment of the pre-initiation complex. Two factors, eIF-4A and eIF-4B, are involved in the scanning process, eIF-4A and possibly eIF-4B having a helicase activity which enables them to break intramolecular base pairs that sometimes form in the mRNA leader region (Figure 8.6). Finally, eIF-5 aids the release of all the other initiation factors at the end of the initiation phase, and eIF-6 plays the same role as IF-3 in bacteria, preventing the large subunit of the ribosome from joining the initiation complex until it is needed.

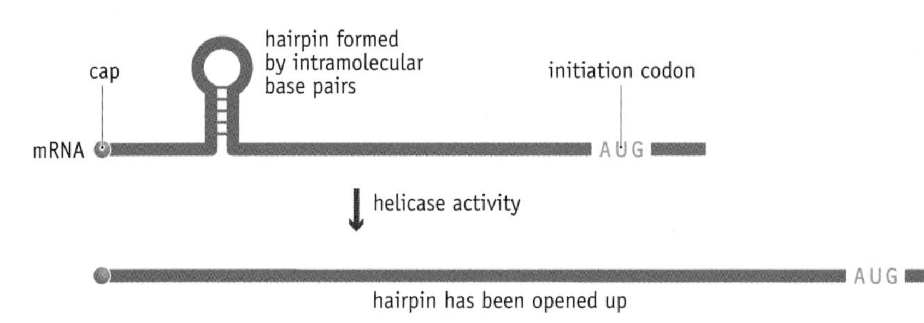

FIGURE 8.6 The helicase activities of eIF-4A and possibly eIF-4B enable hairpin structures resulting from intramolecular base pairing to be broken in the leader region of the mRNA.

Translation of a few eukaryotic mRNAs initiates without scanning

The scanning system for initiation of translation does not apply to every eukaryotic mRNA. This was first recognized with the picornaviruses, a group of viruses with RNA genomes which includes the poliovirus and the human rhinovirus, the latter responsible for the common cold. These viruses are not themselves eukaryotes, but their genes are expressed within the eukaryotic cells that they infect. They make use of at least some of the endogenous noncoding RNAs and proteins, the ones made by the cell, which are subverted by the viruses to their own ends.

Picornavirus mRNAs are not capped but instead have an **internal ribosome entry site** (**IRES**), which is similar in function to the ribosome binding site of bacteria, although the sequences of IRESs and their positions relative to the initiation codon are more variable than the bacterial versions. The presence of IRESs on their mRNAs means that picornaviruses can block protein synthesis in the host cell by inactivating the cap binding complex, without affecting translation of their own mRNAs, although this is not a normal part of the infection strategy of all picornaviruses.

Remarkably, no virus proteins are required for recognition of an IRES by a host ribosome. In other words, the normal eukaryotic cell possesses proteins and/or other factors that enable it to initiate translation by the IRES method. Because of their variability, IRESs are difficult to identify, but it is becoming clear that a few cellular mRNAs possess them and that these are translated, at least under some circumstances, via their IRES rather than by scanning. These cellular mRNAs include a few that are translated when the cell is put under stress, for example, by exposure to heat, radiation, or low oxygen conditions. Under these circumstances, cap-dependent translation is globally suppressed by inactivation of eIF-2, one of the initiation factors present in the pre-initiation complex. The presence of IRESs on the "survival" mRNAs therefore enables these to undergo preferential translation at the time when their products are needed.

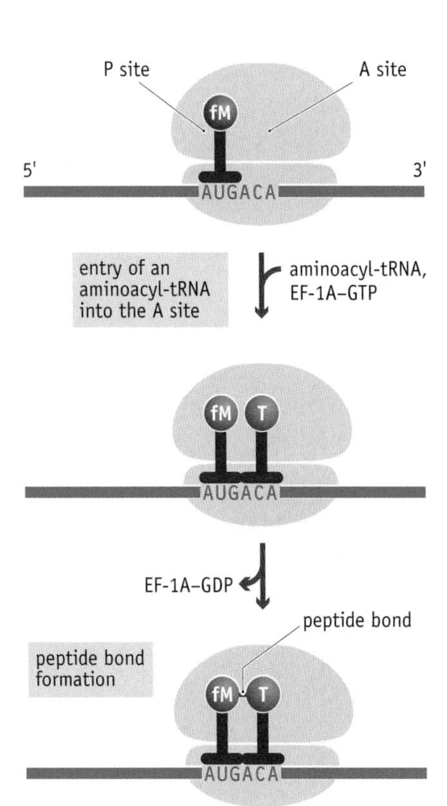

FIGURE 8.7 The first steps in the elongation phase of protein synthesis in *E. coli*. The ribosome contains two sites at which aminoacyl-tRNAs can bind. To begin with, the P site is occupied by the initiator $tRNA_i^{Met}$, and the A site is empty. Elongation begins with entry of the second aminoacyl-tRNA into the A site, accompanied by elongation factor EF-1A and a molecule of GTP. A peptide bond is now formed between the two amino acids. This requires energy released by conversion of the GTP molecule to GDP. fM, *N*-formylmethionine; T, threonine.

Elongation of the polypeptide begins with formation of the first peptide bond

We now move into the elongation phase, during which the protein is synthesized step by step, individual amino acids being attached to the carboxyl terminus of the growing polypeptide. Elongation is similar in bacteria and eukaryotes, so we can deal with the two types of organism together, pointing out the important distinctions as we come to them.

Bringing together the two ribosome subunits creates two sites at which aminoacyl-tRNAs can bind. The first of these, the **peptidyl** (or **P**) **site**, is already occupied by the initiator $tRNA_i^{Met}$, charged with *N*-formylmethionine or methionine, and base-paired with the initiation codon. The second site, the **aminoacyl** (or **A**) **site**, covers the second codon in the open reading frame (Figure 8.7). The structures revealed by X-ray crystallography show that these sites are located in the cavity between the large and small subunits of the ribosome, the codon–anticodon interaction being associated with the small subunit and the aminoacyl end of the tRNA with the large subunit (Figure 8.8).

To begin elongation, the A site becomes filled with the appropriate aminoacyl-tRNA, which in *E. coli* is brought into position by the **elongation factor** EF-1A. This factor helps to ensure that translation is accurate, as it allows only a tRNA that carries its correct amino acid to enter the ribosome,

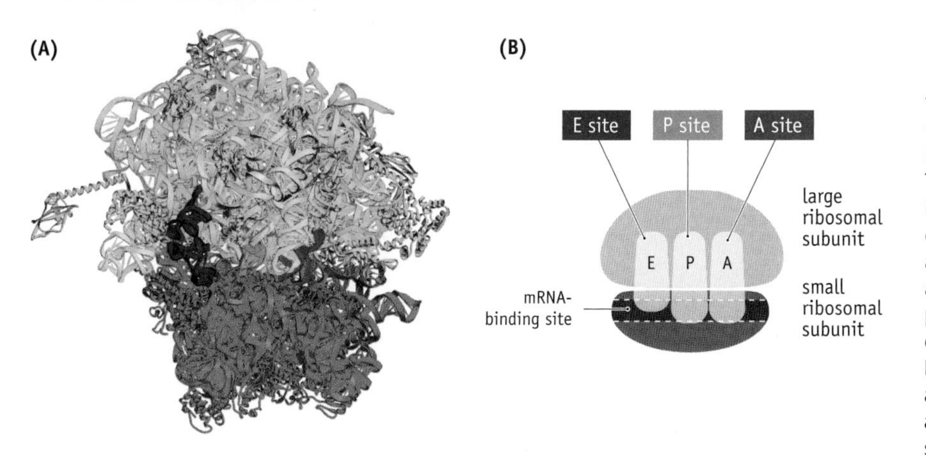

FIGURE 8.8 The structure of a bacterial ribosome during the elongation phase of translation. (A) Detailed structure of the ribosome with the small subunit in dark brown and the large subunit in light brown. The locations of the P and A sites are shown in orange and red, respectively. The E or exit site, which the tRNA passes through as it leaves the ribosome after its amino acid has been attached to the growing polypeptide, is shown in blue. (B) Schematic diagram illustrating the positions of the E, P, and A sites in the ribosome. The mRNA is associated with the small subunit and the aminoacyl end of each tRNA with the large subunit. (Adapted from M. M. Yusupov et al., *Science* 292: 883–896, 2001. With permission from AAAS.)

mischarged tRNAs being rejected at this point. EF-1A also binds a molecule of GTP, which will be hydrolyzed to release energy when the peptide bond is formed. In eukaryotes the equivalent factor is called eEF-1 (Table 8.4).

Once in the A site, the anticodon of the tRNA must form base pairs with the next codon of the mRNA. Again, it is essential that mistakes are not made. The tRNA and mRNA must fit exactly within the A site, and this can happen only if the codon and anticodon are perfectly matched, with each of the three base pairs formed correctly. If there is a mispair at any position in the codon–anticodon interaction, then the tRNA is rejected.

When the correct aminoacyl-tRNA has entered the A site, a peptide bond is formed between the two amino acids. The bond is formed by the enzyme called **peptidyl transferase**. In both bacteria and eukaryotes, peptidyl transferase is an example of a ribozyme—an RNA enzyme. The catalytic activity for peptide bond formation is provided not by a protein but by the largest of the rRNAs present in the large subunit of the ribosome. Peptide bond formation is energy-dependent and requires hydrolysis of the GTP attached to EF-1A (eEF-1 in eukaryotes). This inactivates EF-1A, which is ejected from the ribosome and regenerated by EF-1B. A eukaryotic equivalent of EF-1B has not been identified, and it is possible that one of the subunits of eEF-1 provides the regenerative activity.

TABLE 8.4 FUNCTIONS OF ELONGATION FACTORS IN BACTERIA AND EUKARYOTES

Initiation factor	Function
Bacteria	
EF-1A	Directs the next tRNA to its correct position in the ribosome
EF-1B	Regenerates EF-1A after the latter has yielded the energy contained in its attached GTP molecule
EF-2	Mediates translocation
Eukaryotes	
eEF-1	Complex of four subunits (eEF-1a, eEF-1b, eEF-1d, and eEF-1g); directs the next tRNA to its correct position in the ribosome
eEF-2	Mediates translocation

Note that the bacterial elongation factors have recently been renamed. The older designations were EF-Tu, EF-Ts, and EF-G for EF-1A, EF-1B, and EF-2, respectively.

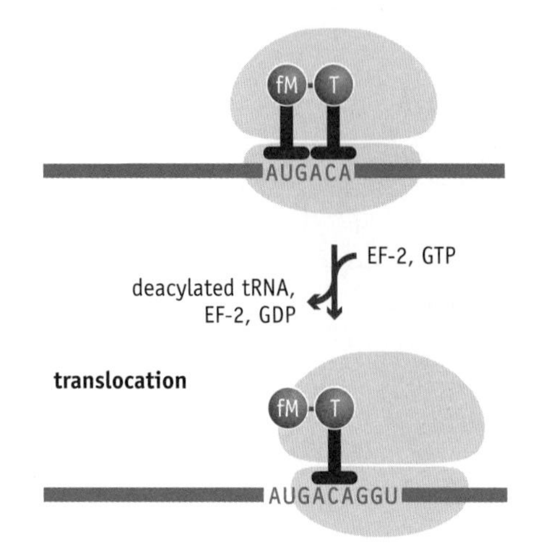

FIGURE 8.9 The first cycle of translocation during protein synthesis in *E. coli*. The ribosome moves three nucleotides along the mRNA, ejecting the deacylated initiator tRNA from the P site, which now becomes occupied by the second tRNA attached to the first two amino acids of the polypeptide. Translocation requires hydrolysis of a molecule of GTP and is mediated by EF-2. fM, *N*-formylmethionine; T, threonine.

Elongation continues until a termination codon is reached

Formation of the first peptide bond results in a dipeptide corresponding to the first two codons of the open reading frame. The attachment between the first amino acid and its tRNA is broken at this stage, leaving the dipeptide attached to the tRNA located in the A site.

The next step is **translocation**, during which the ribosome moves three nucleotides along the mRNA, so a new codon enters the A site (Figure 8.9). This moves the dipeptide-tRNA to the P site, which in turn displaces the deacylated tRNA. In eukaryotes, the deacylated tRNA is simply ejected from the ribosome, but in bacteria the deacylated tRNA departs via a third position, the **exit** (or **E**) **site**. This site was originally looked on as a simple exit point from the ribosome, but it is now known to have an important role in ensuring that translocation moves the ribosome along the mRNA by precisely three nucleotides, thereby ensuring that the ribosome keeps to the correct reading frame.

Translocation requires hydrolysis of a molecule of GTP and is mediated by EF-2 in bacteria and by eEF-2 in eukaryotes. Electron microscopy of ribosomes at different intermediate stages in translocation shows that, in order to move along the mRNA, the ribosome adopts a less compact structure, with the two subunits rotating slightly in opposite directions. This opens up the space between them and enables the ribosome to slide along the mRNA. Translocation results in the A site becoming vacant, allowing a new aminoacyl-tRNA to enter.

The elongation cycle is now repeated, and continues until a termination codon is reached. After several cycles of elongation the start of the mRNA molecule is no longer associated with the ribosome, and a second ribosome can attach and begin to synthesize another copy of the protein. The end result is a **polysome**, an mRNA that is being translated by several ribosomes at once (Figure 8.10). Polysomes have been seen in electron-microscopic images of both prokaryotic and eukaryotic cells.

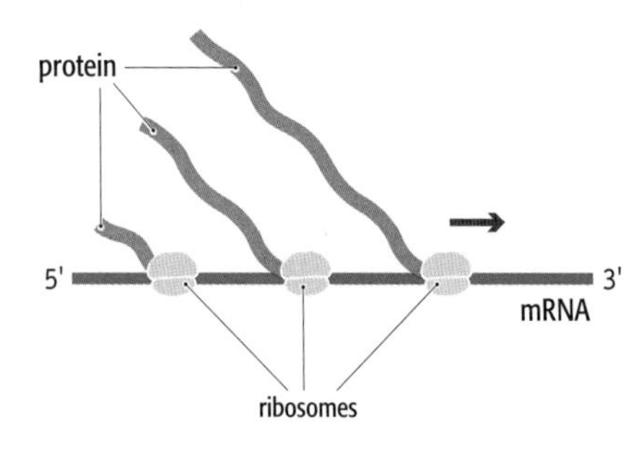

FIGURE 8.10 A polysome. Three ribosomes are attached to a single mRNA. The arrow shows the direction in which the ribosomes are moving along the mRNA.

Termination requires special release factors

Protein synthesis ends when one of the three termination codons enters the A site. There are no tRNA molecules with anticodons able to base-pair with any of the termination codons. Instead a protein **release factor** enters the A site in order to terminate translation.

TABLE 8.5 RELEASE AND RIBOSOME RECYCLING FACTORS IN BACTERIA AND EUKARYOTES

Factor	Function
Bacteria	
RF-1	Recognizes the termination codons 5′-UAA-3′ and 5′-UAG-3′
RF-2	Recognizes 5′-UAA-3′ and 5′-UGA-3′
RF-3	Stimulates dissociation of RF-1 and RF-2 from the ribosome after termination
RRF	Ribosome release factor, responsible for separation of the ribosome subunits after translation has terminated
Eukaryotes	
eRF-1	Recognizes the termination codon
eRF-3	Possibly stimulates dissociation of eRF-1 from the ribosome after termination; possibly causes the ribosome subunits to disassociate after termination of translation

In bacteria there are three release factors (Table 8.5). RF-1 recognizes the termination codons 5′-UAA-3′ and 5′-UAG-3′, and RF-2 recognizes 5′-UAA-3′ and 5′-UGA-3′. RF-3 stimulates release of RF-1 and RF-2 from the ribosome after termination, in a reaction requiring energy from the hydrolysis of GTP (Figure 8.11).

In eukaryotic cells there are just two release factors. These are eRF-1, which recognizes the termination codon, and eRF-3, which may play the same role as RF-3, although this has not been proven. The structure of eRF-1 has been solved by X-ray crystallography, showing that the shape of this protein is very similar to that of a tRNA (Figure 8.12). This immediately suggests that the release factor is able to enter the A site by mimicking a tRNA, but some researchers believe that the structural similarities are deceptive and that there are unresolved complexities about the way in which the release factors work.

The release factors terminate translation, but they do not appear to be responsible for separation of the ribosomal subunits, at least not in bacteria. This is the function of an additional protein called **ribosome recycling factor** (**RRF**), which also has a tRNA-like structure. RRF probably enters the P or A site and "unlocks" the ribosome (see Figure 8.11). Separation of the subunits requires energy, which is released from GTP by EF-2, one of the elongation factors, and also requires the initiation factor IF-3 to prevent the subunits from attaching together again. A eukaryotic equivalent of RRF has not been identified, and this may be one of the functions of eRF-3. The ribosome subunits enter the cytoplasmic pool, where they remain until used again in another round of translation.

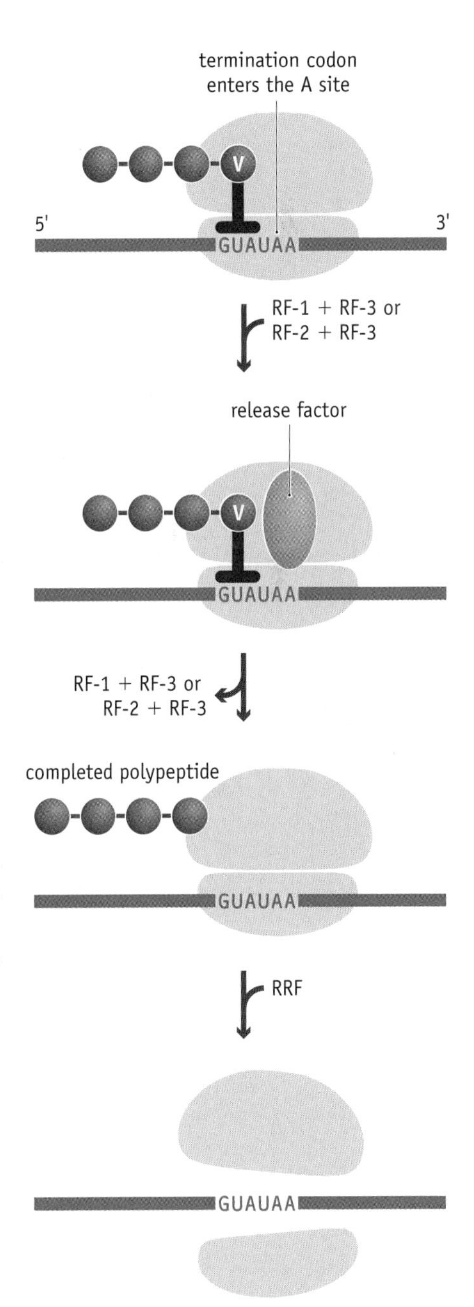

FIGURE 8.11 Termination of translation in *E. coli*. In the top drawing, the ribosome has reached a termination codon. Either RF-1 or RF-2, along with RF-3, enters the A site and releases the completed polypeptide. The ribosome is then separated into its two subunits by the ribosome recycling factor (RRF). V, valine.

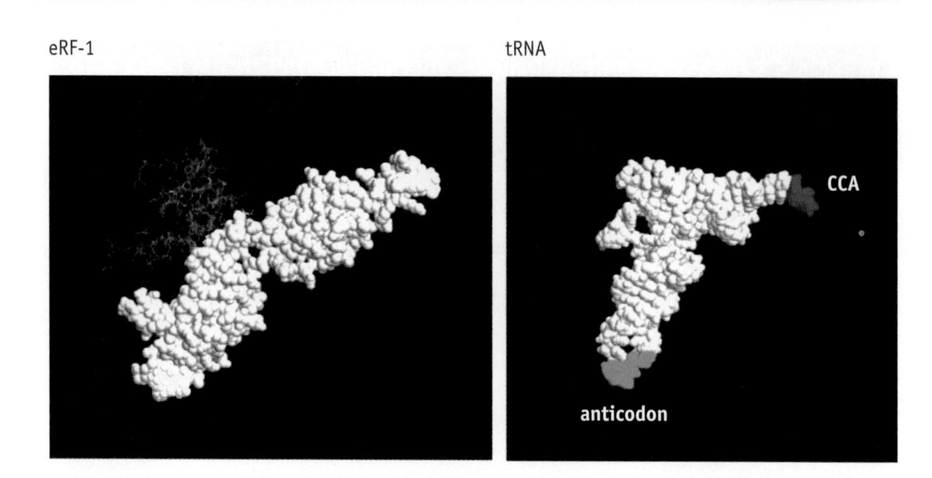

FIGURE 8.12 The structure of the eukaryotic release factor eRF-1 is similar to that of a tRNA. The part of eRF-1 that resembles the tRNA is highlighted in white. The purple segment of eRF-1 interacts with the second eukaryotic release factor, eRF-3. On the tRNA structure, the position of the anticodon is marked, as well as the 3′ end of the molecule to which the amino acid is attached (labeled "CCA"). (From L. L. Kisselev and R. H. Buckingham, *Trends Biochem. Sci.* 25: 561–566, 2000. With permission from Elsevier.)

8.2 POST-TRANSLATIONAL PROCESSING OF PROTEINS

If we were not taking the modern molecular approach to genetics, then we might be tempted to end our study of the gene expression pathway at this point. Our argument would be that "DNA makes RNA makes protein" and that we have reached the stage where a protein has been made, so clearly that is the end of the story. But it is not. Translation results in synthesis of a polypeptide, but this polypeptide is simply a linear chain of amino acids. It is inactive and in such condition is not the final product of the gene expression pathway. Utilization of the biological information contained in a gene requires synthesis of an *active* protein, one able to perform its function in the cell. We must therefore study the events that convert the newly synthesized polypeptide into a functional protein. Only by studying these post-translational processing events will we acquire a complete understanding of how a gene is able to perform its role as a unit of biological information.

The most important part of post-translational processing is protein folding. All polypeptides have to be folded into their correct secondary and tertiary structures before becoming active in the cell. Some proteins also undergo chemical modification, by attachment of new chemical groups, and some also are processed by cutting events carried out by enzymes called **proteases**. These cutting events may remove segments from one or both ends of the polypeptide, resulting in a shortened form of the protein, or they may cut the polypeptide into a number of different segments, all or some of which are then active.

Often these different types of processing occur together, a polypeptide being modified and cut at the same time that it is folded. If this is the case, then the chemical modifications and/or cuts might be needed in order for the polypeptide to take up its correct three-dimensional conformation. Alternatively, a chemical modification or a cutting event might occur after the protein has been folded, possibly as part of a regulatory mechanism that converts a folded but inactive protein into an active form.

Some proteins fold spontaneously in the test tube

In Chapter 7 we examined the four levels of protein structure and learned that all of the information that a polypeptide needs in order to fold into its correct three-dimensional structure is contained within its amino acid sequence. This inviolate link between the amino acid sequence of a protein and its tertiary structure is one of the keys to gene expression, because it

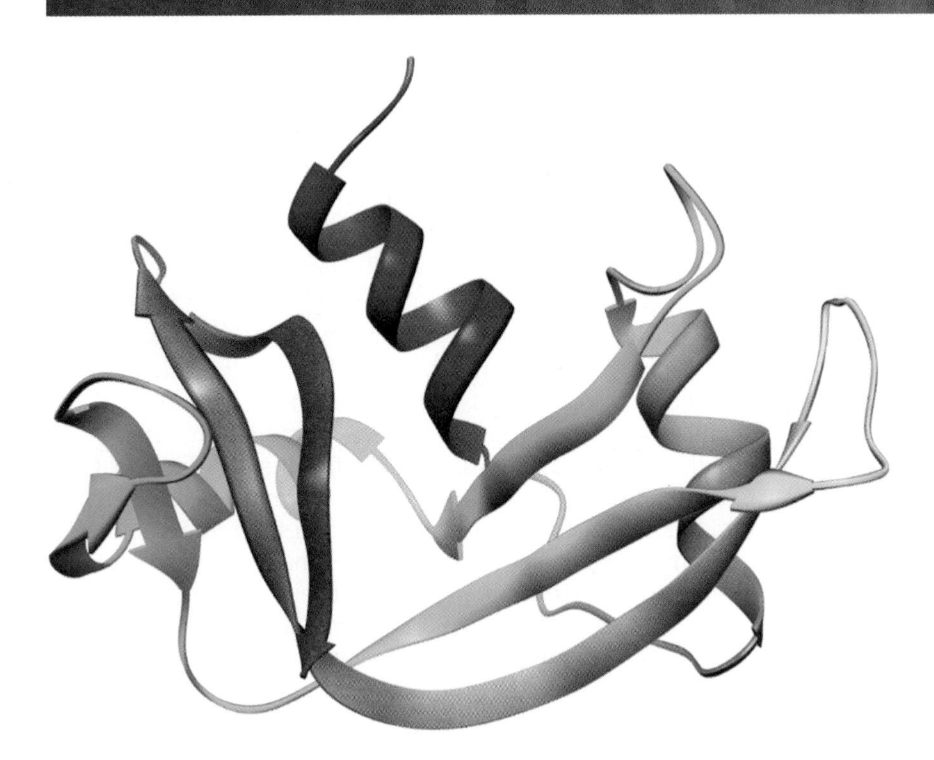

FIGURE 8.13 The tertiary structure of ribonuclease A. The colors indicate each of the various secondary structural components of the protein.

completes the pathway that begins with the biological information contained within a gene and ends with a protein function.

The experiments that showed that the amino acid sequence of a protein specifies its tertiary structure were first carried out with ribonuclease. Ribonuclease is a small protein, just 124 amino acids in length, containing four disulfide bridges and with a tertiary structure that is made up of three α helices and a region of β sheet (Figure 8.13). Studies of its folding can be carried out with purified ribonuclease that has been resuspended in buffer. Addition of urea, a compound that disrupts hydrogen bonding, results in a decrease in the activity of the enzyme, measured by testing its ability to cut RNA. At the same time there is an increase in the viscosity of the solution (Figure 8.14). Both observations indicate that the protein has been denatured by unfolding into an unstructured polypeptide chain. This unfolded protein is no longer able to act as a ribonuclease—hence, the loss of enzymatic activity—and the chains will have a tendency to get tangled, explaining the increase in viscosity.

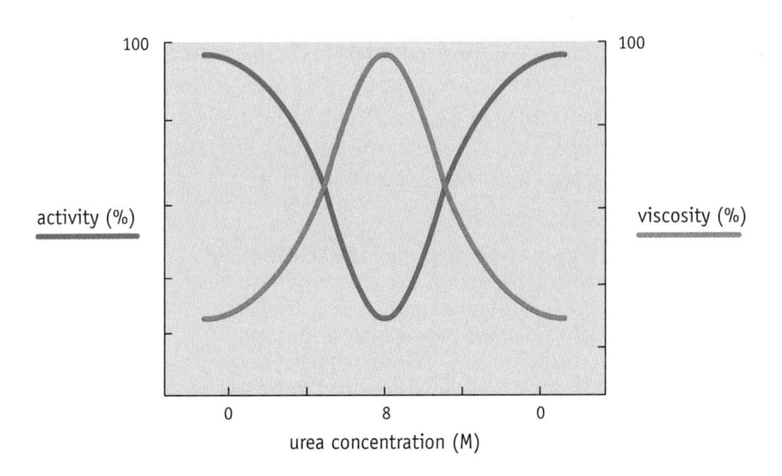

FIGURE 8.14 Denaturation and spontaneous renaturation of ribonuclease in the presence of different amounts of urea. As the urea concentration increases to 8 M, the protein becomes denatured by unfolding. Its activity decreases and the viscosity of the solution increases. When the urea is removed by dialysis, this small protein re-adopts its folded conformation. The activity of the protein increases back to the original level and the viscosity of the solution decreases.

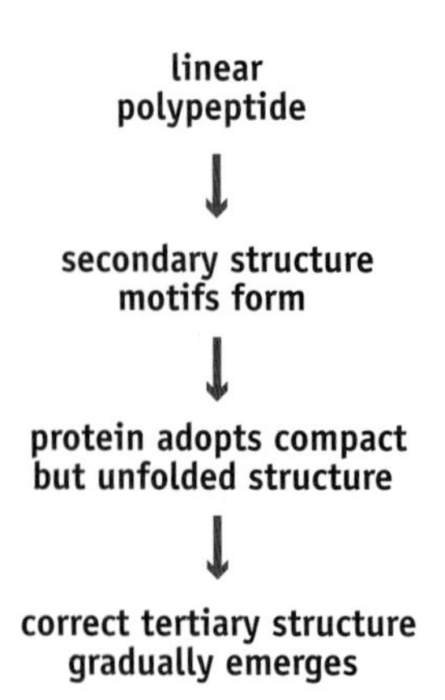

FIGURE 8.15 A typical protein folding pathway.

What if the urea is removed from the solution? Now the viscosity decreases and the enzyme activity reappears. The conclusion is that the protein refolds spontaneously when the denaturant is removed. The same result is obtained when the urea treatment is combined with addition of a reducing agent to break the disulfide bonds. The activity is still regained on renaturation. This shows that the disulfide bonds are not critical to the protein's ability to refold. They merely stabilize the tertiary structure once it has been adopted.

More detailed study of the spontaneous folding of ribonuclease and other small proteins has suggested that the secondary structural motifs along the polypeptide chain form within a few milliseconds of removal of the denaturant. At this stage, the protein adopts a compact, but not folded, organization, with its hydrophobic groups on the inside, shielded from water (Figure 8.15). During the next few seconds or minutes, the secondary structural motifs interact with one another, and the tertiary structure gradually takes shape, often via a series of intermediate conformations that make up the protein's **folding pathway**. There may be more than one possible pathway that a protein can follow to reach its correctly folded structure, and the pathways might also have side branches into which the protein can be diverted, leading to an incorrect structure. If the incorrect structure is sufficiently unstable, then partial or complete unfolding may occur, allowing the protein a second opportunity to pursue a correct folding pathway (Figure 8.16).

Inside cells, protein folding is aided by molecular chaperones

The experiments described above confirm that the amino acid sequence of a protein specifies its tertiary structure, but they do not help us understand how protein folding occurs in the cell. Folding in the cell is probably not an entirely spontaneous process, at least not for all proteins, especially larger ones with structures more complex than that of ribonuclease. When studied in the test tube, these proteins tend to form insoluble aggregates when the denaturant is removed. This is probably because they collapse into interlocked networks when they attempt to protect their hydrophobic groups from water at the start of their folding pathway. Some also get stuck in nonproductive side branches of their folding pathways, taking on an intermediate form that is incorrectly folded but which is too stable to unfold. Experimentally, correct folding can be achieved by diluting the protein, but this is not an option that the cell can take to prevent its unfolded proteins from aggregating or following an incorrect pathway.

The difficulties that larger proteins experience when folding in the test tube indicate that within cells there must be mechanisms that help proteins adopt their correct structures. We now know that the necessary assistance is provided by proteins called **molecular chaperones**. The existence of molecular chaperones does not compromise the underlying principle that the information for folding is carried by a protein's amino acid sequence. Molecular chaperones do not specify the tertiary structure of a protein, they merely help the protein find that correct structure.

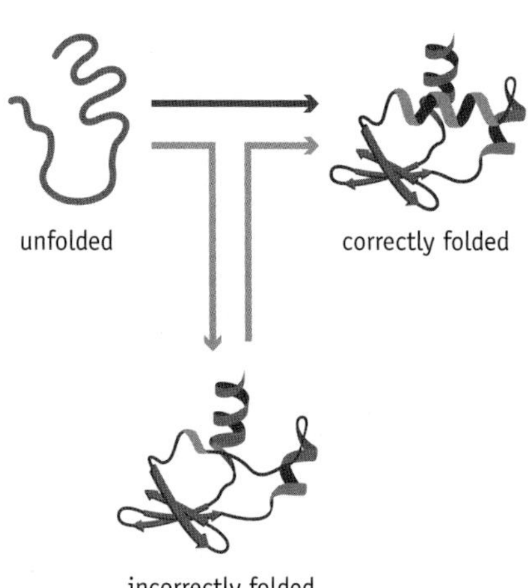

FIGURE 8.16 An incorrectly folded protein might be able to refold into its correct conformation. The black arrow represents the correct folding pathway, leading from the unfolded protein on the left to the active protein on the right. The brown arrow leads to an incorrectly folded conformation, in which one of the α helices has not formed. If this conformation is unstable then the protein will be able to unfold partially, return to its correct folding pathway, and, eventually, reach its active conformation.

There are two types of molecular chaperone. The first are the **Hsp70 chaperones**, which bind to hydrophobic regions of proteins, including proteins that are still being translated (Figure 8.17). The latter is an important point, because if a protein begins to fold before it has been fully synthesized, when only part of the polypeptide is available, then there might be an increased possibility that incorrect branches of the folding pathway will be followed. Hsp70 chaperones are thought to hold the protein in an open conformation and somehow to modulate the association between those parts of the polypeptide that form interactions in the folded protein. Exactly how this is achieved is not understood, but it involves repeated binding and attachment of the Hsp70 protein, each cycle of which requires energy provided by the hydrolysis of ATP. Besides protein folding, the Hsp70 chaperones are also involved in other processes that require shielding of hydrophobic regions in proteins, such as transport through membranes, association of proteins into multisubunit complexes, and disaggregation of proteins that have been damaged by heat stress.

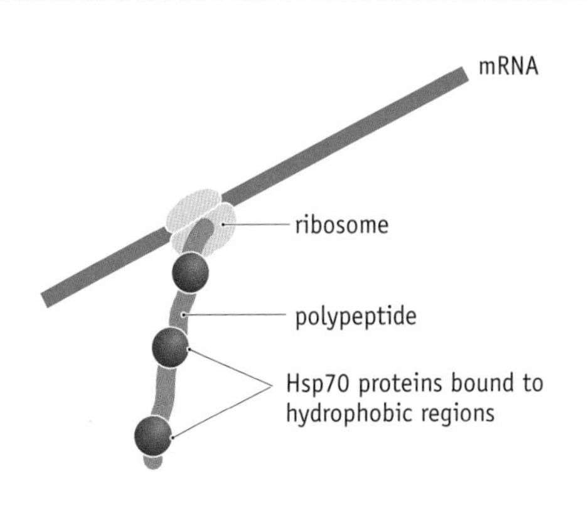

FIGURE 8.17 The role of Hsp70 chaperones. Hsp70 chaperones bind to hydrophobic regions in unfolded polypeptides, including those that are still being translated, and hold the protein in an open conformation to aid its folding.

The second group of molecular chaperones are called the **chaperonins**. These work in a quite different way. The main example is the GroEL/GroES complex, a multisubunit structure that looks like a hollowed-out bullet with a central cavity (Figure 8.18). A single unfolded protein enters the cavity and emerges folded. The mechanism for this is not known, but it is postulated that GroEL/GroES acts as a cage that prevents the unfolded protein from aggregating with other proteins, and that the inside surface of the cavity changes from hydrophobic to hydrophilic in such a way as to promote the burial of hydrophobic amino acids within the protein. This is not the only hypothesis. Another possibility is that the cavity unfolds proteins that have folded incorrectly, passing these unfolded proteins back to the cytoplasm so they can have a second attempt at adopting their correct tertiary structure.

The Hsp70 family of chaperones and the GroEL/GroES chaperonins are present in both bacteria and eukaryotes, but it is thought that in eukaryotes protein folding depends mainly on the action of the Hsp70 proteins. This is probably true also of bacteria, even though the GroEL/GroES chaperonins play a major role in the folding of metabolic enzymes and proteins involved in transcription and translation.

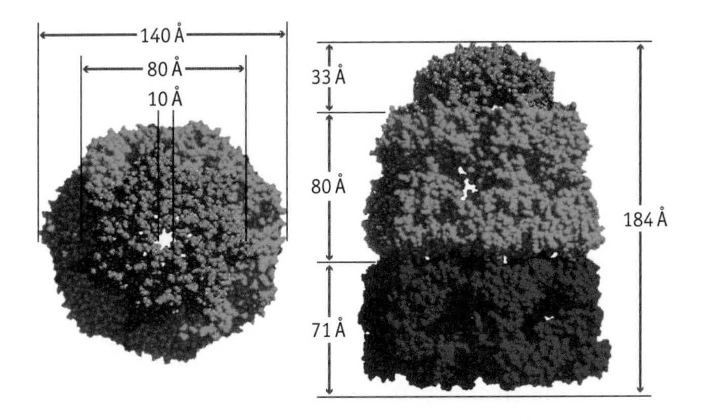

FIGURE 8.18 The GroEL/GroES chaperonin. On the left is the view from the top, and on the right the view from the side. The GroES part of the structure is made up of seven identical protein subunits and is shown in gold. The GroEL components consist of 14 identical proteins arranged into two rings (shown in red and green), each containing seven subunits. The main entrance into the central cavity is through the bottom of the structure shown on the right. 1 Å is equal to 0.1 nm. (Courtesy of Zhaohui Xu, University of Michigan.)

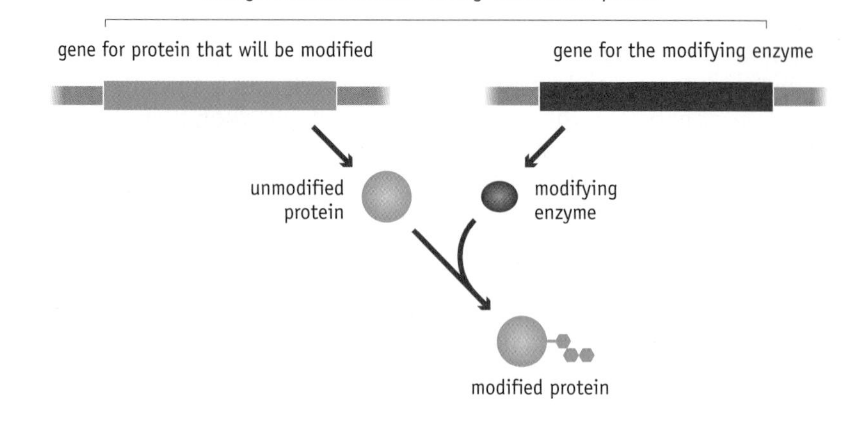

FIGURE 8.19 The biological information needed to synthesize a chemically modified protein resides in both the gene for that protein and the gene for the enzyme that carries out the modification.

Some proteins are chemically modified

The standard genetic code specifies 20 different amino acids, and two others—selenocysteine and pyrrolysine—can be inserted during translation by context-dependent reassignment of 5′-UGA-3′ and 5′-UAG-3′ codons. This repertoire is increased dramatically by post-translational chemical modification of proteins, which results in a vast array of different amino acid types.

Because of chemical modification, the amino acid sequence of a mature, functioning protein might be different from that of the polypeptide coded by the gene. There is still, however, an inviolate link between the gene and the modified protein, because the chemical modifications do not occur at random. Instead they are carried out in a highly specific manner, the same amino acids being modified in the same way in every copy of the protein. In essence, the biological information that specifies the chemical modifications resides not in the gene for the protein that is being modified, but in the genes coding for the enzymes that carry out the modifications (**Figure 8.19**). The structures of these modifying enzymes, as coded by their gene sequences, determine their abilities to make specific chemical modifications at the correct positions on their target proteins.

The simplest types of chemical modification occur in all organisms and involve addition of a small chemical group (e.g., an acetyl, a methyl, or a phosphate group; **Table 8.6**) to an amino acid side chain, or to the amino or carboxyl groups of the terminal amino acids in a polypeptide. Over 150 different modified amino acids have been documented in different proteins. Some proteins undergo an array of different modifications, an example being mammalian histone H3, which can be modified by acetylation, methylation, and phosphorylation at a number of positions along its polypeptide chain (**Figure 8.20**). Histones are components of nucleosomes, the protein structures around which DNA is wound in eukaryotic chromosomes. The

FIGURE 8.20 Post-translational modification of mammalian histone H3. The first 29 amino acids of the protein are shown, along with all of the modifications that can occur in this segment. Ac, acetylation; Me, methylation; P, phosphorylation.

TABLE 8.6 EXAMPLES OF POST-TRANSLATIONAL CHEMICAL MODIFICATIONS		
Modification	**Amino acids that are modified**	**Examples**
Addition of small chemical groups		
Acetylation	Lysine	Histones
Methylation	Lysine	Histones
Phosphorylation	Serine, threonine, tyrosine	Some proteins involved in signal transduction
Hydroxylation	Proline, lysine	Collagen
N-formylation	N-terminal glycine	Melittin
Addition of sugar side chains		
O-linked glycosylation	Serine, threonine	Many membrane proteins and secreted proteins
N-linked glycosylation	Asparagine	Many membrane proteins and secreted proteins
Addition of lipid side chains		
Acylation	Serine, threonine, cysteine	Many membrane proteins
N-myristoylation	N-terminal glycine	Some protein kinases involved in signal transduction
Addition of biotin		
Biotinylation	Lysine	Various carboxylase enzymes

way in which nucleosomes interact with one another is determined by the nature of the chemical modifications on their histone proteins. Chemical modification is therefore a means of modifying a protein's activity, possibly to change its function in a subtle but important way, as with histone modification, or possibly to activate a protein whose function is needed only at particular times.

A more complex type of modification, found predominantly in eukaryotes, is **glycosylation**, the attachment of large carbohydrate side chains to polypeptides. There are two general types of glycosylation (Figure 8.21). ***O*-linked glycosylation** is the attachment of a sugar side chain via the hydroxyl group of a serine or threonine amino acid, and ***N*-linked glycosylation** involves attachment through the amino group on the side chain of asparagine.

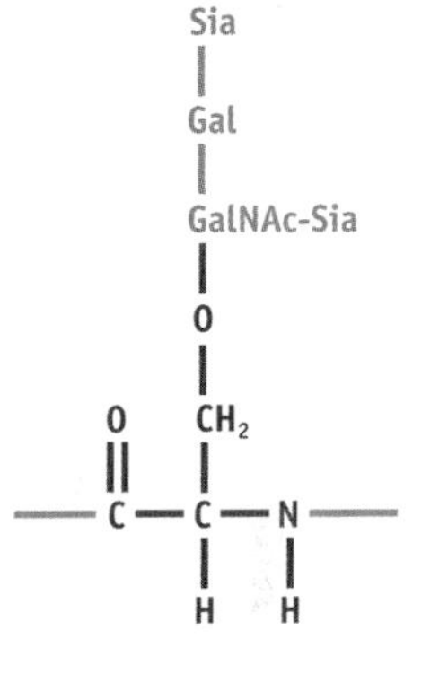

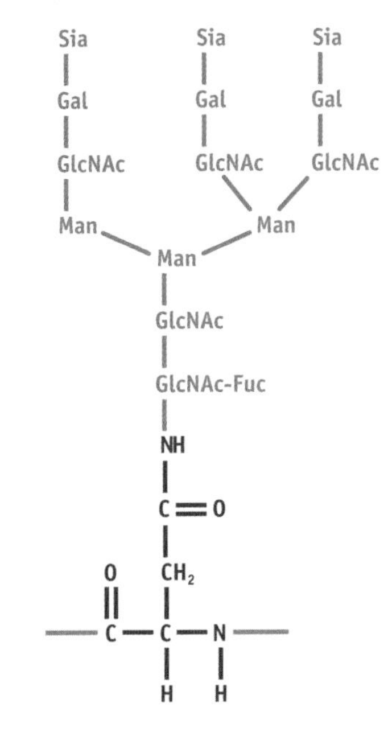

FIGURE 8.21 Glycosylation. (A) *O*-linked glycosylation. The structure shown is found in a number of glycosylated proteins. It is drawn here attached to a serine amino acid, but it can also be linked to a threonine. (B) *N*-linked glycosylation usually results in larger sugar structures than are seen with *O*-linked glycosylation. The drawing shows a typical example of a complex structure attached to an asparagine amino acid. Fuc, fucose; Gal, galactose; GalNAc, *N*-acetylgalactosamine; GlcNAc, *N*-acetylglucosamine; Man, mannose; Sia, sialic acid.

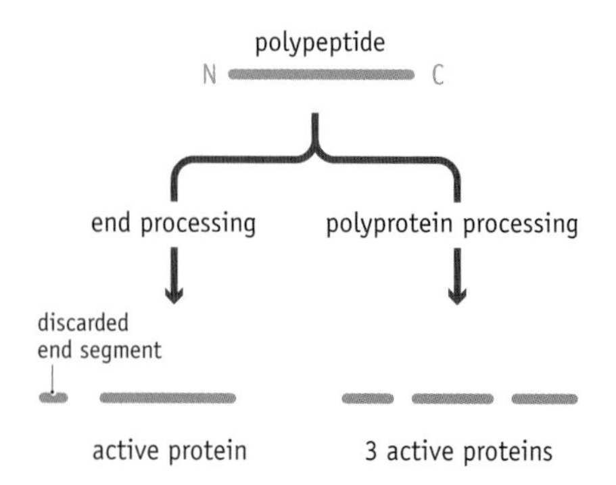

FIGURE 8.22 Two different types of protein processing by proteolytic cleavage. On the left, the protein is processed by removal of the N-terminal segment. C-terminal processing also occurs with some proteins. On the right, a polyprotein is processed to give three different proteins. Not all proteins undergo proteolytic cleavage.

In some eukaryotes, glycosylation can result in attachment to the protein of large structures comprising branched networks of 10 to 20 sugar units of various types. These side chains help to target proteins to particular sites in cells and to increase the stability of proteins circulating in the bloodstream. Another type of large-scale modification involves attachment of long-chain lipids, often to serine or cysteine amino acids. This process is called **acylation** and occurs with many proteins that become associated with membranes.

Some proteins are processed by proteolytic cleavage

Proteolytic cleavage of proteins is relatively common in eukaryotes but less frequent in bacteria. It has two functions (Figure 8.22). The more frequent of these is to remove short pieces from the N- and/or C-terminal regions of polypeptides, leaving a single shortened molecule that folds into the active protein. Less commonly, proteolytic cleavage is used to cut a **polyprotein** into segments, all or some of which are active proteins. As with chemical modification, these cleavage events are carried out in a specific manner and, in many cases, are used to activate a protein or, less frequently, to change its function.

The first type of proteolytic processing—removal of N- or C-terminal segments—is common with secreted polypeptides whose biochemical activities might be deleterious to the cell producing the protein. An example is provided by melittin, the most abundant protein in bee venom and the one responsible for causing cell lysis after injection of the bee sting into the person or animal being stung. Melittin lyses cells in bees as well as animals and so must initially be synthesized as an inactive precursor. This precursor, promelittin, has 22 additional amino acids at its N terminus. The presequence is removed by an extracellular protease that cuts it at 11 positions, releasing the active venom protein. The protease does not cleave within the active sequence, because its mode of action is to release dipeptides with the sequence X–Y, where X is alanine, aspartic acid, or glutamic acid and Y is alanine or proline, motifs that do not occur in the active sequence (Figure 8.23).

A similar type of processing occurs with insulin, the protein made in the islets of Langerhans in the vertebrate pancreas and responsible for controlling blood sugar levels. Insulin is synthesized as preproinsulin, which is 110 amino acids in length (Figure 8.24). The processing pathway involves the removal of the first 24 amino acids to give proinsulin, followed by two

cut sites

APEPEPAPEPEAEADAEADPEAGIGAVLKVLTTGLPALISWIKRKRQQG

FIGURE 8.23 Processing of promelittin, the bee-sting venom. Arrows indicate the cut sites.

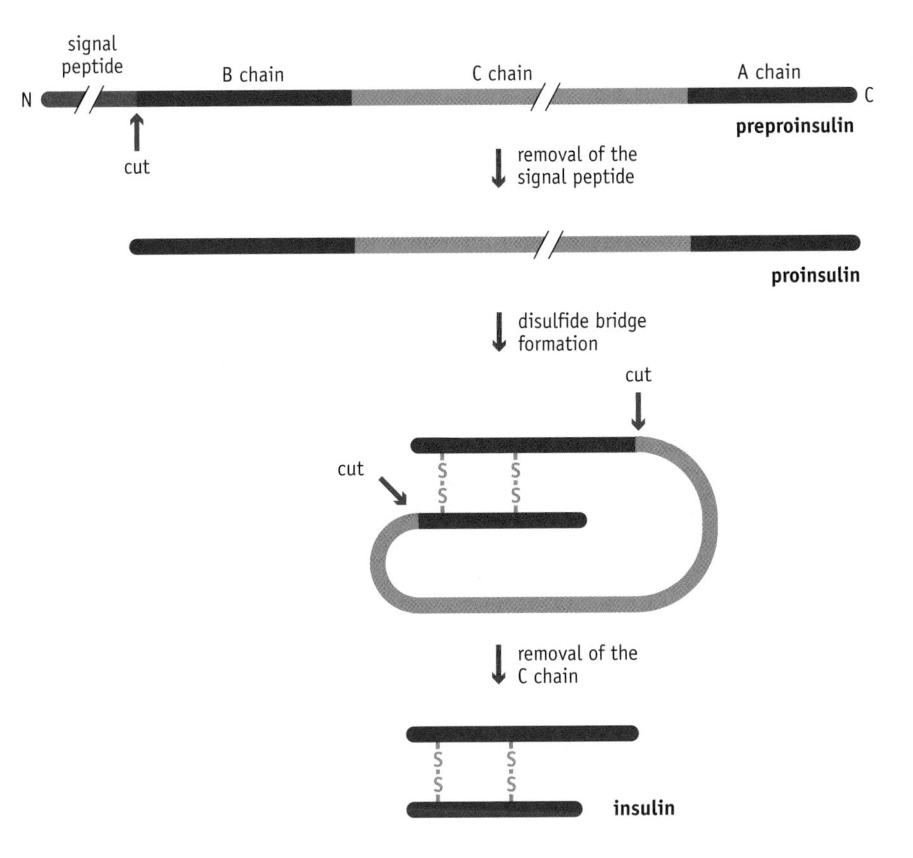

FIGURE 8.24 Processing of preproinsulin. The first 24 amino acids are removed to give proinsulin, which then forms two disulfide bridges. Two additional cuts remove the central segment, to form mature insulin.

additional cuts which excise a central segment. This leaves the two active parts of the protein, the A and B chains, linked together by two disulfide bridges to form mature insulin. The first segment to be removed, the 24 amino acids from the N terminus, is a **signal peptide**, a highly hydrophobic stretch of amino acids that attaches the precursor protein to a membrane prior to transport across that membrane and out of the cell. Signal peptides are commonly found on proteins that bind to and/or cross membranes, in both eukaryotes and prokaryotes.

Polyproteins are long polypeptides that contain a series of mature proteins linked together in head-to-tail fashion. Cleavage of the polyprotein releases the individual proteins, which may have very different functions from one another. Polyproteins are not uncommon in eukaryotes. Several types of virus that infect eukaryotic cells use them as a way of reducing the sizes of their genomes, a single polyprotein gene taking up less space than a series of individual genes. Polyproteins are also involved in the synthesis of peptide hormones in vertebrates. For example, the polyprotein called pro-opiomelanocortin, made in the pituitary gland, contains at least 10 different peptide hormones. These are released by proteolytic cleavage of the polyprotein (Figure 8.25), but not all can be produced at once because of overlaps between individual peptide sequences. Instead, the exact cleavage pattern is different in different cells.

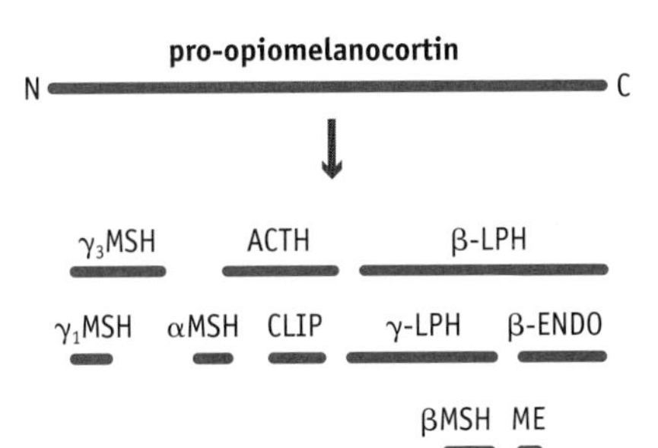

FIGURE 8.25 Processing of the pro-opiomelanocortin polyprotein. ACTH, adrenocorticotropic hormone; CLIP, corticotropin-like intermediate lobe protein; ENDO, endorphin; LPH, lipotropin; ME, met-encephalin; MSH, melanotropin.

8.3 PROTEIN DEGRADATION

The protein synthesis and processing events that we have studied so far in this chapter result in new active proteins that take their place in the cell's proteome. These proteins either replace existing ones that have reached the end of their working lives or provide new protein functions in response to the changing requirements of the cell. But the proteome cannot simply accumulate new protein functions over time. It must also lose those proteins whose functions are no longer needed (Figure 8.26). The removal must be highly selective so that only the correct proteins are degraded. It must also be rapid because under certain conditions the composition of the proteome can change very quickly.

The equivalent need for specific degradation of mRNAs from the transcriptome has been partially explained by the recent discovery of microRNAs. Protein degradation has not yet experienced a similar "eureka" moment when a new discovery helps to explain a long-standing mystery. We do not yet know how specific proteins are targeted for removal from a proteome. Our overview of this topic is therefore restricted to those general processes for protein breakdown that have been described.

There are various protein degradation pathways, but it is not clear how specific proteins are targeted

For many years, protein degradation was an unfashionable subject, and it was not until the 1990s that real progress was made in understanding how protein turnover is linked with changes in cellular activity. Even now, our knowledge centers largely on descriptions of general protein breakdown pathways and less on regulation of those pathways and the mechanisms used to target specific proteins. There appear to be a number of different types of breakdown pathways whose interconnectivities have not yet been traced. This is particularly true in bacteria, which seem to have a range of proteases that work together in controlled degradation of proteins. In eukaryotes, most breakdown involves a single system, involving **ubiquitin** and the **proteasome**.

A link between ubiquitin and protein degradation was first established when it was shown that this abundant 76–amino acid protein is involved in energy-dependent proteolysis reactions in rabbit cells. Subsequent research identified a series of three enzymes that attach ubiquitin molecules, singly or in chains, to lysine amino acids in proteins that subsequently get broken down. Whether or not a protein becomes ubiquitinated depends on the presence or absence within it of amino acid motifs that act as degradation-susceptibility signals. These signals have not been completely characterized, but they include the **N-degron**, a sequence element present at the N terminus of a protein, and **PEST sequences**, internal sequences that are rich in proline (P), glutamic acid (E), serine (S), and threonine (T).

The N-degron and PEST sequences are permanent features of the proteins that contain them and so cannot be straightforward "degradation signals."

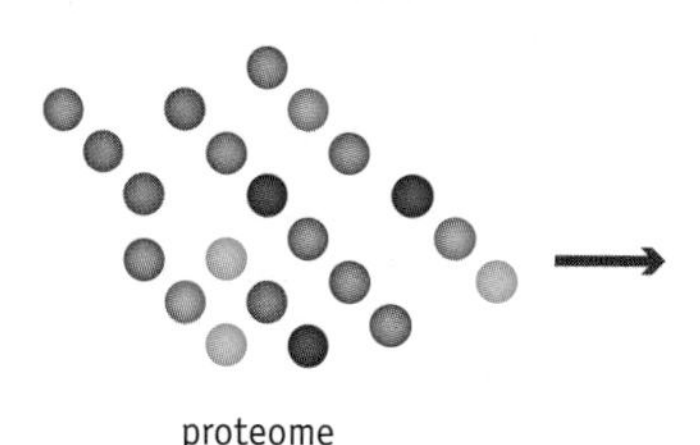

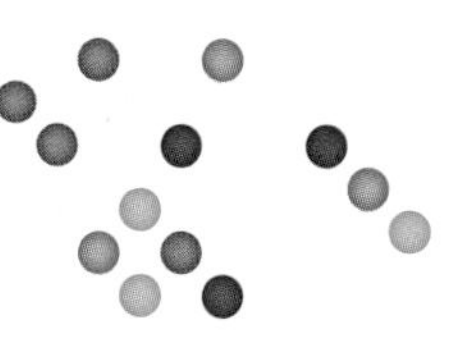

FIGURE 8.26 Protein degradation must be selective so only those proteins whose functions are no longer needed are removed from a proteome. In this example, only the protein represented by the red dots is removed from the proteome.

If they were then these proteins would be broken down as soon as they were synthesized. Instead, they must determine susceptibility to degradation and hence the general stability of a protein in the cell. How this might be linked to the controlled breakdown of selected proteins at specific times is not yet clear.

The second component of the ubiquitin-dependent degradation pathway is the proteasome, the structure within which ubiquitinated proteins are broken down. In eukaryotes the proteasome is a large, multisubunit structure comprising a hollow cylinder and two "caps" (Figure 8.27). The entrance into the cavity within the proteasome is narrow, and a protein must be unfolded before it can enter. This unfolding probably occurs through an energy-dependent process and may involve structures similar to chaperonins, but with unfolding rather than folding activity. After unfolding, the protein can enter the proteasome, within which it is cleaved into short peptides 4 to 10 amino acids in length. These are released back into the cytoplasm, where they are broken down into individual amino acids that can be reutilized in protein synthesis.

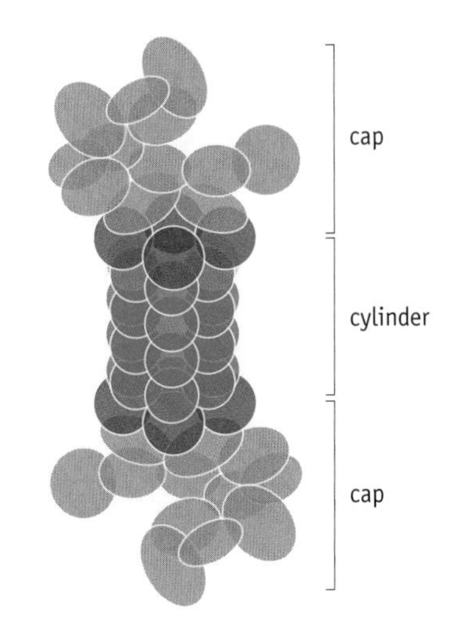

FIGURE 8.27 The eukaryotic proteasome. The protein components of the two caps are shown in brown, orange, and red, and those forming the cylinder in blue.

KEY CONCEPTS

- The collection of proteins in a cell is called the proteome. Within the proteome, the identity and relative abundance of individual proteins represent a balance between the synthesis of new proteins and the degradation of existing ones.
- During translation, the information contained in the nucleotide sequence of an mRNA directs synthesis of a protein. This process follows the rules laid down by the genetic code, with tRNAs enforcing these rules.
- In bacteria the translation initiation complex is built up directly at the initiation codon, the point within the mRNA at which protein synthesis will begin. In eukaryotes the equivalent complex makes its initial attachment at the 5′ end of the mRNA and then scans along the sequence until it locates the initiation codon.
- During the elongation phase of translation, the protein is synthesized step by step, individual amino acids being attached to the carboxyl terminus of the growing polypeptide.
- Protein synthesis ends when the ribosome reaches one of the three termination codons. There are no tRNA molecules with anticodons able to base-pair with any of the termination codons, and instead a protein release factor enters the ribosome in order to terminate translation.
- Utilization of the biological information contained in a gene requires synthesis of an active protein, one able to perform its function in the cell. The initial products of translation are not active proteins. Protein processing is therefore an integral component of the gene expression pathway.
- All proteins must be folded into their correct three-dimensional structures before becoming active in the cell.
- Some proteins also undergo chemical modification, by attachment of new chemical groups. Chemical modification might activate an inactive protein, or change the function of an already active protein.
- Some proteins are also processed by cutting events that may remove segments from one or both ends of the polypeptide, resulting in a

shortened form of the protein, or may break up the polypeptide into a number of different segments, all or some of which are active.

- Pathways for protein breakdown have been identified, but little is known about how those pathways are regulated. The mechanisms used to target specific proteins for degradation have not yet been described.

QUESTIONS AND PROBLEMS (Answers can be found at www.garlandscience.com/introgenetics)

Key Terms

Write short definitions of the following terms:

acylation
aminoacyl site
cap binding complex
chaperonin
elongation factor
exit site
expression proteomics
folding pathway
glycosylation
Hsp70 chaperone
initiation complex
initiation factor
internal ribosome entry site
isoelectric focusing
isoelectric point
isotope-coded affinity tag
Kozak consensus
mass spectrometry
matrix-assisted laser desorption ionization time of flight
molecular chaperone
N-degron
N-linked glycosylation
O-linked glycosylation
peptide mass fingerprinting
peptidyl site
peptidyl transferase
PEST sequence
polyprotein
polysome
pre-initiation complex
protease
proteasome
protein electrophoresis
protein profiling
proteome
proteomics
release factor
ribosome binding site
ribosome recycling factor
scanning
Shine–Dalgarno sequence
signal peptide
translocation
two-dimensional gel electrophoresis
ubiquitin

Self-study Questions

8.1 What is the proteome?

8.2 Describe how the components of a proteome can be studied by protein profiling. What methods are available for making accurate comparisons between the proteomes of two different tissues?

8.3 Outline the events involved in formation of the initiation complex during protein synthesis in *E. coli*.

8.4 What is the importance of *N*-formylmethionine during initiation of protein synthesis in *E. coli*?

8.5 Describe the role of the initiation factors during protein synthesis in *E. coli*.

8.6 Distinguish between the eukaryotic structures called the pre-initiation complex and the initiation complex.

8.7 Explain how the initiation complex is able to ignore certain AUG codons when it scans along a eukaryotic mRNA, and how it identifies the correct initiation codon.

8.8 What is thought to be the role of the poly(A) tail during initiation of protein synthesis in eukaryotes?

8.9 Briefly describe the roles of initiation factors in protein synthesis in eukaryotes.

8.10 How are eukaryotic mRNAs that lack a cap structure translated?

8.11 Give a detailed description of the elongation phase of translation in bacteria and eukaryotes.

8.12 What are the roles of the P, A, and E sites during protein synthesis in *E. coli*? Which of these sites is absent in a eukaryotic ribosome?

8.13 Outline the roles of the three release factors and the ribosome recycling factor during termination of translation in *E. coli*. Which proteins play the equivalent roles during termination in eukaryotes?

8.14 Describe the experiments that showed that a small protein such as ribonuclease can fold spontaneously *in vitro*. Why are larger proteins unable to fold spontaneously?

8.15 Distinguish between the activities of Hsp70 chaperones and chaperonins in protein folding.

8.16 Construct a table that lists, with examples, the various types of post-translational chemical modification that occur with different proteins.

8.17 What is glycosylation? Your answer should distinguish between *O*- and *N*-linked processes.

8.18 Outline the role of proteolytic processing in the synthesis of insulin.

8.19 Give an example of proteolytic processing of a polyprotein.

8.20 Describe the processes thought to be responsible for protein degradation in eukaryotes.

Discussion Topics

8.21 Speculate on the reasons why the poly(A) tail of a eukaryotic mRNA is involved in initiation of translation of the mRNA.

8.22 Bacterial and some eukaryotic mRNAs have internal recognition sequences to which the small subunit of the ribosome attaches in order to initiate protein synthesis. If this relatively straightforward process is suitable for some mRNAs, then why should initiation of most eukaryotic mRNAs have to involve the elaborate scanning mechanism?

8.23 Discuss the extent to which the ribosome is an active or a passive partner in protein synthesis.

8.24 Devise an experiment to demonstrate that peptidyl transferase is a ribozyme.

8.25 To what extent are protein folding studies that are conducted in the test tube good models for protein folding in living cells?

8.26 Using the current information on protein degradation, devise a hypothesis to explain how specific proteins could be individually degraded. Can your hypothesis be tested?

FURTHER READING

Caskey CT (1980) Peptide chain termination. *Trends Biochem. Sci.* 5, 234–237.

Clark B (1980) The elongation step of protein biosynthesis. *Trends Biochem. Sci.* 5, 207–210.

Frydman J (2001) Folding of newly translated proteins *in vivo*: the role of molecular chaperones. *Annu. Rev. Biochem.* 70, 603–649.

Görg A, Weiss W & Dunn MJ (2004) Current two-dimensional electrophoresis technology for proteomics. *Proteomics* 4, 3665–3685.

Hellen CUT & Sarnow P (2001) Internal ribosome entry sites in eukaryotic mRNA molecules. *Genes Dev.* 15, 1593–1612.

Hunt T (1980) The initiation of protein synthesis. *Trends Biochem. Sci.* 5, 178–181.

Jackson RJ, Hellen CUT & Pestova TV (2010) The mechanism of eukaryotic translation initiation and principles of its regulation. *Nat. Rev. Mol. Cell Biol.* 11, 113–127.

Kapp LD & Lorsch JR (2004) The molecular mechanics of eukaryotic translation. *Annu. Rev. Biochem.* 73, 657–704.

Schmeing TM & Ramakrishnan V (2009) What recent ribosome structures have revealed about the mechanism of translation. *Nature* 461, 1234–1242.

Steitz TA & Moore PB (2003) RNA, the first macromolecular catalyst: the ribosome is a ribozyme. *Trends Biochem. Sci.* 28, 411–418. *Describes the evidence that led to the discovery that peptidyl transferase is a ribozyme.*

Varshavsky A (1997) The ubiquitin system. *Trends Biochem. Sci.* 22, 383–387. *Protein degradation.*

Voges D, Zwickl P & Baumeister W (1999) The 26S proteasome: a molecular machine designed for controlled proteolysis. *Annu. Rev. Biochem.* 68, 1015–1068. *More on protein degradation.*

Control of Gene Expression CHAPTER 9

In the previous five chapters we followed the gene expression pathway from the initiation of transcription of a gene through the synthesis, processing, and eventual degradation of its protein product. Now that we understand how the biological information contained in a gene is utilized by the cell, we must look back over the process and examine how gene expression is controlled.

We have already acknowledged that the rate at which a gene is expressed can change and that many genes might be completely switched off at a particular time. Knowing this prompted us to include mRNA and protein degradation within our study of gene expression, as we realized that degradation was needed in order for the composition of the transcriptome and proteome to respond to changes in the rate of expression of individual genes. We have also looked at specific mechanisms for altering the expression pathway of a gene, for example, through use of alternative promoters, by changing the way in which an mRNA is spliced, or by RNA editing. But so far we have only scratched the surface of gene regulation.

First we must ask why it is necessary to control gene expression. In answering this question we will begin to appreciate that the underlying principles of gene regulation are the same in all organisms, but that the control processes are more sophisticated in eukaryotes compared with bacteria. The difference reflects the greater need that eukaryotes, particularly multicellular ones such as humans, have for gene regulation during differentiation and development.

Next we will attempt to distinguish between those regulatory events that determine whether a gene is switched on or off, those that affect the rate at which the RNA or protein product is synthesized once a gene is switched on, and those that have no effect on rate but alter the nature of the gene product. Having understood the distinctions, we will focus on how the initiation of transcription is regulated in bacteria and eukaryotes, as this is looked on as the key step at which the cell controls which genes are active at any one time. We will examine transcriptional regulation in some detail and discover once again that, although the process is more complex in eukaryotes, the underlying principles are the same in all types of organism.

Finally we will survey the most important of the control mechanisms that target steps in gene expression downstream of transcription initiation, asking how these mechanisms work and what influence they have on synthesis of the RNA or protein product.

Throughout the chapter, our attention will be specifically on the regulation of individual genes. This is a narrow focus but one that is necessary in order to deal with what is a complex topic in an accessible way. In Chapter 18 we will place gene regulation in its broader context by examining the genetic basis of differentiation and development.

9.1 THE IMPORTANCE OF GENE REGULATION

The entire complement of genes in a single cell represents a staggering amount of biological information. Some of this information is needed by the cell at all times. For example, most cells continually synthesize ribosomes and so have a continuous requirement for transcription of the rRNA and ribosomal protein genes. Similarly, genes coding for enzymes such as

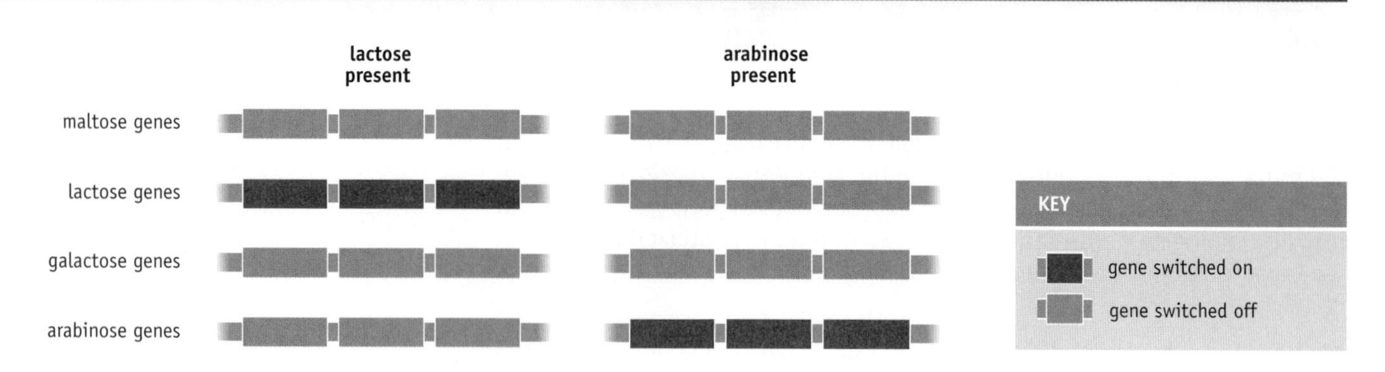

FIGURE 9.1 *E. coli* expresses only the genes coding for those enzymes that it needs to metabolize the sugars present in its environment. For example, if the only sugar present is lactose, then the genes needed for lactose utilization are switched on, and those for utilization of other sugars are switched off. Conversely, when only arabinose is present, the arabinose genes are switched on and the genes for sugars other than arabinose are switched off.

RNA polymerase or those involved in the basic metabolic pathways are active in virtually all cells all of the time. Genes that are active all the time are called **housekeeping genes**, reflecting the role in the cell of the biological information they carry.

Other genes have more specialized roles, and their biological information is needed by the cell only under certain circumstances. All organisms are therefore able to regulate expression of their genes, so that those genes whose RNA or protein products are not needed at a particular time are switched off. This is a straightforward concept but it has extensive ramifications. To illustrate the point we will briefly consider some examples of gene regulation in bacteria and in eukaryotes.

Gene regulation enables bacteria to respond to changes in their environment

Even the simplest bacterium is able to control the expression of its genes. By doing so, it is able to respond to changes in its environment.

The response of bacteria to environmental changes was originally studied with cultures of *Escherichia coli* grown in the laboratory. A culture medium must provide a balanced mixture of nutrients, including compounds that can be metabolized to release the energy that the bacteria need to power their cellular processes. Usually this energy is supplied in the form of sugar. Most bacteria are not particularly fussy about what type of sugar is provided—*E. coli* can metabolize glucose, maltose, lactose, galactose, arabinose, or any of several others. A different set of enzymes is needed to release the energy from each of these compounds, but *E. coli* has genes coding for all of these enzymes.

Exactly which enzymes need to be present in an individual bacterium depends on which sugars are present in the medium. Each bacterium could continuously express all its sugar-utilizing genes, and so have molecules of each enzyme available all the time, but this would waste energy, something all organisms try to avoid. Why synthesize enzymes that act on a sugar that is not available? Instead, *E. coli* expresses only those genes coding for the enzymes it needs to metabolize the sugars that are present. The genes for all the other enzymes are inactive, switched off because their gene products are not required (Figure 9.1).

What if the growth medium is changed by adding a new sugar? Suppose, for example, the original sugar in a medium is glucose. Eventually this is entirely consumed by the bacteria so, to replace it, some lactose is added. Now *E. coli* quickly switches on expression of the genes for lactose metabolism, and switches off those for glucose, which have become redundant. So by regulating expression of its genes, *E. coli* is able to respond quickly to changes in the growth medium without wasting energy by maintaining in an active state genes whose products are of no immediate use.

Gene regulation in eukaryotes must be responsive to more sophisticated demands

At the most basic level, gene regulation achieves the same thing in both bacteria and eukaryotes. It enables a cell to change its biochemical capabilities. The difference is that with eukaryotes we see a greater degree of sophistication with respect to both the signals that influence gene expression and the impact that gene regulation has on the organism.

Eukaryotic cells, just like bacteria, can alter their gene expression patterns in response to changes in their environment. For example, the yeast *Saccharomyces cerevisiae* regulates its genes for sugar utilization enzymes in a manner analogous to *E. coli*. Plant cells switch on genes for photosynthetic proteins in response to light (Figure 9.2A). In a multicellular eukaryote, individual cells, or groups of cells, also respond to stimuli that originate from within the organism. Hormones, growth factors, and other regulatory molecules are produced by one type of cell and cause changes

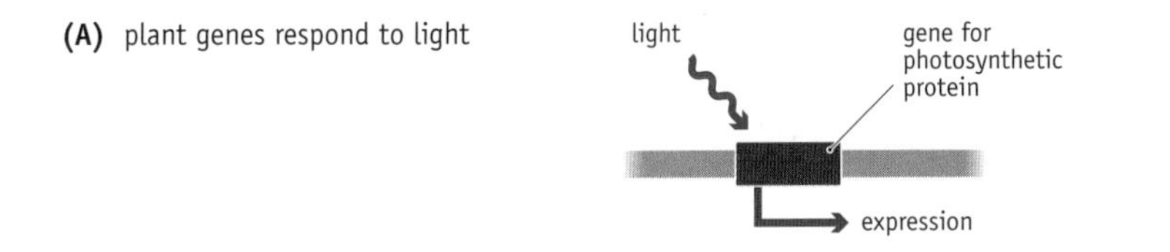

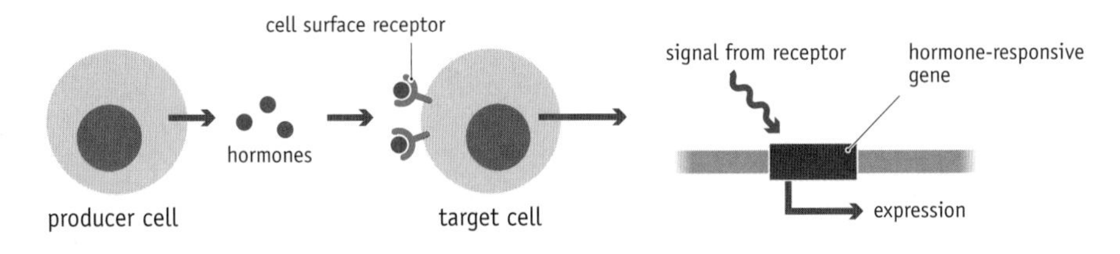

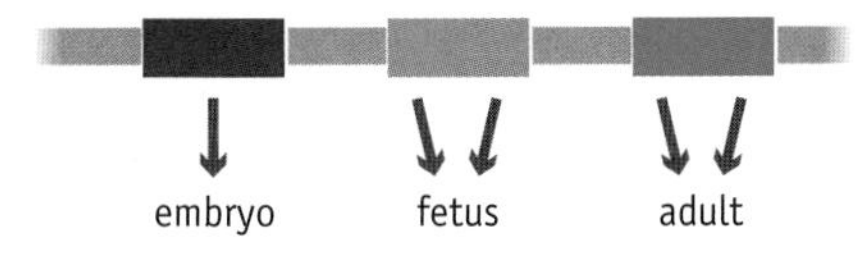

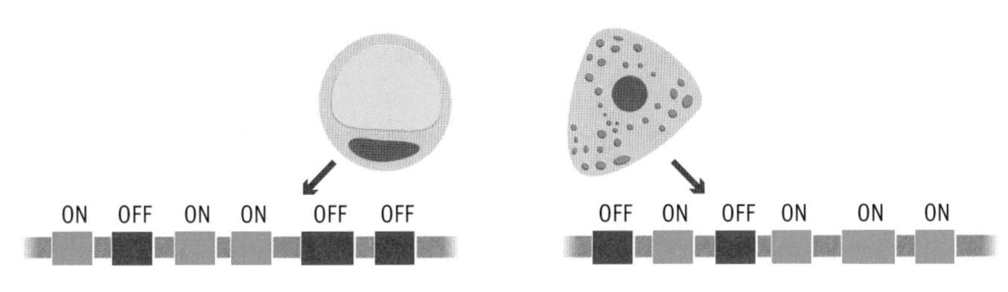

FIGURE 9.2 Examples of gene regulation in multicellular eukaryotes. (A) Some genes in plants are activated by exposure to light. (B) Hormones, growth factors, and other regulatory molecules are produced by one type of cell and cause changes in gene expression in other cells. (C) Some genes are subject to developmental regulation and are expressed at different stages of the life cycle. (D) Different sets of genes are expressed in different types of specialized cell.

in gene expression in other cells (Figure 9.2B). In this way the activities of groups of cells can be coordinated in a manner that is beneficial to the organism as a whole.

Besides these transient changes in gene expression, most eukaryotes (as well as some bacteria) are able to switch genes on and off at different stages of development. An example is provided by the human globin genes, different members of the α and β gene families being expressed in the embryo, fetus, and adult (see Figure 3.20). The distinctive biochemical properties of the resulting globin proteins are matched to the specific physiological roles that hemoglobin plays at these different phases in human development. Developmental regulation therefore enables the biochemistry of an organism to be altered in response to the particular requirements of each stage of its life cycle (Figure 9.2C).

Gene regulation in multicellular eukaryotes is also the key to cell specialization. In humans there are over 400 specialized cell types, each with a different morphology and biochemistry. The distinctions between these cell types are the result of differences in gene expression patterns (Figure 9.2D). A liver cell is very different from a muscle cell because it expresses a different set of genes. These gene expression patterns are permanent—a liver cell can never become a muscle cell—and are laid down very early in development.

The outcomes of gene regulation are therefore more sophisticated in eukaryotes, especially multicellular ones like humans, than they are in bacteria. Now we must move on to consider the various ways in which expression of genes can be controlled. Are these more sophisticated in eukaryotes also?

The underlying principles of gene regulation are the same in all organisms

Although gene expression responds to a wider range of regulatory signals in eukaryotes, and the control processes have more sophisticated outcomes, the underlying principles of gene regulation are the same in all organisms. The difference is in the detail.

Control of gene expression is, in essence, control over the amount of gene product present in the cell. This amount is a balance between the rate of synthesis (how many molecules of the gene product are made per unit time) and the degradation rate (how many molecules are broken down per unit time; Figure 9.3). The result of this balance is a different steady-state concentration for each gene product in the cell. If either the synthesis rate or the degradation rate changes, then the steady-state concentration also changes. Although we have already acknowledged the importance of

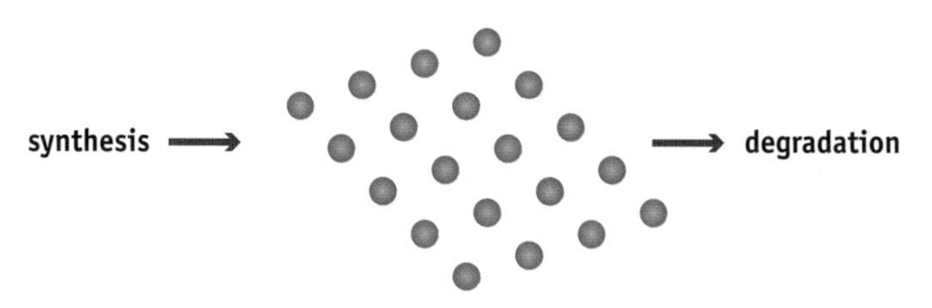

FIGURE 9.3 The amount of a gene product that is present in a cell is a balance between the rate of synthesis and the rate of degradation.

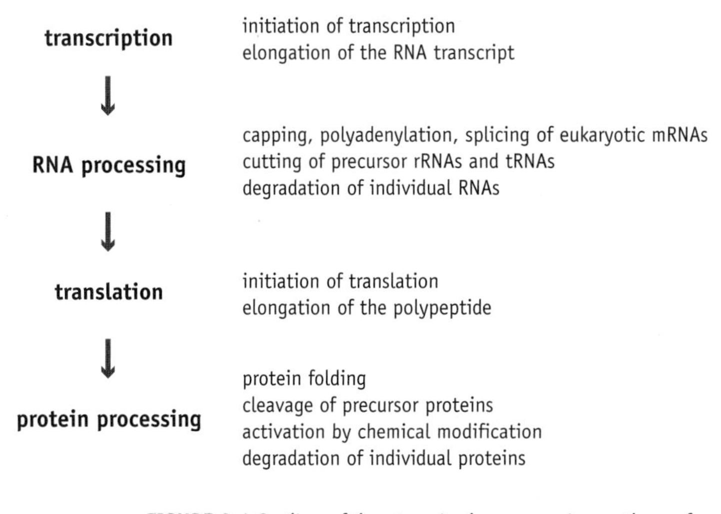

FIGURE 9.4 Outline of the steps in the expression pathway for a eukaryotic gene at which control over synthesis of the gene product can be exerted.

mRNA and protein degradation in the gene expression pathway, geneticists at present believe that the critical variations in the steady state of a gene product arise because of changes in its rate of synthesis. This is true both for proteins and noncoding RNA.

How can the synthesis rate of a gene product be regulated? The answer is by exerting control over any one, or a combination, of the various steps in the gene expression pathway (Figure 9.4). The possibilities include, first of all, transcription. If the number of transcripts synthesized per unit time changes, then the amount of the gene product that is synthesized also changes. Increase the transcription rate and there will be more gene product. Decrease the transcription rate and there will be less gene product. Moving along the pathway, control could also be exerted over RNA processing events. These include not just splicing of mRNA but also capping and polyadenylation, both of which are prerequisites for translation, as well as the various processing events involved in synthesis of functional noncoding RNAs. Slow down one or more of these processing events and product synthesis will fall. Finally, regulation could be exerted during protein synthesis, by controlling either the initial attachment of ribosomes to an mRNA or the rate at which individual ribosomes translocate along an mRNA while making a protein, or by regulating one or more of the various post-translational processing events.

It is becoming clear that organisms use many, possibly all, of the above strategies for controlling expression of their genes. In this regard, eukaryotes do display greater complexity than bacteria simply because the gene expression pathway in eukaryotes has more steps at which regulation could be exerted. Control over mRNA processing, for example, is possible only in eukaryotes because mRNA is not processed in bacteria. But when we consider gene regulation we must be careful to distinguish between the details and the underlying principles. The principle of gene regulation is the need to control synthesis of RNA and protein products by switching genes on and off, and to modulate the rate of expression of those genes that are on. At this primary level, eukaryotes and bacteria are the same because in all organisms the key events that determine whether a gene is on or off take place during the initiation of transcription. Later events in the gene expression pathway are able to change the rate of expression of a gene that is switched on and, in eukaryotes, control the synthesis of

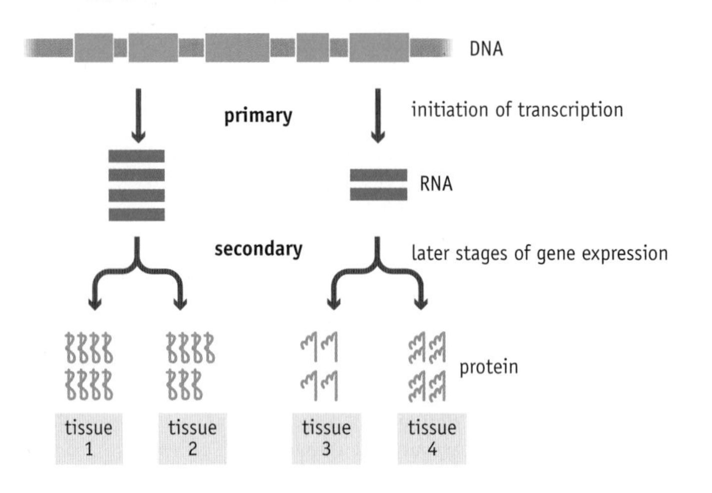

FIGURE 9.5 Initiation of transcription is the primary control point that determines whether a gene is switched on or off. Later events influence the amount of gene product that is made, as illustrated by tissues 1 and 2, or its identity, as in tissues 3 and 4.

alternative products of a single gene (Figure 9.5). But initiation of transcription is the primary control point, and it is this event that we must study in detail in order to understand how gene expression is regulated in bacteria and eukaryotes.

9.2 REGULATION OF TRANSCRIPTION INITIATION IN BACTERIA

The foundation of our understanding of how bacteria regulate transcription of their genes was laid by François Jacob and Jacques Monod in a paper, published in 1961, that is now considered a classic example of experimental analysis and deductive reasoning. Jacob and Monod based their propositions on an intricate genetic analysis of lactose utilization by *E. coli* and described an elegant regulatory system that has subsequently been confirmed in virtually every detail. We will use the lactose system as our central example of how gene regulation works in bacteria.

Four genes are involved in lactose utilization by *E. coli*

Lactose is a disaccharide sugar composed of a single glucose unit attached to a single galactose unit (Figure 9.6A). In order to utilize lactose as an energy source, an *E. coli* cell must first transport lactose molecules from the extracellular environment into the cell and then split the molecules, by hydrolysis, into glucose and galactose (Figure 9.6B). These reactions are catalyzed by three enzymes, each of which has no function other than lactose utilization. These enzymes are lactose permease, which transports lactose into the cell, β-galactosidase, which is responsible for the splitting reaction, and β-galactoside transacetylase, whose precise role in the process is unknown.

In the absence of lactose only a small number of molecules of each enzyme are present in an *E. coli* cell, probably fewer than five of each. When the bacterium encounters lactose, enzyme synthesis is rapidly induced, and within a few minutes levels of up to 5000 molecules of each enzyme per cell are reached. Induction of the three enzymes is coordinate, meaning that each is induced at the same time and to the same extent. This provides a clue to the arrangement of the relevant genes on the *E. coli* DNA molecule.

The three enzymes involved in lactose utilization are coded by genes called *lacZ* (β-galactosidase), *lacY* (permease), and *lacA* (transacetylase). These

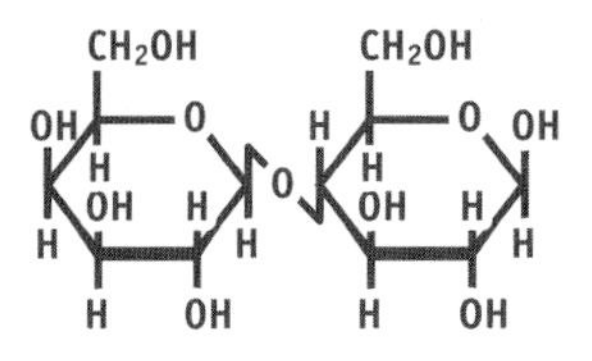

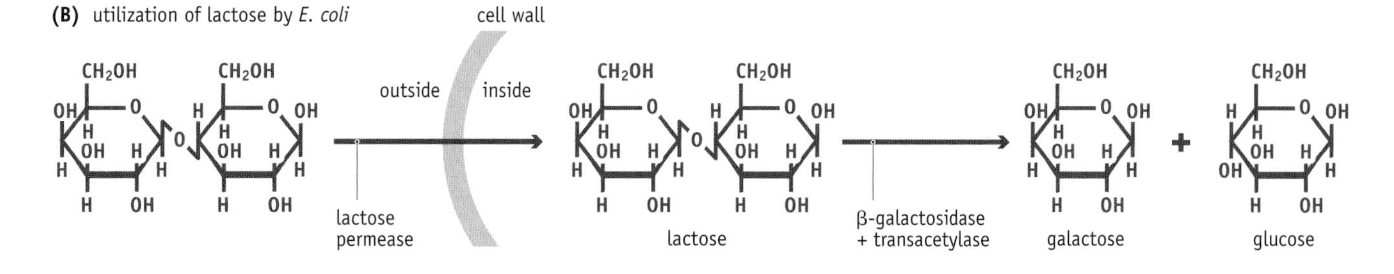

FIGURE 9.6 Lactose and its utilization by *E. coli*. (A) The lactose molecule consists of a glucose unit attached to a galactose unit. (B) In order to utilize lactose as an energy source, the *E. coli* cell must import lactose molecules from the extracellular environment and then split the molecules into their glucose and galactose components.

names follow the convention for genes in *E. coli*, as well as most other organisms, comprising a three-letter abbreviation indicating the function of the gene followed by one or more letters and/or numbers to distinguish between genes of related function. The three genes lie in a cluster with only a very short distance between the end of one gene and the start of the next (Figure 9.7A). There is just one promoter sequence, immediately upstream of *lacZ*—there are no promoters in the small gaps between the genes—and the three genes are transcribed together as a single mRNA molecule. This type of organization is called an **operon**.

Jacob and Monod used genetic analysis techniques to identify *lacZ*, *lacY*, and *lacA*, and to determine their relative positions in the *E. coli* genome. They also discovered an additional gene, which they designated *lacI*. This gene lies just upstream of the lactose gene cluster but is not itself a part of the operon, because it is transcribed from its own promoter and has its own terminator (Figure 9.7B).

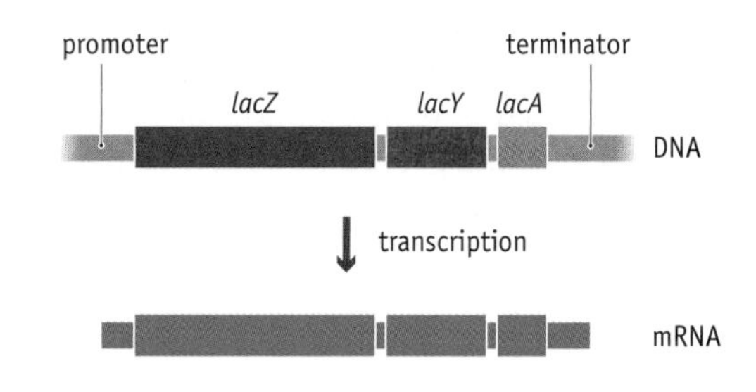

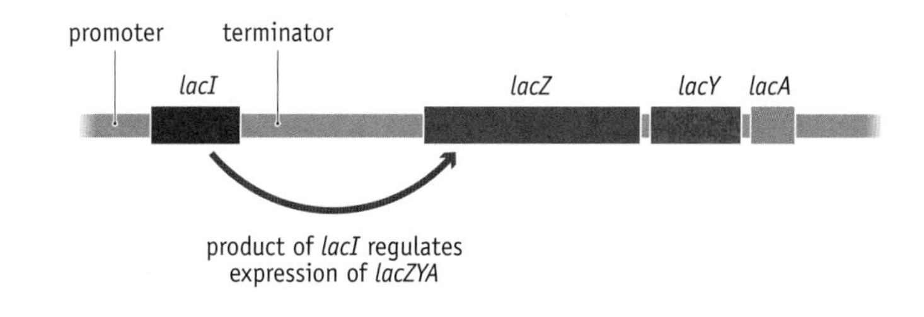

FIGURE 9.7 The genes involved in lactose utilization by *E. coli*. (A) The three genes of the lactose operon are transcribed as a single unit. (B) The fourth gene involved in lactose utilization, *lacI*, has its own promoter and terminator. This gene controls expression of the lactose operon.

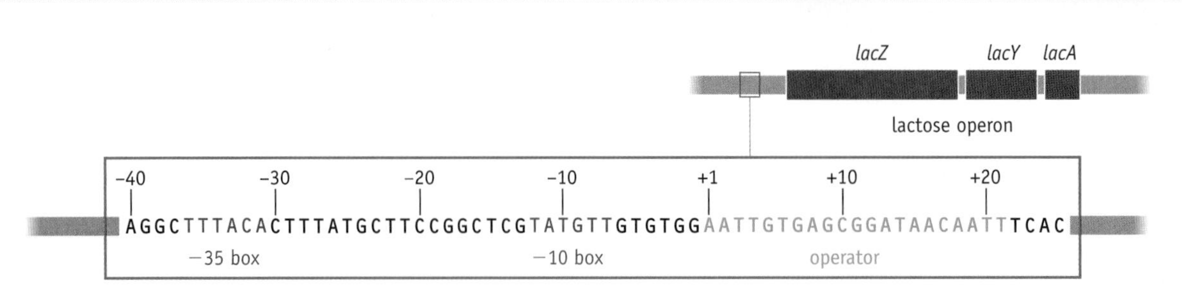

FIGURE 9.8 The DNA sequence immediately upstream of the *lacZ* gene of *E. coli*. The sequence shows the position of the operator relative to the –35 and –10 components of the promoter.

The gene product of *lacI* is intimately involved in lactose utilization but is not an enzyme directly required for the uptake or hydrolysis of the sugar. Instead the *lacI* product regulates the expression of the other three genes. This is evident from the effect of mutations on *lacI*. If *lacI* is inactivated, the lactose operon becomes switched on continuously, even in the absence of lactose. The terminology used by Jacob and Monod is still important today. We refer to *lacZ*, *lacY*, and *lacA* as **structural genes**, because their products contribute to the "biochemical structure" of the cell. In contrast, *lacI* is called a **regulatory gene**, because its function is to control the expression of the other genes.

The regulatory gene codes for a repressor protein

The gene product of *lacI* is a protein that Jacob and Monod called the lactose **repressor**. This protein is able to attach to the *E. coli* DNA molecule at a site between the promoter for the lactose operon and the start of *lacZ*, the first gene in the cluster (Figure 9.8). The attachment site is called the **operator** and was also located by Jacob and Monod by genetic means. The operator in fact overlaps the promoter, so that when the repressor is bound, access to the promoter is blocked so that, in turn, RNA polymerase cannot attach to the DNA and transcription of the lactose operon cannot occur (Figure 9.9). This is what happens if lactose is not available to the cell. To summarize, *if there is no lactose, transcription of the lactose operon does not occur because the promoter is blocked by the repressor.*

Transcription is induced by allolactose. Allolactose is an isomer of lactose and is synthesized as a byproduct during the splitting of lactose into glucose and galactose. When the bacterium encounters a new supply of lactose it takes up a few molecules and splits them. It is able to do this because of the small number of lactose utilization enzymes that are always present. The allolactose that is formed as a byproduct then binds to the repressor, causing a change in the conformation of the protein in such a way that the repressor is no longer able to attach to the operator. The repressor–allolactose complex dissociates from the DNA molecule, enabling RNA polymerase to locate the promoter and begin transcription of the operon (Figure 9.10). This results in synthesis of the much larger numbers of enzyme molecules needed to take up and metabolize the rest of the

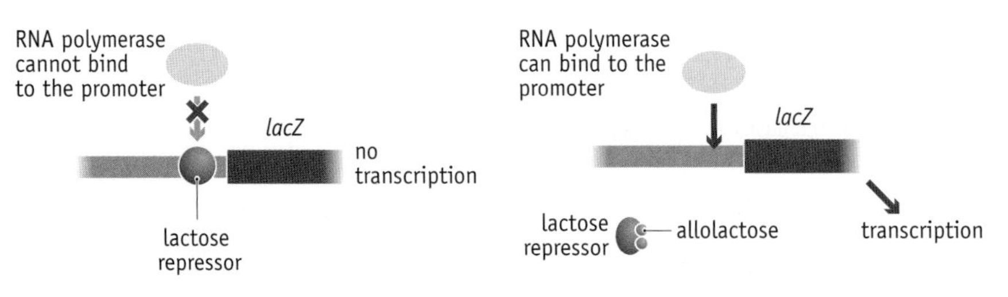

FIGURE 9.9 The RNA polymerase cannot gain access to the promoter when the lactose repressor is bound to the operator.

FIGURE 9.10 The repressor–allolactose complex does not bind to the operator. RNA polymerase can therefore gain access to the promoter.

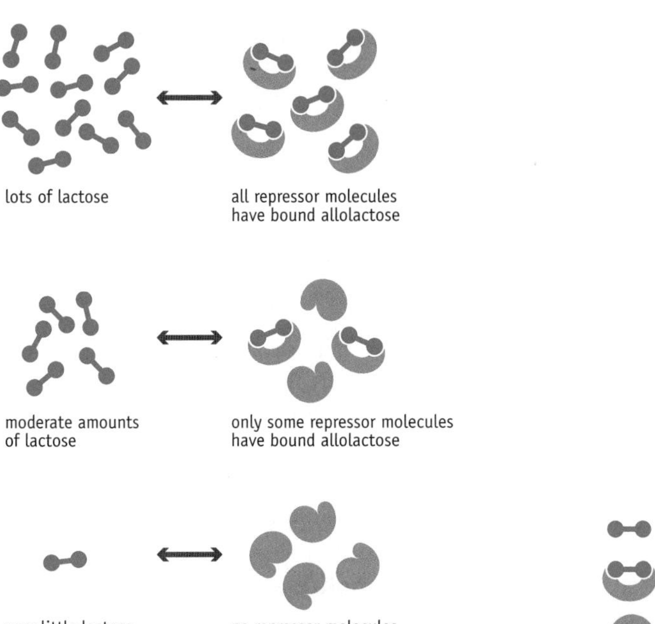

FIGURE 9.11 Repressor–inducer binding is an equilibrium event.

lactose. Allolactose therefore acts as the **inducer** of the operon. *If lactose is present, transcription of the lactose operon occurs as the repressor–inducer complex does not bind to the operator.*

Eventually the lactose utilization enzymes exhaust the available supply of lactose. Repressor–inducer binding is an equilibrium event (Figure 9.11), so when the free lactose concentration decreases, the number of repressor–allolactose complexes also decreases and free repressor molecules start to predominate. These free repressors have regained their original conformation and so can attach once again to the operator. *When the lactose supply is used up, the lactose operon is switched off as the repressor reattaches to the operator.*

Glucose also regulates the lactose operon

The presence or absence of lactose is not the only factor that influences expression of the lactose operon. If a bacterium has a sufficient source of glucose (one of the breakdown products of lactose) for its energy needs, then it does not metabolize lactose even if lactose is also available in the environment. Only when all of the glucose has been used up will the cell start to make use of the lactose (Figure 9.12).

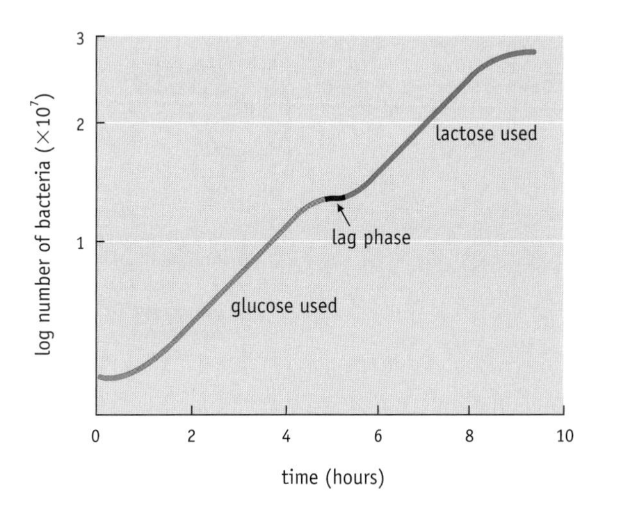

FIGURE 9.12 A typical diauxic growth curve, as seen when *E. coli* is grown in a medium containing a mixture of glucose and lactose. During the first few hours the bacteria divide exponentially, using the glucose as their energy source. When the glucose is used up there is a brief lag period while the lactose genes are switched on before the bacteria return to exponential growth, now using up the lactose. Note that the *y*-axis is a logarithmic scale, and therefore the exponential phases of bacterial growth appear as straight lines on the graph.

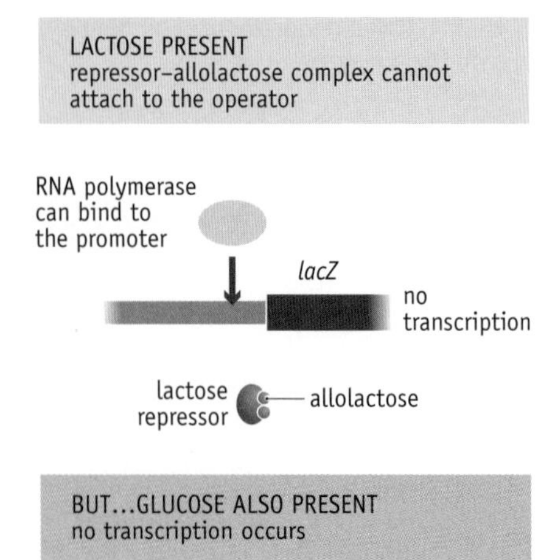

FIGURE 9.13 Glucose overrides the lactose repressor. If lactose is present then the repressor detaches from the operator and the lactose operon should be transcribed, but the operon remains silent if glucose is also present.

This phenomenon, called **diauxie**, was discovered by Monod in 1941. When the details of the lactose operon were worked out some 20 years later it became clear that the diauxie between glucose and lactose must involve a mechanism whereby the presence of glucose can override the inductive effect that lactose usually has on its operon. In the presence of both lactose and glucose, the lactose operon remains switched off, even though some of the lactose in the mixture is converted into allolactose, which binds to the lactose repressor, the situation that under normal circumstances results in transcription of the operon (Figure 9.13).

The explanation for the diauxic response is that glucose prevents expression of the lactose operon, as well as other sugar utilization operons, through an indirect influence on the **catabolite activator protein** (**CAP**). This protein binds to a recognition sequence at various sites in the *E. coli* DNA and activates transcription initiation at downstream promoters, probably by interacting with the σ subunit of the RNA polymerase. Productive initiation of transcription at these promoters is dependent on the presence of bound catabolite activator protein. If the protein is absent then the genes controlled by the promoter are not transcribed.

Glucose does not itself interact with the catabolite activator protein. Instead, glucose controls the level in the cell of the modified nucleotide **cyclic AMP** (**cAMP**; Figure 9.14). It does this by inhibiting the activity of **adenylate cyclase**, the enzyme that synthesizes cAMP from ATP. Inhibition is mediated by a protein called IIA^{Glc}, a component of a multiprotein complex that transports sugars into the bacterium. When glucose is being transported into the cell, IIA^{Glc} becomes modified by removal of phosphate groups that are usually attached to its surface. The modified version of IIA^{Glc} inhibits adenylate cyclase activity. This means that if glucose levels are high, the cAMP content of the cell is low.

The catabolite activator protein can bind to its target sites only in the presence of cAMP, so when glucose is present the protein remains detached and the operons it controls are switched off. In the specific case of diauxie involving glucose plus lactose, the indirect effect of glucose on the catabolite activator protein means that the lactose operon remains inactivated,

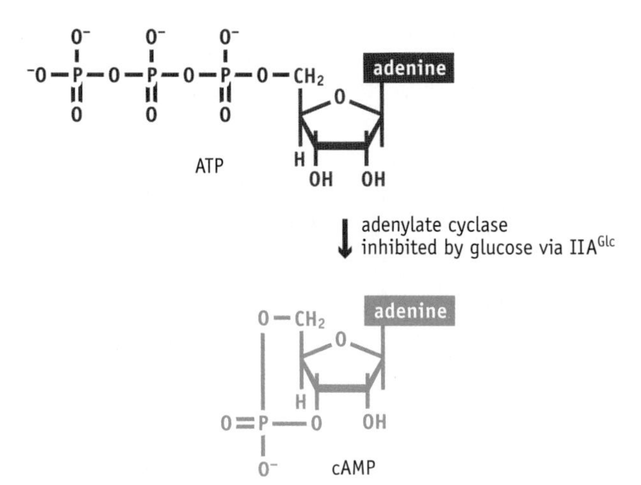

FIGURE 9.14 Glucose controls the level of cAMP in the cell by inhibiting the activity of adenylate cyclase, the enzyme that synthesizes cAMP from ATP. Inhibition is mediated by IIA^{Glc}, a component of the multiprotein complex that transports sugars into the bacterium.

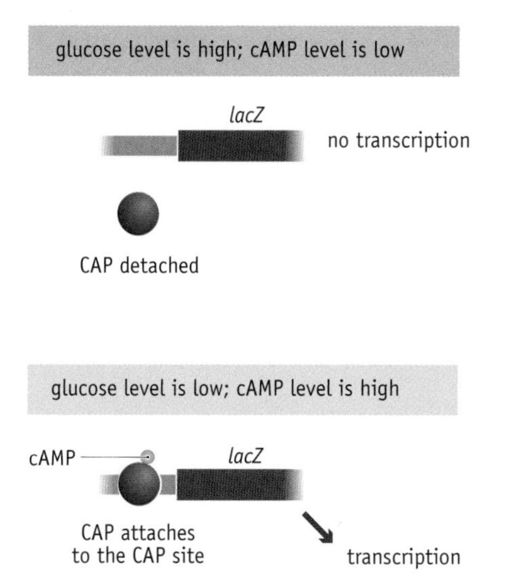

FIGURE 9.15 The catabolite activator protein (CAP) can attach to its DNA binding site only in the presence of cAMP. If glucose is present, the cAMP level is low, so CAP does not bind to the DNA and does not activate the RNA polymerase. Once the glucose has been used up, the cAMP level rises, allowing CAP to bind to the DNA and activate transcription of the lactose operon.

even though the lactose repressor is not bound. The glucose in the medium is therefore used up first. When the glucose is gone, the cAMP level rises and the catabolite activator protein binds to its target sites, including the site upstream of the lactose operon, and transcription of the lactose genes is activated (Figure 9.15).

Operons are common features in prokaryotic genomes

There are almost 600 operons in the *E. coli* genome, each containing two or more genes. Operons are also present in the genomes of most other bacteria and also common in the archaea.

Operons can be broadly divided into two categories. The first type are called **inducible operons** and are typified by the lactose operon. An inducible operon codes for a set of enzymes involved in a metabolic pathway, one that results in breakdown of a substrate compound in the same way that the lactose enzymes break down lactose. The presence of the substrate induces expression of the genes. Other examples of inducible operons are the galactose operon and the arabinose operon.

The second type are called **repressible operons**. These code for enzymes involved in a biosynthetic pathway, one that results in synthesis of a product. The product controls expression of the operon. An example is the tryptophan operon, which is made up of five genes involved in synthesis of the amino acid tryptophan (Figure 9.16). Expression of the operon is controlled by the tryptophan repressor, which attaches to the tryptophan operator and prevents transcription. In this case though, the repressor on

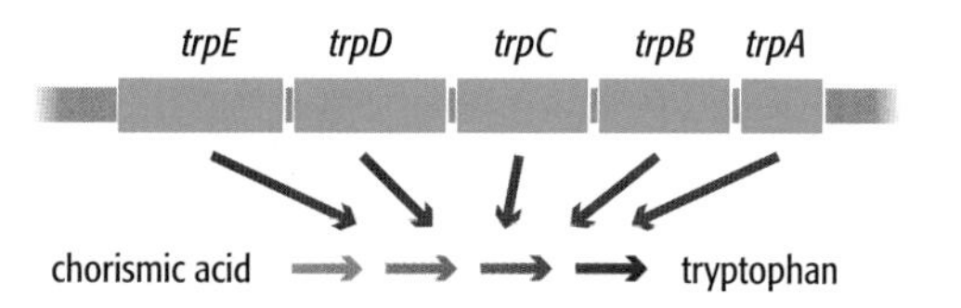

FIGURE 9.16 The tryptophan operon, which contains five genes coding for enzymes involved in the multistep biochemical pathway that converts chorismic acid into the amino acid tryptophan.

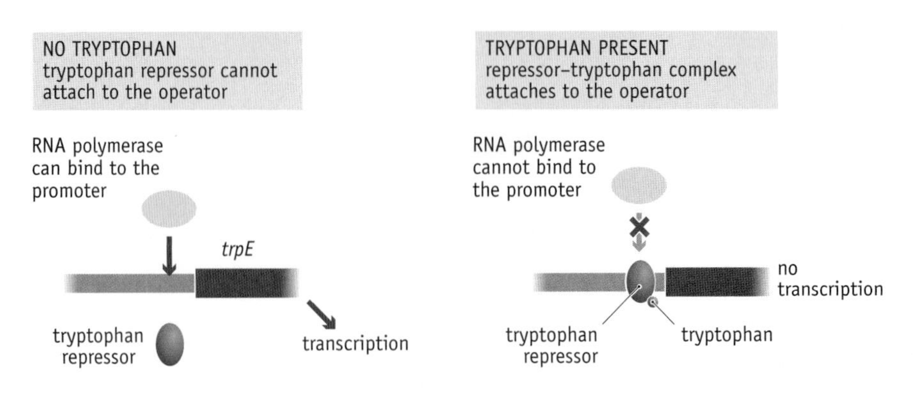

FIGURE 9.17 Regulation of the tryptophan operon of *E. coli*. When tryptophan is present, and so does not need to be synthesized, the operon is switched off because the repressor–tryptophan complex binds to the operator. In the absence of tryptophan the repressor cannot attach to the operator, and the operon is expressed.

its own cannot attach to the operator. Repression of the operon occurs only when the tryptophan repressor binds tryptophan (Figure 9.17). This is of course completely logical, as tryptophan is the product of the biochemical pathway controlled by the operon. If tryptophan is absent, then the enzymes for its synthesis are needed, and the operon must be transcribed. Tryptophan is called the **co-repressor**. A second example of a repressible operon is the phenylalanine operon.

9.3 REGULATION OF TRANSCRIPTION INITIATION IN EUKARYOTES

Now we turn our attention to the regulation of transcription initiation in eukaryotes. We have just learned that in bacteria the initiation of transcription is influenced by DNA-binding proteins, such as the lactose repressor and the catabolite activator protein, which bind to specific sequences located near the attachment site for the RNA polymerase. This is also the basis for transcriptional control in eukaryotes. We will look first at the regulatory sequences.

RNA polymerase II promoters are controlled by a variety of regulatory sequences

RNA polymerase II provides the best illustration of transcription regulation in eukaryotes. Promoters recognized by this polymerase are subject to the most complex control patterns, reflecting the need to ensure that each protein-coding gene is expressed at precisely the level that is appropriate for the prevailing conditions inside and outside of the cell.

We have already examined the promoters for genes transcribed by RNA polymerase II and seen that, in addition to the TATA box and Inr sequence that make up the core promoter, these genes are preceded by a variety of regulatory sequences located in the several kb immediately upstream of the transcription start site (see Figure 4.14). A single gene might be under the control of a number of different regulatory sequences, reflecting the wide range of internal and external signals that a cell must be able to respond to in order to perform its function in a multicellular organism.

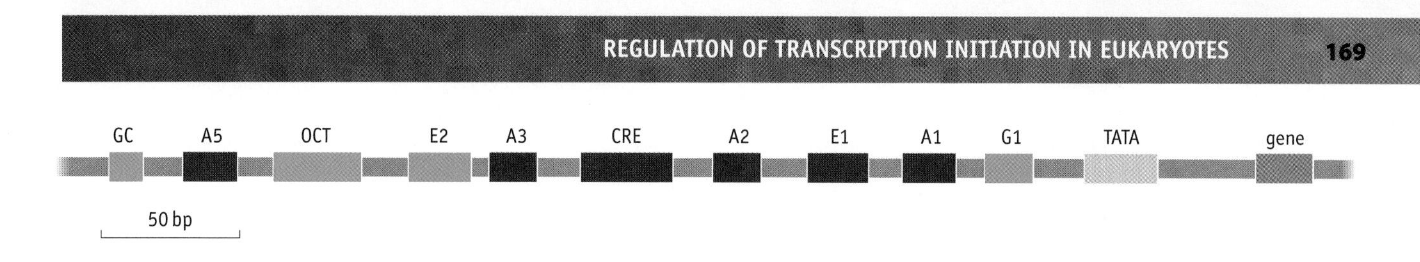

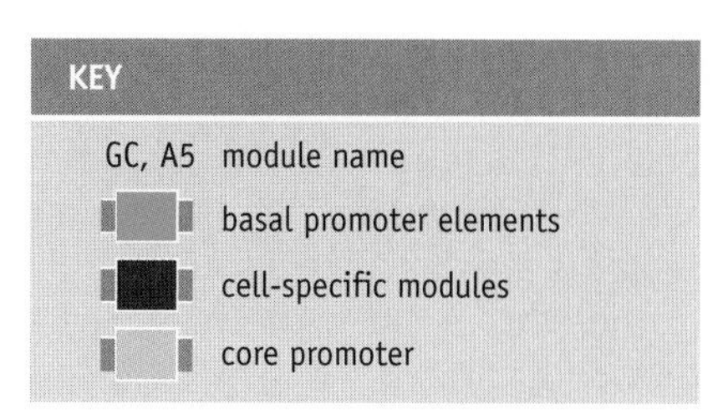

FIGURE 9.18 Regulatory sequences upstream of the human insulin gene. The A5, A3, CRE, A2, E1, and A1 sequences, shown in brown, are cell-specific modules, which ensure that the insulin gene is active only in the β-cells of the pancreas. The GC, OCT, E2, and G1 modules, shown in dark green, are basal promoter elements that control the rate of transcription of the gene when it is switched on. GC, GC box; OCT, octamer sequence.

The human insulin gene, for example, has at least 10 regulatory sequences (Figure 9.18) even though this gene is not subject to a particularly sophisticated control regime.

Some regulatory sequences mediate short-term changes in gene expression, enabling the genes they control to be switched on temporarily when their products are needed. An example in humans is the **heat shock module**, which is recognized by the Hsp70 protein. Hsp70 is thought to detect cellular damage caused by stresses such as heat shock, switching on transcription of genes whose products help repair the damage and protect the cell from further stress. There are also cell-specific modules, whose binding proteins ensure that genes are switched on only in those cell types in which their gene products are needed. The insulin gene has a number of cell-specific modules, as insulin is only made in the β-cells of the pancreas, the gene being switched off in all other tissues. Other regulatory sequences mediate the expression of genes that are active at specific developmental stages.

A final group of regulatory sequences have a rather different function. These are called **basal promoter elements**. They do not respond to any signals from inside or outside the cell but instead control the basal rate of transcription of the gene to which they are attached. The proteins that bind to these sequences therefore ensure that when the gene is switched on, transcription takes place at an appropriate rate. The human insulin gene has four basal promoter elements, including copies of the **GC box** and **octamer sequence**, both of which are found upstream of many protein-coding genes in eukaryotes. Another common basal element, though not present in the insulin control region, is the **CAAT box**.

In addition to the regulatory sequences in the region immediately upstream of a gene, the same and other elements are also contained in **enhancers**. These are longer DNA regions, 200 to 300 bp in length, that are located some distance away from their target genes (Figure 9.19). Often a single enhancer controls expression of more than one gene.

As in bacteria, the regulatory sequences that control expression of a eukaryotic gene are the binding sites for proteins that influence the activity of the RNA polymerase. There is, however, one important difference. The bacterial RNA polymerase has a strong affinity for its promoter, and the basal rate of initiation is relatively high for all but the weakest promoters. With most eukaryotic genes, the reverse is true. The RNA polymerase II initiation complex does not assemble efficiently, and the basal rate of transcript initiation is therefore very low. In order to achieve effective initiation, formation of the complex must be activated by additional proteins.

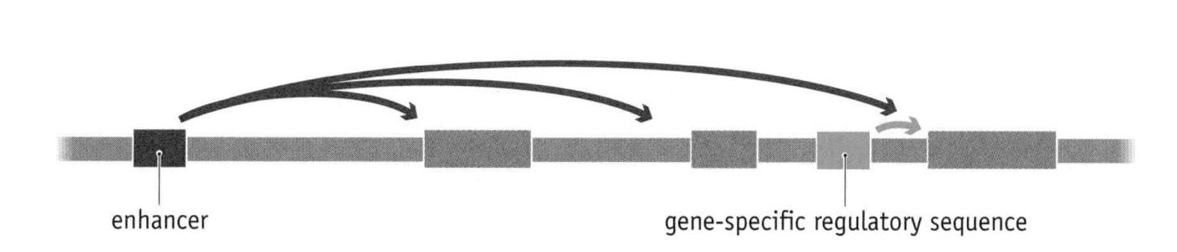

FIGURE 9.19 An enhancer is located some distance away from the genes that it controls. The regulatory sequence (green) is located upstream of its target gene and influences transcription initiation only at that single gene. In contrast, the sequence shown in magenta is an enhancer and influences transcription of all three genes.

RESEARCH BRIEFING 9.1

Proteins that bind to DNA

In both bacteria and eukaryotes, initiation of transcription is regulated by DNA-binding proteins. These proteins attach to specific nucleotide sequences close to the start of the gene or genes whose expression they control.

To bind in this specific fashion a protein must make contact with the double helix in such a way that the nucleotide sequence can be recognized. It does this without breaking the base pairs between the two strands of the helix. This means that part of the protein must penetrate into the major and/or minor grooves. It is only within these parts of the helix that the nucleotide sequence can be read from the arrangement of atoms on the surface of the molecule.

We will explore these DNA-binding proteins and the way they work in a little more detail.

DNA-binding proteins have special structures that enable them to interact with the double helix

The structures of many DNA-binding proteins have been determined by methods such as X-ray crystallography and nuclear magnetic resonance spectroscopy. When these proteins are compared, it becomes clear that there are only a limited number of structures that enable a protein to bind to DNA. Each of these **DNA-binding motifs** is present in a range of proteins, often from very different organisms, and at least some of them probably evolved more than once.

The helix–turn–helix (HTH) motif was the first DNA-binding structure to be identified (Figure 1). We examined this structure in Chapter 7, where we used it as an example of the way in which the amino acid sequence of a protein dictates its function. We learned that one of the two helices is a recognition helix that makes the contacts that enable the DNA sequence to be read. The HTH structure is usually 20 or so amino acids in length and so is just a small part of a protein as a whole. Some of the other parts of the protein also make attachments with the surface of the DNA molecule, primarily to aid the correct positioning of the recognition helix within the major groove. Many bacterial and eukaryotic DNA-binding proteins utilize an HTH motif. In bacteria, HTH motifs are present in some of the best-studied regulatory proteins, including the lactose repressor.

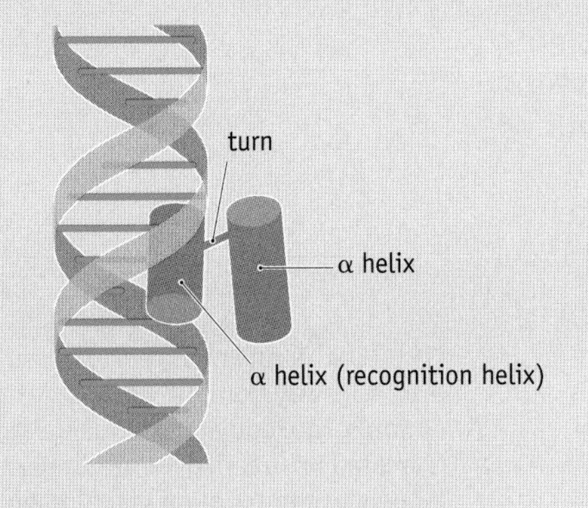

FIGURE 1 The helix–turn–helix motif.

A second important type of DNA-binding motif is the **zinc finger**, which is rare in bacterial proteins but very common in eukaryotes. There are at least six different versions of the zinc finger. The first to be studied in detail was the **Cys_2His_2 finger**, which comprises a series of 12 or so amino acids, including two cysteines and two histidines, which form a segment of β sheet followed by an α helix. These two structures, which form the "finger" projecting from the surface of the protein, hold between them a bound zinc atom, coordinated to the two cysteines and two histidines (Figure 2). The α helix is the part of the motif that makes the critical contacts within

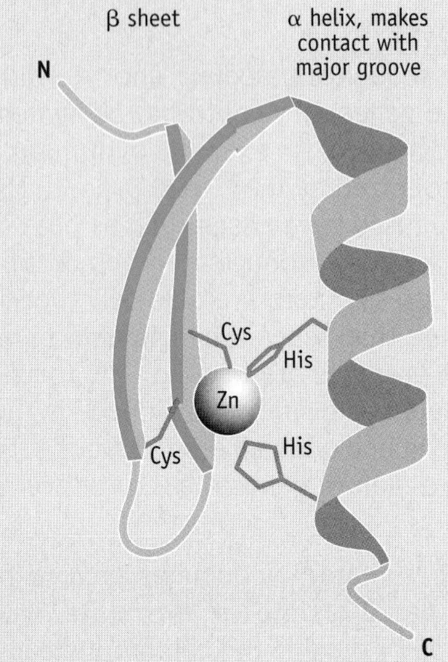

FIGURE 2 A Cys_2His_2 zinc finger.

the major groove. Its positioning within the groove is determined by the β sheet, which interacts with the sugar–phosphate backbone of the DNA, and the zinc atom, which holds the sheet and helix in the appropriate positions relative to one another. Multiple copies of the zinc finger are sometimes found on a single protein. Several have two, three, or four fingers, but there are examples with many more than this, 37 for one toad protein. In most cases, the individual zinc fingers are thought to make independent contacts with the DNA molecule.

Binding proteins can also make nonspecific attachments with DNA molecules

Most proteins that recognize specific sequences are also able to bind nonspecifically to other parts of a DNA molecule. It has been suggested that the amount of DNA in a cell is so large, and the numbers of each binding protein so small, that the proteins spend most, if not all, of their time attached nonspecifically to DNA. The distinction between the nonspecific and specific forms of binding is that the latter is more favorable in thermodynamic terms. As a result, a protein is able to bind to its specific site even though there are literally millions of other sites to which it could attach nonspecifically. To achieve this thermodynamic favorability, the specific binding process must involve the greatest possible number of DNA–protein contacts, which explains in part why the recognition structures of many DNA-binding motifs have evolved to fit snugly into the major groove of the helix, where the opportunity for DNA–protein contacts is greatest.

An intriguing question is whether the specificity of DNA binding can be understood in enough detail for the sequence of a protein's target site to be predicted from examination of the structure of the recognition helix of a DNA-binding motif. To date, this objective has largely eluded us, but some progress has been made with certain types of zinc finger. By comparing the sequences of amino acids in the recognition helices of different zinc fingers with the sequences of nucleotides at their binding sites, it has been possible to identify a set of rules governing the interaction. These enable the nucleotide sequence specificity of a new zinc finger protein to be predicted, admittedly with the possibility of some ambiguity.

Many DNA-binding proteins are dimers

The need to maximize contacts in order to ensure specificity is also one of the reasons why many DNA-binding proteins are dimers, consisting of two proteins attached to one another. This is the case for most HTH proteins and many of the zinc-finger type. Dimerization occurs in such a way that the DNA-binding motifs of the two proteins are both able to access the helix. There may also be some degree of cooperativity between them, so that the resulting number of contacts is greater than twice the number achievable by a monomer.

Besides their DNA-binding motifs, many proteins therefore contain additional characteristic domains that participate in the protein–protein contacts that result in dimer formation. One of these is the **leucine zipper**, which consists of a series of leucines, spaced so they are in a line on the surface of an α helix. These can form contacts with the leucines of the zipper on a second protein, forming the dimer (Figure 3). One interesting possibility is that the protein may unzip, so one subunit can be replaced with another of slightly different type, changing the function of the protein in some subtle way.

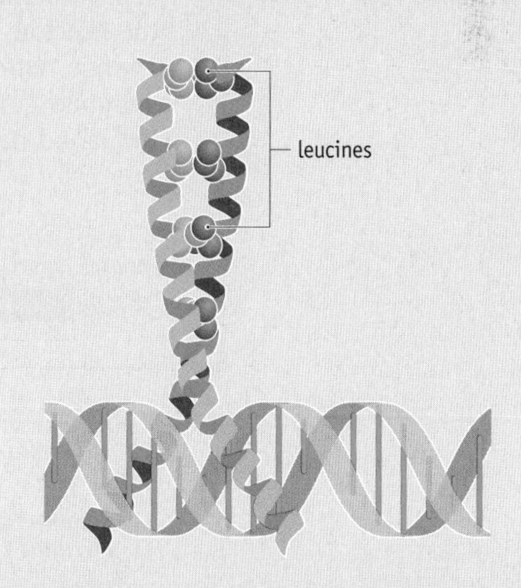

FIGURE 3 A leucine zipper. The orange and red structures are parts of different proteins. Each set of spheres is the R group of a leucine. They are shown interacting with one another. This is a bZIP type of leucine zipper, in which the helices that form the zipper extend to form a pair of DNA-binding motifs. Other zipper proteins have HTH or zinc-finger motifs.

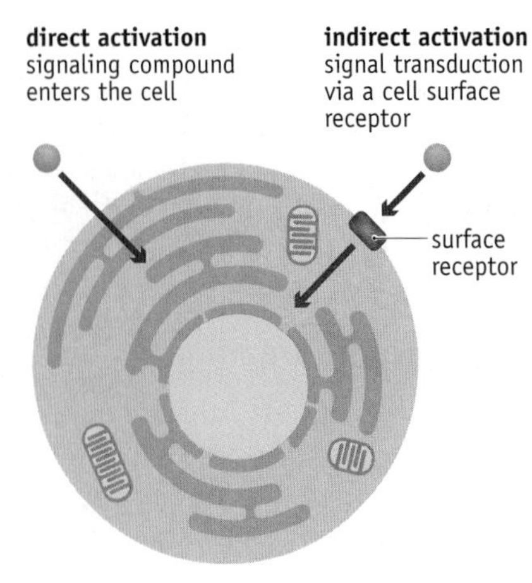

FIGURE 9.20 Two ways in which an extracellular signaling compound can influence events occurring within a cell.

This means that most of the proteins that bind to eukaryotic regulatory sequences are activators of transcription, and repressors are relatively rare.

Signals from outside the cell must be transmitted to the nucleus in order to influence gene expression

In eukaryotes, transcription occurs in the nucleus but must respond to signals from outside the cell. The signal must therefore be transmitted from the cell surface, through the cytoplasm, and into the nucleus (Figure 9.20). There are several ways in which this can occur. The simplest method is for the signaling compound—which might be a hormone or growth factor—to pass through the cell membrane and enter the cell. This is what happens with the **steroid hormones**, which coordinate a range of physiological activities in the cells of higher eukaryotes. They include the sex hormones (estrogens for female sex development, androgens for male sex development), and the glucocorticoid and mineralocorticoid hormones. Once inside the cell, each steroid hormone binds to its own **steroid receptor** protein, which is usually located in the cytoplasm. After binding, the activated receptor migrates into the nucleus, where it attaches to a **hormone response element** upstream of a target gene (Figure 9.21). Response elements for each receptor are located upstream of 50 to 100 genes in humans, often within enhancers, so a single steroid hormone can induce a large-scale change in the biochemical properties of the cell.

Steroid hormones are hydrophobic molecules and so can easily penetrate the cell membrane. This is not the case for many other signaling compounds. Instead, these must bind to a **cell surface receptor**, a protein that spans the cell membrane and is able to transmit the signal into the cell. Attachment of the signaling compound to the part of the receptor on the outside of the cell causes the receptor to undergo a conformational change. Often this change consists of dimerization of the receptor, two subunits combining to form a single structure. This is possible because the liquid nature of the cell membrane allows a limited amount of lateral

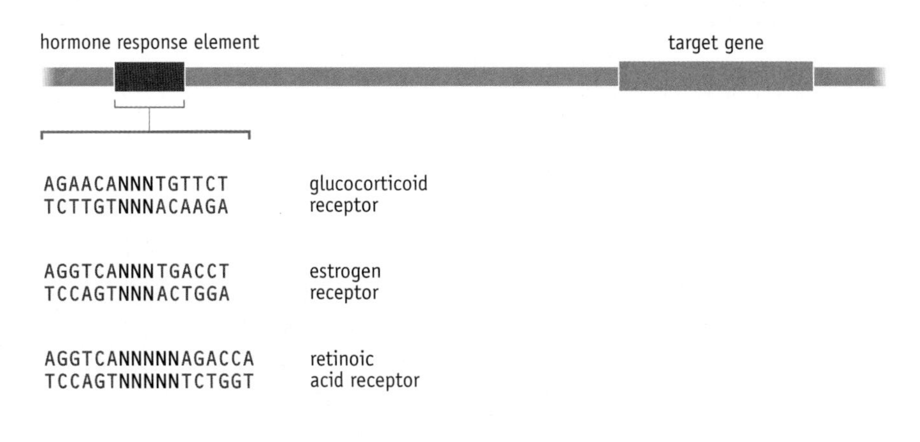

FIGURE 9.21 Hormone response elements. The sequences of three elements are shown. Each element has two parts, which often are exact repeats, possibly in opposite orientations, as is the case with the glucocorticoid and estrogen receptors. N, any nucleotide.

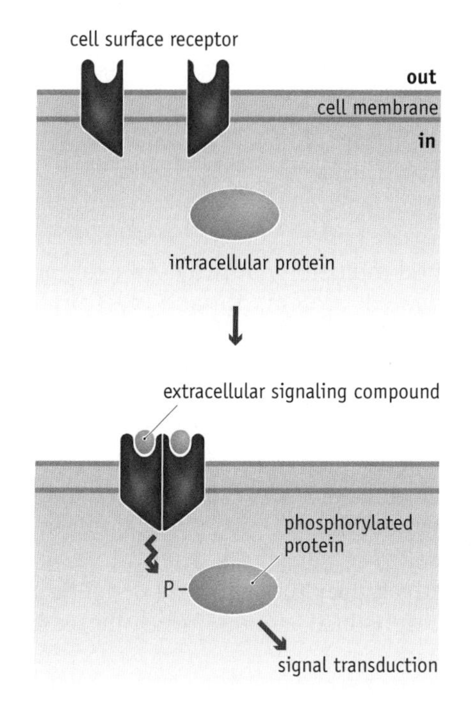

FIGURE 9.22 The role of a cell surface receptor. Binding of the extracellular signaling compound to the outer surface of the receptor protein causes a conformational change in the receptor, often involving dimerization, that results in activation of an intracellular protein, for example by phosphorylation. "P" indicates a phosphate group, PO_3^{2-}.

movement by membrane proteins, enabling the two subunits of a receptor to associate and separate in response to the presence or absence of the extracellular signal (Figure 9.22).

The change in the structure of the receptor induces a biochemical event within the cell, such as attachment of phosphate groups to a cytoplasmic protein. This protein might be a transcription activator that, once phosphorylated, moves into the nucleus and binds to its regulatory sequences. Many cytokines—signaling proteins that control cell growth and division—control gene expression in this way. Their cell surface receptors respond to cytokine binding by phosphorylating transcription activators called **STATs**, which then move to the nucleus and switch on their target genes (Figure 9.23). Many genes can be activated by STATs, but the overall response is modulated by other proteins that interact with different STATs and influence which genes are switched on under a particular set of circumstances. Complexity is entirely expected because the cellular processes that STATs mediate—growth and division—are themselves complex and we anticipate that changes in these processes will require extensive remodeling of the proteome and hence large-scale alterations in gene activity.

The direct phosphorylation of a transcription activator by a cell surface receptor is a straightforward but rather uncommon way of transmitting a signal from the cell surface to the nucleus. With the more prevalent forms of **signal transduction**, the cell surface receptor is just the first component

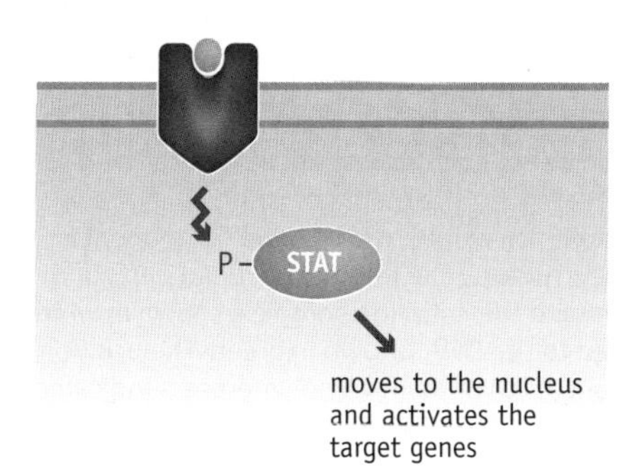

FIGURE 9.23 A STAT is a transcription activator that, once phosphorylated, moves to the nucleus in order to switch on its target genes.

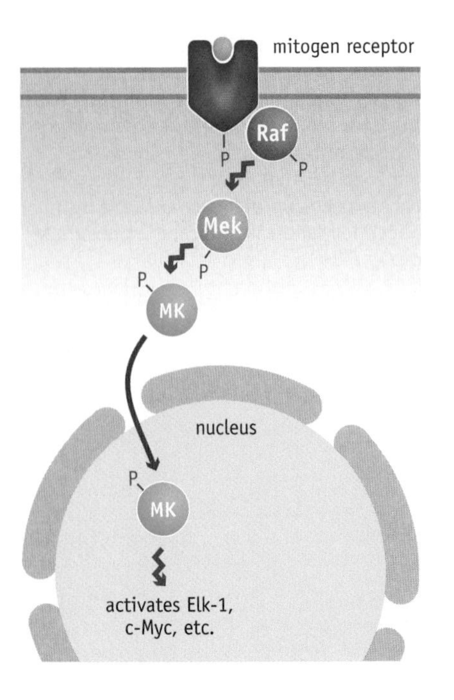

FIGURE 9.24 The MAP kinase signal transduction system. The signaling compound causes the mitogen receptor to dimerize, which results in phosphorylation of the internal signaling protein called Raf. Activation of Raf initiates a cascade that leads via Mek to the MAP kinase. The activated MAP kinase moves into the nucleus, where it switches on various transcription activators, such as Elk-1 and c-Myc. MK, MAP kinase.

of a more complicated multistep pathway. An example is provided by the **MAP** (mitogen-activated protein) **kinase system** (Figure 9.24). This pathway responds to many extracellular signals, including mitogens, compounds with similar effects to cytokines but which specifically stimulate cell division. Binding of the signaling compound results in dimerization of the mitogen receptor, which activates an internal signaling protein called Raf. This protein initiates a cascade of phosphorylation reactions. It phosphorylates Mek, which, in turn, phosphorylates the MAP kinase. The activated MAP kinase now moves into the nucleus, where it switches on, again by phosphorylation, a series of transcription activators. As with STATs, exactly which genes are activated by this pathway depends on interactions with other proteins that transduce signals from other extracellular signaling compounds.

The RNA polymerase II initiation complex is activated via a mediator protein

How can a DNA-binding protein activate transcription of a gene? A physical contact of some kind must be made between the activator protein and the transcription initiation complex. The separation between the control sequence and the transcription start site is not a problem because DNA is flexible and so can bend to enable contact to be made between a bound regulatory protein and the initiation complex. But this does not help us to understand how the regulatory protein influences the activity of the RNA polymerase. How this happens at RNA polymerase II promoters was a mystery for several years, with apparently conflicting evidence coming from work with different organisms. The solution to the problem was found when a large protein called the **mediator** was identified in yeast.

The yeast mediator comprises 21 subunits that form a structure with distinct head, middle, and tail domains. The tail forms a physical contact with an activator protein attached to its regulatory sequence, and the middle and head sections interact with the initiation complex (Figure 9.25). Therefore, rather than direct association of the activator protein with the initiation complex, the association is indirect with the activation signal being transmitted by the mediator.

Mediators in higher eukaryotes are larger than the yeast mediator, with 30 or more subunits making up the human version. One feature of the mammalian mediator is that its subunit composition is variable, raising the possibility that there are several versions, each one responding to a different, although possibly overlapping, set of activators.

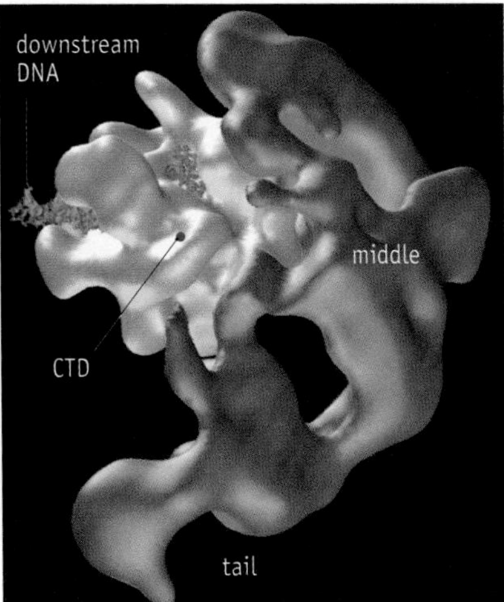

FIGURE 9.25 Interaction between the yeast mediator and the RNA polymerase II initiation complex. The mediator is shown in brown. Its middle and tail sections are visible, and its head component is on the other side of the structure. The middle and head parts of the mediator make contact with the initiation complex, shown in white, which in turn is attached to the DNA, shown in green. The position of the C-terminal domain (CTD) of the RNA polymerase is indicated. (From J.A. Davis et al., *Mol. Cell* 10: 409–415, 2002. With permission from Elsevier.)

How does the mediator affect transcription? The initiation complex is unable to begin transcription until a set of phosphate groups has been added to the C-terminal region of the largest subunit of RNA polymerase II (Figure 9.26). At first it was thought that the mediator activated transcription initiation by adding these phosphates itself. Now we know that the transcription factor TFIIH is responsible for adding the phosphates, although it is still possible that the mediator is also involved in some way. Other experiments suggest that the mediator might activate the very first step in assembly of the initiation complex, when the TATA-binding protein, TBP, attaches to the promoter.

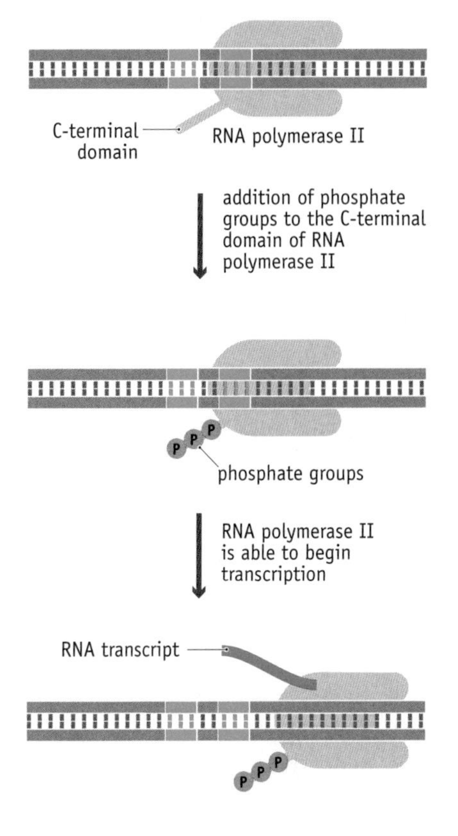

FIGURE 9.26 In order for transcription to begin, phosphate groups must be added to the C-terminal domain of RNA polymerase II. The upper drawing shows RNA polymerase II attached to the core promoter, indicated in green. Phosphorylation of the C-terminal domain activates the polymerase so it is able to begin synthesizing the RNA transcript.

9.4 OTHER STRATEGIES FOR REGULATING GENE EXPRESSION

We have looked closely at how gene expression is controlled by DNA-binding proteins that attach to regulatory sequences and repress or activate initiation of transcription. We have focused on this aspect of gene regulation because geneticists believe that initiation of transcription is the key step that determines whether a gene is switched on or off. But we must not be misled into thinking that this is the only important type of gene regulation. As discussed above, virtually every step in the gene expression pathway is subject to some form of regulation (see Figure 9.4). These other steps might not be primary control points that switch genes on and off, but they have equally important influences on the rate of gene product synthesis.

The problem for the student of genetics attempting to understand gene regulation for the first time is that a broad range of control strategies have evolved in different organisms. It is easy to become swamped in a mass of detail. To avoid this difficulty we will not attempt a comprehensive survey of all of the control mechanisms that are known. Instead, we will study just three examples, which together illustrate the most important of these other ways in which gene expression can be regulated.

Modification of the bacterial RNA polymerase enables different sets of genes to be expressed

The first of these other types of gene regulation also involves transcription initiation, but the control strategy is quite different from any that we have looked at so far. It is known only in bacteria, and it enables sets of genes, not just those present in a single operon, to be switched on or off at the same time. It is used by some bacteria as a means of responding rapidly to emergencies such as heat shock, and by others as the basis for cellular differentiation.

The key to this type of control is the σ subunit of the bacterial RNA polymerase. Although there is just one type of bacterial RNA polymerase, some species possess more than one version of the σ subunit and so can assemble a series of RNA polymerase enzymes with slightly different properties. Because the σ subunit is the part of the polymerase that has the sequence-specific DNA-binding capability, polymerases with different σ subunits will recognize different promoter sequences.

In *E. coli*, the standard σ subunit, which recognizes the consensus promoter sequence shown in Figure 4.12, is called σ^{70} (its molecular mass is approximately 70 kD). *E. coli* also has a second σ subunit, σ^{32}, which is made when

(A) an *E. coli* heat shock gene

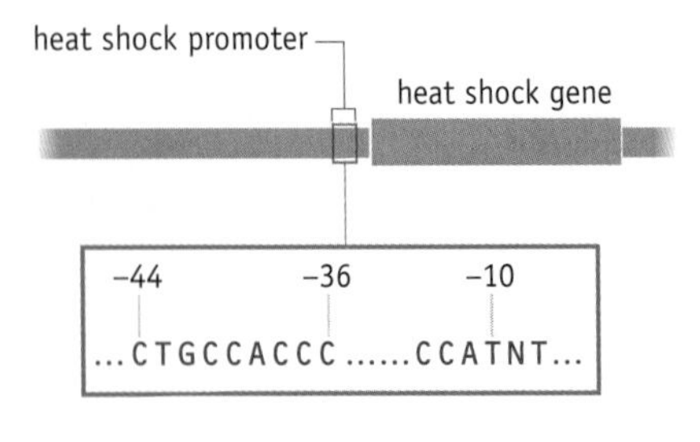

(B) recognition by the σ^{32} subunit

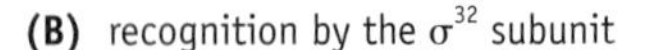

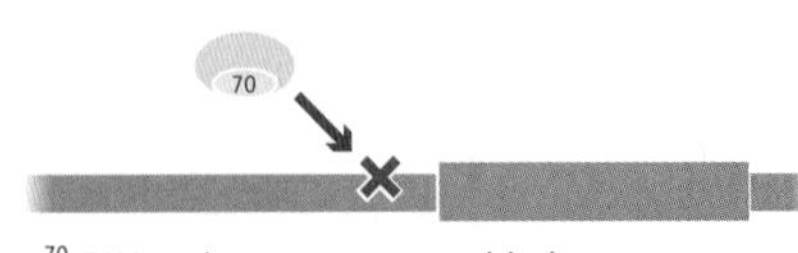

σ^{70} RNA polymerase cannot bind

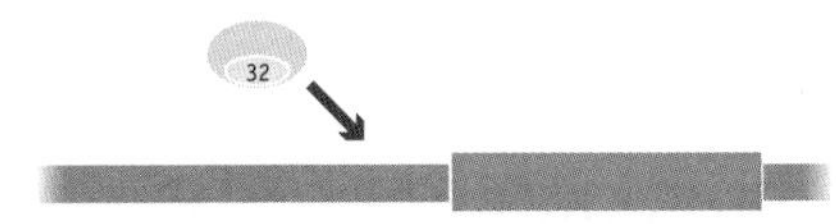

σ^{32} RNA polymerase binds to the heat shock promoter

FIGURE 9.27 Recognition of an *E. coli* heat shock gene by the σ^{32} subunit. (A) The sequence of the heat shock promoter is different from that of the normal *E. coli* promoter. The normal promoter comprises the –35 box, whose consensus sequence is 5′-TTGACA-3′, and the –10 box, 5′-TATAAT-3′. (B) The heat shock promoter is not recognized by the normal *E. coli* RNA polymerase containing the σ^{70} subunit, but is recognized by the σ^{32} RNA polymerase that is active during heat shock. N, any nucleotide.

the bacterium is exposed to a heat shock. This subunit recognizes a different promoter sequence (Figure 9.27), which is found upstream of genes coding for special proteins that help the bacterium withstand heat stress. The bacterium is therefore able to switch on a whole range of new genes by making one simple alteration to the structure of its RNA polymerase.

Alternative σ subunits are also known in other bacteria. For example, *Klebsiella pneumoniae* uses the system to control expression of genes involved in nitrogen fixation, this time with the σ^{54} subunit. The most sophisticated example of this control strategy is provided by *Bacillus* species, which use a whole range of different σ subunits to switch on and off groups of genes during the changeover from normal growth to formation of spores. *Bacillus* is one of several genera of bacteria that produce spores in response to unfavorable environmental conditions (Figure 9.28). These spores are highly resistant to physical and chemical abuse and can survive for decades or even centuries.

The standard *Bacillus subtilis* σ subunits are called σ^A and σ^H. These subunits are synthesized in nonsporulating cells and enable the RNA polymerase to recognize promoters for all the genes it needs to transcribe in order to maintain normal growth and cell division. When sporulation begins, the cell divides into two compartments, one of which will become the spore and the second of which will become the mother cell that eventually dies when the spore is released (Figure 9.29). The two compartments must therefore follow separate differentiation pathways. This is brought about by replacement of the standard σ subunits by two new subunits, σ^F in the prespore and σ^E in the mother cell. Each of these recognizes its own promoter sequence, which is attached to genes specific for prespore or mother cell development, respectively. Later in the sporulation process another change occurs when the σ^F and σ^E subunits are replaced by σ^G and σ^K, respectively. This second change switches on the genes needed in the later stages of spore and mother cell formation. The succession of σ subunits therefore brings about the time-dependent changes in gene activity that underlie differentiation of the original bacterium into a spore.

Transcription termination signals are sometimes ignored

Now we move further into the gene expression pathway and look at regulatory events occurring after the initiation of transcription. Two mechanisms for controlling the synthesis of RNA after transcription has been initiated are known in bacteria.

FIGURE 9.28 Electron micrograph of *Bacillus anthracis* bacteria, showing one cell containing a spore in the final stages of development. (Courtesy of Dr Sherif Zaki and Elizabeth White, Centers for Disease Control and Prevention.)

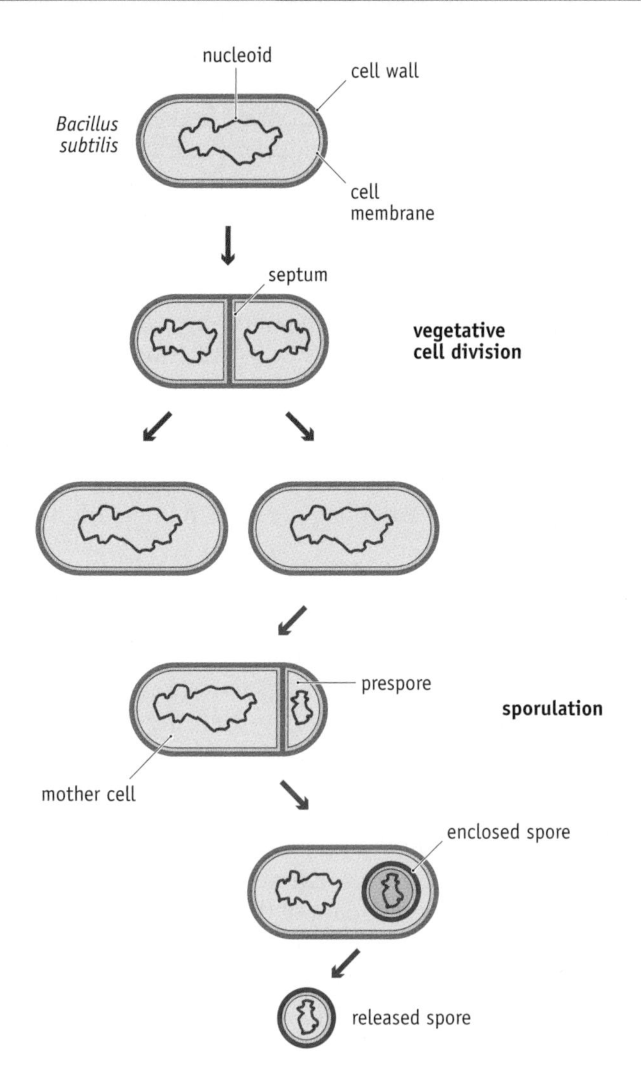

FIGURE 9.29 Sporulation in *Bacillus subtilis*. The top part of the diagram shows the normal vegetative mode of cell division, involving formation of a septum across the center of the bacterium and resulting in two identical daughter cells. The lower part of the diagram shows sporulation, in which the septum forms near one end of the cell, leading to a mother cell and a prespore of different sizes. Eventually the mother cell completely engulfs the prespore. At the end of the process, the mature resistant spore is released.

The first of these is called **antitermination**. This occurs when the RNA polymerase ignores a termination signal and continues elongating its transcript until a second signal is reached (Figure 9.30). Antitermination provides a mechanism whereby one or more of the genes at the end of an operon can be switched off or on as the polymerase either recognizes or fails to recognize a termination signal located upstream of those genes.

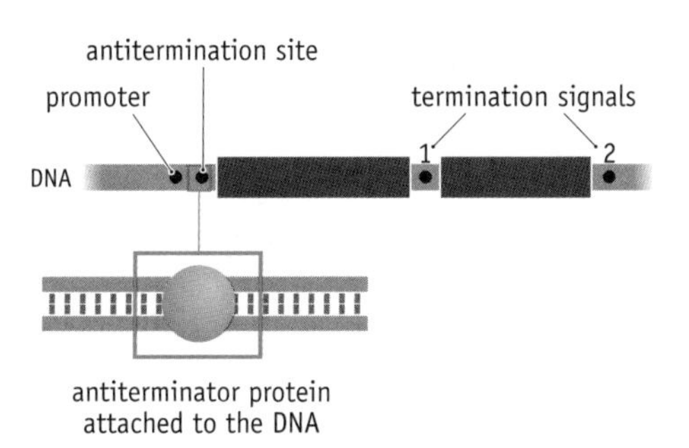

FIGURE 9.30 Antitermination. The antiterminator protein attaches to the DNA at the antitermination site, and transfers to the RNA polymerase as it moves past, subsequently enabling the polymerase to continue transcription through termination signal number 1, so the second of the pair of genes in this operon is transcribed.

Antitermination is controlled by an **antiterminator protein**, which attaches to the DNA near the beginning of the operon and then transfers to the RNA polymerase as it moves past en route to the first termination signal. The presence of the antiterminator protein causes the enzyme to ignore the termination signal, so transcription continues into the downstream region. Exactly how this works is not known, but presumably the antiterminator protein is able to stabilize an intrinsic terminator and/or prevent stalling of the polymerase at a Rho-dependent one.

Although the mechanics of antitermination are unclear, the impact that it can have on gene expression has been described in detail. It is especially important during the infection cycle of bacteriophage λ. This is one of the various viruses that infect *E. coli*. As bacteriophages have relatively few genes—λ has just 48—they were used in the early days of molecular genetics as model systems with which to study the basic principles of gene expression. Studies of bacteriophage λ were particularly instructive with regard to gene regulation.

Successful progress of λ through its infection cycle requires that different sets of bacteriophage genes be switched on and off in the correct order. Immediately after the bacteriophage enters an *E. coli* cell, transcription of the λ genome is initiated when the bacterial RNA polymerase attaches to two promoters, p_L and p_R. Transcription from these promoters results in synthesis of two "immediate early" mRNAs, these terminating at positions t_{L1} and t_{R1} (**Figure 9.31A**). The mRNA transcribed from p_L to t_{L1} codes for the N protein, which is an antiterminator. The N protein attaches to the λ genome at sites *nutL* and *nutR* and transfers to the RNA polymerase as it passes. Now the RNA polymerase ignores the t_{L1} and t_{R1} terminators and continues transcription downstream of these points. The resulting mRNAs encode the "delayed early" proteins (**Figure 9.31B**). Antitermination controlled by the N protein therefore ensures that the immediate early and delayed early proteins are synthesized in the correct order during the λ infection cycle. One of the delayed early proteins, Q, is a second antiterminator that controls the switch to the later stages of the infection cycle.

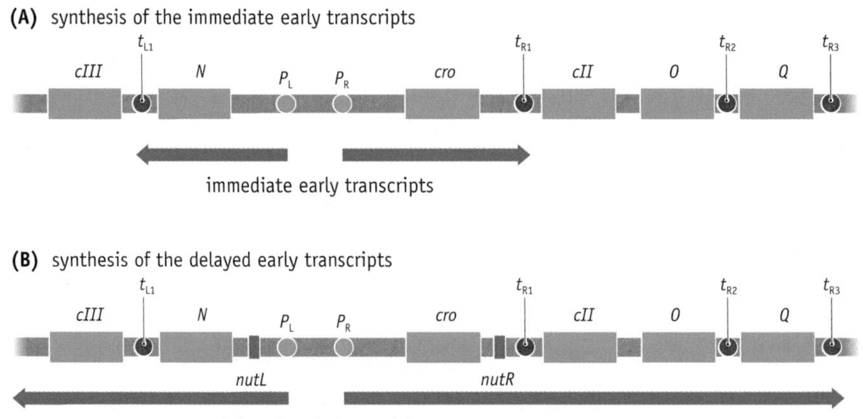

FIGURE 9.31 Antitermination during the infection cycle of bacteriophage λ. (A) Transcription from the promoters p_L and p_R initially results in synthesis of two "immediate early" mRNAs, these terminating at positions t_{L1} and t_{R1}. (B) The mRNA transcribed from p_L to t_{L1} codes for the N protein, which attaches at the antitermination sites *nutL* and *nutR*. Now the RNA polymerase continues transcription downstream of t_{L1} and t_{R1}. Transcription from p_R also ignores terminator t_{R2} and continues until t_{R3} is reached.

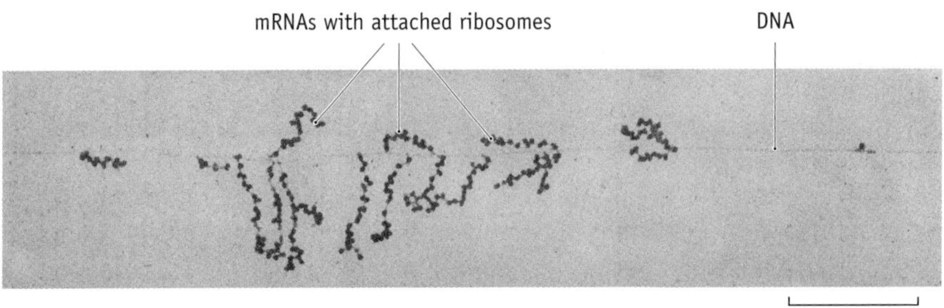

FIGURE 9.32 In bacteria, transcription and translation are often coupled. This electron micrograph shows a part of the *E. coli* DNA molecule. Several mRNAs are being transcribed from the DNA. Each of these mRNAs has ribosomes attached to it—seen as small dark dots. The mRNAs are therefore being translated even though they have not yet been completely transcribed. (From O.L. Miller, B.A. Hamkalo and C.A. Thomas Jr, *Science* 169: 392–395, 1970. With permission from AAAS.)

Attenuation is a second control strategy targeting transcription termination

The second regulatory process used by bacteria to control termination of transcription is called **attenuation**. This occurs primarily with operons that code for enzymes involved in amino acid biosynthesis, although a few other examples are also known.

The basis for attenuation is the coupling between transcription and translation that occurs in bacterial cells, ribosomes attaching to and beginning to translate an mRNA while that molecule is still being synthesized by an RNA polymerase (Figure 9.32). The tryptophan operon of *E. coli* illustrates how attenuation works (Figure 9.33). In this operon, two hairpin loops can form in the region between the start of the transcript and the beginning of *trpE*, the first of the five genes in the operon. The smaller of these loops acts as a termination signal, but the larger hairpin loop, which is closer to the start of the transcript, is more stable. The larger loop overlaps with the termination hairpin, so only one of the two hairpins can form at any one time.

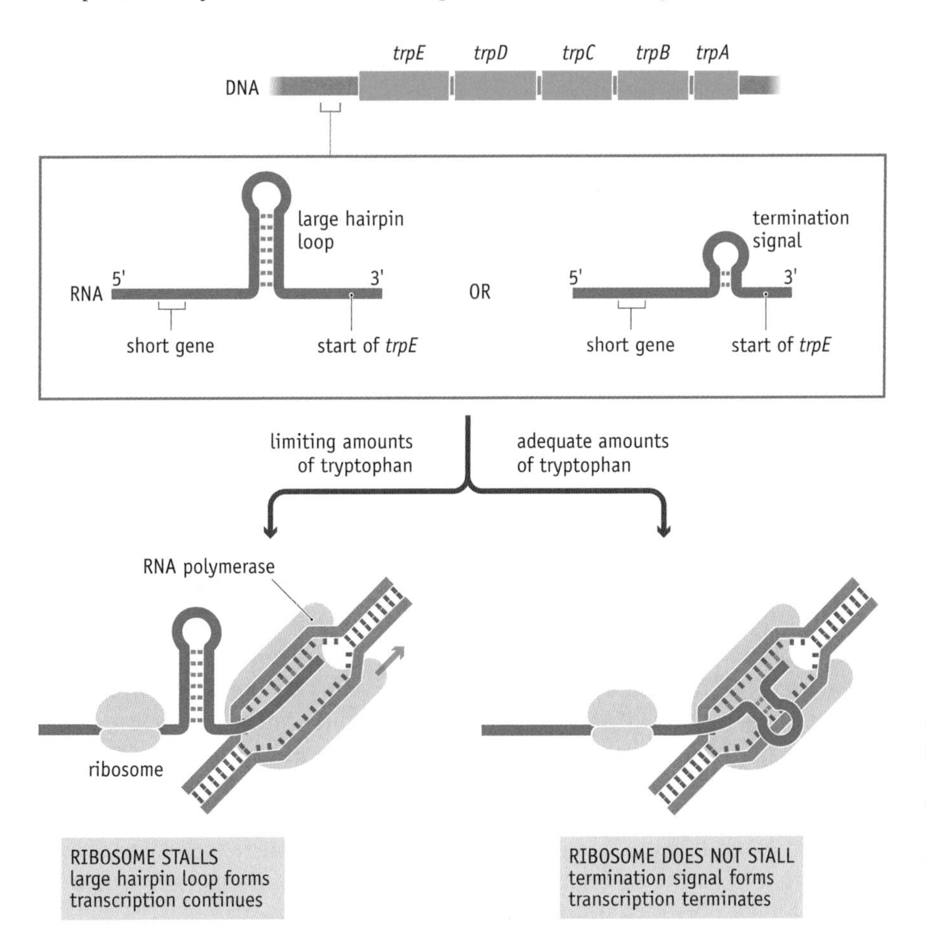

FIGURE 9.33 Attenuation at the tryptophan operon. The two hairpin loops that can form in the region between the start of the transcript and the beginning of *trpE* are shown in the box. The smaller of these hairpin loops acts as a termination signal. The lower part of the diagram shows the different events that occur in the absence or presence of tryptophan. If the amount of tryptophan is limiting, then the ribosome stalls when it reaches the short gene. This enables the polymerase to move ahead so the larger hairpin forms and transcription continues. If the amount of tryptophan is adequate, then the ribosome does not stall but instead keeps up with the polymerase. Now the termination signal forms.

Which loop forms depends on the relative positioning between the RNA polymerase and a ribosome that attaches to the 5′ end of the transcript as soon as it is synthesized. If the ribosome is unable to keep up with the polymerase, then the larger hairpin forms and transcription continues. However, if the ribosome keeps pace with the RNA polymerase, then it disrupts the larger hairpin by attaching to the RNA that makes up part of the stem of this structure. When this happens the termination hairpin is able to form, and transcription stops.

The critical issue is therefore whether or not the ribosome keeps pace with the polymerase. It could fall behind because just upstream of the termination signal is a short gene. This gene codes for a very small protein, just 14 amino acids in length, two of which are tryptophan. If the amount of tryptophan available to the bacterium is low then the ribosome is held up while translating this gene, because it has to wait for the two molecules of tryptophan that it needs. The ribosome therefore falls behind, enabling the polymerase to move ahead so the larger hairpin forms and transcription continues. Because the resulting transcript contains copies of the genes coding for the biosynthesis of tryptophan, its continued elongation addresses the deficiency of this amino acid in the cell. When the amount of tryptophan in the cell reaches a satisfactory level, the attenuation system prevents further transcription of the tryptophan operon, because now the ribosome does not fall behind while making the small protein. Instead it keeps pace with the polymerase, allowing the termination signal to form.

The *E. coli* tryptophan operon is controlled not only by attenuation but also by a repressor. It is thought that repression provides the basic on–off switch and attenuation modulates the precise level of gene expression that occurs. Other *E. coli* operons, such as those for biosynthesis of histidine, leucine, and threonine, are controlled solely by attenuation.

Interestingly, in some bacteria, including *Bacillus subtilis*, the tryptophan operon is one of those that do not have a repressor system and so are regulated entirely by attenuation. In these bacteria, attenuation is mediated not by the speed at which the ribosome tracks along the mRNA, but by an RNA-binding protein called ***trp* RNA-binding attenuation protein** (**TRAP**). In the presence of tryptophan, this protein attaches to the mRNA in the region equivalent to the short gene of the *E. coli* transcript (Figure 9.34). Attachment of TRAP leads to formation of the termination signal and cessation of transcription.

Bacteria and eukaryotes are both able to regulate the initiation of translation

We are now well aware that the initiation of transcription is an important control point at which expression of individual genes can be regulated. The same is true, though to a lesser extent, for initiation of translation.

The best-understood example involves the operons for the ribosomal protein genes of *E. coli* (Figure 9.35). The leader region of the mRNA transcribed from each operon contains a sequence that acts as a binding site for one of the proteins coded by the operon. When this protein is synthesized it can either attach to its position on the ribosomal RNA, or bind to the leader region of the mRNA. The rRNA attachment is favored and occurs if there are free rRNAs in the cell. Once all the free rRNAs have been assembled into ribosomes, the ribosomal protein binds to its mRNA, blocking translation initiation and hence switching off further synthesis of the ribosomal proteins coded by that particular mRNA. Similar events involving other mRNAs ensure that synthesis of each ribosomal protein is coordinated with the amount of free rRNA in the cell.

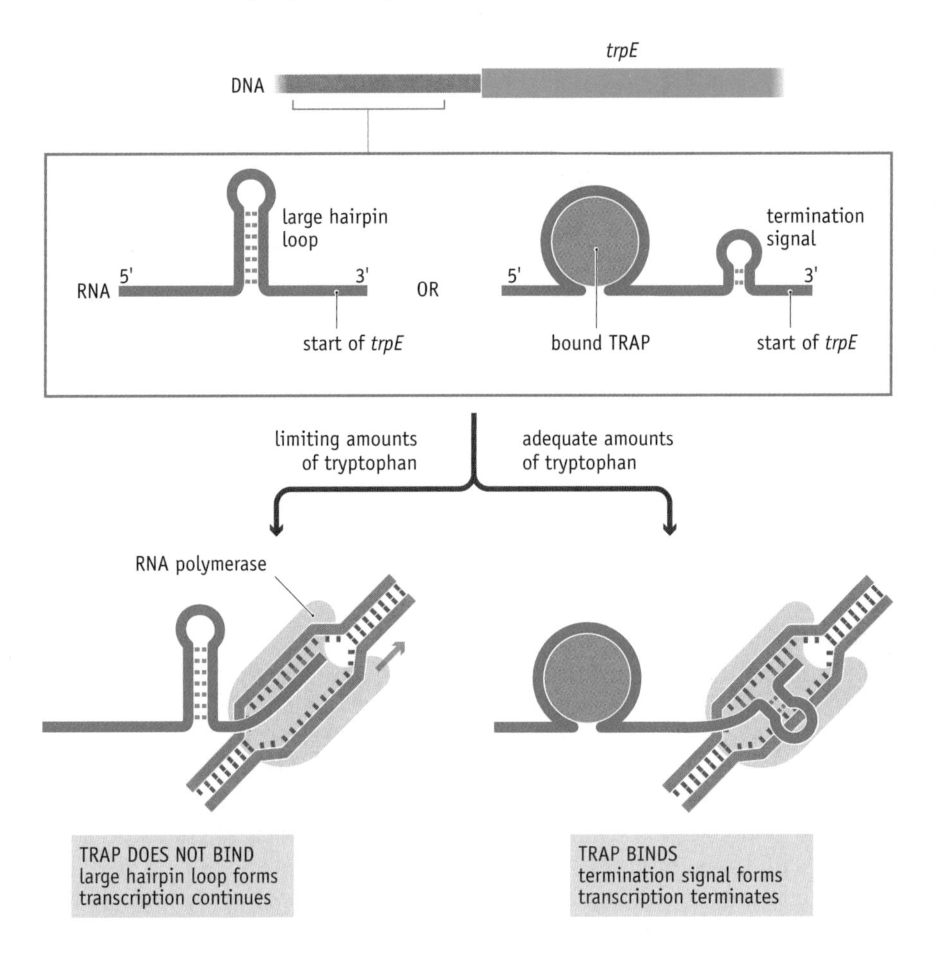

FIGURE 9.34 Regulation of the tryptophan operon of *Bacillus subtilis*. The box shows the two hairpin loops that can form in the region between the start of the transcript and the beginning of *trpE*. The smaller of the hairpin loops, which acts as a termination signal, can form only if the protein called TRAP attaches to the leader region of the transcript. The lower part of the diagram shows the different events that occur in the absence or presence or tryptophan. When the amount of tryptophan is limiting, TRAP does not bind so the larger hairpin forms and transcription continues. When the amount of tryptophan is adequate, TRAP attaches to the mRNA, so that the termination signal forms.

Eukaryotes also have mechanisms for control of translation initiation. In mammals, there is an interesting example with the mRNA for ferritin, an iron-storage protein. In the absence of iron, ferritin is not needed. Its synthesis is inhibited by proteins that bind to sequences called **iron-response elements** located in the leader region of the ferritin mRNA (Figure 9.36). The bound proteins block the ribosome as it attempts to scan along the mRNA in search of the initiation codon. When iron is present, the binding proteins detach and the mRNA is translated. The mRNA for a related protein—the transferrin receptor involved in the uptake of iron by

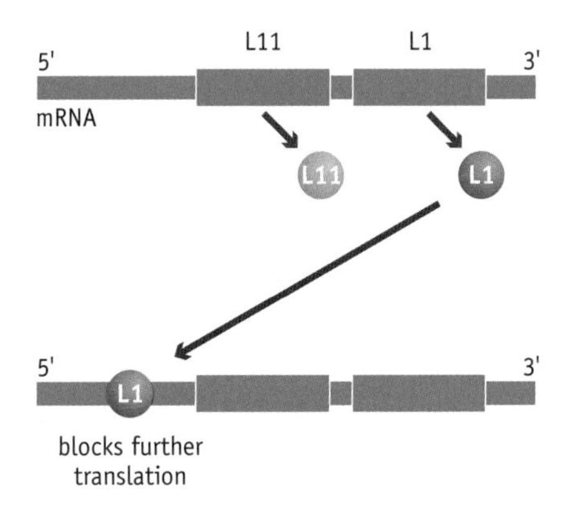

FIGURE 9.35 Regulation of ribosomal protein synthesis in bacteria. The L11 operon of *Escherichia coli* is transcribed into an mRNA carrying copies of the genes for the L11 and L1 ribosomal proteins. When the L1 binding sites on the available 23S rRNA molecules have been filled, L1 binds to the leader region of the mRNA, blocking further initiation of translation.

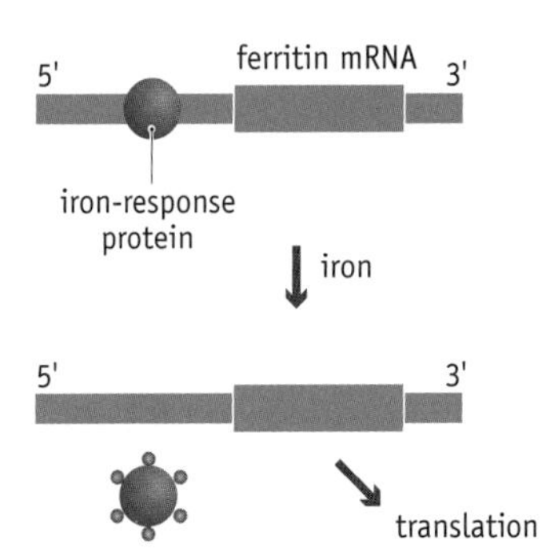

FIGURE 9.36 Regulation of ferritin protein synthesis in mammals. In the absence of iron, the iron-response protein attaches to the ferritin mRNA and prevents its translation. When iron is present, it binds to the iron-response protein, which detaches from the mRNA, enabling ferritin to be synthesized.

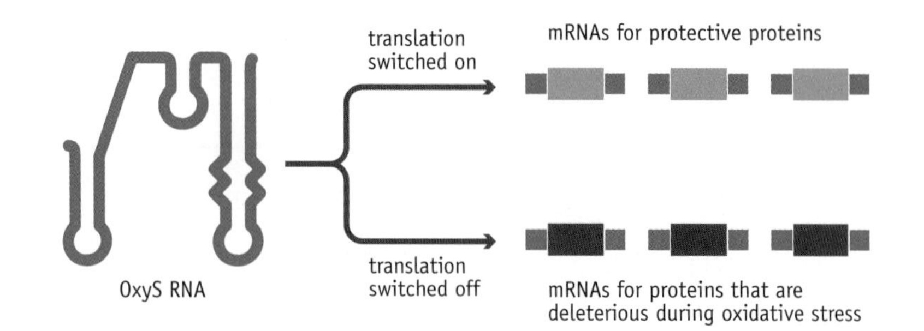

FIGURE 9.37 The role of the OxyS RNA in regulation of gene expression in *E. coli*. OxyS switches on translation of mRNAs for proteins that protect the bacterium from oxidative damage, and switches off translation of other mRNAs whose products would be deleterious during oxidative stress.

the cell—also has iron-response elements. In this case detachment of the binding proteins in the presence of iron results not in translation of the mRNA but in its degradation. This is logical because when iron is present in the cell, there is less requirement for transferrin receptor activity because there is less need to import iron from the outside.

Initiation of translation of some bacterial mRNAs can also be regulated by short RNAs that attach to recognition sequences within the mRNAs. This RNA-mediated mechanism can result in both activation and inhibition of translation of the target mRNAs. One such regulatory RNA in *E. coli* is called OxyS, which is 109 nucleotides in length and regulates translation of 40 or so mRNAs. Synthesis of the OxyS RNA is activated by hydrogen peroxide and other reactive oxygen compounds, which can cause oxidative damage to the cell. Once synthesized, OxyS switches on translation of some mRNAs whose products help protect the bacterium from oxidative damage, and switches off translation of other mRNAs whose products would be deleterious under these circumstances (**Figure 9.37**). We do not yet know how the same RNA can have this differential effect, activating translation of some mRNAs and inhibiting translation of others. Especially perplexing is the discovery that the structure formed by attachment of OxyS to mRNAs that are silenced is not obviously different from the structure formed with mRNAs whose translation is switched on.

KEY CONCEPTS

- Some genes are expressed all the time, but all organisms are able to regulate expression of at least some of their genes. Gene regulation enables bacteria to respond to changes in their environment, and for cells in multicellular organisms to respond to signals from other cells. Gene regulation also underlies differentiation and development.
- Control of gene expression is control over the amount of gene product present in the cell. This amount is a balance between the rate of synthesis (how many molecules of the gene product are made per unit time) and the degradation rate (how many molecules are broken down per unit time).
- The rate of synthesis of a gene product can be regulated by exerting control over any one, or a combination, of the various steps in the gene expression pathway.
- In all organisms the key events that determine whether a gene is on or off take place during the initiation of transcription. Later events in the gene expression pathway are able to change the rate of expression of a gene that is switched on and, in eukaryotes, control the synthesis of alternative products of a single gene.

- The lactose operon of *E. coli* illustrates the basic principles of control over initiation of transcription. Initiation of transcription is influenced by DNA-binding proteins, such as the lactose repressor and the catabolite activator protein, which bind to specific sequences located near the attachment site for the RNA polymerase and influence the ability of the RNA polymerase to access the promoter and initiate RNA synthesis.
- Regulatory sequences and proteins that bind to those sequences also underlie control of transcription initiation in eukaryotes. One difference is that the basal rate of transcription by a eukaryotic RNA polymerase is low, so most eukaryotic regulatory proteins are activators of transcription.
- In eukaryotes, signals from outside the cell must be transmitted from the cell surface, through the cytoplasm, and into the nucleus in order to influence gene expression.
- Later steps in the gene expression pathway, subsequent to initiation of transcription, are also subject to control. Examples are termination of transcription and initiation of translation.

QUESTIONS AND PROBLEMS (Answers can be found at www.garlandscience.com/introgenetics)

Key Terms

Write short definitions of the following terms:

adenylate cyclase
antitermination
antiterminator protein
attenuation
basal promoter element
CAAT box
catabolite activator protein
cell surface receptor
co-repressor
cyclic AMP
Cys_2His_2 finger
diauxie
DNA-binding motif
enhancer
GC box
heat shock module
hormone response element
housekeeping genes
inducer
inducible operon
iron-response element
leucine zipper
MAP kinase system
mediator
octamer sequence
operator
operon
regulatory gene
repressible operon
repressor
signal transduction
STAT
steroid hormone
steroid receptor
structural gene
trp RNA-binding attenuation protein
zinc finger

Self-study Questions

9.1 Using examples, explain why gene regulation is important in bacteria and eukaryotes.

9.2 To what extent are the underlying principles of gene regulation the same in all organisms?

9.3 Distinguish between the terms "structural" and "regulatory" genes, using lactose utilization by *E. coli* as an example.

9.4 Describe how the operator and repressor interact to control expression of the lactose operon. Why is this called an inducible operon?

9.5 In *E. coli*, the galactose operon comprises three structural genes, called *galK*, *galT*, and *galE*. The proteins coded by these genes are involved in metabolism of the sugar galactose. There is also a regulatory gene, *galR*, which codes for a repressor protein that binds to an operator upstream of the *gal* operon. Galactose acts as an inducer. Describe the events that you would expect to occur (a) in the absence of galactose, and (b) in the presence of galactose.

9.6 A mutation is an alteration in nucleotide sequence that may alter the function of a gene, promoter, or regulatory site. Deduce and explain the effects of each of the following mutations on expression of the *lac* operon:

(a) A mutation in the operator, so that the repressor is no longer able to bind

(b) A mutation in *lacI*, so that the repressor is no longer synthesized

(c) A mutation in the promoter, so that RNA polymerase no longer binds

(d) A mutation in *lacI*, so that the repressor no longer binds allolactose

(e) A mutation in *lacY* that causes transcription to terminate in the middle of the gene

(f) A mutation, location unknown, that prevents the *lac* mRNA from being degraded

9.7 What is meant by the term "diauxie"?

9.8 Describe how glucose influences expression of the lactose operon.

9.9 Distinguish between an inducible and a repressible operon.

9.10 Give two examples of DNA-binding motifs found in proteins that attach to DNA molecules.

9.11 What is the importance of leucine zippers in control of gene expression?

9.12 Give examples of regulatory modules found upstream of genes in eukaryotes.

9.13 What are the special features of an enhancer?

9.14 Outline the process by which steroid hormones regulate gene activity in higher eukaryotes.

9.15 Outline the role that STATs play in regulation of gene expression.

9.16 What is meant by signal transduction? Give one example of a signal transduction pathway.

9.17 Describe the differences between the yeast and mammalian mediators, and explain how the mediator is thought to influence the initiation of transcription.

9.18 Give two examples to illustrate the regulation of bacterial gene expression by alternative σ subunits.

9.19 Describe the process called antitermination and explain how it is involved in expression of genes in bacteriophage λ.

9.20 What is attenuation? Your answer should distinguish between the processes occurring in *E. coli* and in *B. subtilis*.

9.21 Giving examples, describe how the initiation of translation is controlled in bacteria and eukaryotes.

Discussion Topics

9.22 To what extent are the underlying principles of gene regulation not the same in all organisms?

9.23 Elaborate on the possibility that the amino acid sequence of a recognition helix can be used to deduce the nucleotide sequence of the DNA binding site for a protein that contains that helix.

9.24 Why is allolactose, rather than lactose, the inducer of the lactose operon?

9.25 To what extent is *E. coli* a good model for the regulation of transcription initiation in eukaryotes? Justify your opinion by providing specific examples of how extrapolations from *E. coli* have been helpful and/or unhelpful in the development of our understanding of equivalent events in eukaryotes.

9.26 Operons are very convenient systems for achieving coordinated regulation of expression of related genes. Operons are common in bacteria, yet they are absent in eukaryotes. Discuss.

9.27 The tryptophan operon of *E. coli* is regulated both by a repressor and by attenuation. Other operons coding for amino acid biosynthetic enzymes are controlled only by attenuation. Discuss.

FURTHER READING

Harrison SC & Aggarwal AK (1990) DNA recognition by proteins with the helix-turn-helix motif. *Annu. Rev. Biochem.* 59, 933–969.

Henkin TM (1996) Control of transcription termination in prokaryotes. *Annu. Rev. Genet.* 30, 35–57. *A detailed account of antitermination and attenuation.*

Horvath CM (2000) STAT proteins and transcriptional responses to extracellular signals. *Trends Biochem. Sci.* 25, 496–502.

Jacob F & Monod J (1961) Genetic regulatory mechanisms in the synthesis of proteins. *J. Mol. Biol.* 3, 318–389. *The original proposal of the operon theory for control of bacterial gene expression.*

Kim Y-J & Lis JT (2005) Interactions between subunits of *Drosophila* mediator and activator proteins. *Trends Biochem. Sci.* 30, 245–249.

Lopez D, Viamakis H & Kolter R (2008) Generation of multiple cell types in *Bacillus subtilis*. *FEMS Microbiol. Lett.* 33, 152–163. *Describes how different σ subunits are involved in sporulation.*

Losick RL & Sonenshein AL (2001) Turning gene regulation on its head. *Science* 293, 2018–2019. *Describes the attenuation systems at the tryptophan operons of* E. coli *and* B. subtilis.

Mackay JP & Crossley M (1998) Zinc fingers are sticking together. *Trends Biochem. Sci.* 23, 1–4.

Ptashne M & Gilbert W (1970) Genetic repressors. *Sci. Am.* 222, 36–44. *The mode of action of repressors and the methods used in isolation of the proteins.*

Tsai M-J & O'Malley BW (1994) Molecular mechanisms of action of steroid/thyroid receptor superfamily members. *Annu. Rev. Biochem.* 63, 451–486. *Control of gene expression by steroid hormones.*

Zubay G, Schwartz D & Beckwith J (1970) Mechanisms of activation of catabolite-sensitive genes: a positive control system. *Proc. Natl. Acad. Sci. USA* 66, 104–110. *An early description of the CAP system.*

GENES AS UNITS OF INHERITANCE

We have learned that biological information is coded within the nucleotide sequence of a gene, and that this information is made available to the cell by the process called gene expression. In the next eight chapters we will focus on the second role of genes, as units of inheritance. We will address three questions. The first is how a complete set of genes is passed to the daughters when the parent cell divides. To answer this question we will examine how individual DNA molecules are replicated and how a daughter cell acquires a full set of these replicated molecules. We must therefore study how DNA molecules are organized inside the chromosomes of eukaryotic cells, and how chromosome division is coordinated with cell division. We must also look at the transmission of DNA molecules when bacteria divide, and we must explore how viruses pass their genes on to their progeny during their infection cycles.

The second question concerns the inheritance of genes during sexual reproduction. Sexual reproduction is preceded by a specialized type of cell division that produces the male and female sex cells. One male and one female cell then fuse to give a fertilized egg that develops into a new version of the organism. We must study how the sex cells are produced, and we must ask how the genes from the male and female parents interact in the new organism that results from sexual reproduction. Answering this question will require us to make a molecular examination of the genetic processes that were first described by Mendel 150 years ago.

The third question concerns the link between the inheritance of genes and the evolution of species. The fact that evolution occurs tells us that genes are not unalterable units and must themselves change over time. We must therefore examine the various processes that alter the sequence of a DNA molecule, as well as the effects that these processes have on the biological information contained in a gene. We will discover that new versions of a gene arise from time to time, and that some of these propagate and spread through a population, the members of the population thereby acquiring an allele that their ancestors lacked. By studying the way that genes are inherited by populations rather than by individuals, we will begin to understand how species can change over time.

PART II

How is a complete set of genes passed to the daughter cells when the parent cell divides? In this chapter we will uncover part of the answer by examining the way in which DNA molecules are replicated. First, we will look at the overall pattern of replication and ask how a single DNA double helix gives rise to two identical daughter helices. We know that complementary base pairing enables polynucleotides to be copied, but there are several ways in which complementary base pairing could lead to replication of the double helix. We must discover which of these is correct. Then we must study the biochemistry and enzymology of DNA replication. What proteins are involved, and how are the new polynucleotides synthesized?

10.1 THE OVERALL PATTERN OF DNA REPLICATION

One of the reasons why the discovery of the double-helix structure was an important breakthrough in genetics is that it immediately suggested a way in which DNA molecules could be copied, and hence solved the great mystery of how genes are able to replicate. Complementary base pairing is the key because it enables each strand of the double helix to act as a template for synthesis of its partner (Figure 10.1). But there appear to be various ways in which template-dependent DNA synthesis could bring about replication of a double helix. We must therefore look at the conceivable replication processes and discover which actually operates in living cells.

DNA replicates semiconservatively, but this causes topological problems

There would seem to be three different ways in which complementary base pairing could be used in order to make a copy of a DNA double helix (Figure 10.2). The first of these is **semiconservative replication**, in which each daughter molecule contains one polynucleotide derived from the original molecule and one newly synthesized strand. The second possibility is **conservative replication**. In this process, one daughter molecule contains both parent polynucleotides and the other daughter contains both newly synthesized strands. The third option is **dispersive replication**, in which each strand of each daughter molecule is composed partly of the original polynucleotide and partly of newly synthesized polynucleotide.

Semiconservative replication is the most straightforward scheme and was the one favored by Watson and Crick when they discovered the double-helix structure in 1953. The conservative model seems less likely because it would require that the two newly synthesized strands detach from their templates and reassociate with one another, which would appear to be a pointless extra step in the replication pathway (Figure 10.3A). Dispersive replication would require repeated template switching, which again would be a seemingly unnecessary elaboration (Figure 10.3B). But science does not progress simply by considering different schemes and judging which is most likely to be true. Experiments were therefore carried out, soon after the discovery of the double helix, to determine which of the three replication models is correct (Research Briefing 10.1). The result of this research was a clear demonstration that DNA replicates semiconservatively. This conclusion raised a problem, because it was widely believed at the time that such a process would be impossible inside a living cell.

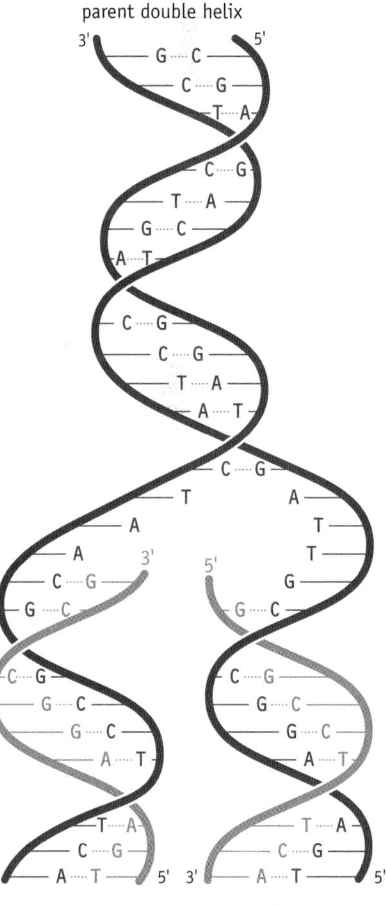

FIGURE 10.1 Complementary base pairing provides a means for making two copies of a double helix. The polynucleotides of the parent double helix are shown in blue, and the newly synthesized strands are in orange.

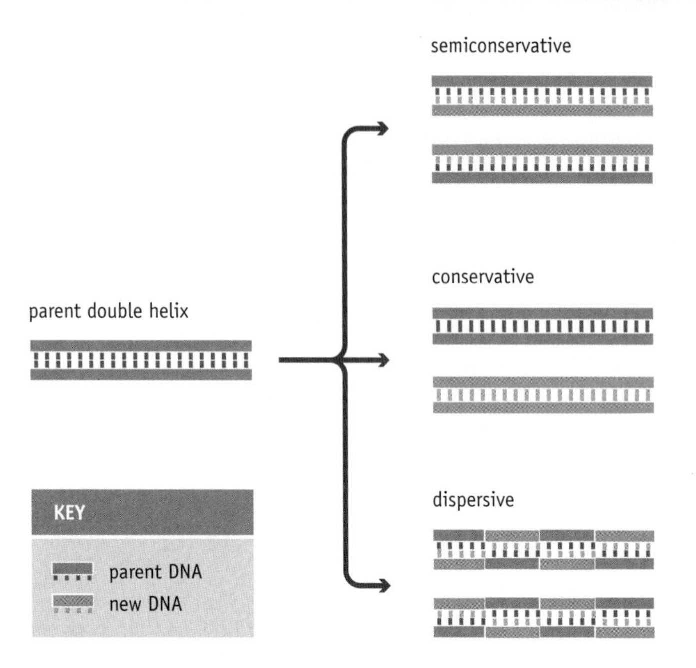

FIGURE 10.2 Three conceivable schemes for DNA replication.

Why should semiconservative replication have been thought impossible? The reason lies in the structure of the double helix. The double helix is **plectonemic**, meaning that the two strands cannot be separated without unwinding. Therefore, during semiconservative replication it would be necessary for the parent double helix to rotate so that its strands could unwind prior to being individually copied. With one turn occurring for every 10 bp of the double helix, complete replication of the DNA molecule in human chromosome 1, which is 225 Mb in length, would require 22.5 million rotations of the chromosomal DNA. It is difficult to imagine how this could occur within the constrained volume of the nucleus, but even so the unwinding of a linear chromosomal DNA molecule is not physically impossible. In contrast, a circular double-stranded molecule, such as those found inside bacteria, would not be able to rotate in the required manner and so, apparently, could not be replicated by the semiconservative process. It took almost 25 years to find a solution to this topological problem.

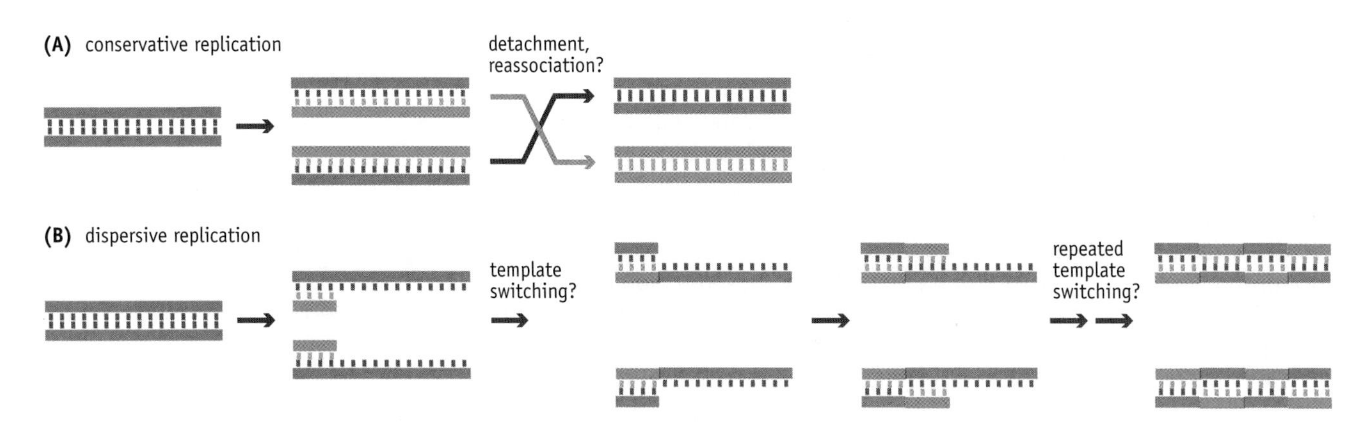

FIGURE 10.3 Complications with conservative and dispersive replication. (A) Conservative replication requires that the two newly synthesized strands detach from their templates and reassociate with one another. (B) Dispersive replication requires repeated template switching.

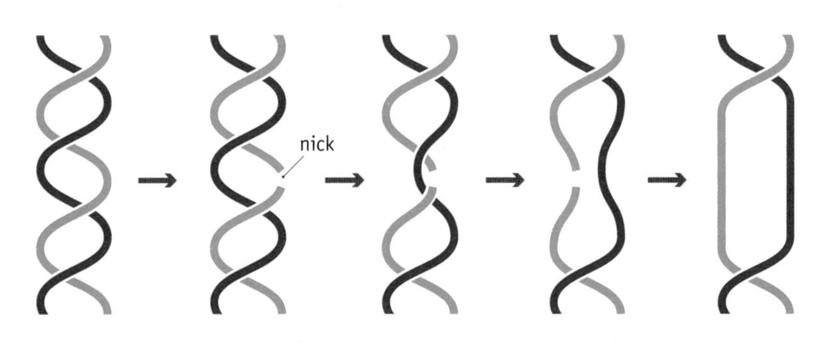

FIGURE 10.4 The mode of action of a type I DNA topoisomerase. A type I topoisomerase introduces a break in one polynucleotide and passes the second polynucleotide through the gap that is formed.

DNA topoisomerases solve the topological problem

This solution to the topological problem was found when the activities of the enzymes called **DNA topoisomerases** were characterized. DNA topoisomerases are enzymes that separate the two strands in a DNA molecule without actually rotating the double helix. They achieve this feat by causing transient breakages in the polynucleotide backbone.

There are two different types of DNA topoisomerase. **Type I topoisomerases** introduce a break in one polynucleotide and pass the second polynucleotide through the gap that is formed (Figure 10.4). The two ends of the broken strand are then re-ligated. This mode of action results in a change in the linking number (the number of times one strand crosses the other in a circular molecule) by 1. **Type II topoisomerases**, on the other hand, break both strands of the double helix, creating a "gate" through which a second segment of the helix is passed (Figure 10.5). This changes the linking number by 2. Despite their different mechanisms, both type I and type II topoisomerases achieve the same end result. They enable the helix to be "unzipped," with the two strands pulled apart sideways so that the molecule does not have to rotate when it is being replicated (Figure 10.6).

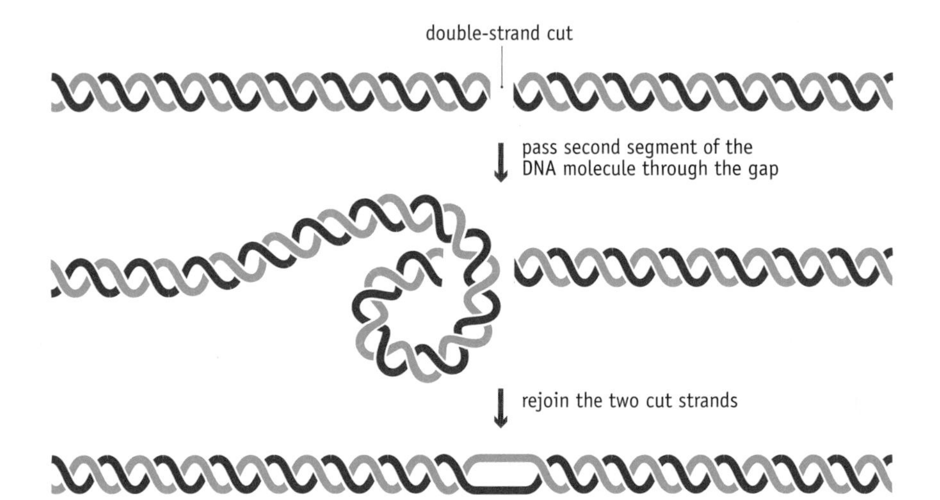

FIGURE 10.5 The mode of action of a type II DNA topoisomerase. A type II topoisomerase breaks both strands of the double helix, creating a gap through which a second segment of the helix is passed.

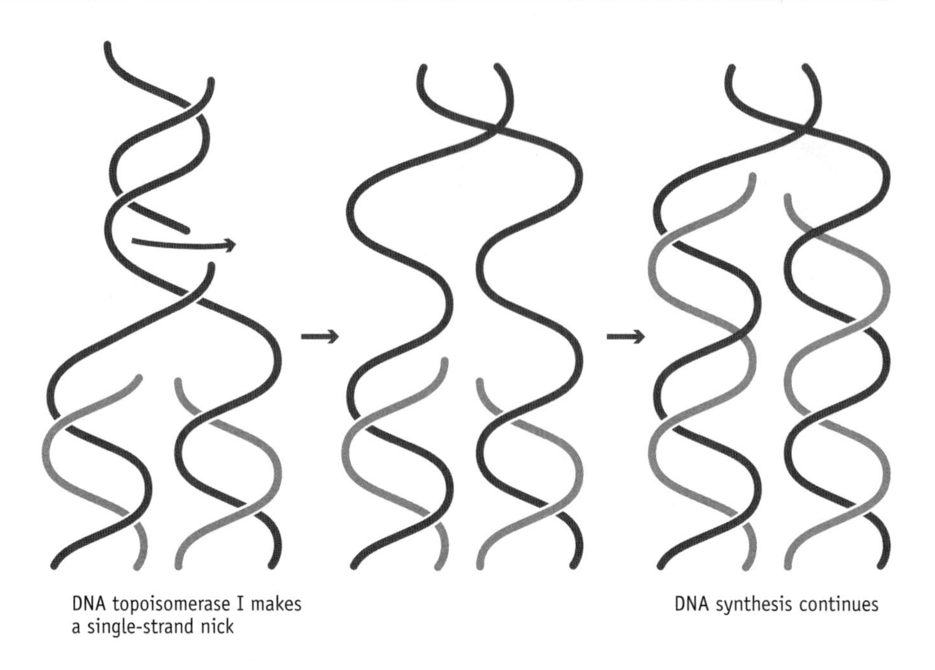

FIGURE 10.6 Unzipping the double helix. During replication, the double helix is "unzipped" as a result of the action of DNA topoisomerases. Replication is therefore able to occur without the helix having to rotate.

Breaking one or both DNA strands might appear to be a drastic solution to the topological problem, leading to the possibility that the topoisomerase might occasionally fail to rejoin a strand and hence inadvertently break a chromosome into two sections. This possibility is reduced by the mode of action of these enzymes. One end of each cut polynucleotide becomes covalently attached to a tyrosine amino acid at the active site of the enzyme, ensuring that these ends of the polynucleotides are held tightly in place while the free ends are being manipulated. Type I and II topoisomerases are subdivided according to the precise chemical structure of the polynucleotide–tyrosine linkage. With type IA and IIA enzymes the link involves a phosphate group attached to the free 5′ end of the cut polynucleotide, and with type IB and IIB enzymes the linkage is via a 3′ phosphate group (Figure 10.7). The A and B topoisomerases probably evolved separately. Both types are present in eukaryotes, but type IB and IIB enzymes are very uncommon in bacteria.

Replication is not the only activity that is complicated by the topology of the double helix. It is becoming increasingly clear that DNA topoisomerases have equally important roles during transcription, and in other processes

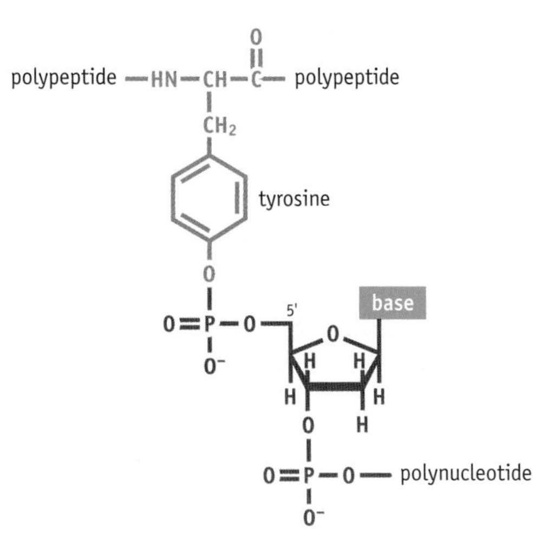

FIGURE 10.7 The linkage between a tyrosine of a type IA or IIA topoisomerase (shown in brown) and the 5′ end of a cut polynucleotide.

that can result in over- or underwinding of DNA. Most topoisomerases are able only to relax DNA, but some bacterial type II enzymes can carry out the reverse reaction and introduce extra turns into DNA molecules. An example is the enzyme called **DNA gyrase**. We will study one of the roles of DNA gyrase in Chapter 13.

Variations on the semiconservative theme

No exceptions to the semiconservative mode of DNA replication are known, but there are several variations on this basic theme. DNA copying via a **replication fork**, as shown in Figure 10.1, is the predominant system, being used by chromosomal DNA molecules in eukaryotes and by the circular DNA molecules of bacteria.

Some smaller circular molecules, such as those present in animal mitochondria, use a slightly different process called **displacement replication**. In these molecules, the point at which replication begins is marked by a **D loop**, a region of approximately 500 bp where the double helix is disrupted by an RNA molecule that is base-paired to one of the DNA strands (Figure 10.8). This RNA molecule acts as the starting point for synthesis of one of the daughter polynucleotides. This polynucleotide is synthesized by continuous copying of one strand of the helix, the second strand being displaced and subsequently copied after synthesis of the first daughter genome has been completed.

The advantage of displacement replication as performed by animal mitochondrial DNA is not clear. In contrast, the special type of displacement process called **rolling-circle replication** is an efficient mechanism for the rapid synthesis of multiple copies of a circular DNA molecule. Rolling-circle

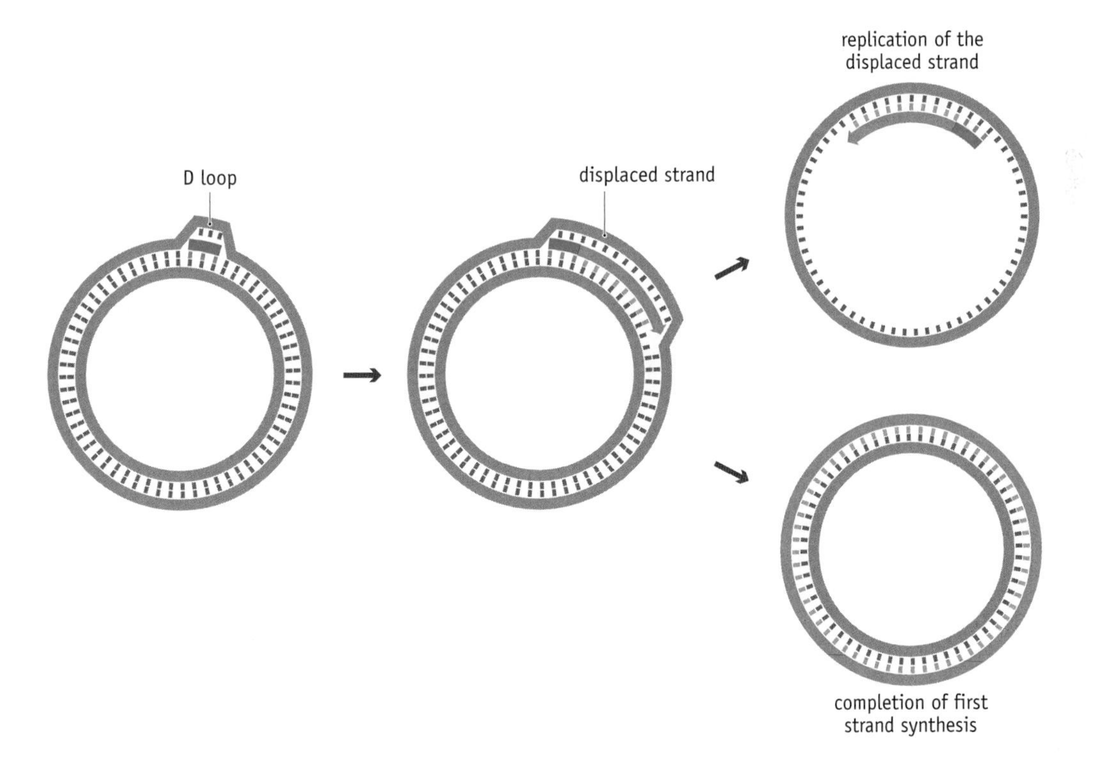

FIGURE 10.8 Displacement replication. The D loop contains a short RNA molecule (shown in blue) that acts as the starting point for synthesis of the first daughter polynucleotide. This polynucleotide is synthesized by continuous copying of one strand of the helix to give the double-stranded molecule at the bottom right. The displaced strand (top right) is then copied by attachment of a second RNA molecule which acts as the starting point for synthesis of the second daughter polynucleotide.

RESEARCH BRIEFING 10.1

DNA replication is semiconservative

The three conceivable ways in which a DNA double helix could be copied are called semiconservative replication, conservative replication, and dispersive replication (see Figure 10.2). The semiconservative scheme seems the most likely to be correct purely on intuitive grounds, but biologists in the 1950s realized that the only way to be certain was to devise an experiment that would distinguish between these three modes of replication and confirm which scheme actually operates. The experiment that settled the question was carried out in 1958 at the California Institute of Technology by Matthew Meselson and Franklin Stahl.

The Meselson–Stahl experiment made use of the heavy isotope of nitrogen

Like many advances in biology, the Meselson–Stahl experiment depended on the use of chemical isotopes, of nitrogen in this case. Nitrogen exists in several isotopic forms. In addition to the "normal" isotope, ^{14}N, which predominates in the environment and has an atomic weight of 14.008, there are a number of other isotopes that occur in much smaller amounts. These include ^{15}N, which because of its greater atomic weight is called "heavy nitrogen." If *Escherichia coli* cells are provided with heavy nitrogen in the form of $^{15}NH_4Cl$, then the bacteria incorporate the isotope into new DNA molecules that they synthesize.

In a mixture of the two, DNA molecules containing only heavy nitrogen can be separated from DNA molecules containing only the normal isotope by density gradient centrifugation. We met this procedure in Chapter 6, where we learned how centrifugation through a sucrose density gradient is used to measure the sedimentation coefficients of ribosomes and other cell components. In a second type of density gradient centrifugation, a solution such as 6 M cesium chloride is used, which is substantially denser than the sucrose solution used to measure sedimentation coefficients. The starting CsCl solution is uniform, the gradient being established during the centrifugation. Cellular components migrate all the way down through the centrifuge tube, but molecules such as DNA and proteins do not reach the bottom. Instead each one comes to rest at a position where its own **buoyant density** equals the cesium chloride density. Meselson and Stahl used this version of density gradient centrifugation because ^{15}N-DNA has a greater buoyant density than ^{14}N-DNA, and therefore forms a band at a lower position in a cesium chloride gradient (Figure 1).

Density gradients reveal the composition of DNA molecules resulting from one and two rounds of replication

Meselson and Stahl grew a culture of *E. coli* in the presence of $^{15}NH_4Cl$ so that the DNA molecules in the cells became labeled with heavy nitrogen. Then the culture was spun in a low-speed centrifuge, the heavy medium discarded, and the bacteria resuspended in medium containing only $^{14}NH_4Cl$. New polynucleotides synthesized after resuspension therefore contained only the normal isotope of nitrogen.

The bacteria were allowed to undergo one round of cell division, which takes roughly 20 minutes for *E. coli*, during which time each DNA molecule replicates just once. Some cells were then taken from the culture, their DNA purified, and a sample analyzed by density gradient centrifugation. The result was a single band of DNA at a position corresponding to a buoyant density intermediate between the values expected for ^{15}N-DNA and ^{14}N-DNA (Figure 2). This shows that after one round of replication each DNA double helix contained roughly equal amounts of ^{15}N-polynucleotide and ^{14}N-polynucleotide.

The culture was then left for another 20 minutes so the bacteria could undergo a second round of cell division. Again, cells were removed and their DNA molecules analyzed in a density gradient. Two bands appeared, one representing the same hybrid molecules as before, and the second corresponding to double helices made entirely of ^{14}N-DNA.

The banding patterns show that replication is semiconservative

How do we interpret the results of the Meselson–Stahl experiment? If we reexamine the three schemes for DNA replication we see that the banding pattern

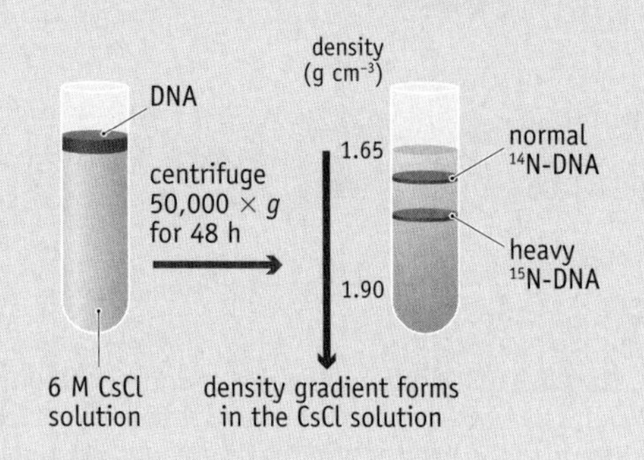

FIGURE 1 The use of density gradient centrifugation to separate ^{14}N- and ^{15}N-DNA molecules.

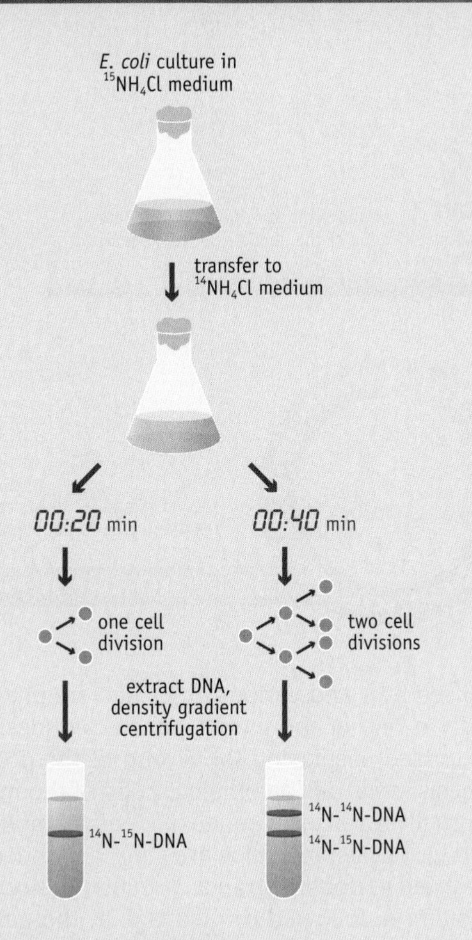

FIGURE 2 The Meselson–Stahl experiment. On the left is the banding pattern seen in the density gradient after a single round of DNA replication, and on the right is the pattern after two rounds of replication.

obtained after a single round of cell division enables conservative replication to be discounted (Figure 3). This scheme predicts that after one round of replication there will be two different types of double helix, one containing just ^{15}N and the other containing just ^{14}N. As just one band is seen, we can rule out conservative replication.

The single ^{14}N-^{15}N-DNA band that was seen after 20 minutes is compatible with both dispersive and semiconservative replication. Under either of these schemes, the daughter molecules will be made up of equal amounts of ^{15}N- and ^{14}N-DNA. If replication is semiconservative then each daughter double helix will comprise one ^{15}N-polynucleotide and one ^{14}N-polynucleotide, and if it is dispersive then both strands will be made up of a mixture of ^{15}N-polynucleotide and ^{14}N-polynucleotide.

To distinguish between semiconservative and dispersive replication, we have to look at the results that Meselson and Stahl obtained after a second round of cell division. The two bands that are now seen indicate that hybrid molecules are still present, but now with a second type of double helix that is made entirely of ^{14}N-DNA. This result is in agreement with the semiconservative mode of replication, because according to this scheme there are now some granddaughter molecules that are made up entirely of ^{14}N-polynucleotides. In contrast, the dispersive mode can be discounted because that method would still produce only hybrid molecules, and in fact would continue to do so for a very large number of cell generations. We can therefore conclude that DNA replication is semiconservative.

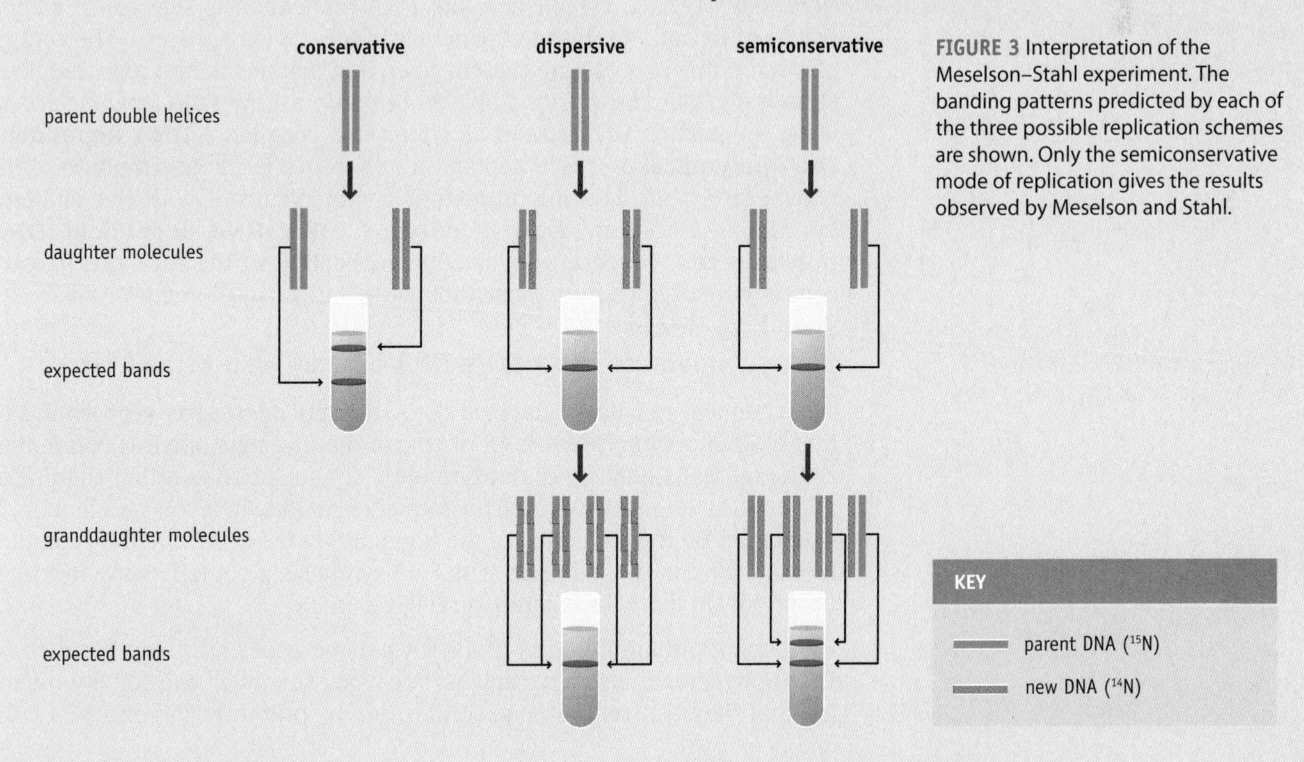

FIGURE 3 Interpretation of the Meselson–Stahl experiment. The banding patterns predicted by each of the three possible replication schemes are shown. Only the semiconservative mode of replication gives the results observed by Meselson and Stahl.

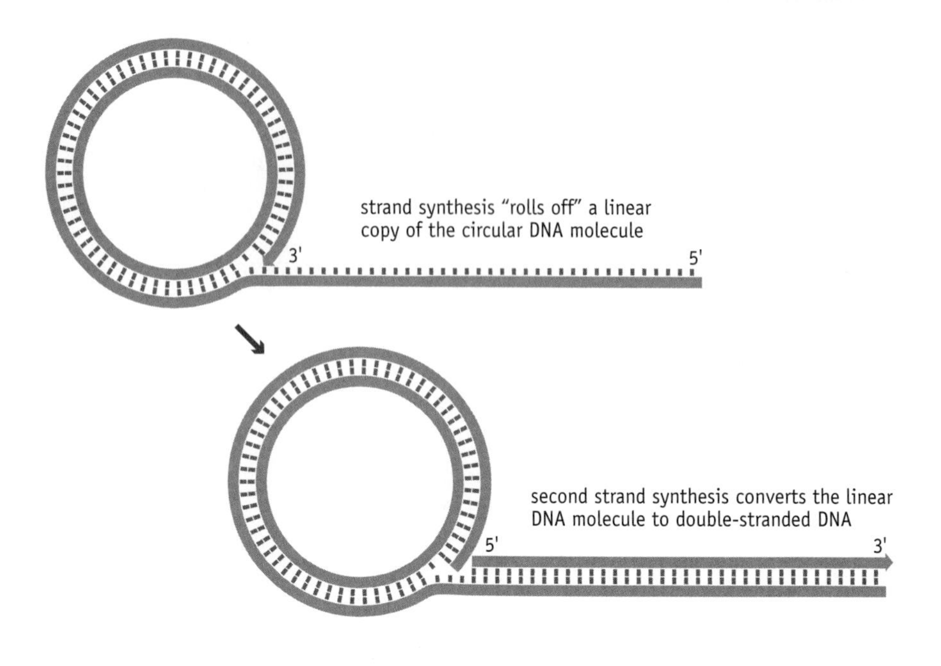

FIGURE 10.9 Rolling-circle replication. The top drawing shows a single strand that has been "rolled off" of the circular molecule by DNA synthesis at its 3′ end. The rolled-off strand is then converted to double-stranded DNA by complementary strand synthesis.

replication, which is used by λ and various other bacteriophages, initiates at a nick that is made in one of the parent polynucleotides. The free 3′ end that results is extended, displacing the 5′ end of the polynucleotide. Continued DNA synthesis "rolls off" a complete copy of the molecule, and further synthesis eventually results in a series of identical molecules linked head to tail (Figure 10.9). These molecules are single-stranded and linear, but can easily be converted to double-stranded circular molecules by complementary strand synthesis, followed by cleavage at the junction points and circularization of the resulting segments.

10.2 DNA POLYMERASES

Now that we have established that DNA replication is semiconservative, we can turn our attention to the details of the copying process. The central players in this process are the enzymes that synthesize the new daughter strands of DNA. An enzyme able to build up a new DNA polynucleotide using an existing DNA strand as a template is called a **DNA-dependent DNA polymerase**. It is acceptable to shorten the name simply to "DNA polymerase," but it is important not to get confused with the different but equally important class of enzymes called **RNA-dependent DNA polymerases**, which are involved in replication of the RNA genomes of certain viruses, including those that cause AIDS.

DNA polymerases synthesize DNA but can also degrade it

The chemical reaction catalyzed by a DNA polymerase is very similar to that of RNA polymerase except of course that the new polynucleotide that is assembled is built up of deoxyribonucleotide subunits rather than ribonucleotides (Figure 10.10). The sequence of the new polynucleotide is dependent on the sequence of the template and is determined by complementary base pairing, and, as with RNA synthesis, DNA polymerization can occur only in the 5′ → 3′ direction (Figure 10.11).

Although the main function of a DNA polymerase is DNA synthesis, most DNA polymerases also have at least one type of exonuclease activity, meaning that they can remove nucleotides one by one from the end of a DNA

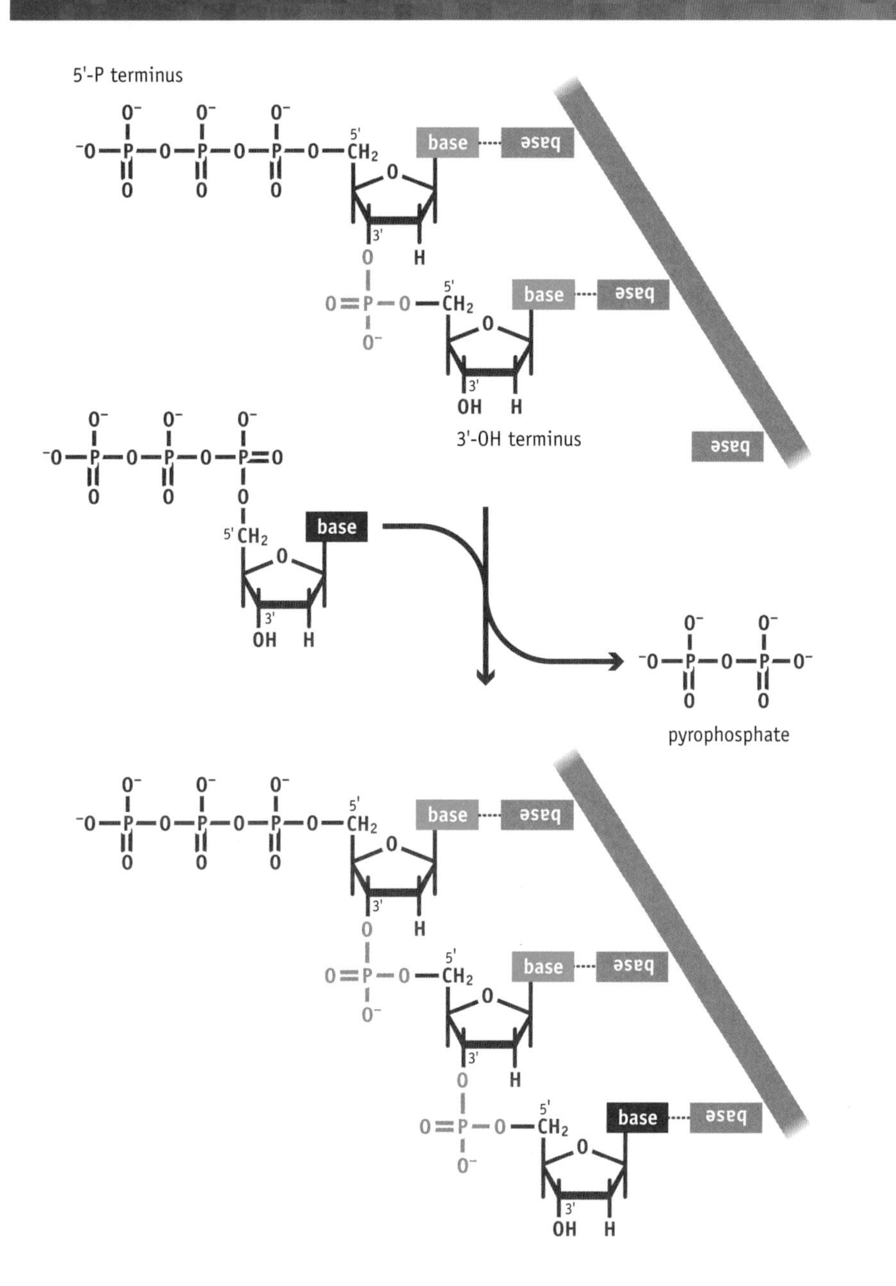

FIGURE 10.10 Template-dependent synthesis of DNA.

polynucleotide. Because the two ends of a polynucleotide are chemically distinct, there are two equally distinct types of exonuclease activity. These are the **3′ → 5′ exonuclease** activity, which enables nucleotides to be removed from the 3′ end of a polynucleotide, and the **5′ → 3′ exonuclease** activity, which removes nucleotides from the 5′ end. Some polymerases are capable of both activities, while others carry out just one or neither of them (Table 10.1).

It may seem odd that DNA polymerases can degrade polynucleotides as well as synthesize them, but when we consider the details of the exonuclease activities the natural logic becomes apparent. The 3′ → 5′ exonuclease

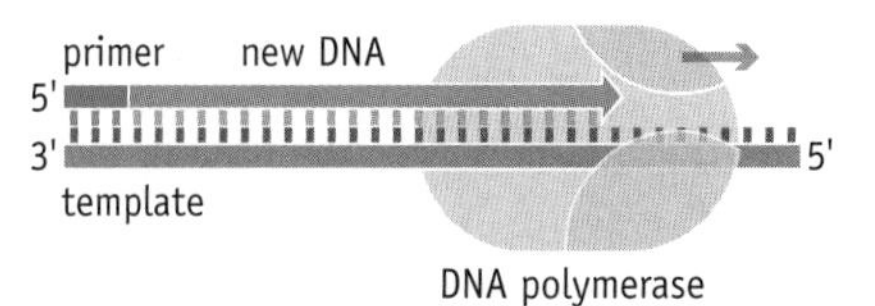

FIGURE 10.11 The action of a DNA polymerase. A DNA polymerase synthesizes DNA in the 5′→3′ direction, with the sequence of the new polynucleotide determined by complementary base pairing. Note that a DNA polymerase cannot initiate DNA synthesis unless there is already a short double-stranded region to act as a primer.

TABLE 10.1 DNA POLYMERASES INVOLVED IN REPLICATION OF BACTERIAL AND EUKARYOTIC GENOMES

Enzyme	Subunits	Exonuclease activities		Function
		3′ → 5′	5′ → 3′	
Bacterial DNA polymerases				
DNA polymerase I	1	Yes	Yes	DNA repair, replication
DNA polymerase III	At least 10	Yes	No	Main replicating enzyme
Eukaryotic DNA polymerases				
DNA polymerase α	4	No	No	Priming during replication
DNA polymerase γ	2	Yes	No	Mitochondrial DNA replication
DNA polymerase δ	2 or 3	Yes	No	Main replicative enzyme
DNA polymerase κ	1	?	?	Required for attachment of cohesin proteins, which hold sister chromatids together

Bacteria and eukaryotes possess other DNA polymerases involved primarily in repair of damaged DNA. These enzymes include DNA polymerases II, IV, and V of *Escherichia coli* and the eukaryotic DNA polymerases β, ε, ζ, η, θ, and ι.

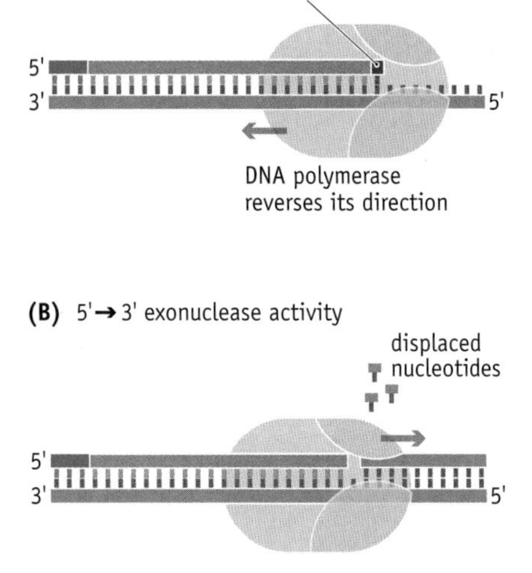

FIGURE 10.12 The exonuclease activities possessed by some DNA polymerases. (A) The 3′ → 5′ exonuclease activity enables a DNA polymerase to remove nucleotides from the 3′ end of a strand that it has just synthesized. (B) The 5′ → 3′ exonuclease activity enables a DNA polymerase to remove part of a polynucleotide that is already attached to the template strand that it is copying.

activity, for example, enables a template-dependent DNA polymerase to remove nucleotides from the 3′ end of a strand that it has just synthesized. This is looked on as a **proofreading** activity because it allows the polymerase to correct errors by removing nucleotides that have been inserted incorrectly (Figure 10.12A).

The 5′ → 3′ exonuclease activity is less common but equally important for the functioning of those polymerases that possess it. It is typically a feature of a polymerase whose role in DNA replication requires that it be able to remove at least part of a polynucleotide that is already attached to the template strand that the polymerase is copying (Figure 10.12B). As we will see later, there is an essential requirement for this activity during DNA replication in bacteria.

Bacteria and eukaryotes possess several types of DNA polymerase

The search for DNA polymerases began in the mid-1950s, as soon as it was realized that DNA synthesis was the key to replication of genes. It was thought that bacteria would probably have just a single DNA polymerase, and when the enzyme now called **DNA polymerase I** was identified in 1957 there was a widespread assumption that this was the main replicating enzyme. The discovery that almost complete inactivation of the *E. coli* *polA* gene, which codes for DNA polymerase I, had no effect on the ability of the cells to replicate their DNA, therefore came as something of a surprise. A second enzyme, **DNA polymerase II**, was then isolated, but this was shown to be involved mainly in repair of damaged DNA rather than DNA replication. It was not until 1972 that the main replicating polymerase of *E. coli*, **DNA polymerase III**, was eventually discovered.

DNA polymerases I and II are single polypeptides, but DNA polymerase III, befitting its role as the main replicating enzyme, is multisubunit, with a molecular mass of approximately 900 kD (Table 10.1). The core enzyme comprises three main subunits, called α, ϵ, and θ. The polymerase activity is specified by the α subunit, and the $3' \rightarrow 5'$ exonuclease activity of the enzyme resides in ϵ. The function of the θ subunit is not clear. It may have a purely structural role in bringing together the other two core subunits and in assembling the various accessory subunits. There are seven of these, the most important being β, which acts as a "sliding clamp" holding the polymerase complex tightly to the DNA.

Eukaryotes have at least ten DNA polymerases, which in mammals are distinguished by Greek suffixes (α, β, γ, δ, etc.), an unfortunate choice of nomenclature as it tempts confusion with the identically named subunits of *E. coli* DNA polymerase III. The main replicating enzyme is **DNA polymerase δ** (Table 10.1), which has two subunits (three according to some researchers) and works in conjunction with an accessory protein called the **proliferating-cell nuclear antigen** (**PCNA**). PCNA is the functional equivalent of the β subunit of *E. coli* DNA polymerase III and holds the enzyme tightly to the DNA that is being copied. **DNA polymerase α** also has an important function in DNA replication, and DNA polymerase γ is responsible for replicating the DNA molecules in mitochondria. Six of the other eukaryotic DNA polymerases are involved in repair of damaged DNA.

The limitations of DNA polymerase activity cause problems during replication

Although DNA polymerases have evolved to replicate DNA, these enzymes have two features that complicate the way in which replication occurs in the cell.

The first difficulty is caused by the limitation that DNA polymerases can synthesize polynucleotides only in the $5' \rightarrow 3'$ direction, which means that the template strands must be read in the $3' \rightarrow 5'$ direction (see Figure 10.11). For one strand of the parent double helix, called the **leading strand**, this is not a problem because the new polynucleotide can be synthesized continuously (Figure 10.13). However, the second or **lagging strand** cannot be copied in a continuous fashion because this would necessitate $3' \rightarrow 5'$ DNA synthesis. Instead the lagging strand has to be replicated in sections. A portion of the parent helix is separated into its two polynucleotides and a short stretch of the lagging strand replicated. A bit more of the helix is separated and another segment of the lagging strand replicated, and so on. At first this process was just a hypothesis, but the isolation in 1969 of **Okazaki fragments**, short pieces of single-stranded DNA associated with DNA replication, confirmed that the suggestion is correct.

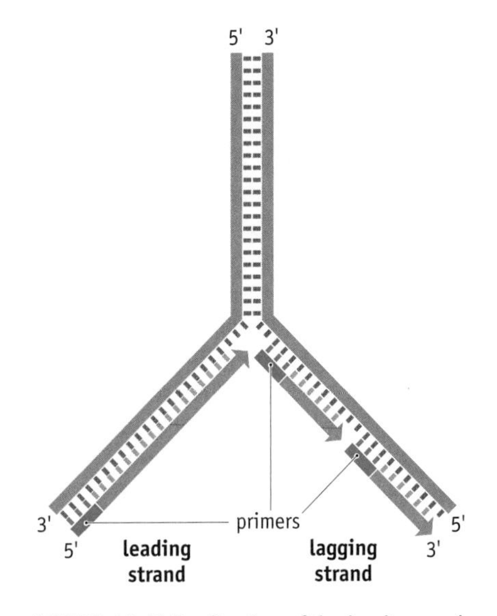

FIGURE 10.13 Replication of the leading and lagging strands of a DNA molecule.

The second difficulty is that, unlike an RNA polymerase, DNA polymerases cannot initiate DNA synthesis unless there is already a short double-stranded region to act as a primer (Figure 10.14). How can this occur during DNA replication? The answer is that the very first few nucleotides attached to either the leading or the lagging strand are not deoxyribonucleotides but ribonucleotides that are put in place by an RNA polymerase enzyme.

RESEARCH BRIEFING 10.2

The polymerase chain reaction

According to DNA folklore, the polymerase chain reaction (PCR) was invented by Kary Mullis during a drive along the coast of California one evening in 1985. Since then it has become one of the central procedures of modern genetics, with applications that extend out into most other areas of biological research.

PCR enables any segment of a DNA molecule, up to about 40 kb in length, to be copied repeatedly so that large quantities are obtained. It can therefore be used to purify a single gene from a mixture of DNA molecules in a cell extract. It is based on DNA replication, so this is a good time to look at how it works and why it has become so important.

In PCR, a thermostable DNA polymerase makes multiple copies of a selected DNA sequence

The PCR procedure results in the repeated copying of a selected region of a DNA molecule. The copying is carried out by the *Taq* polymerase, which is the DNA polymerase I enzyme from the bacterium called *Thermus aquaticus*. This bacterium lives in hot springs where the temperature often exceeds 90°C. In order to survive in such a harsh environment, many of its enzymes are **thermostable**, which means that they remain active at temperatures much higher than those that can be withstood by normal enzymes, such as the ones in *E. coli* and human cells. The biochemical basis of protein thermostability is not fully understood, but it probably centers on structural features that reduce the amount of protein unfolding that occurs at elevated temperatures. In the case of the *Taq* polymerase, its special features, whatever they are, mean that it has an optimal working temperature of 72°C.

To carry out a PCR experiment, a DNA sample is mixed with the *Taq* polymerase, a pair of oligonucleotides, and a supply of nucleotides. The amount of DNA can be very small because PCR is extremely sensitive and will work with just a single starting molecule. The oligonucleotides are needed to prime the DNA synthesis reactions that will be carried out by the *Taq* polymerase. They must attach to the DNA at either side of the segment that is to be copied. The sequences of these attachment sites must therefore be known so that primers that will bind to them can be synthesized.

The reaction is started by heating the mixture to 94°C. At this temperature the hydrogen bonds that hold together the two polynucleotides of the double helix are broken, so the DNA becomes separated, or "denatured," into single-stranded molecules (Figure 1). The temperature is then reduced to 50–60°C, which results in some rejoining of the single strands, but also allows the primers to attach to their annealing positions. DNA synthesis can now begin, so the temperature is raised to 72°C. In this first stage of the PCR, a set of "long products" is synthesized from each strand of the DNA. These polynucleotides have identical 5′ ends but random 3′ ends, the latter representing positions where DNA synthesis has terminated by chance.

The cycle of denaturation–annealing–synthesis is repeated, the long products now acting as templates for new DNA synthesis. This results in a mixture of second-cycle products (Figure 2). Then in the next cycle we see the first "short products" appearing in the reaction mix. These are the key ones because their 5′ and 3′ ends are set by the primer annealing positions (Figure 2). In subsequent cycles, the number of short products accumulates in an exponential fashion, doubling during each cycle, until one of the components of the reaction becomes depleted. This means that after 30 cycles, there will be over 130 million short

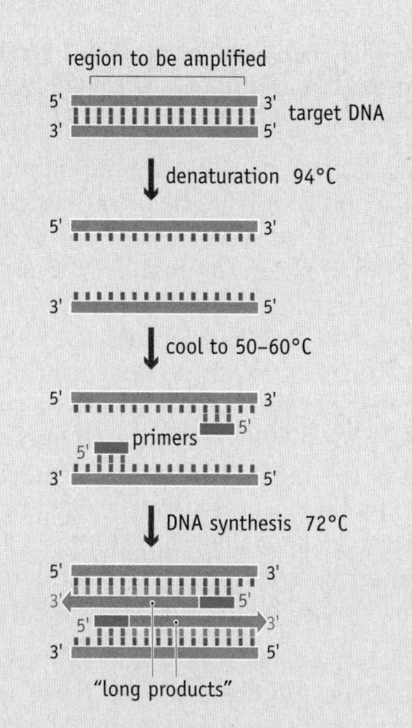

FIGURE 1 The first cycle of a PCR. During this stage of the reaction, the two strands of the target DNA are copied into a pair of "long products", whose 5′ ends are marked by the primers used to initiate the DNA synthesis.

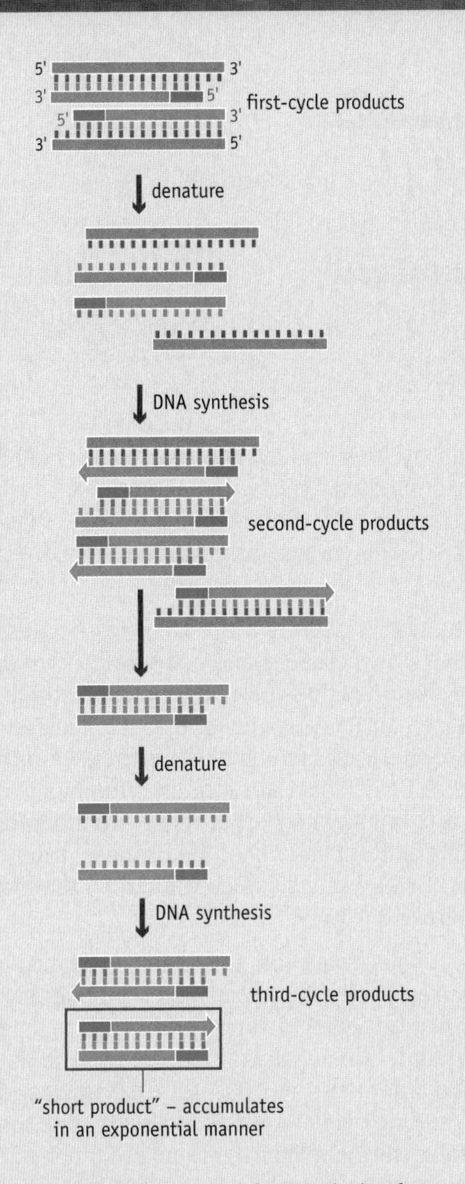

FIGURE 2 The second and third cycles of a PCR. The key feature is the synthesis during the third cycle of the first "short products", ones whose 5′ and 3′ ends are both set by the primer annealing positions. In subsequent cycles, these short products accumulate exponentially in the reaction mixture.

products derived from each starting molecule. In real terms, this equates to several µg of PCR product from a few ng or less of target DNA. The sequence that has been amplified can now be studied in various ways, for example, by DNA sequencing, to obtain information about any genes or other interesting features that it contains.

PCR has applications in many areas of biology

PCR is such a straightforward procedure that it is sometimes difficult to understand how it can have become so important in modern research. First we will deal with its limitations. In order to synthesize primers that will anneal at the correct positions, the sequences of the boundary regions of the DNA to be amplified must be known. This means that PCR cannot be used to purify fragments of genes or other parts of a genome that have never been studied before.

A second constraint is the length of DNA that can be copied. Regions of up to 5 kb can be amplified without too much difficulty, and longer amplifications—up to 40 kb—are possible using modifications of the standard technique. But many genes, especially human ones, are much longer than this, and so cannot be obtained as intact copies by PCR. They can, however, be prepared in segments by carrying out a series of PCRs directed at overlapping regions of their DNA sequence (Figure 3).

What are the strengths of PCR? The need to know the sequences of the primers is not too much of a drawback because there are many applications where the genes that are being studied have already been sequenced. A PCR of the human globin genes, for example, is used to test blood samples for the presence of mutations that might cause the disease called thalassemia. Design of appropriate primers for this PCR is easy because the sequences of the human globin genes are known. After the PCR, the gene copies are sequenced or studied in some other way to determine whether any of the thalassemia mutations are present.

Another clinical application of PCR involves the use of primers specific for the DNA of a disease-causing virus. A positive result indicates that a sample contains the virus and that the person who provided the sample should undergo treatment to prevent onset of the disease. PCR is tremendously sensitive. A carefully set-up reaction yields detectable amounts of DNA even if there is just one DNA molecule in the starting mixture. This means that the technique can detect viruses at the earliest stages of an infection, increasing the chances of treatment being successful. This great sensitivity means that PCR can also be used with DNA from forensic material such as hairs and dried bloodstains or even from the bones of long-dead humans, as we will see in Chapter 21.

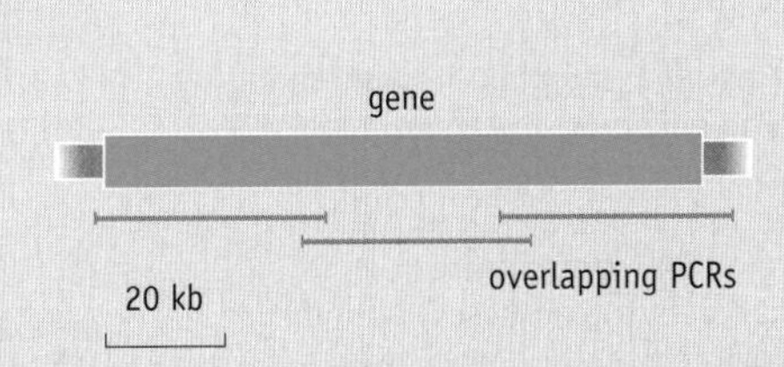

FIGURE 3 Amplification of a gene of approximately 100 kb as a series of overlapping PCRs.

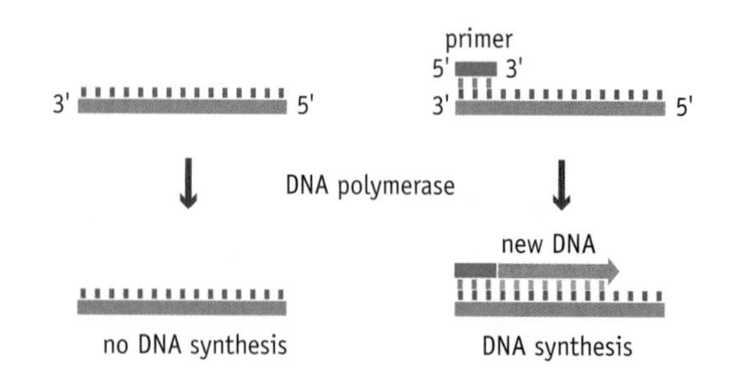

FIGURE 10.14 A primer is needed to initiate DNA synthesis by a DNA polymerase.

In bacteria, primers are synthesized by **primase**, a special RNA polymerase unrelated to the transcribing enzyme, with each primer 4 to 15 nucleotides in length and most starting with the sequence 5′-AG-3′. Once the primer has been completed, strand synthesis is continued by DNA polymerase III (Figure 10.15A).

In eukaryotes the situation is slightly more complex because the primase is tightly bound to DNA polymerase α and cooperates with this enzyme in synthesis of the first few nucleotides of a new polynucleotide. This eukaryotic primase synthesizes an RNA primer of 8 to 12 nucleotides, and then hands over to DNA polymerase α, which extends the RNA primer by adding about 20 nucleotides of DNA. This DNA stretch often has a few ribonucleotides mixed in, but it is not clear whether these are incorporated by DNA polymerase α or by intermittent activity of the primase. After completion of the RNA–DNA primer, DNA synthesis is continued by the main replicative enzyme, DNA polymerase δ (Figure 10.15B).

Priming needs to occur just once on the leading strand, because once primed, the leading-strand copy is synthesized continuously until replication is completed. On the lagging strand, priming is a repeated process that must occur every time a new Okazaki fragment is initiated. In *E. coli*, which makes Okazaki fragments of 1000 to 2000 nucleotides in length, approximately 4000 priming events are needed every time the cell's DNA is replicated. In eukaryotes the Okazaki fragments are much shorter, perhaps less than 200 nucleotides in length, and priming is a highly repetitive event.

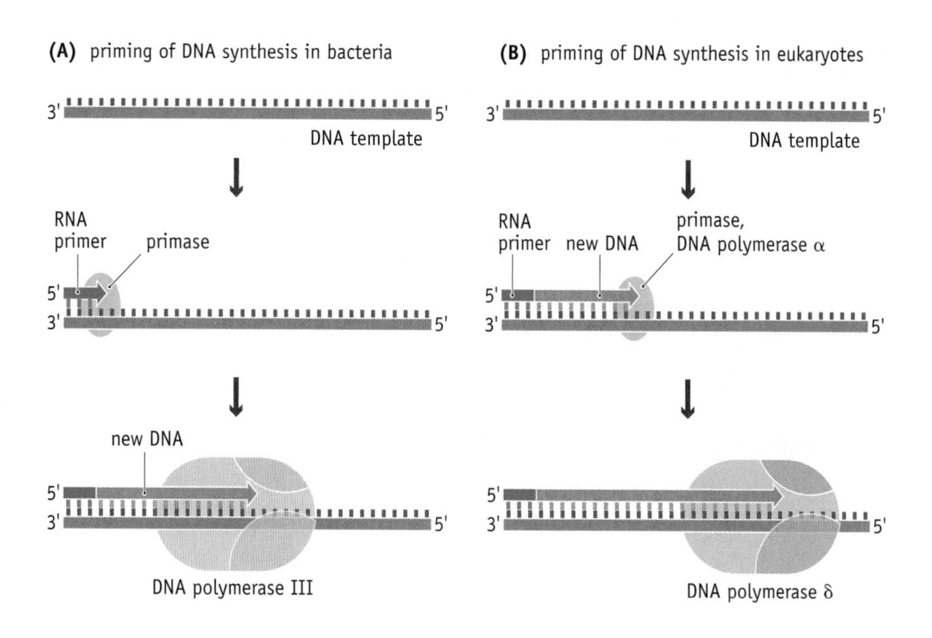

FIGURE 10.15 Priming of DNA synthesis in bacteria and eukaryotes. (A) In bacteria, the primer is synthesized by an RNA polymerase called primase. Strand synthesis is then continued by DNA polymerase III.
(B) In eukaryotes, the primase is bound to DNA polymerase α and works with this enzyme in synthesis of the first 30 or so nucleotides. DNA synthesis is then continued by DNA polymerase δ.

10.3 DNA REPLICATION IN BACTERIA

There are many similarities between DNA replication in bacteria and eukaryotes, and the easiest way to deal with the topic is first to consider the less complex and better understood process occurring in bacteria and then to move on to the special features of replication in eukaryotes.

As with many processes in genetics, we conventionally look on DNA replication as being made up of three phases—initiation, elongation, and termination. Initiation of replication involves recognition of the position(s) on a DNA molecule where replication will begin. Elongation concerns the events occurring at the replication fork, where the parent polynucleotides are copied. And termination, which in general is only vaguely understood, occurs when the parent molecule has been completely replicated. We will consider each of these stages of replication in turn.

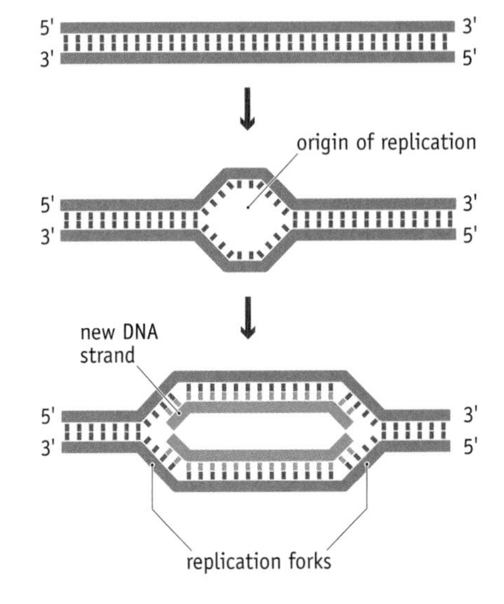

FIGURE 10.16 An origin of replication.

E. coli has a single origin of replication

When a DNA molecule is being replicated, only a limited region is ever in a non-base-paired form. The breakage in base pairing starts at a distinct position, called the **origin of replication** (Figure 10.16), and gradually progresses along the molecule, possibly in both directions, with synthesis of the new polynucleotides occurring as the double helix unzips.

The circular *E. coli* DNA molecule, which is typical of most bacteria, has just one origin of replication. This origin spans approximately 245 bp of DNA and contains two short repeat motifs, one of 9 nucleotides and the other of 13 nucleotides (Figure 10.17). The 9-nucleotide repeat, 5 copies of which are dispersed throughout the origin, is the binding site for a protein called DnaA. With 5 copies of the binding sequence, it might be imagined that 5 copies of DnaA become attached, but in fact bound DnaA proteins cooperate with unbound molecules until some 30 are associated with the origin. The result of DnaA binding is that the double helix opens up ("melts") within the tandem array of three AT-rich, 13-nucleotide repeats located at one end of the origin (Figure 10.18). The exact mechanism is unknown, but DnaA does not appear to possess the enzymatic activity needed to break base pairs, and it is therefore assumed that the helix is melted by torsional stresses introduced by attachment of the DnaA proteins. An attractive model imagines that the DnaA proteins form a barrel-like structure around which the helix is wound.

Melting of the helix initiates a series of events that construct a nascent **replication fork** at either end of the open region. The first step is the attachment of a **prepriming complex** at each of these two positions. Each prepriming complex initially comprises six copies of the DnaB protein and six copies of DnaC. DnaC has a transitory role and is released from the complex soon after it is formed, its function probably being simply to aid the attachment of DnaB. The latter is a **helicase**, an enzyme that can break base pairs.

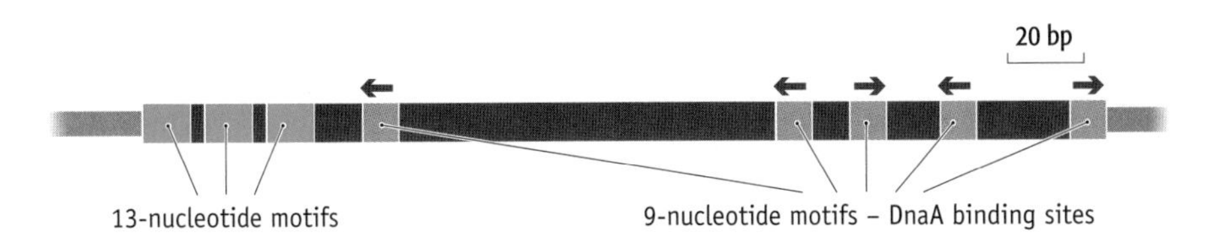

FIGURE 10.17 Sequence motifs within the *E. coli* origin of replication.

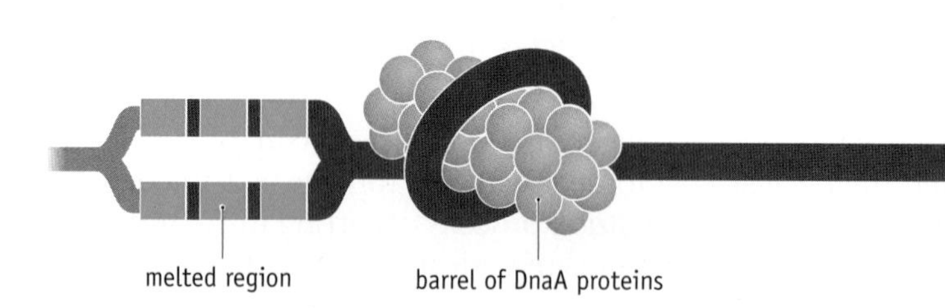

FIGURE 10.18 A model for the attachment of DnaA proteins to the *E. coli* origin of replication, resulting in melting of the helix within the AT-rich 13-nucleotide sequences.

DnaB begins to increase the size of the open region by increasing the lengths of single-stranded DNA that are exposed. These single-stranded regions are naturally "sticky", and the two separated polynucleotides would immediately re-form base pairs after the enzyme has passed, if allowed to. The single strands are also highly susceptible to nuclease attack and are likely to be degraded if not protected in some way. To avoid these unwanted outcomes, **single-strand binding proteins** (**SSBs**) attach to the polynucleotides and prevent them from reassociating or being degraded (Figure 10.19).

As soon as DnaB has begun to increase the size of the open region, the enzymes involved in the elongation phase of DNA replication are able to attach. The replication forks now start to progress away from the origin, and DNA copying begins.

The elongation phase of bacterial DNA replication

The first enzyme involved in DNA copying is recruited soon after the DnaB helicase has bound to the origin. This enzyme is the primase, the special type of RNA polymerase that synthesizes the primers needed before DNA polymerase III can begin to make DNA. Addition of the primase converts the prepriming complex into the structure called the **primosome**, and replication of the leading strand is initiated.

After 1000 to 2000 nucleotides of the leading strand have been replicated, the first round of discontinuous strand synthesis on the lagging strand can begin. The primase, which is still associated with the DnaB helicase in the primosome, makes an RNA primer, which is then extended by DNA polymerase III (Figure 10.20). This means that there are now two DNA polymerase core enzymes attached to the double helix, one copying the leading strand and one copying the lagging strand.

There is also a single **γ complex**, which is made up of five of the accessory subunits of DNA polymerase III (called γ, δ, δ', χ, and ψ). The main function of the γ complex (sometimes called the "clamp loader") is to interact with the β subunit (the "sliding clamp") of each polymerase, and hence control the attachment and removal of the enzymes to and from the template DNA. This function is required primarily during lagging-strand replication, when the polymerase has to attach and detach repeatedly at the start and end of each Okazaki fragment.

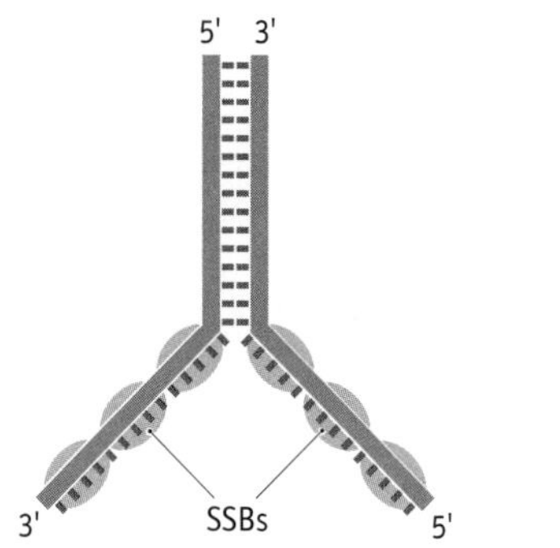

FIGURE 10.19 The role of single-strand binding proteins (SSBs) during DNA replication. SSBs attach to the unpaired polynucleotides produced by helicase action and prevent the strands from base-pairing with one another or being degraded by nucleases.

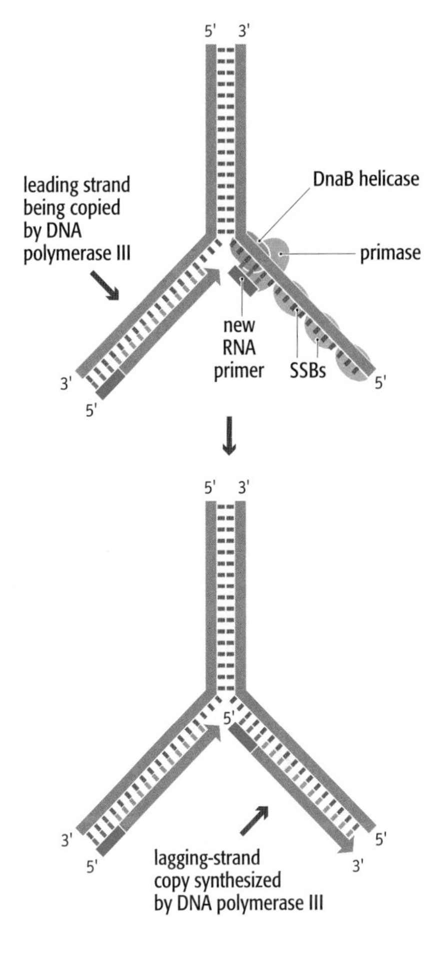

FIGURE 10.20 Priming and synthesis of the lagging-strand copy during DNA replication in *E. coli*. In the top drawing, replication of the leading strand has begun, and the primase has synthesized a primer on the lagging strand. In the lower drawing, this primer has been extended to begin replication of the lagging strand. Note that this is just the first round of discontinuous strand synthesis on the lagging strand. As the replication fork progresses along the parent molecule, additional rounds of lagging-strand replication are needed.

Some models of the replication complex place the two polymerase enzymes in opposite orientations to reflect the different directions in which DNA synthesis occurs, toward the replication fork on the leading strand and away from it on the lagging strand. It is more likely, however, that the pair of enzymes face the same direction and the lagging strand forms a loop. This would enable DNA synthesis on the two strands to proceed in parallel as the complex moves forward, keeping pace with the progress of the replication fork (Figure 10.21). The combination of the two DNA polymerase III enzymes, the γ complex, and the primosome, migrating along the parent DNA and carrying out most of the replicative functions, is called the **replisome**.

After the replisome has passed, the replication process must be completed by joining up the individual Okazaki fragments. This is not a trivial event, because one member of each pair of adjacent Okazaki fragments still has its RNA primer attached at the point where ligation should take place (Figure 10.22). The primer cannot be removed by DNA polymerase III, because this enzyme lacks the required $5' \rightarrow 3'$ exonuclease activity (see Table 10.1). At

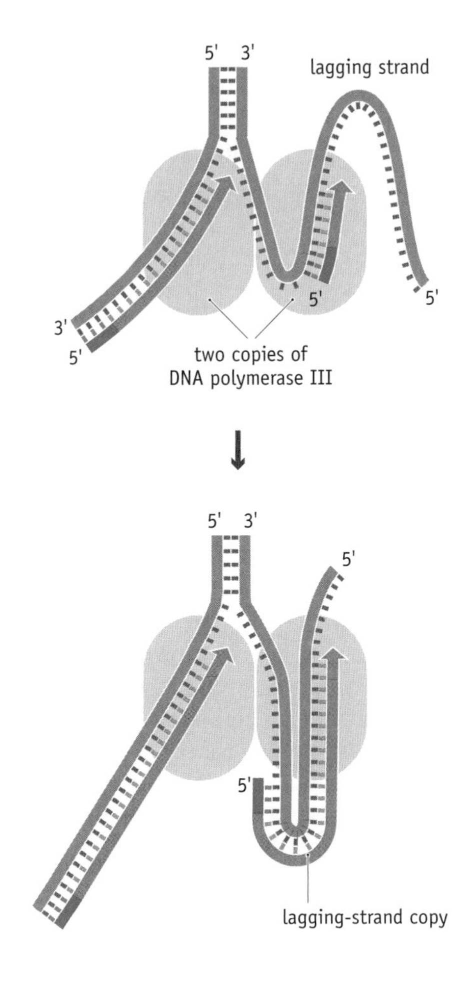

FIGURE 10.21 A model for parallel replication of the leading and lagging strands. There are two copies of DNA polymerase III, one for each strand. It is thought that the lagging strand loops through its copy of the polymerase, in the manner shown. This would enable the leading and lagging strands to be replicated in parallel as the two polymerase enzymes move in the same direction along the parent DNA molecule.

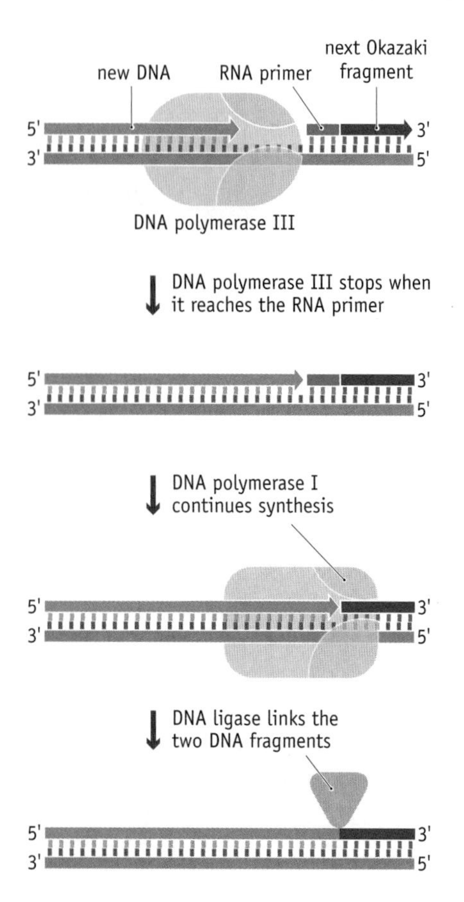

FIGURE 10.22 The series of events involved in joining up adjacent Okazaki fragments during DNA replication in *E. coli*. DNA polymerase III detaches when it reaches the RNA primer of the next Okazaki fragment. DNA polymerase I removes the RNA primer and replaces it with DNA. The final phosphodiester bond is synthesized by DNA ligase.

this point, DNA polymerase III releases the lagging strand and its place is taken by DNA polymerase I, which does have a 5′ → 3′ exonuclease activity and so removes the primer. Usually it also removes the start of the DNA component of the Okazaki fragment, extending the 3′ end of the adjacent fragment into the region of the template that is exposed. The two Okazaki fragments now abut, with the terminal regions of both composed entirely of DNA. All that remains is for the missing phosphodiester bond to be put in place by a **DNA ligase**, linking the two fragments and completing replication of this region of the lagging strand.

Replication of the *E. coli* genome terminates within a defined region

Replication forks proceed along linear DNA molecules, or around circular ones, generally unimpeded except when they encounter a region that is being transcribed. DNA synthesis occurs at approximately 5 times the rate of RNA synthesis, so the replication complex could easily overtake an RNA polymerase, but it probably never does so. Instead it is thought that the replication fork pauses behind the RNA polymerase, proceeding only when the transcript has been completed. Eventually the replication fork reaches the end of a linear molecule or, with a circular one, meets a second replication fork moving in the opposite direction. What happens next is one of the less well understood aspects of DNA replication.

Bacterial DNA molecules are replicated bidirectionally from a single point (**Figure 10.23**), which means that the two replication forks should meet at a position diametrically opposite the origin of replication on the DNA molecule. However, if one fork is delayed, perhaps because it has to replicate extensive regions where transcription is occurring, then it might be possible for the other fork to overshoot the halfway point and continue replication on the "other side" of the molecule (**Figure 10.24**). It is not immediately apparent why this should be undesirable, the daughter molecules presumably being unaffected, but in reality it is not allowed to happen, because of the presence of **terminator sequences**. Six such sequences have been identified in *E. coli* (**Figure 10.25**), each one acting as the recognition site for a sequence-specific DNA-binding protein called **Tus**.

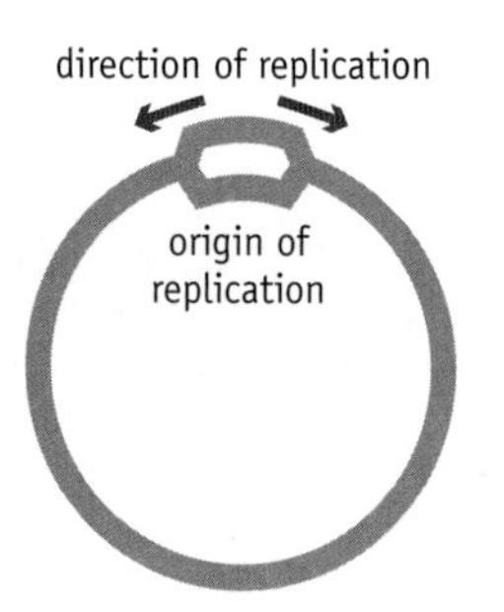

FIGURE 10.23 Bidirectional replication of a circular bacterial DNA molecule.

The mode of action of Tus is quite unusual. When bound to a terminator sequence, a Tus protein allows a replication fork to pass if the fork is moving in one direction, but blocks progress if the fork is moving in the opposite direction. The directionality is set by the orientation of the Tus protein on the double helix. When approaching from one direction, the DnaB helicase, which is responsible for progression of the replication fork, encounters an impenetrable wall of β strands on one side of the Tus protein (**Figure 10.26**). The replication fork cannot get through this wall and so stops. But coming from the other direction, DnaB meets a less rigid part of the Tus structure, and hence is able to displace the protein, enabling the replication fork to continue its progress along the DNA.

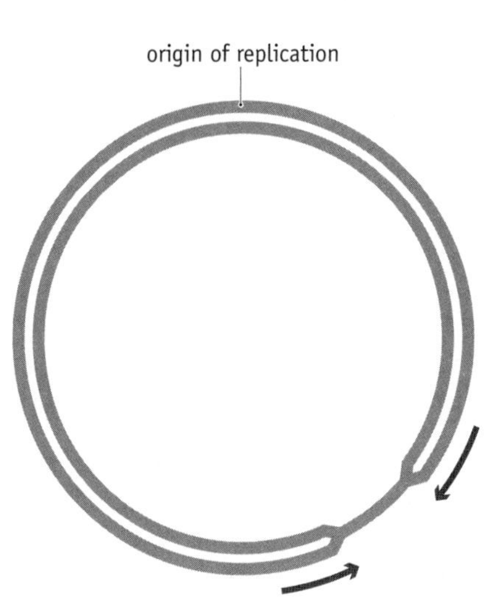

FIGURE 10.24 A situation that is not allowed to occur during replication of the circular *E. coli* genome. One of the replication forks has proceeded some distance past the halfway point.

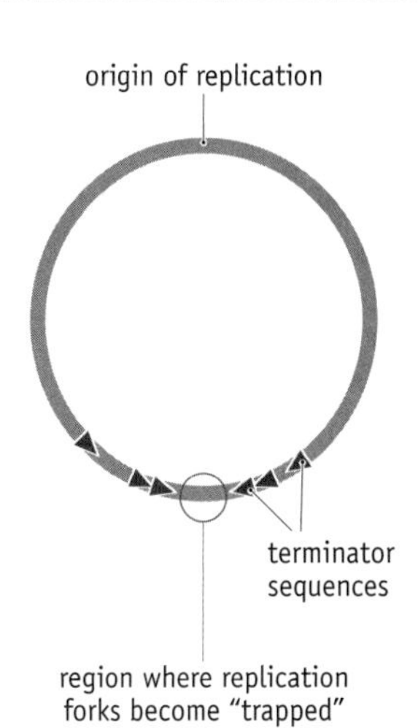

FIGURE 10.25 The positions of the six terminator sequences on the *E. coli* DNA molecule. The arrowheads indicate the direction in which each terminator sequence can be passed by a replication fork.

The orientation of the termination sequences, and hence of the bound Tus proteins, on the *E. coli* DNA molecule is such that both replication forks become trapped within a relatively short region (see Figure 10.25). This ensures that termination always occurs at or near the same position. Exactly what happens when the two replication forks meet is unknown, but meeting is followed by disassembly of the replisomes, either spontaneously or in a controlled fashion. The result is two interlinked daughter molecules, which are separated by topoisomerase IV.

10.4 DNA REPLICATION IN EUKARYOTES

As mentioned above, in outline DNA replication is similar in both bacteria and eukaryotes. Eukaryotic replication does, however, have its own special features. We will now take a look at these.

Eukaryotic genomes have multiple replication origins

Most circular bacterial DNA molecules have a single origin of replication. Eukaryotic chromosomes, in contrast, have multiple origins, 332 in the yeast *Saccharomyces cerevisiae* and some 20,000 in humans. The length of DNA copied by a single replication fork also varies greatly between different organisms. In bacteria, each fork has to travel for several Mb before it meets the second fork progressing in the opposite direction. In humans about 150 kb of DNA is replicated per fork, and in yeast the figure is just 36 kb.

Yeast and other simple eukaryotes have discrete origins of replication similar in some ways to that of *E. coli*. In *S. cerevisiae*, an origin is about 200 bp in length and comprises four segments (A, B1, B2, and B3; Figure 10.27). Two of these (A and B1) make up the **origin recognition sequence**, a stretch of some 40 bp in total that is the binding site for the **origin recognition complex** (**ORC**), a set of six proteins that attach to the origin. The ORC proteins have been described as yeast versions of the *E. coli* DnaA

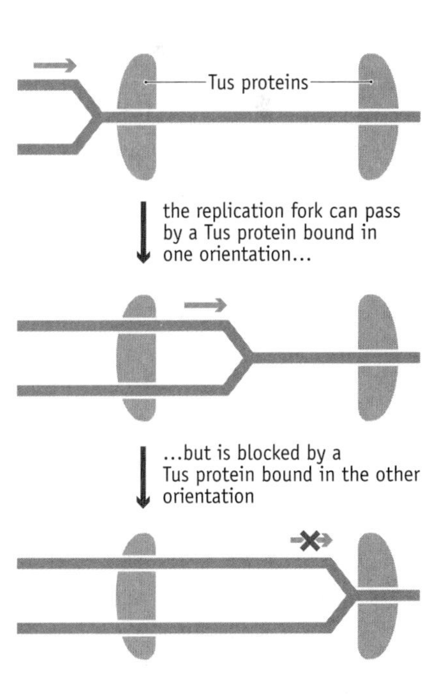

FIGURE 10.26 Bound Tus proteins allow a replication fork to pass when the fork approaches from one direction but not when it approaches from the other direction.

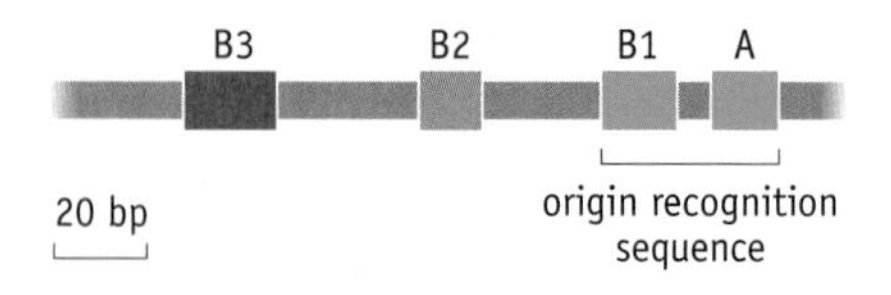

FIGURE 10.27 The structure of a yeast origin of replication.

proteins, but this interpretation is probably not strictly correct because ORCs appear to remain attached to yeast origins even when replication is not being initiated. Rather than being genuine initiator proteins, it is more likely that ORCs are involved in the regulation of DNA replication, acting as mediators between replication origins and the regulatory signals that coordinate the initiation of DNA replication with division of the cell. The real equivalents of DnaA probably bind to a different segment of a yeast origin.

Attempts to identify replication origins in more complex eukaryotes such as humans have been less successful. **Initiation regions** (parts of the chromosomal DNA where replication initiates) have been identified in mammals, suggesting that there are specific regions in mammalian chromosomes where replication begins, but some researchers are doubtful whether these regions contain replication origins equivalent to those in yeast and bacteria. One alternative hypothesis is that replication is initiated by protein structures that have specific positions in the nucleus, the initiation regions simply being those DNA segments located close to these protein structures in the three-dimensional organization of the nucleus.

The eukaryotic replication fork: variations on the bacterial theme

As in bacteria, progress of a eukaryotic replication fork is maintained by helicase activity, although which of the several eukaryotic helicases that have been identified are primarily responsible for DNA melting during replication has not been established. The separated polynucleotides are prevented from reattaching by single-strand binding proteins, the main one of these in eukaryotes being **replication protein A** (**RPA**).

We begin to encounter unique features of the eukaryotic replication process when we examine the method used to prime DNA synthesis. As described above, the eukaryotic DNA polymerase α cooperates with the primase enzyme to put in place the RNA–DNA primers at the start of the leading-strand copy and at the beginning of each Okazaki fragment. DNA polymerase α is not capable of lengthy DNA synthesis, presumably because it lacks the stabilizing effect of a sliding clamp equivalent to the β subunit of *E. coli* DNA polymerase III or the PCNA accessory protein that aids the eukaryotic DNA polymerase δ. This means that although DNA polymerase α can extend the initial RNA primer with about 20 nucleotides of DNA, it must then be replaced by the main replicative enzyme, DNA polymerase δ (see Figure 10.15).

As in *E. coli*, completion of lagging-strand synthesis requires removal of the RNA primer from each Okazaki fragment. There appears to be no eukaryotic DNA polymerase with the 5′ → 3′ exonuclease activity needed for this purpose, and the process is therefore very different from that described for bacterial cells. The central player is the "flap endonuclease," **FEN1**, which associates with DNA polymerase δ at the 3′ end of an Okazaki fragment, in order to degrade the primer from the 5′ end of the adjacent fragment. Understanding exactly how this occurs is complicated by the inability of FEN1 to initiate primer degradation. This is because the ribonucleotide at the extreme 5′ end of the primer has a 5′-triphosphate group which blocks FEN1 activity (Figure 10.28). One possibility is that a helicase breaks the base pairs holding the primer to the template strand, enabling the primer to be pushed aside by DNA polymerase δ as it extends the adjacent Okazaki fragment into the region thus exposed (Figure 10.29). The flap that results can be cut off by FEN1, whose endonuclease activity can cleave the phosphodiester bond at the branch point where the displaced region attaches to the part of the fragment that is still base-paired.

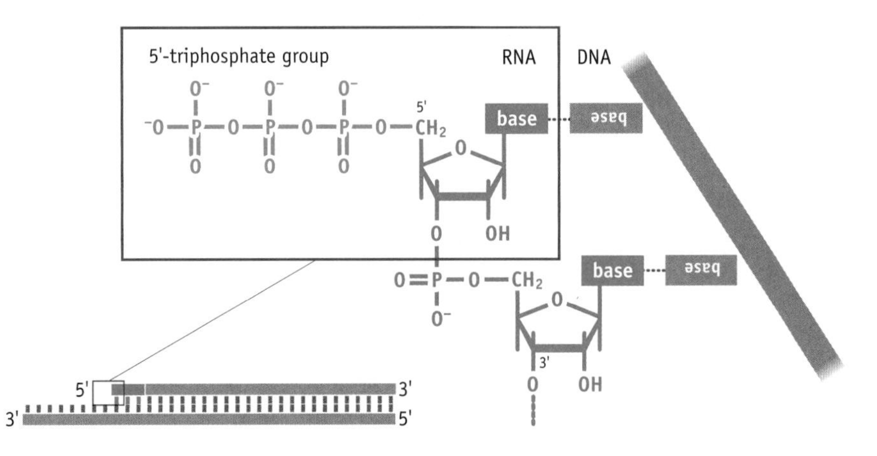

FIGURE 10.28 The "flap endonuclease" FEN1 cannot initiate primer degradation because its activity is blocked by the triphosphate group present at the 5′ end of the primer.

It is thought that both the RNA primer and all of the DNA originally synthesized by DNA polymerase α are removed. This is necessary because DNA polymerase α has no 3′ → 5′ proofreading activity (see Table 10.1) and therefore synthesizes DNA in a relatively error-prone manner. Removal of this region as part of the flap cleaved by FEN1, followed by resynthesis by DNA polymerase δ (which does have a proofreading activity and so makes a highly accurate copy of the template), would prevent these errors from becoming permanent features of the daughter double helix.

The final difference between replication in bacteria and eukaryotes is that in eukaryotes there is no equivalent of the bacterial replisome. Instead, the enzymes and proteins involved in replication form sizable structures within the nucleus, each containing hundreds or thousands of individual replication complexes. These structures are immobile because of attachments with the nuclear matrix, so DNA molecules are threaded through the complexes as they are replicated. The structures are referred to as **replication factories**.

Little is known about termination of replication in eukaryotes

No sequences equivalent to bacterial terminators are known in eukaryotes, and proteins similar to Tus have not been identified. Quite possibly, replication forks meet at random positions and termination simply involves ligation of the ends of the new polynucleotides. We do know that the replication complexes do not break down, because these factories are permanent features of the nucleus.

Rather than concentrating on the molecular events occurring when replication forks meet, attention has been focused on the difficult question of how the daughter DNA molecules produced in a eukaryotic nucleus do not become impossibly tangled up. Although DNA topoisomerases have the ability to untangle DNA molecules, it is generally assumed that tangling is kept to a minimum so that extensive breakage-and-reunion reactions, as catalyzed by topoisomerases, can be avoided. Various models have been proposed to solve this problem. One of these suggests that a eukaryotic chromosome is not randomly packed into its nucleus, but is ordered

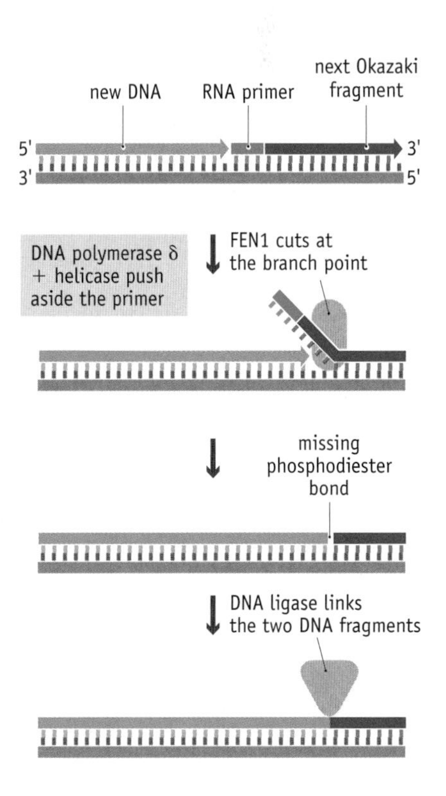

FIGURE 10.29 Completion of lagging-strand replication in eukaryotes. It is thought that a helicase helps DNA polymerase δ to push aside the primer of the next Okazaki fragment. The resulting flap is then cut off by FEN1, and the final phosphodiester bond synthesized by DNA ligase.

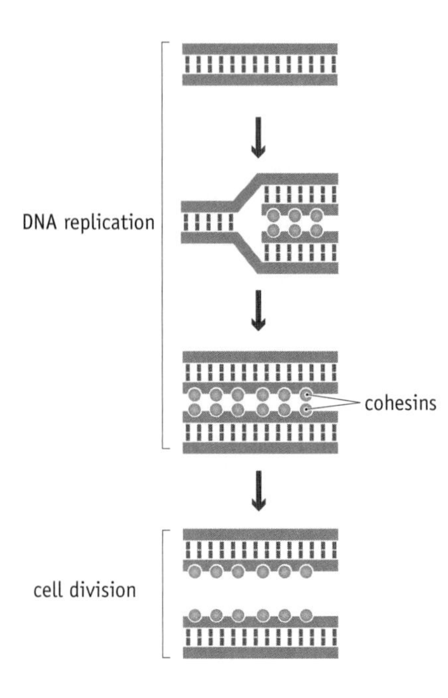

FIGURE 10.30 The role of cohesin proteins. Cohesins attach immediately after passage of the replication fork and hold the daughter molecules together until the cell divides. The cohesins are then cleaved, enabling the replicated chromosomes to separate prior to their distribution into daughter cells.

around the replication factories, which appear to be present in only limited numbers. It is envisaged that each factory replicates a single region of the DNA, maintaining the daughter molecules in a specific arrangement that avoids their entanglement.

Initially, the two daughter molecules are held together by **cohesin** proteins. These are attached to the DNA immediately after passage of the replication fork by a process that appears to involve DNA polymerase κ, an enigmatic enzyme that is essential for replication but whose role does not obviously require a DNA polymerase activity. The cohesins maintain the alignment of the daughter DNA molecules until the cell divides, when they are cleaved by cutting proteins, enabling the daughter chromosomes to separate (Figure 10.30).

KEY CONCEPTS

- The discovery of the double-helix structure suggested a way in which DNA molecules could be copied. Experiments carried out soon after the discovery confirmed that DNA replication occurs in a semiconservative manner.
- Semiconservative replication presents a topological problem, because the two strands of the double helix cannot be separated without unwinding. The problem is solved by the action of DNA topoisomerases.
- An enzyme able to build up a new DNA polynucleotide using an existing DNA strand as a template is called a DNA-dependent DNA polymerase. Most DNA polymerases have exonuclease as well as polymerase activity.
- DNA polymerases can synthesize polynucleotides only in the 5′ → 3′ direction. This means that one strand of the parent double helix has to be replicated in sections.
- DNA polymerases cannot initiate DNA synthesis unless there is already a short double-stranded region to act as a primer. To solve this problem, the first few nucleotides that are attached are ribonucleotides that are put in place by an RNA polymerase enzyme.
- Most bacterial DNA molecules have a single origin of replication, but eukaryotic DNA molecules have many origins.
- Progress of a replication fork along a DNA molecule requires the coordinated activities of several different enzymes and proteins.
- In bacteria, replication forks traveling in opposite directions around a circular DNA molecule become trapped in a region defined by terminator sequences. These sequences prevent one fork or the other from progressing too far around the DNA.
- In eukaryotic nuclei, DNA replication is carried out by large complexes called replication factories. These structures are immobile because of attachments with the nuclear matrix, so DNA molecules are threaded through the complexes as they are replicated.

QUESTIONS AND PROBLEMS (Answers can be found at www.garlandscience.com/introgenetics)

Key Terms

Write short definitions of the following terms:

γ complex
$3' \rightarrow 5'$ exonuclease
$5' \rightarrow 3'$ exonuclease
buoyant density
cohesin
conservative replication
dispersive replication
displacement replication
D loop
DNA-dependent DNA polymerase
DNA gyrase
DNA ligase
DNA polymerase I
DNA polymerase II
DNA polymerase III
DNA polymerase α
DNA polymerase γ
DNA polymerase δ
DNA topoisomerase
flap endonuclease
helicase
initiation region
lagging strand
leading strand
Okazaki fragment
origin of replication
origin recognition complex
origin recognition sequence
plectonemic
prepriming complex
primase
primosome
proliferating-cell nuclear antigen
proofreading
replication factory
replication fork
replication protein A
replisome
RNA-dependent DNA polymerase
rolling-circle replication
semiconservative replication
single-strand binding protein
terminator sequence
thermostable
Tus
type I topoisomerase
type II topoisomerase

Self-study Questions

10.1 Distinguish between the three conceivable ways by which the double helix could be replicated.

10.2 Describe the experiment that showed that DNA replicates in a semiconservative manner.

10.3 Draw the density gradient banding patterns that Meselson and Stahl would have obtained after two rounds of replication if DNA replication had turned out to be (a) conservative, or (b) dispersive.

10.4 What is the topological problem in DNA replication, and how was this problem solved by the discovery of DNA topoisomerases?

10.5 Distinguish between the mode of action of type I and type II DNA topoisomerases.

10.6 With the aid of diagrams, indicate how displacement replication and rolling-circle replication differ from DNA copying via a replication fork.

10.7 What are the roles of the exonuclease activities possessed by some DNA polymerases?

10.8 Describe the structure of *E. coli* DNA polymerase III. How similar are the structures of DNA polymerases I and II?

10.9 Distinguish between the roles of DNA polymerase I and DNA polymerase III during DNA replication in *E. coli*, and between DNA polymerase α and DNA polymerase δ in eukaryotes.

10.10 What impact does the inability of DNA polymerases to synthesize DNA in the $3' \rightarrow 5'$ direction have on DNA replication?

10.11 Explain why DNA replication must be primed and describe how the priming problem is solved by *E. coli* and by eukaryotes.

10.12 Give a detailed description of the structure of the *E. coli* origin of replication and outline the role of each component of the origin in the initiation of replication.

10.13 What are the roles of helicases and single-strand binding proteins during DNA replication?

10.14 Give a detailed description of the events occurring at the replication fork in *E. coli*. Your answer should make a clear distinction between the complexes referred to as the primosome and the replisome.

10.15 How are Okazaki fragments joined together in *E. coli*?

10.16 Why does replication of the *E. coli* DNA molecule always terminate in a defined region?

10.17 Describe the features of replication origins in yeast and humans.

10.18 In what ways do events at the eukaryotic replication fork differ from those occurring in *E. coli*?

10.19 How are Okazaki fragments joined together in eukaryotes?

10.20 What is a replication factory?

10.21 Outline how replication terminates in eukaryotes.

Discussion Topics

10.22 Would it be possible to replicate the DNA molecules present in living cells if DNA topoisomerases did not exist?

10.23 Why is inactivation of the *E. coli polA* gene, coding for DNA polymerase I, not lethal?

10.24 What might be the role of DNA polymerase enzymes in repair of damaged DNA?

10.25 Construct a hypothesis to explain why all DNA polymerases require a primer in order to initiate synthesis of a new polynucleotide. Can your hypothesis be tested?

10.26 Biologists have been aware for many years that repeated rounds of DNA replication ought to result in the gradual shortening of a linear double-stranded molecule. Explain why this should be the case.

10.27 Our current knowledge of DNA replication in eukaryotes is biased toward the events occurring at the replication fork. The next challenge is to convert this description of replication into a model that describes how replication is organized within the nucleus, addressing issues such as the role of replication factories and the processes used to avoid tangling of the daughter molecules. Devise a research plan to address one or more of these issues.

FURTHER READING

Hübscher U, Nasheuer H-P & Syväoja JE (2000) Eukaryotic DNA polymerases: a growing family. *Trends Biochem. Sci.* 25, 143–147.

Johnson A & O'Donnell M (2005) Cellular DNA replicases: components and dynamics at the replication fork. *Annu. Rev. Biochem.* 74, 283–315. *Details of replication in bacteria and eukaryotes.*

Kaplan DL & Bastia D (2009) Mechanisms of polar arrest of a replication fork. *Mol. Microbiol.* 72, 279–285. *Recent ideas on the termination of replication in bacteria.*

Kornberg A (1960) Biologic synthesis of deoxyribonucleic acid. *Science* 131, 1503–1508. *A description of DNA polymerase I.*

Kornberg A (1984) DNA replication. *Trends Biochem. Sci.* 9, 122–124. *An early description of events at the replication fork.*

Meselson M & Stahl F (1958) The replication of DNA in *Escherichia coli*. *Proc. Natl. Acad. Sci. USA* 44, 671–682. *The Meselson–Stahl experiment.*

Mott ML & Berger JM (2007) DNA replication initiation: mechanisms and regulation in bacteria. *Nat. Rev. Microbiol.* 5, 343–354.

Mullis KB (1990) The unusual origins of the polymerase chain reaction. *Sci. Am.* 262, 56–65.

Okazaki T & Okazaki R (1969) Mechanisms of DNA chain growth. *Proc. Natl. Acad. Sci. USA* 64, 1242–1248. *The discovery of Okazaki fragments.*

Pomerantz RT & O'Donnell M (2007) Replisome mechanics: insights into a twin polymerase machine. *Trends Microbiol.* 15, 156–164. *Details of how DNA is synthesized at the replication fork.*

Saiki RK, Gelfand DH, Stoffel S et al. (1988) Primer-directed enzymatic amplification of DNA with a thermostable DNA polymerase. *Science* 239, 487–491. *Description of PCR.*

Wang JC (2002) Cellular roles of DNA topoisomerases: a molecular perspective. *Nat. Rev. Mol. Cell Biol.* 3, 430–440.

Inheritance of Genes During Eukaryotic Cell Division

CHAPTER 11

We are addressing the first of the three big questions that we have asked about the role of genes as units of inheritance. We wish to understand how a complete set of genes is passed on to the daughter cells when the parent cell divides. In the previous chapter we studied the process by which DNA molecules are replicated, but we realize that DNA replication on its own does not explain how genes are inherited. We still have to understand how the DNA molecules resulting from replication are passed on to the daughters when the cell divides. If the cell has more than one DNA molecule, how does it make sure that each daughter cell receives one copy of each of these molecules, and not, by mistake, two copies of one molecule and maybe no copies of another?

To understand how DNA molecules are correctly distributed during cell division we must explore some of the areas of research where genetics and cell biology overlap. First, we must recognize that the problem we are tackling is not the distribution of "naked" DNA molecules from parent to daughter cells, but the transmission of chromosomes. We must therefore understand what chromosomes are and how DNA is organized within them. We must also recognize that cell division is only one stage in the life cycle of a cell. We must therefore examine the cell cycle in its entirety and in particular ask how DNA replication is coordinated with the division phase. We will then be in a position to study the events occurring during cell division and to see how these ensure the correct distribution of chromosomes.

11.1 GENOMES AND CHROMOSOMES

Our study of cell division must be accompanied by an important intellectual step forward. So far in this book our focus has been on the gene, and it will continue to remain so. But we must now begin to consider not just individual genes but the entire genome.

A gene, as we know, is a segment of DNA containing a unit of biological information. The genome, on the other hand, is the entire genetic complement of an organism, made up of one or more lengthy DNA molecules, and comprising all that organism's genes and all of the intergenic DNA between those genes. A gene is a unit of inheritance, but the genome is all of those units added together. We can continue to think about the inheritance of individual genes, but we must recognize that what is actually inherited is the genome. We will therefore use the genome as the starting point for this chapter.

Eukaryotic genomes are contained in chromosomes

Eukaryotic genomes vary greatly in size, the smallest ones being less than 10 Mb in length, and the largest over 50,000 Mb (Table 11.1). This size range coincides to a certain extent with the complexity of the organism, the simplest eukaryotes such as fungi having the smallest genomes. Most of the genome is contained in the nucleus, the exception being the much smaller molecules found in mitochondria and in the chloroplasts of photosynthetic eukaryotes.

All eukaryotic nuclear genomes are made up of linear DNA molecules, each of these molecules contained in a different chromosome. The number of chromosomes varies between species, and appears to be unrelated to the biological features of the organism. Some quite simple eukaryotes have

TABLE 11.1 GENOME SIZES AND GENE NUMBERS FOR VARIOUS EUKARYOTES

Species	Size of genome (Mb)	Approximate gene number
Thalassiosira pseudonana (diatom)	2.5	11,240
Saccharomyces cerevisiae (budding yeast)	12.1	6100
Schizosaccharomyces pombe (fission yeast)	12.5	4900
Caenorhabditis elegans (nematode worm)	97	20,140
Arabidopsis thaliana (plant)	125	25,500
Drosophila melanogaster (fruit fly)	180	13,600
Oryza sativa (rice)	466	40,000
Zea mays (corn)	2300	32,700
Homo sapiens (human)	3200	20,500
Plethodon vandykei (salamander)	70,000	20,000

multiple chromosomes, such as the yeast *Saccharomyces cerevisiae*, which has sixteen. More complex organisms, on the other hand, may have relatively few. The ant *Myrmecia pilosula* has just one chromosome, and the Indian muntjac deer has only three (Figure 11.1).

The human nuclear genome is typical of those of all multicellular animals. It comprises approximately 3200 Mb of DNA, divided into 24 molecules, the shortest 48 Mb in length and the longest 250 Mb. The 24 chromosomes consist of 22 **autosomes** and the two sex chromosomes, X and Y. The vast majority of human cells are **diploid** and so have two copies of each autosome, plus two sex chromosomes, XX for females or XY for males—46 chromosomes in all. These are called **somatic cells**, in contrast to **sex cells** or **gametes**, which are **haploid** and have just 23 chromosomes, comprising one of each autosome and one sex chromosome.

(A)

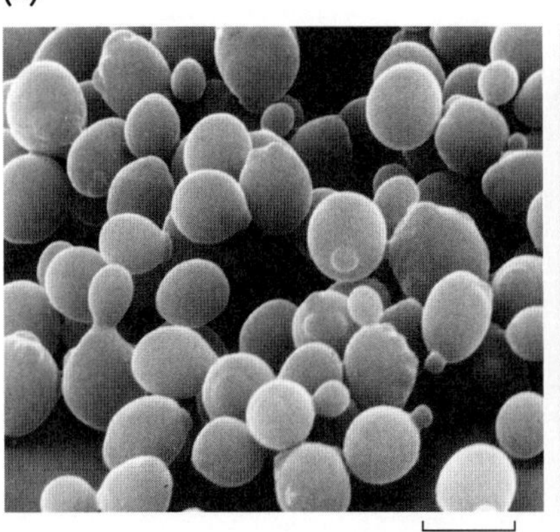

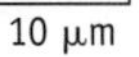

(B) (C)

FIGURE 11.1 Three eukaryotes with different numbers of chromosomes. (A) The yeast *Saccharomyces cerevisiae*, which has 16 chromosomes. (B) The ant *Myrmecia pilosula*, which has just one chromosome. (C) The Indian muntjac deer, which has three chromosomes. (A, Courtesy of Eric Schabtach, University of Oregon; B, Courtesy of Alex Wild; C, Courtesy of Susan Hoffman, Miami University.)

Chromosomes contain DNA and proteins

Chromosomes are present throughout the lifetime of a cell, but they are most apparent during cell division when the familiar structures more accurately referred to as **metaphase chromosomes** reveal themselves (Figure 11.2). Metaphase chromosomes are much shorter than the DNA molecules that they contain, which means fitting a DNA molecule into its chromosome is quite a challenge. The problem is immense. The DNA that makes up the human genome has a total length of about 1 m. This DNA is shared among 24 chromosomes, each of which is only a few thousandths of a millimeter in length in its metaphase form even though it contains, on average, over 4 cm of DNA. There must be a highly organized packaging system to fit such lengthy DNA molecules into such small structures.

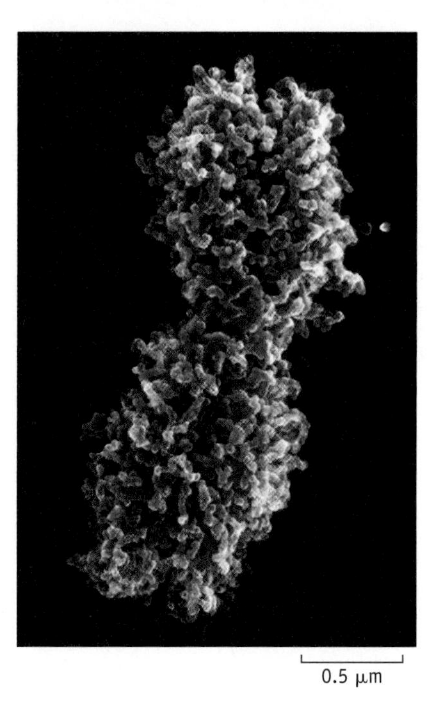

FIGURE 11.2 A scanning electron micrograph of a typical human metaphase chromosome. (From S. Inaga, K. Tanaka, and T. Ushiki (eds), Chromosome Nanoscience and Technology, 2007. With permission from CRC Press.)

The work that led to our current understanding of this packaging system began many years before it was appreciated that DNA is the genetic material. Cytologists toward the end of the nineteenth century coined the term **chromatin** for the component of the nucleus that stains most strongly with basophilic aniline dyes (Figure 11.3). Chromatin is now looked on as the complex association between the DNA of the chromosomes and the proteins to which it binds, the latter including those responsible for packaging the DNA in a regular fashion within the chromosome.

Chromatin is about half protein and half DNA. Of the protein component a portion is made up of a heterogeneous mixture of molecules involved in DNA replication and gene expression, including DNA and RNA polymerases and regulatory proteins. The remainder, associated most intimately with the DNA component of the chromosomes, consists of a group of proteins called **histones** (Table 11.2).

Histones contain a high proportion of basic amino acids, and are remarkable in that they are very similar in all species. If, for example, the H4

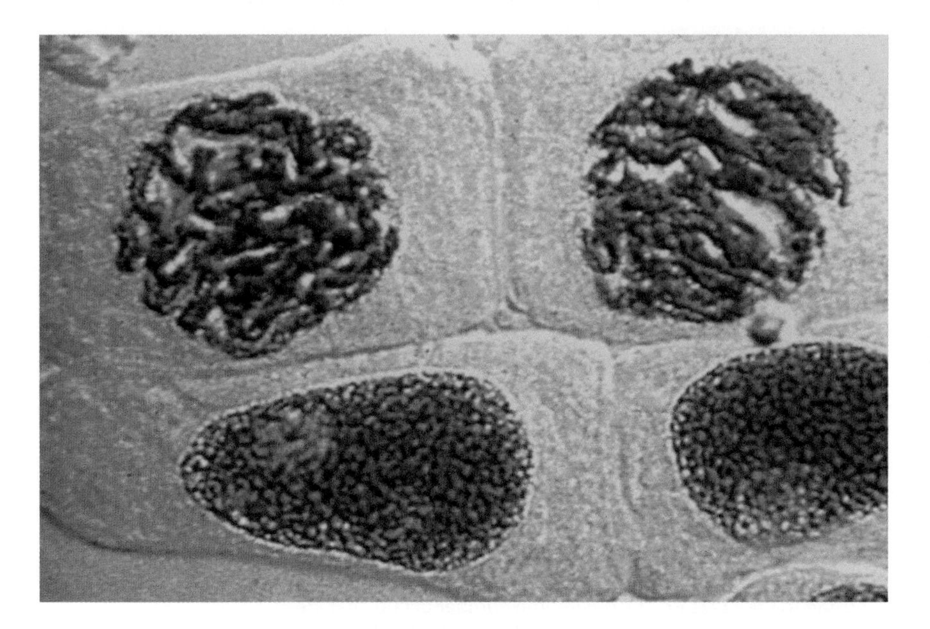

FIGURE 11.3 Light micrograph of cells of the common bluebell, stained so that the DNA content of the nuclei appears red. In the upper two cells the chromosomes are beginning to condense prior to cell division. Microscopic images similar to this one led cytologists in the nineteenth century to use the term "chromatin" to describe the material present in the nucleus. "Chroma" is Greek for color. (Courtesy of G. Gimenez-Martin/Science Photo Library.)

TABLE 11.2 HISTONES

Histone	Molecular mass (kD)	Basic amino acid content
H1	23,000	30%
H2A	14,000	20%
H2B	13,750	22%
H3	15,350	23%
H4	11,300	25%

"Basic amino acid content" refers to the amount of lysine, histidine, and arginine in each protein.
H1 is a family of histones, including H1a–e, $H1^0$, H1t, and H5.

```
cow MSGRGKGGKGLGKGGAKRHRKVLRDNIQGITKPAIRRLARRGGVKRISGLIYEETRGVLKVFLENVIRDAVTYTEHAKRKTVTAMDVVYALKRQGRTLYGFGG
pea MSGRGKGGKGLGKGGAKRHRKVLRDNIQGITKPAIRRLARRGGVKRISGLIYEETRGVLKIFLENVIRDAVTYTEHARRKTVTAMDVVYALKRQGRTLYGFGG
    ************************************************************ **************** *************************
```

FIGURE 11.4 Comparison between the amino acid sequences of histone H4 from cow and pea. The sequences are written using the one-letter abbreviations for the amino acids. Asterisks mark the positions where the same amino acid is present in both sequences.

histone proteins of pea and cow are compared, then we find that only two of the 103 amino acids in the polypeptides are different (Figure 11.4). This is a much greater degree of similarity than we expect for proteins with equivalent functions in such widely divergent organisms. The conservation indicates that histones were among the first proteins to evolve (before the common ancestor of peas and cows lived) and that their structures have not changed for hundreds of millions of years. This in turn suggests that histones play a fundamental role within chromosomes, a role that was set out during the earliest stages of evolution of the eukaryotic cell. It is no surprise to learn that histones are the main components of the DNA packaging system.

Histones are constituents of the nucleosome

The important breakthroughs in understanding DNA packaging were made in the early 1970s by a combination of biochemical analysis and electron microscopy. The biochemical work included **nuclease protection** experiments (Figure 11.5). In this procedure, a DNA–protein complex is treated with an endonuclease, an enzyme that cuts polynucleotides at internal phosphodiester bonds. The endonuclease has to gain access to the DNA in order to cut it, and hence can attack only the phosphodiester bonds

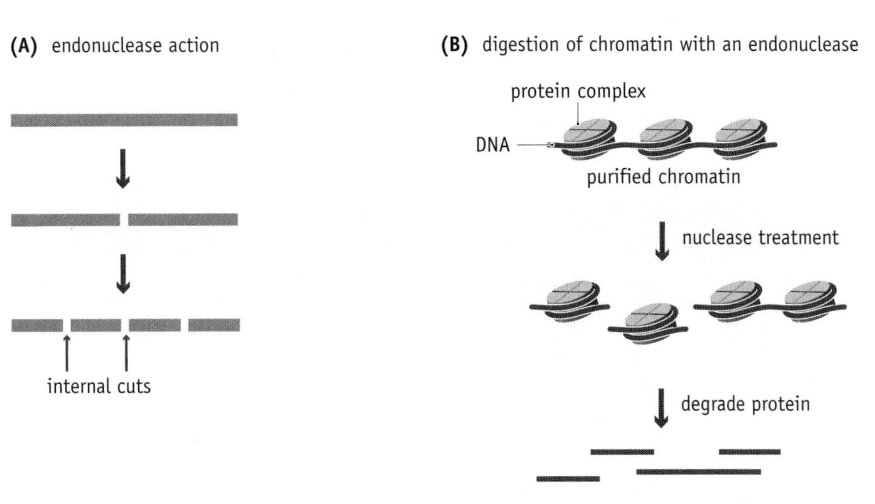

FIGURE 11.5 The nuclease protection technique and its use with purified chromatin. (A) An endonuclease makes cuts at internal phosphodiester bonds within a DNA molecule. The longer the exposure to the nuclease, the greater the number of cuts that are made, until eventually only mononucleotides remain. (B) If proteins are bound to a DNA molecule, then not all phosphodiester bonds can be cut by the endonuclease, as some are protected by the bound protein. The drawing shows purified chromatin being treated with an endonuclease for a short period, so that on average a single cut is made between each bound protein complex. The resulting DNA fragments are 200 bp, and multiples of 200 bp, in length, showing that the protein complexes are about 200 bp apart on the DNA.

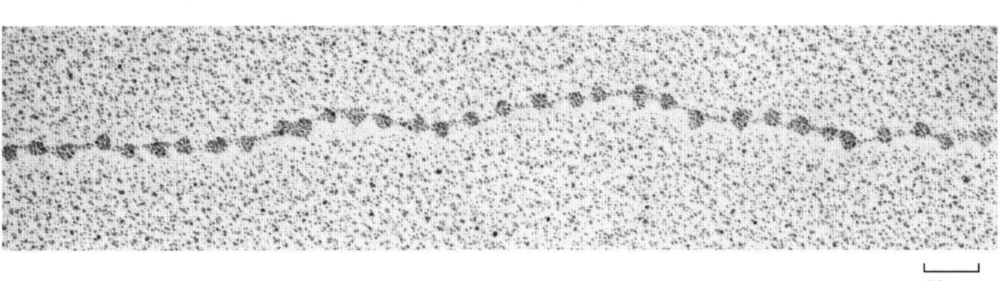

FIGURE 11.6 Electron micrograph showing a strand of DNA with attached nucleosomes, the "beads-on-a-string" structure. (Courtesy of Victoria Foe, University of Washington.)

that are not masked ("protected") by attachment to a protein. These experiments were carried out on chromatin that had been gently extracted from nuclei by methods designed to retain as much of the chromosome structure as possible. The sizes of the resulting DNA fragments were not random as might be expected, but instead had lengths of approximately 200 bp and multiples thereof. The conclusion was that the protein in chromatin is associated with the DNA in a regular fashion, with individual proteins or protein complexes spread out at intervals 200 bp apart.

The biochemical results were complemented by electron microscopic observations of chromatin. The images showed linear arrays of spherical structures referred to as **beads on a string** (Figure 11.6). These are generally believed to be the direct visualization of the protein complexes attached to a DNA molecule.

The spherical particles are called **nucleosomes** and contain equal amounts of each histone except H1. The histones form a barrel-shaped **core octamer** consisting of two molecules each of H2A, H2B, H3, and H4, with the DNA molecule wound twice around each nucleosome (Figure 11.7A). In humans, exactly 146 bp of DNA is associated with the nucleosome and hence protected from endonuclease digestion. The only parts of the DNA molecule available to the endonuclease are the 50- to 70-bp stretches of **linker DNA** that join the individual nucleosomes to one another. A single cleavage in each linker region therefore gives rise to fragments of DNA that are approximately 200 bp in length.

The one histone not found in the core octamer of the nucleosome is H1. We now know that histone H1 is not a single protein but a group of proteins, all closely related to one another and now collectively called the **linker histones**. In vertebrates these include histones H1a–e, $H1^0$, H1t, and H5. A single linker histone is attached to each nucleosome, to form the **chromatosome**, but the precise positioning of this linker histone is not known. The most popular model places the linker histone on the outer surface of the nucleosome, possibly keeping the DNA in place in the same way that you would use your finger to stop a knot from coming undone before tying the bow (Figure 11.7B).

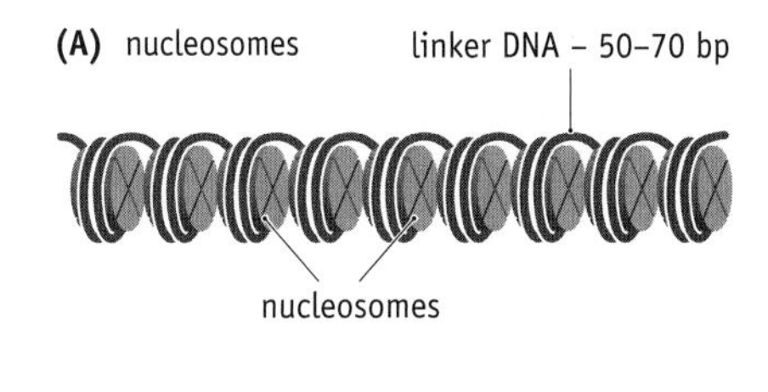

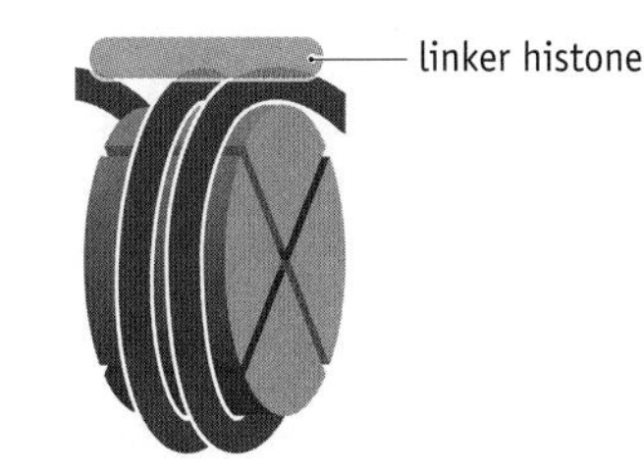

FIGURE 11.7 Nucleosomes and the chromatosome. (A) The model for the "beads-on-a-string" structure in which each bead is a barrel-shaped nucleosome with the DNA wound twice around the outside. (B) In the chromatosome, the precise position of the linker histone relative to the nucleosome is not known but, as shown here, the linker histone may act as a clamp, preventing the DNA from detaching from the outside of the nucleosome.

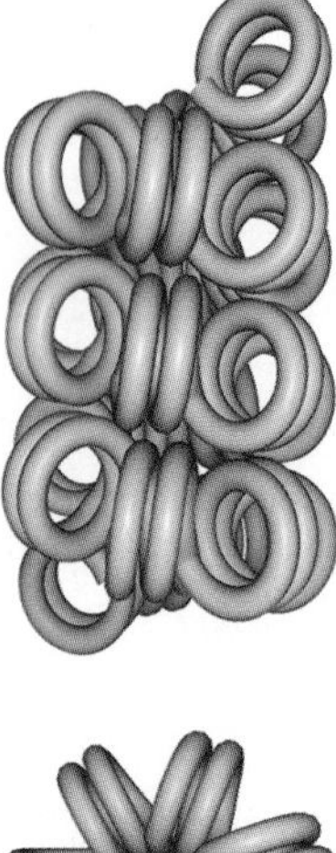

FIGURE 11.8 The solenoid model for the 30-nm chromatin fiber. The nucleosomes have been left out of the drawing in order to show the way in which the DNA molecule is coiled within the fiber. The upper drawing shows the view from the side, and the lower drawing is the view along the axis of the fiber.

Nucleosomes associate to form the 30-nm chromatin fiber

A DNA molecule of 6 cm, a bit above average for a single human chromosome, would take up 1 cm when in the beads-on-a-string form. This is still a million times greater than the length of a metaphase chromosome. We must therefore look for additional, higher levels of chromatin structure.

The next level of chromatin structure was discovered by electron microscopic studies, which revealed a chromatin fiber, less dispersed than the beads-on-a-string form, about 30 nm in diameter. This fiber is formed by packing together individual nucleosomes. Exactly how the nucleosomes are arranged is not understood, but the original "solenoid" model is still favored by most biologists (Figure 11.8). The **30-nm chromatin fiber** reduces the length of the beads-on-a-string structure by about 7 times, so our original 6-cm piece of DNA is now about 1.4 mm in length.

Between cell divisions, the 30-nm fiber is probably the most organized form of chromatin for those parts of a chromosome that contain genes that are being expressed. The DNA in these regions cannot be too condensed, as it must remain accessible to the proteins involved in the transcription process. We refer to these regions as **euchromatin**.

With the electron microscope it is possible to see loops of DNA within the euchromatin regions, each loop between 40 and 100 kb in length and predominantly in the form of the 30-nm chromatin fiber. The loops are attached to the **nuclear matrix**, the complex network of protein and RNA fibrils that permeate the entire nucleus. The attachments to the matrix are made by AT-rich segments of DNA called **matrix-associated regions** (**MARs**) or **scaffold attachment regions** (**SARs**) (Figure 11.9).

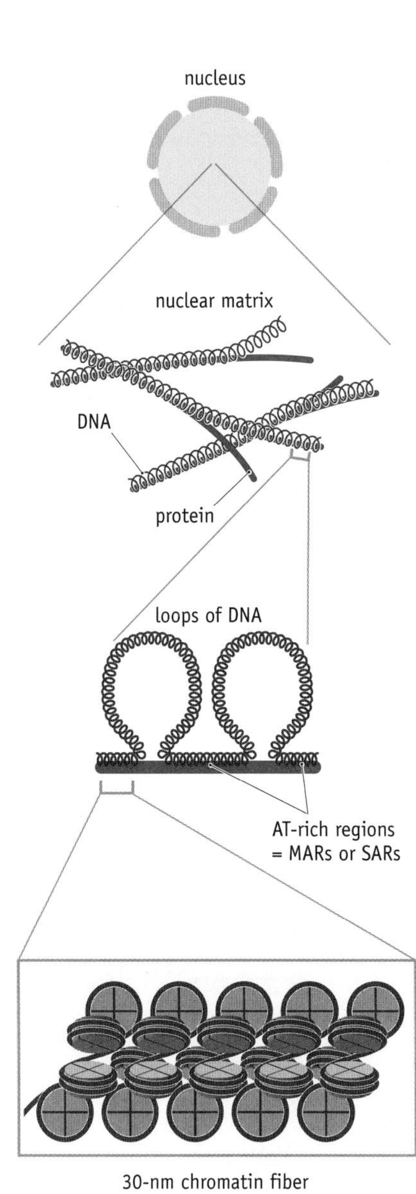

FIGURE 11.9 A scheme for the organization of DNA in euchromatin. The DNA in euchromatin is attached to the nuclear matrix, a fibrous protein-based structure that permeates the nucleus. Loops of DNA are attached to the matrix by AT-rich sequences called matrix-associated or scaffold attachment regions (MARs or SARs). The DNA in these loops is in the form of the 30-nm chromatin fiber.

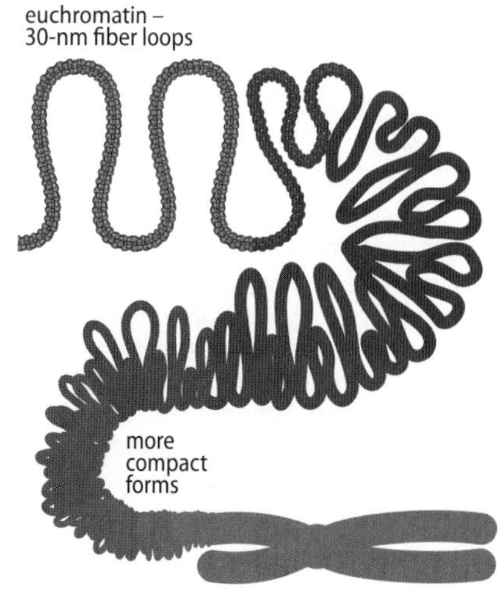

FIGURE 11.10 A scheme for the more compact conformations of the 30-nm fiber thought to be present in a metaphase chromosome.

How DNA is packaged into more compact chromosome structures is poorly understood

Nucleosomes, the 30-nm fiber, and euchromatin can be looked on as the first three levels of DNA packaging. Moving up the hierarchy we eventually reach the metaphase chromosomes, the most compact form of chromatin in eukaryotes. Exactly how these higher levels of packaging are organized is not yet understood. It is thought that they are more condensed versions of the structure postulated for euchromatin, one possibility being that histones in different loops of the 30-nm fiber interact with one another to draw the euchromatin structure into more compact conformations (Figure 11.10).

One reason for the difficulty in understanding these higher levels of chromatin structure is that chromosomes are visible only in dividing cells. Between cell divisions, the chromosomes become less compact and cannot be distinguished as individual structures unless specialized techniques are used. When nondividing nuclei are examined by microscopy, all that can be seen is a mixture of light- and dark-staining areas (Figure 11.11). The light areas are the euchromatin and the dark areas are called **heterochromatin**, the latter made up of DNA that is more condensed than in euchromatin.

Two types of heterochromatin are recognized. The first is **constitutive heterochromatin**, which is a permanent feature of all cells and represents DNA that contains no genes and so can always be retained in a compact organization. In contrast, **facultative heterochromatin** is not a permanent feature but is seen in some cells some of the time. Facultative heterochromatin is thought to contain genes that are inactive in some cells or at some periods of the cell's life cycle. When these genes are inactive, their DNA regions are compacted into heterochromatin.

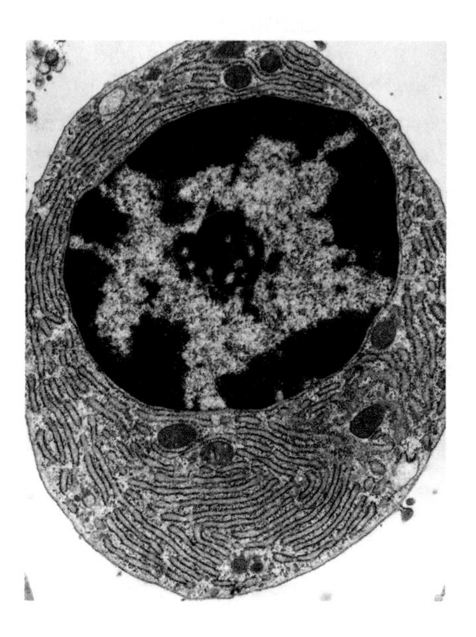

FIGURE 11.11 Electron micrograph of a plasma cell from guinea pig. Within the circular nucleus, the light areas are euchromatin and the dark areas are heterochromatin. (Courtesy of Don Fawcett/Science Photo Library.)

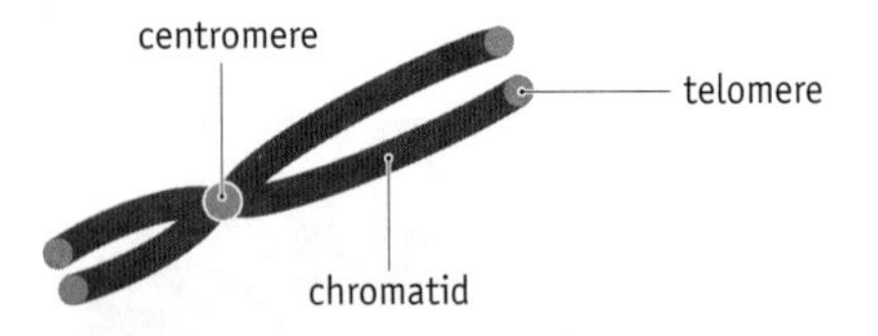

FIGURE 11.12 The typical appearance of a metaphase chromosome. Metaphase chromosomes are formed after DNA replication has taken place, so each one is, in effect, two chromosomes linked together at the centromere. The arms are called the chromatids. A telomere is the extreme end of a chromatid.

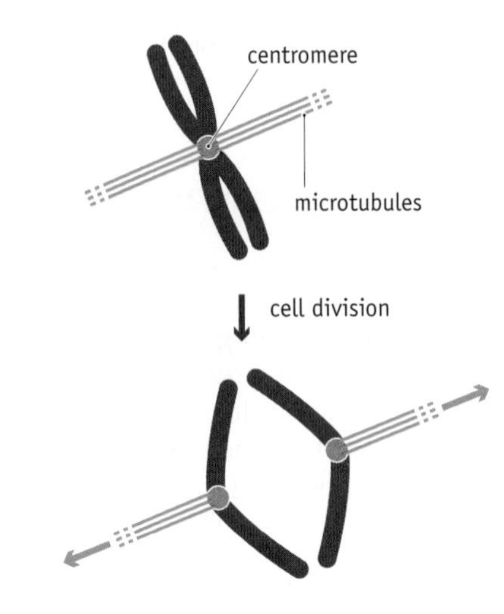

FIGURE 11.13 The role of the centromeres during cell division. During cell division, individual chromosomes are drawn apart by the contraction of microtubules attached to the centromere.

11.2 THE SPECIAL FEATURES OF METAPHASE CHROMOSOMES

Although metaphase chromosomes are transient structures that are formed only during cell division, they are important because their visual appearance enables individual chromosomes to be distinguished and their important morphological features to be recognized.

Individual metaphase chromosomes have distinct morphologies

The typical appearance of a metaphase chromosome is shown in Figure 11.12. This is in reality two chromosomes joined together because by the stage of the cell cycle when chromosomes condense and become visible by light microscopy, DNA replication has already occurred. A meta-phase chromosome therefore contains the two daughter chromosomes linked together at their **centromeres**. The centromere is also the position at which the chromosome attaches to the microtubules that draw the daughters into their respective nuclei during cell division (Figure 11.13).

The position of the centromere is characteristic for a particular chromosome (Figure 11.14). The centromere can be in the middle of the chromosome (**metacentric**), a little off center (**submetacentric**), toward one end of the chromosome (**acrocentric**), or located very close to one end (**telocentric**). The regions on either side of the centromere are called the chromosome arms or **chromatids**.

The differing positions of the centromere mean that individual chromosomes can be recognized. Further distinguishing features are revealed

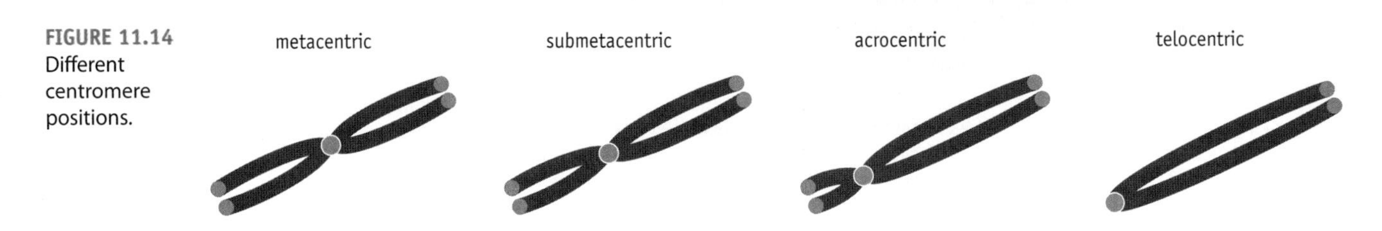

FIGURE 11.14 Different centromere positions.

TABLE 11.3 STAINING TECHNIQUES USED TO PRODUCE CHROMOSOME BANDING PATTERNS

Technique	Procedure	Banding pattern
G banding	Mild proteolysis followed by staining with Giemsa	Dark bands are AT-rich
		Pale bands are GC-rich
R banding	Heat denaturation followed by staining with Giemsa	Dark bands are GC-rich
		Pale bands are AT-rich
Q banding	Staining with quinacrine	Dark bands are AT-rich
		Pale bands are GC-rich
C banding	Denaturation with barium hydroxide and then staining with Giemsa	Dark bands contain constitutive heterochromatin

when chromosomes are stained. There are a number of different staining techniques (Table 11.3), each resulting in a banding pattern that is characteristic for a particular chromosome. The set of chromosomes possessed by an organism can therefore be represented as a **karyogram**, in which the banded appearance of each one is depicted. The human karyogram is shown in Figure 11.15.

Centromeres contain repetitive DNA and modified nucleosomes

DNA is present along the entire length of a chromosome including within the centromeres. This centromeric DNA is made up largely of repeat sequences. In humans, these repeats are 171 bp in length and are called **alphoid DNA**. There are 1500 to 30,000 copies of the alphoid DNA per chromosome, spanning a region of between 0.9 and 5.2 Mb.

A few genes are also present in the centromeric regions, but not many. For a long time it was thought that there were none at all, and the centromeric DNA tended to be left out when "entire" genome sequences were obtained for different organisms. Now we know that there are genes within the centromeres of most species, though the frequency is usually less than 100 genes per Mb of DNA, compared with 250 or more genes per Mb for the noncentromeric regions.

The centromeres of most eukaryotes also contain nucleosomes, similar to those in other regions of the chromosome but some of them containing the protein CENP-A instead of histone H3. CENP-A-containing nucleosomes are more compact and structurally rigid than those containing H3. It has been suggested that the arrangement of CENP-A and H3 nucleosomes along the DNA is such that the CENP-A versions are located on the surface of the centromere. Here they form an outer shell on which the **kinetochores** are constructed. These are the structures that act as the attachment points for the microtubules which draw the divided chromosomes into the daughter nuclei. Kinetochores form on the metaphase chromosomes, each of the daughter chromosomes constructing its own kinetochore on the conjoined centromere. One kinetochore is on one side of the centromere, and one on the other (Figure 11.16).

Telomeres protect chromosome ends

The ends of a chromosome are called the **telomeres**. These are specialized structures with a number of important roles. The first is simply to protect

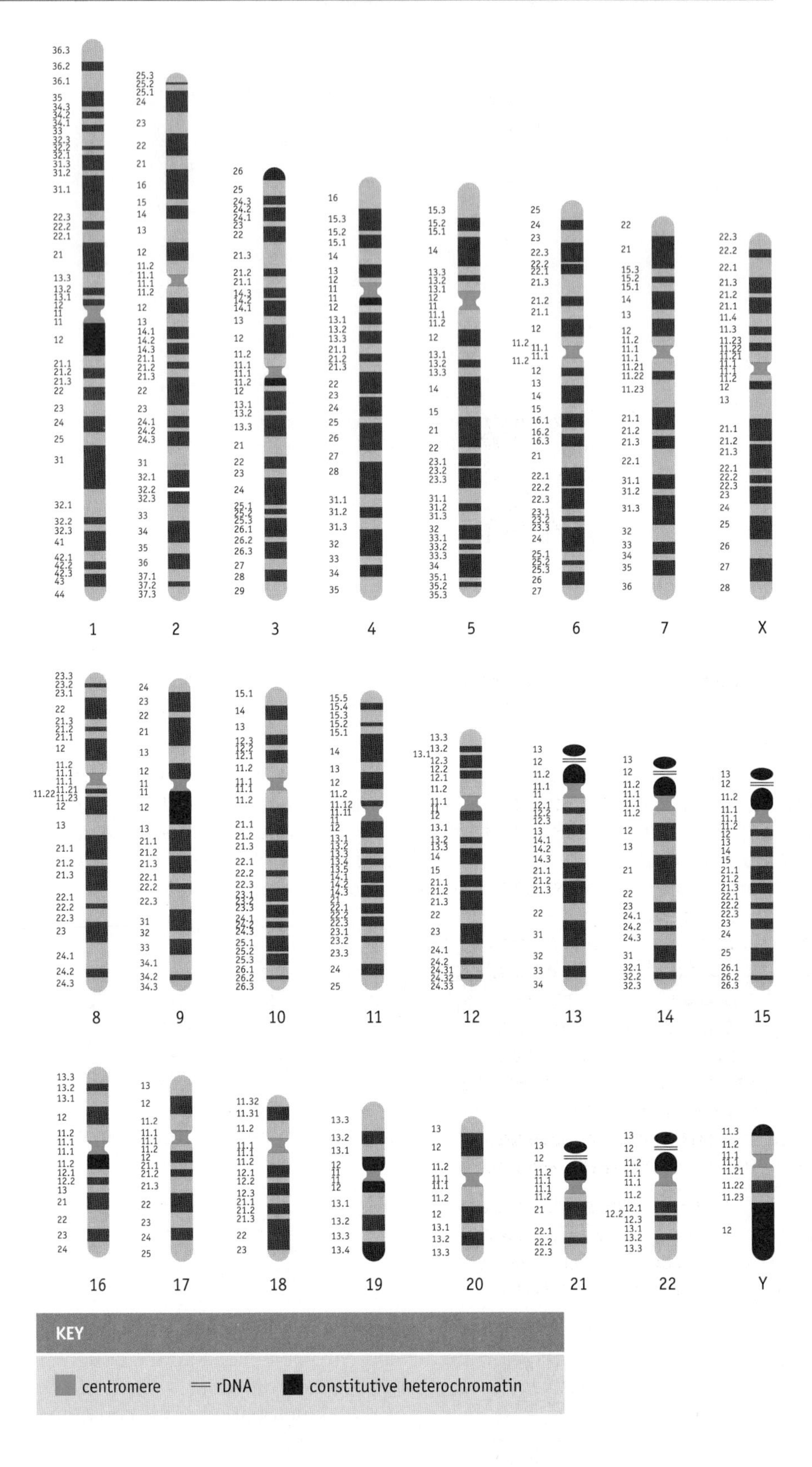

FIGURE 11.15 The human karyogram. The chromosomes are shown with the G-banding pattern obtained after Giemsa staining. Chromosome numbers are given below each structure and the band numbers to the left. "rDNA" is a region containing a cluster of repeat units for the ribosomal RNA genes.

FIGURE 11.16 The positions of the kinetochores on a metaphase chromosome.

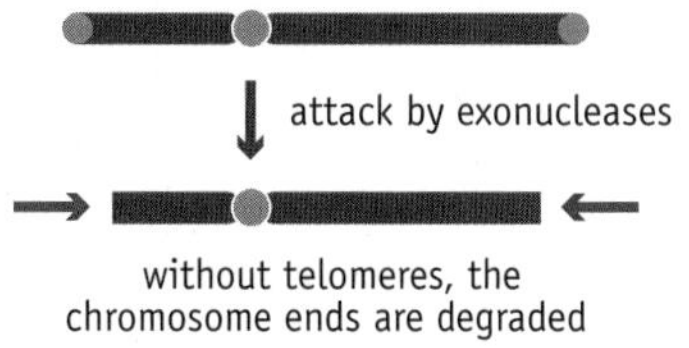

attack by exonucleases

without telomeres, the chromosome ends are degraded

FIGURE 11.17 One of the roles of the telomeres is to prevent exonucleases from degrading the ends of a chromosome.

the ends of the DNA molecule contained within the chromosome from attack by exonuclease enzymes. If these enzymes gain access to the chromosomal DNA then they will degrade some of the terminal region (Figure 11.17). The natural ends of the chromosomal DNA also need protection from the cell's DNA repair systems. These include very efficient mechanisms for repairing a broken chromosome, driven by proteins that bind to the ends of the DNA molecule that are exposed at the break point. If these repair proteins confuse a natural chromosome end for a break then they might join two chromosomes together. The telomeres prevent this from happening.

Like centromeres, the telomeres contain special DNA sequences. Telomeric DNA is made up of hundreds of copies of a short sequence, 5′-TTAGGG-3′ in humans (Figure 11.18). On the other strand the repeat sequence is 5′-CCCTAA-3′, which means that one of the strands in the telomeric region is rich in G nucleotides while the other is rich in Cs. This is true for all eukaryotes, even though the actual repeat sequence varies. The G-rich strand, which contributes the 3′ end of the DNA molecule, extends for up to 200 nucleotides beyond the terminus of the C-rich strand, giving a single-stranded overhang.

Nucleosomes are not present in the telomeric regions of chromosomes. Instead, two special proteins bind to the repeat sequences. In humans, these are called TRF1, which helps to regulate the length of the telomere, and TRF2, which maintains the single-strand extension. If TRF2 is inactivated then this extension is lost and the two polynucleotides fuse together in a covalent linkage. Other telomeric proteins are thought to form a linkage between the telomere and the periphery of the nucleus, the area in which the chromosome ends are localized.

Chromosomes should get progressively shorter during multiple rounds of replication

Telomeres must also overcome a problem posed by DNA replication. In theory, every round of DNA replication should result in a slight decrease in the length of a chromosome, something that does not happen in practice, at least not in all cells.

To understand why chromosomes should gradually get shorter we must recall one of the basic features of DNA replication. DNA synthesis can occur only in the 5′ → 3′ direction, which means that the lagging strand must be copied discontinuously, as a series of Okazaki fragments (see Figure 10.13). This presents two problems at the ends of a linear DNA molecule such as those found in eukaryotic chromosomes. The first is that the extreme 3′ end of the lagging strand might not be copied because the final Okazaki fragment cannot be primed, the natural position for the priming site being

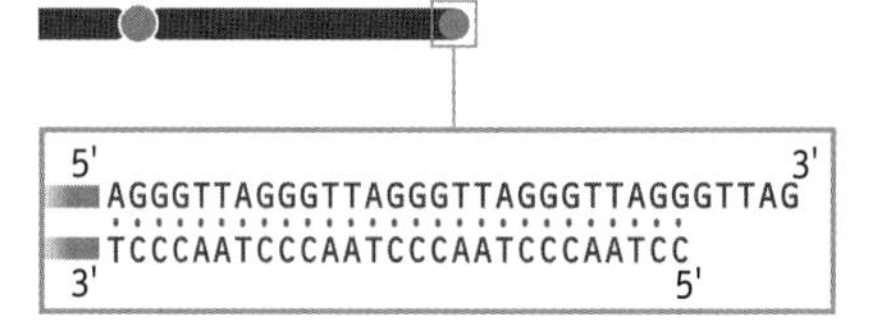

FIGURE 11.18 The DNA sequence at the end of a human telomere. The single-strand extension is up to 200 nucleotides in length.

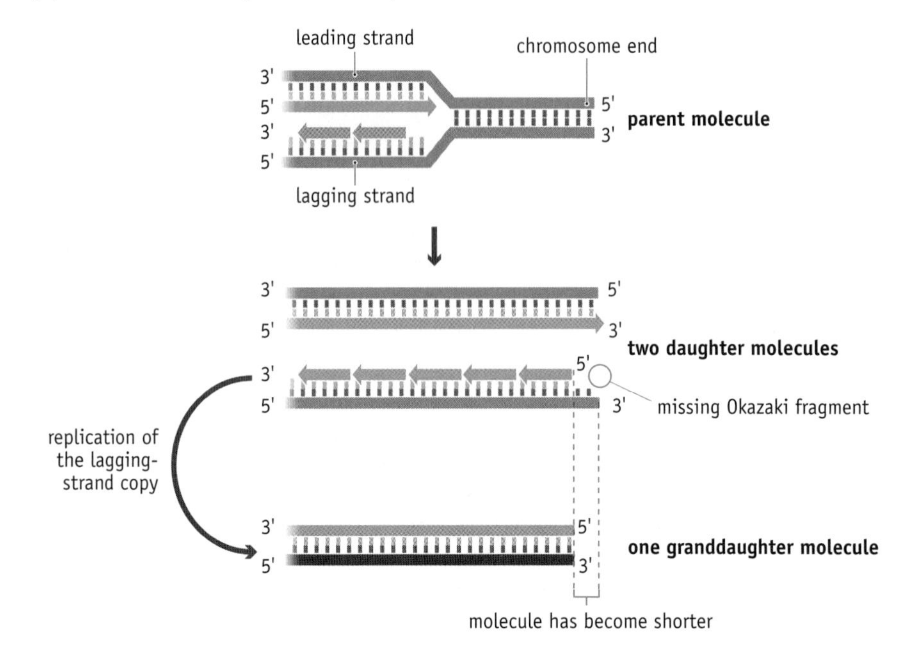

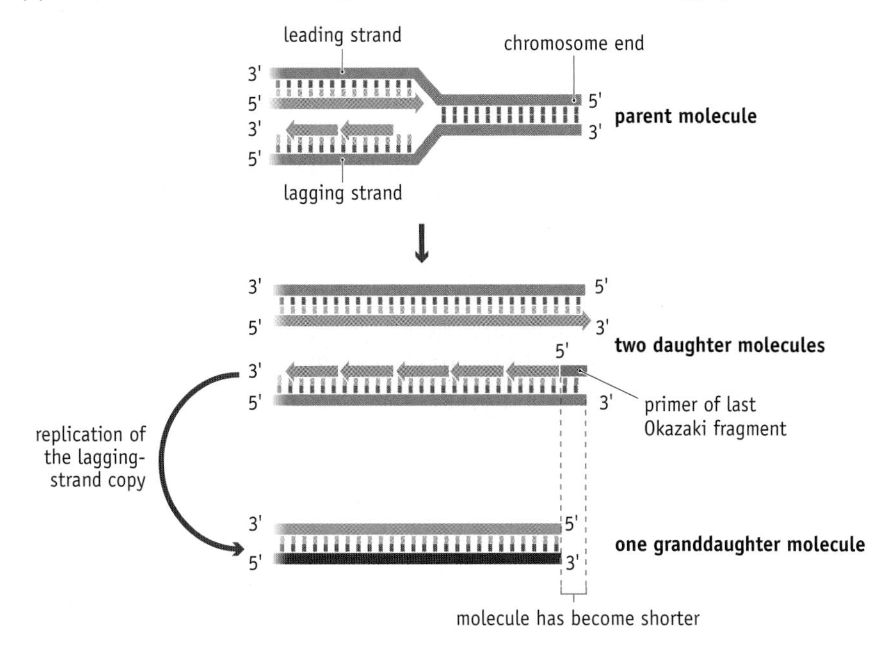

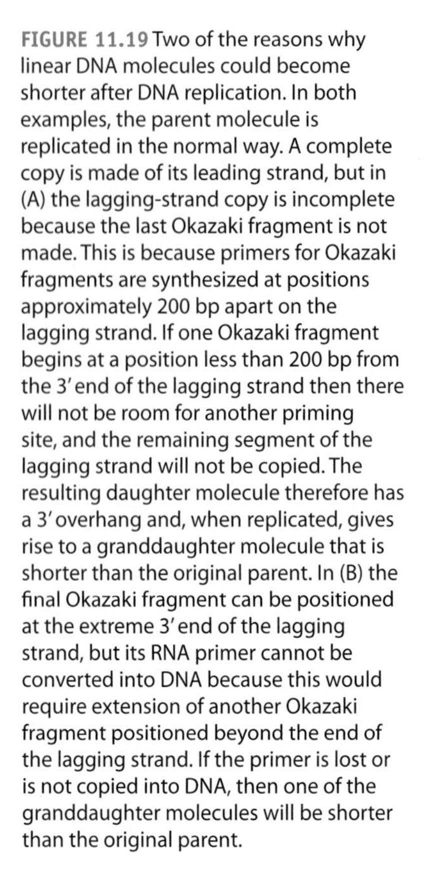
FIGURE 11.19 Two of the reasons why linear DNA molecules could become shorter after DNA replication. In both examples, the parent molecule is replicated in the normal way. A complete copy is made of its leading strand, but in (A) the lagging-strand copy is incomplete because the last Okazaki fragment is not made. This is because primers for Okazaki fragments are synthesized at positions approximately 200 bp apart on the lagging strand. If one Okazaki fragment begins at a position less than 200 bp from the 3′ end of the lagging strand then there will not be room for another priming site, and the remaining segment of the lagging strand will not be copied. The resulting daughter molecule therefore has a 3′ overhang and, when replicated, gives rise to a granddaughter molecule that is shorter than the original parent. In (B) the final Okazaki fragment can be positioned at the extreme 3′ end of the lagging strand, but its RNA primer cannot be converted into DNA because this would require extension of another Okazaki fragment positioned beyond the end of the lagging strand. If the primer is lost or is not copied into DNA, then one of the granddaughter molecules will be shorter than the original parent.

beyond the end of the lagging strand (Figure 11.19A). The absence of this Okazaki fragment means that the lagging-strand copy is shorter than it should be. If the copy remains this length, then when it acts as a parent polynucleotide in the next round of replication the resulting daughter molecule will be shorter than its grandparent.

A second problem arises if the primer for the last Okazaki fragment is placed at the extreme 3′ end of the lagging strand. Shortening will still occur, although to a lesser extent, because this terminal RNA primer cannot be converted into DNA by the standard processes for primer removal (Figure 11.19B). This is because the methods for primer removal (see Figure 10.29) require extension of the 3′ end of an adjacent Okazaki fragment,

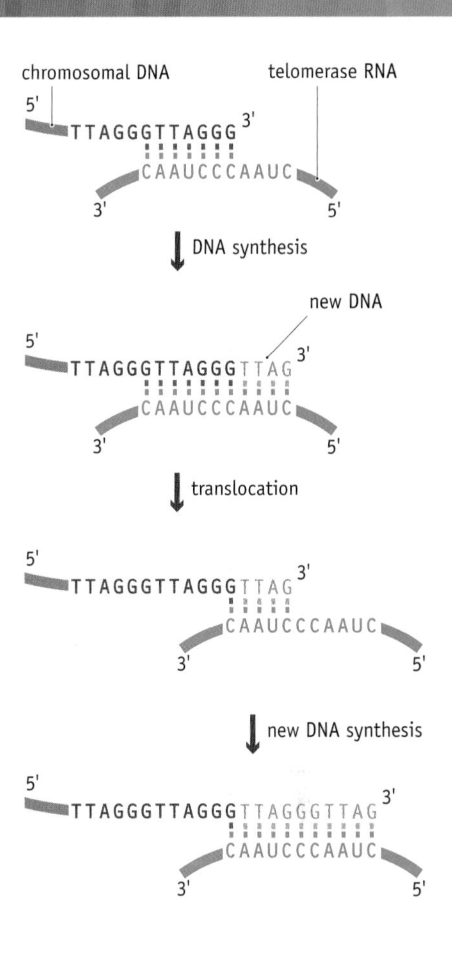

FIGURE 11.20 Extension of the end of a human chromosome by telomerase. The 3′ end of a human chromosomal DNA molecule is shown. The sequence comprises repeats of the human telomere motif 5′-TTAGGG-3′. The telomerase RNA base-pairs to the end of the DNA molecule which is then extended a short distance. The telomerase RNA translocates to a new base-pairing position slightly further along the DNA polynucleotide and the molecule is extended by a few more nucleotides.

which cannot occur at the very end of the molecule because the necessary Okazaki fragment is absent.

Telomeres are designed to solve these problems in the following way. Most of the telomeric DNA is copied in the normal fashion during DNA replication, but this is not the only way in which it can be synthesized. To compensate for the limitations of the replication process, telomeres can be extended by an independent mechanism catalyzed by the enzyme **telomerase**. This is an unusual enzyme in that it consists of both protein and RNA. In the human enzyme the RNA component is 450 nucleotides in length and contains near its 5′ end the sequence 5′-CUAACCCUAAC-3′. Note that the central region of this sequence is the reverse complement of the human telomere repeat 5′-TTAGGG-3′. Telomerase RNA can therefore base-pair to the single-stranded DNA overhang that is present at the end of the telomere (Figure 11.20). Base pairing provides a template that enables the 3′ end of the DNA to be extended by a few nucleotides. The telomerase RNA then translocates to a new base-pairing position slightly further along the DNA polynucleotide and the molecule is extended by a few more nucleotides. The process can be repeated until the chromosome end has been extended by a sufficient amount.

Telomerase can add nucleotides to the end of only the G-rich strand. It is not clear how the other polynucleotide—the C-rich strand—is extended, but it is presumed that when the G-rich strand is long enough, the primase–DNA polymerase α complex attaches to the end of the C-rich strand and initiates synthesis of a new Okazaki fragment (Figure 11.21). This requires the use of a new RNA primer, which explains why the C-rich strand is always shorter than the G-rich one. The important point is that the overall length of the chromosomal DNA has not been reduced.

11.3 THE CELL CYCLE

We have seen how a eukaryotic genome is packaged into one or more chromosomes. Now we must examine how chromosomes are passed on to the daughter cells when the cell divides. This means that we must become familiar with the different stages in the cell cycle.

Human cells growing in culture divide about once every 24 hours. In the body the time taken to pass through a single cell cycle is usually longer but the events that take place are the same in all cases. These events are coordinated and controlled to ensure that the chromosomes are replicated at the appropriate time, and that cell division, when it occurs, results in the orderly partitioning of the replicated chromosomes into the daughter cells.

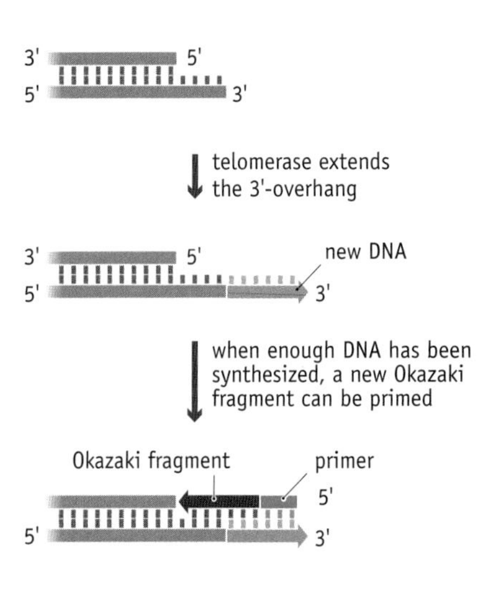

FIGURE 11.21 Completion of the extension process at the end of a chromosome. It is believed that after telomerase has extended the 3′ end of the chromosome, a new Okazaki fragment is primed and synthesized, resulting in a double-stranded end.

RESEARCH BRIEFING 11.1

Telomerase, senescence, and cancer

Telomerase is the enzyme that maintains the ends of the linear DNA molecules present in eukaryotic chromosomes. It is needed to ensure that these DNA molecules do not gradually get shorter every time they are replicated. It is therefore surprising to discover that telomerase is not active in all mammalian cells. The enzyme is functional in the early embryo, but after birth is active only in the reproductive and **stem cells**. The latter are progenitor cells that divide continually throughout the lifetime of an organism, producing new cells to maintain organs and tissues in a functioning state.

Cells that lack telomerase activity undergo chromosome shortening every time they divide. Eventually, after many cell divisions, the telomeric repeat sequences are completely removed. This results in loss of the protein "cap" normally present at each chromosome end. These caps protect the ends from exonuclease attack, and prevent the ends from being mistaken for internal breaks. The proteins that form the protective cap recognize the telomere repeats as their binding sequences, and so have no attachment points after the telomeres have been deleted.

The absence of telomerase activity might be linked with senescence

For several years biologists have attempted to link the gradual shortening of chromosome ends with **cell senescence**. Senescence was originally studied in cell cultures (Figure 1). All normal cell cultures have a limited lifetime. After 50 or so rounds of division the cells enter a senescent state in which they remain alive but cannot divide (Figure 2). Senescence is thought to be a protective mechanism that counters the tendency of a cell lineage to accumulate defects such as broken or rearranged chromosomes. Eventually, these defects could lead to the cells becoming cancerous. The senescence process prevents this by ensuring that the lineage is terminated before the danger point is reached.

A link between senescence and telomere shortening was first made purely on intuitive grounds. If a cell lineage becomes senescent after a certain number of cell divisions, then there must be a mechanism by which those cells are able to count the number of divisions that they have been through. The incremental shortening of the telomeres is one of a very limited number of cellular processes that enable the number of divisions that have taken place to be recorded. Experimental proof that telomere shortening does indeed have a role in senescence was obtained when it was shown that with some mammalian cell lines, notably fibroblast cultures (connective tissue cells), senescence can be delayed by engineering the cells so that they synthesize active telomerase.

Cellular senescence might be linked with aging

Is there a similar link between cell senescence and aging? The latter is the process that affects the whole organism and leads eventually to its death. In human cells, there appears to be a correlation between the average telomere length and the age of a person, and some diseases that lead to early aging are associated with relatively short telomere lengths. Some scientists have even reported that stress and obesity, both of which can reduce an individual's life expectancy, can accelerate the rate at which a person's telomeres shorten.

Some of the diseases that result in early aging have been linked directly with the activity of telomerase. With these, telomerase activity is reduced, either because the enzyme itself is defective, or because other proteins that usually regulate telomerase activity are not functioning correctly. Telomerase defects are likely to have their major impact on the stem cells, which depend on continued activity of the enzyme in order to stay viable so that they can initiate new cell lineages throughout an organism's lifetime.

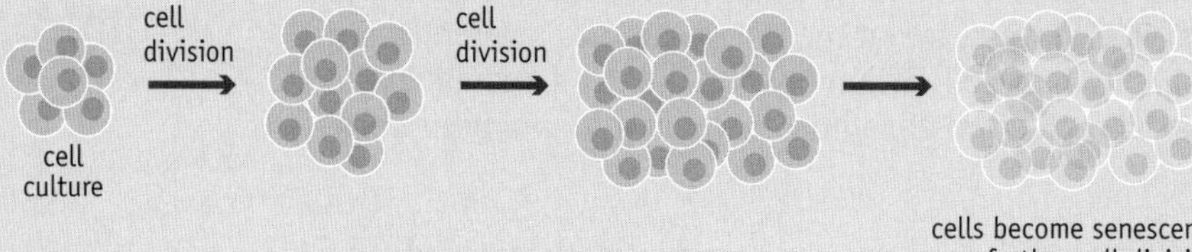

FIGURE 1 Cultured cells become senescent after multiple cell divisions.

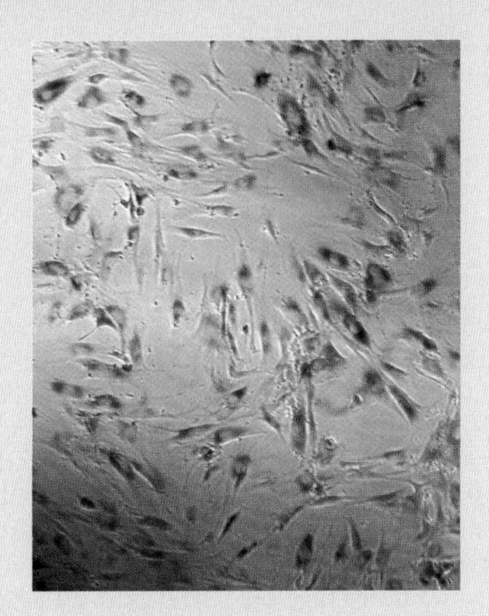

FIGURE 2 These blue cells are human mesenchymal stem cells that have become senescent or lost the ability to divide after X-ray irradiation. This image, obtained using senescence-associated β-galactosidase (SA-β-gal) staining, helped Berkeley Lab scientists better understand the process that triggers senescence in mesenchymal stem cells. SA-β-gal is an enzyme that is active only in senescent cells. An assay that detects the activity of this enzyme, such as the staining procedure used here, enables senescent cells to be distinguished from nonsenescent ones. (Adapted from D. Wang and D. J. Jang, *Cancer Res.* 69: 8200–8207, 2009. With permission from the American Association for Cancer Research.)

Any attempt to establish a link between telomeres and aging must therefore make a careful distinction between senescence caused by telomere shortening in somatic cells, and a loss of regenerative ability due to defective telomerase activity in stem cells. Making any kind of extrapolation from these cellular activities to aging of an organism is fraught with difficulties.

The presence of telomerase might be linked with cancer

Not all somatic cell lines display senescence. Cancerous cells are able to divide continuously in culture, their immortality being looked on as analogous to tumor growth in an intact organism. With several types of cancer, this absence of senescence is associated with activation of telomerase. Sometimes this results in the telomeres retaining their normal length through multiple cell divisions, but often the telomeres become longer than normal because the telomerase is overactive.

It is not clear whether telomerase activation is a *cause* or an *effect* of cancer, although the former seems more likely because at least one type of cancer, dyskeratosis congenita, appears to result from a mutation in the gene specifying the RNA component of human telomerase. The question is critical to understanding why a cancer arises, but is less relevant to the therapeutic issue, which centers on whether telomerase could be a target for drugs designed to combat the cancer. Such a therapy could be successful even if telomerase activation is an effect of the cancer, because inactivation by drugs would induce senescence of the cancer cells and hence prevent their proliferation.

Attempts to inactivate telomerase in cancer cells have focused on both the protein and RNA components of the enzyme. The protein has been used to prepare vaccines that contain antibodies that should bind to and inactivate any telomerase proteins that they encounter. In clinical trials, these vaccines have been able to reduce the number of cancerous cells circulating in the bloodstream of patients, reducing the chances that their cancers might spread to other parts of their body. Whether the vaccines can reduce the growth of existing tumors, or prevent a cancer from becoming established in the first place, is not yet clear. A second approach makes use of a short oligonucleotide that is complementary to part of the RNA component of telomerase. The idea is that this oligonucleotide will inhibit a telomerase enzyme by binding to its RNA molecule (Figure 3).

Vaccines and oligonucleotides targeted at telomerase are among the most promising of the various strategies being used to combat cancer. The major stumbling block at the moment is the ability of some cells to maintain their telomeres, and hence avoid senescence, by an alternative process that does not require telomerase. This involves the transfer of telomere repeats from DNA molecules that have not yet reached a critically short length to others that are close to the danger point. This process is switched on in some cancer cells, counteracting the effects of telomerase inactivation.

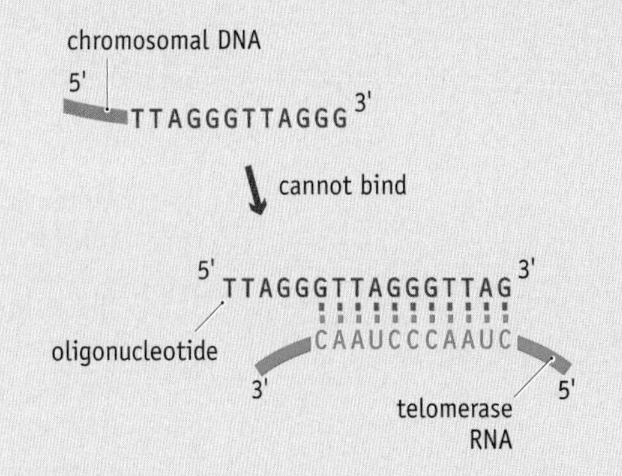

FIGURE 3 The use of an oligonucleotide to inhibit telomerase activity.

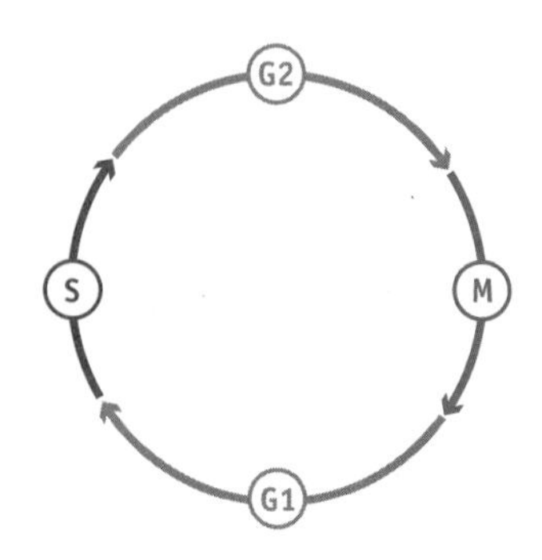

FIGURE 11.22 The cell cycle. The lengths of the individual phases vary in different cells. Abbreviations: G1 and G2, gap phases; M, mitosis; S, synthesis phase.

There are four phases within the cell cycle

The concept of a cell cycle emerged from light microscopy studies carried out by the early cell biologists. Their observations showed that dividing cells pass through repeated cycles of **mitosis**—the period when nuclear and cell division occur—and **interphase**, a less dramatic period when few dynamic changes can be detected with the light microscope. It was understood that chromosomes divide during interphase, so when DNA was identified as the genetic material, interphase took on a new importance as the period when genome replication must take place. This led to a reinterpretation of the cell cycle as a four-stage process (Figure 11.22).

The first of these four stages is called the **G1 phase**. This is generally the longest phase of the cell cycle and for some cells may last for weeks or even months. "G" stands for "gap," implying that this is an unimportant period, which is misleading, as it is the time when the cell is growing, metabolizing, and, within a multicellular organism, performing its specialized function. During G1 the cell has its standard complement of chromosomes, 46 for a diploid human cell.

The second stage is **S phase** or "synthesis" phase. This is the period when the chromosomal DNA molecules are replicated. At the start of S phase the replication origins are activated and bidirectional DNA synthesis progresses along the linear molecules until replication is complete. Despite the presence of multiple origins on each chromosome, S phase can take 6 to 8 hours to complete.

There then follows a second gap period called the **G2 phase**. The cell is now committed to division, so G2 rarely lasts longer than 3 to 4 hours. During G2 the cell is effectively tetraploid, as each chromosome has been replicated and so contains two DNA molecules. At the end of G2 the chromosomes begin to condense to form the metaphase structures. This leads into **M phase**, when nuclear and cell division take place.

Two distinct types of nuclear division can occur, called mitosis and **meiosis**. Usually, each takes less than 1 hour to complete. Mitosis gives rise to two daughter nuclei, each containing the same chromosome complement as the parent, and is the standard type of division for somatic cells. We will therefore be focusing on mitosis in this chapter. Meiosis is the specialized type of cell division that gives rise to the male and female reproductive cells. We will be dealing with the inheritance of DNA molecules during sexual reproduction in Chapter 14, and so will study meiosis then.

DNA replication must be coordinated with the rest of the cell cycle

It is clearly important that DNA replication be coordinated with the other phases of the cell cycle so that the genome is completely replicated, but replicated only once, before mitosis occurs. Coordination of the cell cycle involves a variety of regulatory proteins, collectively called **cyclins**. Some of these are active at the beginning of S phase and so are thought to switch on DNA replication. Others prevent a second round of DNA replication from occurring during G2 phase, while the cell is preparing itself for mitosis.

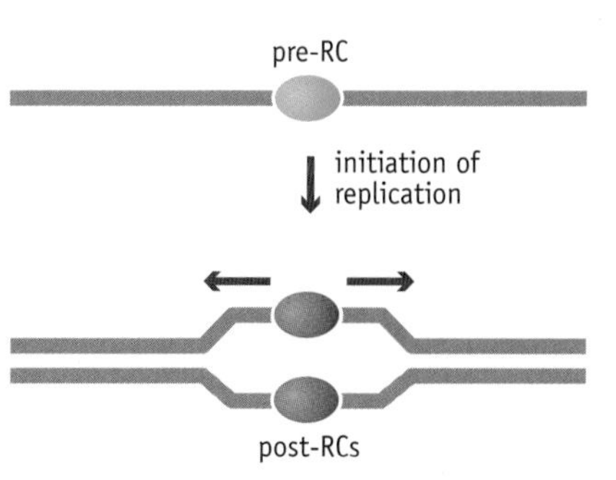

FIGURE 11.23 The relationship between a pre-RC and a post-RC. A pre-RC is converted into two post-RCs when the replication forks move away from the origin.

Cyclins are thought to influence DNA replication by regulating events at the replication origin. The initiation of DNA replication at an origin requires construction of a **pre-replication complex** (**pre-RC**). A pre-RC is converted to **post-replication complexes** (**post-RCs**) when the replication forks move away from the origin (Figure 11.23). A post-RC is unable to initiate replication, which means that segments of the genome that have been replicated cannot be accidentally recopied before mitosis has occurred.

The key protein involved in construction of a pre-RC is Cdc6p, which was originally identified in yeast and subsequently also shown to be present in higher eukaryotes. The first indication of the role of Cdc6p was provided by experiments in which the gene for this protein was inactivated. The resulting cells were unable to synthesize pre-RCs. Conversely, when the Cdc6p gene is overexpressed, so that large amounts of the protein are made, the genome undergoes multiple rounds of replication in the absence of mitosis. Cdc6p is synthesized during G1 phase and recruits the MCM helicase protein into the pre-RC (Figure 11.24). This helicase is probably responsible for breaking the base pairs within the origin, allowing DNA synthesis to begin.

A second important group of pre-RC proteins are the **replication licensing factors** (**RLFs**). These become bound to the chromosomal DNA toward the end of M phase and remain in place until the start of S phase, after which they are gradually removed from the DNA as it is replicated. Their removal may be the key event that converts a pre-RC into a post-RC and so prevents reinitiation of replication at an origin that has already directed a round of replication.

Control can also be exerted during S phase

The G1 to S phase transition is probably the major control point for genome replication, but it is not the only one. The specific events occurring during S phase are also subject to control.

Initiation of replication does not occur at the same time at all replication origins, nor is "origin firing" an entirely random process. Some parts of the genome are replicated early in S phase and some later, the pattern of replication being consistent from one round of the cell cycle to the next. The general pattern is that actively transcribed genes and the centromere are replicated early in S phase, and telomeres and nontranscribed regions of each chromosome later on (Figure 11.25). Early-firing origins are therefore tissue-specific and reflect the pattern of gene expression occurring in a particular cell. The migration rate of individual replication forks also shows variability. The mean speed in yeast is 2.9 kb per minute, but some forks move much more quickly, up to 11 kb per minute for the most active ones.

Understanding what determines the firing time of a replication origin is proving quite difficult. It is not simply the sequence of the origin, because transfer of a DNA segment from its normal position to another site in the same or a different chromosome can result in a change in the firing pattern

(A) Cdc6p is synthesized during G1 phase

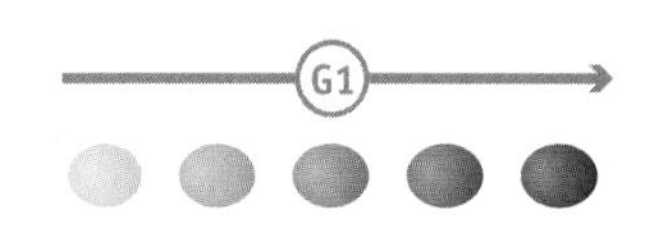

(B) Cdc6p recruits the MCM helicase to the pre-RC

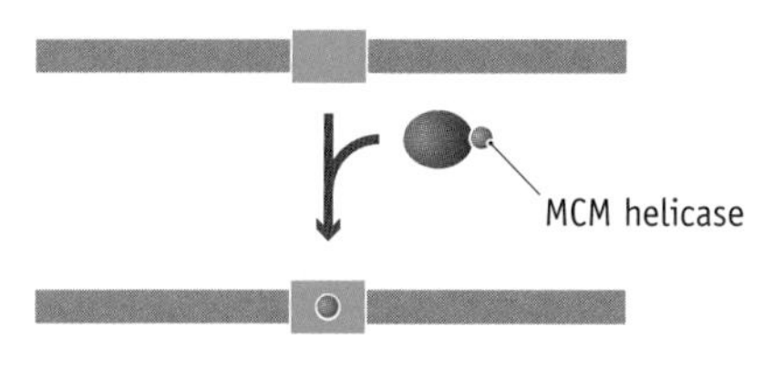

FIGURE 11.24 (A) Cdc6p is synthesized during G1 phase, and (B) recruits the MCM helicase to the pre-RC.

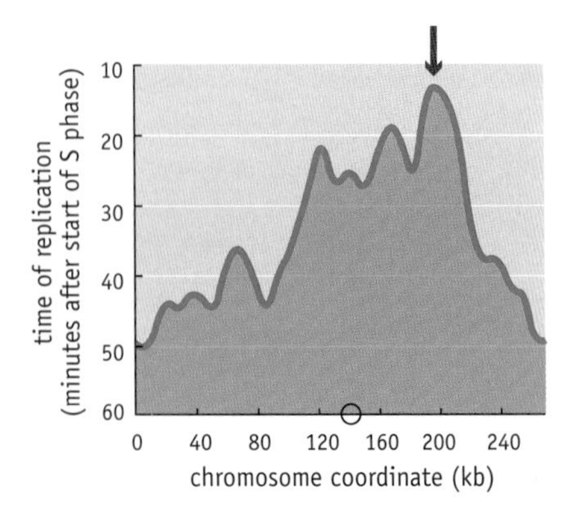

FIGURE 11.25 Timing of origin firing along chromosome VI of *Saccharomyces cerevisiae*. The arrow indicates the point where replication of this chromosome begins, and the circle on the *x*-axis is the centromere.

of origins contained in that segment (Figure 11.26). This positional effect may be linked with the way that the DNA is packaged in different parts of a chromosome. The position of the origin in the nucleus may also be important as origins that become active at similar periods within S phase appear to be clustered together, at least in mammals.

DNA replication can also be brought to a temporary or permanent halt during S phase. This phenomenon was first identified when it was shown that one of the responses of yeast cells to DNA damage is a slowing down and possibly a complete cessation of genome replication. This response is linked with the activation of genes whose products are involved in DNA repair. A set of S-phase cyclins respond to signals from proteins that detect the presence of damage in the template DNA at a replication fork. The signals from these damage-sensing proteins can elicit different types of cellular response. The replication process can be arrested by repressing the firing of origins of replication that are usually activated at later stages in S phase or by slowing the progression of existing replication forks. If the damage is not excessive then DNA repair processes are activated and replication is subsequently able to recommence.

Alternatively the cell may be shunted into the pathway of programmed cell death called **apoptosis**. The death of a single somatic cell as a result of DNA damage is usually less dangerous than allowing that cell to replicate its mutated DNA and possibly give rise to a tumor or other cancerous

(A) normal location in the genome

origins of replication

1 2 3

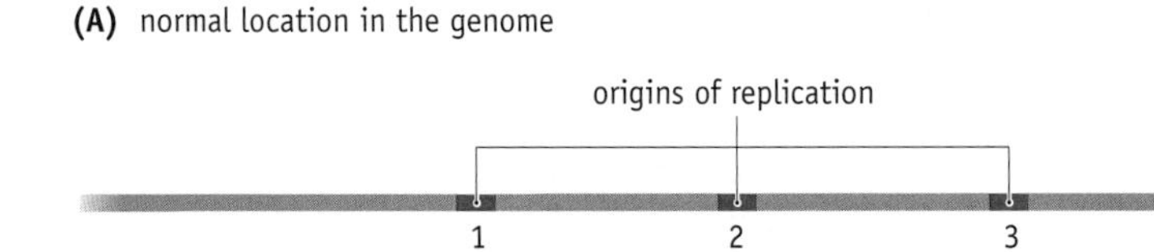

firing pattern: 1, then 2, then 3

(B) move segment to new location

firing pattern: 2, then 1, then 3

FIGURE 11.26 Transfer of a segment of DNA from (A) its normal location to (B) a new position in a chromosome can result in a change in the order in which replication origins contained within the segment are fired.

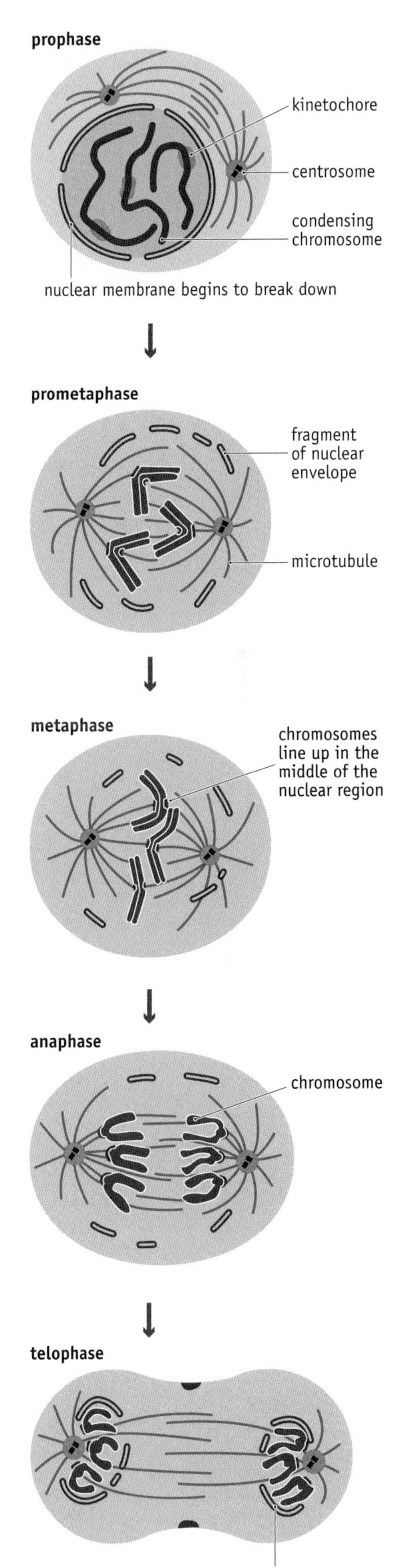

FIGURE 11.27 Mitosis. (Adapted from B. Alberts et al., Essential Cell Biology, 3rd ed. New York: Garland Science, 2010.)

growth. In mammals, a central player in induction of cell cycle arrest and apoptosis is the protein called p53. This is classified as a **tumor-suppressor protein**, because when this protein is defective, cells with damaged genomes can avoid the S-phase checkpoints and possibly proliferate into a cancer. p53 is a sequence-specific DNA-binding protein that activates a number of genes thought to be directly responsible for apoptosis, and it also represses expression of other genes that must be switched off to facilitate these processes.

Mitosis ensures the correct partitioning of chromosomes

Following a successful S phase, the cell is able to begin mitosis, which will result in two daughter cells, each containing a complete set of chromosomes from the parent. The chromosomes were replicated during S phase, so at the start of mitosis each one contains two daughter DNA molecules attached to one another at the centromere. The chromosomes are still in their interphase conformations, comprising regions of euchromatin and heterochromatin.

Mitosis itself is conventionally looked on as comprising five phases (Figure 11.27). The first of these is **prophase**. This is the period when the chromosomes condense and become visible under the light microscope. It is also the period when, in most eukaryotes, the nuclear membrane breaks down and the two **centrosomes** of the cell move to positions on either side of the nuclear region.

Prophase is followed by **prometaphase**. The chromosomes, now fully condensed, begin to attach to the microtubules radiating out from the centrosomes. In **metaphase**, the chromosomes line up in the middle of the nuclear region, and then, during **anaphase**, each pair of replicated chromosomes separates. One member of each pair is drawn toward one of the centrosomes. During the later part of anaphase, the cell begins to constrict around its middle, prior to division.

Telophase completes a round of mitosis. By now the chromosome pairs are fully separated and new nuclear membranes are being formed around each set. The middle of the cell continues to constrict until division is complete and the two daughters are formed.

How is the appropriate partitioning of chromosomes assured? The answer lies in the structure of the metaphase chromosome, specifically the positioning of the kinetochores on the surface of the conjoined centromeres. As indicated in Figure 11.16, two kinetochores form, one on each centromere. The two kinetochores are therefore on opposite sides of the chromosome pair.

The kinetochores are the attachment points for the microtubules that radiate out from the centrosomes. The centrosomes are themselves located on the opposite sides of the **mitotic spindle**, the microtubular arrangement that now occupies the region of the cell previously taken up by the nucleus (Figure 11.28). During prometaphase, the microtubules of the spindle begin to attach to the chromosomes, migrating along the chromatids until they locate one or the other of the kinetochores. The first microtubule to locate a kinetochore will orient that kinetochore toward the centrosome to which

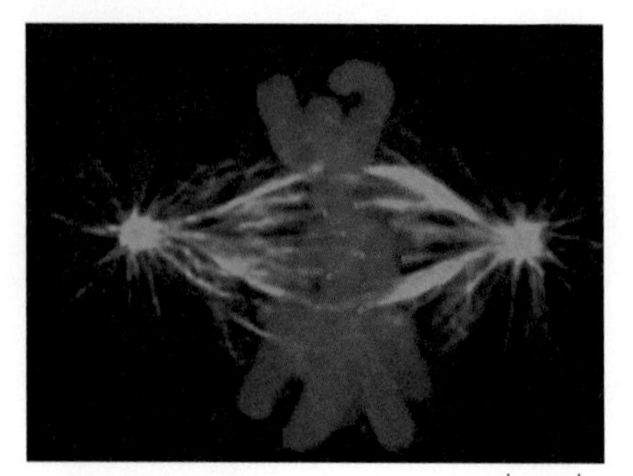

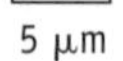

FIGURE 11.28 Fluorescence micrograph of a cell undergoing mitosis, showing the mitotic spindle in green, with chromosomes in blue and the kinetochores visible as red dots. (From A. Desai, *Curr. Biol.* 10: R508, 2000. With permission from Elsevier.)

the microtubule is attached (**Figure 11.29**). The other kinetochore is now more likely to be caught by a microtubule from the other centrosome.

Eventually, up to 40 microtubules will attach to either kinetochore. The two sets of microtubules begin to pull in opposite directions, the tension that is set up confirming that the correct attachments have been made. If both kinetochores have been captured by microtubules from the same centrosome, then this tension is not generated and mitosis cannot continue.

At the start of anaphase, the cohesin proteins that hold the daughter chromosomes together are degraded by protease enzymes, and the microtubules begin to draw the freed daughters toward the centrosomes. The natural outcome is that one of the daughter chromosomes of each pair

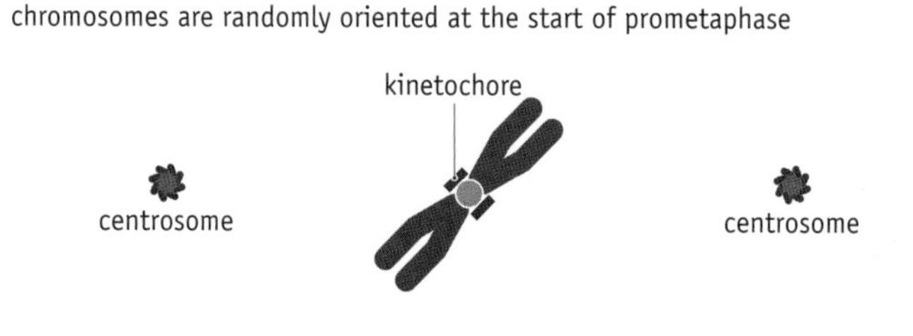

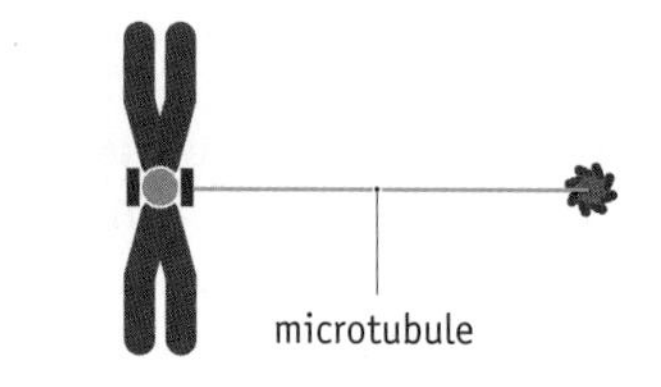

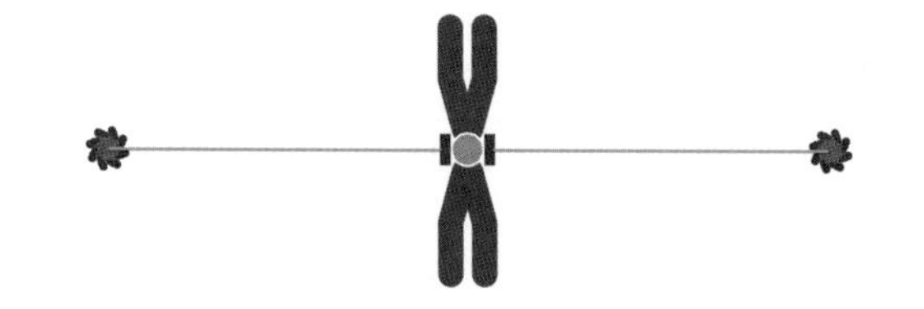

FIGURE 11.29 The way in which attachment of microtubules results in movement of the daughter chromosomes toward different centrosomes.

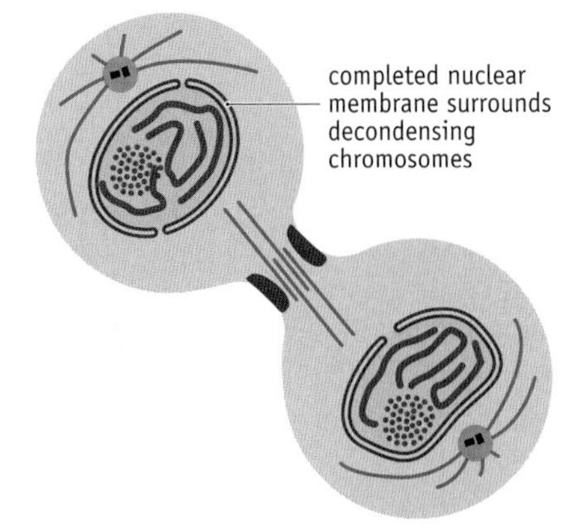

FIGURE 11.30 Cytokinesis. (Adapted from B. Alberts et al., Essential Cell Biology, 3rd ed. New York: Garland Science, 2010.)

moves toward one centrosome, and the other toward the other centrosome. The result is that a complete set of chromosomes is assembled at each of the poles of the mitotic spindle. During telophase, each set becomes enclosed within a new nucleus. Division of the cell—**cytokinesis**—occurs across the plane of the mitotic spindle, ensuring that each daughter cell forms around one of the divided nuclei (Figure 11.30). Each daughter cell therefore inherits a complete set of chromosomes and hence a complete copy of the parent cell's genome.

KEY CONCEPTS

- The genome is the entire genetic complement of an organism, made up of one or more lengthy DNA molecules, and comprising all that organism's genes and all of the intergenic DNA between those genes. A gene is a unit of inheritance and the genome is all of those units added together.
- All eukaryotic genomes are made up of linear DNA molecules, each of these molecules contained in a different chromosome. The number of chromosomes varies between species.
- A packaging system is needed to fit a DNA molecule into its chromosome. This packaging system is hierarchical. The lowest level of packaging involves winding the DNA around octamers of histone proteins to form nucleosomes. Nucleosomes then pack together to form the 30-nm fiber.
- Chromosomes are visible only in dividing cells. Between cell divisions, the nucleus contains a mixture of light- and dark-staining areas. The light areas are euchromatin, made up mainly of 30-nm fiber. The dark areas are called heterochromatin, and are made up of DNA that is more condensed than in euchromatin.
- A metaphase chromosome contains two replicated daughter chromosomes linked together at the centromere. The centromere is also the position at which the chromosome attaches to the microtubules that draw the daughters into their respective nuclei during cell division.
- The ends of a chromosome are called the telomeres. Telomeres have various roles. Importantly, they prevent the ends of the chromosome from becoming progressively shorter.
- Events during the cell cycle are coordinated and controlled to ensure that the chromosomes are replicated at the appropriate time, and that cell division, when it occurs, results in the orderly partitioning of the replicated chromosomes into the daughter cells. Coordination of the cell cycle involves a variety of regulatory proteins, collectively called cyclins.
- Mitosis ensures the correct partitioning of chromosomes during cell division.

QUESTIONS AND PROBLEMS (Answers can be found at www.garlandscience.com/introgenetics)

Key Terms

Write short definitions of the following terms:

30-nm chromatin fiber
acrocentric
alphoid DNA
anaphase
apoptosis
autosome
beads-on-a-string
cell senescence
centromere
centrosome
chromatid
chromatin
chromatosome
constitutive heterochromatin
core octamer
cyclin
cytokinesis
diploid
euchromatin
facultative heterochromatin
G1 phase
G2 phase
gamete
haploid
heterochromatin
histone
interphase
karyogram
kinetochore
linker DNA
linker histone
M phase
matrix-associated region
meiosis
metacentric
metaphase
metaphase chromosome
mitosis
mitotic spindle
nuclear matrix
nuclease protection
nucleosome
post-replication complex
pre-replication complex
prometaphase
prophase
replication licensing factor
S phase
scaffold attachment region
sex cell
somatic cell
stem cell
submetacentric
telocentric
telomerase
telomere
telophase
tumor-suppressor protein

Self-study Questions

11.1 When chromosome number is considered, how typical are humans of eukaryotes in general?

11.2 Explain why a packaging system is needed to fit a DNA molecule into its chromosome.

11.3 Describe the role of histones in packaging DNA into eukaryotic chromosomes.

11.4 What does the treatment of chromatin with nucleases reveal about the packaging of eukaryotic DNA?

11.5 Distinguish between the beads-on-a-string structure and the 30-nm chromatin fiber.

11.6 How is the 30-nm chromatin fiber attached to the nuclear matrix?

11.7 Explain the difference between euchromatin and heterochromatin, and between the two different types of heterochromatin.

11.8 Describe the main morphological features of a human metaphase chromosome.

11.9 Draw four diagrams to show the differences between metacentric, submetacentric, acrocentric, and telocentric chromosomes.

11.10 What are the special features of the DNA sequences and the nucleosomes present in centromeres?

11.11 Summarize the roles that telomeres play. What are the special features of the DNA present in telomeres?

11.12 Explain why the ends of a chromosomal DNA molecule could become shortened after repeated rounds of DNA replication, and show how telomerase prevents this from occurring.

11.13 Discuss the links between telomeres, cell senescence, and cancer.

11.14 Explain what is meant by the "cell cycle," and distinguish between the different stages of the cell cycle.

11.15 Outline our current knowledge regarding the way in which DNA replication is coordinated with the rest of the cell cycle.

11.16 What is meant by "origin firing," and how does origin firing vary along a chromosome?

11.17 Describe what happens to a cell that has accumulated too much DNA damage for its chromosomes to be replicated.

11.18 Draw a series of diagrams to illustrate the main stages of mitosis.

11.19 What roles do kinetochores play during mitosis?

11.20 Describe how the correct partitioning of chromosomes is ensured during mitosis.

Discussion Topics

11.21 Although the size of a eukaryotic genome coincides to a certain extent with the complexity of the organism, the relationship is only approximate. For example, the salamander genome is over 20 times the length of the human genome. Few humans would accept that we are only one-twentieth as complex as a salamander. What factors might be responsible for the large genomes of certain organisms?

11.22 Why are chromosome numbers so different in different organisms, and why does the number not relate to the complexity of the organism?

11.23 Discuss the impact that the presence of nucleosomes is likely to have on the expression of individual genes.

11.24 Discuss the implications that the existence of facultative heterochromatin has for the organization of genes within a eukaryotic genome.

11.25 Devise an experiment that would enable the time of firing of each of the replication origins on a yeast chromosome to be worked out.

FURTHER READING

Cech TR (2004) Beginning to understand the end of the chromosome. *Cell* 116, 273–279. *Reviews all aspects of telomerase.*

Collado M, Blasco MA & Serrano M (2007) Cellular senescence in cancer and aging. *Cell* 130, 223–233.

Kornberg RD & Klug A (1981) The nucleosome. *Sci. Am.* 244, 48–60.

Mazia D (1961) How cells divide. *Sci. Am.* 205, 100–120.

Nurse P (2000) A long twentieth century of the cell cycle and beyond. *Cell* 100, 71–78. *Reviews progress in understanding the cell cycle.*

Robinson PJJ & Rhodes D (2006) Structure of the "30 nm" chromatin fibre: a key role for the linker histone. *Curr. Opin. Struct. Biol.* 16, 336–343. *Reviews models for the structure of the 30-nm fiber.*

Schueler MG, Higgins AW, Rudd MK et al. (2001) Genomic and genetic definition of a functional human centromere. *Science* 294, 109–115. *Details the sequence features of human centromeres.*

Shay JW & Wright WE (2006) Telomerase therapeutics for cancer: challenges and new directions. *Nat. Rev. Drug Discov.* 5, 577–584. *Methods for inhibiting telomerase in order to treat cancer.*

Stillman B (1996) Cell cycle control of DNA replication. *Science* 274, 1659–1664.

Travers A (1999) The location of the linker histone on the nucleosome. *Trends Biochem. Sci.* 24, 4–7.

van Steensel B, Smogorzewska A & de Lange T (1998) TRF2 protects human telomeres from end-to-end fusions. *Cell* 92, 401–413.

Verheijen R, van Venrooij W & Ramaekers F (1988) The nuclear matrix: structure and composition. *J. Cell Sci.* 90, 11–36.

Inheritance of Genes in Bacteria CHAPTER 12

As humans we tend to look on the human genome as the most important. As geneticists we must take a broader view and consider the genomes of all types of organism. This means that we must ask how genes are inherited not only during eukaryotic cell division, but also in bacteria.

We must therefore examine the structures of bacterial genomes and ask how these are replicated and how complete copies are passed on to daughter cells when a bacterium divides. In doing this, we will make two important discoveries that will challenge our understanding of genomes and inheritance. First, we will learn that some bacteria possess a variety of DNA molecules and defining exactly which ones constitute the organism's genome can be difficult. Some of these DNA molecules, called **plasmids**, are mobile and can move easily between bacteria, even between members of different species. Second, we will see that linear descent—repeated passage of the genome from parent to daughter cells—is not the only way in which bacteria obtain genes. Some of the genes in bacterial genomes have been acquired from other bacteria, possibly other species, by **horizontal gene transfer** (Figure 12.1).

Most of the early research in bacterial genetics was carried out with *Escherichia coli*. This is a common bacterium that normally lives in the lower intestine of humans and other mammals (Figure 12.2). Most strains are harmless, but a few types have genes that code for proteins that are toxic to the host. These pathogenic types are often responsible for outbreaks of food poisoning.

E. coli was first isolated in 1885. The ease with which it can be grown on solid agar medium or in a liquid broth culture led to the species gradually being adopted as a model bacterium for all types of microbiology research, including bacterial genetics. This was a lucky choice, because *E. coli* has a relatively uncomplicated genome. Although it is host to a number of plasmids, it does not suffer from the ambiguities presented by the additional DNA molecules found in some other bacteria. We will therefore begin our

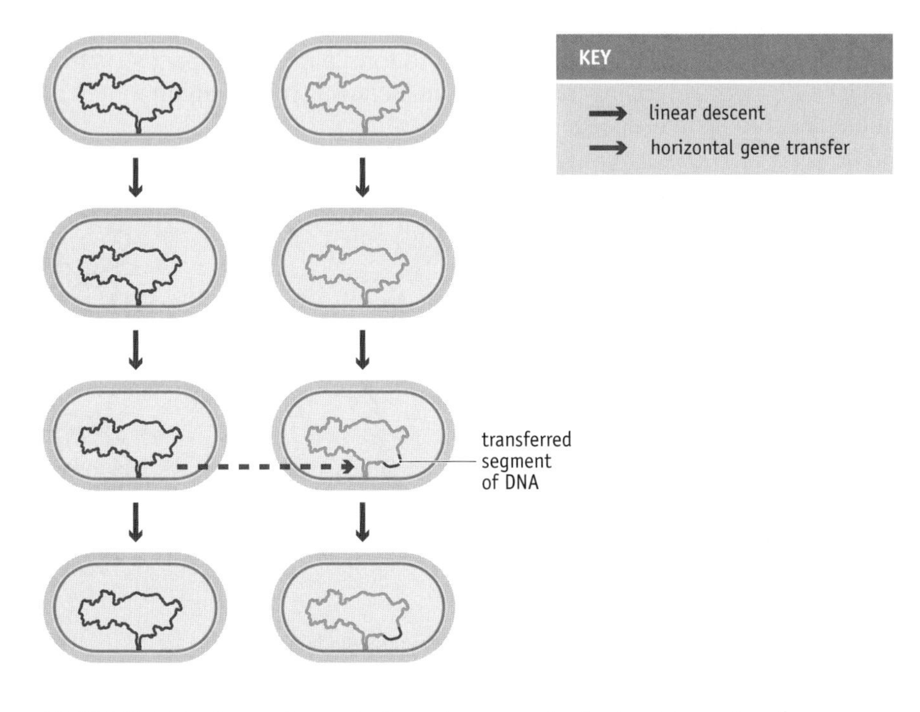

FIGURE 12.1 Inheritance of genes by linear descent and by horizontal gene transfer.

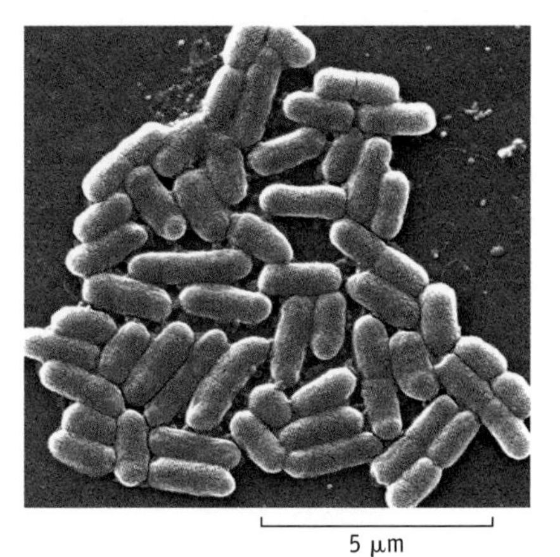

FIGURE 12.2 Scanning electron micrograph of *Escherichia coli* cells. (Courtesy of Janice Carr, Centers for Disease Control and Prevention.)

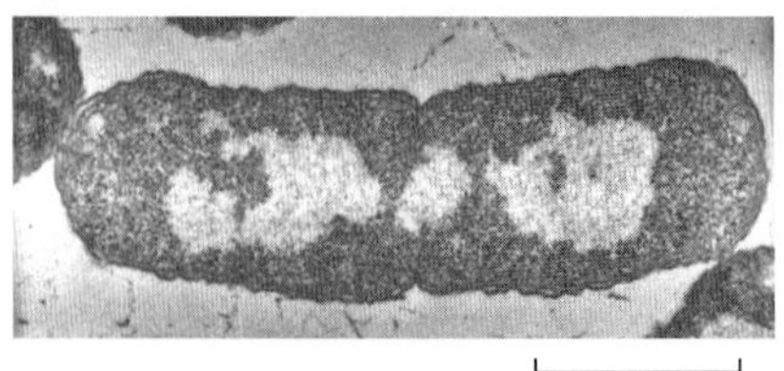

FIGURE 12.3 The *Escherichia coli* nucleoid. This transmission electron micrograph shows the cross-section of a dividing *E. coli* cell. The nucleoid is the lightly staining area in the center of the cell. (Courtesy of Conrad Woldringh, University of Amsterdam.)

study of the inheritance of bacterial genes with *E. coli*, and then look at the complications posed by the more complex types of genomes. Finally, we will examine the various processes that can bring about horizontal gene transfer and look at the impact that these processes have had on the composition of bacterial genomes.

12.1 INHERITANCE OF GENES IN *E. COLI*

Although bacterial cells are not divided into membrane-bound compartments, their internal structure is not entirely featureless. The electron microscope reveals two distinct regions within an *E. coli* cell (Figure 12.3). There is a central area called the **nucleoid**, which takes up about one-third of the volume of the cell. The nucleoid is surrounded by a peripheral region that is usually referred to simply as the cytoplasm. Most of the DNA in an *E. coli* cell is contained in the nucleoid, so it is here that we must go in order to study the *E. coli* genome.

The *E. coli* nucleoid contains supercoiled DNA attached to a protein core

The *E. coli* nucleoid is made up of DNA and protein. The DNA is a single, circular molecule of 4639 kb, corresponding to a contour length (i.e., circumference) of approximately 1.6 mm. This molecule is often referred to as the **bacterial chromosome**, an unfortunate use of terminology because it has none of the features of a eukaryotic chromosome.

The *E. coli* chromosome has to be squeezed into a cell that is about 1 μm by 2 μm. This is not impossible, as a DNA molecule is very thin and so does not take up much space when it is folded up tightly. This tight folding can be achieved by **supercoiling**, which occurs when additional turns are introduced into the DNA double helix (positive supercoiling) or if turns are removed (negative supercoiling). With a linear molecule, the torsional stress introduced by over- or underwinding is immediately released by rotation of the ends of the DNA molecule, but a circular molecule, having no ends, cannot reduce the strain in this way. Instead the circular molecule responds by winding around itself to form a more compact structure (Figure 12.4). Supercoiling is therefore an ideal way of packaging a circular molecule into a small space.

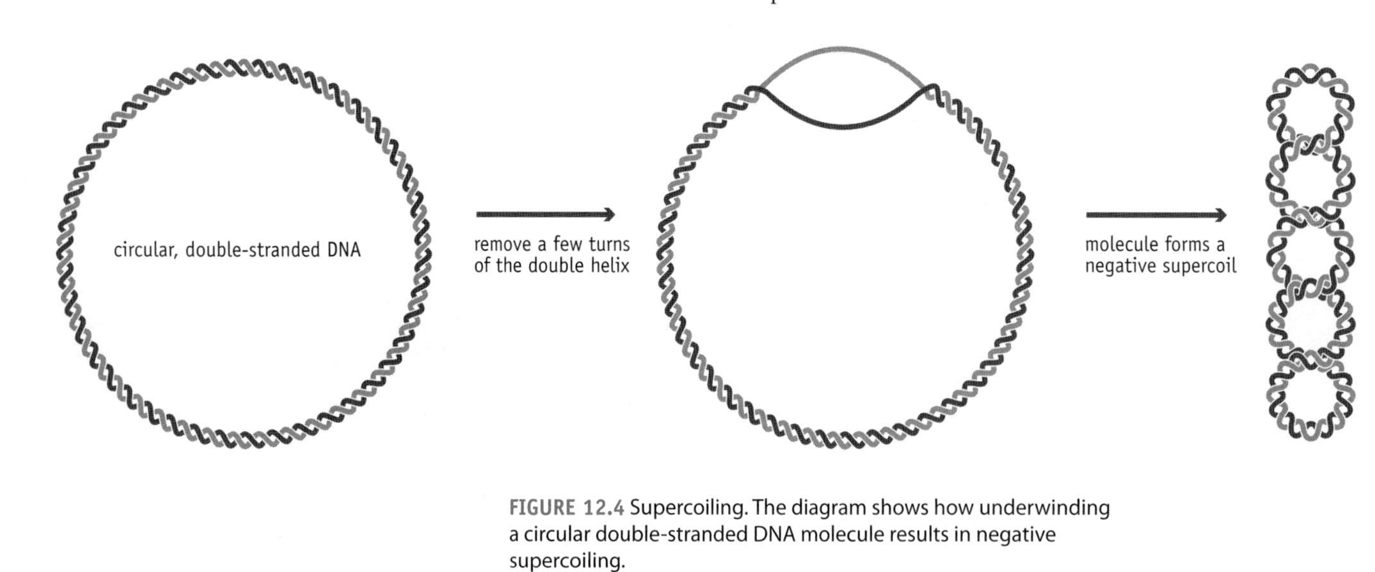

FIGURE 12.4 Supercoiling. The diagram shows how underwinding a circular double-stranded DNA molecule results in negative supercoiling.

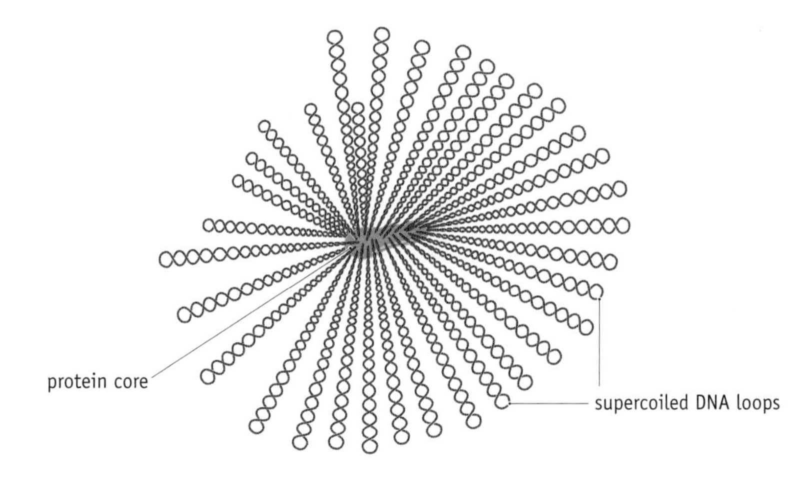

FIGURE 12.5 A model for the structure of the *Escherichia coli* nucleoid.

Evidence that supercoiling is involved in packaging the circular *E. coli* chromosome was first obtained in the 1970s from examination of isolated nucleoids. In 1981, supercoiling was confirmed as a feature of DNA in living cells. In *E. coli*, the supercoiling is thought to be generated and controlled by two enzymes, DNA gyrase and DNA topoisomerase I.

The supercoiling introduced into the *E. coli* chromosome must be organized in such a way that the genes within the molecule are still accessible, so they can be transcribed as and when necessary. It must also be possible to replicate the DNA and separate the daughter molecules without everything getting tangled up. This means that the DNA molecule must be folded up in a very ordered manner. Currently, we believe that the *E. coli* DNA is attached to a protein core from which supercoiled loops, each containing 10 to 100 kb of DNA, radiate out into the cell (Figure 12.5). The protein component of the nucleoid includes DNA gyrase and DNA topoisomerase I, the two enzymes that are primarily responsible for maintaining the supercoiled state, as well as a set of at least four proteins believed to have a more specific role in packaging the bacterial DNA. The most abundant of these packaging proteins is HU, which forms a tetramer around which approximately 60 bp of DNA becomes wound (Figure 12.6). There are some 60,000 HU proteins per *E. coli* cell, enough to cover about one-fifth of the DNA molecule, but it is not known whether the tetramers are evenly spaced along the loops of supercoiled DNA that radiate out into the cell, or are restricted to the protein core of the nucleoid.

Plasmids are independent DNA molecules within a bacterial cell

In addition to the chromosome in its nucleoid, the cytoplasm of an *E. coli* cell might contain other DNA molecules called plasmids (Figure 12.7). Most

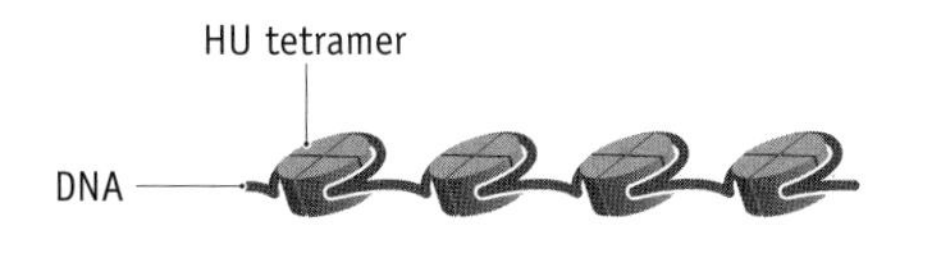

FIGURE 12.6 The role of HU proteins in packaging bacterial DNA. HU forms a tetramer around which approximately 60 bp of DNA becomes wound.

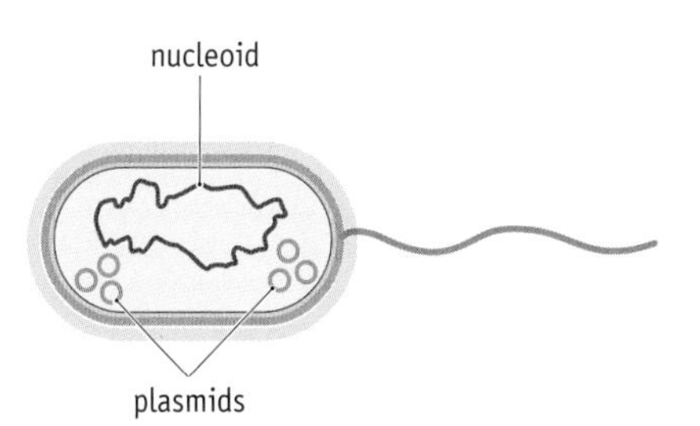

FIGURE 12.7 Plasmids are small, usually circular DNA molecules that are found inside some bacterial cells.

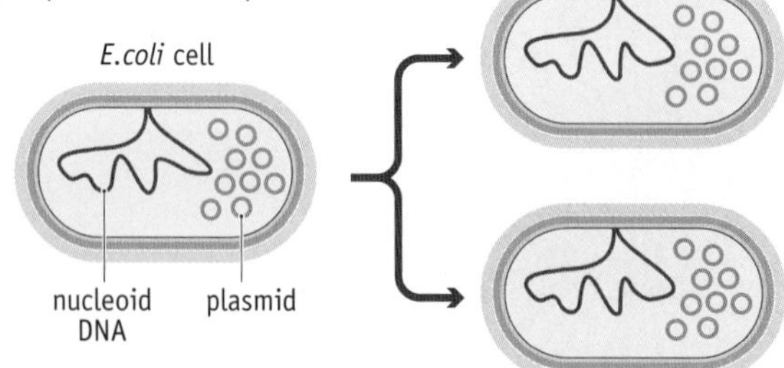

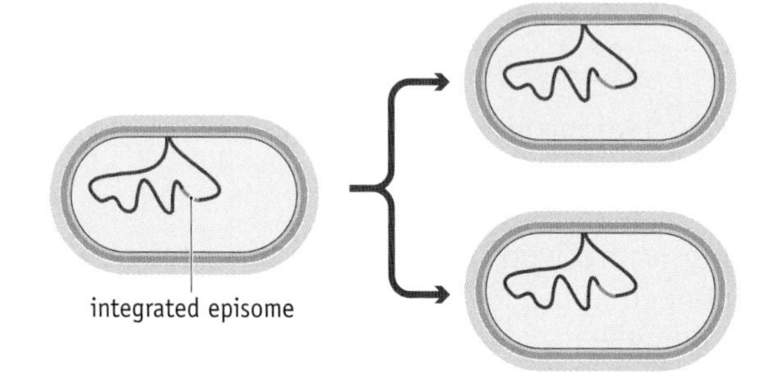

FIGURE 12.8 Two mechanisms for plasmid replication. (A) Some plasmids replicate as independent entities. (B) Episomes are able to insert into the bacterial chromosome.

plasmids are circular, though some linear ones are known. Plasmids almost always carry one or more genes, and often these genes are responsible for a useful characteristic. For example, the ability of some *E. coli* strains to survive in normally toxic concentrations of antibiotics such as chloramphenicol or ampicillin is due to the presence in the bacterium of a plasmid that carries genes conferring resistance to these antibiotics.

All plasmids possess at least one DNA sequence that can act as an origin of replication, so they are able to multiply in the cell independently of the bacterial chromosome (Figure 12.8A). The smaller plasmids make use of the cell's own DNA replicative enzymes to make copies of themselves, whereas some of the larger ones carry genes that code for special enzymes that are specific for plasmid replication. A few types of plasmid are also able to replicate by inserting themselves into the chromosome (Figure 12.8B). These integrative plasmids or **episomes** may be stably maintained in this form through numerous cell divisions, but always at some stage exist as independent elements.

Plasmids can be classified according to the genes they carry

Virtually all species of bacteria harbor plasmids. Their sizes vary from approximately 1 kb for the smallest to over 250 kb for the larger ones. Some plasmids are restricted to just a few related species and are found in no other bacteria, but others have a broad host range and can exist in numerous species, although possibly displaying a preference for a few kinds of bacteria in which they are more common.

Plasmids as a whole are most usefully classified according to the genes that they carry and the characteristics that those genes confer on the host bacterium. According to this classification there are five main types of plasmid (Table 12.1).

The most common type are **resistance** or **R plasmids**. These carry genes conferring on the host bacterium resistance to one or more antibacterial agents, such as chloramphenicol, ampicillin, or mercury. R plasmids are very important in clinical microbiology because their spread through

TABLE 12.1 FEATURES OF TYPICAL PLASMIDS

Type of plasmid	Gene functions	Examples
Resistance	Resistance to antibacterial agents	Rbk of *Escherichia coli* and other bacteria
Fertility	Conjugation and DNA transfer between bacteria	F of *E. coli*
Killer	Synthesis of toxins that kill other bacteria	Col of *E. coli*, for colicin production
Degradative	Enzymes for metabolism of unusual molecules	TOL of *Pseudomonas putida*, for toluene metabolism
Virulence	Pathogenicity	Ti of *Agrobacterium tumefaciens*, conferring the ability to cause crown gall disease in dicotyledonous plants

natural populations can have profound consequences for the treatment of bacterial infections. An example of an R plasmid is Rbk, which is commonly found in *E. coli* but also occurs in other bacteria.

Some species of bacteria, including *E. coli*, often have copies of a second type of plasmid called a **fertility** or **F plasmid**. These are able to direct **conjugation** between different bacteria of the same, or related, species. Conjugation is a process that enables two bacteria to join together so that plasmids, and possibly parts of the chromosome, can be passed from one cell to another. Conjugation is therefore one of the processes that can lead to horizontal gene transfer, and we will look at it in more detail when we deal with this topic later in the chapter.

Some plasmids carry genes that code for toxic proteins which kill other bacteria. We assume that bacteria carrying these plasmids gain an advantage in the competition for scarce resources. Plasmids of this type are not common in bacterial species as a whole, but are present in some strains of *E. coli*. These are the **Col plasmids**, coding for toxins called **colicins**. In species other than *E. coli*, we also find **degradative plasmids** and **virulence plasmids**. Degradative plasmids allow the host bacterium to metabolize unusual molecules such as toluene and salicylic acid. Several examples occur in the *Pseudomonas* genus of bacteria, such as the TOL plasmid of *Pseudomonas putida*. Virulence plasmids confer pathogenicity on the host bacterium. The best known example is the Ti plasmid of *Agrobacterium tumefaciens*, which induces crown gall disease in dicotyledonous plants.

A single bacterium can have multiple copies of the same or different plasmids

Each type of plasmid has its own characteristic **copy number**, this being the number of copies of the plasmid that are present in a single bacterial cell. Some plasmids, especially the larger ones, are **stringent** and have a low copy number of perhaps just one or two per cell. Others, called **relaxed** plasmids, are present in multiple copies of 50 or more.

There can also be more than one type of plasmid in a single cell—cells of *E. coli* have been known to contain up to seven different plasmids at once. To be able to coexist in the same cell, different plasmids must be **compatible**. If two plasmids are incompatible, then one or the other will be quite rapidly lost from the cell. Different types of plasmid can therefore

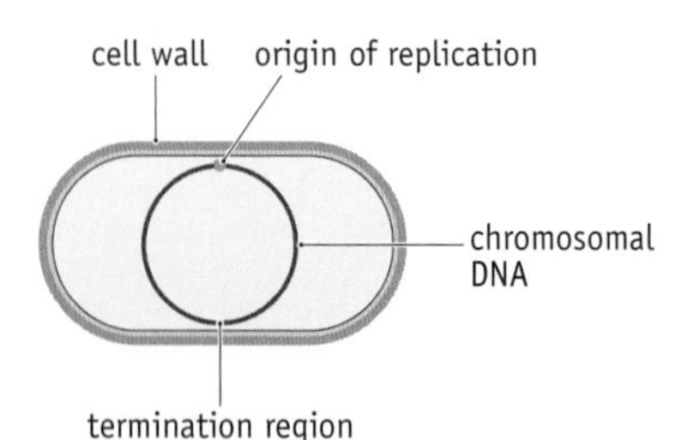

FIGURE 12.9 The orientation of the chromosomal DNA inside an *E. coli* cell. Note that, in reality, the DNA molecule is much too long to be laid out inside the cell as a circle, as shown here, and instead is in a supercoiled conformation as illustrated in Figure 12.5. The important point is that the replication origin and termination region are located on opposite sides of the cell.

be assigned to different **incompatibility groups** on the basis of whether or not they can coexist. Plasmids from a single incompatibility group (which cannot coexist) are often related to one another in various ways.

The factors determining the copy number and compatibility relationships between plasmids are not well understood, but events during plasmid replication are thought to underlie both phenomena. One theory holds that copy number is determined by an inhibitor molecule that prevents further replication of the plasmid once the characteristic value is reached. It has also been suggested that replication of incompatible plasmids is controlled by the same inhibitor molecule. A second theory suggests that plasmid replication requires attachment to a specific binding site on the cell membrane and that incompatible plasmids compete for the same attachment sites.

How daughter cells acquire copies of the bacterial chromosome and plasmids is not well understood

Although it has just a single chromosome, *E. coli* still must have a mechanism that ensures that the replicated copies of this molecule are passed to the daughter cells so each receives a single, intact copy. Inheritance of plasmids is not such an absolute requirement, because these DNA molecules do not carry any essential genes, but during the vast majority of cell divisions the plasmids present in the parent cell become evenly distributed between the daughters. This means that mechanisms must also exist for partitioning of plasmids.

We know relatively little about the way in which the replicated bacterial chromosomes are passed on to the daughter cells. What we do know suggests that the arrangement of the chromosome within the nucleoid plays an important role. This arrangement is not random. In different cells the equivalent parts of the DNA molecule are always located at equivalent positions. In *E. coli*, for example, the chromosomal DNA is located in the middle part of the rod-shaped cell. The origin of replication is always adjacent to the cell wall, with the replication termination region diametrically opposite, adjacent to the wall on the other side of the cell (Figure 12.9).

When the chromosome begins to replicate, the origins of replication of the two daughter DNA molecules move apart from one another, each toward a different end of the cell (Figure 12.10). How this occurs is not known. A cytoskeletal protein called MreB is required, but may play a passive role, simply providing a framework along which the origins travel. One theory is that the origins are pushed apart by RNA polymerase enzymes as these gain access to the newly replicated strands in order to transcribe the genes close to the origin.

Whatever the process, the outcome is that the two daughter chromosomes move toward either end of the bacterium as the parent molecule is being replicated. The replicated DNA molecules therefore take up the appropriate positions needed for them to become enclosed within the daughter cells that are formed when cytokinesis divides the bacterium into two. In a rich growth medium, one in which the bacteria are able to divide every 20 minutes or so, the septum begins to form across the middle of the parent cell before chromosome replication has been completed (Figure 12.11). The last parts of the replicated DNA molecules are "pumped" across the septum by special translocase enzymes.

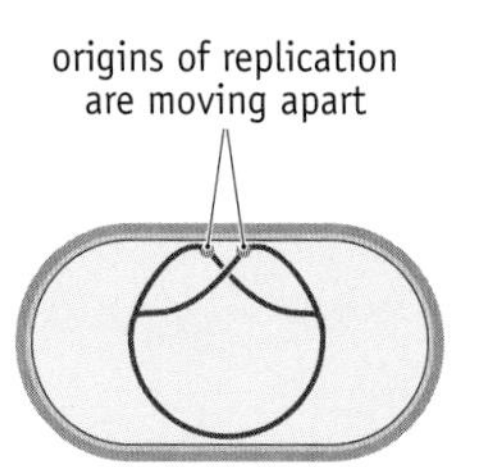

FIGURE 12.10 During replication of the *E. coli* chromosome, the origins of replication of the two daughter molecules move apart from one another.

Plasmid partitioning is understood better, although there are still gaps in our knowledge. There are at least three different partitioning mechanisms, each used by different types of plasmid. The "type I" system, which is used by several of the clinically important R plasmids of *E. coli*, is best understood. In this system, the two members of a pair of daughter plasmids are pushed apart from one another by growth of a microfilament made up of ParM proteins (Figure 12.12). These proteins attach to each plasmid at a special sequence called *parC*, which is the binding site for another protein, ParR. The latter forms a platform for attachment of the first of the ParM proteins in the growing microfilament. Sadly, *parC* is sometimes called the plasmid "centromere," even though it has no similarity to a eukaryotic centromere other than being the attachment site for a microfilament (which is quite different from a eukaryotic microtubule). The ParM filament grows along the axis of the bacterium, so the replicated plasmids become pushed to opposite ends. The other plasmid partitioning systems work in similar ways, the main distinction being in the identity of the protein that forms the filament that pushes the daughter plasmids to the poles.

septum

chromosome replication not yet completed

FIGURE 12.11 In rapidly dividing *E. coli* cells, septum formation might occur before the chromosomal DNA has been completely replicated.

12.2 VARIATIONS ON THE *E. COLI* THEME

How typical is *E. coli* of bacteria in general? In the early days of bacterial genetics there tended to be an assumption that most if not all bacteria would prove to have very similar genomes, and therefore that discoveries made with *E. coli* would also apply to other species. We now know that this view was incorrect, and that in some respects *E. coli* is quite an unusual bacterium.

Bacterial genomes vary greatly in size, and some are linear DNA molecules

One feature of the *E. coli* genome that is "average" is its size. At 4639 kb, the *E. coli* genome is in the midrange for bacterial genomes. Most bacterial genomes are less than 5 Mb, but the variation among those genomes that have been completely sequenced is from 160 kb for *Carsonella ruddii* to 13.0 Mb for *Sorangium cellulosum* (Table 12.2). A few unsequenced genomes are substantially larger than this. *Bacillus megaterium*, for example, has a huge genome of 30 Mb, 2½ times larger than the smallest eukaryotic ones. These sizes reflect the number of genes that the bacterium possesses, which in turn reflects lifestyle. *C. ruddii* is an endosymbiont that lives inside some types of insect. In this protective environment it does not need many of the genes required by a free-living bacterium such as *S. cellulosum*, a cellulose-degrading bacterium that lives in soil. *C. ruddii* has about 180 genes, whereas *S. cellulosum* has almost 9400, twice as many as *E. coli*.

TABLE 12.2 GENOME SIZES FOR VARIOUS PROKARYOTES

Species	Size of genome (Mb)
Carsonella ruddii	0.16
Mycoplasma genitalium	0.58
Mycobacterium tuberculosis H37Rv	4.41
Escherichia coli K12	4.64
Sorangium cellulosum So ce56	13.0

The strain designation (e.g., K12) is given if specified by the group who sequenced the genome. With many bacterial species, different strains have different genome sizes.

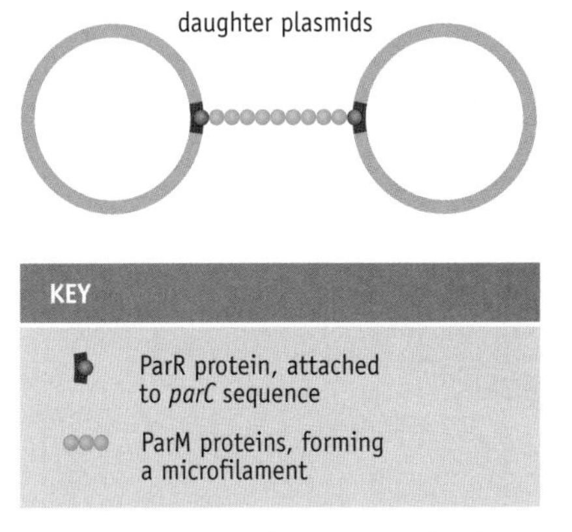

FIGURE 12.12 Plasmid partitioning by the type I system.

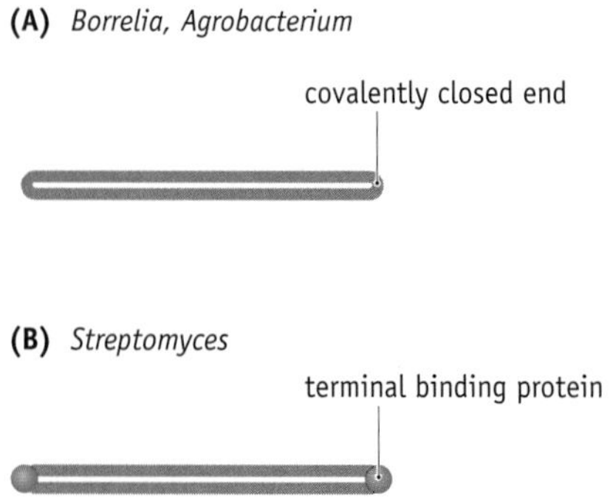

FIGURE 12.13 The ends of the linear chromosomes found in (A) *Borrelia* and *Agrobacterium*, and (B) *Streptomyces*.

The *E. coli* genome, as described above, is a single, circular DNA molecule. This is also the case with the vast majority of bacterial genomes that have been studied, but an increasing number of linear versions are being found. The first of these, for *Borrelia burgdorferi*, the organism that causes Lyme disease, was described in 1989, and during the following years similar discoveries were made for *Streptomyces coelicolor* and *Agrobacterium tumefaciens*.

Linear molecules have free ends, which must be distinguishable from DNA breaks, so these chromosomes require terminal structures equivalent to the telomeres of eukaryotic chromosomes. In *Borrelia* and *Agrobacterium*, the real chromosome ends are distinguishable because a covalent linkage is formed between the 5′ and 3′ ends of the polynucleotides in the DNA double helix, and in *Streptomyces* the ends appear to be marked by special binding proteins (**Figure 12.13**).

Some bacteria have multipartite genomes

A second, more widespread variation on the *E. coli* theme is the presence in some bacteria of multipartite genomes—genomes that are divided into two or more DNA molecules. This feature is more complicated than it might at first appear, because there can sometimes be difficulties in distinguishing a genuine part of a genome from a plasmid. Because the genes carried by a plasmid are nonessential, the bacterium being able to survive without them, a plasmid is generally looked on as an accessory molecule rather than a genuine part of a bacterium's genome.

When we find a bacterium that contains more than one DNA molecule, we must therefore establish that the additional molecules are not plasmids before concluding that the species has a multipartite genome. This is easy enough with many bacteria, including *E. coli*, which harbors various combinations of plasmids, none of which is more than a few kb in size and none of which carry genes essential for survival of the bacterium.

With other bacteria the distinction is not so easy (**Table 12.3**). *Vibrio cholerae*, the pathogenic bacterium that causes cholera, is a case in

TABLE 12.3 EXAMPLES OF GENOME ORGANIZATION IN BACTERIA

Species	DNA molecules	Size (Mb)
Escherichia coli K12	One circular molecule	4.639
Vibrio cholerae El Tor N16961	Two circular molecules	
	Main chromosome	2.961
	Megaplasmid	1.073
Deinococcus radiodurans R1	Four circular molecules	
	Chromosome 1	2.649
	Chromosome 2	0.412
	Megaplasmid	0.177
	Plasmid	0.046

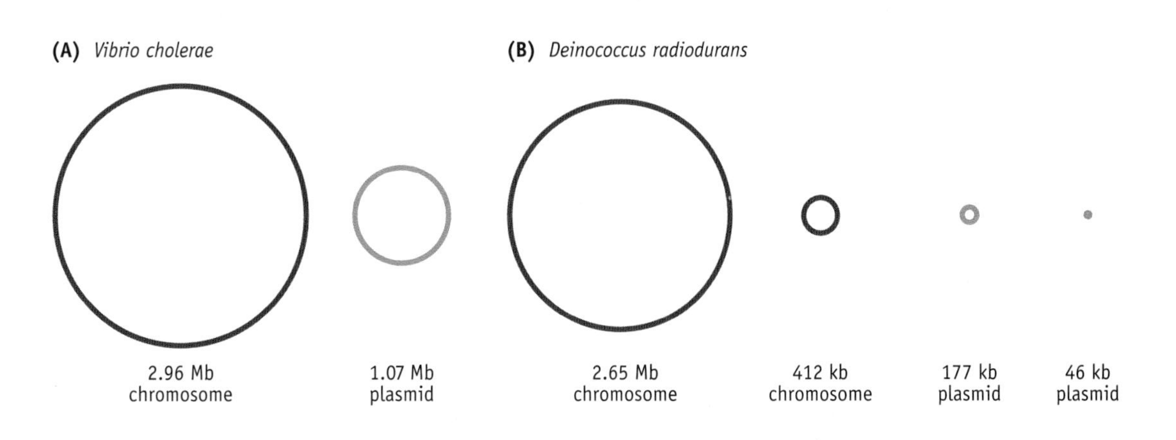

FIGURE 12.14 The multipartite genomes of (A) *Vibrio cholerae*, and (B) *Deinococcus radiodurans*.

point. Members of this species have two circular DNA molecules, one of 2.96 Mb and the other of 1.07 Mb, with 71% of the organism's 3878 protein-coding genes on the larger of these (Figure 12.14A). Both molecules carry essential genes, although most of those specifying the central cellular activities, such as gene expression and energy generation, as well as the genes that confer pathogenicity, are located on the larger molecule. It would appear obvious that these two DNA molecules together constitute the *Vibrio* genome. The complications arise when the smaller of the two molecules is examined more closely. This molecule contains an **integron**, a set of genes and other DNA sequences that enable plasmids to capture genes from bacteriophages and other plasmids. It therefore appears that the smaller molecule is a plasmid rather than a chromosome, even though some of the genes it carries are essential to the bacterium.

Deinococcus radiodurans, whose genome is of particular interest because it contains many genes that help this bacterium resist the harmful effects of radiation, is constructed on similar lines. Its essential genes are distributed among two circular DNA molecules, considered to be genuine chromosomes, and two plasmids (Figure 12.14B).

The complications posed by bacteria such as *Vibrio* and *Deinococcus* have prompted microbial geneticists to invent a new term—**chromid**—to describe a plasmid that carries essential genes. This means that we now distinguish between three, rather than just two, types of DNA molecule that might be found in a bacterium (Figure 12.15). First, there are one or more bacterial chromosomes, carrying essential genes and located in the nucleoid. The second are genuine plasmids, which are distinct from a bacterial chromosome because of their special plasmid partitioning system and whose genes are nonessential to the bacterium. And third are the chromids, which use a plasmid partitioning system but which carry genes that the bacterium needs to survive. According to this nomenclature, *Vibrio cholerae* has one chromosome and one chromid, and *Deinococcus radiodurans* has two chromosomes and two chromids.

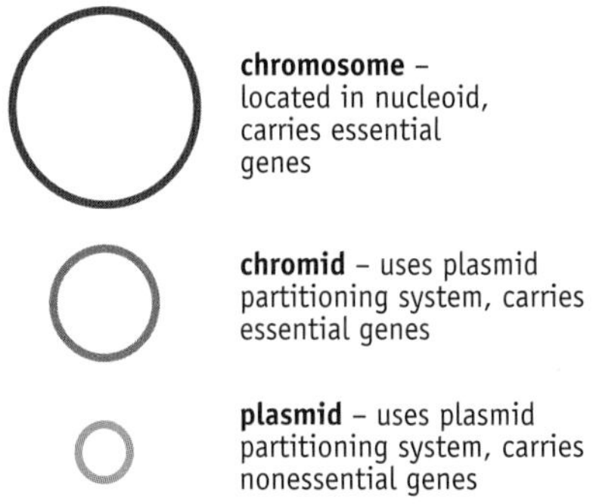

chromosome – located in nucleoid, carries essential genes

chromid – uses plasmid partitioning system, carries essential genes

plasmid – uses plasmid partitioning system, carries nonessential genes

FIGURE 12.15 The relationship between chromids, chromosomes, and plasmids.

12.3 TRANSFER OF GENES BETWEEN BACTERIA

So far, we have considered only the way in which bacterial genes, whether on a chromosome, chromid, or plasmid, are inherited by linear descent—by daughter cells when a parent cell divides. Bacteria can also acquire genes directly from other bacteria. This is called horizontal or lateral gene transfer.

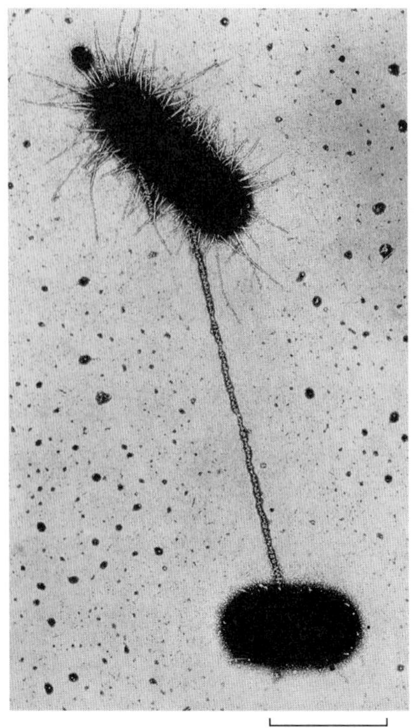

FIGURE 12.16 Electron micrograph showing two bacteria joined together by a pilus. One of the two bacteria also has a number of shorter pili radiating from its surface. (Courtesy of the late Charles Brinton.)

It has been known since the 1940s that plasmids and occasionally chromosomal genes can move horizontally, but it was thought that the transfer could take place only between members of the same species, or occasionally between closely related species. This assumption was overturned in the 1990s when complete genome sequences were obtained for a variety of bacterial species. Comparisons of these sequences revealed that the same genes are present in quite distantly related species, suggesting that the barriers to horizontal gene transfer are much less rigid than previously thought. It also became clear that although we know a great deal about the processes by which plasmids and chromosomal genes move within species, we know much less about how genes are transferred from one species to another. We will look first at our understanding of these within-species processes, and then examine the implications of the much broader examples of horizontal gene transfer revealed by genome sequencing.

Plasmids can be transferred between bacteria by conjugation

One of the major breakthroughs in twentieth-century microbiology was made in 1946 when Joshua Lederberg, a 21-year-old graduate student at Yale University, discovered that bacteria can exchange genes by a process that was subsequently called conjugation. During the next decade, Lederberg and other microbial geneticists who studied conjugation showed that the transfer of genes is unidirectional from donor to recipient cells. The donor cells are referred to as **F$^+$** ("F-plus," F for fertility), and the recipient cells are called **F$^-$** ("F-minus") cells.

Conjugation involves a tubelike structure, called the **pilus**, which F$^+$ cells are able to construct (Figure 12.16). An F$^+$ cell forms a connection with an F$^-$ cell via its sex pilus, and because the pilus is hollow it has been assumed that DNA transfer occurs through it. This has never been demonstrated, and it is more likely that the pilus simply brings the F$^+$ and F$^-$ cells into close contact (Figure 12.17). It may be that the contact is so close that the outer membranes on the surfaces of the two bacteria fuse, enabling DNA to be transferred directly from one cell to the other.

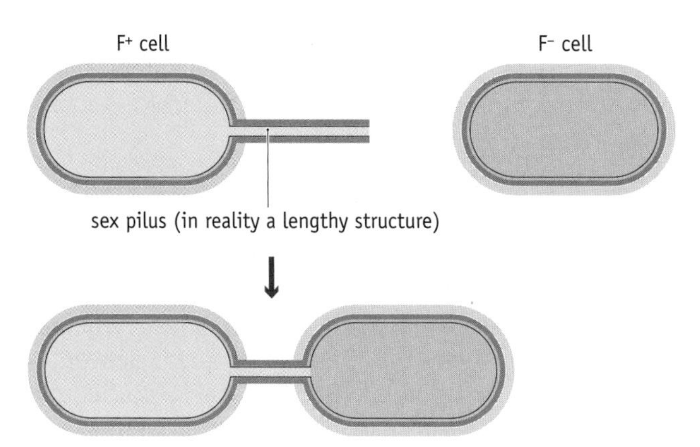

FIGURE 12.17 The role of the sex pilus in bringing two bacteria together.

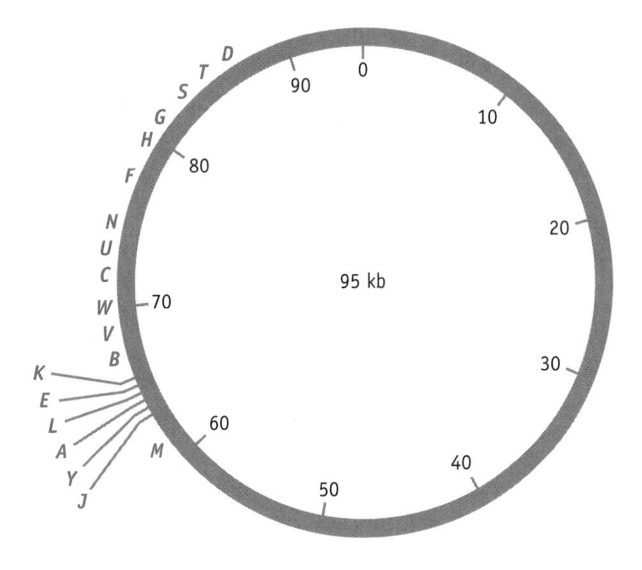

FIGURE 12.18 The F plasmid. The numbers indicate the distance in kb in a clockwise direction from the replication origin. The letters indicate the positions of the *tra* genes (*traA*, *traB*, etc.).

The difference between an F^+ and an F^- cell is that the former contains a fertility plasmid. The best studied of these is the F plasmid of *E. coli*. The F plasmid is 95 kb in size and carries a large operon that contains the *tra* (for transfer) genes (Figure 12.18). These genes code for proteins involved in synthesis and assembly of the pilus and in the DNA transfer process itself.

Conjugation is always between an F^+ cell and an F^- one. The F^+ cell, the one that contains an F plasmid, initiates conjugation by making a pilus, which subsequently attaches to an F^- cell (Figure 12.19). The two cells become drawn together, and a copy of the F plasmid is transferred from the F^+ to the F^- bacterium. Replication is by the rolling-circle mechanism, with the parent plasmid remaining in the F^+ cell and the copy transferring to the recipient F^- cell as it is rolled off of the parent. The F^- cell therefore becomes F^+. Theoretically, this should mean that, over time, all *E. coli* bacteria will become F^+. This is not the case, so presumably the F plasmid is occasionally lost from an F^+ bacterium, so there are always F^- cells in a population.

A large number of bacterial species are able to conjugate in the same way as *E. coli*, and in all of these the process is controlled by a plasmid analogous to the F plasmid. Fertility plasmids of this type are called self-transmissible, which means that they can set up conjugation and mobilize themselves into the recipient cell. Conjugation (i.e., setting up the initial contact) and

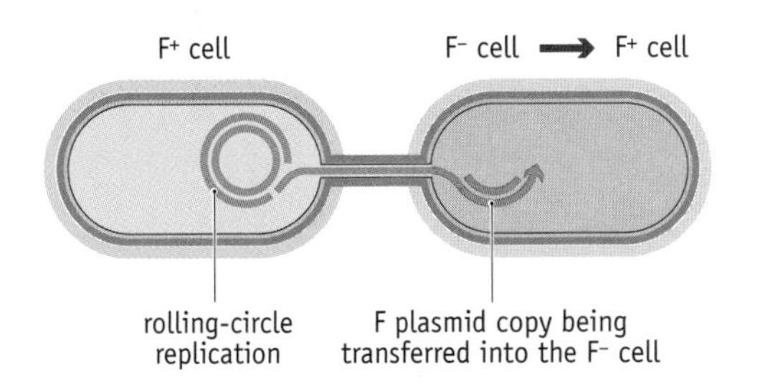

FIGURE 12.19 Transfer of the F plasmid during conjugation between an F^+ and an F^- cell. The transfer is shown occurring through the sex pilus, although it may be that the transfer occurs directly across the walls of the two cells.

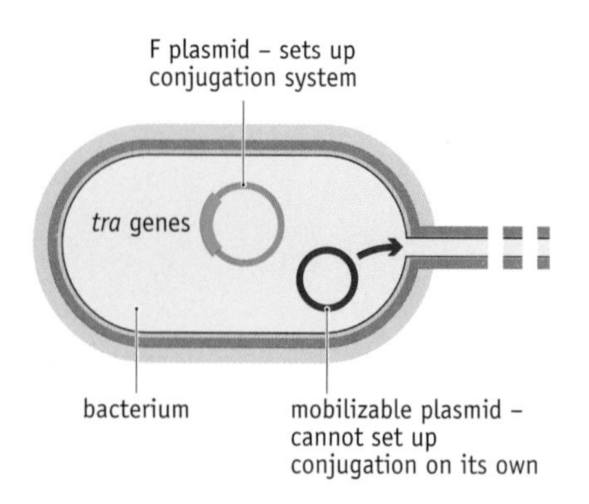

FIGURE 12.20 Some plasmids cannot set up conjugation but can be mobilized if they share a cell with a conjugative plasmid.

mobilization (passage of plasmid DNA from donor to recipient) are two distinct functions, and only self-transmissible plasmids are able to direct both. A few other plasmids can only set up the conjugation contact between cells and not mobilize on their own, and some can mobilize but only if they are coexisting in the cell with a second plasmid that can set up the initial contact (Figure 12.20). Other plasmids are totally nonfertile and can neither conjugate nor mobilize.

Chromosomal genes can also be transferred during conjugation

Bacterial conjugation is important not only because it enables plasmids to be transferred between bacteria. Chromosomal genes can also be passed from donor cell to recipient. How does this come about?

There are several possible ways, the first being simply that a small random piece of the donor cell's genome is transferred along with the F plasmid (Figure 12.21). This probably occurs fairly infrequently.

More important are the gene transfer properties of two special types of donor cell, called **Hfr** and **F′** ("F-prime"). In Hfr cells, the F plasmid has become integrated into the *E. coli* chromosomal DNA. The integrated form of the F plasmid can still direct conjugal transfer, but in this case, in addition to transferring itself, it also carries into the F^- cell a copy of at least some of the *E. coli* DNA molecule to which it is attached (Figure 12.22). It takes approximately 100 minutes for the entire *E. coli* chromosome to be transferred in this way, but conjugation rarely continues for this long. Termination of conjugation interrupts the DNA transfer, so usually only a part of the chromosomal DNA is passed to the recipient cell.

F′ cells, the second type of donor cell regularly associated with transfer of chromosomal genes, occasionally arise from Hfr cells when the integrated F plasmid becomes excised from the genome. Normally this event results in an F^+ cell, but sometimes excision of the F plasmid is not entirely accurate, and a small segment of the adjacent bacterial DNA is also snipped out. This leads to an F′ plasmid that carries a segment of the bacterium's genome, possibly including a few genes (Figure 12.23). Conjugation involving an F′ cell always results in transfer of these plasmid-borne bacterial genes.

Bacterial genes can also be transferred without contact between donor and recipient

Bacteria can also acquire segments of chromosomal DNA from other cells by processes that do not involve direct cell-to-cell contact. There are two such processes, called **transformation** and **transduction**.

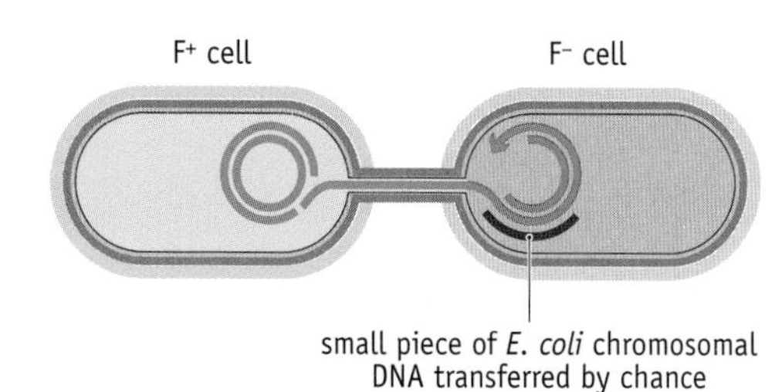

FIGURE 12.21 Chance transfer of a segment of the bacterial chromosome during conjugation between an F^+ and an F^- cell.

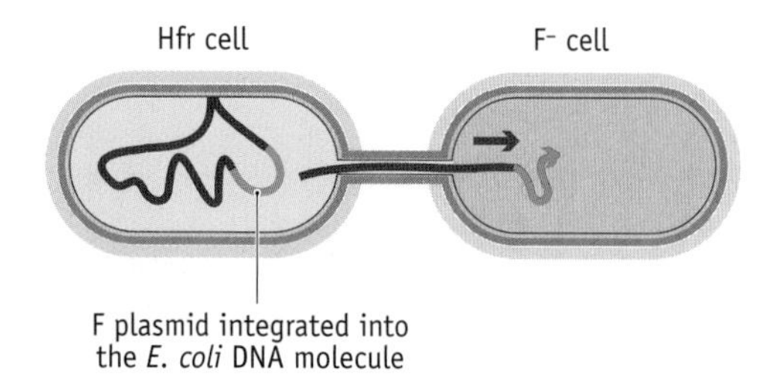

FIGURE 12.22 Transfer of chromosomal DNA during conjugation between an Hfr and an F^- cell.

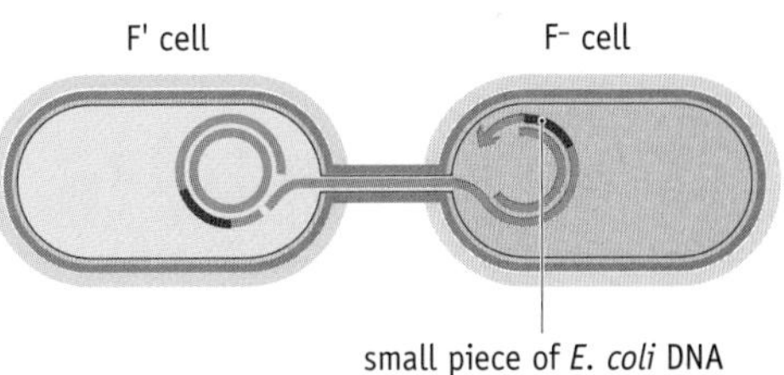

FIGURE 12.23 Transfer of chromosomal DNA during conjugation between an F′ and an F⁻ cell.

Transformation is the more straightforward of these processes. During transformation, a bacterium simply takes up DNA that it encounters in its local environment. Some bacteria, such as *Bacillus* and *Haemophilus* species, have efficient mechanisms for DNA uptake. The outer membranes of these species contain proteins that bind DNA and transfer it into the cell (Figure 12.24). Other species, including *E. coli*, do not have DNA uptake proteins, but some DNA can still penetrate the outer surface of the bacterium and enter the cell. In the laboratory, artificial treatments can improve the transformation efficiency of *E. coli* cells by rendering them more competent for DNA uptake.

The second type of DNA transfer that does not require cell-to-cell contact is transduction. In transduction, the transfer is mediated by a bacteriophage, one of the viruses that infect bacterial cells. Some bacteriophages do not immediately kill the host cell. Instead the virus genome becomes integrated into the host bacterium's chromosome, just like an F plasmid in an Hfr cell (Figure 12.25). The integrated virus genome is called a **prophage**. At some stage the prophage will become excised from the host chromosome and replicate, and new virus particles will be constructed around the replicated virus genomes (Figure 12.26). This is accompanied by breakdown of the bacterial chromosome into small fragments, some of which, by chance, will be about the same size as the bacteriophage genome. By mistake, one of these small chromosome fragments might be packaged into a virus particle. The resulting particle is still infective, as infection is solely a function of the proteins within which the DNA is enclosed. The fragment of bacterial DNA can therefore be transferred to a new cell when that recipient becomes "infected" with the transducing virus particle.

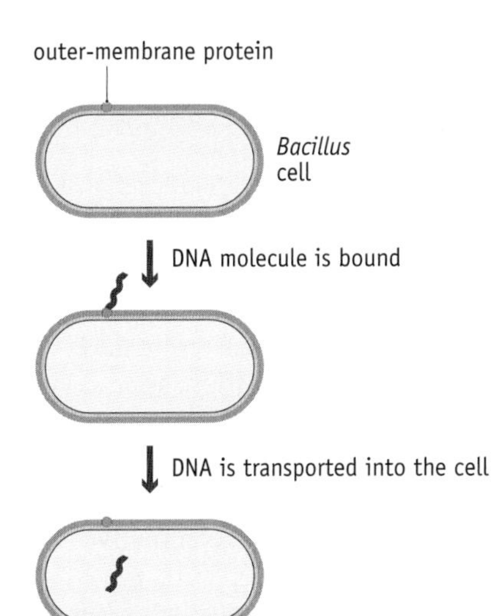

FIGURE 12.24 The role of outer-membrane proteins in DNA uptake during transformation.

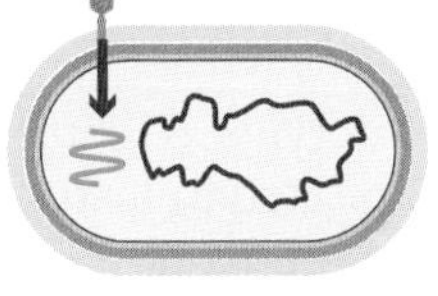

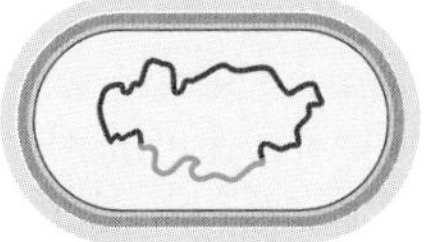

FIGURE 12.25 Integration of a bacteriophage DNA molecule into a bacterial chromosome.

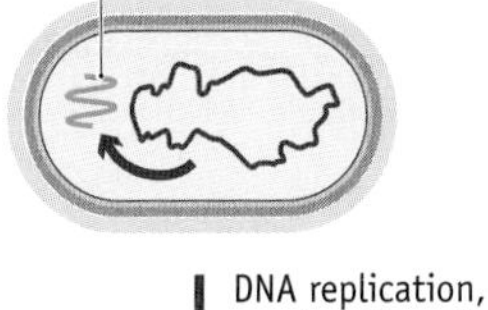

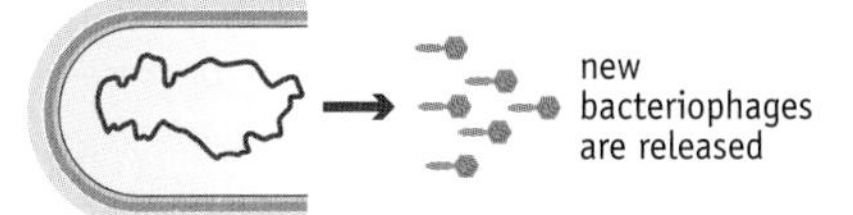

FIGURE 12.26 Excision of a prophage from a bacterial chromosome, followed by synthesis of new bacteriophages.

RESEARCH BRIEFING 12.1

Using conjugation to map genes in bacteria

The discovery of conjugation was important not just for what it revealed about the transfer of DNA between individual bacteria. It also led to a technique that enables the relative positions of genes to be mapped on the *E. coli* chromosome. Before the days of genome sequencing, such maps were the only information that geneticists had about the organization of genes within bacterial chromosomes. When genome sequences were first obtained, maps such as the one for *E. coli* were used to identify the genes that were revealed in the DNA sequence.

Conjugation can be used in gene mapping because of the transfer of chromosomal genes that occurs when the donor bacteria are Hfr cells. The possibilities were first recognized by Elie Wollman and François Jacob at the Pasteur Institute early in 1955, as a result of what has become known as the **interrupted mating experiment**.

Interrupted mating shows that, during conjugation, not all genes are transferred at the same time

Wollman and Jacob's aim was to chart the transfer of bacterial genes from an Hfr cell into an F^- cell. To do this they needed an F^- strain that carried several mutated genes and an Hfr strain that possessed correct or **wild-type** copies of these genes. Transfer of each wild-type gene from the Hfr cell to the F^- cell would be signaled by acquisition by the recipient cells of the characteristic specified by that gene.

The set of genes studied by Wollman and Jacob is listed in Table 1. The experiment was carried out by mixing the Hfr and F^- cells together and then removing samples from the culture at various time intervals (Figure 1). Each sample was immediately agitated to disrupt the pili linking Hfr and F^- cells and so interrupt the mating. The cells were then spread onto an agar medium containing streptomycin. Only the recipient cells would be able to grow on this medium because the original F^- cells were streptomycin-resistant, whereas the Hfr bacteria were sensitive to this antibiotic. The recipient cells therefore gave rise to colonies. The biochemical characteristics of the bacteria in each colony were tested by transferring them to other agar plates. Whether or not the cells still required threonine (see Table 1) was tested

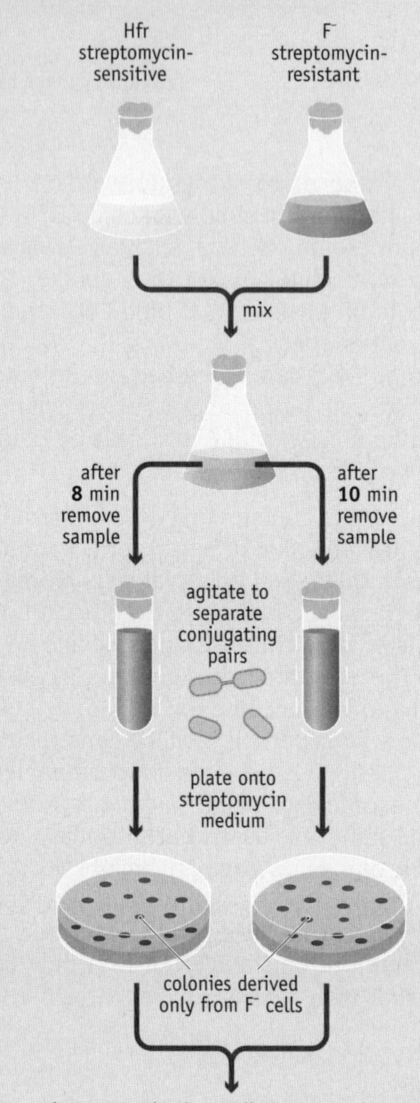

FIGURE 1 The interrupted mating experiment.

TABLE 1 GENES STUDIED BY WOLLMAN AND JACOB

Gene	Characteristic
thr	Ability to synthesize threonine
leu	Ability to synthesize leucine
azi	Sensitivity to azide
ton	Resistance to colicin and T1 bacteriophage
lac	Ability to use lactose as an energy source
gal	Ability to use galactose as an energy source

time after mixing (min)	genes transferred
8	*thr, leu*
10	*thr, leu, azi*
15	*thr, leu, azi, ton*
20	*thr, leu, azi, ton, lac*
25	*thr, leu, azi, ton, lac, gal*

FIGURE 2 The order of transfer of genes during the interrupted mating experiment.

by placing some on an agar medium that lacked threonine and seeing if they would grow. Their ability to use lactose as an energy source was tested with a medium that contained lactose but no other sugars. And so on, for all the genes listed in Table 1.

Wollman and Jacob were not sure exactly what to expect from the interrupted mating experiment because, at that time, the precise details of Hfr conjugation had not been worked out. Although they had their own ideas, the actual results came as a complete surprise. Rather than all the genes being transferred at once, they were passed from the Hfr to the F⁻ cells at different times. First, after about 8 minutes of mating, recipient cells that had acquired the ability to synthesize threonine and leucine began to appear (Figure 2). As the mating continued, the proportion of cells able to make these two amino acids increased until, after a few more minutes, almost all the recipient cells possessed the wild-type genes for these characteristics. Next, after about 10 minutes, azide sensitivity began to appear in the recipient cells, gradually increasing in frequency until about 80% of the population were azide-sensitive. Eventually, each of the wild-type characteristics had been transferred to the recipient cells. Wollman and Jacob portrayed the results in the form of a graph illustrating the kinetics of transfer of each gene (Figure 3).

The time of transfer of a gene indicates its position on the *E. coli* chromosome

The interrupted mating experiment is easily explained once we remember that Hfr cells transfer a portion, or possibly all, of the host DNA as a linear molecule into F⁻ cells. During this transfer the F region is the first to enter the recipient cell, followed by the rest of the *E. coli* chromosome. When mating is interrupted, the recipient cells will have had time to acquire only a few of the wild-type genes from the donor bacteria. By measuring the times at which different characteristics appeared in the recipient cells, Wollman and Jacob were in effect mapping the relative positions of the genes on the *E. coli* chromosome. Indeed, map distances on *E. coli* DNA are now expressed in minutes, with 1 minute representing the length of DNA that takes 1 minute to transfer. There are 100 minutes on the map because transfer of the entire *E. coli* chromosome takes about 100 minutes.

Transfer of the entire chromosome is not very likely during a single conjugation. Many bacteria naturally abandon mating well before 100 minutes have passed. To map the entire *E. coli* DNA it has been necessary to carry out a series of interrupted mating experiments, with a variety of Hfr strains, each with the F plasmid inserted at a different position. The parts of the *E. coli* chromosome that can be studied with each of these strains overlap, enabling the complete map to be built up. These experiments show that although genes are transferred in a linear fashion, the *E. coli* chromosome is actually circular.

Two additional points about the kinetics of transfer should be noted. The first is that there is a slope to each line on the graph shown in Figure 3. This indicates that conjugation is not absolutely synchronous in the population of cells, and that a particular gene is transferred in some matings before it is in others. For this reason, conjugation mapping is not completely accurate and is unable to distinguish the relative positions of two genes less than about 2 minutes apart, equivalent to about 90 kb of the chromosome. The second point to note is that each gene has a plateau value that indicates the proportion of the population of F⁻ cells that eventually obtains the gene. This plateau value is always less than 100% and becomes gradually smaller for genes transferred later in the mating. The plateau indicates that not all F⁻ bacteria become involved in conjugation and that many matings terminate naturally during the course of the experiment.

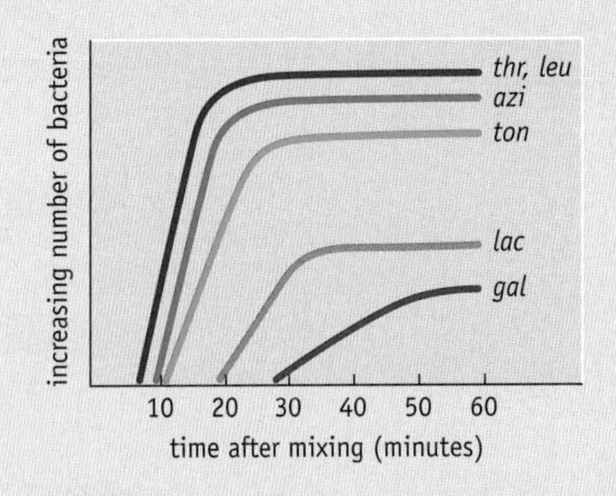

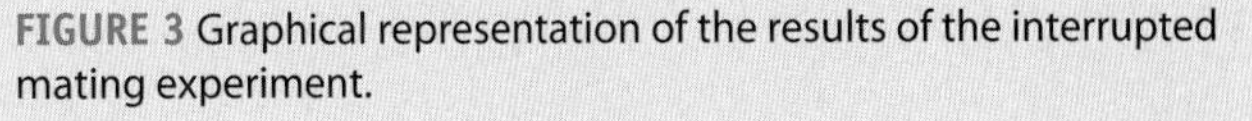
FIGURE 3 Graphical representation of the results of the interrupted mating experiment.

Transferred bacterial DNA can become a permanent feature of the recipient's genome

What happens to the DNA that a recipient cell acquires by conjugation, transformation, or transduction? If the DNA is an intact plasmid, then often it will be able to replicate in its new host, adding to the plasmid complement of the recipient cell. This is the outcome of most conjugation events and might also arise from transformation, as it is quite possible for plasmid DNA to be taken up from the environment in this way. Horizontal transfer of plasmids is immensely important in microbiology because it is responsible for the spread of antibiotic resistance genes. This has brought about the evolution of strains of pathogenic bacteria that can no longer be controlled with the antibiotics that previously were effective against them. An important example is methicillin-resistant *Staphylococcus aureus* (MRSA). There is a common belief that the "M" in MRSA stands for "multiple," but it does not. It is bad enough that MRSA has become resistant to the methicillin antibiotics usually used to prevent *S. aureus* from invading wounds in hospital patients. Should it genuinely become "multiply resistant," then the problems in controlling it will be substantially worse.

What about the chromosomal DNA that is transferred into a recipient cell by transformation, transduction, and occasionally conjugation? This DNA might be broken down and used as a source of carbon, nitrogen, and phosphorus for construction of new molecules within the recipient bacterium. But sometimes the DNA can survive in the cell. For this to happen, the DNA must become integrated into the bacterial chromosome, or into a chromid. This is possible if the transferred DNA has the same or a similar sequence to a segment of the recipient cell's chromosome, as will be the case if it comes from the same species. The segment of transferred DNA might then replace the equivalent stretch of DNA in the chromosome (Figure 12.27). If this happens then the recipient bacterium will inherit the genes from the donor. As the genes are the same as the ones that it had before, the replacement might appear to be immaterial. But bacterial genes, just like eukaryotic genes, have variant forms called alleles, each one with a slightly different DNA sequence. Transfer might therefore result in the recipient acquiring a different allele for one or more of the genes that it obtains. Its biological characteristics might therefore change in a small way.

Genome sequencing projects have told us that transfer of DNA also occurs between quite different species. Exactly how this happens is not known. We presume that the transfer is by transformation, because unrelated species

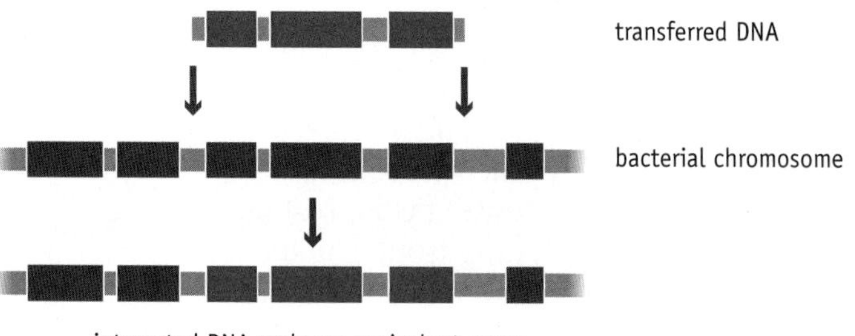

FIGURE 12.27 Integration of transferred DNA might result in replacement of a segment of the bacterial chromosome containing one or more genes.

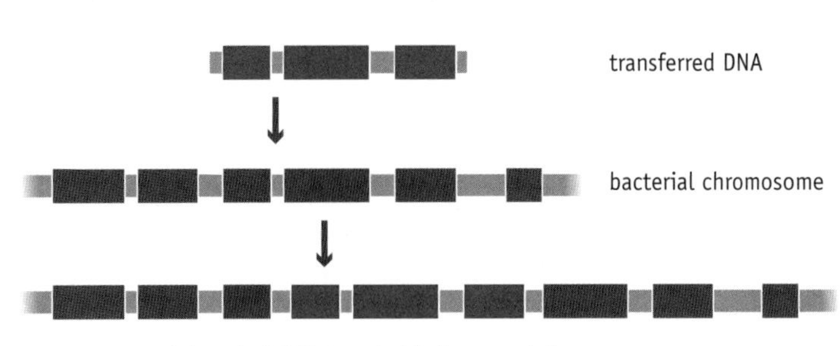

FIGURE 12.28 Random integration of transferred DNA into a bacterial chromosome, without replacement of any of the existing genes.

are unable to conjugate, and transduction can take place only within the host range of the transducing bacteriophage, which is always quite narrow. The incoming DNA is unlikely to have extensive sequence similarity with any part of the recipient's chromosome, so integration is assumed to occur randomly (Figure 12.28). Once it becomes a part of the chromosome, the new DNA, and any genes it contains, will become a permanent part of the recipient's genome. The genes will be passed on to daughter cells along with the rest of the genome when the parent cell divides.

Transfer across a species boundary might not happen very frequently, but it occurs often enough for bacteria to acquire, from time to time, new genes from other species. Most bacterial genomes contain a few hundred kb of DNA acquired by horizontal gene transfer, and in some cases the figure is higher. In the common laboratory strain of *E. coli*, 12.8% of the genome, corresponding to 0.59 Mb, is estimated to have been obtained in this way (Figure 12.29). The genome of the thermophilic bacterium *Thermotoga maritima* has 1877 genes, 451 of which appear to have been obtained from other species, including members of the second prokaryotic kingdom, the archaea. Transfer in the other direction, from bacteria to archaea, is equally prevalent. The picture that is emerging is one in which prokaryotes living in similar ecological niches exchange genes with one another in order to increase their individual fitness for survival in their particular environment. Many of the *Thermotoga* genes that have been obtained from archaea have probably helped this bacterium acquire its ability to tolerate high temperatures.

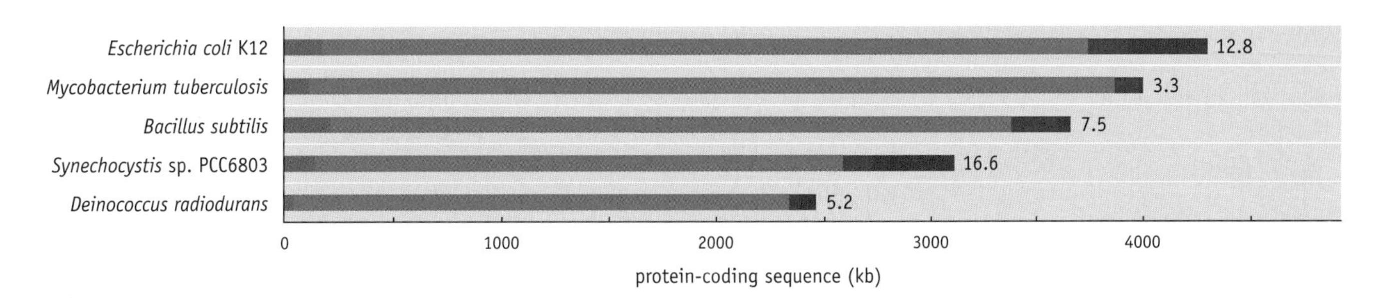

FIGURE 12.29 Most bacterial genomes contain DNA acquired by horizontal gene transfer. The chart shows the DNA that is unique to a particular species in blue and the DNA that has been acquired by horizontal gene transfer in red. The number at the end of each bar indicates the percentage of the genome that derives from horizontal transfer. The intergenic DNA content of each genome has been ignored, so the figures refer only to the genes.

KEY CONCEPTS

- Most of the DNA in a bacterium is contained in the nucleoid. The nucleoid makes up about one-third of the volume of the cell.
- The *E. coli* nucleoid contains the bacterial chromosome attached to a protein core from which supercoiled loops, each containing 10 to 100 kb of DNA, radiate out into the cell.
- During division of a bacterium, the two daughter DNA molecules move toward either end of the cell as the parent chromosome is being replicated. The mechanism by which this occurs is poorly understood.
- Plasmids are DNA molecules, usually circular, that lead an independent existence within a bacterium. Plasmids carry genes, but these are not essential for the survival of the bacterium. The plasmid genes may, however, confer a useful characteristic such as resistance to an antibiotic.
- Some bacteria have multipartite genomes. In some of these species it is difficult to distinguish between chromosomal DNA molecules, which carry essential genes, and plasmids. A new term, chromid, has been introduced to describe molecules that have plasmidlike features but carry essential genes.
- Conjugation, transformation, and transduction are processes that can result in the transfer of DNA between bacteria. Conjugation and transduction can result in transfer only between members of the same or related species. Transformation can, in theory, occur between members of any two species.
- Horizontal gene transfer between members of different species is evident from comparisons between bacterial genome sequences.
- Prokaryotes, both bacteria and archaea, living in similar ecological niches exchange genes with one another in order to increase their individual fitness for survival in their particular environment.

QUESTIONS AND PROBLEMS (Answers can be found at www.garlandscience.com/introgenetics)

Key Terms

Write short definitions of the following terms:

bacterial chromosome
chromid
Col plasmid
colicin
compatible
conjugation
copy number
degradative plasmid
episome
F^+ cell
F^- cell
F′ cell
fertility plasmid
Hfr cell
horizontal gene transfer
incompatibility group
integron
interrupted mating experiment
nucleoid
pilus
plasmid
prophage
relaxed
resistance plasmid
stringent
supercoiling
transduction
transformation
virulence plasmid
wild-type

Self-study Questions

12.1 Describe the structure of the *E. coli* nucleoid.

12.2 What is supercoiling, and what effect does it have on a circular DNA molecule?

12.3 What proteins are present in the *E. coli* nucleoid?

12.4 Explain what a plasmid is, and provide examples of the different types of plasmid that are known.

12.5 What is the link between plasmids and horizontal gene transfer?

12.6 Distinguish between the terms "stringent" and "relaxed" as applied to plasmids.

12.7 Explain what an incompatibility group is, and outline the proposed link between plasmid incompatibility and copy number.

12.8 Describe how the *E. coli* genome is replicated and partitioned between the daughter cells.

12.9 Outline the type I system for plasmid partitioning.

12.10 The *E. coli* genome is a single, circular DNA molecule. What other types of genome structure are found among prokaryotes?

12.11 Describe the genome organizations of *Vibrio cholerae* and *Deinococcus radiodurans*.

12.12 What are the distinguishing features of bacterial chromosomes, plasmids, and chromids?

12.13 Describe how plasmids are transferred between bacterial cells by conjugation.

12.14 Plasmid R100 carries a gene for resistance to streptomycin. You have an F$^+$ strain of *E. coli* that contains R100. Design an experiment to test whether R100 is able to mobilize on its own.

12.15 Distinguish between F$^+$, F$^-$, Hfr, and F′ cells.

12.16 Describe how Wollman and Jacob carried out the first interrupted mating experiment.

12.17 Explain how interrupted mating experiments proved that the *E. coli* DNA molecule is circular.

12.18 A series of interrupted mating experiments with different Hfr strains produces the following linear gene maps:

1. *A J F C*

2. *D H E B*

3. *I B E H*

4. *J F C I*

5. *A G D H*

Draw a circular map of the bacterial DNA molecule showing the positions of these genes.

12.19 Describe two ways in which genes can be transferred without contact between bacteria.

12.20 How can transferred DNA become incorporated into the genome of the recipient bacterium?

12.21 Outline the evidence for the occurrence of horizontal gene transfer between prokaryotes of different species.

Discussion Topics

12.22 Genetic elements that reproduce within or along with a host genome, but confer no benefit on the host, are called "selfish." Discuss this concept, in particular as it applies to plasmids.

12.23 Should the traditional view of the bacterial genome as a single, circular DNA molecule be abandoned? If so, what new definition of "bacterial genome" should be adopted?

12.24 The obligate intracellular parasitic bacterium *Mycoplasma genitalium* has just 470 genes. Why does this organism require so few genes?

12.25 Discuss possible ways in which plasmid copy number could be controlled.

12.26 Research Briefing 12.1 describes how conjugation can be used to map the relative positions of genes on the *E. coli* chromosome. Because of the lack of synchrony within a population of bacteria, it is impossible to use conjugation mapping to distinguish the relative positions of two genes that are less than about 90 kb apart. Discuss ways in which the other processes that result in DNA transfer between bacteria—transformation and transduction—might be used to map the positions of genes that are closer than 90 kb.

12.27 The biological species concept states that a species is a group of interbreeding individuals that are reproductively isolated from other such groups. Does this definition of species hold for prokaryotes? If not, then how might a prokaryotic species be defined?

FURTHER READING

Boucher Y, Douady CJ, Papke RT et al. (2003) Lateral gene transfer and the origins of prokaryotic groups. *Annu. Rev. Genet.* 37, 283–328.

Harrison PW, Lower RPJ, Kim NKD & Young JP (2010) Introducing the bacterial "chromid": not a chromosome, not a plasmid. *Trends Microbiol.* 18, 141–148.

Heidelberg JF, Eisen JA, Nelson WC et al. (2000) DNA sequence of both chromosomes of the cholera pathogen *Vibrio cholerae*. *Nature* 406, 477–483.

Holloway BW (1979) Plasmids that mobilize bacterial chromosome. *Plasmid* 2, 1–19. *Includes a description of F plasmids from bacteria other than E. coli.*

Novick RP (1980) Plasmids. *Sci. Am.* 243, 76–90. *A general review.*

Rocha EPC (2008) The organization of the bacterial genome. *Annu. Rev. Genet.* 42, 211–233. *Includes details of nucleoid structure and chromosome partitioning.*

Thanbichler M & Shapiro L (2008) Getting organized—how bacterial cells move proteins and DNA. *Nat. Rev. Microbiol.* 6, 28–40. *Chromosome and plasmid partitioning.*

White O, Eisen JA, Heidelberg JF et al. (1999) Genome sequence of the radioresistant bacterium *Deinococcus radiodurans* R1. *Science* 286, 1571–1577.

Willetts N & Skurray R (1980) The conjugation system of F-like plasmids. *Annu. Rev. Genet.* 14, 47–76.

Wollman EL & Jacob F (1955) Sur la mécanisme du transfert de matériel génétique au cours de la recombinaison chez *Escherichia coli* K12. *Comptes Rendus de l'Académie des Sciences* 240, 2449–2451. *The first interrupted mating experiment.*

Inheritance of Genes During Virus Infection Cycles CHAPTER 13

We have studied the inheritance of genes during eukaryotic and bacterial cell division and also examined how genes can be transferred between bacteria of the same and different species. To complete the picture we must now look at how genes are inherited during virus life cycles.

The viruses are the simplest form of life on the planet. In fact viruses are so simple in biological terms that we have to ask ourselves whether they can really be thought of as living organisms. Doubts arise partly because viruses are constructed differently from all other forms of life—viruses are not cells—and partly because of the nature of the virus life cycle. Viruses are obligate parasites of the most extreme kind. They reproduce only within a host cell, and in order to replicate and express their genomes they must subvert at least part of the host's genetic machinery to their own ends. Some viruses possess genes coding for their own DNA polymerase and RNA polymerase enzymes, but many depend on the host enzymes for genome replication and transcription. All viruses make use of the host's ribosomes and translation apparatus for synthesis of the polypeptides that make up the protein coats of their progeny. This means that virus genes must be matched to the host genetic system. Viruses are therefore quite specific for particular organisms, and individual types cannot infect a broad spectrum of species.

There are a multitude of different types of virus, and planning a strategy for studying them can be quite difficult. The best approach is to follow history. Geneticists in the 1930s chose bacteriophages—the viruses that infect bacteria—as model organisms with which to study virus infection cycles. This was because it is relatively easy to grow cultures of bacteria that are infected with bacteriophages. In later years, attention turned to viruses that infect eukaryotes, including the **retroviruses**, the group to which the human immunodeficiency viruses that cause AIDS belong.

We will follow this same approach. First we will become familiar with the basic features of virus life cycles by examining the bacteriophages. Then we will move on to the viruses of eukaryotes, with our focus largely on the retroviruses, not because these are the only important eukaryotic viruses, but because they display features that we do not encounter among the bacteriophages. Our study of retroviruses will then lead us to the very edge of what we understand to be life.

13.1 BACTERIOPHAGES

Bacteriophages (sometimes called "phages") are common but frequently overlooked members of the natural environment. From time to time, research has been directed at the possibility of using them to control or treat bacterial infections. As the antibiotic age draws to a close, with many bacteria acquiring resistance due to the spread of plasmids, bacteriophages may once again be looked on as potential benefactors of the human race.

We are mainly interested in how genes are inherited during bacteriophage infection cycles, but before studying this topic we must become familiar with the bacteriophages themselves.

Bacteriophages have diverse structures and equally diverse genomes

Bacteriophages are constructed primarily from two components, protein and nucleic acid. The protein forms a coat, or **capsid**, within which the nucleic acid genome is contained.

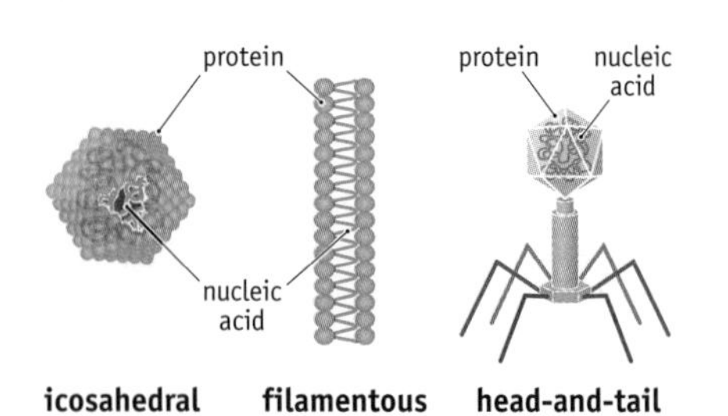

FIGURE 13.1 The three types of capsid structures commonly displayed by bacteriophages.

Three basic bacteriophage types can be distinguished from their capsid structures (Figure 13.1). The first of these have **icosahedral** capsids, in which the individual protein subunits (**protomers**) are arranged into a 20-faced geometric structure that surrounds the nucleic acid. Examples of icosahedral bacteriophages are MS2, which infects *Escherichia coli*, and PM2, which infects *Pseudomonas aeruginosa*. The second type have **filamentous** capsids, in which the protomers are arranged in a helix, producing a rod-shaped structure. The *E. coli* bacteriophage called M13 is an example. The **head-and-tail** bacteriophages combine the features of the other two types. Their capsid is made up of an icosahedral head, containing the nucleic acid, and a filamentous tail, which facilitates entry of the nucleic acid into the host cell. They may also have other structures, such as the "legs" possessed by the *E. coli* bacteriophage T4.

The term "nucleic acid" has to be used when referring to bacteriophage genomes because in some cases these molecules are made of RNA. Viruses are the one form of "life" in which the genetic material is not always DNA. Bacteriophages and other viruses also break another rule. Their genomes, whether of DNA or RNA, can be single-stranded as well as double-stranded.

A whole range of different genome structures are known among the bacteriophages, as summarized in Table 13.1. With most types of bacteriophage there is a single DNA or RNA molecule that comprises the entire genome. A few RNA bacteriophages have **segmented genomes**, meaning that their genes are carried by a number of different RNA molecules. The sizes of

TABLE 13.1 FEATURES OF SOME TYPICAL BACTERIOPHAGES AND THEIR GENOMES

Phage	Host	Capsid structure	Genome structure	Genome size (kb)	Number of genes
λ	*E. coli*	Head-and-tail	Double-stranded linear DNA	49.5	48
ϕX174	*E. coli*	Icosahedral	Single-stranded circular DNA	5.4	11
f6	*Pseudomonas phaseolica*	Icosahedral	Double-stranded segmented linear RNA	2.9, 4.0, 6.4	13
M13	*E. coli*	Filamentous	Single-stranded circular DNA	6.4	10
MS2	*E. coli*	Icosahedral	Single-stranded linear RNA	3.6	3
PM2	*Pseudomonas aeruginosa*	Icosahedral	Double-stranded linear DNA	10.0	approx. 21
SP01	*Bacillus subtilis*	Head-and-tail	Double-stranded linear DNA	150	100+
T2, T4, T6	*E. coli*	Head-and-tail	Double-stranded linear DNA	166	150+
T7	*E. coli*	Head-and-tail	Double-stranded linear DNA	39.9	55+

The genome structure is that in the phage capsid. Some genomes exist in different forms within the host cell.

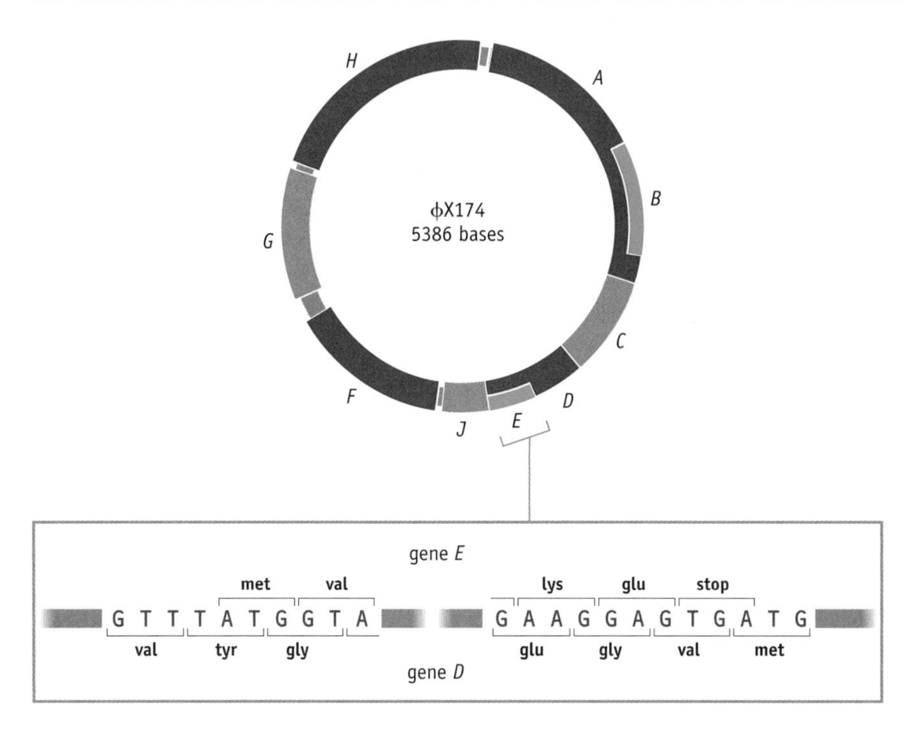

FIGURE 13.2 The ϕX174 genome contains overlapping genes. The expanded region shows the DNA sequence at the start and end of the overlap between genes *E* and *D*, showing that the reading frames used by these two genes are different.

bacteriophage genomes vary enormously, from about 1.6 kb for the smallest bacteriophages to over 150 kb for large ones such as T2, T4, and T6.

Bacteriophage genomes, being relatively small, were among the first to be studied comprehensively by the rapid and efficient DNA sequencing methods that were developed in the late 1970s. Gene numbers vary from just 3 in the case of MS2 to over 100 for the more complex head-and-tail bacteriophages (see Table 13.1). The smaller bacteriophage genomes, of course, contain relatively few genes, but these can be organized in a very complex manner. Phage ϕX174, for example, manages to pack into its genome "extra" biological information because several of its genes overlap (Figure 13.2). These **overlapping genes** share nucleotide sequences (gene *B*, for example, is contained entirely within gene *A*), but code for different gene products because the transcripts are translated from different start positions and in different reading frames. Overlapping genes are not uncommon in viruses.

The larger bacteriophage genomes contain more genes, reflecting the more complex capsid structures of these bacteriophages and a dependence on a greater number of bacteriophage-encoded enzymes during the infection cycle. The T4 genome, for example, includes some 50 genes involved solely in construction of the bacteriophage capsid (Figure 13.3). Despite their complexity, even these large bacteriophages still require at least some host-encoded proteins and RNAs in order to carry through their infection cycles.

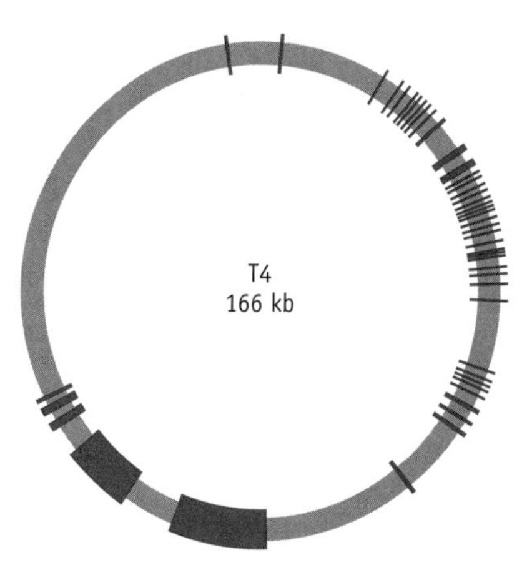

FIGURE 13.3 The T4 genome. Only those genes coding for components of the bacteriophage capsid are shown (red). About 100 additional genes involved in other aspects of the bacteriophage life cycle are omitted.

The lytic infection cycle of bacteriophage T4

Bacteriophages are classified into two groups according to their life cycle. The two types are called **lytic** and **lysogenic**. The fundamental difference between these groups is that a lytic bacteriophage kills its host bacterium

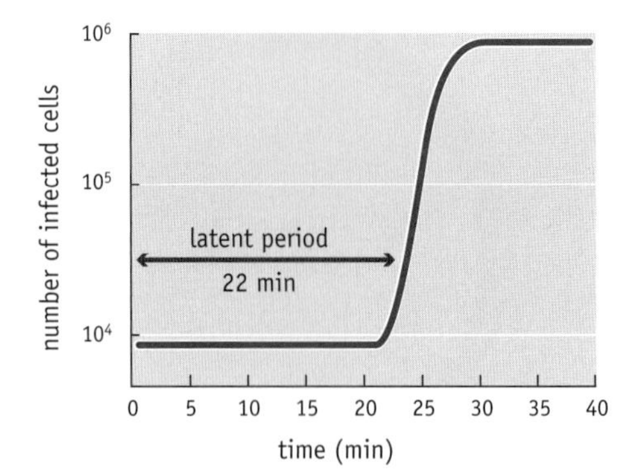

FIGURE 13.4 The time course of the T4 infection cycle. There is no change in the number of infected cells during the first 22 minutes of infection. Then the number of infected cells starts to increase, showing that lysis of the original hosts is occurring, and that the new bacteriophages that are being produced are infecting other cells.

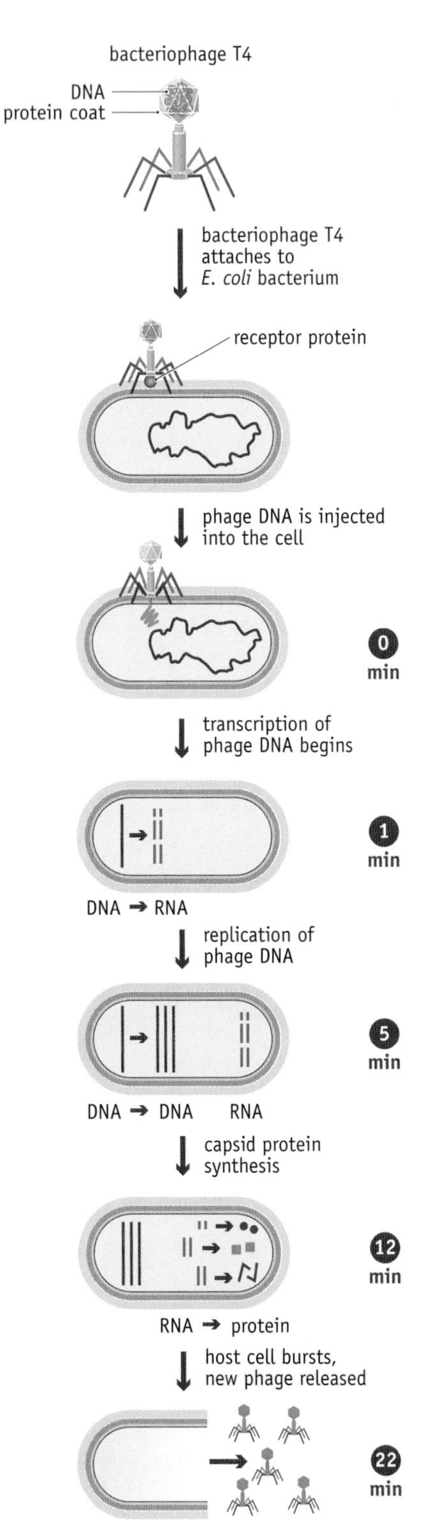

FIGURE 13.5 Events occurring during the T4 infection cycle.

very soon after the initial infection, usually within 30 minutes, whereas a lysogenic bacteriophage can remain quiescent within its host for a substantial period of time, even throughout numerous generations of the host cell. These two life cycles are typified by two *E. coli* bacteriophages, the lytic (or **virulent**) T4 and the lysogenic (or **temperate**) λ.

The T4 infection cycle can be followed by adding T4 bacteriophages to a culture of *E. coli*, waiting 3 minutes for the bacteriophages to attach to the bacteria, and then measuring the number of infected cells over a period of 40 minutes (Figure 13.4). The first interesting feature of the infection cycle is that there is no change in the number of infected cells during the first 22 minutes of infection. This **latent period** is the time needed for the bacteriophages to reproduce within their hosts. After 22 minutes the number of infected cells starts to increase, showing that lysis of the original hosts is occurring, and that the new bacteriophages that have been produced are now infecting other cells in the culture. This infection cycle is referred to as the **one-step growth curve**.

The molecular events occurring during the T4 infection cycle are as follows. The initial event is attachment of the bacteriophage to a receptor protein on the outside of the bacterium (Figure 13.5). Different bacteriophages have different receptors. For T4, the receptor is a protein called OmpC. OmpC is a type of outer-membrane protein called a porin, which forms a channel through the membrane and facilitates the uptake of nutrients. After attachment, the bacteriophage injects its DNA genome into the cell through its tail structure.

The latent period now begins. The name is, in fact, a misnomer. Although to the outside observer nothing very much seems to happen, inside the cell there is frenzied activity directed at synthesis of new bacteriophage particles. Immediately after entry of the bacteriophage DNA, the synthesis of host DNA, RNA, and protein stops and transcription of the bacteriophage genome begins. Within 5 minutes the bacterial DNA molecule has been broken down and the resulting nucleotides are being utilized in replication of the T4 genome. After 12 minutes new bacteriophage capsid proteins start to appear and the first complete bacteriophage particles are assembled. Finally, at the end of the latent period, the cell bursts and the new bacteriophages are released. A typical infection cycle produces 200 to 300 T4 bacteriophages per cell, all of which can go on to infect other bacteria.

The lytic infection cycle is regulated by expression of early and late genes

With all bacteriophages, but especially those with larger genomes, the question arises as to how gene expression is regulated in order to ensure that the correct activities occur at the right time during the infection cycle. With most bacteriophages, genome replication precedes synthesis of capsid proteins. Similarly, synthesis of lysozyme, the enzyme that causes the bacterium to burst, must be delayed until the very end of the infection cycle. Individual bacteriophage genes must therefore be expressed at different times in order for the infection cycle to proceed correctly.

One of the simplest strategies for regulating a lytic infection cycle is displayed by ϕX174. This bacteriophage exerts no control over transcription of its genes. All 11 genes are transcribed by the host RNA polymerase as soon as the bacteriophage DNA enters the cell. Genome replication and capsid synthesis occur more or less at the same time, but lysozyme synthesis is delayed because the mRNA for this enzyme is translated slowly.

With most other bacteriophages there are distinct phases of gene expression. Traditionally, two groups of genes are recognized. These are **early genes**, whose products are needed during the early stages of infection, and **late genes**, which remain inactive until toward the end of the cycle. There may, however, be other divisions within these groups, some bacteriophages having "very early" genes, for example. A number of strategies are employed by bacteriophages to ensure that these groups of genes are expressed in the correct order. Many of these schemes utilize a **cascade system**, meaning that the appearance in the cell of the translation products from one set of genes switches on transcription of the next set of genes (Figure 13.6).

With T4, for example, the very first genes to be expressed are transcribed by the *E. coli* RNA polymerase from a few standard *E. coli* promoter sequences present on the bacteriophage genome. The very-early-gene products include proteins that modify the σ subunit of the host RNA polymerase so it no longer recognizes *E. coli* promoters, thereby switching off host gene expression. Instead, the RNA polymerase now specifically transcribes a second set of bacteriophage genes (Figure 13.7). One of these genes specifies

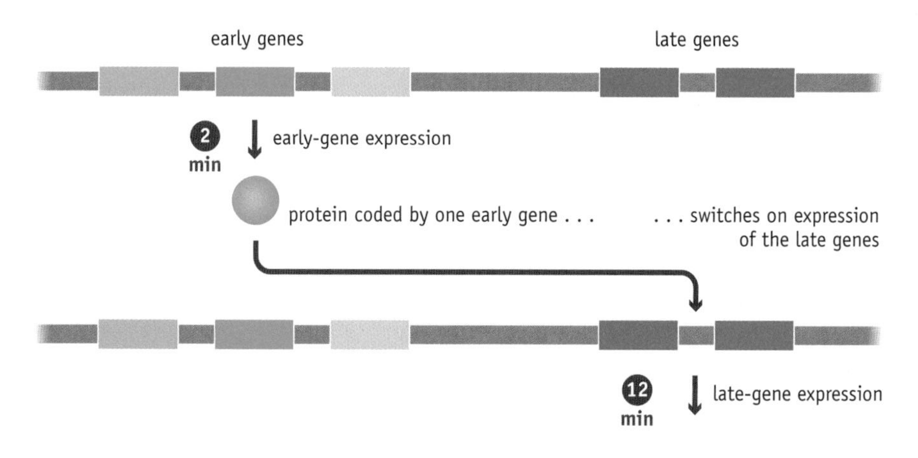

FIGURE 13.6 A cascade system for ensuring the correct timing of early- and late-gene expression during a bacteriophage infection cycle. The proteins coded by the early genes include one that, after it has been synthesized, switches on expression of the late genes.

first set of T4 genes are transcribed by the *E. coli* RNA polymerase

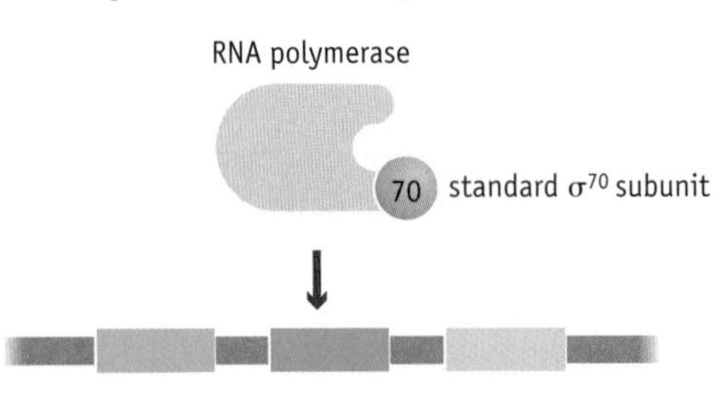

modification of the σ^{70} subunit leads to transcription of a second set of T4 genes

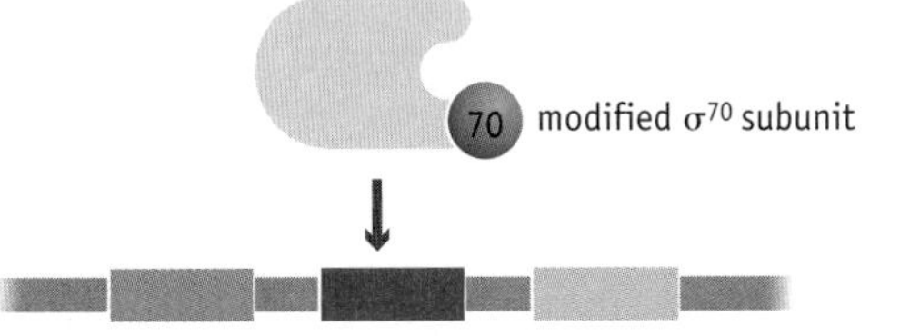

replacement of σ^{70} with σ^{55} leads to transcription of a third set of T4 genes

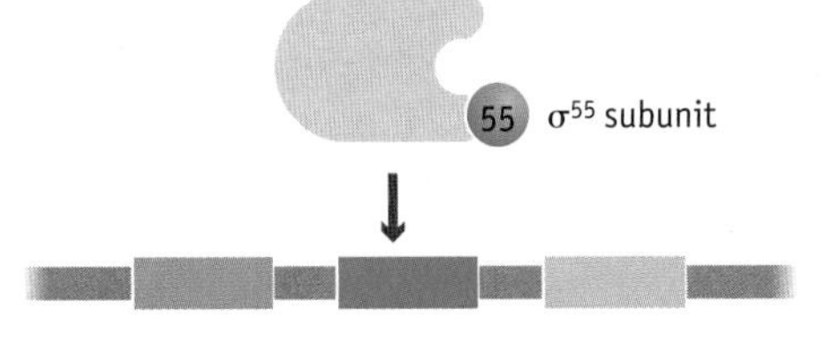

FIGURE 13.7 Sequential expression of T4 genes resulting from modifications to the *E. coli* RNA polymerase.

a new σ subunit, σ^{55}, which replaces the host's σ^{70} version, so the RNA polymerase now transcribes a third set of bacteriophage genes. The individual groups of genes are therefore expressed in the correct order, the products of one set switching on expression of the next set.

The lysogenic infection cycle of bacteriophage λ

Bacteriophage λ, like most temperate bacteriophages, can follow a lytic infection cycle but is more usually associated with the alternative lysogenic cycle. The distinction is that during a lysogenic cycle the bacteriophage genome becomes integrated into the host DNA.

Integration occurs immediately after entry of the bacteriophage DNA into the cell, and results in a quiescent form of the bacteriophage, called the **prophage** (Figure 13.8). Integration occurs by a site-specific mechanism involving identical 15-bp sequences present in the λ and *E. coli* genomes. This means that the λ genome always integrates at the same position within the *E. coli* DNA molecule.

The integrated prophage can be retained in the host DNA molecule for many cell generations, being replicated along with the bacterial genome and passed on with it to the daughter cells. However, the switch to the lytic mode of infection occurs if the prophage is **induced** by any one of several chemical or physical stimuli. Each of these appears to be linked to DNA damage and possibly therefore signals the imminent death of the host by natural causes.

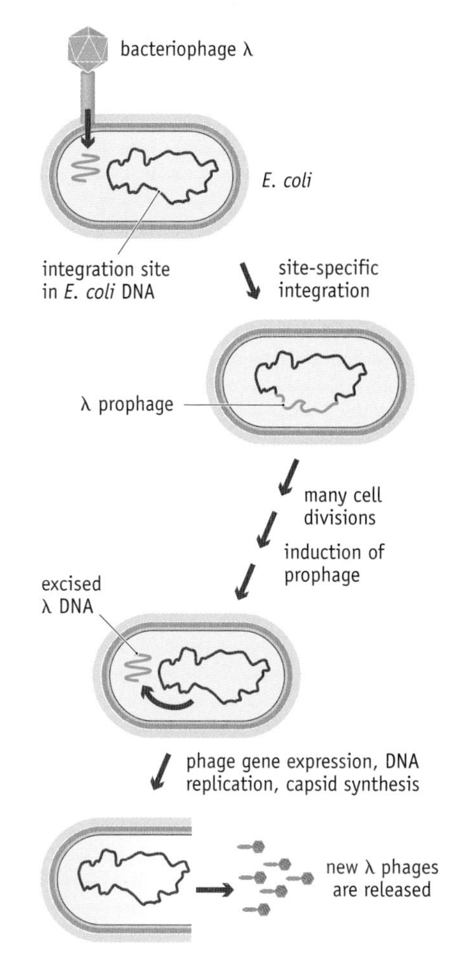

FIGURE 13.8 Events occurring during the lysogenic infection cycle of bacteriophage λ. Integration of the λ DNA into the host chromosome results in the quiescent prophage, which can be retained in the host DNA molecule for many cell divisions. Following induction, the λ genome is excised from the host DNA and directs synthesis of new bacteriophage particles.

In response to these stimuli, the bacteriophage genome is excised from the host DNA and converted into a circular molecule. The λ genome then replicates by the rolling-circle mechanism. This produces a series of **concatamers** (linear genomes linked end to end) that are cleaved by an endonuclease (coded by λ gene *A*) at a 12-bp recognition sequence called the ***cos*** site. The linear genomes that result are packaged into λ bacteriophage particles, and these new bacteriophages are released from the cell. The genomes recircularize immediately after injection into a new host.

Lysogeny adds an additional level of complexity to the bacteriophage life cycle, and ensures that the bacteriophage is able to adopt the particular infection strategy best suited to the prevailing conditions.

Many genes are involved in establishment and maintenance of lysogeny

Lysogeny adds an additional layer of complexity to the pattern of bacteriophage gene expression that occurs during the infection cycle. In particular, it raises three questions. How does the bacteriophage "decide" whether to follow the lytic or the lysogenic cycle, how is lysogeny maintained, and how is the prophage induced to break lysogeny? A considerable amount is known about gene expression during λ infection, so much so that fairly complete answers can be given for each of these three questions. What follows is a summary.

We will deal with the second question—how is lysogeny maintained—before questions 1 and 3. The first step in the lytic infection cycle is expression of the early λ genes. These are transcribed from two promoters, p_L and p_R, located on either side of a regulatory gene called *c*I (Figure 13.9). During lysogeny, p_L and p_R are switched off because the *c*I gene product, which is a repressor protein, is bound to operators adjacent to these promoters. As a result, the early genes are not expressed and the bacteriophage cannot enter the lytic cycle. Lysogeny is maintained for numerous cell divisions because the *c*I gene is continuously expressed, albeit at a low level, so that the amount of cI repressor present in the cell is always enough to keep p_L

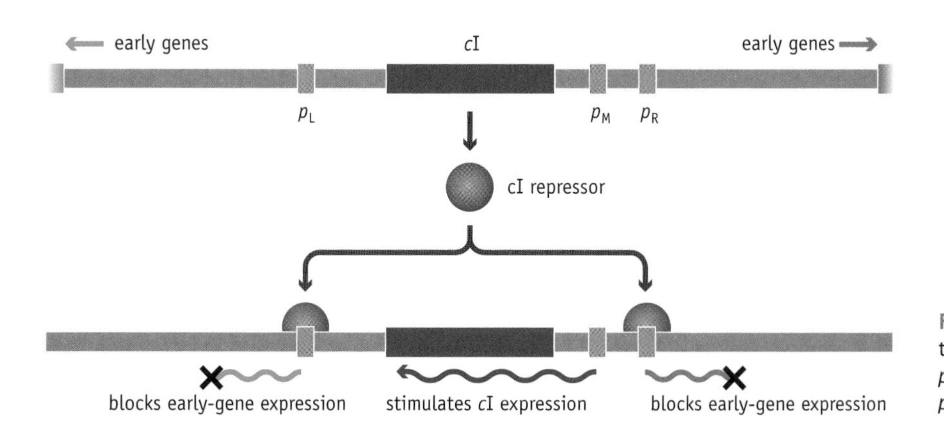

FIGURE 13.9 The cI repressor blocks transcription from the early gene promoters p_L and p_R, but stimulates transcription from p_M, the promoter for the cI gene.

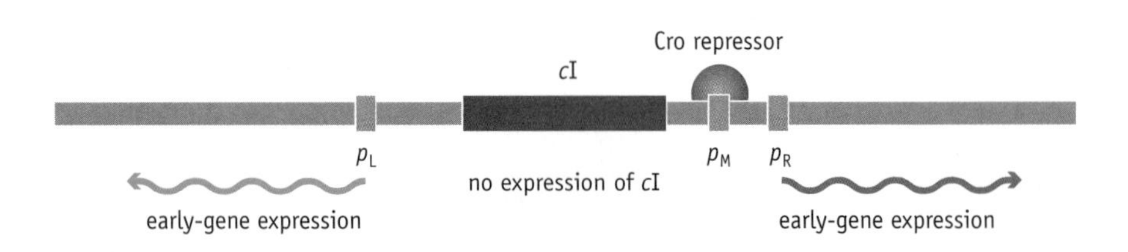

FIGURE 13.10 The Cro repressor blocks transcription of *c*I.

and p_R switched off. This continued expression of *c*I occurs because the cI repressor not only blocks transcription from p_L and p_R, but also stimulates transcription from p_M, the promoter for the *c*I gene. The dual role of the cI repressor is therefore the key to lysogeny.

How does the bacteriophage "decide" whether to follow the lytic or the lysogenic cycle? This depends on the outcome of a race between the cI and cro proteins. When a λ DNA molecule enters an *E. coli* cell, the host's RNA polymerase enzymes attach to the various promoters on the molecule and start transcribing the λ genes. Once the *c*I gene is expressed the cI repressor blocks expression of the early genes, preventing entry into the lytic cycle and enabling lysogeny to be established. But lysogeny is not always the outcome of a λ infection. This is because a second gene, *cro*, also codes for a repressor, but in this case one that prevents transcription of *c*I (Figure 13.10). Both the *c*I and *cro* genes are expressed immediately after the λ DNA molecule enters the cell. If the cI repressor is synthesized more quickly than the Cro repressor, then early-gene expression is blocked and lysogeny follows. However, if the Cro repressor wins the race it blocks expression of the *c*I gene before enough cI repressor has been synthesized to switch the early genes off. As a result, the bacteriophage enters the lytic infection cycle. The decision between lysis and lysogeny therefore depends on which of the two gene products, cI and Cro, accumulates the quickest. The decision is influenced by the products of other λ genes, which are able to assess the physiological state of the host cell, and hence ensure that the appropriate choice is made between lysogeny and immediate lysis.

Finally, how is lysogeny ended? This requires inactivation of the cI repressor. Lysogeny is maintained as long as the cI repressor is bound to the operators adjacent to p_L and p_R. The prophage will therefore be induced if the levels of active cI repressor decline below a certain point. This may happen by chance, leading to spontaneous induction, or may occur in response to physical or chemical stimuli. These stimuli activate a general protective mechanism in *E. coli*, the **SOS response**. Part of the SOS response is expression of an *E. coli* gene, *recA*, coding for the RecA protein. RecA inactivates the cI repressor by cutting it in half (Figure 13.11). This switches on expression of the early genes, enabling the bacteriophage to enter the lytic cycle. Inactivation of the cI repressor also means that transcription of the *c*I gene is no longer stimulated, avoiding the possibility of lysogeny being

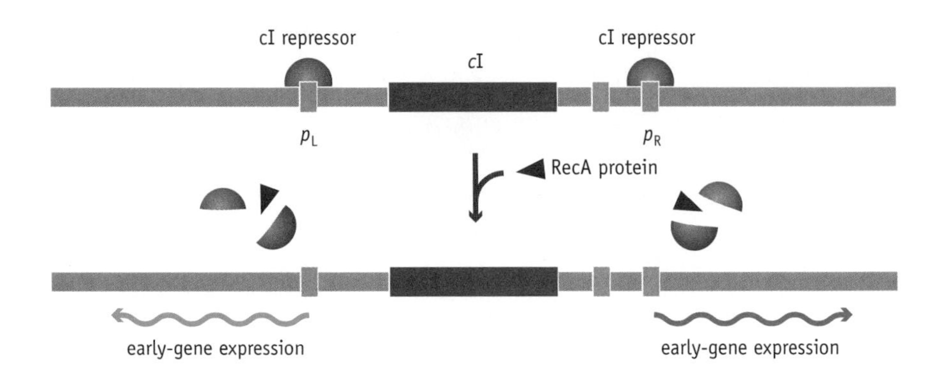

FIGURE 13.11 The RecA protein cleaves the cI repressor, enabling the early genes to be expressed.

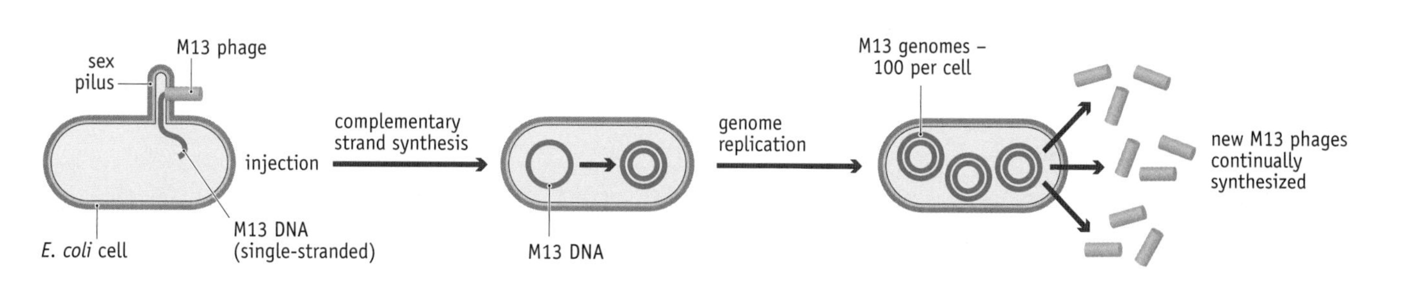

FIGURE 13.12 The unusual infection cycle of bacteriophage M13. This bacteriophage attaches to the sex pilus in order to inject its DNA into the host cell. This means that it cannot infect F^- bacteria, and is sometimes called a "male-specific" bacteriophage.

reestablished through the synthesis of more cI repressor. Inactivation of the cI repressor therefore leads to induction of the prophage.

There are some unusual bacteriophage life cycles

Although lysis and lysogeny are the two most typical bacteriophage life cycles, they are not the only ones. One or two other bacteriophages display unusual infection cycles that are neither truly lytic nor truly lysogenic. An example is provided by a third *E. coli* bacteriophage, M13, which is unusual in that its genome is a *single*-stranded, circular DNA molecule, 6400 nucleotides in length.

After injection into the bacterium, the M13 genome is replicated by synthesis of the complementary strand, producing a double-stranded, circular DNA molecule (Figure 13.12). This molecule then undergoes further replication until there are over 100 copies of it in the cell. At this stage the infection cycle takes on characteristics of both lytic and lysogenic bacteriophages. As with lytic bacteriophages, M13 coat proteins are synthesized, and new bacteriophage particles are assembled and released from the cell. However, as with lysogenic bacteriophages, cell bursting does not occur and the infected bacteria continue to grow and divide. Copies of the bacteriophage genome are passed on to daughter bacteria during cell division, and M13 assembly and release continues. The M13 infection cycle is therefore partly lytic and partly lysogenic.

13.2 VIRUSES OF EUKARYOTES

All eukaryotes act as hosts for viruses of one kind or another. Indeed, most eukaryotes are susceptible to infection by a broad range of virus types. Just think of the number of viral diseases that humans can catch. Because of the medical relevance, the viruses of humans and animals have received most research attention, with plant viruses, capable of destroying crops, a rather distant second.

In many respects eukaryotic viruses resemble bacteriophages, but the fact that their hosts are eukaryotes rather than bacteria forces some differences upon them. Their genes have to be expressed within eukaryotic cells and so must resemble eukaryotic genes. They therefore need the complex upstream sequences required to activate transcription by RNA polymerase II, and they may contain introns.

Eukaryotic viruses have diverse structures and infection strategies

The capsids of eukaryotic viruses are either icosahedral or filamentous—the head-and-tail structure is unique to bacteriophages. A second distinct feature of eukaryotic viruses, especially those with animal hosts, is that

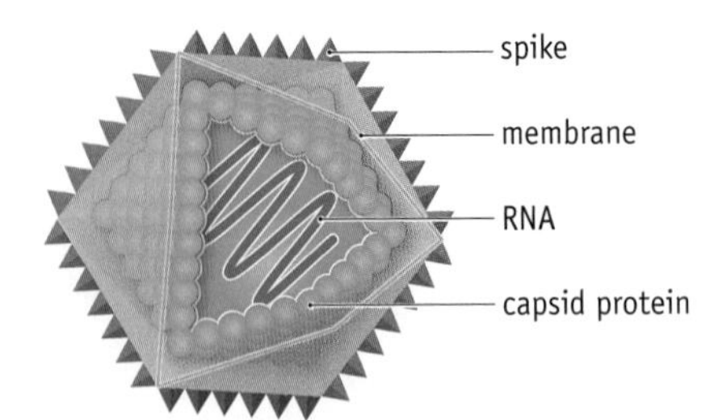

FIGURE 13.13 The structure of a eukaryotic retrovirus. The capsid is surrounded by a lipid membrane to which additional virus proteins are attached. The proteins form the "spikes" on the outer surface of the virus.

the capsid may be surrounded by a lipid membrane, forming an additional component to the virus structure (Figure 13.13). This membrane is derived from the host when the new virus particle leaves the cell, and may subsequently be modified by insertion of virus-specific proteins.

Virus genomes display a great variety of structures. They may be DNA or RNA, single- or double-stranded (or partly double-stranded with single-stranded regions), linear or circular, segmented or nonsegmented. For reasons that no one has ever understood, the vast majority of plant viruses have RNA genomes. Genome sizes cover approximately the same range as seen with bacteriophages, although the largest viral genomes (e.g., vaccinia virus, at 240 kb) are rather larger than the largest bacteriophage genomes. Eukaryotic virus structure is summarized in Table 13.2.

Most eukaryotic viruses follow only the lytic infection cycle, but few of these viruses take over the host cell's genetic machinery in the way that a lytic bacteriophage takes over a host bacterium. Many viruses coexist with their host cells for long periods, possibly years. The host-cell functions cease only toward the end of the infection cycle, when the virus progeny that have been stored in the cell are released (Figure 13.14A). Other viruses have infection cycles similar to M13 in *E. coli*, continually synthesizing new virus particles, which are extruded from the cell (Figure 13.14B).

TABLE 13.2 FEATURES OF SOME TYPICAL EUKARYOTIC VIRUSES AND THEIR GENOMES

Virus	Host	Genome structure	Genome size (kb)	Number of genes
Adenovirus	Mammals	Double-stranded linear DNA	36.0	30
Hepatitis B	Mammals	Partly double-stranded circular DNA	3.2	4
Influenza virus	Mammals	Single-stranded segmented linear RNA	13.5	12
Parvovirus	Mammals	Single-stranded linear DNA	1.6	5
Poliovirus	Mammals	Single-stranded linear RNA	7.6	8
Reovirus	Mammals	Double-stranded segmented linear RNA	22.5	22
Retroviruses	Mammals, birds	Single-stranded linear RNA	6.0–9.0	3
SV40	Monkeys	Double-stranded circular DNA	5.0	5
Tobacco mosaic virus	Plants	Single-stranded linear RNA	6.4	6
Vaccinia virus	Mammals	Double-stranded circular DNA	240	240

The genome structure is that in the virus capsid. Some genomes exist in different forms within the host cell.

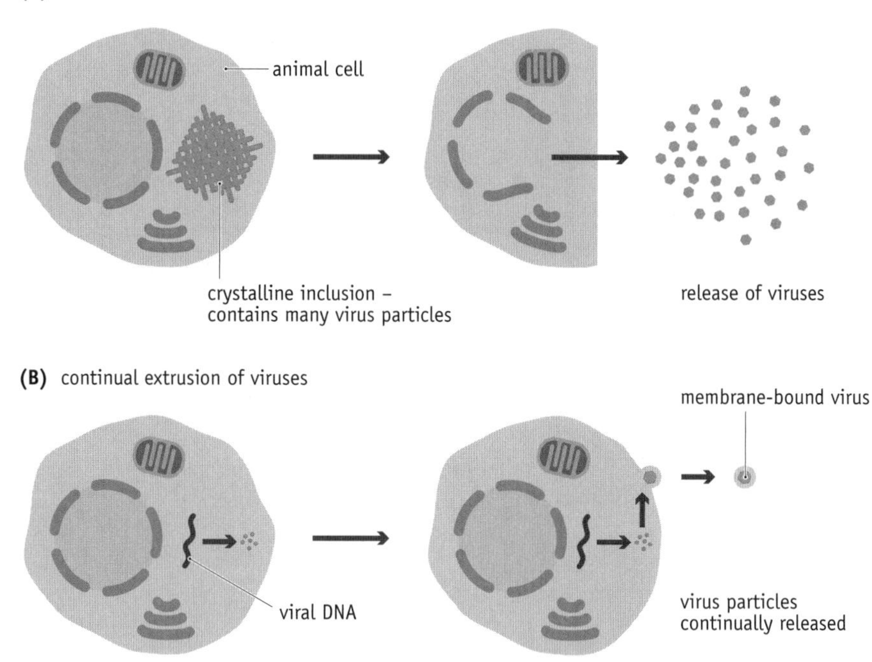

FIGURE 13.14 The progeny of a eukaryotic virus might be (A) stored in the host cell prior to release, or (B) continually extruded from the infected cell. In (A), the stored viruses form a crystalline inclusion inside the cell. In (B), the virus particles become coated with a small part of the outer cell membrane as they are released from the cell.

Viral retroelements integrate into the host-cell DNA

Many eukaryotic viruses can set up long-term infections without integrating into the host-cell DNA, but this does not mean that there are no eukaryotic equivalents to lysogenic bacteriophages. A number of DNA and RNA viruses are able to integrate into the genomes of their hosts, sometimes with drastic effects on the host cell. **Viral retroelements** are important examples of integrative eukaryotic viruses. These viruses are also interesting because their replication pathways include a novel step in which an RNA version of the genome is converted into DNA.

There are two kinds of viral retroelement. These are the **retroviruses**, whose capsids contain the RNA version of the genome, and the **pararetroviruses**, whose encapsidated genome is made of DNA. The ability of viral retroelements to convert RNA into DNA requires the enzyme called **reverse transcriptase**, which is capable of making a DNA copy of an RNA template (Figure 13.15).

The typical retroviral genome is a single-stranded RNA molecule, 6000 to 9000 nucleotides in length. After entry into the cell the genome is copied into double-stranded DNA by a few molecules of reverse transcriptase that the virus carries in its capsid. The double-stranded version of the genome

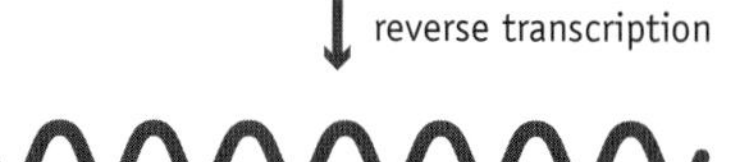

FIGURE 13.15 Reverse transcription of an RNA template into DNA. The single-stranded DNA resulting from reverse transcription can be copied into double-stranded DNA by a DNA-dependent DNA polymerase.

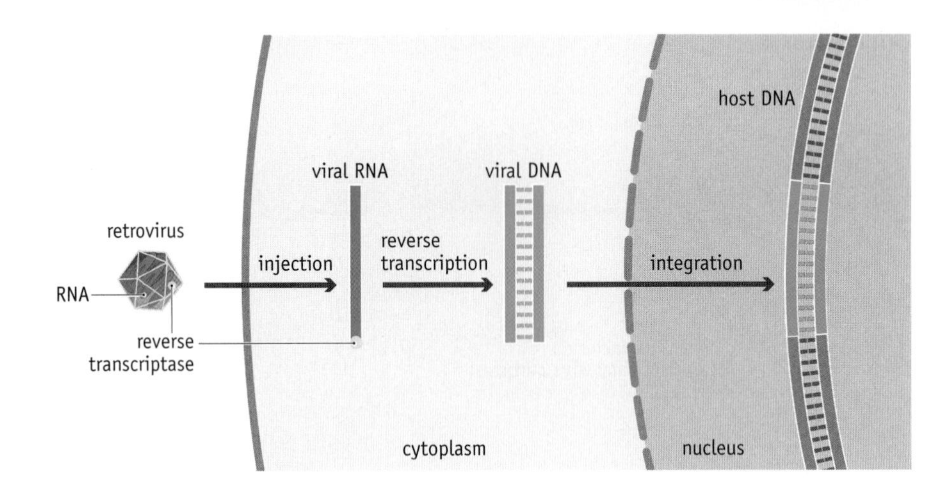

FIGURE 13.16 Insertion of a retroviral genome into a host chromosome.

then integrates into the host DNA (Figure 13.16). Unlike λ, the retroviral genome has no sequence similarity with its insertion site in the host DNA, and appears to insert at more or less random positions.

Integration of the viral genome into the host DNA is a prerequisite for expression of the retrovirus genes. There are three of these, called *gag*, *pol*, and *env* (Figure 13.17). Each codes for a polyprotein that is cleaved, after translation, into two or more functional gene products. These products include the virus coat proteins (from *env*) and the reverse transcriptase (from *pol*). The protein products combine with full-length RNA transcripts of the retroviral genome to produce new virus particles.

The causative agents of AIDS were shown in 1983–1984 to be retroviruses. The first AIDS virus to be isolated is called human immunodeficiency virus 1, or HIV-1, and is responsible for the most prevalent and pathogenic form of AIDS. A related virus, HIV-2, is less widespread and causes a milder form of the disease. The HIVs attack certain types of lymphocyte in the bloodstream, thereby depressing the immune response of the host. These lymphocytes carry on their surfaces multiple copies of a protein called CD4, which acts as a receptor for the virus. An HIV particle binds to a CD4 protein and then enters the lymphocyte after fusion between its lipid envelope and the cell membrane.

Some retroviruses cause cancer

The human immunodeficiency viruses are not the only retroviruses capable of causing diseases. Several retroviruses can induce **cell transformation**, possibly leading to cancer. Cell transformation involves changes in cell morphology and physiology. In cell cultures, transformation results in a loss of control over growth, so that transformed cells grow as a disorganized mass, rather than as a monolayer (Figure 13.18). In whole animals, cell transformation is thought to underlie the development of tumors.

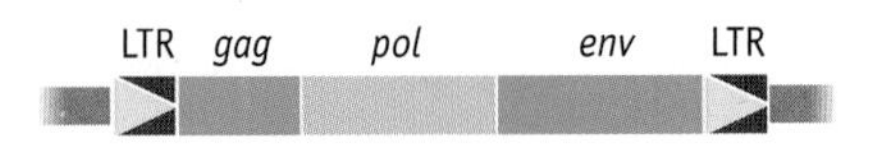

FIGURE 13.17 A retrovirus genome. LTR, long terminal repeat.

There appear to be two distinct ways in which retroviruses can cause cell transformation. With some retroviruses, such as the leukemia viruses, cell transformation is a natural consequence of infection, although it may be induced only after a long latent period during which the integrated form of the virus lies quiescent within the host genome. Other retroviruses cause

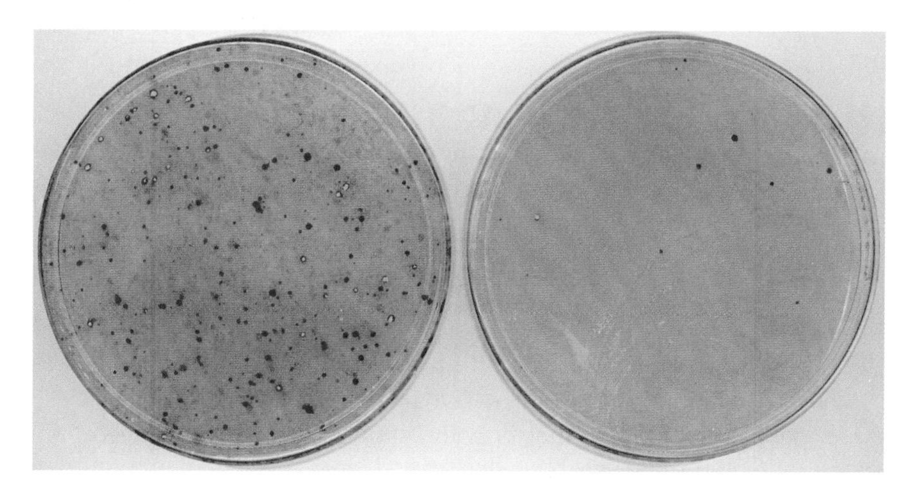

FIGURE 13.18 Transformation of cultured human cells. In the Petri dish on the right, normal human cells are growing in a monolayer. On the left, the cells have been transformed. Clumps of cells can be seen, showing that some of the processes that normally control cell growth have been disrupted. (Courtesy of Klaus Bister, University of Innsbruck.)

cell transformation because of abnormalities in their genome structures. These **acute transforming viruses** carry cellular genes that they have captured from previous cells that they have infected. In at least one transforming retrovirus (Rous sarcoma virus) this cellular gene is present side by side with the standard retroviral genes (Figure 13.19A). In others the cellular gene replaces part of the retroviral gene complement (Figure 13.19B). In the latter case the retrovirus may be **defective**, meaning that it is unable to replicate and produce new viruses, as it has lost genes coding for vital replication enzymes and/or capsid proteins. These defective retroviruses are not always inactive, however, as they can make use of proteins provided by other retroviruses in the same cell (Figure 13.20).

The ability of an acute transforming virus to cause cell transformation lies in the nature of the cellular gene that has been captured. Often this captured gene is a v-*onc* gene, with "v" standing for "viral" and "onc" standing for **oncogene**, a gene that codes for a protein involved in cell proliferation. The normal cellular version of such a gene is subject to strict regulation and expressed only in limited quantities when needed. It is thought that expression of the v-*onc* follows a different, less controlled pattern, either because of changes in the gene structure or because of the influence of promoters within the retrovirus. One result of this altered expression pattern could be a loss of control over cell division, leading to the transformed state.

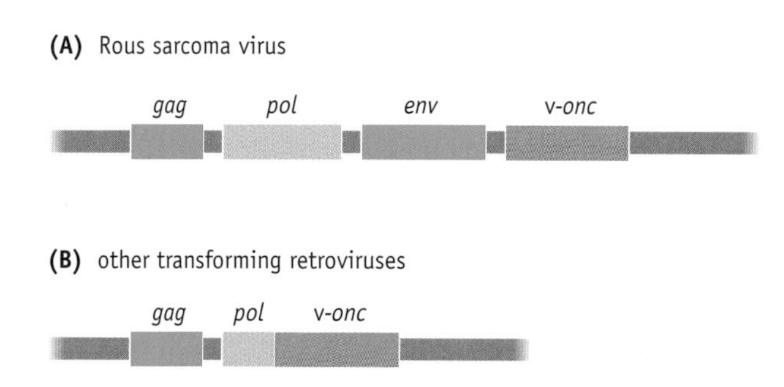

FIGURE 13.19 The genomes of (A) Rous sarcoma virus, and (B) an acute transforming virus. The gene labeled v-*onc* is the cellular gene that has been captured by the virus.

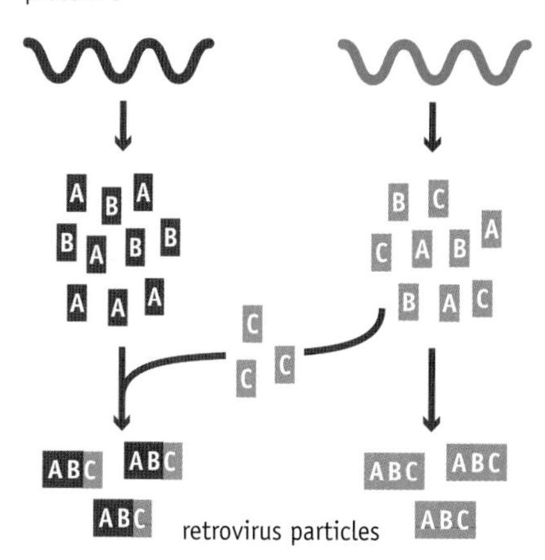

FIGURE 13.20 A defective retrovirus may be able to give rise to infective virus particles if it shares the cell with a nondefective retrovirus. The latter acts as a "helper," providing the proteins that the defective virus is unable to synthesize.

RESEARCH BRIEFING 13.1

Reconstruction of the 1918 influenza virus

Research into the ways in which viruses cause disease takes many forms. An illustration of one of the more unconventional of these approaches is the reconstruction of the genome of the virus responsible for the 1918 influenza epidemic. Studies of the virus genome have provided insights into the reasons why this strain of influenza was so virulent.

Influenza epidemics occur every year. Most people who become infected with the virus recover after about a week's illness, but up to half a million die every year. These are mostly the very young or elderly, or people already suffering from other conditions such as heart disease. Once exposed to the virus, a healthy individual becomes immune to further infection with that strain. The newfound immunity leads to an "arms race," the virus genome continually undergoing mutations that lead to small changes in the proteins in the virus capsid. The mutations give rise to a new strain of the virus that is not recognized so well by the antibodies that gave protection against the previous strain (Figure 1). A new cycle of infection therefore occurs.

Occasionally, the virus undergoes a more dramatic change that results in a much more serious epidemic. This can happen in various ways. One possibility is that a strain of the virus that has evolved in birds or an animal such as pigs jumps into the human population. For humans, this is an entirely new strain, not a new version of a strain from a previous year. So there is no human immunity at all and many more people are likely to die (Figure 2). Such a strain is what the popular media refer to as "bird flu" and "swine flu." Another possibility is that two different strains of the virus recombine to produce a hybrid genome. This might be two human strains or one human virus and a version from another species (Figure 3). The new variant is again so different from any that humans have encountered in the past that many people are likely to die before immunity develops in the population as a whole. The 2009 "swine flu" virus was an example, this being a reassortment of three previous influenza strains, one from humans, one from pigs, and one from birds. Fortunately, it did not lead to the widespread epidemic that was feared when it first emerged.

There was no such good fortune with the 1918 virus. This strain gave rise to the most severe influenza epidemic in recorded history. It lasted from 1918 until 1920 and resulted in 20 million to 100 million deaths—so many that nobody is sure of the exact number. Clinicians have debated whether this high death rate occurred because the strain was particularly virulent, or whether the mortality reflected a decreased level of resistance among the stressed and, in some cases, starving populations of Europe immediately after the First World War. There have also been questions about the origins of the strain. Alternative theories are that it resulted from recombination between human and swine flu, or that it was a form of avian flu that jumped to humans. We are overdue another large flu epidemic, and a better knowledge of the origins and virulence of the 1918 virus might help us prepare for its inevitable arrival.

Genes from the 1918 influenza virus have been recovered from ancient specimens

How can we study the 1918 influenza strain if the virus is continually evolving by mutation and recombination? No examples of the strain will survive among the human or animal populations alive today. The answer is to recover remnants of the virus genome from individuals who died during the 1918 epidemic. It is possible to obtain these remnants from preserved tissue sections that were taken from flu victims. Material has also been recovered from the lungs of a female who died in Alaska and whose body became preserved after burial in the permafrost.

These ancient samples do not contain intact copies of the virus genome, but sometimes small fragments are preserved. Like many viruses, the influenza genome is made of RNA. The RNA fragments recovered from the ancient samples were therefore converted to DNA by treating with reverse transcriptase, and then studied

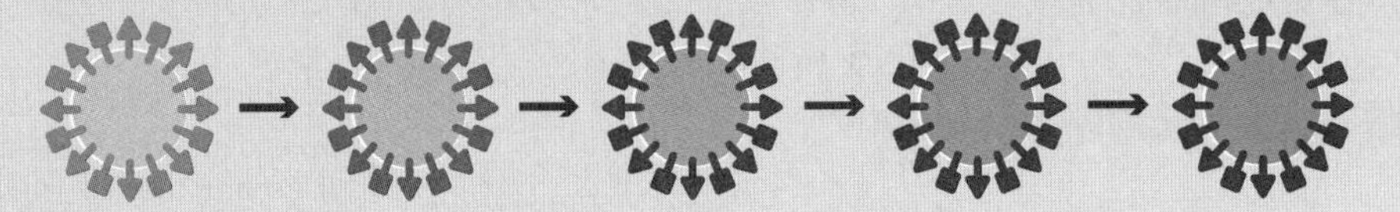

FIGURE 1 The influenza virus evolves gradually over time, giving rise to annual outbreaks of the disease.

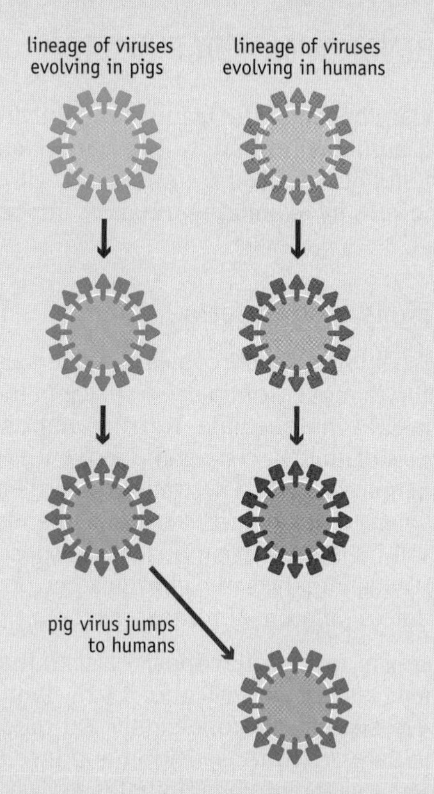

FIGURE 2 A more severe epidemic is likely to result if a strain of influenza jumps from pigs or birds to the human population.

by PCR. In the first set of experiments, the PCRs were directed at the genes for the more important of the virus proteins. These included the hemagglutinin and neuraminidase proteins, which lie on the surface of the virus capsid. They are the proteins that are usually recognized by the host immune system, and also the ones that determine how effectively the virus is able to invade human tissues and cause disease.

Comparisons of the sequences of the 1918 genes with the equivalent genes of modern influenza viruses have suggested that 1918 flu was caused by an avian virus that adapted to humans, rather than a reassorted human–swine virus. The hemagglutinin PCRs showed that in the 1918 virus this protein had largely retained its avian structure, but with small alterations that enabled the virus to bind to surface receptors on human cells. These are unusual features not seen in modern influenza strains, and could well explain the high virulence of the 1918 virus.

The complete 1918 virus genome has been reconstructed

The information obtained from preserved specimens has also enabled partial and complete copies of the 1918 virus genome to be reconstructed. Initially 1918 sequences were inserted into a modern virus genome and mice inoculated with the resulting hybrids. The mice suffered severe lung damage when the 1918 hemagglutinin gene was present in the virus construct, much more so than when the virus carried a modern hemagglutinin. This confirmed the suspicions, based on the gene sequence, that the structure of this particular version of the hemagglutinin protein was one of the underlying factors responsible for the severity of the 1918 epidemic.

Further studies showed that the virulence of 1918 flu was not just due to the hemagglutinin protein. Experiments with macaque monkeys, whose immune system is very similar to that of humans, showed that the 1918 virus elicits a stronger immune response than contemporary strains of flu. Surely a stronger immune response should reduce virulence by destroying more of the infecting viruses? Paradoxically, the enhanced response might, in fact, benefit the virus. This is because the resulting tissue inflammation facilitates entry of the virus particles into the host's lung cells. Here we have an explanation for one of the most tragic features of the 1918 epidemic. There was very high mortality among younger people of 15 to 45 years of age, whose relatively robust immune systems should have made them better able to withstand the effects of the virus. It turns out that the strength of their immune response was the cause of their death.

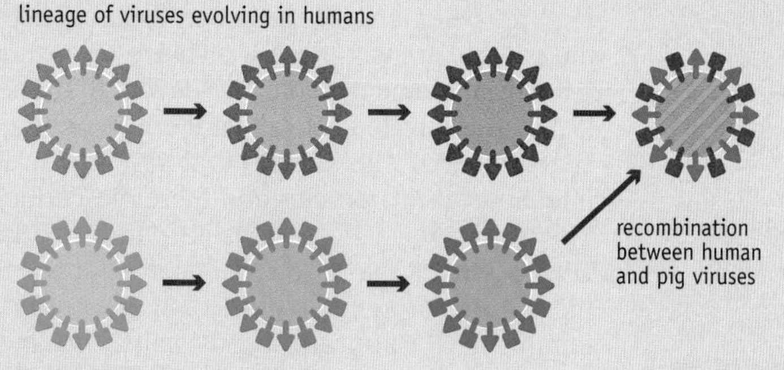

FIGURE 3 A severe epidemic can also result from reassortment between two strains of influenza virus, to produce a new variant with a combination of features not previously encountered.

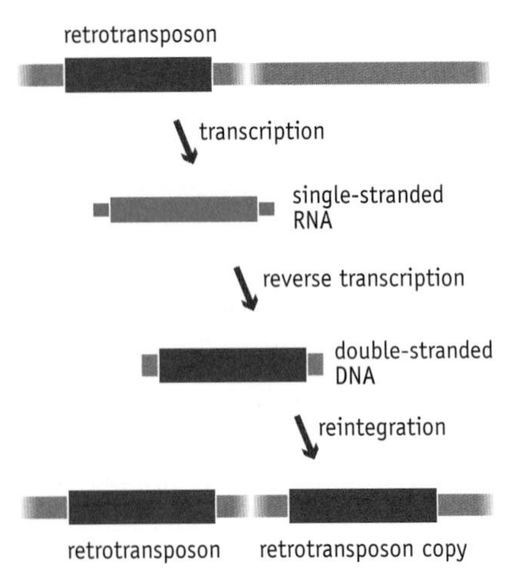

FIGURE 13.21 Retrotransposition. An RNA copy of the retrotransposon is converted into double-stranded DNA, which reintegrates into the genome.

13.3 TOWARD AND BEYOND THE EDGE OF LIFE

Viruses occupy the boundary between the living and nonliving worlds. At the very edge of life, and arguably beyond it, lie a variety of genetic elements that might or might not be classified as viruses. We will complete our survey of virus infection cycles by examining the most important of these enigmatic entities.

There are cellular versions of viral retroelements

Many eukaryotic genomes contain multiple copies of sequences, called **RNA transposons**, that are able to move from place to place in the genome by a process that involves an RNA intermediate. **Retrotransposition**, as it is called, begins with synthesis of an RNA copy of the sequence by the normal process of transcription (Figure 13.21). The transcript is then copied into double-stranded DNA, which initially exists as an independent molecule outside the genome. Finally, the DNA copy of the transposon integrates into the genome, possibly back into the same chromosome occupied by the original unit, or possibly into a different chromosome.

If we compare the mechanism for retrotransposition with that for replication of a viral retroelement, as shown in Figure 13.16, then we see that the two processes are very similar. The one significant difference is that the RNA molecule that initiates retrotransposition is transcribed from an endogenous sequence, whereas the one that initiates replication of a viral retroelement comes from outside the cell. This close similarity alerts us to the possibility that a relationship exists between RNA transposons and viral retroelements.

RNA transposons can be broadly classified into two types, those that have **long terminal repeats** (**LTRs**) and those that do not. Long terminal repeats are also possessed by viral retroelements (see Figure 13.17), and it is now clear that these viruses and the endogenous LTR retrotransposons are members of the same superfamily of elements.

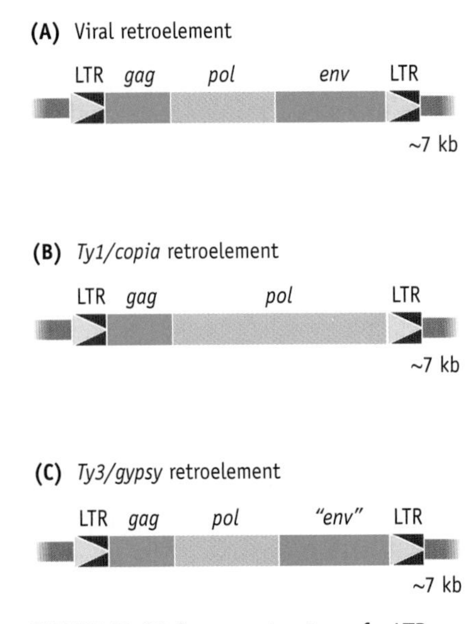

FIGURE 13.22 Genome structures for LTR retroelements.

The first LTR retrotransposon to be discovered was the *Ty* sequence of yeast, which is 6.3 kb in length and has a copy number of 25 to 35 in most *Saccharomyces cerevisiae* genomes. There are several types of *Ty* element in yeast genomes. The most abundant of these, *Ty1*, is similar to the *copia* retrotransposon of the fruit fly. These elements are therefore now called the *Ty1/copia* family. If we compare the structure of a *Ty1/copia* retrotransposon with that of a viral retroelement, then we see clear family relationships (Figure 13.22). Each *Ty1/copia* element contains two genes, called *TyA* and *TyB* in yeast, which are similar to the *gag* and *pol* genes of a viral retroelement. In particular, *TyB* codes for a polyprotein that includes the reverse transcriptase that plays the central role in transposition of a *Ty1/copia* element. Note, however, that the *Ty1/copia* retrotransposon lacks an equivalent of the viral *env* gene, the one that codes for the viral coat proteins. This means that *Ty1/copia* retrotransposons cannot form infectious virus particles and therefore cannot escape from their host cell. They do, however, form viruslike particles (VLPs) consisting of the RNA and DNA copies of the retrotransposon attached to core proteins derived from the TyA polyprotein (Figure 13.23). In contrast, the members of the second family of LTR retrotransposons, called *Ty3/gypsy* (again after the yeast and fruit fly versions), do have an equivalent of the *env* gene (Figure 13.22C), and at least some of these can form infectious viruses. Although classed as endogenous retrotransposons, these infectious versions should be looked upon as viral retroelements.

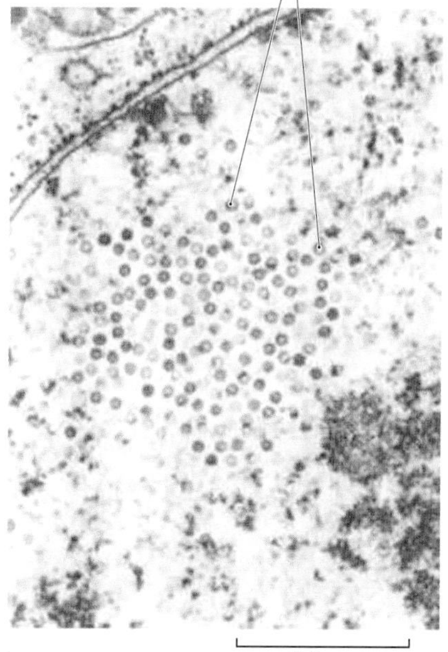

FIGURE 13.23 Electron micrograph showing viruslike particles of the *copia* retrotransposon in the nucleus of a *Drosophila melanogaster* cell. (From K. Yoshioka et al., *EMBO J.* 9: 535–541, 1990. With permission from Macmillan Publishers Ltd.)

LTR retrotransposons make up substantial parts of many eukaryotic genomes, and are particularly abundant in the larger plant genomes, especially those of grasses such as maize. They also make up an important component of invertebrate and some vertebrate genomes, but in the genomes of humans and other mammals all the LTR elements appear to be decayed viral retroelements rather than true retrotransposons. These sequences are called **endogenous retroviruses** (**ERVs**), and with a copy number of approximately 240,000 they make up 4.7% of the human genome. Human ERVs are 6 to 11 kb in length and have copies of the *gag*, *pol*, and *env* genes (Figure 13.24). Although most contain mutations or deletions that inactivate one or more of these genes, a few members of the HERV-K group have functional sequences. By comparing the positions of the HERV-K elements in the genomes of different individuals it has been inferred that at least some of these are active transposons, unlike the majority of human ERVs, which are inactive sequences that are not capable of additional proliferation.

Satellite RNAs, virusoids, viroids, and prions are all probably beyond the edge of life

When we move to the very edge of life we meet a variety of infectious molecules related to, but different from, viruses. The **satellite RNAs**, also called **virusoids**, are examples. These are RNA molecules some 320 to 400 nucleotides in length, each containing a single gene or a very small number of genes. Satellite RNAs and virusoids cannot construct their own capsids, and instead move from cell to cell within the capsids of helper viruses. The distinction between the two groups is that a satellite virus shares the capsid with the genome of the helper virus whereas a virusoid RNA molecule becomes encapsidated on its own. Both are generally looked on as parasites of their helper viruses, although there appear to be a few cases where the helper cannot replicate without the satellite RNA or virusoid, suggesting that at least some of the relationships are symbiotic.

Satellite RNAs and virusoids are both found predominantly in plants, as is a more extreme group called the **viroids**. These are RNA molecules of 240 to 375 nucleotides that contain no genes. Viroids never become encapsidated, spreading instead from cell to cell as naked RNA. They include some economically important pathogens, such as the citrus exocortis viroid, which inhibits the growth of citrus fruit trees.

Finally, there are the **prions**. These are infectious, disease-causing particles that contain no nucleic acid. Prions are responsible for scrapie in sheep and goats, and their transmission to cattle has led to the new disease called BSE—bovine spongiform encephalopathy. Whether their further transmission to humans causes a variant form of Creutzfeldt–Jakob disease (CJD) is controversial but accepted by many biologists. At first prions were thought to be viruses but it is now clear that they are made solely of protein.

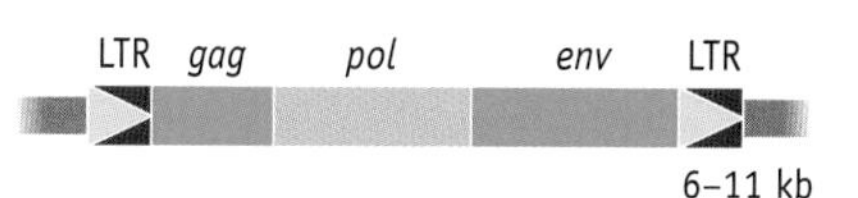

FIGURE 13.24 An intact human endogenous retrovirus sequence. LTR, long terminal repeat.

The normal version of the prion protein, called PrP^C, is coded by a mammalian nuclear gene and synthesized in the brain, although its function is unknown. PrP^C is easily digested by proteases, whereas the infectious

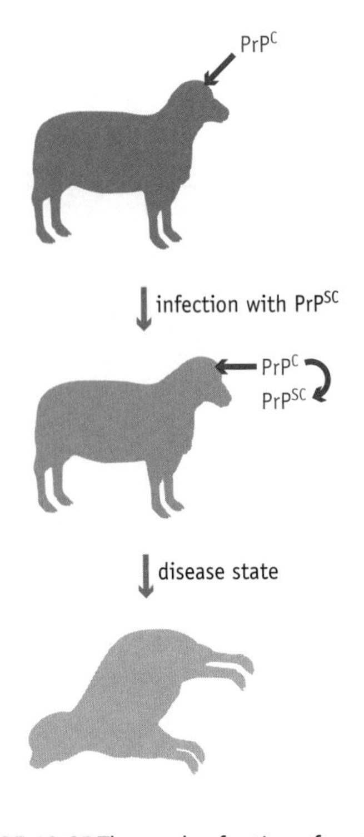

FIGURE 13.25 The mode of action of a prion.

version, PrP^{SC}, has a more highly β-sheeted structure that is resistant to proteases and forms fibrillar aggregates that are seen in infected tissues. Once inside a cell, PrP^{SC} molecules are able to convert newly synthesized PrP^{C} proteins into the infectious form, by a mechanism that is not yet understood, resulting in the disease state. Transfer of one or more of these PrP^{SC} proteins to a new animal results in accumulation of new PrP^{SC} proteins in the brain of that animal, transmitting the disease (**Figure 13.25**). Infectious proteins with similar properties are known in lower eukaryotes, examples being the Ure3 and Psi^{+} prions of *Saccharomyces cerevisiae*. It is clear, however, that prions are *gene products* rather than genetic material, and despite their infectious properties, which led to the initial confusion regarding their status, they are unrelated to viruses or to subviral particles such as viroids and virusoids.

KEY CONCEPTS

- Viruses are the simplest form of life on the planet. They are so simple in biological terms that it is doubtful whether they can really be thought of as living organisms.
- The progeny resulting from a virus infection cycle inherit genes from the parent virus.
- Bacteriophages are classified into two groups according to their life cycle. The two types are called lytic and lysogenic. The fundamental difference between these groups is that a lytic bacteriophage kills its host bacterium very soon after the initial infection, whereas a lysogenic bacteriophage can remain quiescent throughout numerous generations of the host cell.
- Expression of bacteriophage genes is regulated so that the correct genes are expressed at the appropriate times in the infection cycle.
- Most eukaryotic viruses follow only the lytic infection cycle, but few take over the host cell's genetic machinery in the way that a bacteriophage does. Many viruses coexist with their host cells for long periods, possibly years.
- Viral retroelements are examples of eukaryotic viruses whose genomes become integrated into the host DNA. Their replication pathways include a novel step in which an RNA version of the genome is converted into DNA.
- The causative agents of AIDS are retroviruses, as are some of the viruses that cause cell transformation, which may lead to cancer.
- Many eukaryotic genomes contain multiple copies of RNA transposons, which are able to move from place to place in the genome by a process that involves an RNA intermediate. Some of these transposons are closely related to viral retroelements.
- Satellite RNAs, virusoids, and viroids are subviral particles that are probably beyond the edge of life.
- Prions are infectious, disease-causing particles that contain no nucleic acid. They are made of protein and so are gene products rather than genes.

QUESTIONS AND PROBLEMS (Answers can be found at www.garlandscience.com/introgenetics)

Key Terms

Write short definitions of the following terms:

acute transforming virus
capsid
cascade system
cell transformation
concatamer
cos site
defective retrovirus
early gene
endogenous retrovirus
filamentous bacteriophage
head-and-tail bacteriophage
icosahedral
induction of prophage
late gene
latent period
long terminal repeat
lysogenic infection cycle
lytic infection cycle
oncogene
one-step growth curve
overlapping genes
pararetrovirus
prion
prophage
protomer
retrotransposition
retrovirus
reverse transcriptase
RNA transposon
satellite RNA
segmented genome
SOS response
temperate bacteriophage
viral retroelement
viroid
virulent bacteriophage
virusoid

Self-study Questions

13.1 Describe the different capsid structures seen with bacteriophages.

13.2 Describe the different types of bacteriophage genome.

13.3 What are overlapping genes, as found in some viral genomes?

13.4 List the key differences between the infection cycles of lytic and lysogenic bacteriophages.

13.5 Draw and annotate a graph that illustrates the one-step growth curve. Explain how this graph relates to events occurring during infection of *E. coli* with T4 bacteriophage.

13.6 Describe what is meant by the terms "early genes" and "late genes." Outline a typical strategy by which a bacteriophage ensures that its genes are expressed at the correct time.

13.7 Describe the process by which a bacteriophage λ prophage is excised from the *E. coli* genome and is packaged into new bacteriophage particles.

13.8 Explain how bacteriophage λ makes the decision to enter lysogeny and how the state is maintained.

13.9 Outline the link between the *E. coli* SOS response and induction of a λ prophage.

13.10 Explain in what respects the M13 infection cycle is lytic and in what respects it is lysogenic.

13.11 Describe in what ways the structures of eukaryotic viruses are similar to and distinct from the structures of bacteriophages.

13.12 Distinguish between a retrovirus and a pararetrovirus.

13.13 Describe the infection cycle of a retrovirus. Your answer should explain the roles of the three genes in the retrovirus genome.

13.14 What is the distinctive feature of an acute transforming virus, and how does this relate to its ability to cause cell transformation?

13.15 Explain why an oncogene can be harmless when present in the human genome but result in cancer when carried by a retrovirus.

13.16 Using examples, describe the relationships between RNA transposons and viral retroelements.

13.17 Describe the differences between the *Ty1/copia* and *Ty3/gypsy* families of LTR retrotransposons.

13.18 What are the features of the LTR elements present in the human genome?

13.19 Describe the key features of, and differences between, virusoids and viroids.

13.20 What is a prion, and how does it cause disease?

13.21 Describe how studies of influenza virus remnants in preserved specimens have aided our understanding of the causes of the 1918 influenza epidemic.

Discussion Topics

13.22 To what extent can viruses be considered a form of life?

13.23 Bacteriophages with small genomes (for example, ϕX174) are able to replicate very successfully in their hosts. Why, then, should other bacteriophages, such as T4, have large and complicated genomes?

13.24 Bacteriophages were extensively used in the middle decades of the twentieth century as model systems with which to study genes. Many of the basic principles of gene structure and regulation were discovered through this work. Discuss the advantages, and disadvantages, of bacteriophages for this type of research.

13.25 Devise a research project aimed at discovering whether HIV-1 was originally an animal virus that jumped into the human population.

13.26 Why do LTR retroelements have long terminal repeats?

13.27 At what point in the series virus–virusoid–viroid–prion do we reach the stage where we are no longer studying genetics?

FURTHER READING

Belshe RB (2005) The origins of pandemic influenza—lessons from the 1918 virus. *N. Engl. J. Med.* 353, 2209–2211.

Bishop JM (1983) Cellular oncogenes and retroviruses. *Annu. Rev. Biochem.* 52, 301–354.

Butler PJG & Klug A (1978) The assembly of a virus. *Sci. Am.* 239, 52–59. *Describes the basic features of virus structure.*

Diener TO (1984) Viroids. *Trends Biochem. Sci.* 9, 133–136.

Herskowitz I (1973) Control of gene expression in bacteriophage lambda. *Annu. Rev. Genet.* 7, 289–324.

Marvin DA (1998) Filamentous phage structure, infection and assembly. *Curr. Opin. Struct. Biol.* 8, 150–158. *Bacteriophage M13.*

Miller ES, Kutter E, Mosig G et al. (2003) Bacteriophage T4 genome. *Microbiol. Mol. Biol. Rev.* 67, 86–156.

Patience C, Wilkinson DA & Weiss RA (1997) Our retroviral heritage. *Trends Genet.* 13, 116–120. *Endogenous retroviruses.*

Prusiner SB (1996) Molecular biology and pathogenesis of prion diseases. *Trends Biochem. Sci.* 21, 482–487.

Song SU, Gerasimova T, Kurkulos M, Boeke JD & Corces VG (1994) An env-like protein encoded by a *Drosophila* retroelement: evidence that *gypsy* is an infectious retrovirus. *Genes Dev.* 8, 2046–2057.

Inheritance of DNA Molecules During Eukaryotic Sexual Reproduction CHAPTER 14

In the next two chapters we will address the second of the three big questions that we have asked about the role of the gene as a unit of inheritance. How are genes inherited during sexual reproduction?

Sexual reproduction begins with a specialized type of cell division, called **meiosis**, which produces the male and female sex cells, or **gametes**. The cell that undergoes meiosis is diploid—like all somatic cells, it has two copies of the genome. The gametes, on the other hand, are haploid. Each gamete has just one copy of the genome (Figure 14.1). Fusion of a pair of gametes, one from the male parent and one from the female, results in a fertilized egg cell. The egg cell is therefore diploid, having inherited one copy of the genome from the male parent, and one copy from the female parent. Repeated division of the egg cell and its progeny by mitosis gives rise to the somatic cells in the new organism.

In order to understand how genes are inherited during this process we must examine sexual reproduction from two different angles. The first angle, you will not be surprised to learn, is the molecular one. We must follow through the steps in sexual reproduction and understand what happens to the DNA molecules at each stage. We already know what happens during mitosis, so we must focus on the novel type of cell division that produces the gametes. We will discover that, unlike mitosis, meiosis does not involve simply the replication of DNA molecules and the passage of the daughters into the cells that are formed when the parent cell divides. Meiosis is more complicated because it also provides an opportunity for DNA molecules to exchange segments by recombination. In this chapter we will focus on the inheritance of DNA during meiosis with emphasis on the molecular basis of recombination.

The second approach that we must take to sexual reproduction is the one taken by the early geneticists such as Mendel. The new organism that results from sexual reproduction inherits a complete complement of genes from each parent. Sometimes the two copies of an individual gene will be identical—they will both be the same allele—but sometimes they will not. We must therefore explore how pairs of alleles interact to define the

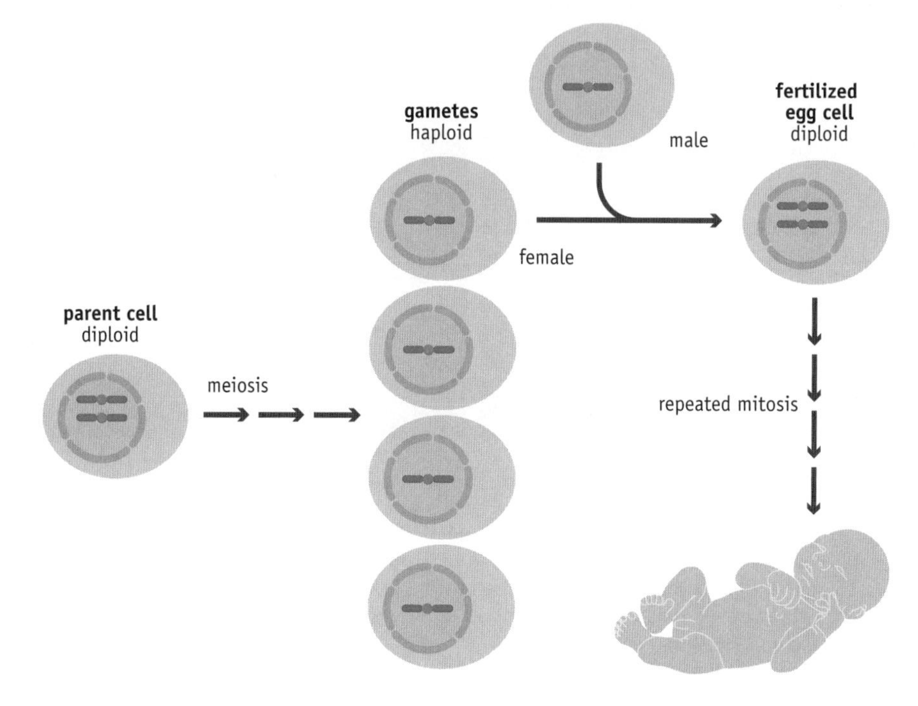

FIGURE 14.1 The cellular events occurring during sexual reproduction.

biological characteristics of the new organism. This aspect of sexual reproduction—which concerns the inheritance of genes rather than just DNA molecules—will be the subject of the next chapter.

14.1 INHERITANCE OF DNA MOLECULES DURING MEIOSIS

The vast majority of the cell divisions that take place in a multicellular eukaryote occur by mitosis. Approximately 10^{17} mitoses are needed to produce all the cells required during a human lifetime. A much smaller number of cell divisions follow the alternative pathway, called meiosis, that gives rise to the sex cells, or gametes. One of the key features of meiosis is that the gametes are haploid, each containing just a single copy of the genome. We will begin our study of meiosis by focusing on how the diploid parent cell is converted into haploid daughters.

Meiosis requires two successive cell divisions

In order to understand meiosis we must first take a step back and remind ourselves what it means when we say a cell is diploid. Each somatic cell has two copies of each chromosome. In humans this means 46 chromosomes in all—two copies of each of the 22 autosomes and a pair of sex chromosomes, XX for a female cell and XY for a male (Figure 14.2). We refer to the two copies of each chromosome as a **homologous** pair or as **homologous chromosomes**.

At the start of cell division, whether meiosis or mitosis, every homologous chromosome has itself replicated, the pairs of daughter chromosomes remaining attached at their centromeres. The cell at this point is therefore tetraploid, possessing *four* copies of each chromosome (Figure 14.3).

Meiosis results in haploid cells because it is, in effect, two successive cell divisions of the tetraploid cell, with no DNA replication occurring during either of these divisions. The two divisions are called meiosis I and meiosis

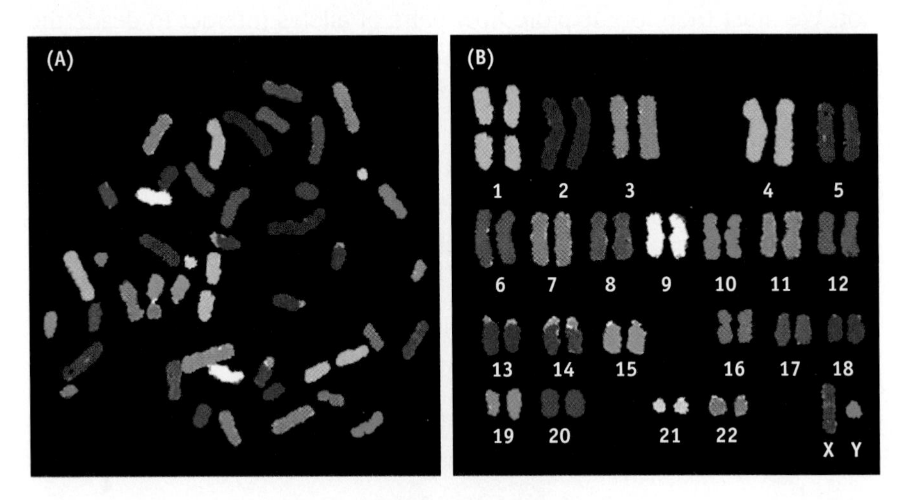

FIGURE 14.2 Chromosomes from a human somatic cell. Each chromosome has been "painted" by labeling with a different fluorescent marker. (A) The chromosomes as originally obtained from the cell. (B) The same image manipulated so that pairs of homologous chromosomes are lined up next to one another. Note that the cell was male. (From E. Schröck et al., *Science* 273: 494–497, 1996. With permission from AAAS.)

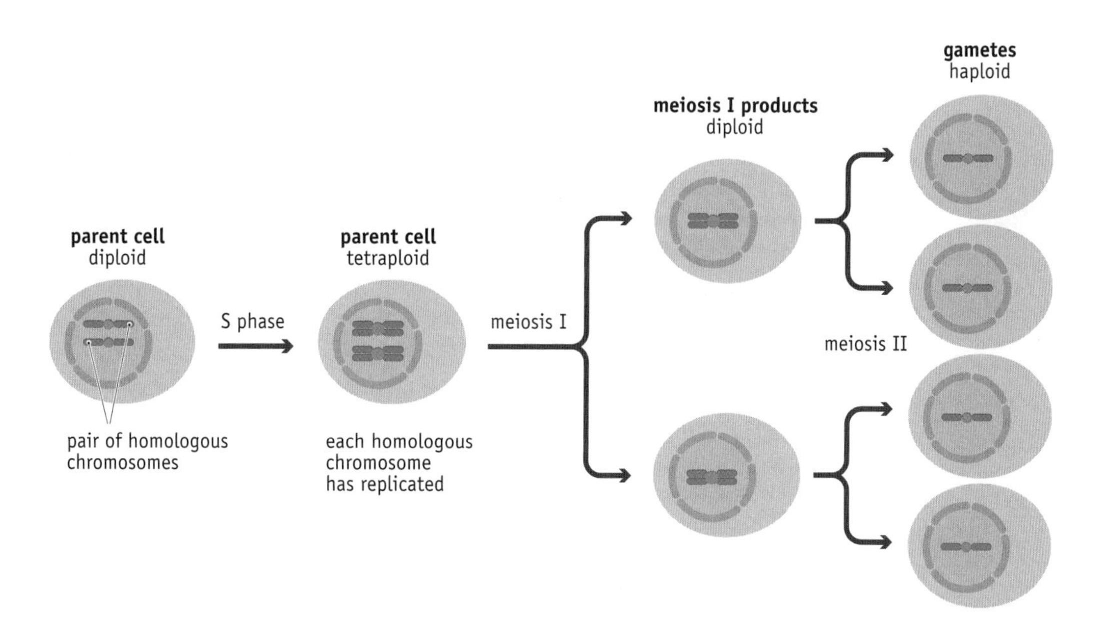

FIGURE 14.3 Outline of the events occurring during meiosis, indicating the changes in ploidy that occur during the process. During S phase of the cell cycle, each chromosome is replicated, the daughters remaining attached to one another at their centromeres. At this stage, the cell is therefore tetraploid.

II. Meiosis I results in two cells, each containing one copy of each of the replicated homologous chromosomes. The cells resulting from meiosis I are therefore diploid. During the second meiotic division, the attachment that holds each pair of daughter chromosomes together breaks. One daughter chromosome passes into one gamete, and the second daughter into the second gamete. The gametes are therefore haploid.

During meiosis I, bivalents are formed between homologous chromosomes

As meiosis I begins, the replicated chromosomes condense and migrate to the middle of the nuclear region (Figure 14.4). It is at this stage that we see the most important difference between meiosis and mitosis. During mitosis, homologous chromosomes remain separate from one another. In meiosis I, however, the pairs of homologous chromosomes are by no

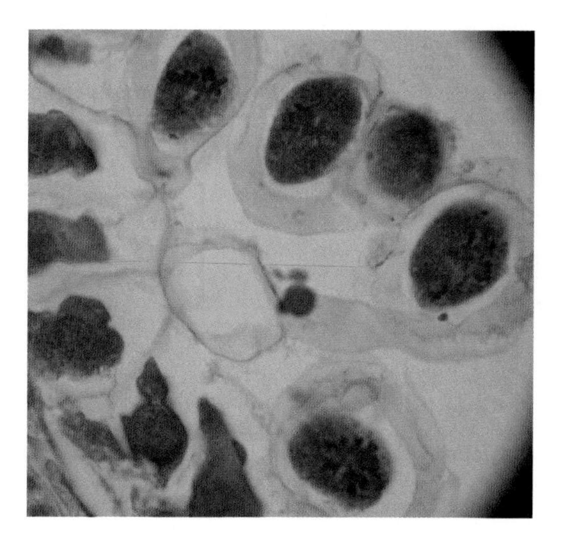

FIGURE 14.4 Early events during meiosis I in the lily. In this cross-section of a lily anther the first stages in production of pollen—the male gametes—can be seen. DNA is stained red, and the remainder of each nucleus appears green. In several nuclei, the chromosomes have condensed and have migrated to the middle part of the nucleus. (Courtesy of Professor David B. Fankhauser, University of Cincinnati Clermont College. http://biology.clc.uc.edu/Fankhauser)

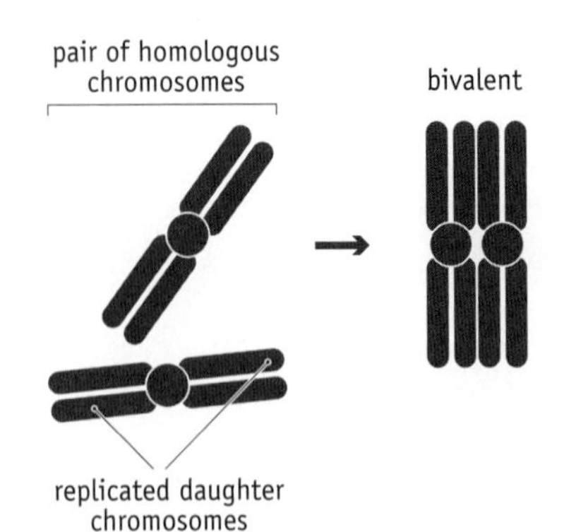

FIGURE 14.5 Formation of a bivalent between a pair of homologous chromosomes.

means independent. Instead, each chromosome finds its homolog and forms a **bivalent** (Figure 14.5). It is these bivalents, not the independent chromosomes, that line up in the middle of the nucleus.

After formation of the bivalents, meiosis I continues in a similar manner to a mitotic cell division (Figure 14.6). Microtubules radiate out from the centrosomes, attach to the kinetochores, and begin to pull in opposite directions. The tension exerted on a bivalent breaks it apart, and the two homologous chromosomes are pulled in opposite directions. A complete set of chromosomes is therefore assembled at each of the poles of the mitotic spindle. Meiosis I is then completed by cytokinesis.

Meiosis II begins almost immediately (Figure 14.7). In each of the cells, the chromosomes again migrate to the middle of the nucleus. Each cell has only one member of each pair of homologous chromosomes, so the chromosomes remain independent of one another. Microtubules attach to the kinetochores and pull the chromosomes apart, separating the two daughters, which again move in opposite directions. A second round of cytokinesis creates the four gametes.

Formation of bivalents ensures that siblings are not identical to one another

To a geneticist, the most exciting thing about sexual reproduction is the formation of bivalents. One reason why formation of bivalents is so exciting is that it is the reason why siblings—children who have the same parents—are not exactly the same. This is true for all siblings except identical twins, who derive from a single fertilized egg cell.

The difference between siblings is such an obvious fact of life that it is easy to forget to ask why it should be the case. We should ask this question because we know that mitosis results in identical daughter cells, each of which contains the same set of DNA molecules. If meiosis had the same outcome, then all the gametes produced by an individual would be identical to one another (Figure 14.8). The mother's gametes would all be the same, as would those of the father. Fusion of maternal and paternal gametes would produce identical fertilized egg cells that would develop into identical siblings.

Why are siblings different from one another, each a blend of the parents' biological characteristics, but each with its own individuality? There are two reasons, both of them outcomes of the formation of bivalents. The first results from the separation of the bivalents during the anaphase period of

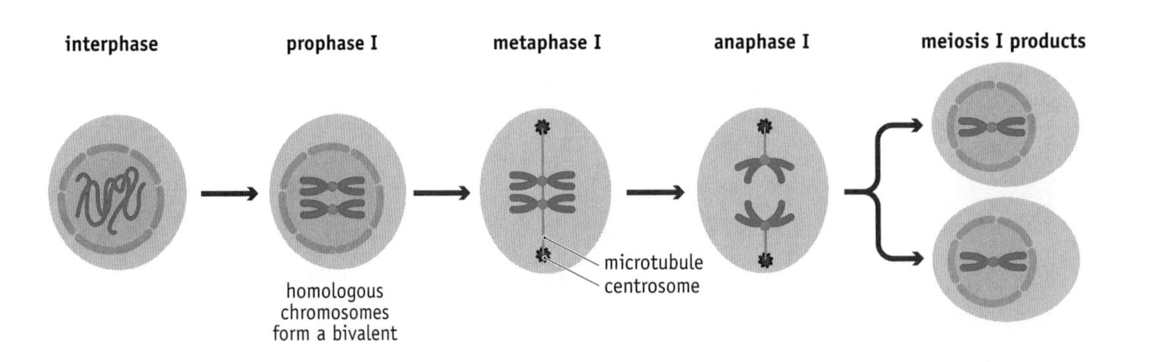

FIGURE 14.6 Meiosis I. The homologous chromosomes form bivalents during prophase I, and microtubules attach to the kinetochores during metaphase I. The chromosomes are then pulled toward opposite centrosomes during anaphase I, with meiosis I being completed by cytokinesis. Compare with events occurring during mitosis, as shown in Figure 11.27.

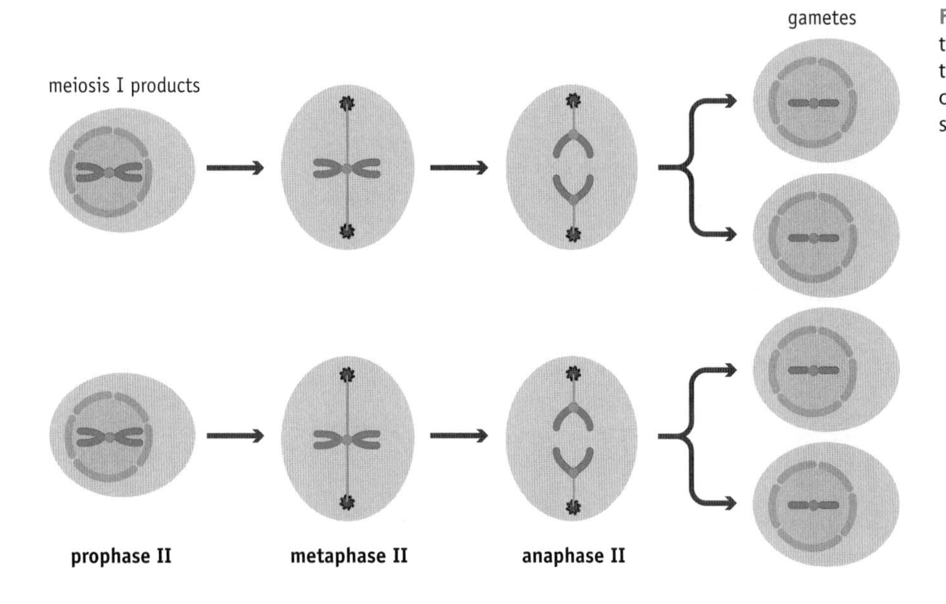

FIGURE 14.7 Meiosis II. The events are similar to those occurring during meiosis I, except that each starting cell has only one member of each pair of homologous chromosomes, so no bivalents are formed.

meiosis I. The two homologous chromosomes in a bivalent are not identical. They contain the same set of genes but will possess different alleles of many of those genes (Figure 14.9). When the bivalents break apart, the two chromosomes are pulled toward opposite centrosomes. Which

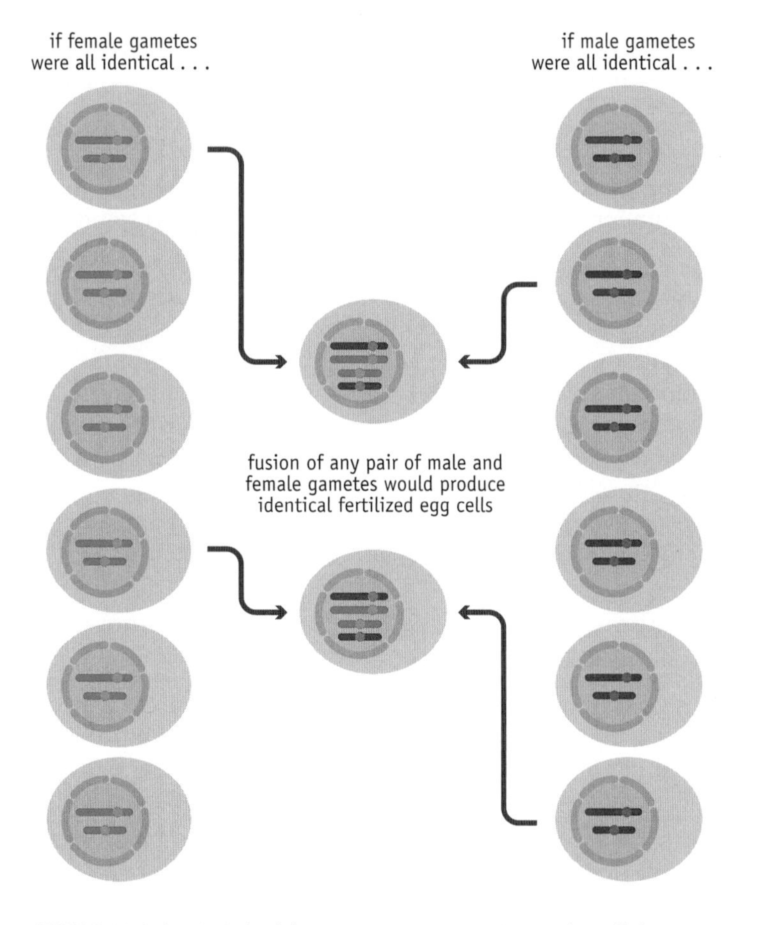

FIGURE 14.8 If meiosis had the same outcome as mitosis, then all the gametes produced by an individual would be identical. Fusion of gametes from female and male partners would always result in identical fertilized egg cells. Siblings developing from these egg cells would be genetically identical.

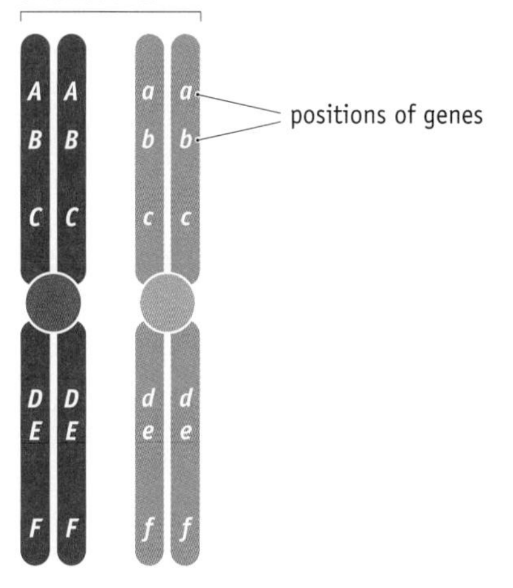

FIGURE 14.9 The pair of homologous chromosomes in a bivalent are not identical. In this example, *A* and *a*, etc., are different alleles of the same gene.

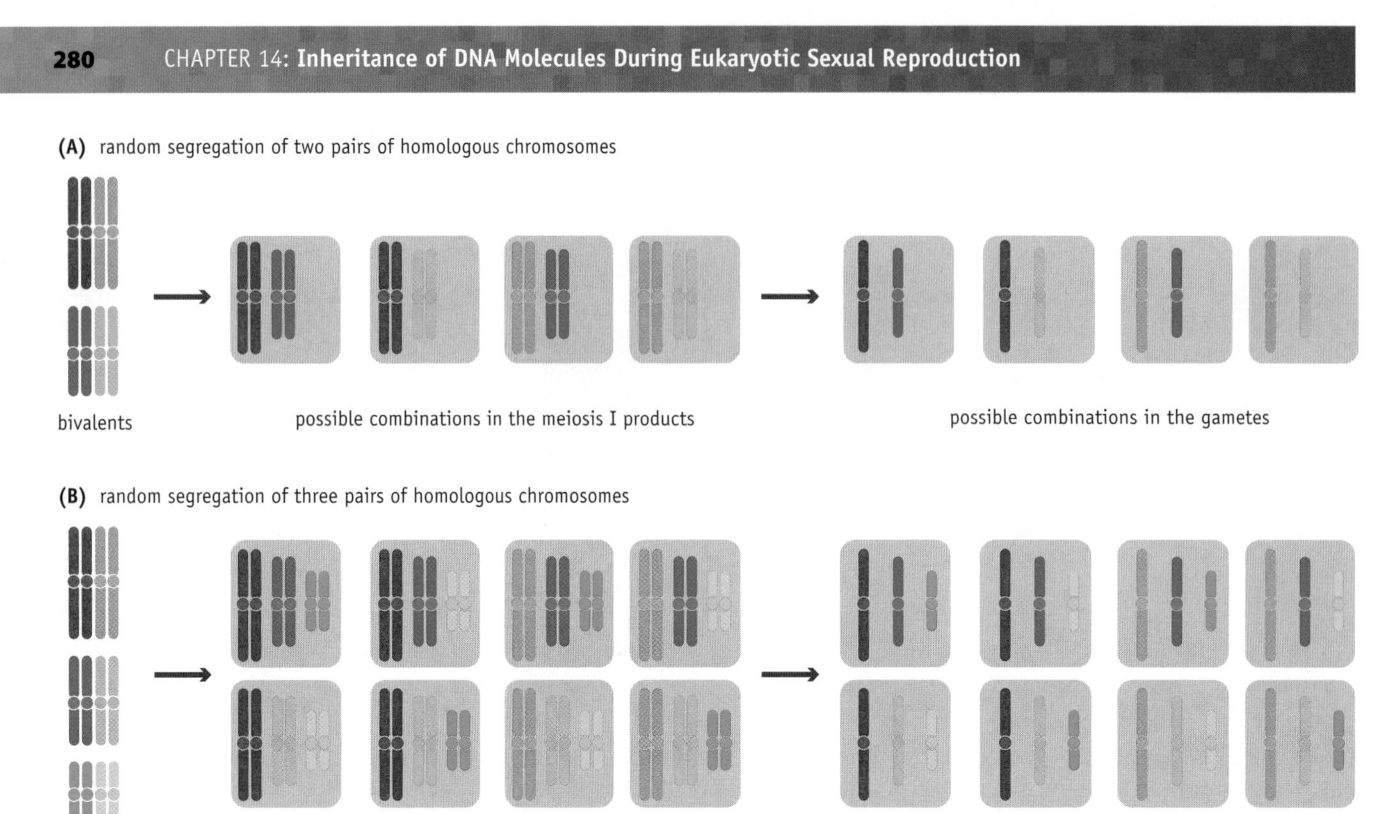

FIGURE 14.10 Random segregation of homologous chromosomes during anaphase I means that the gametes resulting from meiosis are not all identical. The individual members of each pair of homologous chromosomes are given dark and light shading. The various combinations that can arise in the diploid meiosis I products and in the gametes are shown.

chromosome goes in which direction is entirely random, depending only on the orientation of the bivalent with respect to the two spindle poles.

Imagine a cell with two pairs of homologous chromosomes. Meiosis can result in gametes with any of four different chromosome combinations, depending on the directions in which the members of each homologous pair segregate during anaphase I (Figure 14.10A). Each of these chromosome combinations corresponds to a gamete with a different set of alleles. For a cell with three chromosomes, there are $2^3 = 8$ possible combinations (Figure 14.10B). For human cells, with 46 chromosomes in the diploid set, there are 2^{23} possible ways in which the pairs of homologous chromosomes can be distributed during meiosis I. That is more than 8 million combinations, each combination giving a gamete with a different set of alleles.

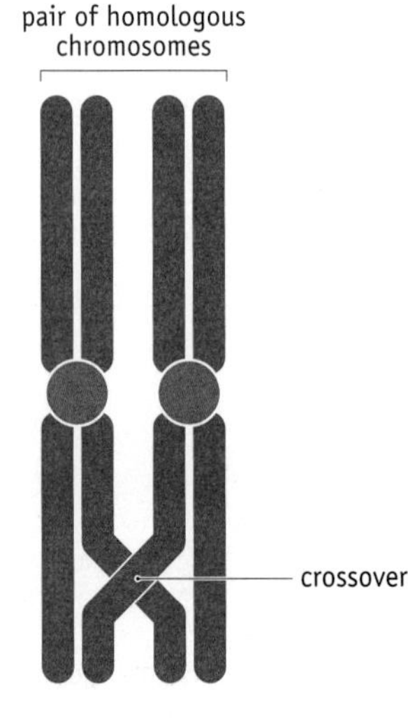

FIGURE 14.11 A crossover between two chromatids in a bivalent.

Recombination occurs between homologous chromosomes within a bivalent

Random segregation of chromosomes during anaphase I is not the only factor influencing the variability of the gametes resulting from meiosis. The bivalent contributes to the resulting variability in a second, even more important way.

Within the bivalent, the chromosome arms—the chromatids—can exchange segments of DNA (Figure 14.11). This exchange is called **crossing over** or **recombination**. It was first discovered by a Belgian cell biologist, F. A. Janssens, in 1909. Janssens had made a very close examination of the bivalents formed during meiosis in salamanders. He did not actually see crossovers between the chromatids in a bivalent, as this was not possible with the types of microscope available at that time. But in a flash

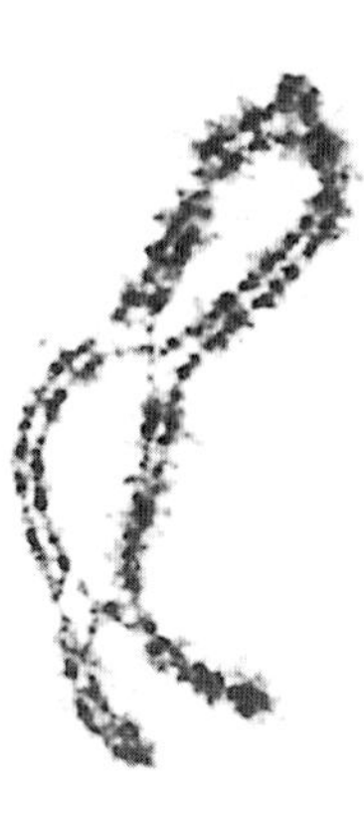

FIGURE 14.12 Electron micrograph of a bivalent in a grasshopper nucleus showing three crossovers. (Courtesy of the late Professor Bernard John.)

of insight he suggested that the close proximity of the chromatids could lead to breakage and transfer of segments between chromosomes. Later, with more sophisticated microscopes, cell biologists were able to see the actual crossover points, or **chiasmata**. The micrograph shown in Figure 14.12 is fairly typical, with three separate chiasmata along the length of a chromosome.

What impact does crossing over have on the genetic variability of the gametes resulting from meiosis? If the pair of homologous chromosomes in a bivalent exchange a segment of DNA, then any genes present in that segment will also be exchanged. The alleles present in one chromosome will be transferred to the other, and vice versa. Furthermore, the two chromosomes in each member of the homologous pair can participate in different exchanges. The outcome could be that after crossing over the four chromosomes in the bivalent are all different from one another (Figure 14.13).

Recombination increases the variability of the gametes resulting from meiosis, by changing the allele combinations within individual chromosomes prior to the random segregation of those chromosomes during anaphase I. Random segregation on its own can give rise to any of 8 million different chromosome combinations in the gametes resulting from human meiosis. The number of possible *allele* combinations in those gametes is infinitely higher than 8 million, because of the randomness of the DNA exchanges that are possible within each bivalent.

14.2 THE MOLECULAR BASIS OF RECOMBINATION

The term "recombination" was originally used by geneticists to describe the outcome of crossing over between pairs of homologous chromosomes. In this context it refers specifically to the generation of new allele combinations during meiosis. In the 1960s it was realized that crossing over involves the breakage and subsequent rejoining of DNA molecules, and "recombination" gradually became the name given to any event that involves breakage and reunion of DNA molecules. Today, we know that recombination is important not only in crossing over, but also in the repair of damaged DNA molecules, one of the topics that we will study in

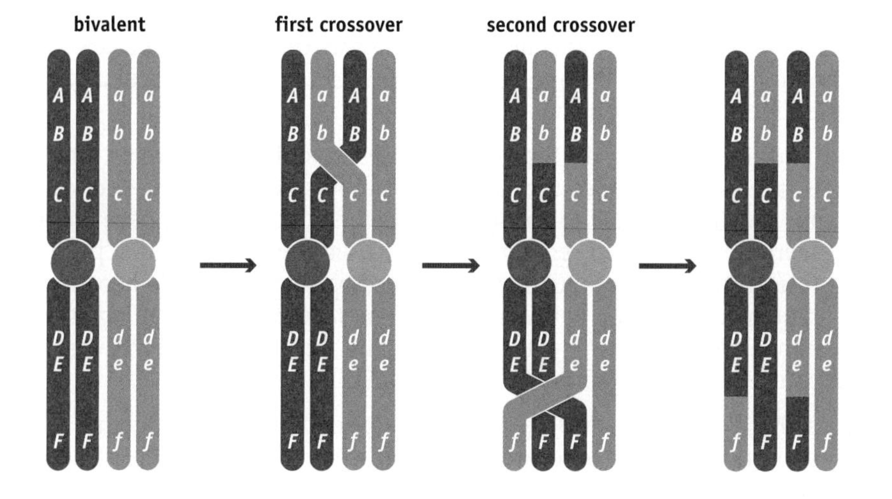

FIGURE 14.13 One possible outcome of two crossovers in a bivalent. At the end of the process, the four chromosomes are all different from one another. Many other results are possible, depending on the number of crossovers, the positions where they occur, and the members of the bivalent that participate in these crossovers.

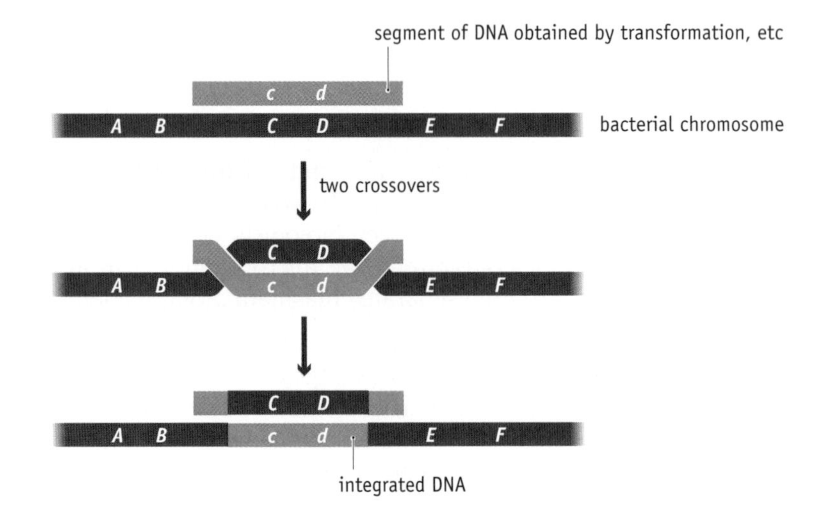

FIGURE 14.14 One of the outcomes of recombination. Integration into a bacterial chromosome of a piece of DNA acquired by transformation, conjugation, or transduction requires a pair of crossovers. In the example shown, integration results in a change in the alleles present at genes C and D.

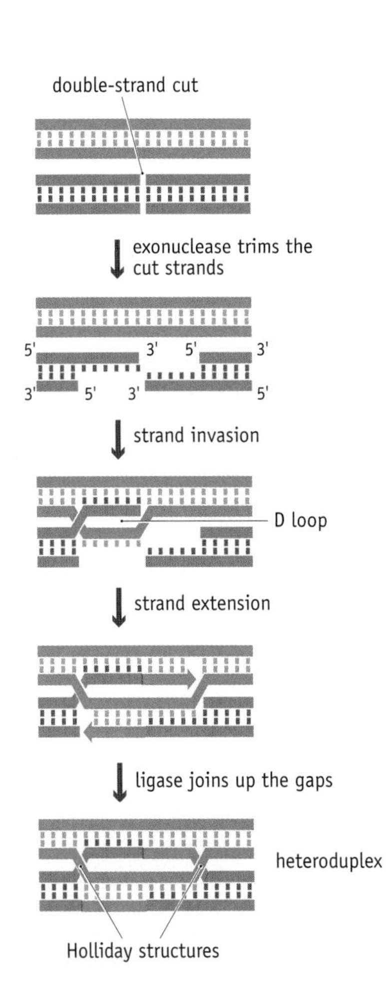

FIGURE 14.15 The initial steps in homologous recombination, resulting in formation of a heteroduplex.

Chapter 16. Recombination is also involved in the integration into bacterial genomes of DNA acquired by transformation, conjugation, or transduction, as well as the integration of episomes and lysogenic bacteriophage genomes (Figure 14.14).

Because it is such an important process in genetics, we must spend some time understanding exactly how recombination takes place. We must examine the way in which DNA segments are exchanged during crossing over, and we must become familiar with the enzymes and other proteins that catalyze and control the process.

Homologous recombination begins with formation of a DNA heteroduplex

Homologous recombination is the name given to the type of recombination that occurs during crossing over between homologous chromosomes. It occurs between two double-stranded DNA molecules that have regions where the nucleotide sequences are the same or at least very similar.

Homologous recombination begins when the two double-stranded molecules line up adjacent to one another (Figure 14.15). A double-strand cut is made in one of the molecules, breaking this one into two pieces. One strand in each half of this molecule is then shortened by removal of a few nucleotides, giving each end a 3′ overhang.

The partnership between the two chromosomes is set up when one of the 3′ overhangs invades the uncut DNA molecule, displacing one of its strands and forming a **D loop**. Strand invasion is stabilized by base pairing between the transferred segment of polynucleotide and the intact polynucleotide of the recipient molecule. This base pairing is possible because of the sequence similarity between the two molecules. The invading strand is extended by new DNA synthesis, enlarging the D loop, and at the same time the other broken strand is also extended.

After strand extension, the free polynucleotide ends are joined together. This gives a structure called a **heteroduplex**, in which the two double-stranded molecules are linked together by a pair of **Holliday structures**. These are named after the scientist Robin Holliday, who proposed the first model for recombination and realized that the process must involve formation of a heteroduplex. Each Holliday structure is dynamic and can move along the heteroduplex. This **branch migration** results in the exchange of longer segments of DNA (Figure 14.16).

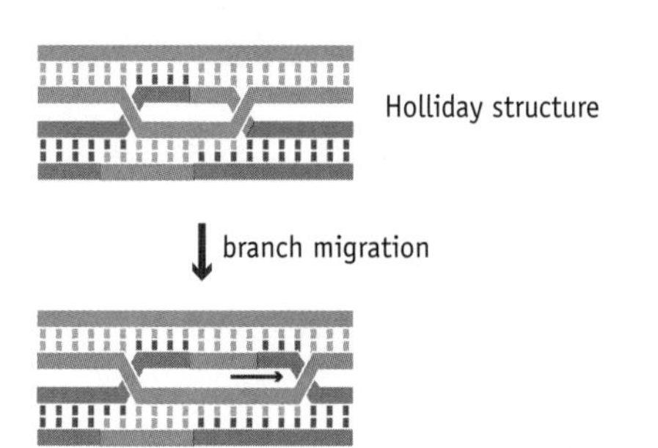

FIGURE 14.16 Branch migration. In this example, one of the two Holliday structures resulting from the recombination process shown in Figure 14.15 migrates in such a way that the heteroduplex is extended.

Cleavage of the Holliday structures results in recombination

Separation, or **resolution**, of the heteroduplex back into individual double-stranded molecules occurs by cleavage of the Holliday structures. The two-dimensional representation used in Figures 14.15 and 14.16 is rather unhelpful in this regard because it obscures the true topology of the Holliday structures. In particular, it fails to reveal that each Holliday structure can be cleaved in either of two orientations, with very different outcomes. To understand this point, we must depict a Holliday structure in its true three-dimensional configuration, called the **chi form**, as revealed by electron-microscopic observation of recombining DNA molecules (Figure 14.17).

Now the two possible cleavages become much clearer. First, the cut can be made left to right across the chi form. This is equivalent to cleavage directly across the Holliday structures as drawn at the top in Figure 14.17. Cleavage of a Holliday structure in this way, followed by joining of the cut strands, gives two molecules that have exchanged short segments of polynucleotide. As the exchanged strands have similar sequences, the effect on the genetic constitution of each molecule is relatively minor, and this outcome is thought not to be relatively common during meiosis.

More frequently, a Holliday structure is resolved by cutting the chi structure in the second of the two possible orientations, up–down in Figure 14.17. This is the type of cleavage that is not readily apparent from the two-dimensional representation of the Holliday structure. But it is vitally important

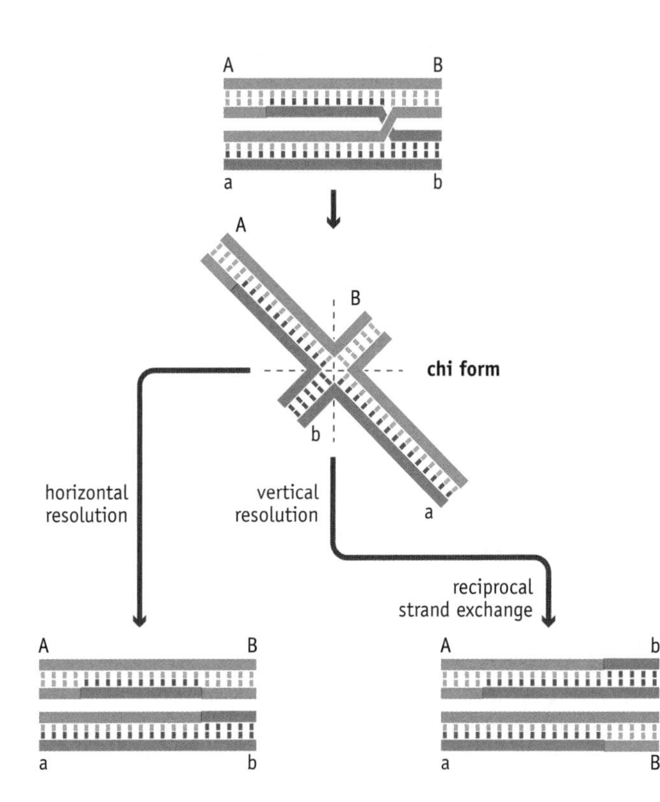

FIGURE 14.17 The two possible ways of resolving a Holliday structure.

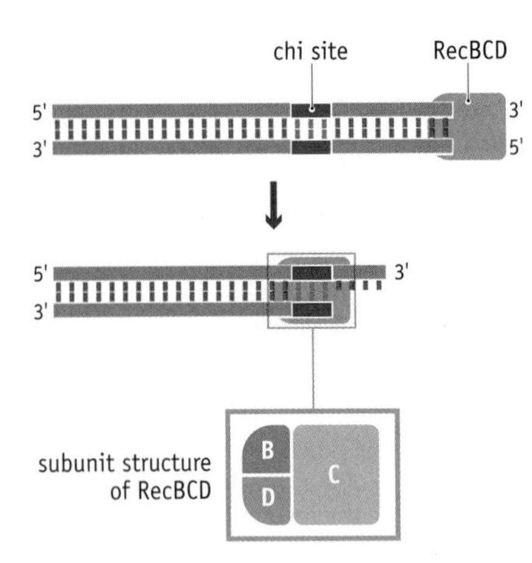

FIGURE 14.18 The role of the RecBCD enzyme in homologous recombination in *E. coli*.

because up–down cleavage results in **reciprocal strand exchange**, double-stranded DNA being transferred between the two molecules so that the end of one molecule is exchanged for the end of the other molecule. This is the DNA transfer that occurs during crossing over.

The biochemical pathways for homologous recombination have been studied in *E. coli*

Now that we have explored the way in which the DNA molecules behave during homologous recombination we can turn our attention to the enzymes and other proteins that are responsible for carrying out the process. Here we are faced with a difficulty. Research into the biochemistry of recombination in *Escherichia coli* is well advanced, but much less is known about the equivalent events in eukaryotes such as humans. This is a problem because there appear to be some important differences between the recombination processes in bacteria and eukaryotes, so we cannot be certain that everything that has been learned from studies of *E. coli* is directly applicable to eukaryotes. With that caveat in mind, we will begin by examining what we know about the enzymes and proteins involved in homologous recombination in bacteria.

At the biochemical level, there are three distinct recombination systems in *E. coli*, each involving a different set of proteins. These are called the RecBCD, RecE, and RecF pathways. The RecBCD pathway is the best studied and illustrates the key points of all three pathways.

The central player in the RecBCD pathway is the **RecBCD enzyme**, which, as its name implies, is made up of three different proteins. Two of these—RecB and RecD—are helicases, enzymes capable of breaking the base pairs that hold polynucleotides together. To initiate homologous recombination, one copy of the RecBCD enzyme attaches to the free ends of a chromosome at a double-strand break. Using its helicase activity, RecBCD then progresses along the DNA molecule at a rate of approximately 1 kb per second until it reaches the first copy of the eight-nucleotide consensus sequence 5′-GCTGGTGG-3′. This is called the **chi site**. The sequence occurs on average once every 6 kb in *E. coli* DNA. Exactly what happens at a chi site is not fully understood, but the outcome is that the RecBCD enzyme produces a double-stranded molecule with a 3′ overhang, as is required for the initiation of recombination (Figure 14.18).

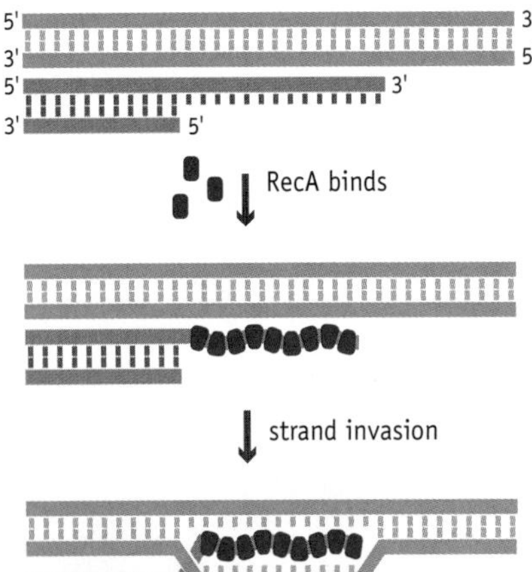

FIGURE 14.19 The role of the RecA protein in formation of the D loop during homologous recombination in *E. coli*.

The next step is establishment of the heteroduplex. This stage is mediated by the RecA protein, which forms a protein-coated DNA filament that is able to invade the intact double helix and set up the D loop (Figure 14.19). An intermediate in formation of the D loop is a **triplex** structure, a three-stranded DNA helix in which the invading polynucleotide lies within the major groove of the intact helix and forms hydrogen bonds with the base pairs it encounters.

Once the heteroduplex has been established, the subsequent events are probably common to all three recombination systems. If branch migration occurs, then it is catalyzed by the RuvA and RuvB proteins, both of which attach to a Holliday structure. Four copies of RuvA bind directly to the branch point, forming a core to which two RuvB rings, each consisting of eight proteins, attach, one to either side (Figure 14.20). The resulting structure might act as a "molecular motor," rotating the helices in the required manner so that the branch point moves.

Branch migration does not appear to be a random process. Instead it stops preferentially at the sequence 5′-(A/T)TT(G/C)-3′. In this sequence, (A/T) and (G/C) denote that either of the two nucleotides can be present at the

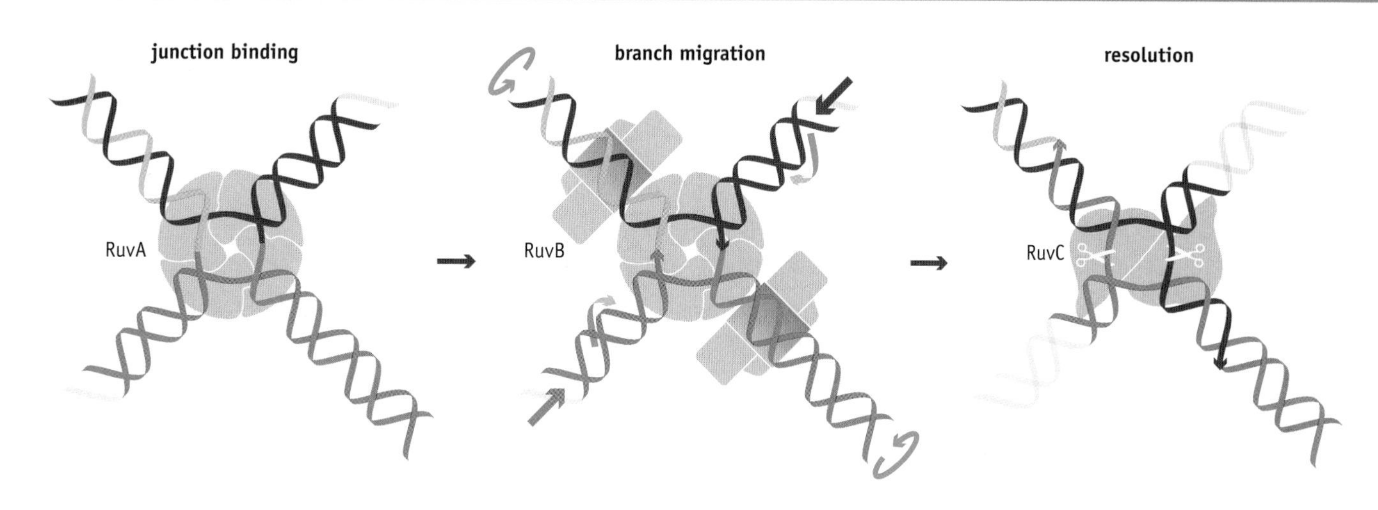

FIGURE 14.20 The role of the Ruv proteins in homologous recombination in *E. coli*. In the central drawing the orange and black arrows indicate the direction in which the double helices are being driven through the RuvAB complex, and the green arrows show how the strands rotate during branch migration.

positions indicated. The sequence occurs frequently in the *E. coli* genome, so presumably migration does not halt at the first instance of the motif that is reached. When branch migration has ended, the RuvAB complex detaches and is replaced by two RuvC proteins. RuvC is a nuclease and so can carry out the cleavage that resolves the Holliday structure. The cuts are made between the second T and the (G/C) components of the recognition sequence.

The biochemical basis of recombination in eukaryotes is less well understood

The biochemical events underlying recombination appear to be similar in all organisms, so in a general sense the *E. coli* RecBCD pathway describes the process by which recombination occurs in eukaryotes. It is the details that are tantalizingly different.

Studies of the yeast *Saccharomyces cerevisiae* have identified proteins with functions equivalent to those of the participants in the RecBCD pathway. In particular, two yeast proteins called RAD51 and DMC1 are the homologs of RecA of *E. coli*. Although RAD51 and DMC1 are thought to have their own individual roles, the two proteins are found together at the same locations within nuclei that are undergoing meiosis, and they are thought to work together in many homologous recombination events. Proteins homologous to RAD51 and DMC1 are also known in several other eukaryotes including humans.

One puzzling aspect of homologous recombination in eukaryotes is the mechanism by which Holliday structures are resolved. For many years, eukaryotic proteins similar to RuvC of *E. coli* were sought but not found. Other types of nuclease that can cut Holliday structures in the test tube were discovered, but the cleaved ends that they leave cannot be ligated to one another easily. It therefore seems unlikely that they have any role in recombination in living cells. During this search it was also realized that, even in bacteria, it is possible to resolve a Holliday structure without the use of a nuclease enzyme. A combination of helicases and topoisomerases can separate the heteroduplex back into independent chromosomes.

It began to look possible that there is no eukaryotic equivalent of RuvC, but recently the puzzle has been solved with isolation of the GEN1 protein from human cells. This protein cuts the Holliday structure in almost exactly the same way as RuvC, and, importantly, the cut DNA ends resulting from GEN1 cleavage can be joined together easily.

KEY CONCEPTS

- Meiosis is the type of cell division that gives rise to the sex cells, or gametes. One of the key features of meiosis is that the gametes are haploid, each containing just a single copy of the genome.
- Meiosis results in haploid cells because it is, in effect, two successive cell divisions, with no DNA replication occurring during either of them.
- During meiosis I, the two members of a pair of homologous chromosomes line up alongside one another to form a bivalent.
- The random segregation of chromosomes during anaphase I, when the bivalents separate, gives rise to a vast range of possible combinations in the resulting gametes.
- The variability of the gametes is increased even further by crossing over between homologous chromosomes within a bivalent.
- Recombination involves formation of a heteroduplex, a structure within which two double-stranded DNA molecules are linked together by a pair of crossovers.
- Resolution of a heteroduplex can lead to exchange of chromosome segments.
- The biochemistry of the recombination pathway in *E. coli* is well understood. The equivalent events in eukaryotes are less well studied, but proteins similar to those involved in bacterial recombination are known in yeast and humans.

QUESTIONS AND PROBLEMS (Answers can be found at www.garlandscience.com/introgenetics)

Key Terms

Write short definitions of the following terms:

bivalent
branch migration
chi form
chi site
chiasmata
crossing over
D loop
gamete
heteroduplex
Holliday structure
homologous chromosomes
meiosis
RecBCD enzyme
reciprocal strand exchange
recombination
resolution of a heteroduplex
triplex helix

Self-study Questions

14.1 Explain how meiosis results in haploid cells whereas mitosis gives rise to diploid cells.

14.2 Explain what is meant by the term "homologous chromosomes." How many pairs of homologous chromosomes are present in a diploid human cell?

14.3 Describe how bivalents are formed and how they separate during meiosis I.

14.4 How does microtubule attachment ensure the correct partitioning of chromosomes during meiosis?

14.5 Describe the link between formation of bivalents and the genetic differences displayed by siblings.

14.6 How many chromosome combinations can result from meiosis of a cell containing four pairs of homologous chromosomes?

14.7 How was crossing over between homologous chromosomes first discovered?

14.8 Describe the impact that crossing over has on the genetic variability of the gametes resulting from meiosis.

14.9 What is a heteroduplex, and how is it formed during recombination?

14.10 Draw a series of diagrams that illustrate the events occurring during homologous recombination, up to the stage when branch migration occurs.

14.11 What occurs as a result of branch migration during homologous recombination?

14.12 Draw a diagram showing the chi form of a Holliday structure.

14.13 Describe the different ways in which the chi form of a Holliday structure can be cleaved, and explain the effect that each of these cleavages has on the structures of the resulting DNA molecules.

14.14 What is the role of the RecBCD enzyme during homologous recombination in *E. coli*?

14.15 Explain how the triplex intermediate is formed during homologous recombination in *E. coli*.

14.16 Describe the molecular basis of branch migration in *E. coli*.

14.17 Describe the molecular basis of cleavage of the Holliday structure in *E. coli*.

14.18 What are the roles of the RAD51 and DMC1 proteins in yeast?

14.19 Explain why the molecular mechanism behind resolution of Holliday structures in eukaryotes was a puzzle, and describe how this puzzle has been solved.

Discussion Topics

14.20 What complications arise during meiosis of a male human cell that do not arise during meiosis of a female cell?

14.21 Crossing over between homologous chromosomes is the basis of methods used to map the positions of genes on eukaryotic chromosomes. How might crossing over be used in this way?

14.22 Some biologists believe that the main function of recombination is in DNA repair, the process by which mutations and other forms of DNA damage are corrected. Explain how recombination could be used to repair a damaged DNA molecule.

14.23 What would be the result of recombination between the *E. coli* chromosome and a small circular plasmid? Assume that the plasmid has a region where the nucleotide sequence is identical to part of the chromosome.

FURTHER READING

Chen Z, Yang H & Pavietich NP (2008) Mechanism of homologous recombination from the RecA-ssDNA/dsDNA structures. *Nature* 453, 489–494. *Details of how the triplex structure is set up by the RecA protein.*

Dillingham MS & Kowalczykowski SC (2008) RecBCD enzyme and the repair of double-stranded DNA breaks. *Microbiol. Mol. Biol. Rev.* 72, 642–671. *The mode of action of the RecBCD enzyme during homologous recombination and other events in which it is involved.*

Murray AW & Szostak JW (1985) Chromosome segregation in mitosis and meiosis. *Annu. Rev. Cell Biol.* 1, 289–315.

Nasmyth K (2001) Disseminating the genomes: joining, resolving and separating sister chromatids during mitosis and meiosis. *Annu. Rev. Genet.* 35, 673–745.

Pyle AM (2004) Big engine finds small breaks. *Nature* 432, 157–158. *The structure of the RecBCD complex.*

Rafferty JB, Sedelnikova SE, Hargreaves D et al. (1996) Crystal structure of DNA recombination protein RuvA and a model for its binding to the Holliday junction. *Science* 274, 415–421.

Rass U, Compton SA, Natos J et al. (2010) Mechanism of Holliday junction resolution by the human GEN1 protein. *Genes Dev.* 24, 1559–1569.

Roeder GS (1997) Meiotic chromosomes: it takes two to tango. *Genes Dev.* 11, 2600–2621. *Describes the formation of bivalents and exchange of DNA by recombination.*

Shinagawa H & Iwasaki H (1996) Processing the Holliday junction in homologous recombination. *Trends Biochem. Sci.* 21, 107–111.

Szostak JW, Orr-Weaver TK, Rothstein RJ & Stahl EW (1983) The double-strand-break repair model for recombination. *Cell* 33, 25–35. *The molecular basis of recombination.*

Inheritance of Genes During Eukaryotic Sexual Reproduction CHAPTER 15

From our study of meiosis we have gained a clear understanding of the behavior of chromosomes, and the DNA molecules that they contain, during sexual reproduction. In particular we have learned that formation of bivalents between homologous chromosomes is the key step that leads to genetic variability among the gametes resulting from meiosis. New allele combinations are generated by recombination within bivalents and by random segregation of homologous chromosomes when the bivalents break down.

Now that we have a firm grasp of how DNA molecules behave during meiosis, we can turn our attention to the second of the two angles from which we must view sexual reproduction. Now we must study the inheritance of genes rather than DNA molecules. This takes us back to the beginnings of genetics, as the questions that we will ask are the same ones that were studied by Gregor Mendel and the other early geneticists.

The first question that we must address is what happens when the new organism resulting from sexual reproduction inherits two different alleles of a particular gene. In Chapter 3 we learned that alleles specify different variants of a biological characteristic, an example being the pair of alleles that specify the round and wrinkled variants of pea shape. If an organism contains two different alleles for a particular gene, then which of the two variants of the biological characteristic does it display? This question was almost completely answered by Mendel, a few final embellishments being added by his successors in the early part of the twentieth century.

The second question concerns interactions between groups of genes. Many biological characteristics are specified by groups of genes working together. These include relatively simple characteristics such as the color of a flower as well as more complex ones such as susceptibility to certain diseases in humans. How do the allele combinations that are inherited for these groups of genes influence the biological characteristic specified by the genes as a whole? Mendel did not himself study this question, but the discoveries that he made laid the foundation for the extensive work on gene interactions that has taken place since his time.

The final question asks what effect the crossings over that occur between pairs of homologous chromosomes during meiosis have on inheritance of the genes that are located on a single chromosome. This is an important applied question, because measurement of the frequency at which crossovers occur between two genes is used as a way of identifying the positions of those genes on a chromosome. These gene mapping techniques were developed 100 years ago by Thomas Hunt Morgan, the second-most important geneticist after Mendel, and are still used today. Their applications include the identification of human genes involved in disease.

15.1 RELATIONSHIPS BETWEEN PAIRS OF ALLELES

If an organism inherits two different alleles of a gene, then which of the two variants of the biological information specified by that gene does it display? To begin to answer this question we will return to one of the biological characteristics studied by Mendel. This is pea shape, which we examined in Chapter 3 when we were exploring the role of genes as units of biological information.

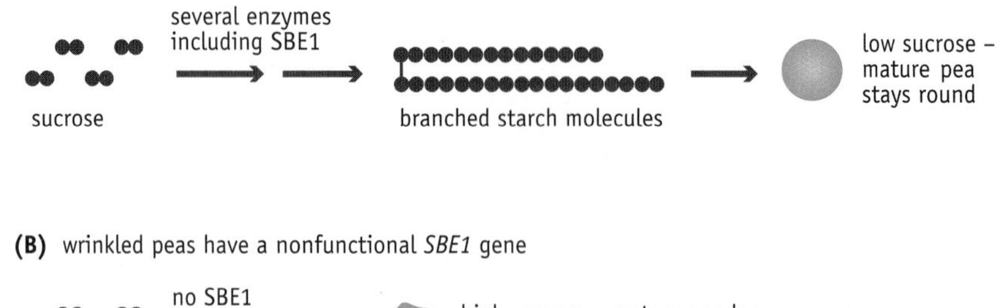

FIGURE 15.1 The basis of the round and wrinkled pea phenotypes. (A) Round peas have a functional *SBE1* gene, so convert sucrose into starch. This means that the pea stays round when it matures. (B) Wrinkled peas have a nonfunctional *SBE1* gene and convert less sucrose into starch. They lose water when they mature and become wrinkled.

The allele for round peas is dominant over the one for wrinkled peas

Pea shape illustrates the simplest type of relationship that exists between a pair of alleles. Remember that the difference between a round pea and a wrinkled one is that a round pea has a functional gene for the starch branching enzyme SBE1 and is therefore able to convert the sucrose it produces by photosynthesis into starch (Figure 15.1). The presence of starch reduces water uptake as the pea develops, so the pea stays round when it reaches maturity and dehydrates. In plants that produce wrinkled peas, the *SBE1* gene is nonfunctional because it contains an extra piece of DNA that disrupts the sequence of codons so the mRNA cannot be translated into protein. No SBE1 enzyme is made, and the peas have a high sucrose content, because less sucrose is converted into starch. These peas absorb much more water as they develop, and so dehydrate more when they reach maturity, causing the pea to collapse and become wrinkled.

The biological characteristic described as pea shape is therefore specified by the *SBE1* gene. This gene has two alleles. The first is the functional version that codes for round peas. We will use the same notation as Mendel and refer to this as the *R* allele. The second allele is the nonfunctional gene that gives rise to wrinkled peas, which we will call *r*.

Now consider a pea plant in which both copies of the *SBE1* gene are functional (Figure 15.2A). We would describe the **genotype** of this plant as *RR* and its **phenotype** as "round peas." "Genotype" refers to the genetic constitution—the combination of alleles that are present—and "phenotype" is the observable characteristic resulting from that combination of alleles. Because this plant has two identical alleles—*RR* in this case—we say that it is a **homozygote**.

A plant that has two nonfunctional copies of the *SBE1* gene is also a homozygote, but this time with the genotype *rr*. As no SBE1 enzyme is produced, the phenotype displayed by this plant is "wrinkled peas" (Figure 15.2B).

What if a plant has inherited an *R* allele from one parent and an *r* allele from another? Its genotype will be *Rr*, and the plant will be a **heterozygote**. But

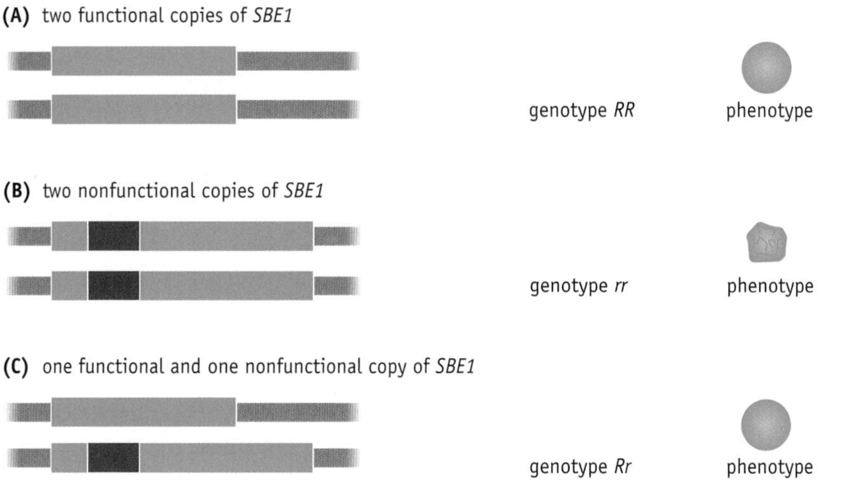

(A) two functional copies of *SBE1*

genotype *RR* phenotype

(B) two nonfunctional copies of *SBE1*

genotype *rr* phenotype

(C) one functional and one nonfunctional copy of *SBE1*

genotype *Rr* phenotype

FIGURE 15.2 The relationship between genotype and phenotype for round and wrinkled peas. The allele for round peas, *R*, is dominant to the allele for wrinkled peas, *r*. This means that peas with the genotypes *RR* and *Rr* are both round (parts A and C). Only peas with the genotype *rr* are wrinkled (part B).

it still has a functional copy of the *SBE1* gene and is able to synthesize the starch debranching enzyme. It can therefore convert sucrose into starch. Its peas will be round (Figure 15.2C). The terminology that we use is the same as that invented by Mendel. We say that the *R* allele, the one that determines the phenotype of the heterozygote, is **dominant**. The *r* allele, whose phenotypic effect is masked in the heterozygote, is **recessive**.

Some pairs of alleles display incomplete dominance

Round and wrinkled peas illustrate the most simple type of relationship between a pair of alleles, the relationship in which one is dominant and the other recessive and the dominant allele contributes the phenotype of the heterozygote. Often, as with pea shape, the dominant allele is the functional version of a gene and the recessive allele is a nonfunctional variant. But the relationship between the functional and nonfunctional alleles of a gene is not always so straightforward.

When the *SBE1* gene is heterozygous, the phenotype is round pea not just because the functional copy of the gene directs synthesis of the starch branching enzyme. The functional copy of the gene must be expressed at a rate sufficient to maintain the required levels of the SBE1 enzyme in the plant's cells. In the case of SBE1, this is what happens, and the pea stays nice and round.

What if the single functional copy of a gene in a heterozygote is *not* able to direct enough protein synthesis to maintain the required levels of the gene product in the cell? Now the phenotype of the heterozygote might not be the same as that of a homozygote with two functional copies of the gene. As an example we will consider synthesis of the flower pigment anthocyanin in carnations. Anthocyanin is a type of flavonoid that gives petals a deep red color. It is synthesized from a white precursor by the enzyme anthocyanin synthase, which is coded by the *ANS* gene. This gene has two alleles: *A*, corresponding to the functional copy of the gene, and *a*, the inactive version (Figure 15.3).

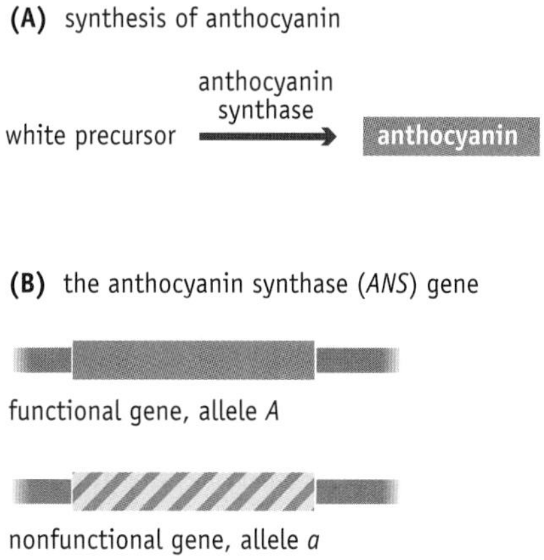

(A) synthesis of anthocyanin

white precursor —anthocyanin synthase→ anthocyanin

(B) the anthocyanin synthase (*ANS*) gene

functional gene, allele *A*

nonfunctional gene, allele *a*

FIGURE 15.3 The *ANS* gene of carnations. (A) The product of the *ANS* gene is anthocyanin synthase, which synthesizes anthocyanin from a white precursor. (B) The functional and nonfunctional alleles of the *ANS* gene.

RESEARCH BRIEFING 15.1

Mendel's discovery of the First Law of Genetics

The dominant–recessive relationship between alleles was first discovered by Gregor Mendel over 150 years ago. It is sometimes difficult to appreciate, in our modern molecular age, that Mendel and the geneticists who followed him were able to make quite detailed deductions about relationships between alleles simply by studying the inheritance of physical characteristics such as the shapes of peas. Mendel was not, of course, the first to recognize that children resemble their parents and that the same principle of inheritance applies throughout the natural world. But Mendel was the first person to show how inheritance could be studied in an informative, scientific manner. His work is quite rightly looked on as the first real experiments in genetics. We will therefore examine how he discovered the First Law of Genetics.

Mendel's experiments were carefully planned

Mendel's aim was to find a pattern to the way in which inherited characteristics are passed from parents to succeeding generations. In the mid-nineteenth century the fact that a child displays a mixture of characteristics from both its mother and father was ascribed to vague events such as "cytoplasmic mixing." The accepted wisdom was that inheritance could not be studied in a scientific or statistical manner. Mendel was not so sure.

Mendel decided to work with the garden pea, *Pisum sativum*. Pea plants are easy to grow, and their life cycle is relatively short, so several generations of plant can be obtained in a single growing season. They are therefore ideal experimental organisms for studying the way in which inherited characteristics are passed from parents to offspring.

The first step was to choose the characteristics to study. Mendel decided to work with characteristics that exist as alternative forms, with a plant able to display one version of the characteristic or another, but not both at once. Pea shape is a good example, because a pea can be round or wrinkled, but it cannot be both at the same time. Mendel also realized that he might need to type these characteristics in several hundred plants in order to be able to distinguish a pattern to inheritance, should one exist. He therefore chose characteristics that are easy to score simply by looking at the plants, rather than needing dissection or chemical analysis (Table 1).

Mendel's crosses revealed a regular pattern to the inheritance of characteristics

In his first set of experiments, Mendel allowed his pea plants to **self-fertilize**, the female carpels within the flowers being fertilized with pollen from the same plant. Mendel discovered that the progeny resulting from self-fertilization always displayed the characteristics of the parent plant. So, for example, self-fertilization of a plant with wrinkled peas gave rise to many progeny plants, but all of these had wrinkled peas. Nowadays we would say that the plants are **pure-breeding**.

Next, Mendel set up experiments in which plants were **cross-fertilized**, the pollen from one plant fertilizing the carpels of a second plant. One of these crosses was between plants with round peas and plants with wrinkled peas. All of the progeny—the plants we refer to as the F_1 generation—had round peas. These plants were then allowed to self-fertilize, giving rise to the F_2 generation. Now plants with wrinkled peas reappeared, the F_2 generation yielding a total of 5474 round peas and 1850 wrinkled peas. This is a ratio of 2.96 to 1. Mendel performed the same type of cross for each of his seven characteristics. In each cross only one of the versions of the characteristic was displayed in the F_1 generation, but both appeared in the F_2 generation, always with a ratio of approximately 3 to 1 (Table 2).

Why do all the F_1 plants have the same phenotype?

To explain the consistent pattern of inheritance that he had discovered, Mendel proposed that each

TABLE 1 THE CHARACTERISTICS STUDIED BY MENDEL

Characteristic	Alternative forms
Pea color	Yellow, green
Stem height	Tall, short
Pod location	Axial, terminal
Pod morphology	Full, constricted
Pod color	Green, yellow
Pea morphology	Round, wrinkled
Flower color	Violet, white

TABLE 2 THE RESULTS OF MENDEL'S CROSSES

Cross	F_1 generation	F_2 generation	
		Numbers	Ratio
Round × wrinkled peas	All round	5474 round, 1850 wrinkled	2.96 : 1
Tall × short stems	All tall	787 tall, 277 short	2.84 : 1
Axial × terminal pods	All axial	651 axial, 207 terminal	3.15 : 1
Full × constricted pods	All full	882 full, 299 constricted	2.95 : 1
Green × yellow pods	All green	428 green, 152 yellow	2.82 : 1
Yellow × green peas	All yellow	6022 yellow, 2001 green	3.01 : 1
Violet × white flowers	All violet	705 violet, 224 white	3.15 : 1

characteristic is controlled by a **unit factor**, and that each unit factor can exist in more than one form. These unit factors are exactly the same things that we now call genes, the two alternative forms being the alleles. Mendel was even so perceptive as to imagine that his unit factors are physical structures present inside cells—a century before genes were shown to be made of DNA.

Mendel realized that the most important aspect of his crosses was that one of the parental characteristics disappeared in the F_1 generation and then reappeared in the F_2 generation. In the crosses between plants with round peas and plants with wrinkled peas, for example, the wrinkled characteristic was absent in the F_1 generation, but present again in the F_2 plants. This means that the F_1 plants must contain two alleles, one for round peas and one for wrinkled peas. If this were not the case then how could both alleles be transmitted to the F_2 plants? This led Mendel to conclude that each gene exists as a pair of alleles, and that the genotype of the F_1 plants could be written as *Rr*. These plants are therefore heterozygotes, and *R* is dominant to *r*. Remember that this was a long time before it was shown that genes lie on chromosomes and that a diploid cell has two copies of each chromosome. Mendel deduced all this simply by examining the results of the crosses listed in Table 2.

Why is there a 3:1 ratio in the F_2 generation?

Having established how one of the parental characteristics could disappear in the F_1 generation and reappear in the F_2 plants, Mendel went on to explain how the 3:1 ratios arise in the F_2 generation. Given that the F_1 plants have the genotype *Rr*, the F_1 cross can be written as:

$$Rr \times Rr$$

This notation is acceptable even though the F_1 cross is a self-fertilization, meaning that in reality it does not involve two individual plants, but instead different parts of the same plant.

We know that the F_1 cross produces some plants with wrinkled peas, whose genotypes must be *rr*. These F_2 plants must have obtained an *r* allele from each F_1 "parent." There is no other way in which they could arise. This suggests that during sexual reproduction the alleles of each parent separate (we use the term **segregate**), producing intermediate structures that contain just one allele. These intermediate structures are the gametes that were discovered, some 30 years later, by the first cell biologists to study meiosis. Two gametes, one from each parent, fuse to bring together the pair of alleles carried by a member of the next generation.

How do these events result in the 3:1 ratio? The useful construction called the **Punnett square** (Figure 1) helps us to see that the 3:1 ratio arises naturally, so long as the F_1 gametes are able to segregate in an entirely random fashion. If segregation is random, then the genotypes displayed by the F_2 plants will be 1*RR* : 2*Rr* : 1*rr*. As both *RR* and *Rr* plants have round peas, there will be three round peas to every wrinkled pea, exactly as Mendel observed. The 3:1 ratio therefore led Mendel to his First Law of Genetics, which states that *alleles segregate randomly*.

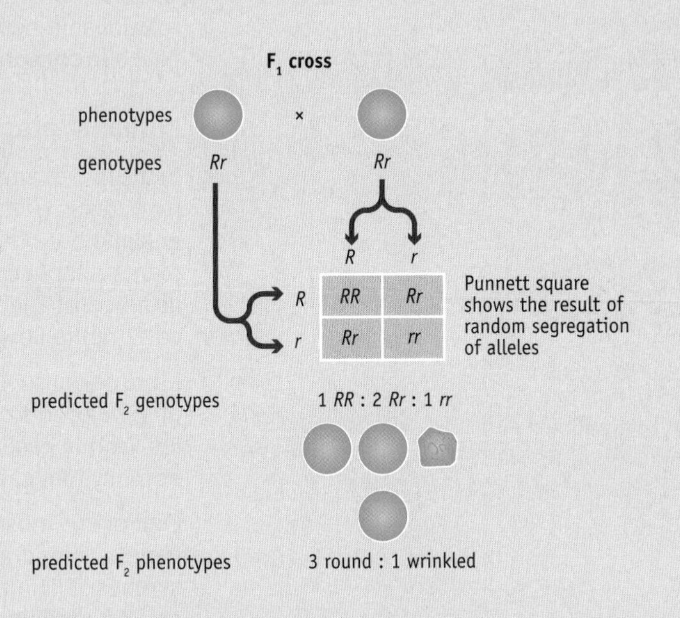

FIGURE 1 The explanation of the 3:1 ratio in the F_2 generation.

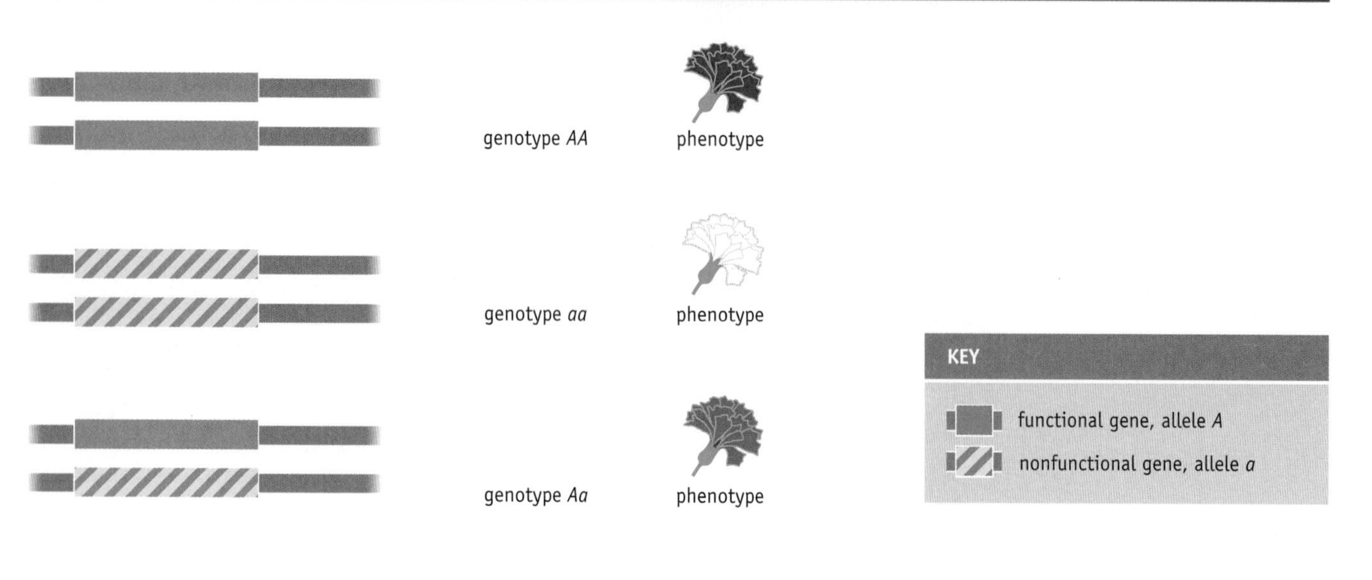

FIGURE 15.4 Incomplete dominance as displayed by flower color in carnations. The phenotype of the heterozygote is intermediate between those of the two homozygotes.

A plant that is homozygous for allele *A*, and hence has the genotype *AA*, is able to make anthocyanin synthase and so produces lots of anthocyanin and has red flowers (Figure 15.4). The other homozygote has the genotype *aa* and makes no anthocyanin synthase and hence no pigment. Its flowers are white.

If the *A* and *a* alleles had a simple dominant–recessive relationship, like that displayed by the *SBE1* gene, then a heterozygous plant, *Aa*, would have red flowers. But in this case the 50% reduction in overall gene expression that occurs in the heterozygote—because there is only one functional copy of the gene—is insufficient to maintain the amount of anthocyanin synthase in the cells at the required level. Only a limited amount of anthocyanin is made, and the flowers are a rather attractive pink color. This type of relationship between the functional and nonfunctional alleles of a gene is called **incomplete dominance**.

Lethal alleles result in death of a homozygote

In the two examples that we have studied so far, a homozygote that contains two copies of the nonfunctional allele is unable to synthesize a particular enzyme but otherwise is healthy. The absence of the SBE1 enzyme in pea plants simply means that the peas are wrinkled rather than round, and absence of the pigment synthesis enzyme in carnations affects only the color of the flowers.

More serious consequences arise if the organism cannot tolerate the loss of gene product that occurs in a homozygote with two nonfunctional alleles. In this case the nonfunctional allele is said to be **lethal**, because the homozygote cannot survive.

One of the most interesting examples of a lethal allele concerns the *aurea* gene of snapdragon. This gene codes for a protein involved in chlorophyll synthesis. The plant must make chlorophyll in order to photosynthesize, and if it cannot photosynthesize then it cannot survive. All is well in a plant that has two functional copies of the *aurea* gene. Its genotype is usually

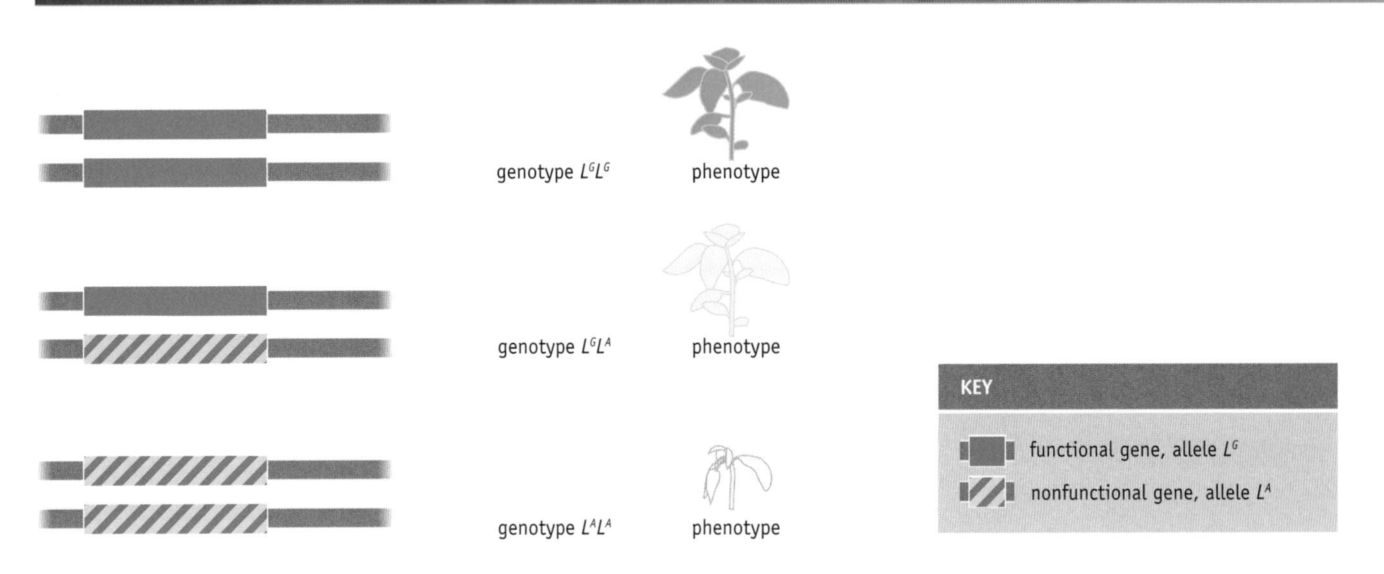

FIGURE 15.5 The L^A allele of the snapdragon *aurea* gene is lethal. The homozygote L^AL^A cannot survive. The heterozygote L^GL^A produces enough chlorophyll to survive but its leaves are pale yellow.

denoted L^GL^G. It has two functional alleles and so makes plenty of chlorophyll, appearing green in color and able to grow healthily (Figure 15.5).

In the heterozygote, genotype L^GL^A, the single active L^G allele is able to direct synthesis of some chlorophyll, enough for the plant to survive even though its leaves are pale yellow rather than green. The name *aurea* in fact refers to this "golden" variety. But a homozygote with two inactive copies of the gene is unable to synthesize chlorophyll and cannot photosynthesize. Homozygous L^AL^A seeds are able to germinate, but the seedlings are pure white and do not survive for long. So L^A and L^G display incomplete dominance and L^A is a lethal allele.

The *aurea* gene provides an example of a lethal allele which, in the homozygous form, allows the organism to begin to develop, but prevents it from reaching maturity. Other lethal alleles have more drastic effects and prevent a fertilized egg cell from surviving, or possibly even prevent fertilization from occurring at all. From our studies of gene expression we can think of many examples of such aggressively lethal alleles. An allele specifying a nonfunctional version of RNA polymerase, or of a ribosomal protein or one of the enzymes required for DNA replication, would clearly be lethal in the homozygous condition.

Some alleles are codominant

The final type of allele relationship that we will study is **codominance**. To display codominance, both members of a pair of alleles must be functional. Codominance then occurs when both alleles are active in the heterozygous state.

Normally, to observe codominance it is necessary to examine the biochemistry of the organism. This is the case with one of the best-understood examples, the MN blood group series in humans. A person's blood group depends on the identity of certain protein molecules in the blood. The M and N proteins are coded by a pair of alleles designated L^M and L^N (Figure 15.6). A person who has the homozygous L^ML^M genotype is able to make

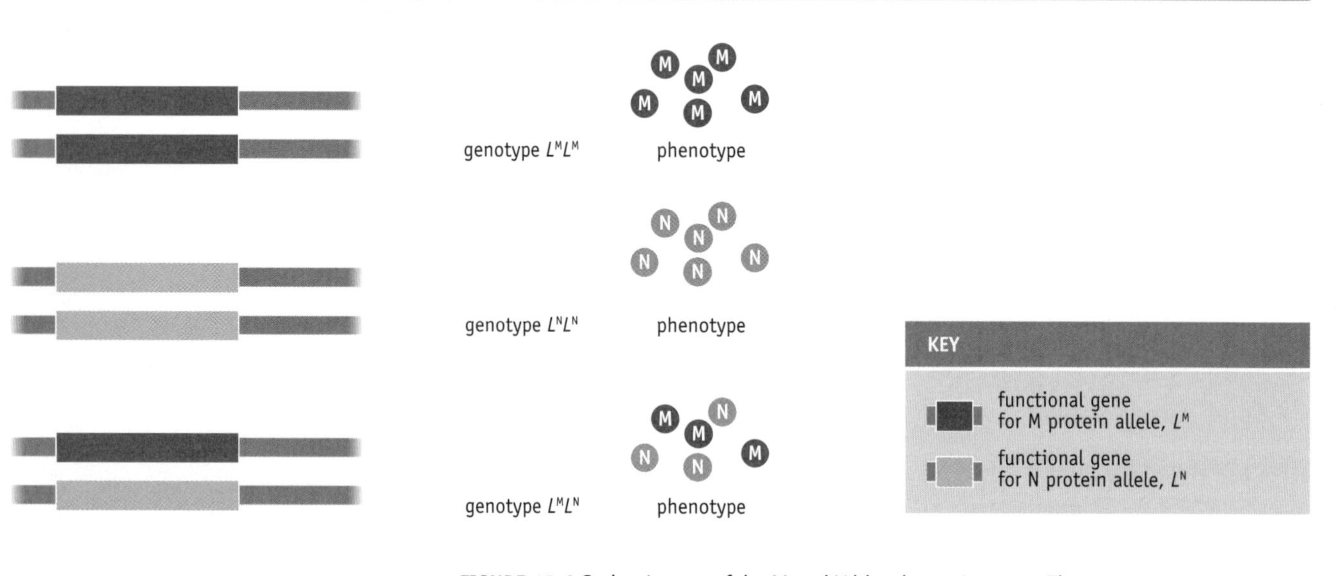

FIGURE 15.6 Codominance of the M and N blood protein genes. The heterozygote L^ML^N is able to make both the M and N proteins and so has blood type MN.

just M proteins, and so has blood group M. Similarly, the homozygous L^NL^N genotype produces N proteins and gives rise to blood group N. However, in the heterozygote, both alleles are present (L^ML^N) and both still direct synthesis of proteins. The blood type is MN. Neither allele is dominant.

Codominance is displayed by many allele pairs in which both members are functional versions of a gene. But codominance is by no means the only type of relationship that a pair of functional alleles can have. Two functional alleles can also form a dominant–recessive pair, and although quite unusual, they might also exhibit incomplete dominance. An important point to note is that identification of the relationship displayed by any pair of alleles is often the starting point for understanding the molecular basis of those alleles, for discovering, for example, whether one is a nonfunctional version of the gene or whether both are active but in different ways. In Chapter 20 we will see how understanding such relationships is important in an applied setting, in the study of human genetic disease.

15.2 INTERACTIONS BETWEEN ALLELES OF DIFFERENT GENES

Although many phenotypes can be linked to the activity of just a single gene, there are numerous others that result from interactions between two or more separate genes. Understanding these interactions is crucial to our awareness of how the genome as a whole specifies the biological characteristics of an organism.

The question that we must ask is how the more complex characteristics that are controlled by groups of genes are influenced by the presence of different alleles of those genes (Figure 15.7). We will begin by looking at relatively simple interactions that involve just two genes working together, and then examine the greater complexity presented by **quantitative traits**, which are specified by the combined activities of large numbers of genes.

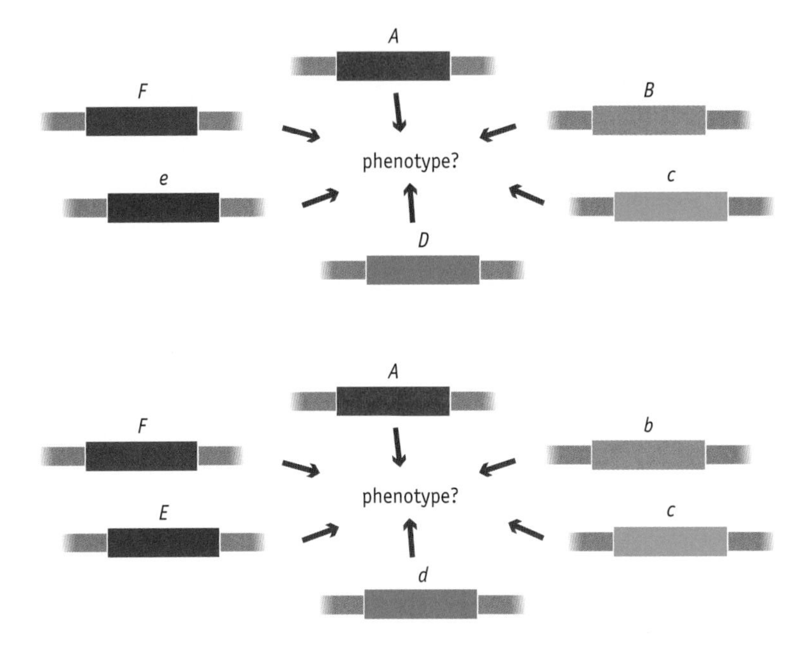

FIGURE 15.7 How are complex phenotypes that are specified by more than one gene affected by the presence of different alleles of those genes? In this example, the phenotype is specified by six genes, A–F, each of which exists as two alleles. Two of the possible $2^6 = 64$ allele combinations are shown. Each combination could result in a slightly different phenotype.

Functional alleles of interacting genes can have additive effects

Flower pigmentation illustrates various ways in which different genes can interact to produce a single phenotype. Some plant species are able to synthesize several different flower pigments of the same or similar colors. An example is the daylily of the genus *Hemerocallis* (Figure 15.8), which makes two types of anthocyanin pigment, called cyanidin and pelargonidin. Both are red in color. Cyanidin synthase is specified by gene C and pelargonidin synthase by gene P. Each gene has a functional allele, *C* or *P*, and a nonfunctional version, *c* or *p*. Each pair of alleles—*C* and *c*, and *P* and *p*—displays incomplete dominance. You will realize that these allele relationships are the same as those displayed by the anthocyanin synthase gene of carnation that we studied earlier.

FIGURE 15.8 A daylily of the genus *Hemerocallis*.

Armed with our knowledge of the phenotypes resulting from different combinations of pigment synthesis alleles in carnations, we can begin to deduce the effect that the two interacting genes, C and P, will have on flower color in the daylilies. A double homozygote with two copies of each functional allele, genotype *CCPP*, will synthesize cyanidin and pelargonidin at their maximum levels and have deep red flowers (Figure 15.9A). At the other end of the scale, a second double homozygote *ccpp* will make neither pigment and have pure white flowers.

So far the outcome is the same as if there were just one pigment synthesis gene. The differences arise when we consider the other possibilities (Figure 15.9B). Besides the double homozygotes, we can have a variety of other genotypes when the two genes are considered together. These are *CcPP*, *CcPp*, *Ccpp*, *CCPp*, *CCpp*, *ccPP*, and *ccPp*. What will be the phenotype of plants with each of these genotypes?

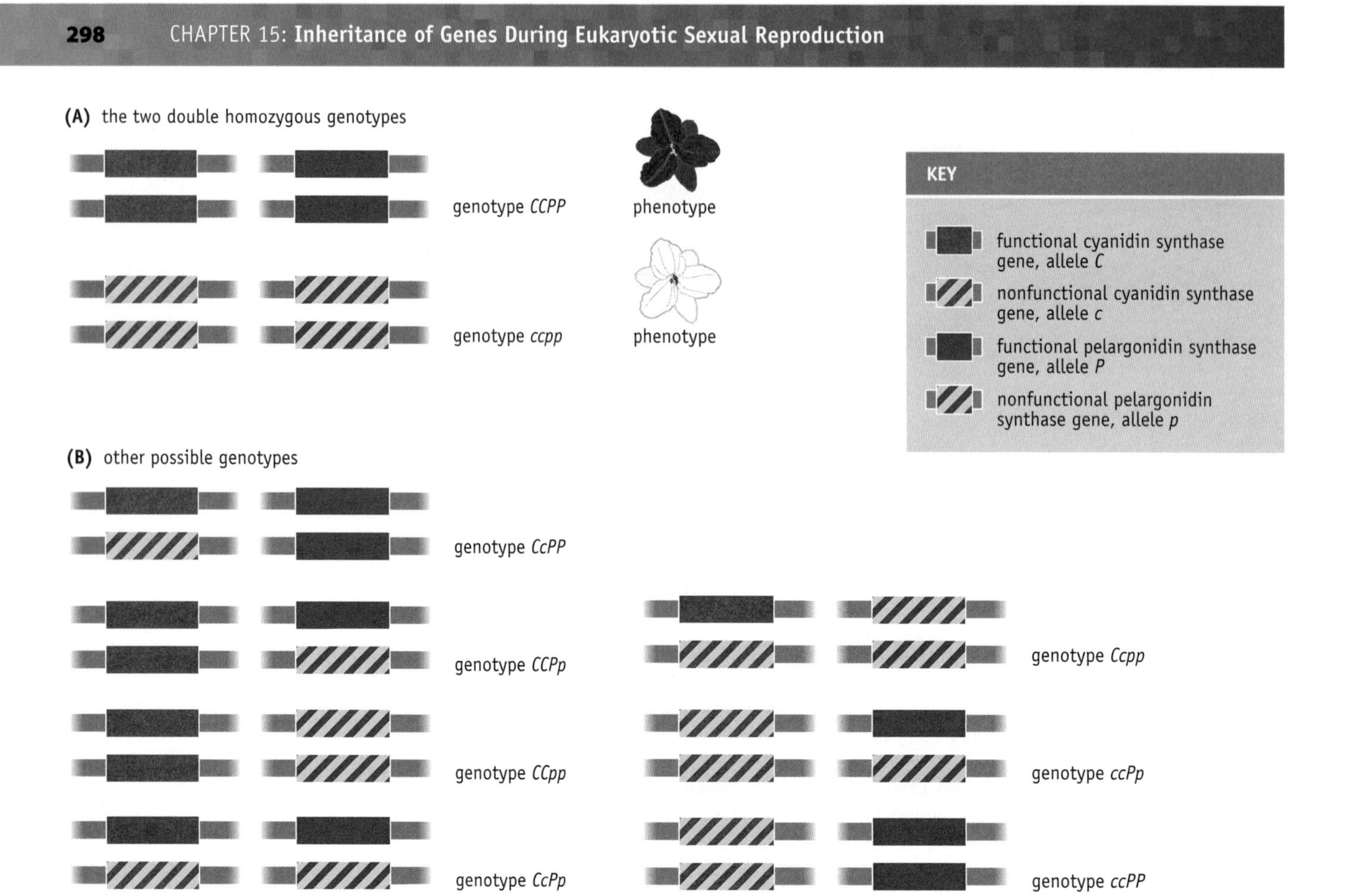

FIGURE 15.9 The genetic basis of flower color in daylilies. (A) The phenotypes of the two double homozygotes can be predicted. (B) There are seven additional genotypes in which one or both genes are heterozygous.

The easiest way to address this problem is to look on each functional allele copy as contributing a unit of pigment (Table 15.1). According to this interpretation, the double homozygote *CCPP* makes 4 units of pigments, resulting in a deep red flower. The double homozygote *ccpp* makes 0 units of pigments and so the flower is white. In between, we have genotypes that have 1, 2, or 3 functional alleles, contributing 1, 2, or 3 units of pigments. This means there will be three intermediate shades of flower color between red and white. These would be a light pink, pink, and light red, though they would probably be called something more adventurous by the company selling daylilies, perhaps seashell, baby pink, and salmon.

Table 15.1 makes two important points. The first is that, as we have seen, alleles of different genes can combine in an additive fashion to determine the nature of a phenotype. The second point is that the number of genotypes is not the same for each of the five possible phenotypes. There is only one way of making "red" and one way of making "white," but there are

TABLE 15.1 PIGMENT PRODUCTION BY TWO INTERACTING SYNTHESIS GENES

Number of pigment units	Genotypes	Phenotype
4	*CCPP*	Red
3	*CCPp, CcPP*	Light red
2	*CCpp, ccPP, CcPp*	Pink
1	*Ccpp, ccPp*	Light pink
0	*ccpp*	White

two possible genotypes each for "light pink" and "light red," and three for "pink." In a large population of the plants, we would therefore expect relatively few to have red or white flowers, and we would expect pink to be the most common color. We would go further and predict that the numbers of each flower color would form a **normal**, or **Gaussian**, **distribution**, the distribution commonly called a bell-shaped curve (Figure 15.10). This is an important observation that we will return to when we study quantitative traits specified by multiple genes.

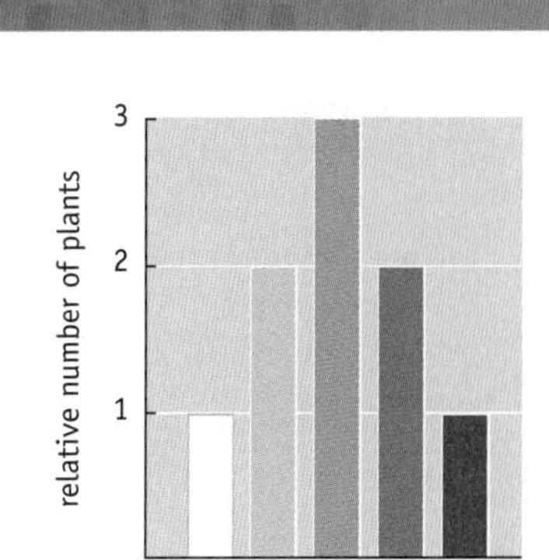

FIGURE 15.10 The expected distribution of flower colors in a large population of daylilies.

Important interactions occur between genes controlling different steps in a biochemical pathway

The example of flower color that we have just considered involves two genes that are responsible for synthesis of different types of red pigment. Their functional alleles therefore have an additive effect on the phenotype—on the color of the flowers that are produced.

Now we will consider the type of interaction that would be displayed by a pair of genes that work together in controlling synthesis of a single flower pigment. To start with, we will use a hypothetical example where the pigment results from a two-step biochemical pathway that converts a precursor compound into an intermediate and then into the pigment (Figure 15.11). Genes X and Y code for the enzymes that catalyze the two steps of this pathway, and as usual, alleles *X* and *Y* are functional, and *x* and *y* are nonfunctional.

When two genes work together in a single pathway, at least one functional allele of each gene must be present in order for the product to be synthesized. In our example, a functional allele of gene X is needed for the precursor to be converted into the intermediate compound, and a functional allele of gene Y is needed to convert the intermediate into the pigment. This means that there are four genotypes that give rise to pigment synthesis: *XXYY*, *XxYY*, *XXYy*, and *XxYy*. Each of the other five possible genotypes—*xxYY*, *xxYy*, *XXyy*, *Xxyy*, *xxyy*—lacks a functional allele of either gene X or gene Y and so cannot make the pigment.

This type of relationship is common in biology. It holds for all cases where two or more genes specify enzymes responsible for different steps in a biochemical pathway. We have used pigment synthesis as our example, but it could equally well be a pathway for synthesis of an amino acid or an essential vitamin. The relationship is sometimes called **complementary gene action**, as a particular combination of alleles of two separate genes is needed in order to produce the phenotype.

Epistasis is an interaction in which alleles of one gene mask the effect of a second gene

Although often referred to as complementary gene action, it is more accurate to describe the relationship we have just studied between genes X and Y as **epistasis**. Epistasis is where certain alleles of one gene mask or cancel the effects of alleles of a second gene. In the example shown in Figure 15.11, the *xx* genotype masks the genotype for gene Y. It makes no

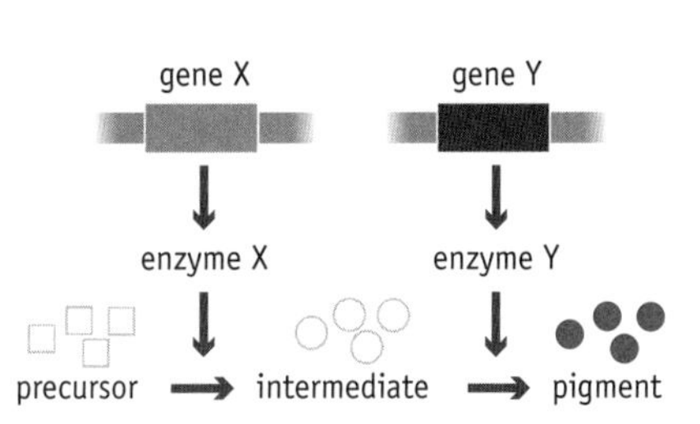

FIGURE 15.11 A pair of genes coding for enzymes involved in different steps in a biochemical pathway leading to synthesis of a pigment. The gene X product converts the colorless precursor compound into a colorless intermediate. The product of gene Y converts this intermediate into the pigment.

difference whether this genotype is *YY*, *Yy*, or *yy*. The same phenotype—no pigment synthesis—always occurs. Similarly, *yy* masks the genotype for gene X. The two genes are epistatic toward one another.

In the example that we have just considered, the masking effect is conferred by a recessive, nonfunctional allele of the epistatic gene. It is also possible for a dominant, functional allele to be epistatic toward a second gene. Continuing our plant pigment theme, an example is provided by fruit color in squash. The orange color of the mesocarp, or fleshy part, of squashes is due to the presence of carotenoid pigments (Figure 15.12), which also increase the nutritional value of the fruit. The carotenoid pigment is synthesized from a green precursor by an enzyme coded by gene G (Figure 15.13). The functional allele *G* displays dominance over the nonfunctional allele *g*, so genotypes *GG* and *Gg* give orange fruits, and *gg* gives green ones.

FIGURE 15.12 A butternut squash showing the orange fleshy mesocarp. (Courtesy of Andy Roberts, UK.)

The epistatic gene is called W. When this gene is present as one or more of its dominant alleles, *WW* or *Ww*, the fruit is white. The biochemical explanation for this example of epistasis is that the *W* allele codes for a protein that inhibits synthesis of the green precursor. As the green compound is not made, the mesocarp is nonpigmented and appears white. The genotype of gene G is immaterial because there is no green precursor for its enzyme to convert into the carotenoid pigment. Gene W therefore displays dominant epistasis toward gene G.

Interactions between multiple genes result in quantitative traits

So far we have considered only the ways in which pairs of genes can interact. As there are 20,500 genes in the human genome we might anticipate that a simple pairwise interaction would be a relatively uncommon phenomenon, with multiple gene interactions being much more prevalent. We must therefore examine situations where many genes contribute to a single phenotype.

An indication of the effect that multiple genes might have was provided by the additive effects of the daylily pigmentation genes that we considered above. We recognized five different levels of pigmentation from white to red, these different degrees of color resulting from interaction between just two genes. The five levels of pigmentation are expected to occur at different frequencies in a random population of plants, the distribution as a whole forming a bell curve (see Figure 15.10).

The bell curve is a characteristic description of phenotypes resulting from interactions between multiple genes. This is true whether there are three, four, or many hundred interacting genes, the only difference being that the curve becomes smoother as more genes are added (Figure 15.14). A continuous Gaussian distribution pattern is precisely what is found if a quantitative trait such as height, which is thought to be specified by many genes, is examined in the human population (Figure 15.15). Many quantitative traits, including height, are also affected by environmental factors such as diet, complicating attempts to work out their genetic basis.

Today, quantitative traits are extensively studied because of their applied importance. For example, the productivity of most crops and farm animals is determined by quantitative traits such as seed size and the meat-to-weight

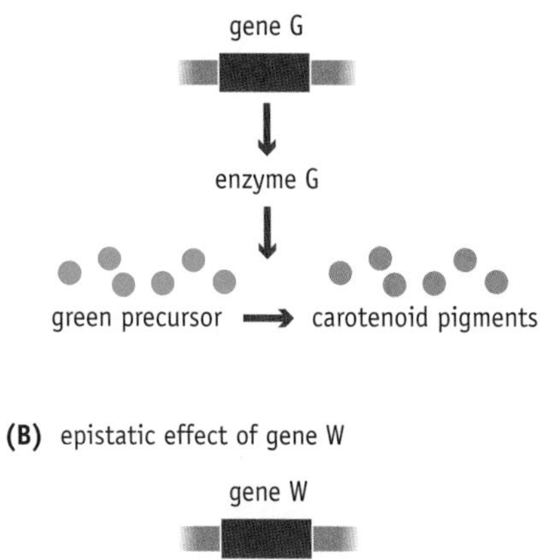

FIGURE 15.13 Carotenoid pigment synthesis in squash.
(A) Synthesis of carotenoid pigments by the product of gene G.
(B) The epistatic effect of gene W.

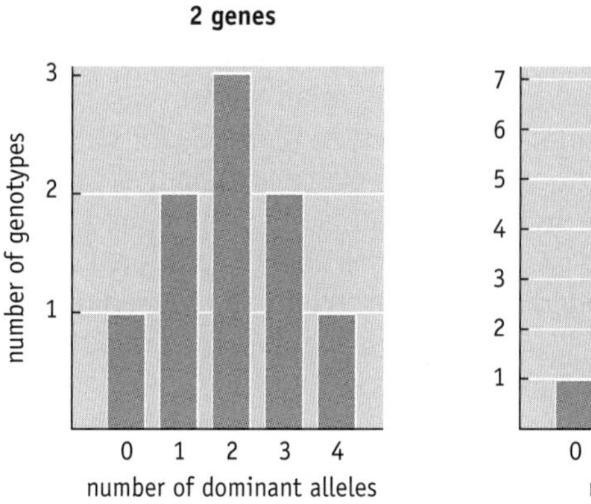

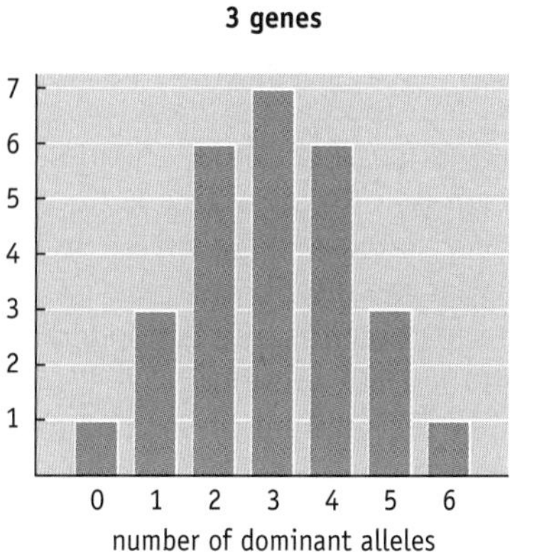

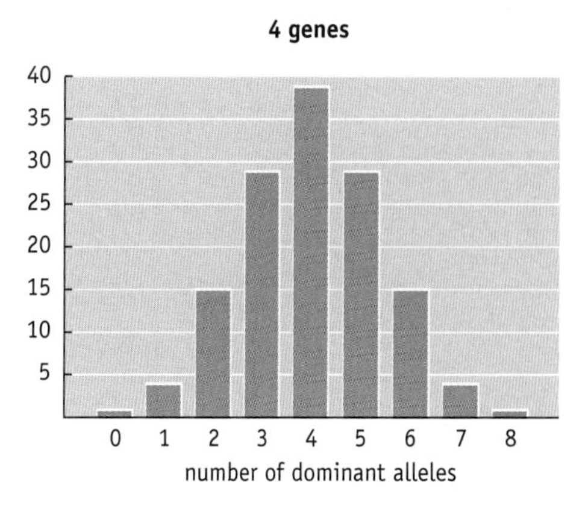

FIGURE 15.14 The predicted distribution of phenotypes for two, three, and four genes acting in an additive fashion and displaying incomplete dominance.

ratio. In medicine, quantitative traits are important because susceptibility to certain diseases depends on interactions between multiple genes. Susceptibility therefore displays a Gaussian distribution in the population as a whole. Typically, those individuals falling at one extreme of the distribution are likely to develop the disease (Figure 15.16).

15.3 INHERITANCE OF GENES LOCATED ON THE SAME CHROMOSOME

The final aspect of sexual reproduction that we must study is the way in which genes that are located on a single chromosome are inherited. Of course, those genes are always inherited together, but the allele combination present on a chromosome might change during meiosis, because of the exchange of segments between homologous chromosomes that occurs as a result of recombination.

When viewed solely from the point of view of allele inheritance during sexual reproduction, the effects of recombination are fairly trivial. Some alleles may be exchanged for other ones, and this may affect the phenotype of the organism resulting from sexual reproduction. We have already covered all of the important points in the earlier parts of this chapter. The reason

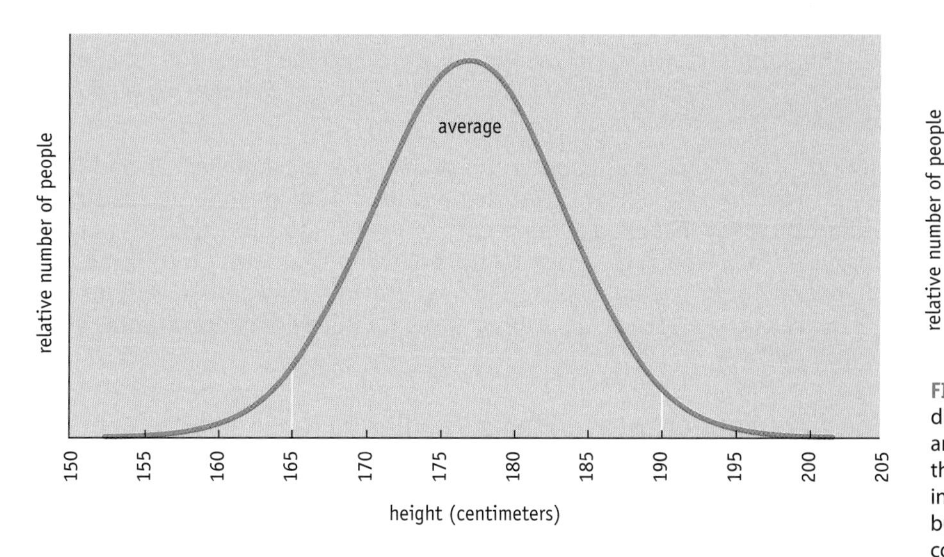

FIGURE 15.15 The distribution pattern for height among male adults from the United States.

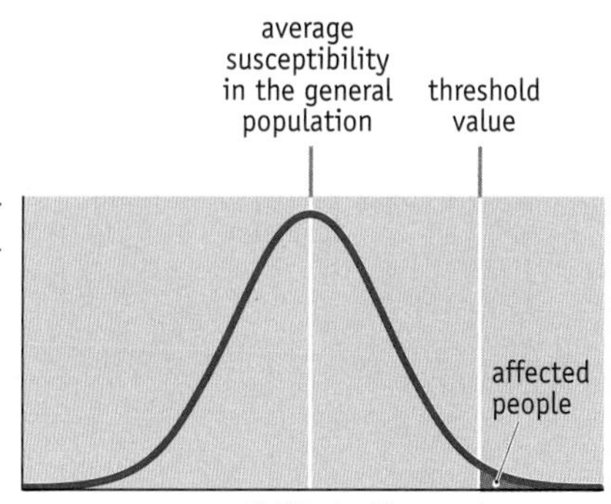

FIGURE 15.16 Susceptibility to certain diseases is a quantitative trait in humans, and displays a Gaussian distribution when the population as a whole is considered. All individuals will display some susceptibility, but only those whose particular combination of alleles places them to the right of the threshold value are likely to develop the disease.

why we must study the effect of recombination on allele inheritance is that recombination forms the basis for the method used to map the positions of genes on chromosomes. Even today, when it is relatively easy to obtain the DNA sequence of an entire chromosome, we need methods for working out the positions of genes. This is because it can be very difficult to locate a gene simply by examining a DNA sequence, especially if the gene is split into exons that are distributed over many kb. Some idea of its approximate position in the chromosome is needed so the search can be focused on the right part of the DNA sequence. We must therefore spend some time understanding how recombination is used to map genes.

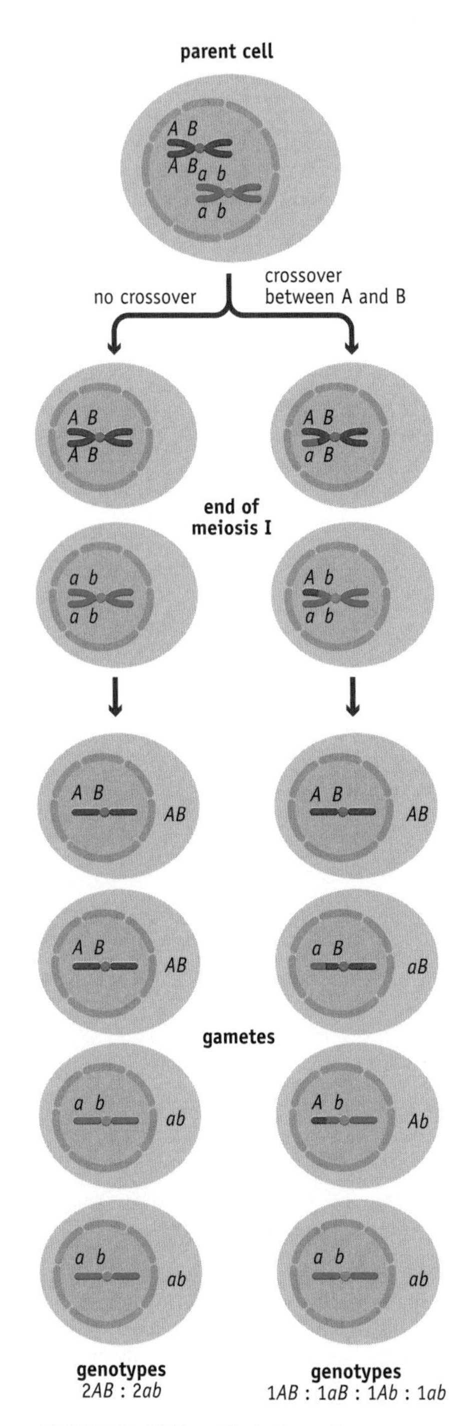

FIGURE 15.17 The effect of crossing over on the genotypes of the gametes resulting from a meiosis.

Crossing over results in gametes with recombinant genotypes

How can recombination be used to map the positions of genes on a chromosome? To answer this question we need to think about the effect that crossing over can have on the inheritance of genes.

Let us consider two genes, each of which has two alleles. We will call the first gene A and its alleles *A* and *a*, and the second gene B with alleles *B* and *b*. Imagine that the two genes are located on chromosome number 1 of the fruit fly *Drosophila melanogaster*. We will use fruit flies in this example because the techniques used to map genes were first applied to them, by Thomas Hunt Morgan and his students way back in the 1910s.

We are going to follow the meiosis of a diploid nucleus in which one copy of chromosome 1 has alleles *A* and *B* and the second has *a* and *b*. This situation is illustrated in Figure 15.17. There are two different scenarios. In the first, a crossover does not occur between genes A and B. If this is what happens, then two of the resulting gametes will contain copies of chromosome 1 with alleles *A* and *B* and the other two will contain copies of this chromosome with alleles *a* and *b*. In other words, two of the gametes will have the genotype *AB* and two will have the genotype *ab*.

The second scenario is that a crossover does occur between genes A and B. In this instance, segments of DNA containing gene A are exchanged between homologous chromosomes. The eventual result is that each gamete has a different genotype. The proportion will therefore be 1*AB* : 1*aB* : 1*Ab* : 1*ab*.

Now think about what would happen if we looked at the results of meiosis in 100 identical cells (Figure 15.18A). If crossovers never occur, 200 of the resulting gametes will have the genotype *AB* and the other 200, the genotype *ab*. This is called complete **linkage**, the two genes behaving as if they were a single unit during meiosis.

Now consider what will happen if crossovers occur between A and B in some of the meioses. Now the allele pairs are not being inherited as single units. Let us say that crossovers occur during 40 of the 100 meioses (Figure 15.18B). This would give rise to the following gametes: 160*AB*, 160*ab*, 40*Ab*, and 40*aB*. The linkage is not complete, it is only partial. Besides the two **parental** genotypes (*AB*, *ab*) we see gametes with **recombinant** genotypes (*Ab*, *aB*).

The frequency of recombinants enables the map positions of genes to be worked out

Now that we appreciate the effect that crossing over has on the genotypes of the gametes resulting from meiosis, we only have to take a small step in order to understand how the positions of genes are mapped on a chromosome.

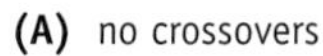
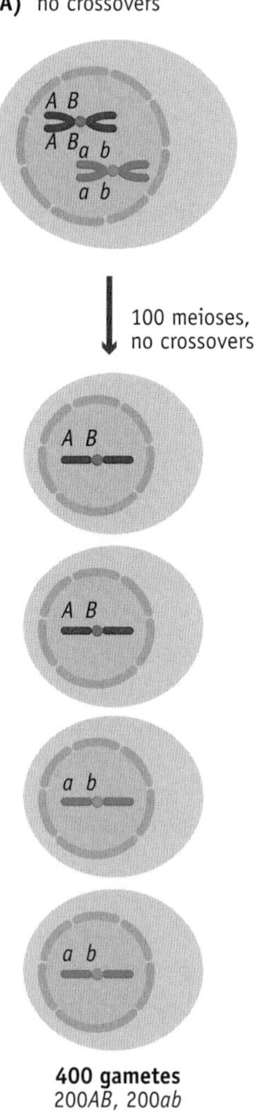
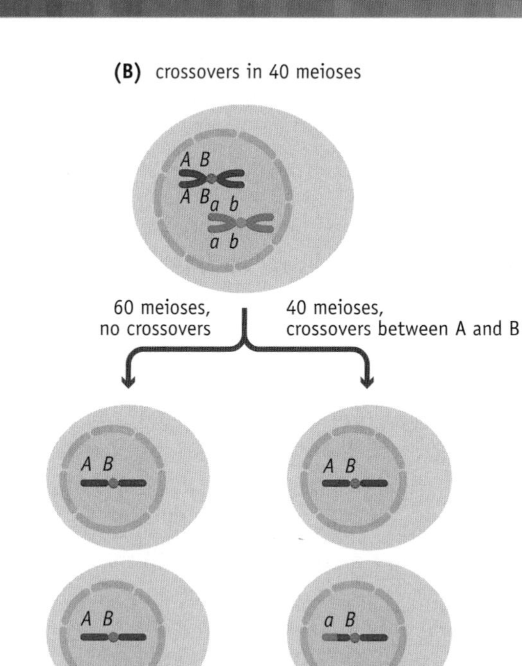

FIGURE 15.18 Linkage. The gametes resulting from 100 meioses are examined. (A) Genes A and B display complete linkage, as no crossovers have occurred between them. (B) The genes display partial linkage, as crossovers have occurred between them during some of the meioses.

Let us assume that crossing over is a random event, there being an equal chance of it occurring at any position along a pair of lined-up chromosomes. If this assumption is correct then two genes that are close together will be separated by crossovers less frequently than two genes that are more distant from one another. Furthermore, the frequency with which the genes are unlinked by crossovers will be directly proportional to how far apart they are on their chromosome. The **recombination frequency** is therefore a measure of the distance between two genes. If you work out the recombination frequencies for different pairs of genes, you can construct a map of their relative positions on the chromosome (Figure 15.19).

The very first genetic map was constructed for chromosome 1 of the fruit fly by Arthur Sturtevant, an undergraduate who was working in Morgan's laboratory in 1913. Two of the genes that he studied specified eye color, gene V for vermilion eyes and gene W for eyes that were white. A third, gene M, gave rise to miniature wings, and the fourth, Y, gave a yellow body. Sturtevant worked out the recombination frequencies between each pair of genes, obtaining values between 1.3% and 33.7%. From the data he was able to draw the map shown in Figure 15.20.

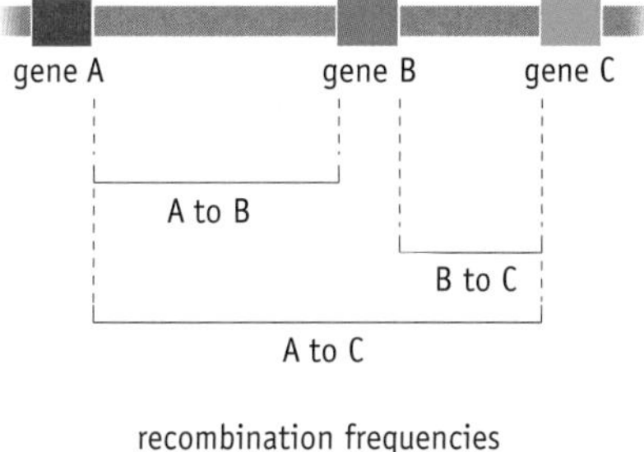

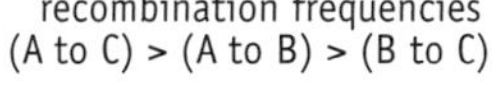

FIGURE 15.19 The principle behind the use of recombination frequencies to map the positions of genes on a chromosome.

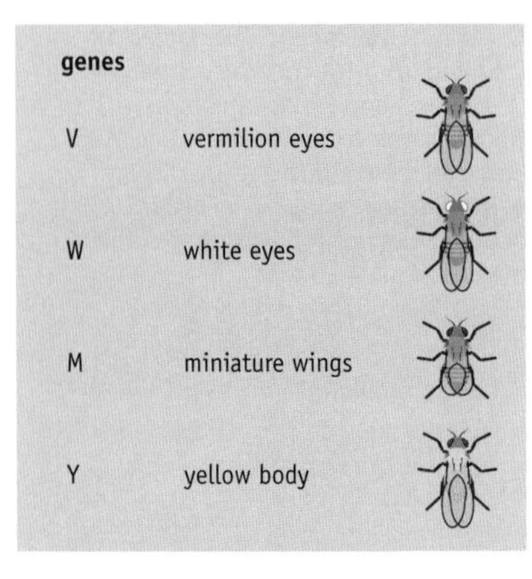

recombination frequencies

between M and V = 3.0%
between M and Y = 33.7%
between V and W = 29.4%
between W and Y = 1.3%

deduced map positions

Y W V M
0 1.3 30.7 33.7

FIGURE 15.20 Sturtevant's map showing the positions of four genes on chromosome 1 of the fruit fly. Recombination frequencies between the genes are shown, along with their deduced map positions.

Morgan's group then set about mapping as many fruit fly genes as possible and by 1915 had assigned locations for 85 of them. These genes fall into four **linkage groups**, corresponding to the four pairs of chromosomes seen in the fruit fly nucleus. The distances between genes are expressed in **map units**, one map unit being the distance between two genes that recombine with a frequency of 1%. According to this notation, the distance between the genes for white eyes and yellow body, which recombine with a frequency of 1.3%, is 1.3 map units. More recently the name **centimorgan** (**cM**) has begun to replace the map unit.

When we compare Sturtevant and Morgan's map, which, remember, was put together a century ago, with the actual locations of these genes, taken from the *D. melanogaster* genome sequence, we see that the positions are more or less correct, but not precisely so (Figure 15.21). This is because the assumption about the randomness of crossovers is not entirely justified. Crossovers are less frequent near the centromere of a chromosome, and some regions, called **recombination hotspots**, are more likely to be involved in crossovers than others. This means that a genetic map distance does not precisely indicate the actual distance between two genes. Also, we now realize that a single chromosome can participate in more than one crossover at the same time, but that there are limitations on how close together these crossovers can be, leading to more inaccuracies in the mapping procedure. Despite these qualifications, **linkage analysis** usually makes correct deductions about gene order, and distance estimates are sufficiently accurate for a genetic map to be valuable when searching for genes in a genome sequence.

In practice, gene mapping requires planned breeding experiments or pedigree analysis

Working out a genetic map requires that we obtain information on the recombination frequencies between pairs of genes. How do we do this in practice?

Most gene mapping projects make use of the procedures first devised by Morgan. These involve analysis of the progeny of experimental crosses set up between parents of known genotypes, as illustrated in Research Briefing 15.2. This approach is applicable, at least in theory, to all eukaryotes, though practical considerations sometimes make it difficult if not impossible to carry out. The organism must have a relatively short life cycle in order for progeny to be obtained in a reasonable amount of time. It is therefore appropriate for fruit flies, mice, and the like, but less so for elephants.

Gene mapping by planned breeding experiments is also not applicable to humans, not just because of our long life cycle but because factors other than the interests of geneticists determine the genotypes of the partners that participate in human breeding events. But we can still map genes in humans. Indeed, of all the genomes in existence, our own is the one for

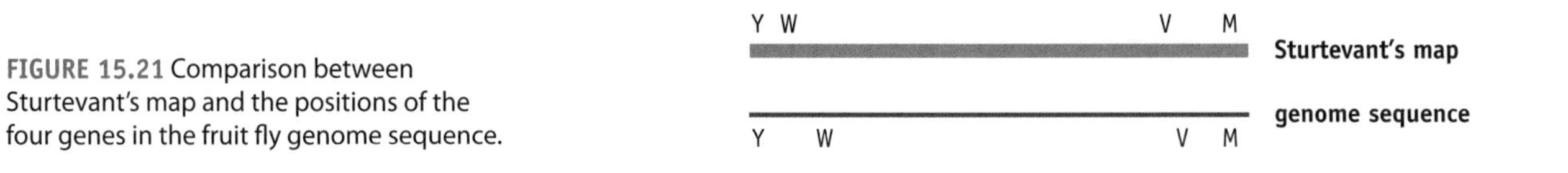

FIGURE 15.21 Comparison between Sturtevant's map and the positions of the four genes in the fruit fly genome sequence.

which we would like to have the most detailed gene maps, as this is one of the ways of identifying genes that are involved in disease. Identification of the causative gene is often the first step in designing a cure or treatment for a disease.

Gene mapping in humans is carried out by pedigree analysis

With humans, rather than carrying out planned experiments, data for the calculation of recombination frequencies have to be obtained by examining the genotypes of the members of successive generations of existing families. This means that only limited data are available, and their interpretation is often difficult. The limitation arises because genotypes of one or more family members are often unobtainable because those individuals are dead or unwilling to cooperate.

We will follow through a typical human **pedigree analysis** in order to explore what this method can achieve and to become familiar with some of the problems that can arise along the way.

Imagine that we are studying a genetic disease present in a family of two parents and six children (Figure 15.22A). Genetic diseases are frequently

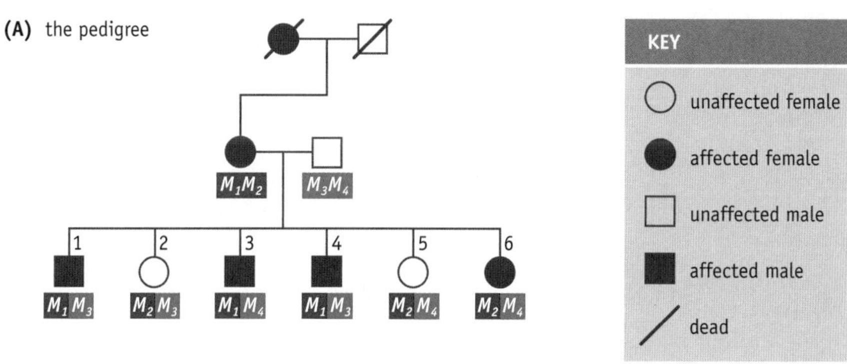

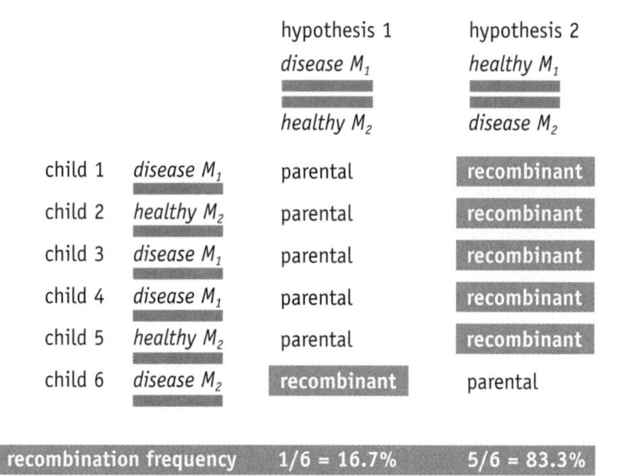

(C) resurrection of the maternal grandmother

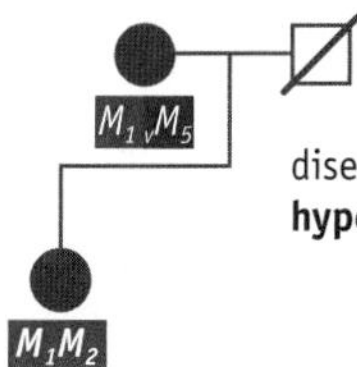

disease allele must be linked to M_1
hypothesis 1 is correct

FIGURE 15.22 An example of human pedigree analysis. (A) The pedigree shows inheritance of a genetic disease in a family of two living parents and six children, with information about the maternal grandparents available from family records. The disease allele (closed symbols) is dominant over the healthy allele (open symbols). (B) The pedigree can be interpreted in two different ways. Hypothesis 1 gives a low recombination frequency and indicates that the disease gene is tightly linked to gene M. Hypothesis 2 suggests that the disease gene and gene M are much less closely linked. In (C), the issue is resolved by the reappearance of the maternal grandmother, whose genotype is consistent only with hypothesis 1.

RESEARCH BRIEFING 15.2

Mapping genes in eukaryotes

The first methods for mapping the positions of genes on eukaryotic chromosomes were devised by Thomas Hunt Morgan and his colleagues roughly a century ago. These methods have stood the test of time and are still used today with organisms with which planned breeding experiments can be carried out. Here we will study how these experiments are performed and how the resulting data are converted into a gene map.

The test cross is central to gene mapping when breeding experiments are possible

The key to gene mapping is being able to determine the genotypes of the gametes resulting from meiosis (see Figure 15.18). Then it is possible to deduce the crossover events that have occurred, and from these to calculate the recombination frequency for a pair of genes. The complication with a genetic cross is that the resulting diploid progeny are the product not of one meiosis but of two—one in each parent. In most organisms crossover events are equally likely to occur during production of both the male and female gametes. Somehow we have to be able to disentangle from the genotypes of the diploid progeny the crossover events that occurred in each of these two meioses.

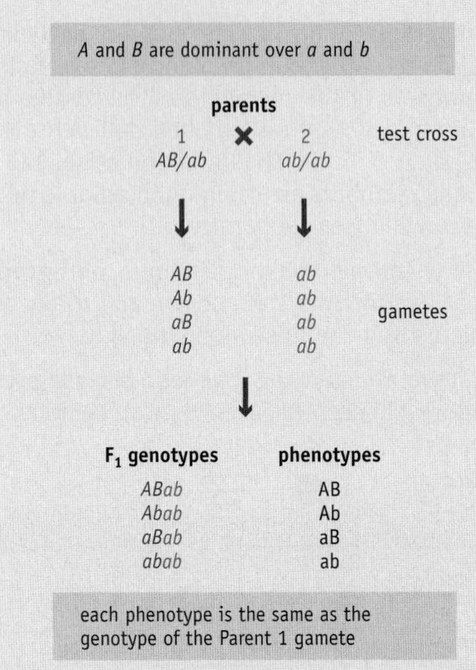

FIGURE 1 A test cross. Parent 1 can give rise to gametes with either of four possible genotypes, depending on whether or not crossing over takes place between genes A and B. Parent 2 can only give gametes with the double recessive genotype, regardless of whether there is crossing over. Parent 2 therefore makes no contribution to the phenotypes of the progeny. The phenotype of each individual in the F_1 generation is the same as the genotype of the gamete from Parent 1 that gave rise to that individual.

The solution to this conundrum is to use a **test cross**, as illustrated in Figure 1. Here we have set up a test cross to map the positions of genes A (alleles *A* and *a*) and B (alleles *B* and *b*). The critical feature of a test cross is the genotypes of the two parents. One of these parents has to be a **double heterozygote**. This means that all four alleles are present in this parent. Its genotype is therefore *AB/ab*. This notation indicates that one pair of the homologous chromosomes has alleles *A* and *B*, and the other has *a* and *b*. Double heterozygotes can be obtained by crossing two pure-breeding strains, for example, *AB/AB* × *ab/ab*.

The second parent has to be a **double homozygote**. In this parent both homologous copies of the chromosome are the same. The alleles must be the recessive ones, which means that in the example shown in Figure 1 both homologous chromosomes have alleles *a* and *b*, and the genotype of the parent is *ab/ab*.

Note that all the gametes from the second parent (the double homozygote) will have the genotype *ab* regardless of whether crossovers occur during their production. Alleles *a* and *b* are both recessive, so meiosis in this parent is, in effect, invisible when the genotypes of the progeny are examined. This means that, as shown in Figure 1, the genotypes of the diploid progeny can be unambiguously converted into the genotypes of the gametes from the double heterozygous parent. The test cross therefore enables us to make a direct examination of a single meiosis and hence to calculate a recombination frequency and map distance for the two genes being studied.

Multipoint crosses—mapping the positions of more than two genes at once

The power of gene mapping is enhanced if more than two genes are followed in a single cross. Not only does this generate recombination frequencies more quickly, it also enables the relative order of genes on a chromosome to be determined by simple inspection of the data. This is because two crossover events are required to unlink the central gene from the two outer

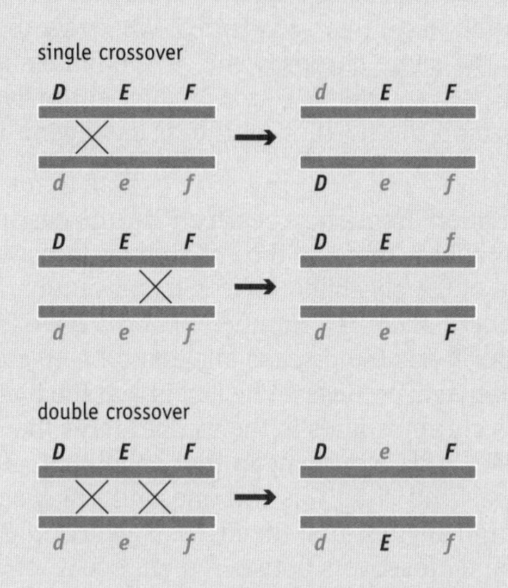

FIGURE 2 A single crossover is enough to unlink one of the outer genes, but two crossovers are needed to unlink the central gene.

genes in a series of three, whereas either of the two outer genes can be unlinked by just a single crossover (Figure 2). Two crossover events are less likely than a single one, so unlinking of the central gene will occur relatively infrequently.

If more than two genes are mapped at once then the breeding experiment is called a **multipoint cross**. A set of typical data from a three-point cross is shown in Table 1. A test cross has been set up between a triple heterozygote (*ABC/abc*) and a triple homozygote (*abc/abc*). The most frequent progeny are those with one of the two parental genotypes, resulting from an absence of recombination events in the region containing the genes A, B, and C. Two other classes of progeny are relatively frequent, comprising 51 and 63 progeny, respectively, in the example shown. Both of these are presumed to arise from a single recombination. Inspection of their genotypes shows that in the first of these two classes, gene A has become unlinked from B and C, and in the second class, gene B has become unlinked from A and C. The implication is that A and B are the outer genes. This is confirmed by the number of progeny in which marker C has become unlinked from A and B. There are only two of these, showing that a double recombination is needed to produce this genotype. Gene C is therefore between A and B.

TABLE 1 SET OF TYPICAL DATA FOR A MULTIPOINT CROSS BETWEEN A TRIPLE HETEROZYGOTE (*ABC/abc*) AND A TRIPLE HOMOZYGOTE (*abc/abc*)

Genotypes of progeny	Number of progeny	Inferred recombination events
ABC/abc, abc/abc	987	None (parental genotypes)
aBC/abc, Abc/abc	51	One, between A and B/C
AbC/abc, aBc/abc	63	One, between B and A/C
ABc/abc, abC/abc	2	Two, one between C and A and one between C and B

Having established the gene order we can use the frequencies of the genotypes to infer the distances between genes A and C, and C and B. Table 1 shows that there are 51 recombinants generated by a single recombination between genes A and C. We must also include the progeny generated by double recombination events, because with each of these one of the events was located between genes A and C. So the recombination frequency for genes A and C is calculated as:

$$\text{RF} = \frac{sr\ (\text{A} \leftrightarrow \text{C}) + dr}{t} \times 100$$

In this equation, RF is the recombination frequency, sr (A ↔ C) the number of progeny resulting from single recombination events between genes A and C, dr the number of progeny resulting from double recombination events, and t the total number of progeny. Putting in the appropriate numbers we discover that the recombination frequency for genes A and C is

$$\text{RF} = \frac{51 + 2}{1103} \times 100 = 4.8\%$$

and for genes C and B the recombination frequency is

$$\text{RF} = \frac{63 + 2}{1103} \times 100 = 5.9\%$$

The deduced map is therefore as shown in Figure 3.

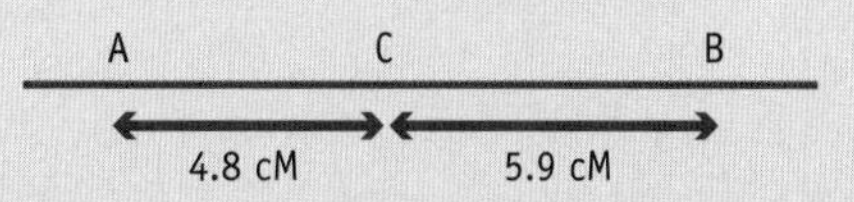

FIGURE 3 The map deduced from the multipoint cross.

used as gene markers in humans, the disease state being one allele and the healthy state being a second. In our pedigree, the mother is affected by the disease, as are four of her children. We know from family accounts that the maternal grandmother also suffered from this disease, but both she and her husband—the maternal grandfather—are now dead. We can include them in the pedigree, with slashes indicating that they are dead, but we cannot obtain any further information on their genotypes. We know that the disease gene is present on the same chromosome as a multiallelic gene M, four alleles of which—M_1, M_2, M_3, and M_4—are present in the living family members. Our aim is to map the position of the disease gene relative to gene M.

To establish a recombination frequency between the disease gene and gene M we must determine how many of the children are recombinants. If we look at the genotypes of the six children we see that numbers 1, 3, and 4 have the disease allele and allele M_1. Numbers 2 and 5 have the healthy allele and M_2, and number 6 has the disease allele and M_2. We can therefore construct two alternative hypotheses. The first is that the two copies of the relevant homologous chromosomes in the mother have the genotypes *Disease-*M_1 and *Healthy-*M_2. This would mean that children 1, 2, 3, 4, and 5 have parental genotypes and child 6 is the one and only recombinant (Figure 15.22B). This would suggest that the disease gene and gene M are relatively closely linked and crossovers between them occur infrequently.

The alternative hypothesis is that the mother's chromosomes have the genotypes *Healthy-*M_1 and *Disease-*M_2. According to this hypothesis, children 1 through 5 are recombinants, and child 6 has the parental genotype. This would mean that the two genes are relatively far apart on the chromosome. We cannot determine which of these hypotheses is correct. The data are frustratingly ambiguous.

The most satisfying solution to the problem posed by the pedigree in Figure 15.22 would be to know the genotype of the grandmother. Let us pretend that this is a soap opera family and it turns out that the grandmother is not really dead. Her genotype for gene M is found to be M_1M_5 (Figure 15.22C). This tells us that the chromosome inherited from her by the mother must have had the genotype *Disease-*M_1. It could not have been *Disease-*M_2 because the grandmother does not possess the M_2 allele. We can now distinguish between our two hypotheses and conclude with certainty that only child 6 is a recombinant.

Data from human pedigrees are analyzed statistically, using a measure called the **lod score**. This stands for logarithm of the odds that the genes are linked and is used primarily to determine whether the two genes being studied lie on the same chromosome. A lod score of 3 or more corresponds to odds of 1000:1 and is usually taken as the minimum for confidently concluding that this is the case. If it can be established that the two genes are on the same chromosome then lod scores can be calculated for each of a range of recombination frequencies, in order to identify the frequency most likely to have given rise to the data obtained by pedigree analysis.

Lod scores are substantially more reliable if data are available for more than one pedigree, and the analysis is less ambiguous for families with larger numbers of children, and, as we saw above, it is important that the members of at least three generations can be genotyped. For this reason, family collections have been established, such as the one maintained by the Centre d'Études du Polymorphisme Humaine (CEPH) in Paris. The CEPH collection contains cultured cell lines from families in which all four grandparents as well as at least eight grandchildren could be sampled. This collection is available for gene mapping by any researcher who agrees to submit the resulting data to the central CEPH database.

KEY CONCEPTS

- Knowing the relationship displayed by a pair of alleles is often the starting point for understanding the molecular basis of those alleles. Often one allele is the functional version of the gene and another is nonfunctional.
- Many pairs of alleles display a relationship in which one allele is dominant and the other recessive. The phenotypic effect of the recessive allele is masked in a heterozygote.
- Some pairs of alleles display incomplete dominance. The phenotype of the heterozygote is intermediate between those of the two homozygotes.
- Some alleles are lethal when present in the homozygous condition.
- Many phenotypes can be linked to the activity of just a single gene, but there are numerous others that result from interactions between two or more separate genes. Understanding these interactions is crucial to our awareness of how the genome as a whole specifies the biological characteristics of an organism.
- Sometimes an allele of one gene will mask or cancel the alleles of a second gene. This is called epistasis.
- The phenotypes resulting from a quantitative trait, one specified by many genes acting together, display a Gaussian distribution when examined in the population as a whole.
- The frequency of recombination between a pair of genes that lie on the same chromosome can be used to map the relative positions of those two genes.
- Gene mapping in many organisms is carried out by examining the genotypes of the offspring resulting from planned breeding experiments.
- With humans, rather than carrying out planned experiments, data for the calculation of recombination frequencies are obtained by pedigree analysis. This involves examination of the genotypes of the members of successive generations of existing families.

QUESTIONS AND PROBLEMS (Answers can be found at www.garlandscience.com/introgenetics)

Key Terms

Write short definitions of the following terms:

centimorgan
codominance
complementary gene action
cross-fertilization
dominant
double heterozygote
double homozygote
epistasis
Gaussian distribution
genotype
heterozygote
homozygote
incomplete dominance
lethal allele
linkage
linkage analysis
linkage group
lod score
map unit
multipoint cross
normal distribution
parental genotype
pedigree analysis
phenotype

Punnett square
pure-breeding
quantitative trait
recessive
recombinant genotype
recombination frequency
recombination hotspot
segregation
self-fertilization
test cross
unit factor

Self-study Questions

15.1 Distinguish between the terms "genotype" and "phenotype."

15.2 Carefully explain how Mendel was able to deduce that alleles occur in pairs and that they display dominance and recessiveness.

15.3 Describe the molecular basis for the dominant–recessive relationship between the round and wrinkled pea alleles.

15.4 Explain how Mendel derived the First Law of Genetics.

15.5 Green pods are dominant to yellow pods in pea plants. What will be the phenotypes and genotypes of the F_1 and F_2 plants from a green pod (*GG*) and yellow pod (*gg*) cross?

15.6 What would be the outcome of crossing an F_1 plant from the cross described in Question 15.5 with the yellow pod parent (this is called a backcross).

15.7 What would be the ratio of phenotypes in the F_1 and F_2 generations resulting from a cross between two parent plants with the genotypes *AA* and *aa*, if *A* and *a* display incomplete dominance?

15.8 Explain what is meant by the term "lethal allele." What is the molecular basis of the lethality of the *aurea* allele of snapdragon?

15.9 What would be the ratio of phenotypes resulting from a cross between a pair of heterozygotes, genotype *Aa*, if *a* was a lethal allele?

15.10 What is codominance, and how is it usually detected?

15.11 What can flower pigmentation tell us about interactions between genes?

15.12 Explain why the phenotypes resulting from a quantitative trait display a Gaussian distribution.

15.13 Give an example of complementary gene action.

15.14 Describe how fruit color in squash illustrates the principle of epistasis.

15.15 Distinguish between a parental and a recombinant genotype. How do gametes with recombinant genotypes arise during meiosis?

15.16 Why does the frequency of recombinant gametes resulting from meiosis enable the map positions of genes to be worked out?

15.17 How were genes first mapped on chromosome 1 of the fruit fly?

15.18 Explain why the positions of genes deduced by genetic mapping are slightly different from their actual locations on a chromosome.

15.19 Describe the factors that limit the amount of data that are available from pedigree analysis.

15.20 What is a lod score, and what information does it provide about the relative positions of two genes in the human genome?

15.21 Why is a double homozygote used for test crosses in linkage analysis? Why must the homozygous alleles be recessive for the characteristics being studied?

Discussion Topics

15.22 What features are desirable for an organism that is to be used for extensive studies of gene inheritance?

15.23 The genes for pod color and pea shape are on different chromosomes in peas. What would be the ratio of phenotypes in the F_1 and F_2 generations resulting from a cross between a parent plant with green pods and round peas, genotype *GGRR*, and one with yellow pods and wrinkled peas, *ggrr*?

15.24 Most wild mice have black hairs with fine yellow bands, giving them the coat color referred to as agouti, which has evolved to provide the best camouflage in the mouse's natural environment. The black coloration is coded by gene C and the yellow banding by gene A. The following observations have been made:

- Mice with the genotype *C_A_* (this notation indicates that each mouse possesses at least one dominant allele for each of the two genes) have black hairs (coded by C) and yellow bands (coded by A) and so have agouti coats.

- Mice that lack the dominant allele of gene A (genotypes *C_aa*) still have black hairs but this time without the yellow bands. They have black coats.
- Mice with the double recessive genotype, *ccaa*, have white hairs without yellow bands and their coats are white.
- We would anticipate that mice that lack the dominant allele of gene C but do have at least one dominant allele for gene A (genotypes *ccA_*) would have white hairs with yellow bands. But, in fact, these mice have pure white hairs exactly the same as the double recessive ones.

Suggest a hypothesis that takes account of each of these observations. How might your hypothesis be tested?

15.25 Vermilion eyes and rudimentary wings are specified by two genes on chromosome 1 of the fruit fly. A cross between *VVRR* (red eyes and normal wings) and *vvrr* (vermilion eyes and rudimentary wings) produced the following F_2 generation:

359 flies with red eyes and normal wings

381 flies with vermilion eyes and rudimentary wings

131 flies with red eyes and rudimentary wings

139 flies with vermilion eyes and normal wings

What is the map distance between the genes for vermilion eyes and rudimentary wings?

15.26 Genes A, B, C, and D lie on the same fruit fly chromosome. In a series of crosses the following recombination frequencies were observed:

Genes	Recombination frequency (%)
A, C	40
A, D	25
B, D	5
B, C	10

Draw a map of the chromosome showing the positions of these genes.

15.27 Discuss the reasons why gene mapping is still important today even though it is now relatively easy to obtain a complete DNA sequence for an organism's genome.

FURTHER READING

Baur E (1908) Die Aurea-Sippen von Antirrhinum majus. *Mol. Gen. Genet.* 1, 124–125. *The original publication of the lethal aurea allele of snapdragon.*

Cordell HJ (2002) Epistasis: what it means, what it doesn't mean, and statistical methods to detect it in humans. *Hum. Mol. Genet.* 20, 2463–2468.

Fincham JRS (1990) Mendel – now down to the molecular level. *Nature* 343, 208–209. *Describes the molecular basis of round and wrinkled peas.*

Gudbjartsson DF, Walter GB, Thorleifsson G et al. (2008) Many sequence variants affecting diversity of adult human height. *Nat. Genet.* 40, 609–615.

Mackay TFC (2001) Quantitative trait loci in *Drosophila*. *Nat. Rev. Genet.* 2, 11–20. *Explains how quantitative traits are studied in fruit flies and indicates how they might be studied more effectively in humans.*

Morton NE (1955) Sequential tests for the detection of linkage. *Am. J. Hum. Genet.* 7, 277–318. *The use of lod scores in human pedigree analysis.*

Orel V (1995) Gregor Mendel: The First Geneticist. Oxford: Oxford University Press.

Ott J (1999) Analysis of Human Genetic Linkage. Baltimore: Johns Hopkins University Press. *Describes all aspects of human pedigree analysis.*

Shine I & Wrobel S (1976) Thomas Hunt Morgan: Pioneer of Genetics. Lexington, KY: University Press of Kentucky.

Sturtevant AH (1913) The linear arrangement of six sex-linked factors in *Drosophila* as shown by mode of association. *J. Exp. Zool.* 14, 39–45. *Construction of the first linkage map for the fruit fly.*

Mutation and DNA Repair CHAPTER 16

At the beginning of Part II we posed three big questions regarding the gene as a unit of inheritance. We have dealt with the first two of these questions, concerning the inheritance of genes during cell division and during sexual reproduction. The third question, which we will explore in the next two chapters, concerns the link between the inheritance of genes and the evolution of species. The fact that evolution occurs tells us that the biological characteristics of a species are able to change over time. We must therefore ask how the passage of genes from generation to generation enables a species to acquire genetic diversity and eventually to evolve into a new species.

To understand how genetic diversity is acquired we must study the various events that can result in a change in a DNA sequence. These changes are called **mutations**, and they can arise in two ways. The first is from an error in DNA replication. Although DNA replication is virtually error-free, mistakes do occasionally occur. An error in replication results in a daughter molecule that contains a **mismatch**, a position at which a base pair does not form because the nucleotides opposite each other in the double helix are not complementary (Figure 16.1A). When the mismatched daughter molecule is itself replicated the two granddaughter molecules that are produced are not identical. One has the correct nucleotide sequence, but the second contains a mutation.

The second way in which mutations can arise is through the action of a chemical or physical **mutagen**. Various chemicals react with DNA molecules and change the structure of one or more nucleotides within the double helix. Heat, ultraviolet radiation, and other physical mutagens have similar effects. Changing the structure of a nucleotide may affect its base-pairing properties, resulting in a mutation when the DNA molecule is replicated (Figure 16.1B).

As we will see, mutations can be harmless, or they may have a serious, even lethal, effect on the organism. Because they can be harmful, all organisms have **DNA repair** mechanisms which correct the vast majority of the mutations that occur in their DNA molecules. Despite these measures a few mutations slip through. If a mutation that has not been repaired is present in a cell that undergoes meiosis and gives rise to gametes, then the mutation might be inherited by one of that organism's progeny. It might then become a new and permanent feature of the genetic diversity of the species (Figure 16.2).

In this chapter we will focus on the generation of diversity. We will look first at the processes that give rise to mutations and the mechanisms that are used to repair mutated DNA molecules. We will then examine the various possible effects that a mutation can have on a cell or organism. In the next chapter we will explore the possible fate of new alleles that arise as a result of mutation.

16.1 THE CAUSES OF MUTATIONS

A mutation is a change in the nucleotide sequence of a DNA molecule. The simplest type of sequence change is a **point mutation**, in which one nucleotide is replaced by another (Figure 16.3). Point mutations are also called simple mutations, or single-site mutations. A point mutation is classified as a **transition** if it is a purine to purine (A $\rightleftharpoons$ G) or a pyrimidine to

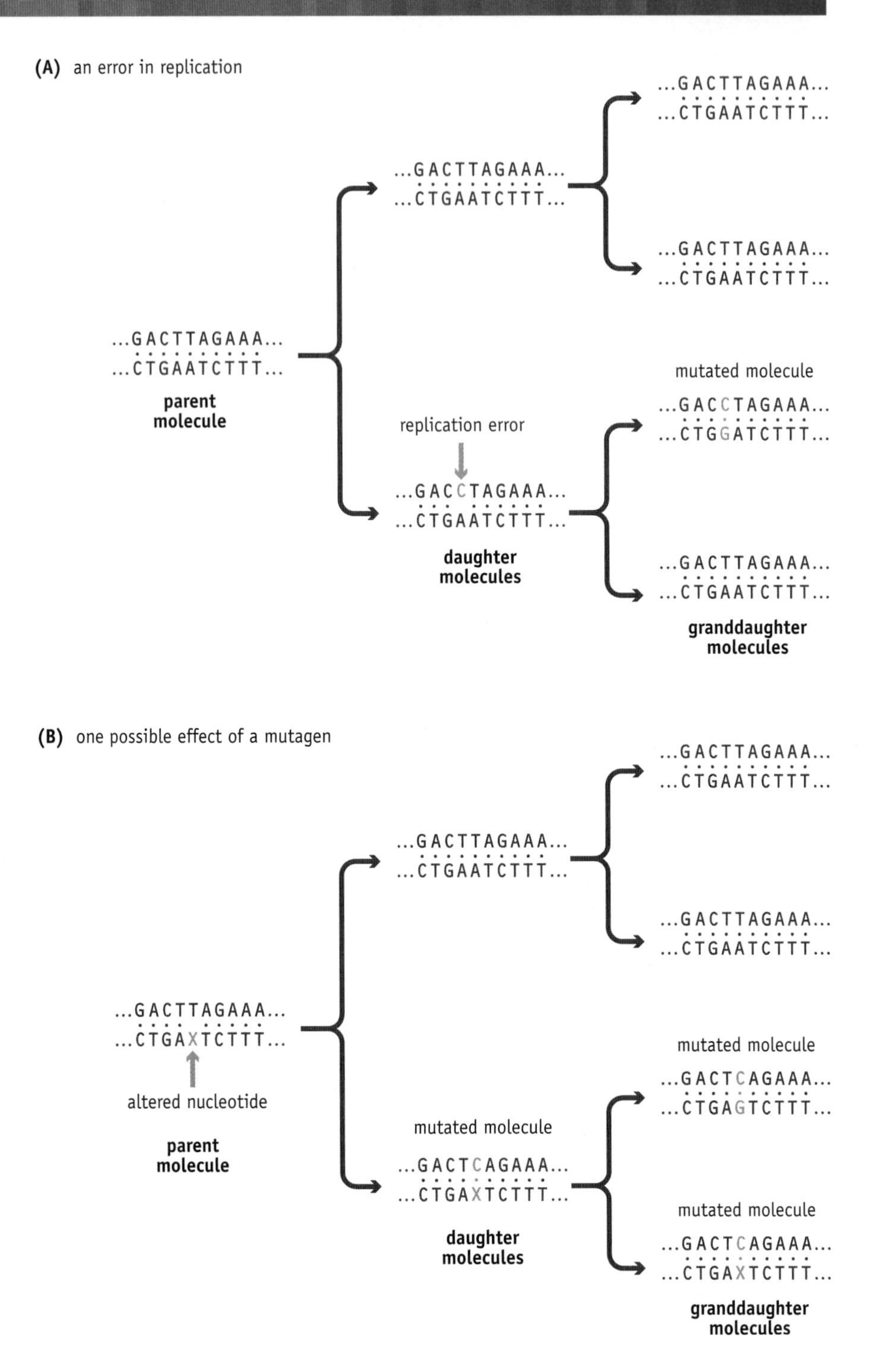

FIGURE 16.1 Examples of mutations. (A) An error in replication leads to a mismatch in one of the daughter double helices. When the mismatched molecule is itself replicated it gives one double helix with the correct sequence and one with a mutated sequence. (B) A mutagen has altered the structure of an A in the lower strand of the parent molecule, giving nucleotide X, which does not base-pair with the T in the other strand. When the parent molecule is replicated, X base-pairs with C, giving a mutated daughter molecule. When this daughter molecule is replicated, both granddaughters inherit the mutation.

pyrimidine (T ⇌ C) change, or a **transversion** if the alteration is purine to pyrimidine or vice versa (A or G ⇌ T or C).

Other types of mutation include **insertions** and **deletions**, which are the addition or removal of anything from one base pair up to quite extensive pieces of DNA, and **inversions**, which involve excision of a portion of the double helix followed by its reinsertion at the same position but in the reverse orientation.

Some mutations arise from errors in DNA replication that escape the proof-reading functions of the replicating enzymes, and others are due to the

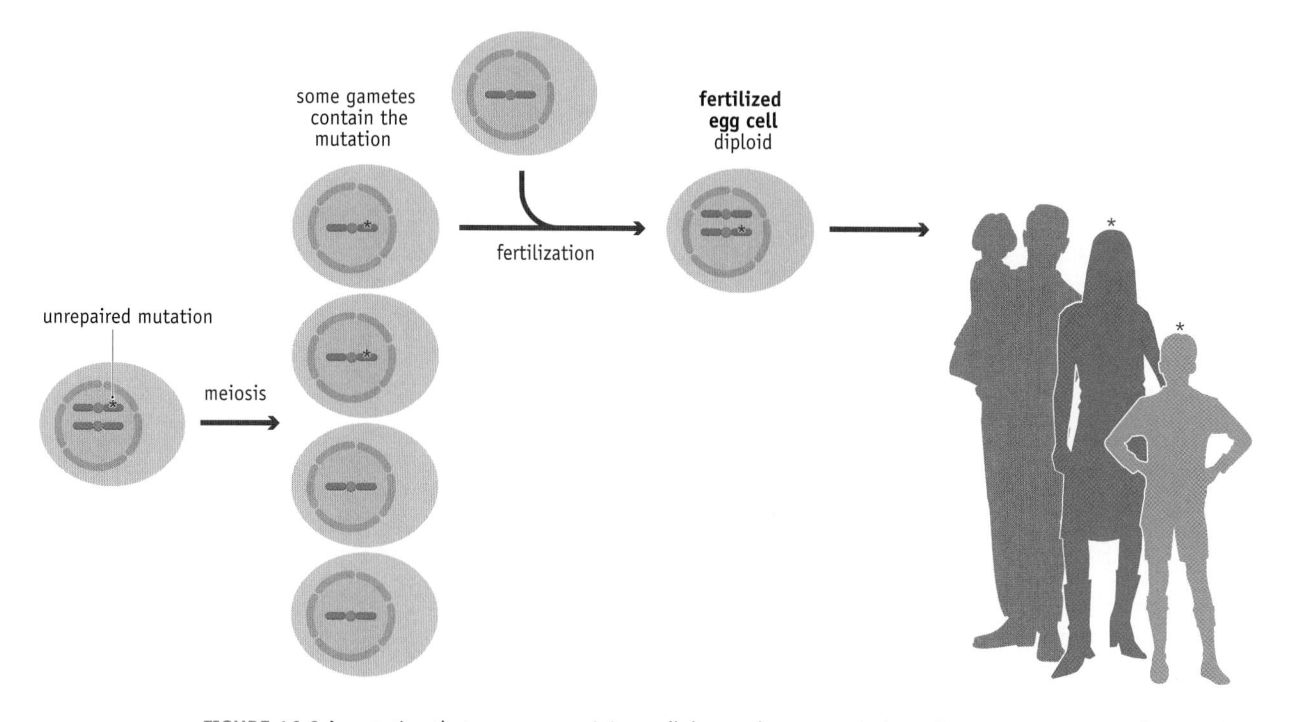

FIGURE 16.2 A mutation that escapes repair in a cell that undergoes meiosis can become a permanent feature of the genetic diversity of the species. The unrepaired mutation that occurred in the cell on the left has been inherited by the woman and passed on by her to her elder child.

action of mutagens that occur naturally in the environment and cause structural changes in DNA molecules. We begin this chapter by examining the molecular basis of these two types of event.

Errors in replication are a source of point mutations

During DNA replication, the sequence of the new strand of DNA that is being made is determined by complementary base pairing with the template polynucleotide (Figure 16.4). When considered purely as a chemical

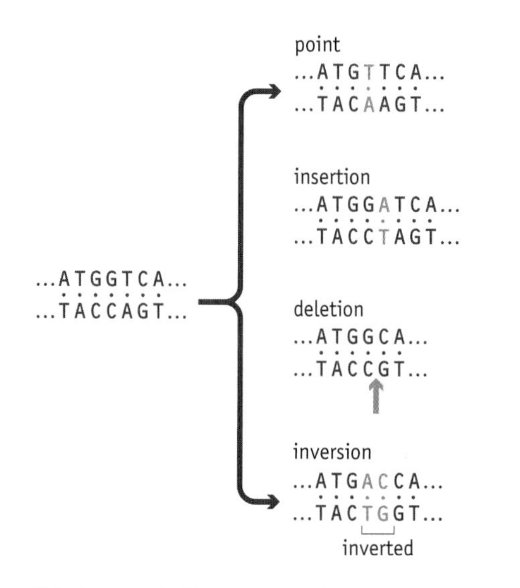

FIGURE 16.3 Different types of DNA sequence alteration that can occur by mutation. Here we are simply comparing the original and mutated DNA molecules, taking no account of the steps needed to go from one to another.

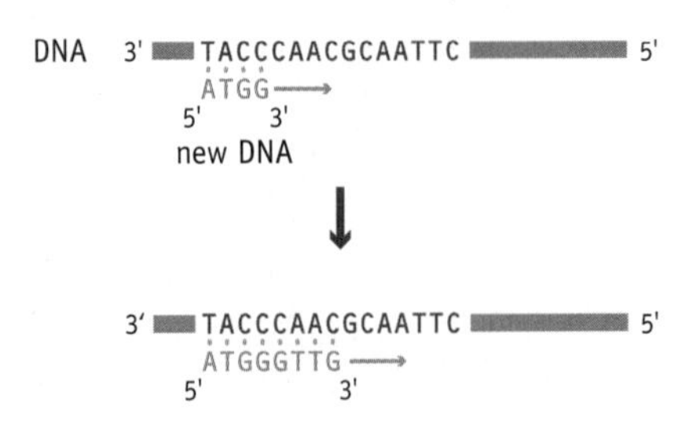

FIGURE 16.4 Template-dependent DNA replication.

reaction, complementary base pairing is not particularly accurate. Nobody has yet devised a way of carrying out the template-dependent synthesis of DNA without the aid of enzymes, but if the process could be carried out simply as a chemical reaction in a test tube then the resulting polynucleotide would probably have point mutations at 5 to 10 positions out of every hundred. This represents an error rate of 5–10%, which would be completely unacceptable during DNA replication in a living cell.

The accuracy of DNA replication in the cell is increased by a checking process that is operated by the DNA polymerase enzyme and reduces the possibility of an incorrect nucleotide being incorporated into the growing strand (Figure 16.5). This selection process acts at different stages during the polymerization reaction. The identity of the nucleotide is checked when it is first bound to DNA polymerase and checked again when the nucleotide is moved to the active site of the enzyme. At either of these stages, the enzyme is able to reject the nucleotide if it recognizes that it is not the right one.

Should an incorrect nucleotide evade this surveillance system and become attached to the 3′ end of the polynucleotide that is being synthesized, then all is not lost. Most DNA polymerase enzymes possess a 3′ → 5′ exonuclease activity and so are able to go back over their work and remove a stretch of newly synthesized polynucleotide that contains one or more incorrect pairings (Figure 16.6). This is called proofreading.

Despite these precautions, some errors do creep through, each one causing a mismatch in the daughter molecule and a permanent point mutation in one of the granddaughter molecules, as shown in Figure 16.1A. But not all of these point mutations can be blamed on the polymerase enzymes. Sometimes an error occurs even though the enzyme adds the "correct"

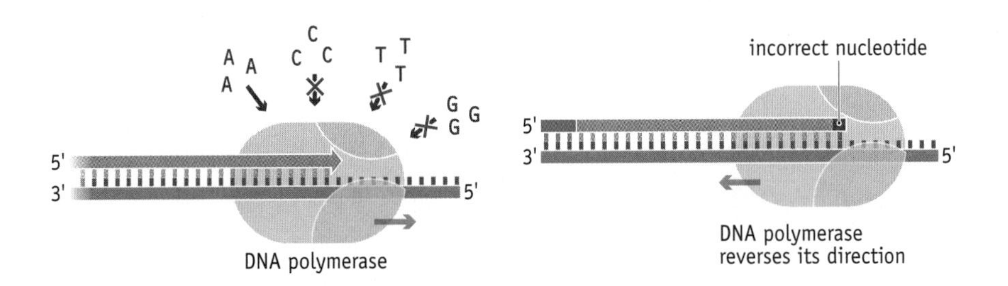

FIGURE 16.5 During DNA replication, the DNA polymerase selects the correct nucleotide to insert at each position.

FIGURE 16.6 The proofreading activity possessed by many DNA polymerases.

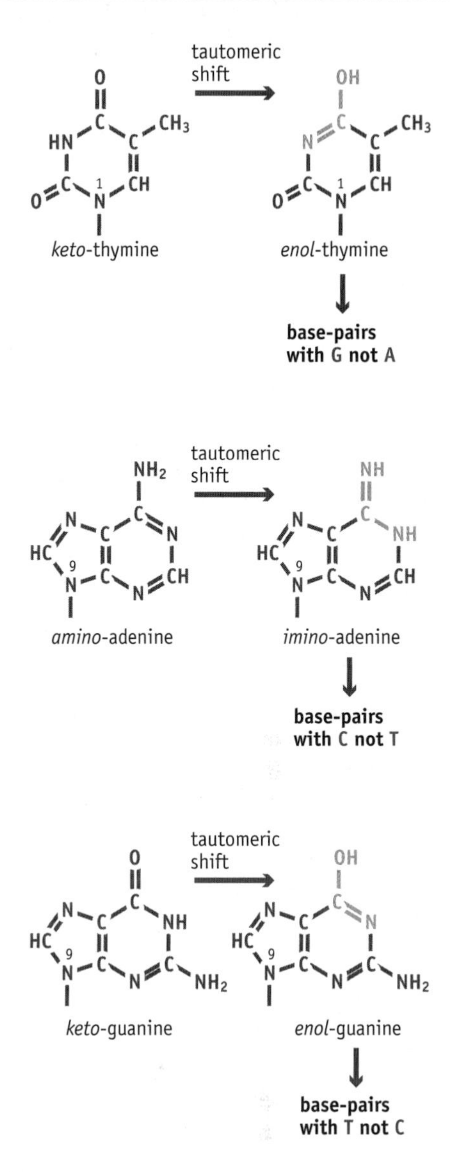

FIGURE 16.7 The effects of tautomerism on base pairing. In each of these three examples, the two tautomeric forms of the base have different pairing properties. Cytosine also has *amino* and *imino* tautomers but both pair with guanine.

nucleotide, the one that base-pairs with the template. This is because each nucleotide base can exist as either of two alternative structural forms called **tautomers**. For example, thymine exists as two tautomers, the *keto* and *enol* forms, with individual molecules occasionally undergoing a shift from one tautomer to the other (Figure 16.7). The equilibrium is biased very much toward the *keto* form, but every now and then the *enol* version of thymine occurs in the template DNA at the precise time that the replication fork is moving past. This will lead to an error, because *enol*-thymine base-pairs with G rather than A. The same problem can occur with adenine, the rare *imino* tautomer of this base preferentially forming a pair with C, and with guanine, *enol*-guanine pairing with thymine. After replication, the rare tautomer will inevitably revert to its more common form, leading to a mismatch in the daughter double helix.

Replication errors can also lead to insertion and deletion mutations

Not all errors in replication are point mutations. Aberrant replication can also result in small numbers of extra nucleotides being inserted into the polynucleotide that is being synthesized, or some nucleotides in the template not being copied. An insertion or deletion that occurs within the coding region of a gene might result in a frameshift mutation, which changes the reading frame used for translation of the protein specified by the gene (Figure 16.8). There is a tendency to use "frameshift" to describe all insertions and deletions, but this is inaccurate because inserting or deleting three nucleotides, or multiples of three, simply adds or removes codons or parts of adjacent codons without affecting the reading frame. Also, of course, many insertions and deletions occur outside of genes, within the intergenic regions of a DNA molecule.

Insertion and deletion mutations can affect all parts of the genome but are particularly prevalent when the template DNA contains short repeated sequences. This is because repeated sequences can induce **replication slippage**, in which the template strand and its copy shift their relative positions so that part of the template is either copied twice or missed. The result is that the new polynucleotide has a larger or smaller number, respectively, of the repeat units (Figure 16.9).

Replication slippage is probably responsible for the **trinucleotide repeat expansion diseases**. These are human neurodegenerative diseases that

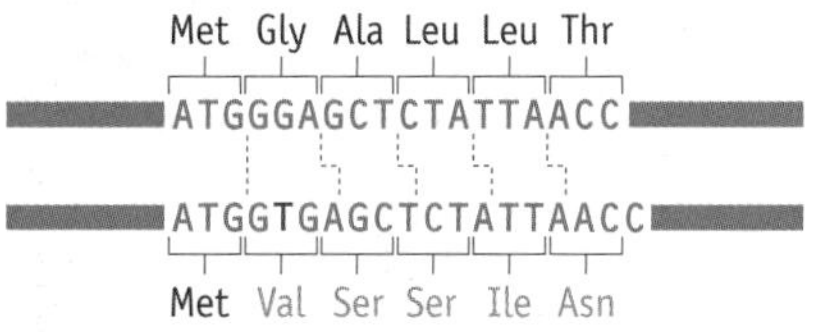

Met Gly Ala Leu Leu Thr
ATGGGAGCTCTATTAACC
ATGGTGAGCTCTATTAACC
Met Val Ser Ser Ile Asn

FIGURE 16.8 Insertion of a nucleotide changes the reading frame used for translation of the protein coded by a gene.

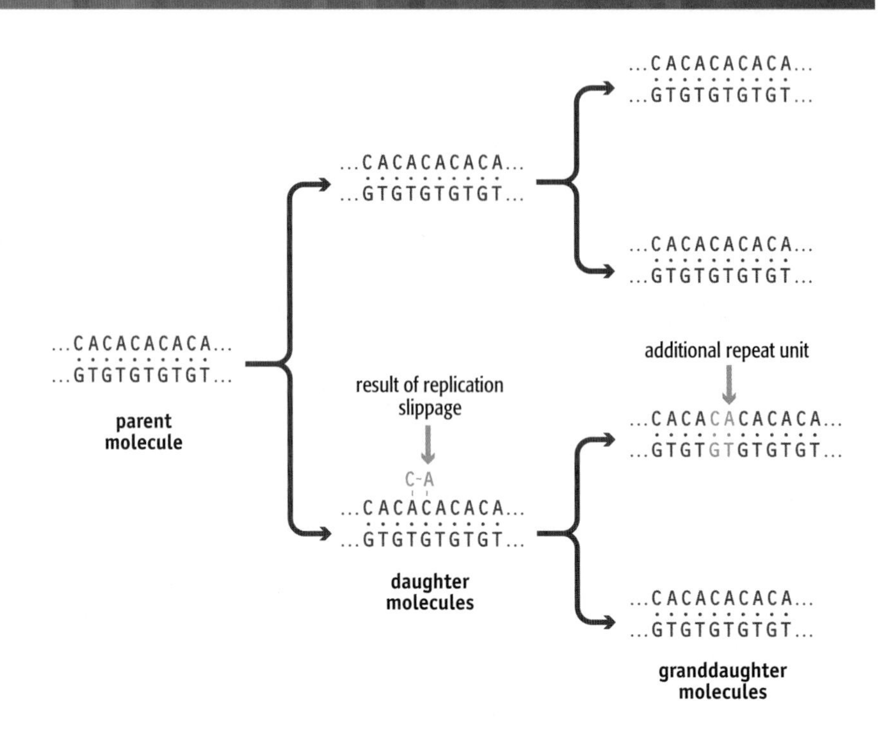

FIGURE 16.9 Replication slippage. The diagram shows replication of a short repeat sequence. Slippage has occurred during replication of the parent molecule, inserting an additional repeat unit into the newly synthesized polynucleotide of one of the daughter molecules. When this daughter molecule replicates it gives a granddaughter molecule whose repeat sequence is one unit longer than that of the original parent.

have long been known but whose molecular basis has been discovered only in recent years. Each of these diseases is caused by a relatively short series of trinucleotide repeats becoming elongated to two or more times its normal length. For example, the human *HD* gene contains the sequence 5′-CAG-3′ repeated between 6 and 35 times in tandem, coding for a series of glutamines in the protein product. In Huntington's disease this repeat expands to a copy number of 36 to 121, increasing the length of the polyglutamine tract and resulting in a dysfunctional protein (Figure 16.10).

Mutagens are one type of environmental agent that causes damage to cells

Many chemicals that occur naturally in the environment have mutagenic properties, and these have been supplemented in recent years with other chemical mutagens that result from human industrial activity. Physical agents such as radiation are also mutagenic. Most organisms are exposed to greater or lesser amounts of these various mutagens, their genomes suffering damage as a result.

Mutagens cause mutations in three different ways (Figure 16.11). Some act as **base analogs** and are mistakenly used as substrates when new DNA is synthesized at the replication fork. Others react directly with DNA, causing structural changes that lead to miscopying of the template strand when the DNA is replicated. These structural changes are diverse, as we will see

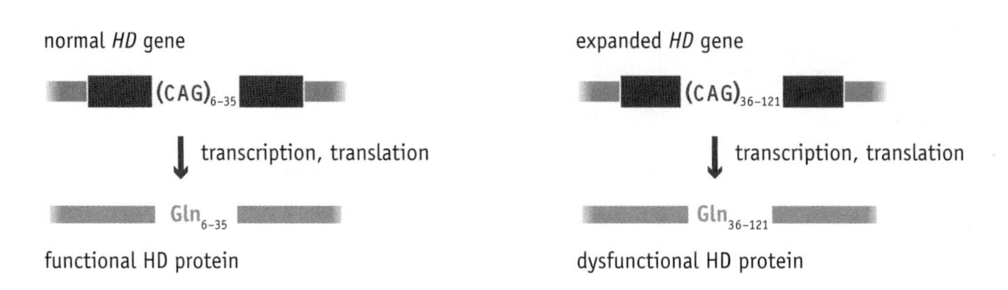

FIGURE 16.10 The genetic basis of Huntington's disease.

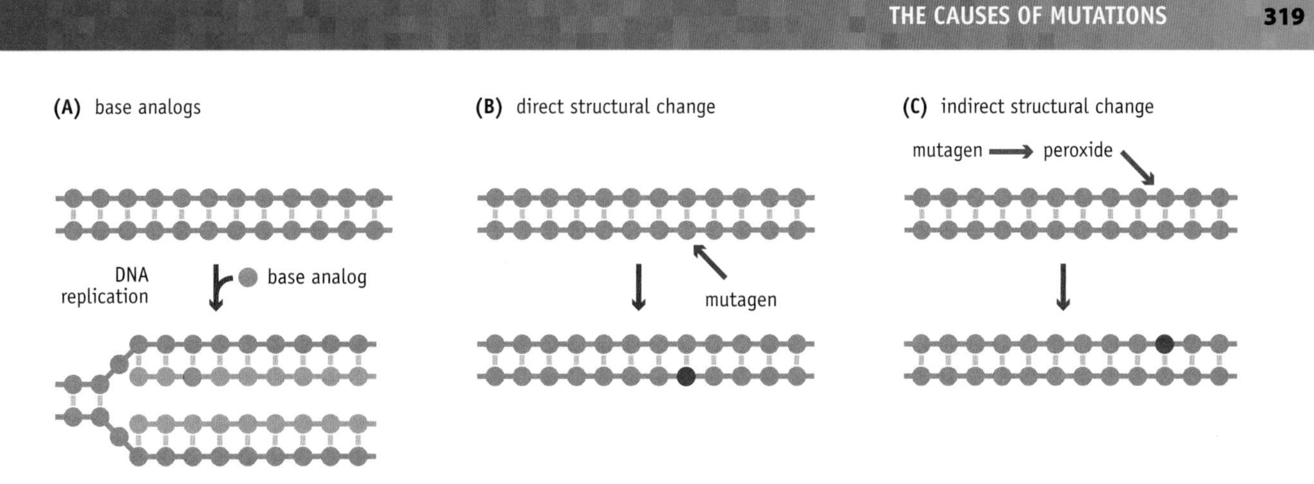

FIGURE 16.11 Three ways in which mutagens can cause mutations. (A) A base analog can be mistakenly used as a substrate during DNA replication. (B) Some mutagens react directly with DNA, causing a structural change that might be miscopied during DNA replication. (C) Other mutagens act indirectly on DNA, by causing the cell to synthesize chemicals that have a direct mutagenic effect.

when we look at individual mutagens. And finally, some mutagens act indirectly on DNA. They do not themselves affect DNA structure, but instead cause the cell to synthesize chemicals such as peroxides that have a direct mutagenic effect.

It is important to understand that a mutagen, defined as a chemical or physical agent that causes mutations, is distinct from other types of environmental agent that cause damage to cells in ways other than by causing mutations. These include **carcinogens**, which cause cancer (the neoplastic transformation of cells), **oncogens**, which cause tumor formation, and **teratogens**, which cause developmental abnormalities. There are overlaps between these categories (for example, some mutagens are also carcinogens), but each type of agent has a distinct biological effect. The definition of "mutagen" also makes a distinction between true mutagens and other agents that damage DNA without causing mutations, such as **clastogens**, which cause breaks in DNA molecules and subsequent fragmentation of chromosomes. This type of damage might block replication and cause the cell to die, but it is not a mutation in the strict sense of the term and the causative agents are therefore not mutagens.

There are many types of chemical mutagen

The range of chemical mutagens is so vast that it is difficult to devise an all-embracing classification. We will therefore restrict our study to the most important types (Table 16.1).

First, there are base analogs. These are purine and pyrimidine bases that are similar enough to the standard bases to be incorporated into nucleotides

TABLE 16.1 IMPORTANT TYPES OF CHEMICAL MUTAGEN

Type	Examples
Base analogs	5-Bromouracil, 2-aminopurine
Deaminating agents	Nitrous acid, sodium bisulfite
Alkylating agents	Ethylmethane sulfonate, dimethylnitrosamine, methyl halides
Intercalating agents	Ethidium bromide

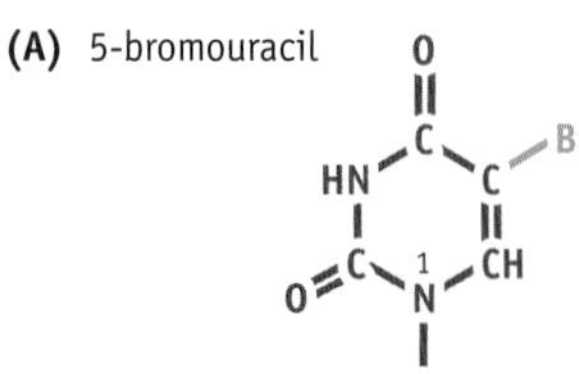

(B) base pairing with 5-bromouracil

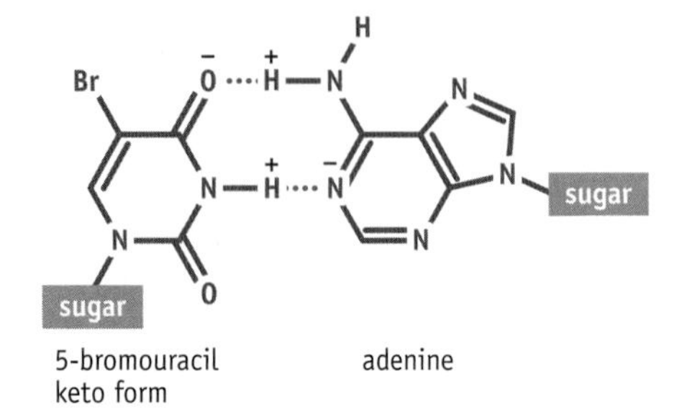

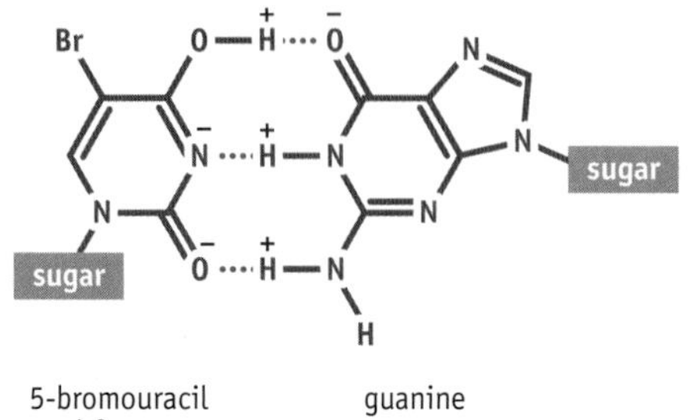

(C) the mutagenic effect of 5-bromouracil

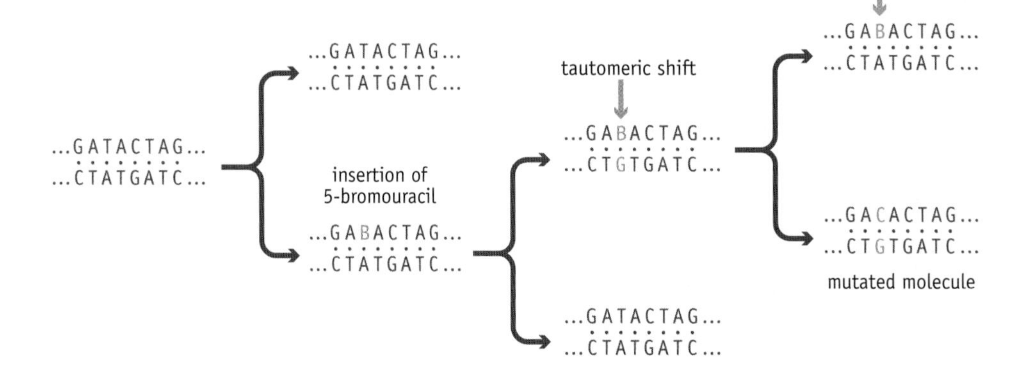

FIGURE 16.12 5-Bromouracil and its mutagenic effect. (A) 5-Bromouracil is a modified form of thymine. (B) The *keto* form of 5-bromouracil base-pairs with adenine, and the *enol* form with guanine. (C) Three cycles of DNA replication are shown. During the first cycle, 5-bromouracil acts as a base analog and replaces a thymine nucleotide in one of the daughter molecules. During the second cycle of replication, the relatively common *enol* tautomer of 5-bromouracil is present, leading to a base-pairing change. The third round of replication converts this error into a point mutation.

when these are synthesized by the cell. The resulting unusual nucleotides can then be used as substrates for DNA synthesis during genome replication. For example, **5-bromouracil** (**5-bU**) has the same base-pairing properties as thymine, and nucleotides containing 5-bU can be added to the daughter polynucleotide at positions opposite As in the template. The mutagenic effect arises because the equilibrium between the two tautomers of 5-bU is shifted more toward the rarer *enol* form than is the case with thymine. This means that during the next round of replication there is a relatively high chance of the polymerase encountering *enol*-5-bU, which, like *enol*-thymine, pairs with G rather than A. This results in a point mutation (Figure 16.12). **2-Aminopurine** acts in a similar way. It is an analog of adenine with an *amino* tautomer that pairs with thymine and an *imino* tautomer that pairs with cytosine. The *imino* form is more common than *amino*-adenine and induces T-to-C transitions during DNA replication.

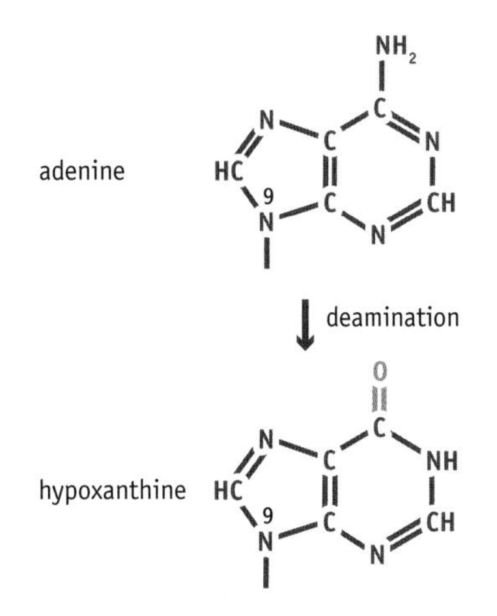

FIGURE 16.13 Hypoxanthine is a deaminated version of adenine.

Deaminating agents also cause point mutations. A certain amount of base deamination (removal of an amino group) occurs spontaneously in cellular DNA molecules, with the rate being increased by chemicals such as nitrous acid, which deaminates adenine, cytosine, and guanine, and sodium bisulfite, which acts only on cytosine. Thymine, of course, has no amino group and so cannot be deaminated. Deamination of adenine gives hypoxanthine (Figure 16.13), which pairs with C rather than T, and deamination of cytosine gives uracil, which pairs with A rather than G. Deamination of these two bases therefore results in point mutations when the template strand is copied. Deamination of guanine is bad for the cell because the resulting base, xanthine, blocks replication when it appears in the template polynucleotide, but this is not a mutagenic effect according to our definition of mutation.

Alkylating agents are a third type of mutagen that can give rise to point mutations. Chemicals such as **ethylmethane sulfonate** (**EMS**) and dimethylnitrosamine add alkyl groups to nucleotides in DNA molecules, as do methylating agents such as methyl halides, which are present in the atmosphere, and the products of nitrite metabolism. The effect of alkylation depends on the position at which the nucleotide is modified and the type of alkyl group that is added. Methylations, for example, often result in modified nucleotides with altered base-pairing properties and so lead to point mutations. Other alkylations block replication by forming crosslinks between the two strands of a DNA molecule, or by adding large alkyl groups that prevent progress of the replication complex.

The final type of chemical mutagen that we will look at is the **intercalating agents**. The best-known mutagen of this type is **ethidium bromide**, which is sometimes used as a stain for DNA because it fluoresces when exposed to ultraviolet radiation. Ethidium bromide and other intercalating agents are flat molecules that can slip between base pairs in the double helix, slightly unwinding the helix and hence increasing the distance between adjacent base pairs (Figure 16.14). It has been assumed that this is likely to lead specifically to an insertion or deletion, but exposure to an intercalating agent can also lead to other types of mutation.

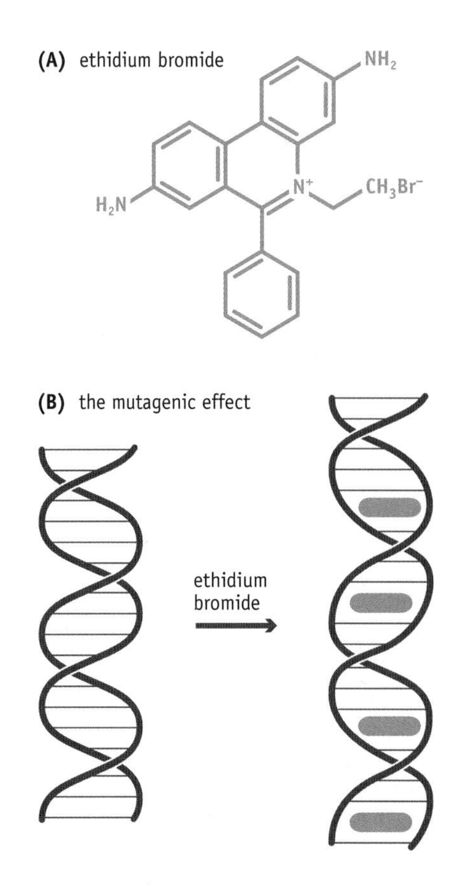

FIGURE 16.14 The mutagenic effect of ethidium bromide. (A) Ethidium bromide is a flat plate-like molecule that is able to slot in between the base pairs of the double helix. (B) Ethidium bromide molecules, viewed sideways on, are shown intercalated into a double helix, increasing the distance between adjacent base pairs.

There are also several types of physical mutagen

It is not only the chemical composition of the environment that can cause mutations in DNA molecules. There are also a variety of physical mutagens. An example is ultraviolet (UV) radiation, which induces dimerization of adjacent pyrimidine bases, especially if these are both thymines. This results in a **cyclobutyl dimer** (Figure 16.15A). Other pyrimidine combinations also form dimers, the order of frequency being 5′-CT-3′ > 5′-TC-3′ > 5′-CC-3′. Purine dimers are much less common. UV-induced dimerization usually results in a deletion mutation when the modified strand is copied. Another type of UV-induced **photoproduct** is the **(6–4) lesion**, in which carbons number 4 and 6 of adjacent pyrimidines become covalently linked (Figure 16.15B).

Ionizing radiation is also mutagenic, having various effects on DNA depending on the type of radiation and its intensity. Point, insertion, and/or deletion mutations might arise, as well as more severe forms of DNA damage that prevent subsequent replication of the genome. Some types of ionizing radiation act directly on DNA, and others act indirectly by stimulating the formation of reactive molecules such as peroxides in the cell.

Heat is mutagenic. Heat stimulates the water-induced cleavage of the β-*N*-glycosidic bond that attaches the base to the sugar component of the nucleotide (Figure 16.16A). This occurs more frequently with purines than

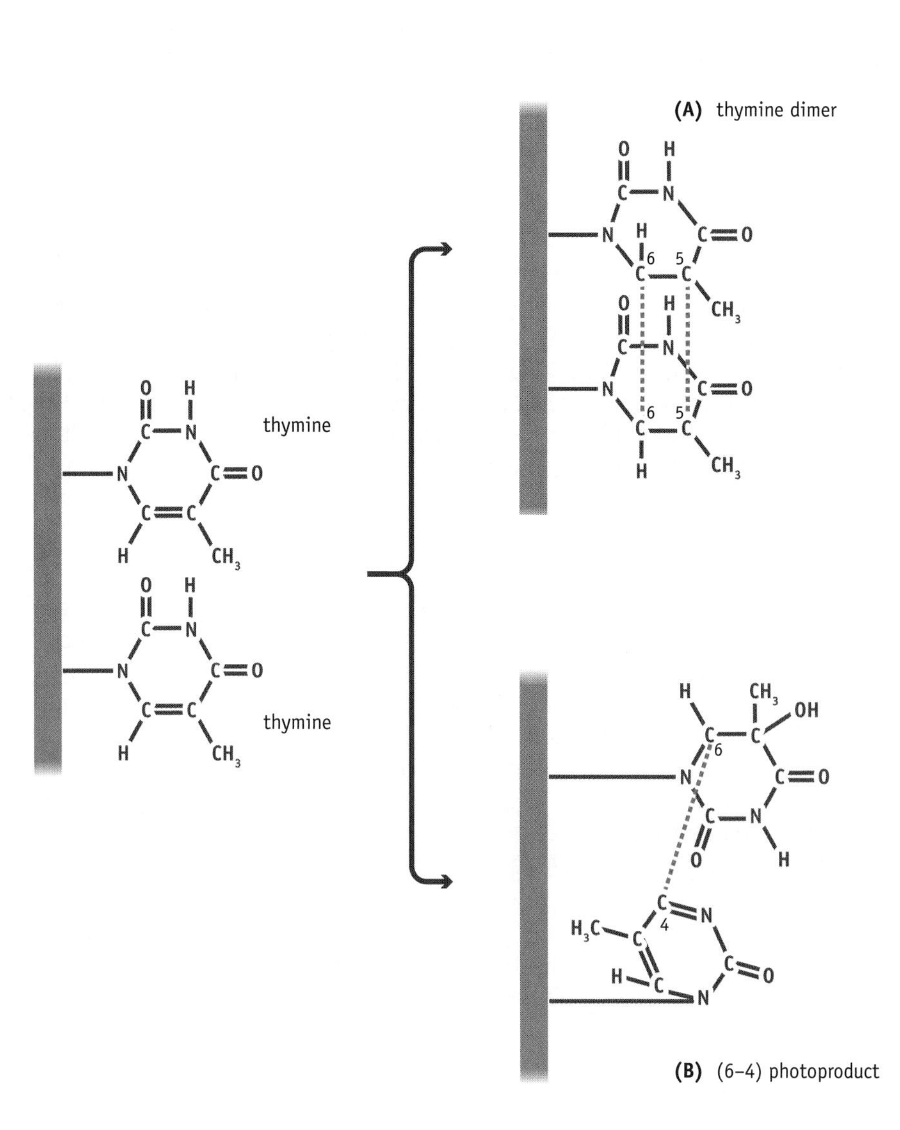

FIGURE 16.15 Photoproducts induced by UV irradiation. A segment of a polynucleotide containing two adjacent thymine bases is shown. (A) A thymine dimer contains two UV-induced covalent bonds, one linking the carbons at position 6 and the other linking the carbons at position 5. (B) The (6–4) lesion involves formation of a UV-induced covalent bond between carbons 4 and 6 of the adjacent nucleotides. In (A) and (B) the UV-induced covalent bonds are shown as dotted lines.

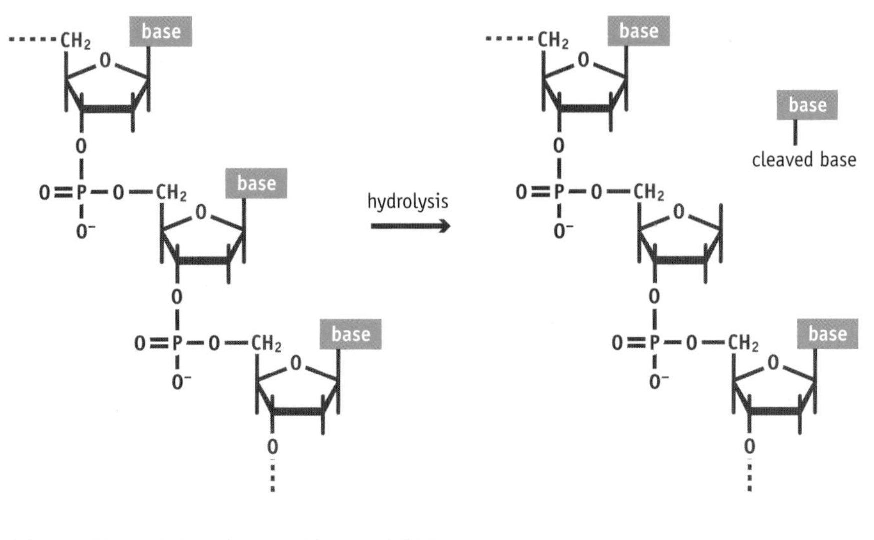

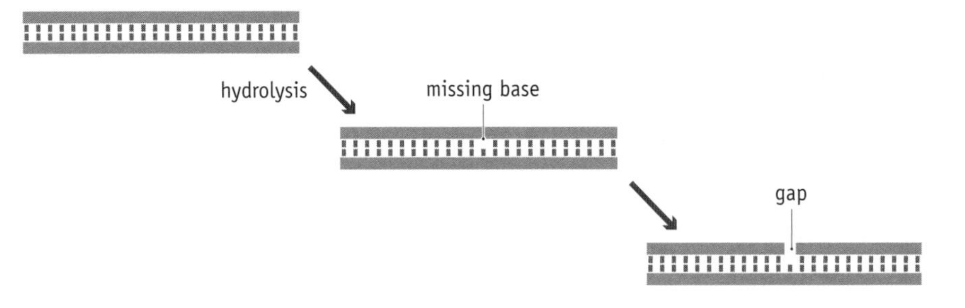

FIGURE 16.16 The mutagenic effect of heat. (A) Heat induces hydrolysis of β-*N*-glycosidic bonds, resulting in a baseless site in a polynucleotide. (B) Schematic representation of the effect of heat-induced hydrolysis on a double-stranded DNA molecule. The baseless site is unstable and degrades, leaving a gap in one strand.

with pyrimidines and results in an **AP** (apurinic/apyrimidinic) or **baseless site**. The sugar–phosphate that is left is unstable and rapidly degrades, leaving a gap if the DNA molecule is double-stranded (Figure 16.16B). This reaction is not normally mutagenic because cells have effective systems for repairing gaps, which is reassuring when one considers that 10,000 AP sites are generated in each human cell per day.

16.2 DNA REPAIR

In view of the thousands of damage events that genomes suffer every day, coupled with the errors that occur when the genome replicates, it is essential that cells possess efficient repair systems. Without these repair systems it would be only a few hours before key genes became inactivated by DNA damage. Similarly, cell lineages would accumulate replication errors at such a rate that their genomes would become dysfunctional after a few thousand cell divisions.

Most cells possess four different types of DNA repair systems (Figure 16.17). The first of these are the **direct repair systems**, which, as the name suggests, act directly on damaged nucleotides, converting each one back to its original structure. The second type is **excision repair**. This involves excision of a segment of the polynucleotide containing a damaged site, followed by resynthesis of the correct nucleotide sequence by a DNA polymerase. The segment removed may be just one nucleotide in length or

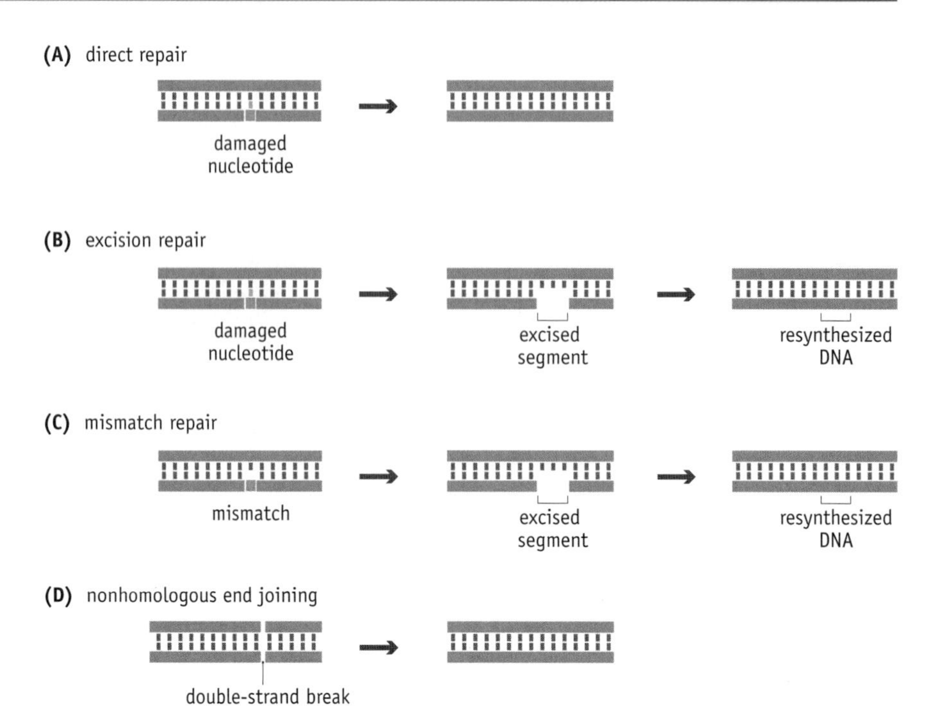

FIGURE 16.17 Four different types of DNA repair system. (A) Direct repair converts a damaged nucleotide back to its original structure. (B) Excision repair involves removal of a segment of the polynucleotide containing a damaged nucleotide, followed by resynthesis of the correct DNA sequence. (C) Mismatch repair is similar to excision repair, but is initiated by detection of a base-pairing error. (D) Nonhomologous end joining repairs double-strand breaks.

much longer. A special type of excision repair is **mismatch repair**, which corrects errors of replication, again by excising a stretch of single-stranded DNA containing the offending nucleotide and then repairing the resulting gap. Finally, **nonhomologous end joining** (**NHEJ**) is used to mend double-strand breaks.

The impact of these pathways cannot be overstated. The error rate for DNA synthesis in *E. coli* is 1 mistake for every 10^7 bp that are replicated, but the overall error rate for replication of the *E. coli* genome is only 1 in 10^{11} to 1 in 10^{10}. The improvement compared with the polymerase error rate is due entirely to the mismatch repair system. The repair enzymes scan newly replicated DNA for positions where the bases are unpaired and correct the mistakes that the replication enzymes make. Because of the efficiency of this system, on average only one uncorrected replication error occurs every 1000 times that the *E. coli* genome is copied.

The repair processes are not particularly complicated, but we need to spend some time examining how each one works and what they achieve.

Direct repair systems fill in nicks and correct some types of nucleotide modification

Most of the types of DNA damage that are caused by chemical or physical mutagens can be repaired only by excision of the damaged nucleotide followed by resynthesis of a new stretch of DNA, as shown in Figure 16.17B. Only a few types of damage can be repaired directly without removal of the nucleotide. These include some of the products of alkylation, which are directly reversible by enzymes that transfer the alkyl group from the nucleotide to their own polypeptide chains. Enzymes capable of doing this are present in many different organisms and include the **Ada enzyme** of *E. coli*, which is involved in an adaptive process that this bacterium is able to activate in response to DNA damage. Ada removes alkyl groups attached to the oxygen atoms at positions 4 and 6 of thymine and guanine, respectively, and can also repair phosphodiester bonds that have become methylated. Other alkylation repair enzymes have more restricted specificities, an example being human **MGMT** (O^6-methylguanine-DNA methyltransferase), which, as its name suggests, removes alkyl groups only from position 6 of guanine.

Cyclobutyl dimers resulting from UV damage can also be repaired directly, by a light-dependent system called **photoreactivation**. In *E. coli*, the process involves the enzyme called **DNA photolyase**. When stimulated by light with a wavelength between 300 and 500 nm this enzyme binds to cyclobutyl dimers and converts them back to the original nucleotides. Photoreactivation is a widespread but not universal type of repair. It is known in many but not all bacteria and also in quite a few eukaryotes, including some vertebrates, but is absent in humans and other placental mammals. A similar type of photoreactivation involves the **(6–4) photoproduct photolyase** and results in repair of (6–4) lesions. Neither *E. coli* nor humans have this enzyme but it is possessed by a variety of other organisms.

Many types of damaged nucleotide can be repaired by base excision

Base excision is the least complex of the various repair systems that involve removal of one or more damaged nucleotides followed by resynthesis of DNA to span the resulting gap. It is used to repair many modified nucleotides whose bases have suffered relatively minor damage resulting from, for example, exposure to alkylating agents or ionizing radiation.

The process is initiated by a **DNA glycosylase**, which cleaves the β-*N*-glycosidic bond between a damaged base and the sugar component of the nucleotide (Figure 16.18A). Each DNA glycosylase has a limited specificity, the specificities of the glycosylases possessed by a cell determining the range of damaged nucleotides that can be repaired in this way. Most organisms are able to deal with deaminated bases such as uracil (deaminated cytosine) and hypoxanthine (deaminated adenine), oxidation products such as 5-hydroxycytosine and thymine glycol, and methylated bases such as 3-methyladenine, 7-methylguanine, and 2-methylcytosine.

A DNA glycosylase removes a damaged base by "flipping" the structure to a position outside of the helix and then detaching it from the polynucleotide. This creates an AP, or baseless, site that is converted into a single-nucleotide gap in the second step of the repair pathway, which is outlined in Figure 16.18B. This step can be carried out in a variety of ways. The standard method makes use of an **AP endonuclease**, which cuts the phosphodiester bond on the 5′ side of the AP site. Some AP endonucleases can also remove the sugar from the AP site, this being all that remains of the damaged nucleotide, but others lack this ability and so work in conjunction with a separate **phosphodiesterase**, which cuts out the sugar. An alternative pathway for converting the AP site into a gap utilizes the endonuclease activity possessed by some DNA glycosylases, which can make a cut at the 3′ side of the AP site, probably at the same time that the damaged base is removed, followed again by removal of the sugar by a phosphodiesterase.

The single-nucleotide gap is filled by a DNA polymerase, using base pairing with the undamaged base in the other strand of the DNA molecule to ensure that the correct nucleotide is inserted. In *E. coli* the gap is filled by DNA polymerase I and in mammals by DNA polymerase β. After gap filling, the final phosphodiester bond is put in place by a DNA ligase.

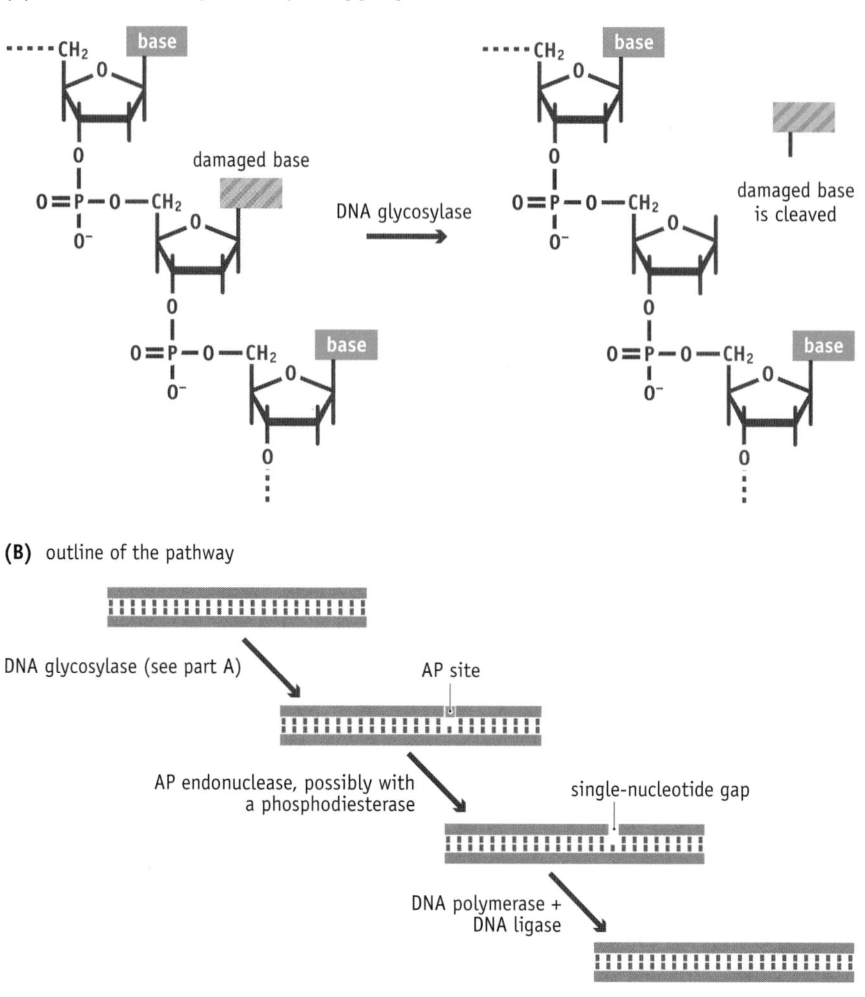

FIGURE 16.18 Base excision repair. (A) Excision of a damaged nucleotide by a DNA glycosylase. (B) Schematic representation of the base excision repair pathway.

Nucleotide excision repair is used to correct more extensive types of damage

Nucleotide excision repair is a second type of excision repair system, able to deal with more extreme forms of damage such as intrastrand crosslinks and bases that have become modified by attachment of large chemical groups. It is also able to correct cyclobutyl dimers by a **dark repair** process, providing those organisms that do not have the photoreactivation system (such as humans) with a means of repairing this type of damage.

In nucleotide excision repair, a segment of single-stranded DNA containing the damaged nucleotide(s) is excised and replaced with new DNA. The process is therefore similar to base excision repair except that it is not preceded by selective base removal, and a longer stretch of polynucleotide is cut out. The best studied example of nucleotide excision repair is the **short-patch** process of *E. coli*, so called because the region of polynucleotide that is excised and subsequently "patched" is relatively short, usually 12 nucleotides in length.

Short-patch repair is initiated by a multienzyme complex called the **UvrABC endonuclease**. In the first stage of the process a trimer comprising two UvrA proteins and one copy of UvrB attaches to the DNA at the damaged site (Figure 16.19). How the site is recognized is not known, but the broad specificity of the process indicates that individual types of

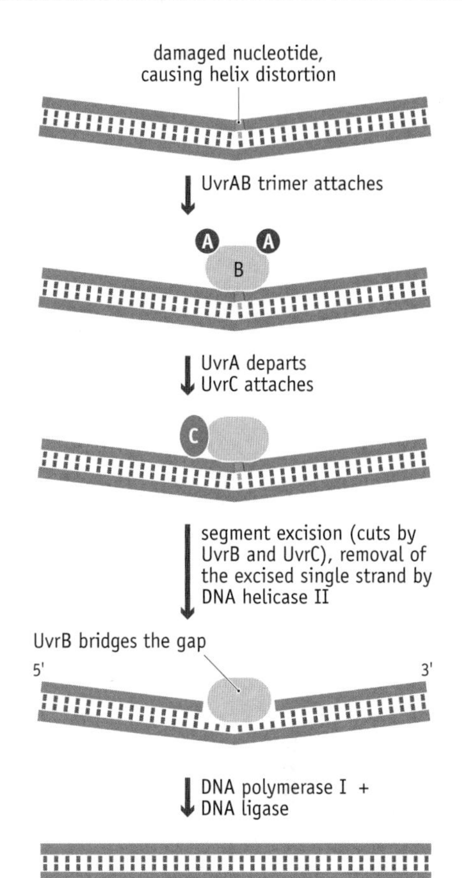

FIGURE 16.19 Short-patch nucleotide excision repair in *E. coli*. The process begins with recognition of the damaged site by the UvrAB trimer. UvrA leaves the complex and UvrC joins, causing a cut to be made on either side of the damaged site. UvrB bridges the gap while it is repaired by DNA polymerase I and DNA ligase. In the top part of the drawing, the damaged nucleotide is shown distorting the helix because this is thought to be one of the recognition signals for the UvrAB trimer.

damage are not directly detected and that the complex must search for a more general attribute of DNA damage such as distortion of the double helix. UvrA may be the part of the complex most involved in damage location because it dissociates once the site has been found and plays no further part in the repair process.

Departure of UvrA allows UvrC to bind, forming a UvrBC dimer that cuts the polynucleotide on either side of the damaged site. The first cut is made by UvrB at the fifth phosphodiester bond downstream of the damaged nucleotide, and the second cut is made by UvrC at the eighth phosphodiester bond upstream. This results in a 12-nucleotide excision, although there is some variability, especially in the position of the UvrB cut site.

The excised segment is then removed, usually as an intact piece of DNA, by DNA helicase II, which detaches the segment by breaking the base pairs holding it to the second strand. UvrC also detaches at this stage, but UvrB remains in place and bridges the gap produced by the excision. The bound UvrB is thought to prevent the single-stranded region that has been exposed from base-pairing with itself. Alternatively, the role of UvrB could be to prevent this strand from becoming damaged, or possibly to direct the DNA polymerase to the site that needs to be repaired. As in base excision repair, the gap is filled by DNA polymerase I and the last phosphodiester bond is synthesized by DNA ligase.

E. coli also has a **long-patch** nucleotide excision repair system that involves Uvr proteins but differs in that the piece of DNA that is excised can be anywhere up to 2 kb in length. Long-patch repair has been less well studied and the process is not understood in detail, but it is presumed to work on more extensive forms of damage, possibly regions where groups of nucleotides, rather than just individual ones, have become modified. Eukaryotes have just one type of nucleotide excision repair pathway, resulting in replacement of only 24 to 29 nucleotides of DNA, and seemingly unrelated to either of the excision pathways in bacteria.

Mismatch repair corrects errors of replication

Each of the repair systems that we have looked at so far—direct, base excision, and nucleotide excision repair—recognize and act upon DNA damage caused by mutagens. This means that they search for abnormal chemical structures such as modified nucleotides, cyclobutyl dimers, and intrastrand crosslinks. They cannot, however, correct mismatches resulting from errors in replication, because the mismatched nucleotide is not abnormal in any way. It is simply an A, C, G, or T that has been inserted at the wrong position. As these nucleotides look exactly like any other nucleotide, the mismatch repair system that corrects replication errors has to detect not the mismatched nucleotide itself but the absence of base pairing between the parent and daughter strands. Once it has found a mismatch, the repair system excises part of the daughter polynucleotide and fills in the gap, in a manner similar to base and nucleotide excision repair.

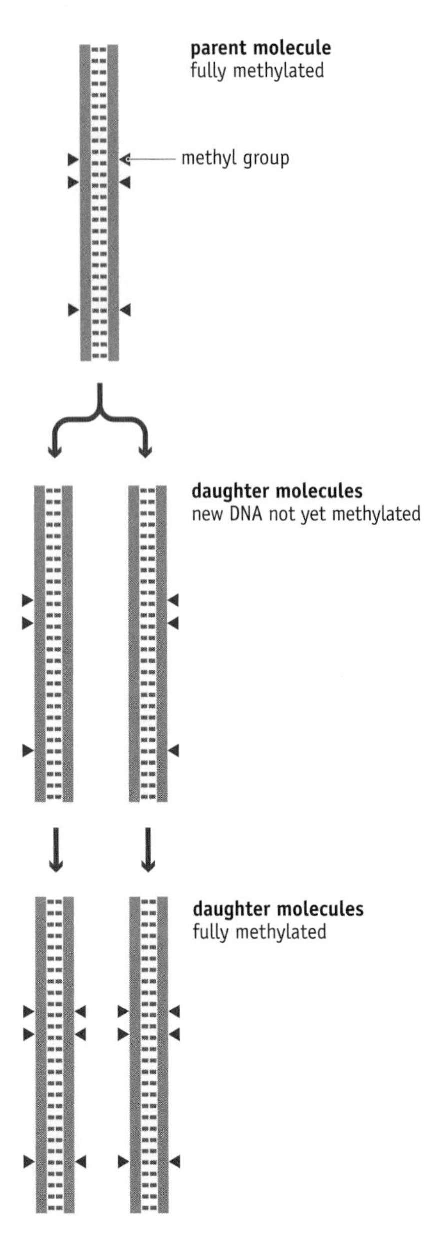

FIGURE 16.20 Methylation of newly synthesized DNA in *E. coli* does not occur immediately after replication, providing a window of opportunity for the mismatch repair proteins to recognize the daughter strands and correct replication errors.

There is one difficulty. The repair must be made in the daughter polynucleotide, because it is in this newly synthesized strand that the error has occurred. The parent polynucleotide, on the other hand, has the correct sequence. How does the repair process know which strand is which? In *E. coli* the answer is that the parent strand is methylated and so can be distinguished from the newly synthesized polynucleotide, which is not methylated.

E. coli DNA is methylated because of the activities of two enzymes. These are **DNA adenine methylase** (**Dam**), which converts adenines to 6-methyladenines in the sequence 5′-GATC-3′, and **DNA cytosine methylase** (**Dcm**), which converts cytosines to 5-methylcytosines in 5′-CCAGG-3′ and 5′-CCTGG-3′. These methylations are not mutagenic, the modified nucleotides having the same base-pairing properties as the unmodified versions. There is a delay between DNA replication and methylation of the daughter strand, and it is during this window of opportunity that the repair system scans the DNA for mismatches and makes the required corrections in the unmethylated, daughter strand (Figure 16.20).

E. coli has at least three mismatch repair systems, called long-patch, short–patch, and very-short-patch, the names indicating the relative lengths of the excised and resynthesized segments. The long-patch system replaces up to 1 kb or more of DNA and requires the MutH, MutL, and MutS proteins, as well as the DNA helicase II that we met during nucleotide excision repair. MutS recognizes the mismatch, and MutH distinguishes the two strands by binding to unmethylated 5′-GATC-3′ sequences (Figure 16.21). After binding, MutH cuts the phosphodiester bond immediately upstream of the G in the methylation sequence, and DNA helicase II detaches a segment of the single strand starting from this point. There does not appear to be an enzyme that cuts the strand downstream of the mismatch. Instead, the detached single-stranded region is degraded by an exonuclease that follows the helicase and continues beyond the mismatch site. The gap is then filled in by DNA polymerase I and DNA ligase.

The mismatch repair processes of eukaryotes probably work in a similar way, although methylation might not be the method used to distinguish between the parent and daughter polynucleotides. Methylation has been implicated in mismatch repair in mammalian cells, but the DNA of some eukaryotes, including fruit flies and yeast, is not extensively methylated. It is therefore thought that in these organisms a different method must be used to identify the daughter strand. Possibilities include an association between the repair enzymes and the replication complex, so that repair is coupled with DNA synthesis, or use of single-strand binding proteins to mark the parent strand.

DNA breaks can also be repaired

Finally, we must study how breaks in DNA molecules are repaired. A single-stranded break in a double-stranded DNA molecule, such as is produced by some types of oxidative damage, does not present the cell with a critical problem, as the double helix retains its overall intactness. In humans, the exposed strand is coated with PARP1 proteins, which protect it from damage. The break can then be filled in by DNA polymerase and ligase enzymes (Figure 16.22).

A double-stranded break is more serious because this converts the original double helix into two separate fragments that have to be brought back together again in order for the break to be repaired. The two broken ends must be protected in some way, because loss of nucleotides would result

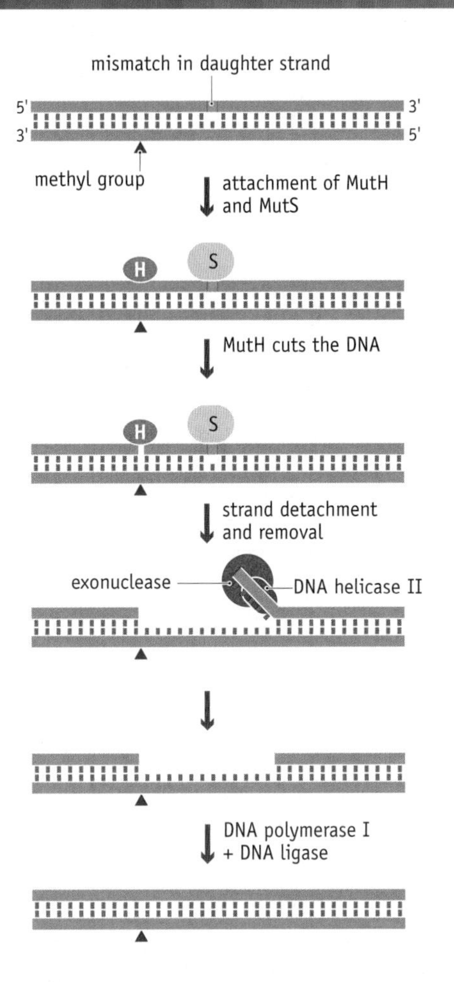

FIGURE 16.21 Long-patch mismatch repair in *E. coli*. The process begins with attachment of MutS, which recognizes the mismatch, and MutH, which distinguishes the methylated and unmethylated polynucleotides. MutH cuts the unmethylated polynucleotide at the methylation sequence, and DNA helicase II detaches the single strand around the mismatch. The detached single strand is degraded by an exonuclease and the gap is filled by DNA polymerase I and DNA ligase.

in a deletion mutation appearing at the repaired breakpoint. The repair processes must also ensure that the correct ends are joined. If there are two broken chromosomes in the nucleus, then the correct pairs must be brought together so that the original structures are restored. Experimental studies indicate that achieving this outcome is difficult and if two chromosomes are broken then misrepair resulting in hybrid structures occurs relatively frequently. Even if only one chromosome is broken, there is still a possibility that a natural chromosome end could be confused as a break and an incorrect repair made.

Double-strand breaks are generated by exposure to ionizing radiation and some chemical mutagens, and breakages can also occur during DNA replication. These breaks can be repaired by the nonhomologous end joining process, which involves a pair of proteins, called Ku, which bind the DNA ends on either side of the break (Figure 16.23). The individual Ku proteins have an affinity for one another, which means that the two broken ends of the DNA molecule are brought into proximity and can be joined back together by a DNA ligase.

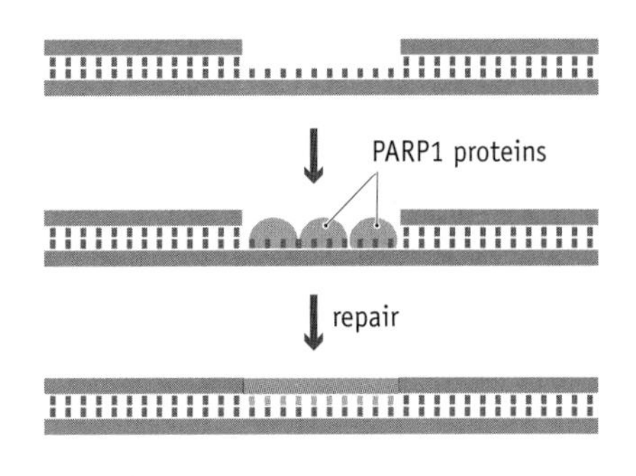

FIGURE 16.22 Single-strand break repair.

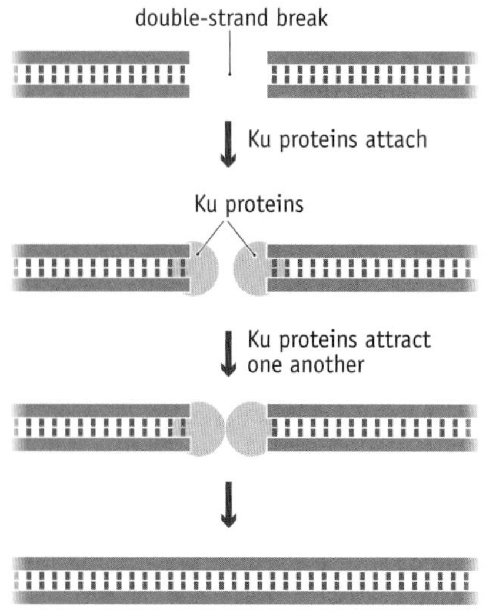

FIGURE 16.23 Repair of a double-strand break by nonhomologous end joining. Ku proteins attach to the DNA ends on either side of the break. The affinity between these proteins draws the two ends together so the break can be repaired by a DNA ligase.

NHEJ was originally thought to be restricted to eukaryotes, but searches of the protein databases have uncovered bacterial versions of the Ku proteins. Experimental studies have indicated that these act in conjunction with bacterial ligases in a simplified type of double-strand break repair.

In an emergency, DNA damage can be bypassed during genome replication

If a region of the genome has suffered extensive damage, then it is conceivable that the repair processes will be overwhelmed. The cell then faces a stark choice between dying and attempting to replicate the damaged region even though this replication may be error-prone and result in mutated daughter molecules. When faced with this choice, *E. coli* cells invariably take the second option, by inducing one of several emergency procedures for bypassing sites of major damage.

The best studied of these bypass processes occurs as part of the **SOS response**. This system enables an *E. coli* cell to replicate its DNA even though the template polynucleotides contain AP sites and/or photoproducts that would normally block, or at least delay, the replication complex. Bypass of these sites requires construction of a **mutasome**, comprising DNA polymerase V (also called the UmuD'$_2$C complex) and several copies of the RecA protein. RecA is a single-stranded DNA-binding protein that we met previously when we examined its role in recombination. In this bypass system, RecA coats the damaged strands, enabling DNA polymerase V to carry out its own, error-prone DNA synthesis until the damaged region has been passed (Figure 16.24). The main replicating enzyme, DNA polymerase III, then takes over once again.

The SOS response is primarily looked on as the last best chance that the bacterium has to replicate its DNA and hence survive under adverse conditions. However, the price of survival is an increased mutation rate, because the mutasome does not repair damage, it simply allows a damaged region of a polynucleotide to be replicated. When it encounters a damaged position, DNA polymerase V selects a nucleotide more or less at random, although with some preference for placing an A opposite an AP site.

Defects in DNA repair underlie human diseases, including cancers

The importance of DNA repair is emphasized by the number and severity of inherited human diseases that have been linked with defects in one of the repair processes. One of the best characterized of these is xeroderma pigmentosum, which results from a mutation in any one of several genes for proteins involved in nucleotide excision repair. Nucleotide excision is the only way in which human cells can repair cyclobutyl dimers and other photoproducts, so it is no surprise that the symptoms of xeroderma pigmentosum include hypersensitivity to UV radiation, patients suffering more mutations than normal on exposure to sunlight, and often suffering from skin cancer.

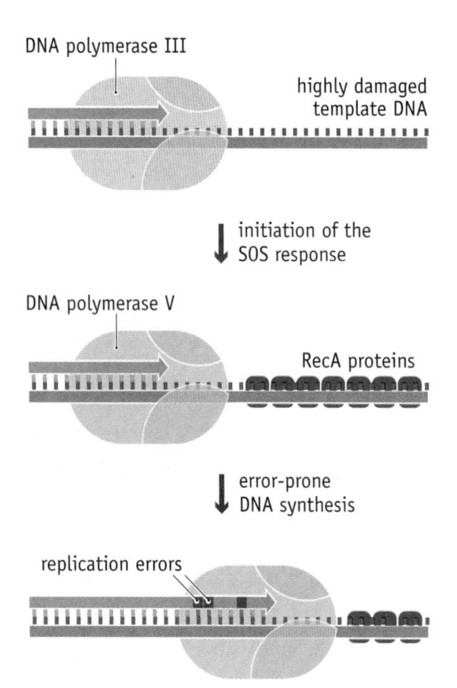

FIGURE 16.24 The SOS response of *E. coli*. During DNA replication, a highly damaged region can be "bypassed" by the SOS response. DNA polymerase III, the main replicating enzyme, detaches and is replaced by the mutasome, a combination of DNA polymerase V and the RecA protein. RecA coats the damaged polynucleotides, enabling DNA polymerase V to carry out error-prone DNA synthesis through the damaged region. DNA polymerase III then takes over again.

Defects in nucleotide excision repair are also linked with breast and ovarian cancers, and Cockayne syndrome, a complex disease manifested by growth and neurological disorders. It is also the underlying cause of the cancer-susceptibility syndrome called HNPCC (hereditary nonpolyposis colorectal cancer), although this disease was originally identified as a defect in mismatch repair. Ataxia telangiectasia, the symptoms of which include sensitivity to ionizing radiation, results from defects in the *ATX* gene, which is involved in the damage detection process. Other diseases that are associated with a breakdown in DNA repair are Bloom syndrome and Werner syndrome, which are caused by inactivation of a DNA helicase that may have a role in NHEJ, Fanconi anemia, which confers sensitivity to chemicals that cause crosslinks in DNA but whose biochemical basis is not yet known, and spinocerebellar ataxia, which results from defects in the pathway used to repair single-strand breaks.

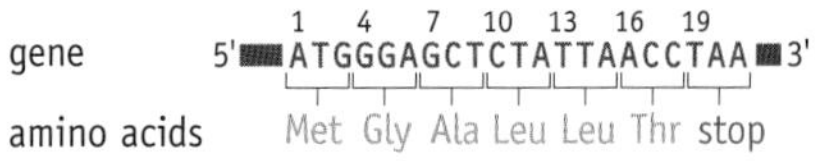

gene 5' ATGGGAGCTCTATTAACCTAA 3' (positions 1 4 7 10 13 16 19)
amino acids Met Gly Ala Leu Leu Thr stop

FIGURE 16.25 A short imaginary gene.

16.3 THE EFFECTS OF MUTATIONS ON GENES, CELLS, AND ORGANISMS

Now that we understand how mutations arise in DNA molecules, and appreciate that many of them are repaired before they can be propagated by DNA replication, we must consider the effect of those few mutations that slip through.

Mutations have various effects on the biological information contained in a gene

Many mutations are **silent** because they occur in intergenic regions. They result in changes in the nucleotide sequence but have no effect on the genome's biological information. But what are the likely effects if a mutation occurs in the coding region of a gene? To answer this question we will invent a short imaginary gene (Figure 16.25) which, although coding for a protein only six amino acids in length, will show us most of the possible consequences.

The first possibility is that a point mutation takes place at the third nucleotide position of a codon and changes the codon but, owing to the degeneracy of the genetic code, not the amino acid. In the example shown in Figure 16.26, nucleotide 15 is changed from A to G. This changes the fifth codon of our gene from 5′-TTA-3′, coding for leucine, to 5′-TTG-3′, which also codes for leucine. This is called a **synonymous mutation**, and as it has no effect on the amino acid sequence of the gene product it is silent as far as the cell is concerned.

A **nonsynonymous mutation**, on the other hand, is a point change that does change the amino acid sequence of the protein coded by the gene. Most point changes at the first or second nucleotide positions of a codon result in nonsynonymous mutations, as do a few third-position changes. In our gene, changing nucleotide 4 from G to A produces an arginine codon instead of a glycine codon (Figure 16.27). Similarly, changing nucleotide 15 from A to T specifies phenylalanine rather than leucine. A nonsynonymous mutation gives rise to a protein with a single amino acid change. Whether or not this changes the phenotype of the cell depends on the precise role of the mutated amino acid in the protein. Most proteins can tolerate some changes in their amino acid sequence, although a nonsynonymous mutation that alters an amino acid essential for structure or function will inactivate the protein and lead to a mutant phenotype.

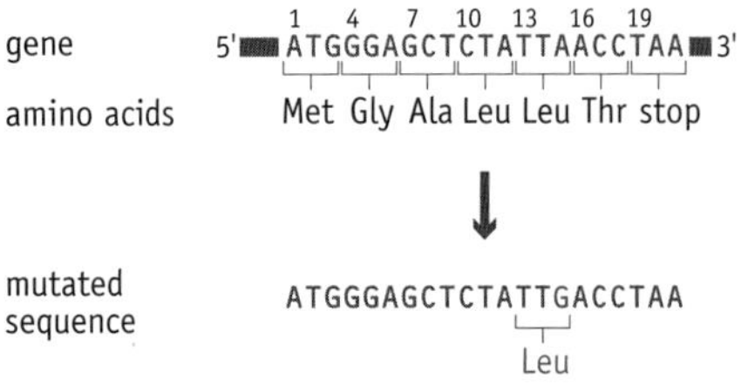

gene 5' ATGGGAGCTCTATTAACCTAA 3' (positions 1 4 7 10 13 16 19)
amino acids Met Gly Ala Leu Leu Thr stop
↓
mutated sequence ATGGGAGCTCTATTGACCTAA
Leu

FIGURE 16.26 A silent mutation within a gene. Nucleotide 15 has been changed from A to G, but because of the degeneracy of the genetic code, the amino acid sequence of the protein remains the same.

1 4 7 10 13 16 19
ATGGGAGCTCTATTAACCTAA
Met Gly Ala Leu Leu Thr stop
ATGAGAGCTCTATTAACCTAA
Arg
ATGGGAGCTCTATTTACCTAA
Phe

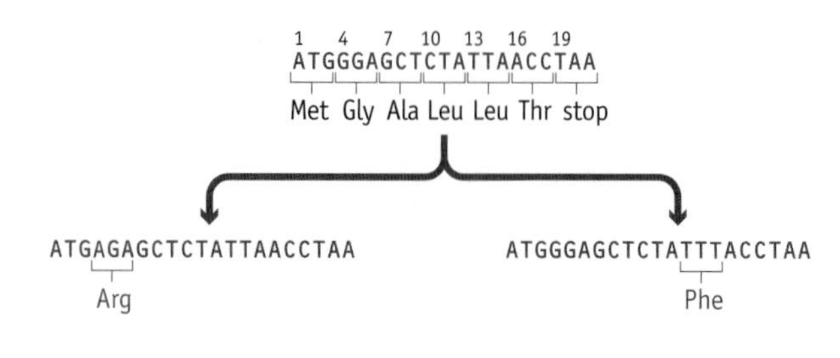

FIGURE 16.27 Two nonsynonymous mutations. Changing nucleotide 4 from G to A produces an arginine codon instead of a glycine codon, and changing nucleotide 15 from A to T gives phenylalanine rather than leucine.

Other mutations are much more likely to result in a mutant phenotype. A **nonsense mutation** (Figure 16.28) is a point mutation that changes a codon specifying an amino acid into a termination codon. The result is a gene that codes for a truncated protein, one that has lost a segment at its carboxyl terminus. In many cases, although not always, this segment will include amino acids essential for the protein's activity. A **frameshift mutation** (Figure 16.29) is the usual consequence of an insertion or deletion event. This is because the addition or removal of any number of base pairs that is not a multiple of three causes the ribosome to read a completely new set of codons downstream from the mutation. Finally, a **readthrough mutation** (Figure 16.30) converts a termination codon into one specifying an amino acid. This results in readthrough of the stop signal so the protein is extended by an additional series of amino acids at its carboxyl terminus. Most proteins can tolerate short extensions without an effect on function, but longer extensions might interfere with folding of the protein and so result in reduced activity.

Mutations have various effects on multicellular organisms

In order to produce a mutant phenotype the nucleotide sequence alteration must give rise to a gene product that is unable to fulfill its function in the cell. In many cases the cell will be unable to tolerate the loss of the function and will die. Sequence alterations that cause cell death are called **lethal mutations**. However, not all mutations that produce a change in phenotype are so drastic in their effect. Some mutations inactivate proteins that are not essential to the cell, and others result in proteins with reduced or modified activities. What effects would these nonlethal mutations have on a multicellular organism?

1 4 7 10 13 16 19
ATGGGAGCTCTATTAACCTAA
Met Gly Ala Leu Leu Thr stop
ATGGGAGCTCTATGAACCTAA
stop

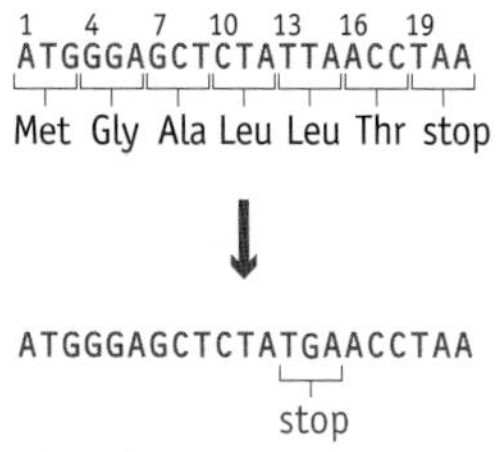

FIGURE 16.28 A nonsense mutation. Changing nucleotide 14 from T to G converts the leucine codon into a termination codon.

1 4 7 10 13 16 19 22 25
ATGGGAGCTCTATTAACCTAATTTGA
Met Gly Ala Leu Leu Thr stop
insertion
deletion
ATGGGGAGCTCTATTAACCTAATTTGA
Met Gly Ser Ser Ile Asn Leu Ile stop
ATGGGGCTCTATTAACCTAATTTGA
Met Gly Leu Tyr stop

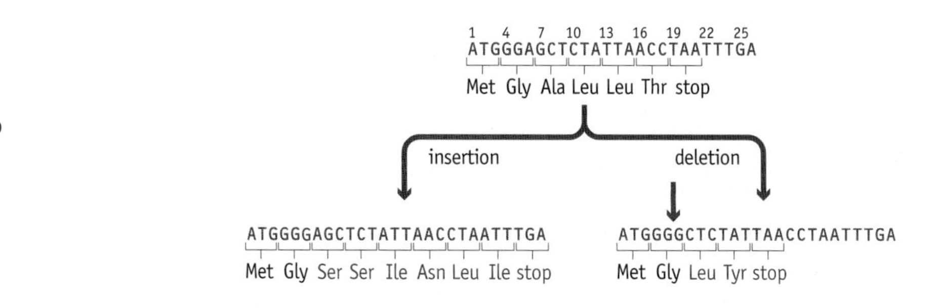

FIGURE 16.29 Two frameshift mutations, one caused by inserting a G between nucleotides 5 and 6, and the other by deleting nucleotide 6.

We must distinguish between mutations that occur in somatic cells, the ones that make up the bulk of the organism but are not passed on during reproduction, and those in sex cells, which give rise to gametes. Because somatic cells do not pass on copies of their genomes to the next generation, a somatic cell mutation is important only for the organism in which it occurs. It has no potential evolutionary impact. In fact, most somatic cell mutations have no significant effect, even if they result in cell death, because there are many other identical cells in the same tissue and the loss of one cell is immaterial. An exception is when a mutation causes a somatic cell to malfunction in a way that is harmful to the organism, for instance by inducing tumor formation or other cancerous activity.

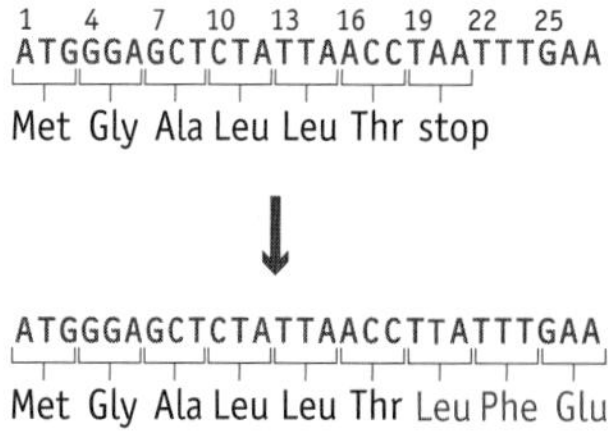

1 4 7 10 13 16 19 22 25
ATGGGAGCTCTATTAACCTAATTTGAA
Met Gly Ala Leu Leu Thr stop
ATGGGAGCTCTATTAACCTTATTTGAA
Met Gly Ala Leu Leu Thr Leu Phe Glu

FIGURE 16.30 A readthrough mutation. Changing nucleotide 20 from A to T converts the termination codon into one specifying leucine.

Mutations in germ cells are more important because they can be transmitted to members of the next generation and will then be present in all the cells of any individual who inherits the mutation. Most mutations, including all silent ones and many in coding regions, will still not change the phenotype of the organism in any significant way. Those that do have an effect can be divided into two categories, **loss of function** and **gain of function**.

Loss of function is the normal result of a mutation that reduces or abolishes a protein activity. Because most eukaryotes are diploid, meaning that they have two copies of each chromosome and hence two copies of each gene, a loss-of-function mutation might not result in a change of phenotype, as the second copy of the gene is still active and compensates for the loss (Figure 16.31). There are some exceptions, one example being **haploinsufficiency**, where the organism is unable to tolerate the approximately 50% reduction in protein activity resulting from the loss of one of the gene copies.

Gain-of-function mutations are much less common. The mutation must be one that confers an abnormal activity on a protein. Examples are known with proteins that act as cell surface receptors. These proteins respond to external stimuli by sending into the cell signals that switch genes on or off as appropriate. A mutation could result in the receptor protein sending signals into the cell even though the external stimulus is absent (Figure 16.32). The cell therefore acts as if it is continuously being affected by the stimulus, resulting in the gain in function. Gain-of-function mutations cannot be compensated by the presence of the second, unmutated gene copy in a diploid organism.

loss-of-function mutation
no protein product
pair of homologous chromosomes
protein product

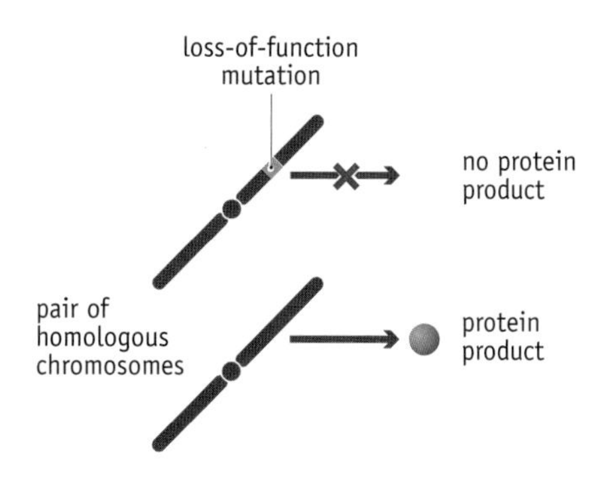

FIGURE 16.31 A loss-of-function mutation is usually recessive because a functional version of the gene is present on the second chromosome copy.

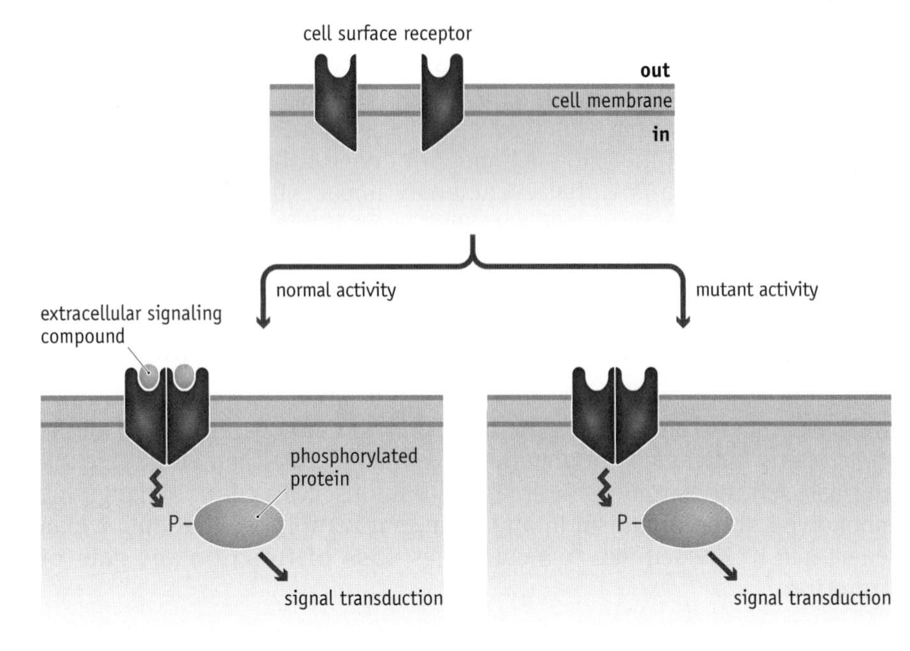

FIGURE 16.32 One way in which a gain-of-function can arise. The top drawing shows a cell surface receptor. The normal activity is shown on the left. The presence of the external signaling compound causes the receptor to dimerize and activate an internal protein. This protein initiates a signal transduction pathway that switches on a particular set of genes whose products are needed only when the external stimulus is present. On the right is the activity of the mutant receptor protein. Now the receptor protein dimerizes even though there is no external signaling compound. The genes targeted by the signal transduction pathway are therefore switched on all the time, and the mutant cell "gains" the function specified by these genes.

Assessing the effects of mutations on the phenotypes of multicellular organisms can be difficult. Not all mutations have an immediate impact. Some display **incomplete penetrance** and others have **delayed onset**. Typically, incomplete penetrance occurs when a mutation merely provides a predisposition to developing the phenotype, which appears only in response to a later trigger (Figure 16.33). Incomplete penetrance is displayed by a number of human genetic cancers, with the triggering factor possibly linked to the environment or to the person's lifestyle. If the trigger is never encountered during an individual's lifetime then the person remains healthy until dying of some other cause.

Delayed-onset phenotypes are expressed only late in an individual's lifetime, possibly again due to the need for a trigger but in this case one that results from the accumulation of some factor over time (Figure 16.34). Many of the debilitating human afflictions of old age, such as Huntington's disease, fall into this category, with the accumulating trigger thought possibly to be toxins or some kind of cellular damage. As with incomplete penetrance, some individuals with a predisposing genotype might never express the phenotype simply because they do not accumulate the trigger to a critical level during their lifetime. The difference between the two phenomena is that with incomplete penetrance the phenotype may never appear however long the person lives, but with delayed onset it will inevitably appear if the person lives long enough.

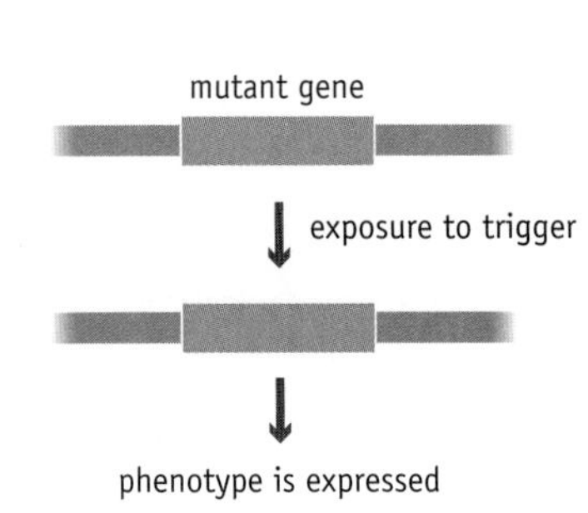

FIGURE 16.33 Incomplete penetrance.

A second mutation may reverse the phenotypic effect of an earlier mutation

There are various ways in which the effect of a mutation can be reversed by a second mutation, the simplest being when the second mutation restores the original nucleotide sequence of the DNA molecule. A point mutation can be reversed by a second point mutation, an insertion event by a subsequent deletion, and so on. These events are called **back mutations** and are not very likely unless the site at which the original mutation occurred has some natural predisposition toward mutation.

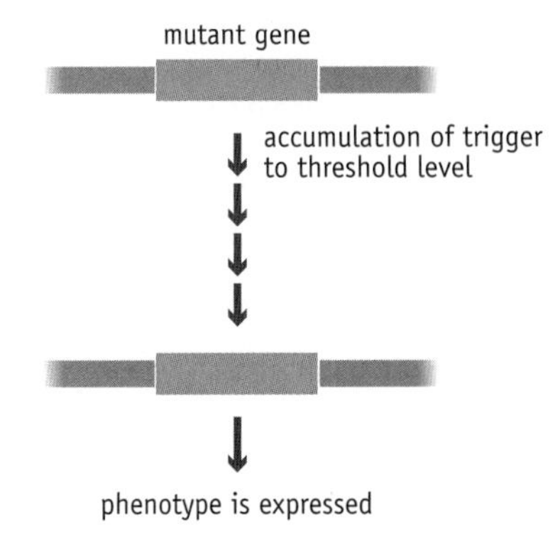

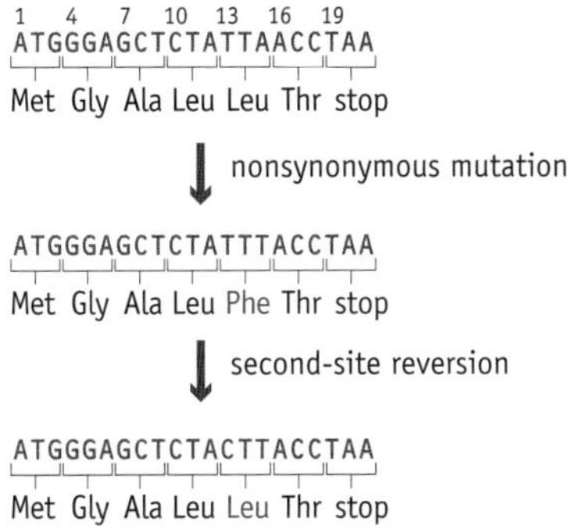

Many mutations can also be corrected by a **second-site reversion**, a second mutation that restores the original phenotype but does not return the DNA sequence to its precise unmutated form. To illustrate one of the ways in which this might occur we will return to the made-up gene that we used earlier to illustrate some of the effects of mutations. One of the nonsynonymous mutations shown in Figure 16.27 involved changing nucleotide 15 from A to T, changing a leucine codon (5′-TTA-3′) into a codon for phenylalanine (5′-TTT-3′). If we now introduce a second mutation, changing nucleotide 13 from T to C, the codon becomes 5′-CTT-3′, which codes for leucine once more (Figure 16.35). The nonsynonymous mutation has been corrected and the gene product returned to normal, even though the nucleotide sequence remains different from the original.

A second-site reversion occurs within the *same* gene as the original mutation. The final way of reversing the effect of a mutation is by **suppression**, in which the second mutation occurs in a *different* gene. There are several examples of this phenomenon but the most common is the reversal of a nonsense mutation by a suppressive mutation in the anticodon of a tRNA. For instance, the anticodon for $tRNA^{Trp}$ is 5′-CCA-3′, to decode the tryptophan codon 5′-UGG-3′. If the first nucleotide of the anticodon in the tRNA is changed by mutation from C to U, then the anticodon becomes 5′-UCA-3′ and now decodes the termination codon 5′-UGA-3′ (Figure 16.36A). This termination codon is therefore read as a tryptophan codon during protein synthesis. The shortened polypeptide that we produced by a nonsense mutation in Figure 16.28 would now be restored to full length (Figure 16.36B). Note, however, that the amino acid inserted into the protein when the suppressed termination codon is read is not the one found at the corresponding position in the unmutated version. As with a nonsynonymous mutation, such a small alteration in the amino acid sequence can probably be tolerated by the protein, so the activity is restored.

mutant gene
accumulation of trigger to threshold level
phenotype is expressed

FIGURE 16.34 Delayed onset.

1 4 7 10 13 16 19
ATGGGAGCTCTATTAACCTAA
Met Gly Ala Leu Leu Thr stop
nonsynonymous mutation
ATGGGAGCTCTATTTACCTAA
Met Gly Ala Leu Phe Thr stop
second-site reversion
ATGGGAGCTCTACTTACCTAA
Met Gly Ala Leu Leu Thr stop

FIGURE 16.35 An example of second-site reversion. The first mutation changes nucleotide 15 from A to T, converting a leucine codon into one for phenylalanine. The second mutation changes nucleotide 13 from T to C, causing a reversion back to a leucine codon, without restoring the original nucleotide sequence.

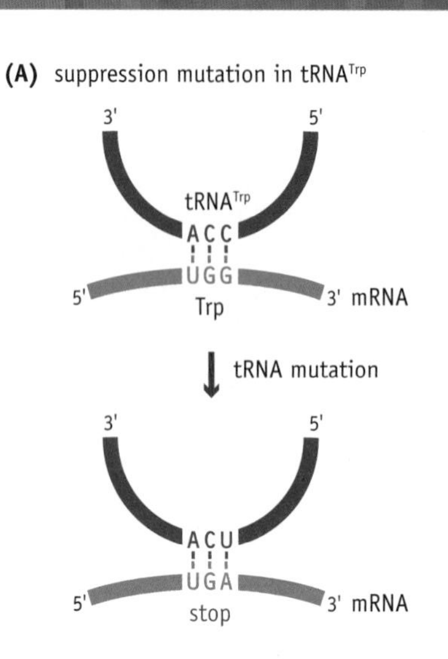

(B) suppression of a nonsense mutation

1 4 7 10 13 16 19
ATGGGAGCTCTATTAACCTAA
Met Gly Ala Leu Leu Thr stop

nonsense mutation

ATGGGAGCTCTATGAACCTAA
Met Gly Ala Leu stop

suppressive mutation in a tRNA gene

ATGGGAGCTCTATGAACCTAA
Met Gly Ala Leu Trp Thr stop

incorrect amino acid but polypeptide is the correct length

FIGURE 16.36 Suppression. (A) If the first nucleotide (reading in the 5′→3′ direction) of the tRNA^Trp anticodon is changed from C to U, then the new anticodon will be able to base-pair with the termination codon 5′-UGA-3′. (B) A nonsense mutation in our imaginary gene has converted a leucine codon into a termination codon. This mutation could be suppressed by the mutant tRNA^Trp, which will read the termination codon as one specifying tryptophan.

KEY CONCEPTS

- A mutation is a change in the nucleotide sequence of a DNA molecule. Mutations can arise as errors in replication or from the effects of chemical or physical mutagens.
- All organisms have DNA repair mechanisms that correct the vast majority of the mutations that occur in their DNA molecules.
- If a mutation that has not been repaired is present in a cell that undergoes meiosis and gives rise to gametes, then the mutation might be inherited by one of that organism's progeny. It might then become a new and permanent feature of the genetic diversity of the species.
- Most cells possess four different types of DNA repair system. Direct repair systems act directly on damaged nucleotides, converting each one back to its original structure. Excision repair involves excision of a segment of the polynucleotide containing a damaged site, followed by resynthesis of the correct nucleotide sequence. Mismatch repair corrects errors of replication. Nonhomologous end joining is used to mend double-strand breaks.
- In bacteria and mammals, mismatches resulting from errors in replication can be repaired correctly because the parent strand, which contains the correct sequence, is methylated. Some eukaryotes use a different method to distinguish the parent and daughter strands.
- The importance of DNA repair is emphasized by the number and severity of inherited human diseases that have been linked with defects in one of the repair processes.
- Mutations can have various effects on the biological information contained in a gene. Some have no effect. At the other extreme, some mutations result in loss of a protein function.
- Some mutations alter the protein product in such a way that a gain of function results. The mutated cell possesses an abnormal protein activity that is absent from the unmutated cell.
- The effects of a mutation can be reversed by back mutation, second-site reversion, or suppression.

QUESTIONS AND PROBLEMS (Answers can be found at www.garlandscience.com/introgenetics)

Key Terms

Write short definitions of the following terms:

(6–4) lesion
(6–4) photoproduct photolyase
Ada enzyme
alkylating agent
2-aminopurine
AP endonuclease
AP site
back mutation
base analog
baseless site
5-bromouracil
carcinogen
clastogens
cyclobutyl dimer
dark repair
deaminating agent
delayed onset
deletion
direct repair
DNA adenine methylase
DNA cytosine methylase
DNA glycosylase
DNA photolyase
DNA repair
ethidium bromide
ethylmethane sulfonate
excision repair
frameshift mutation
gain of function
haploinsufficiency
incomplete penetrance
insertion
intercalating agent
inversion
lethal mutation
long-patch repair
loss of function
MGMT
mismatch
mismatch repair
mutagen
mutasome
mutation
nonhomologous end joining
nonsense mutation
nonsynonymous mutation
oncogen
phosphodiesterase
photoproduct
photoreactivation
point mutation
readthrough mutation
replication slippage
second-site reversion
short-patch repair
silent mutation
SOS response
suppression
synonymous mutation
tautomer
teratogens
transition
transversion
trinucleotide repeat expansion disease
UvrABC endonuclease

Self-study Questions

16.1 What is the difference between a transition and a transversion mutation?

16.2 Distinguish between the following terms: (a) point mutation, (b) insertion mutation, (c) deletion mutation, (d) inversion mutation.

16.3 Describe how tautomerism of nucleotide bases can lead to errors occurring during the replication process.

16.4 What is replication slippage, and how can it lead to an error in replication?

16.5 What is the link between replication slippage and a trinucleotide repeat expansion disease?

16.6 Describe the difference between a mutagen and a clastogen.

16.7 Using examples, outline how base analogs result in mutations.

16.8 Distinguish between the mutagenic effects of deaminating agents, alkylating agents, and intercalating agents.

16.9 What is the difference between a cyclobutyl dimer and a (6–4) lesion? How does each type of structure arise?

16.10 Why is heat mutagenic?

16.11 Distinguish between the direct, excision, and mismatch systems for DNA repair.

16.12 Give examples that illustrate the direct repair of DNA damage in bacteria and eukaryotes.

16.13 Describe the base excision repair process, paying particular attention to the mode of action of DNA glycosylases.

16.14 Explain how nucleotide excision repair results in correction of mutations in *E. coli*.

16.15 Describe the importance of DNA methylation in the mismatch repair system of *E. coli*.

16.16 Outline how double-strand breaks are repaired by nonhomologous end joining.

16.17 What is a mutasome, and why is the error-prone replication that it directs sometimes of value to an *E. coli* cell?

16.18 Give examples of human diseases that have been linked to defects in DNA repair.

16.19 Explain why nonsense and frameshift mutations are more likely to result in an altered phenotype than are nonsynonymous mutations.

16.20 Which of the following types of mutation could be caused by a point change in a nucleotide sequence? (a) Nonsynonymous, (b) nonsense, (c) frameshift, (d) back, (e) second-site reversion, (f) suppression. Which could be caused by an insertion?

16.21 Here is the nucleotide sequence of a short gene:

5'-ATGGGTCGTACGACCGGTAGTTACTGGTTCAGTTAA-3'

Write out the amino acid sequence of the polypeptide coded by this gene. Now introduce the following mutations and describe their effects on the polypeptide: (a) a silent mutation, (b) a nonsynonymous mutation, (c) a nonsense mutation, (d) a frameshift mutation.

16.22 Describe the effect that haploinsufficiency has on the phenotype of an individual who is heterozygous for a loss-of-function mutation. What is the molecular basis for haploinsufficiency?

16.23 What is the difference between incomplete penetrance and delayed onset?

16.24 Give one example of the way in which the effects of a mutation can be reversed by suppression.

Discussion Topics

16.25 Explain why a purine-to-purine or pyrimidine-to-pyrimidine point mutation is called a transition, whereas a purine-to-pyrimidine (or vice versa) change is called a transversion.

16.26 What would be the anticipated ratio of transitions to transversions in a large number of mutations?

16.27 In some eukaryotes, the mismatch repair process is able to recognize the daughter strand of a double helix even though the two strands lack distinctive methylation patterns. Propose a mechanism by which the daughter strand can be recognized in the absence of methylation. How would you test your hypothesis?

16.28 Explain how point mutations could result in each of the following codon changes:
(a) a glycine codon to an alanine codon
(b) a tryptophan codon to a termination codon
(c) a cysteine codon to an arginine codon
(d) a serine codon to an isoleucine codon
(e) a serine codon (either AGT or AGC) to a threonine codon
(f) the previous mutation, back to serine but without recreating the original nucleotide sequence.

16.29 The example of suppression described in the text ought to result in all UGA termination codons being read through. Clearly this would result in a lot of elongated polypeptides. Would you expect this to be lethal? If not, why not? If so, then what might be the precise nature of this type of suppressive mutation?

16.30 Propose a scheme for a suppressive mutation that does not involve changing a tRNA gene.

FURTHER READING

David SS, O'Shea VL & Kundu S (2007) Base-excision repair of oxidative DNA damage. *Nature* 447, 941–950.

Drake JW, Glickman BW & Ripley LS (1983) Updating the theory of mutation. *Am. Sci.* 71, 621–630. *There are very few good general reviews of mutation, so this one is worth searching out.*

Fisher E & Scambler P (1994) Human haploinsufficiency—one for sorrow, two for joy. *Nat. Genet.* 7, 5–7.

Hearst JE (1995) The structure of photolyase: using photon energy for DNA repair. *Science* 268, 1858–1859.

Kunkel TA & Bebenek K (2000) DNA replication fidelity. *Annu. Rev. Biochem.* 69, 497–529. *Covers the processes that ensure that the minimum number of errors are made during DNA replication.*

Lehmann AR (1995) Nucleotide excision repair and the link with transcription. *Trends Biochem. Sci.* 20, 402–405.

Li G-M (2008) Mechanisms and functions of DNA mismatch repair. *Cell Res.* 18, 85–98.

Lieber MR (2010) The mechanism of double-strand break repair by the nonhomologous DNA end-joining pathway. *Annu. Rev. Biochem.* 79, 181–211.

Mirkin AM (2007) Expandable DNA repeats and human disease. *Nature* 447, 932–940. *Trinucleotide repeat expansion diseases.*

Subba Rao K (2007) Mechanisms of disease: DNA repair defects and neurological disease. *Nat. Clin. Pract. Neurol.* 3, 162–172.

Sutton MD, Smith BT, Godoy VG & Walker GC (2000) The SOS response: recent insights into *umuDC*-dependent mutagenesis and DNA damage tolerance. *Annu. Rev. Genet.* 34, 479–497.

Inheritance of Genes in Populations CHAPTER 17

Only a very small number of the mutations occurring in a multicellular organism are passed on to the next generation. To be inherited in this way a mutation must occur in a sex cell, evade the cell's repair enzymes, and be passed on to a gamete that participates in formation of a fertilized egg cell (Figure 17.1). Working out how many such mutations occur per generation is very difficult, but it is thought that each human being possesses over 100 new mutations that are absent from the somatic cells of their mother and father, and which therefore arose in the sex cells of those two parents prior to the individual's conception. This might sound like a minute degree of change, bearing in mind that the human genome contains 3200 Mb and only 4% of this comprises coding DNA. Most, if not all, of those 100 unique mutations possessed by an individual will therefore be in intergenic regions, but occasionally one will occur in the coding part of a gene, creating a new allele.

In this chapter we will explore the possible fate of the new alleles that arise in this way. This means that we must investigate how genes are inherited not by individuals but by populations. Geneticists use the term "population" to describe a group of organisms whose members are likely to breed with one another. A population could therefore be all the members of a species. Alternatively, a species might be split into two or more populations with little interbreeding between them. Usually this is because the populations have become separated from one another by geographical barriers (Figure 17.2).

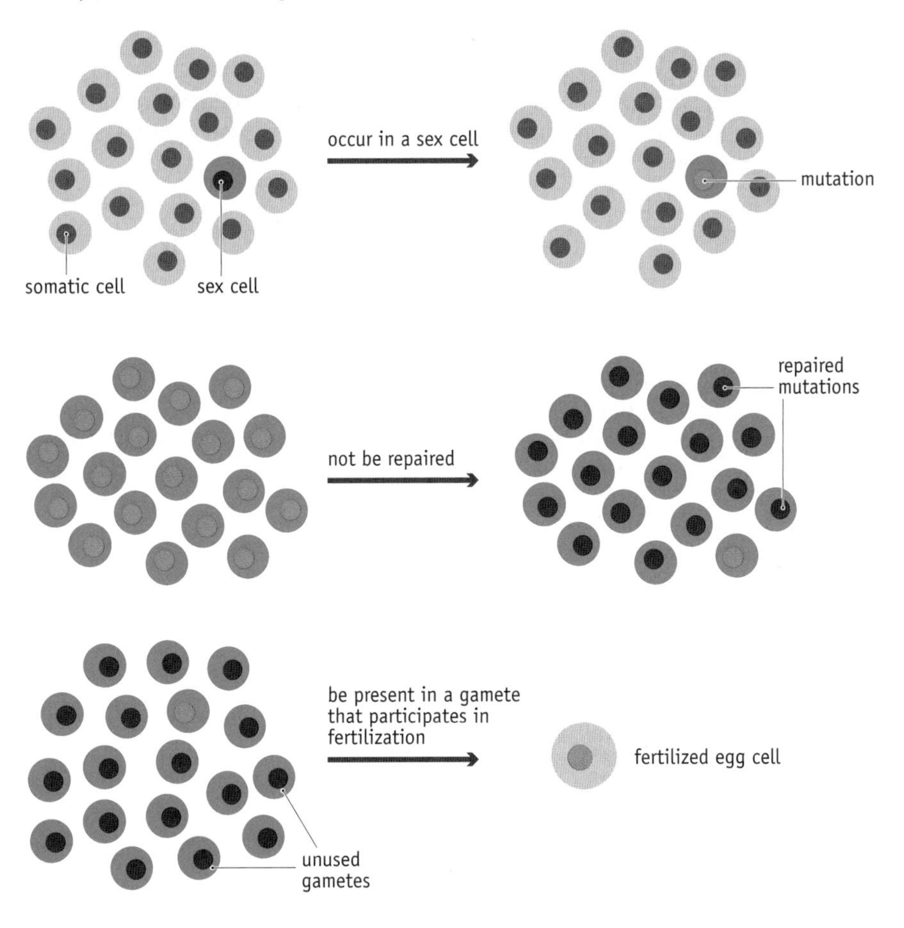

FIGURE 17.1 Only a small number of mutations are passed on to the next generation.

(A) all members of a species can form a single population

(B) a species can be split into populations separated by a geographical barrier

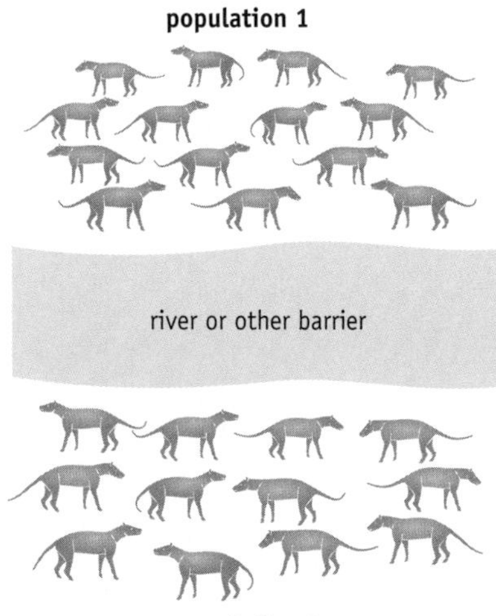

FIGURE 17.2 Populations. (A) All the members of a species could be a single population. (B) A species may be split into two or more populations.

Within a population, the number of individuals that possess a particular allele can change over time. These changes in allele frequencies are referred to as **microevolution**. They occur as a result of random processes called **genetic drift** and of more directed processes called **selection**. We will examine these processes in order to understand how they might affect the fate of a new allele created by mutation.

Populations exist in an environment that is constantly changing. We must therefore study the impact that the environment can have on the allele frequencies displayed by a population. We must also examine what happens if there is a sudden reduction in the size of the population, or if a new geographical barrier splits a population into two independent subpopulations. This will lead us to the final topic that we will consider in our study of the gene as a unit of inheritance, the process by which new species arise.

17.1 FACTORS AFFECTING THE FREQUENCIES OF ALLELES IN POPULATIONS

Imagine a gene called A, which has two alleles referred to, naturally, as *A* and *a*. Imagine also that within a population, 95% of all the alleles for gene A are *A*, which means that 5% will be *a* (Figure 17.3A). The frequency of allele *A* is therefore 0.95 and that for *a* is 0.05. We usually denote these frequencies as p and q, so in this case $p = 0.95$ and $q = 0.05$. Now imagine that 1000 years pass and we again determine the frequencies of *A* and *a* in the same population. We find that things are different because the value for p is now 0.85 and that of q is 0.15 (Figure 17.3B). The allele frequencies have changed and the population as a whole has undergone microevolution. How has this come about?

Allele frequencies do not change simply as a result of mating

In the early 1900s, when genetics first became a major topic of conversation for scientists as a whole, a common misconception was that recessive alleles should gradually be lost from a population. This notion is most famously (and probably unfairly) credited to Udny Yule, a Scottish statistician, who observed that in humans, abnormally short fingers are dominant to fingers of normal length. The number of people with normal fingers should therefore decrease over time until everybody has short fingers. Is this what geneticists are saying must happen?

Yule's argument was based on the expected phenotypes of the children of parents who were heterozygous for the finger length gene. If we call this the F gene, then the two alleles will be *F*, for short fingers, and *f*, for normal fingers. If the two parents are heterozygotes, then both will have the genotype *Ff* and both will produce equal numbers of gametes carrying the *F* and *f* alleles. A child is therefore equally likely to inherit an *F* or *f* allele

(A) population in which $p = 0.95$, $q = 0.05$

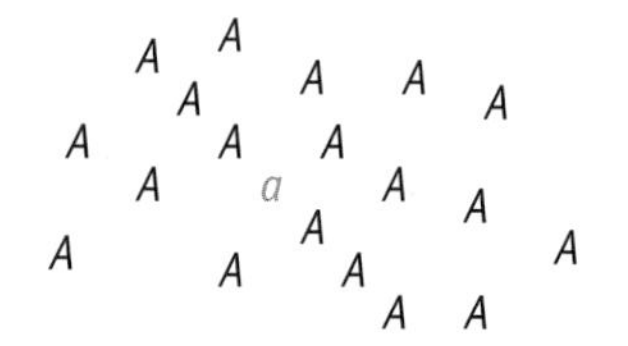

(B) population in which $p = 0.85$, $q = 0.15$

FIGURE 17.3 Populations displaying different allele frequencies. Gene A has two alleles, *A* and *a*, whose frequencies are p and q, respectively.

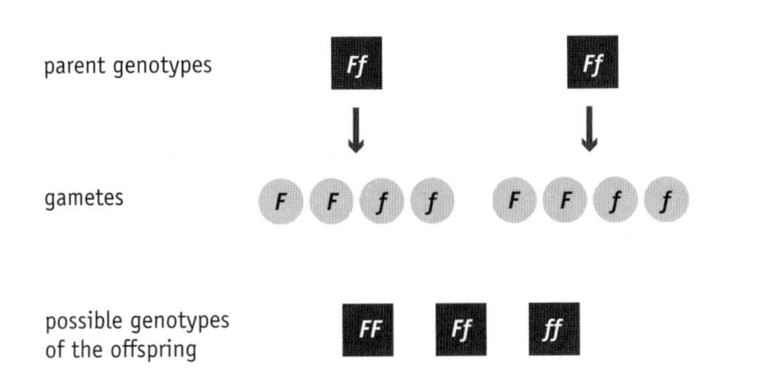

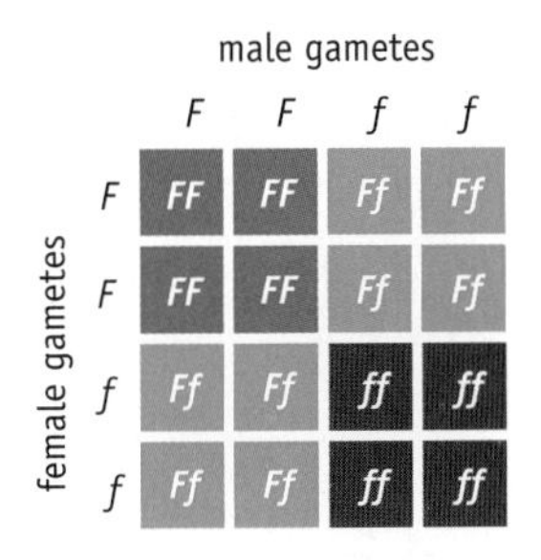

FIGURE 17.4 The outcome of a cross between two individuals who are heterozygous for the same gene. (A) Gametes from the male and female parents combine in a random fashion, so all three possible genotypes will be present among the offspring. (B) The ratio of genotypes among the offspring can be predicted statistically. The "Punnett square," shown here, is the simplest way of doing this. The genotypes of the four male gametes are listed along the top of the square and those for the four female gametes on the left. The boxes indicate the genotypes of the offspring arising from each pairwise combination of gametes. Counting the genotypes in the boxes gives the predicted ratio among the offspring.

from either parent, and so could have any of the three possible genotypes, *FF*, *Ff*, or *ff*. Mendel had shown experimentally with his pea plants that if enough such crosses occur, then the ratio of genotypes in the next generation will be 1*FF* : 2*Ff* : 1*ff* (Figure 17.4). The point that Yule seized on was that when rewritten in terms of the phenotype, this outcome is three short-fingered children to every one with normal fingers. This is because in the heterozygote the dominant allele, in this case short fingers, determines the phenotype that is observed. If every mating between heterozygotes results in three children with the dominant phenotype and only one with the recessive phenotype, then surely the number of people displaying the recessive phenotype would decrease over time?

Well, no, actually. Yule would be correct only if heterozygotes sought out one another in order to raise families. Within a population, matings occur much more randomly between individuals of different genotypes. Yule's argument was shown to be a fallacy in 1908 by Godfrey Hardy and Wilhelm Weinberg, who independently proved that the allele frequencies in a population will not change simply as a result of mating. The proof of what we now call the **Hardy–Weinberg equilibrium** is described in Research Briefing 17.1. Besides dispelling the notion that recessive alleles will gradually become less common in a population, the work of Hardy and Weinberg was important because it showed that the rules governing the inheritance of alleles by individuals, as had recently been described by Mendel, applied also to populations.

The Hardy–Weinberg equilibrium therefore shows that mating, by itself, has no effect on the allele frequencies present in a population. But in Figure 17.3 we considered a scenario in which allele frequencies *did* change over time, and we have learned that such changes are considered important by biologists because they underlie microevolution. We must therefore consider the factors that disrupt the Hardy–Weinberg equilibrium for a set of alleles within a population and lead to changes in the frequencies of those alleles over time.

Genetic drift occurs because populations are not of infinite size

A central assumption of the Hardy–Weinberg equilibrium is that all possible matings within a population occur at their predicted statistical frequencies. This assumption is true only if the population is of infinite size, because any finite population will be subject to **stochastic** effects—the deviations from statistical perfection that occur with real events in the real world. Statistics predicts the way events are most likely to turn out, but we are all aware

RESEARCH BRIEFING 17.1

The Hardy–Weinberg equilibrium

The misconception that recessive alleles should gradually be lost from a population was laid to rest by Godfrey Hardy, an English mathematician, and Wilhelm Weinberg, a German doctor. Hardy and Weinberg independently used Mendel's First Law of Genetics to show that the allele frequencies in a population will not change simply as a result of mating. Proof of this principle requires that we use a little algebra, but the analysis is not too difficult to follow.

	pA	qa
pA	AA $p \times p = p^2$	Aa $p \times q$
qa	Aa $p \times q$	aa $q \times q = q^2$

FIGURE 1 A Punnett square showing the frequencies of genotypes that are expected as a result of random matings for a gene whose alleles, *A* and *a*, have frequencies of *p* and *q*, respectively.

If mating is random, genotype frequencies can be calculated

First, we will calculate the frequency of different genotypes in a population in which mating is occurring at random. To do this, we will consider an imaginary gene A. This gene has two alleles *A* and *a*, which have frequencies *p* and *q*, respectively. What are the frequencies of the three possible genotypes, *AA*, *Aa*, and *aa*, in the population as a whole?

Mendel's First Law tells us that alleles segregate randomly. We can therefore imagine gametes containing the *A* and *a* alleles forming a pool, from which pairs of gametes are taken, at random, to produce a new individual. We can use a Punnett square to calculate the frequencies of each of the three genotypes (Figure 1). This analysis shows us that the frequencies of these genotypes in the population will be

$$AA = p \times p = p^2$$

$$Aa = (p \times q) + (p \times q) = 2pq$$

$$aa = q \times q = q^2$$

The total genotype frequency must equal 1, so

$$p^2 + 2pq + q^2 = 1$$

To show that this relationship holds true, and the sum of the three frequencies always equals one, in Figure 2, we do the relevant calculations for two populations, one in which $p = 0.95$ and $q = 0.05$, and the other where $p = 0.85$ and $q = 0.15$.

Genotype frequencies do not change in successive generations

A population in which the frequencies of the genotypes *AA*, *Aa*, and *aa* are p^2, $2pq$, and q^2 is said to display the Hardy–Weinberg equilibrium. The important point that Hardy and Weinberg established is that these genotype frequencies are maintained in successive populations.

There are various ways of proving that the equilibrium is maintained. The simplest is to consider all the possible matings that can occur within the population, and to work out the genotypes of the offspring produced by those matings. This analysis is depicted in Table 1.

First, we consider matings in which both parents have the genotype *AA*. These matings will, of course, result in offspring that also have the *AA* genotype. The matings will occur at a frequency of $p^2 \times p^2$, which equals p^4. This means that these matings result in a set of individuals with the *AA* genotype, the frequency of these individuals in the next generation being p^4.

Now, consider matings in which one parent is *AA* and the other *Aa*. The frequencies of these genotypes are p^2 and $2pq$, respectively. If we are not thinking too clearly, then we might conclude that the mating frequency is therefore $p^2 \times 2pq$. This would be incorrect because we have to take account not only of matings in which the male parent is *AA* and the female *Aa*, but also ones in which the parental genotypes are the other way around, where the male is *Aa* and the female *AA*. So the correct frequency is $2 \times (p^2 \times 2pq) = 4p^3q$.

Matings between *AA* and *Aa* parents will produce equal numbers of *AA* and *Aa* offspring. These matings will therefore contribute *AA* and *Aa* individuals to the

(A) $p = 0.95$ and $q = 0.05$:

$$p^2 + 2pq + q^2 = 0.95^2 + (2 \times 0.95 \times 0.05) + 0.05^2$$
$$= 0.9025 + 0.0950 + 0.0025$$
$$= 1$$

(B) $p = 0.85$ and $q = 0.15$:

$$p^2 + 2pq + q^2 = 0.85^2 + (2 \times 0.85 \times 0.15) + 0.15^2$$
$$= 0.7225 + 0.2550 + 0.0225$$
$$= 1$$

FIGURE 2 Calculations of the expected genotype frequencies in two populations. (A) Population in which $p = 0.95$ and $q = 0.05$. (B) Population in which $p = 0.85$ and $q = 0.15$.

next generation, each at a frequency of one-half of $4p^3q$, which is $2p^3q$.

The next set of matings in Table 1 are between *AA* and *aa* parents. The frequencies of the genotypes are p^2 and q^2, respectively, and once again we must account for two sets of matings depending on which parent is male and which female. So the frequency of these matings is $2p^2q^2$. All these matings give rise to heterozygous offspring, so we place $2p^2q^2$ in the column for *Aa*.

Now that we are into our stride the rest of Table 1 becomes less daunting. Matings between two heterozygotes will occur with a frequency of $2pq \times 2pq = 4p^2q^2$. This mating was illustrated in Figure 17.4, so we know that the ratio of offspring will be 1 *AA* : 2 *Aa* : 1 *aa*. The entries in Table 1 are therefore p^2q^2 for *AA*, $2p^2q^2$ for *Aa*, and p^2q^2 for *aa*.

Matings between *Aa* and *aa* parents will occur at a frequency of $2(2pq \times q^2) = 4pq^3$. Half the offspring will have the *Aa* genotype and half *aa*. And, finally, the frequency of matings between two *aa* parents will be $q^2 \times q^2 = q^4$, and all the offspring will be *aa*.

The final proof that frequencies do not change

Now that we have completed the Table, we can add up the individual entries in order to calculate the total frequency of each genotype. The calculations are

$$AA = p^4 + 2p^3q + p^2q^2$$
$$= p^2(p^2 + 2pq + q^2)$$
$$Aa = 2p^3q + 2p^2q^2 + 2p^2q^2 + 2pq^3$$
$$= 2p^3q + 4p^2q^2 + 2pq^3$$
$$= 2pq(p^2 + 2pq + q^2)$$
$$aa = p^2q^2 + 2pq^3 + q^4$$
$$= q^2(p^2 + 2pq + q^2)$$

We know that $(p^2 + 2pq + q^2) = 1$, so these formulas can be simplified to $AA = p^2$, $Aa = 2pq$, and $aa = q^2$. These are the frequencies of the three genotypes in the new generation that we have created.

If we go back to the beginning of the analysis, we see that $AA = p^2$, $Aa = 2pq$, and $aa = q^2$ are also the frequencies that we had in the parent population. In other words, the Hardy–Weinberg equilibrium has not been disturbed by creation of the new generation.

The notion that recessive alleles will gradually become less common in the population is therefore shown to be a fallacy. Mating, by itself, has no effect on the allele frequencies present in the population as a whole, so long as those matings occur at random.

TABLE 1 THE OUTCOMES OF ALL POSSIBLE MATINGS THAT CAN OCCUR IN A POPULATION IN WHICH A PAIR OF ALLELES, *A* AND *a*, ARE AT HARDY–WEINBERG EQUILIBRIUM

		Contribution to the next generation		
Mating	Mating frequency	*AA*	*Aa*	*aa*
AA × *AA*	$p^2 \times p^2 = p^4$	p^4		
AA × *Aa*	$2(p^2 \times 2pq) = 4p^3q$	$2p^3q$	$2p^3q$	
AA × *aa*	$2(p^2 \times q^2) = 2p^2q^2$		$2p^2q^2$	
Aa × *Aa*	$2pq \times 2pq = 4p^2q^2$	p^2q^2	$2p^2q^2$	p^2q^2
Aa × *aa*	$2(2pq \times q^2) = 4pq^3$		$2pq^3$	$2pq^3$
aa × *aa*	$q^2 \times q^2 = q^4$			q^4

(A) sampling does not affect allele frequencies

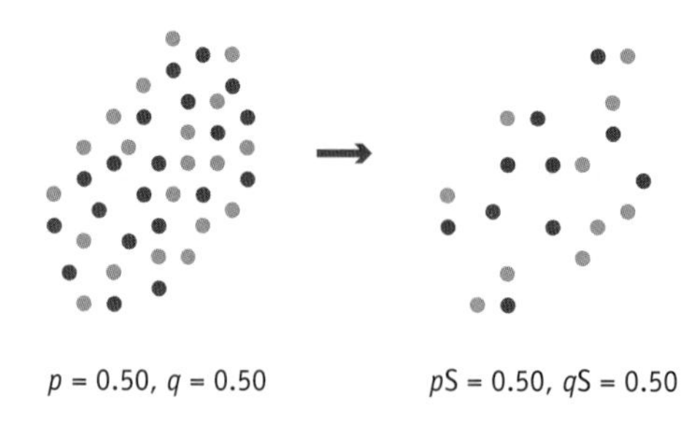

(B) sampling changes allele frequencies

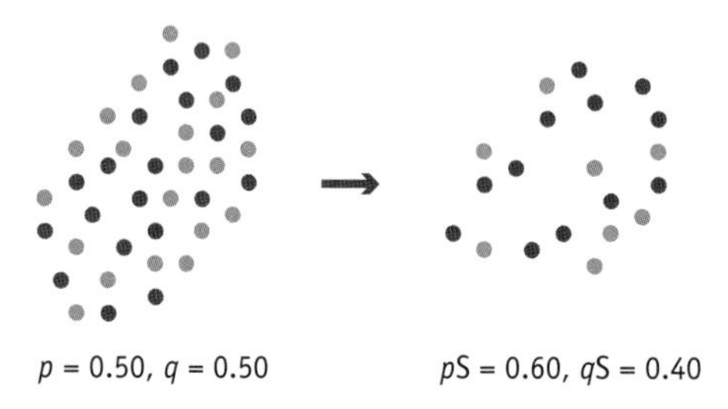

FIGURE 17.5 The sampling that occurs when each new generation arises might affect the allele frequencies in a population. These examples consider a single gene with two alleles *A* and *a*. The frequencies of *A* and *a* in the original population are *p* and *q*, respectively. The frequencies in the samples are *p*S and *q*S. In (A) the allele frequencies in the sample are the same as the frequencies in the starting population. In (B) the allele frequencies in the sample are different.

that deviations from statistical predictions occur. This is true with a simple process such as spinning a coin, and is equally true with a more complex process such as the inheritance of alleles by a population.

Stochastic effects are important in populations because, for each new generation to arise, a finite number of alleles are sampled from the preceding generation. We can refer to the allele frequencies in these samples as *p*S and *q*S, for alleles *A* and *a*, respectively. The allele frequencies in the sample *might* be precisely equal to the allele frequencies in the population as a whole, so that $pS = p$ and $qS = q$, but this is by no means certain (Figure 17.5). Any deviation from $pS = p$ and $qS = q$ will lead to a change in the allele frequencies of the succeeding generation. This process is called **genetic drift**.

No real population can be of infinite size, but larger populations are closer to infinite size than are smaller ones. We would therefore expect that samples taken from larger populations will, on average, deviate from $pS = p$ and $qS = q$ less than samples taken from smaller populations (Figure 17.6). If this assumption is true then genetic drift should have a more dramatic effect on the allele frequencies in smaller populations. Computer simulations have shown that this prediction is indeed correct (Figure 17.7). In these simulations the effects of genetic drift are modeled for pairs of alleles that begin with equal frequency ($p = 0.5$, $q = 0.5$) in populations of different sizes. When the population number, *N*, is set at 20, the allele frequencies fluctuate dramatically, whereas when $N = 1000$ the fluctuations, although still occurring, are less extreme.

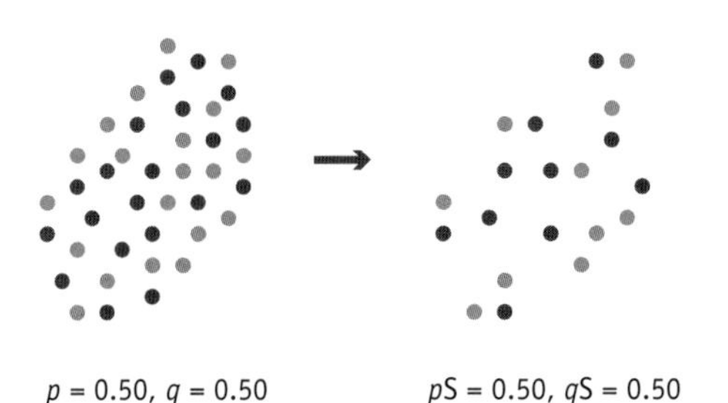

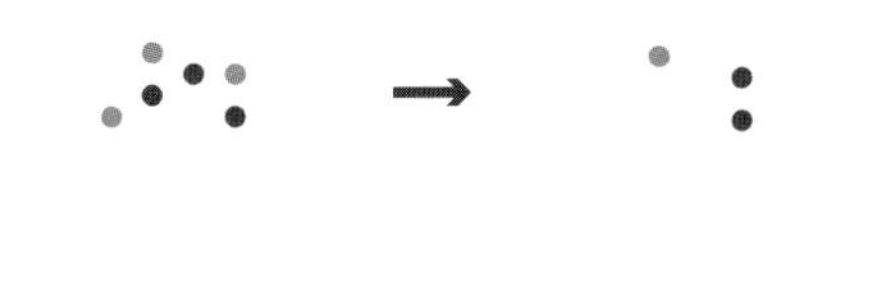

FIGURE 17.6 Sampling effects are greater with a small population. Again, we consider a single gene with two alleles *A* and *a*, whose frequencies are *p* and *q*, respectively, in the original population, and *pS* and *qS* in the samples. (A) and (B) illustrate the expectation that a sample taken from a larger population will display a smaller deviation in allele frequencies than a sample taken from a smaller population.

Alleles become fixed at a rate determined by the effective population size

The simulations shown in Figure 17.7 illustrate an important outcome of genetic drift. Over time allele frequencies will drift so that at some point one will become lost from the population. The corollary is that the other allele reaches **fixation**, the situation when all members of the population are homozygous for that allele. In Figure 17.7 this occurred after 26 generations for a population size of 20. It did not occur at all during the time span of the simulation for a population size of 1000, but the statistical prediction is that one member of a pair of alleles will always become fixed as a result of genetic drift. This has been confirmed by the results of simulations that have been allowed to run for a greater number of generations.

On average, the time to fixation is longer in larger populations. We might imagine that this time to fixation would be *proportional* to the population

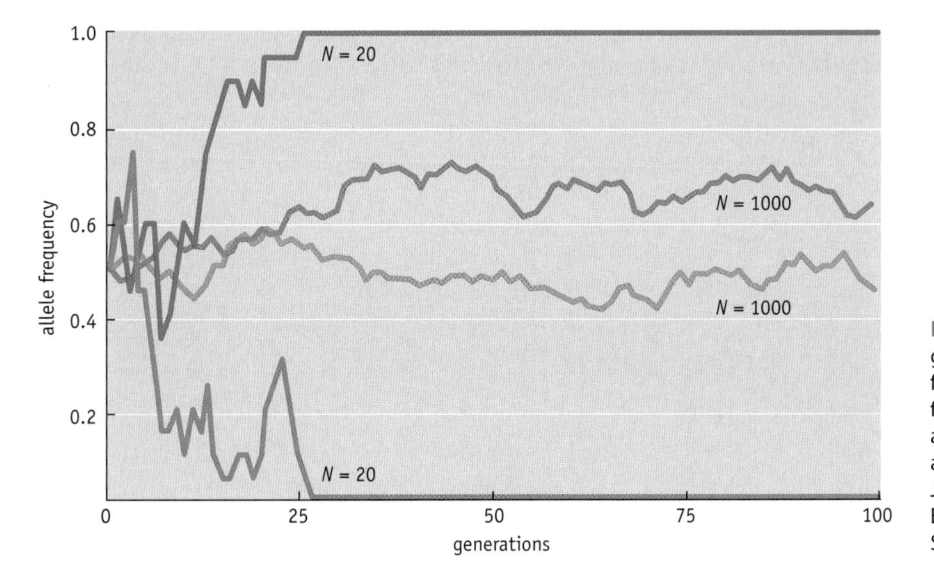

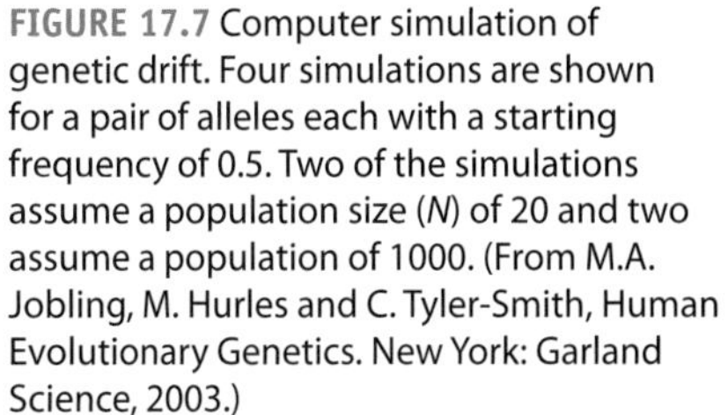
FIGURE 17.7 Computer simulation of genetic drift. Four simulations are shown for a pair of alleles each with a starting frequency of 0.5. Two of the simulations assume a population size (*N*) of 20 and two assume a population of 1000. (From M.A. Jobling, M. Hurles and C. Tyler-Smith, Human Evolutionary Genetics. New York: Garland Science, 2003.)

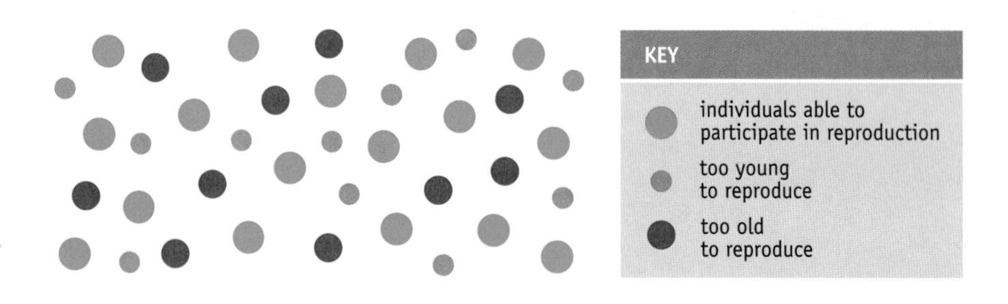

FIGURE 17.8 Only some members of a population can participate in sexual reproduction.

size, but this is not strictly correct. All of the issues that we have so far considered regarding inheritance of alleles within a population assume that all members of the population participate in reproduction so that their alleles contribute to the next generation. This is certainly not true of human populations and it is equally untrue of most other animal populations. In any population there will be some members who have not yet reached reproductive age and others who are too old to contribute their alleles to the next generation (Figure 17.8).

The part of the population that does participate in reproduction is called the **effective population size**, N_e. This figure is not simply the number of individuals of reproductive age, because it is influenced by a variety of factors, including the relative numbers of males and females. It is almost always a number smaller than the total population size, and often is substantially smaller.

The more rapid loss and fixation of alleles in smaller populations is one of the reasons why the population size of an endangered species is so critical. We become concerned not simply that the last remaining individuals will die but also that a small population might quickly lose genetic variability. Without variability the population might not be able to respond to a disease which, in a more variable population, might not have such a devastating effect because some individuals will have a greater tolerance and hence be able to survive (Figure 17.9). Understanding the effective population size of an endangered species, and its degree of genetic variability, is essential for designing planned breeding programs aimed at rescuing the species from impending extinction.

New alleles are continually created by mutation

If genetic drift leads inevitably to the fixation and elimination of alleles, then why do populations display any genetic variability at all? We might expect that, over time, all variation will be lost because all but one of the

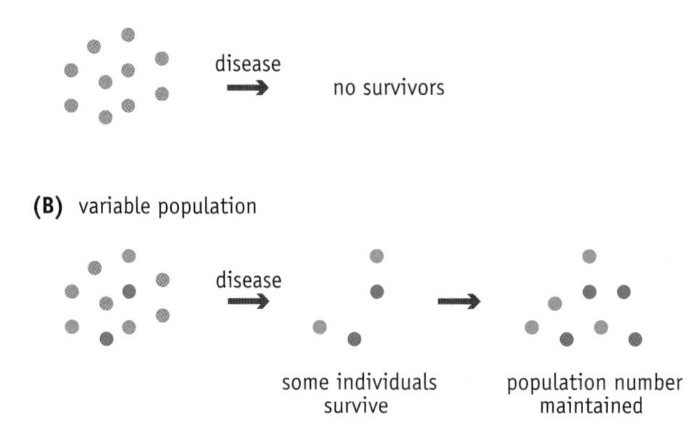

FIGURE 17.9 A possible problem for a population with low genetic variability. Individuals with the orange genotype are susceptible to disease. (A) The population has low variability. When disease strikes, there are no survivors. (B) The population has a greater degree of variability. Disease kills the individuals with the orange genotype, but other individuals survive and are able to re-establish the original population number.

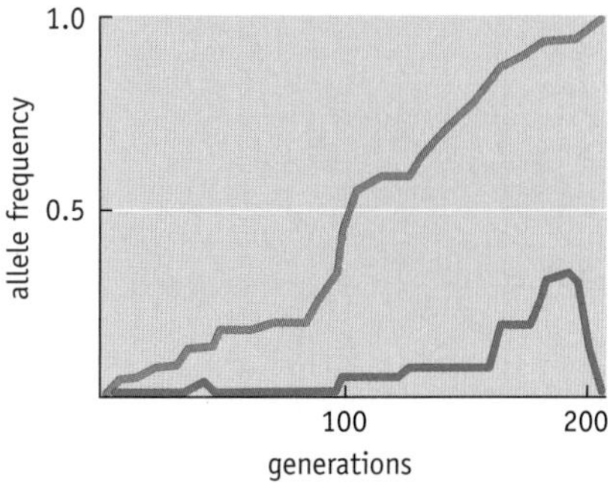

FIGURE 17.10 Possible fates of a new allele. The orange line shows an allele that eventually becomes fixed in its population. The blue line shows the more common situation where the allele is lost.

alleles for any particular gene will become eliminated. The answer is, of course, that new alleles are continually created by mutation.

Initially a new allele will have a very low frequency in the population as a whole—indeed it will be possessed by just one individual. The chances it will be eliminated by drift are therefore relatively high. Elimination is not, however, inevitable, and if it does occur it might not occur immediately. Computer simulations show that some new alleles increase in frequency to such an extent as a result of drift that eventually they in turn become fixed in the population (Figure 17.10).

The creation of new alleles by mutation, and the subsequent drift of those alleles to fixation or elimination, means that over time the genetic features of a population will change. These molecular events enable a population to undergo microevolution, but in the absence of any other forces the rate of microevolutionary change will be very slow. In real populations, additional factors contribute to the rate of microevolution by bringing about changes in allele frequencies. The most important of these is **natural selection**.

Natural selection favors individuals with high fitness for their environment

Darwin's evolutionary theory holds that natural selection favors individuals with characteristics that provide them with a reproductive advantage over other members of a population. This reproductive advantage can take a number of different forms. There can be an increased likelihood that an individual survives to reproductive age, an increased likelihood that the individual finds a mate, or increased fertility so that a greater proportion of an individual's matings are productive (Figure 17.11).

The outcome of natural selection is that advantageous traits increase in a population over time, because the individuals possessing those traits, being more successful at reproduction, make a greater contribution to later generations than do those individuals with less advantageous traits. A new

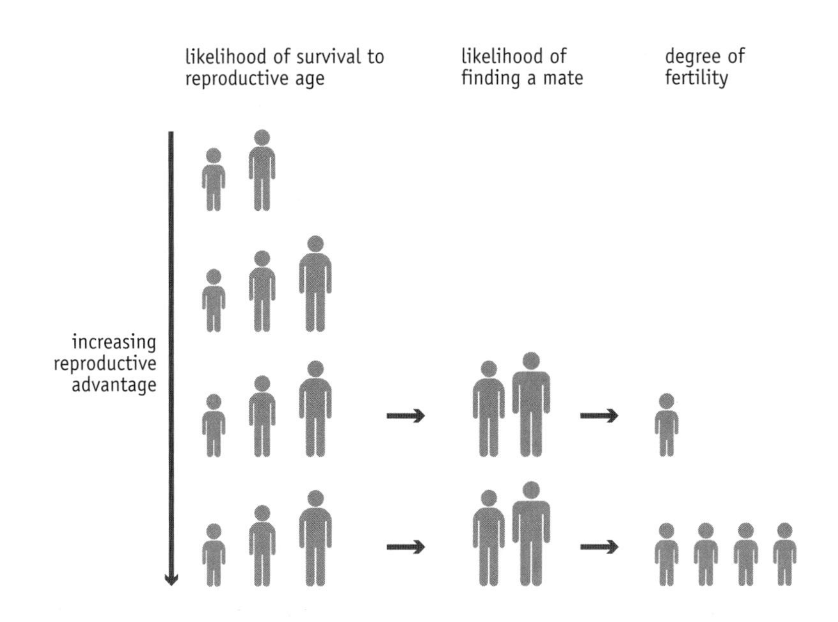

FIGURE 17.11 Various factors that influence the reproductive advantage of an individual.

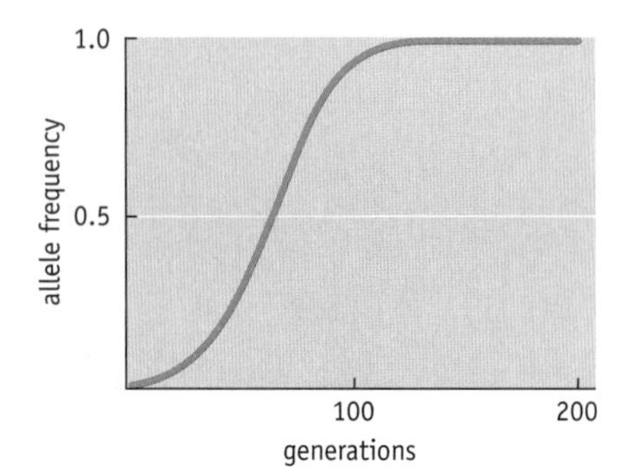

FIGURE 17.12 An allele that specifies an advantageous trait can display a relatively rapid increase in frequency due to positive natural selection. Compare this plot with those in Figure 17.10, where the new alleles were not subject to selection.

allele that specifies an advantageous trait can therefore display a relatively rapid increase in frequency (Figure 17.12). The degree of reproductive advantage possessed by an individual is called its **fitness**. A key feature of fitness is that it varies according to the environment. An individual who has high reproductive success under one set of conditions might have lower reproductive success if the environment changes.

The importance of the environment in determining the fitness of an individual is illustrated by one of the classic examples of natural selection, melanism of the peppered moth, *Biston betularia* (Figure 17.13). This species is common throughout Britain and until the mid-nineteenth century existed predominantly as a light-colored form, the variety called *typica*. In 1848 a dark-colored variety, *carbonaria*, was first observed near Manchester, which at that time was one of the most industrialized cities in Europe. The dark body and wings of *carbonaria* moths are caused by increased synthesis of the dark pigment melanin in the skin cells (Figure 17.14).

During the second half of the nineteenth century, the *carbonaria* version of the peppered moth became more frequent, to the extent that by 1900 over 90% of the moths in the Manchester area had dark bodies and wings (Figure 17.15). This increase is attributed to natural selection on the grounds that the dark coloration provides a camouflage when the moth alights on a tree trunk or other structure that has become blackened due to the airborne pollution typical of industrial areas in the nineteenth century. In contrast, the lighter *typica* version of the peppered moth is more conspicuous against a blackened background, and so is more likely to be seen and eaten by a passing bird. The dark-winged moths are therefore more likely to survive until reproductive age, and their progeny will therefore increase in frequency over time. Similar increases in the frequencies of dark versions of other species of moth were recorded in industrial areas elsewhere in Europe and in North America.

We do not know when the *carbonaria* mutation occurred in peppered moths, but we suspect that it was some time before the Industrial Revolution. In the pre-industrial environment, these dark moths had relatively low fitness and hence were uncommon in the population as a whole. Industrialization, however, resulted in a change in the environment, such that the fitness of the *carbonaria* moths increased and that of the *typica* version decreased. Consequently, the dark moths increased in numbers and became more frequent in the population. In rural parts of Britain, where there was no industrial pollution and hence the environment did not change, the fitness of *carbonaria* moths remained low and their numbers did not increase during the nineteenth century. During the twentieth century, when air pollution became reduced in Britain due to stricter controls on factory emissions, *carbonaria* moths became less frequent again, their fitness declining as the environment changed back to its less blackened, pre-industrial state.

Natural selection acts on phenotypes but causes changes in the frequencies of the underlying genotypes

Natural selection is a biological rather than a genetic process. It acts on the phenotypes of individuals, because it is the phenotype that determines the fitness of the individual for its environment. Natural selection does not act

FIGURE 17.13 Pale and dark (melanic) forms of the peppered moth, *Biston betularia cognataria*. (Courtesy of Bruce Grant. This work was published in B.S. Grant and L.L. Wiseman, *J. Hered.* 93: 86–90, 2002.)

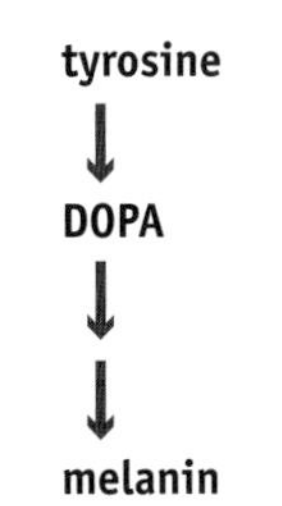

FIGURE 17.14 The pathway for synthesis of melanin pigments from tyrosine. DOPA, 3,4-dihydroxyphenylalanine.

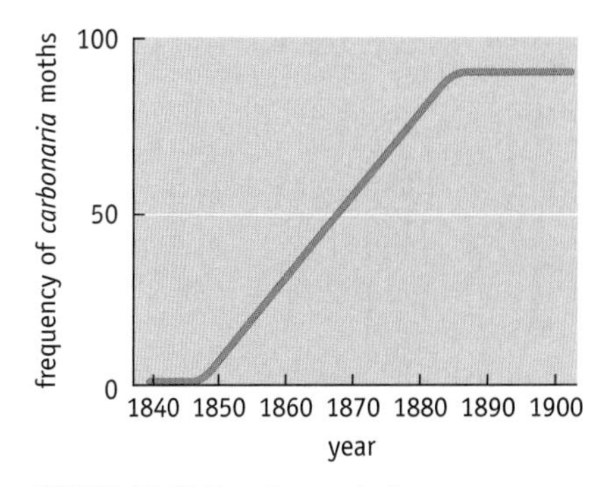

FIGURE 17.15 The change in frequency of *carbonaria* moths that occurred in the vicinity of Manchester, England, during the nineteenth century.

directly on genes, but the changes that it brings about within a population have important effects on allele frequencies. To explore this point we will look at the genetic changes occurring in peppered moth populations as a result of selection for the *carbonaria* variety.

Crosses between *carbonaria* and *typica* moths show that the *carbonaria* allele is dominant and the *typica* allele recessive. From what we have learned about the molecular basis to dominance and recessiveness, this result will not be surprising. We would anticipate that an allele specifying synthesis of a dark pigment would dominate over one specifying either synthesis of a light-colored pigment or synthesis of no pigment at all.

If we denote the pigment synthesis gene as M, for melanin, and its alleles *M* and *m*, then *carbonaria* would have the *MM* and *Mm* genotypes, and *typica* moths would be *mm*. Selection for the *carbonaria* form in industrial areas therefore increases the frequencies of the *MM* and *Mm* genotypes in the population, and decreases the frequency of *mm*. This in turn means that the *M* allele becomes more frequent, and the *m* allele less so (Figure 17.16).

Continuing selection will eventually drive one allele to fixation and the other to extinction. Exactly how long this takes depends on the relative fitness of the different phenotypes specified by the alleles (Figure 17.17). Fitness is very difficult to measure with any degree of accuracy, especially in natural populations, as it requires a knowledge of the average number of offspring produced by individuals of each genotype. Most studies of the influence of fitness on allele frequencies have therefore made use of computer simulations.

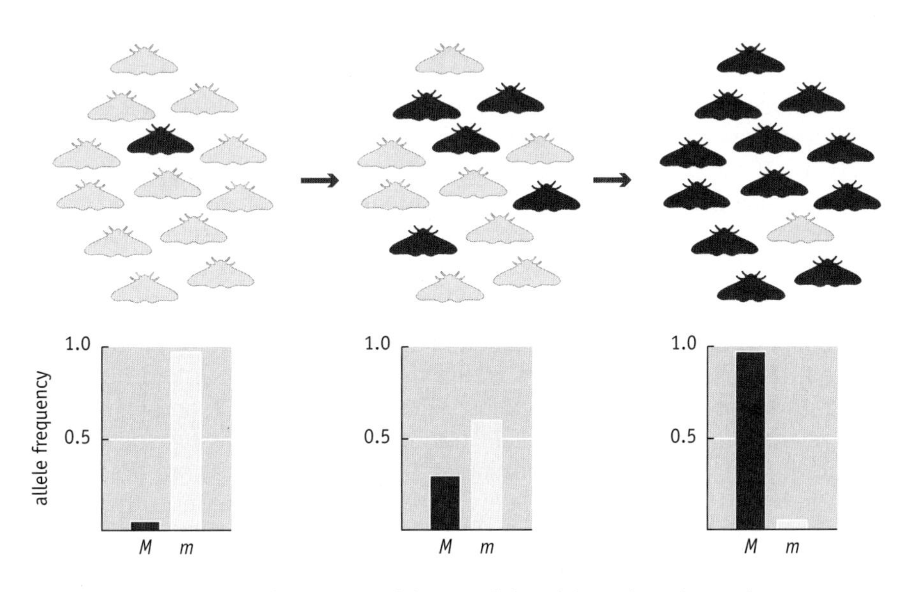

FIGURE 17.16 Changes in frequencies of the two alleles of the melanin biosynthesis gene resulting from selection for the *carbonaria* phenotype of the peppered moth.

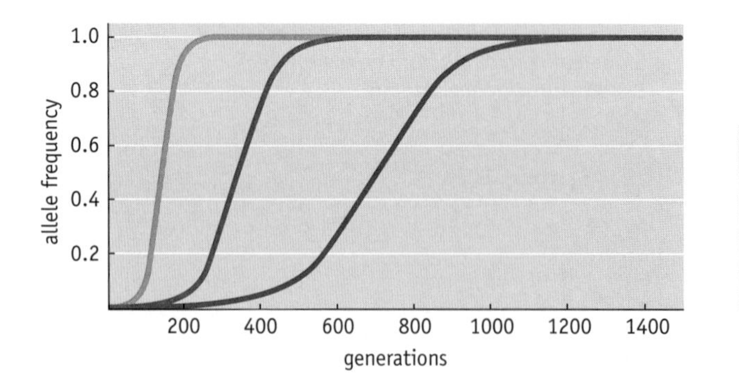

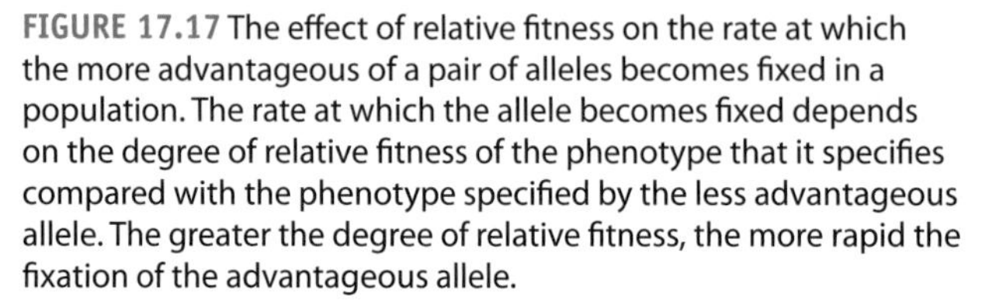
FIGURE 17.17 The effect of relative fitness on the rate at which the more advantageous of a pair of alleles becomes fixed in a population. The rate at which the allele becomes fixed depends on the degree of relative fitness of the phenotype that it specifies compared with the phenotype specified by the less advantageous allele. The greater the degree of relative fitness, the more rapid the fixation of the advantageous allele.

In these simulations, relative fitness is expressed as the **selection coefficient**, the most successful genotype being assigned a selection coefficient of 0, and the least successful given a value of 1. Individuals with the latter genotype would never produce progeny, perhaps because the genotype is lethal. Figures in between represent intermediate degrees of fitness. A selection coefficient of 0.05, for example, would be given to a genotype that confers a fitness level that is 95% that of the most successful genotype.

Alleles not subject to selection can hitchhike with ones that are

We must not forget that genes are contained within DNA molecules. This means that selection results not simply in a change in the frequency of an individual allele, but a change in frequency of the segment of DNA containing that allele (Figure 17.18A). Over time, the segment that increases in frequency will become shorter in length. This is because of the effect of crossing-over events that occur during the production of gametes in individuals who are heterozygous for the mutation. Some of these crossovers will replace parts of the segment containing the mutation with the equivalent stretches of DNA from chromosomes that lack the mutation (Figure 17.18B). If the frequency of the segment containing the mutation is rapidly increasing due to a strong selective pressure, then fixation could occur while the DNA segment is still relatively long. Under these circumstances, what becomes fixed might not just be the allele responding to selection, but also a substantial length of DNA to either side of this allele (Figure 17.18C).

All of this would merely be of academic interest were it not for the possibility that other genes are located within the segment of DNA that becomes fixed. Imagine three adjacent genes, A, B, and C. Genes A and C have no impact on fitness and their allele frequencies are changing solely as a result of genetic drift. In one particular DNA molecule they are present as alleles *a* and *c*, which, by chance, are relatively uncommon in the population as a whole. Now imagine that, within this DNA molecule, gene B undergoes mutation, creating a new allele, which we call b^*, which confers an increase in fitness and so becomes subject to positive selection. The resulting increase in the frequency of allele b^* in the population will be

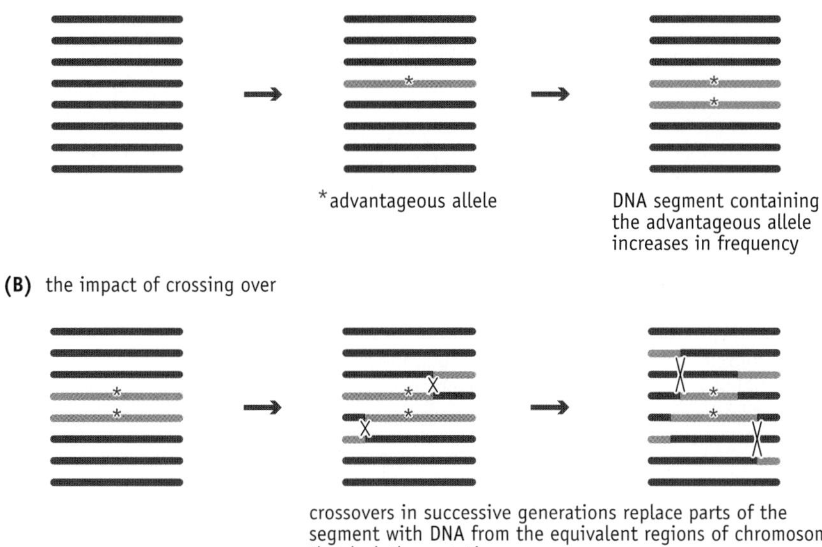

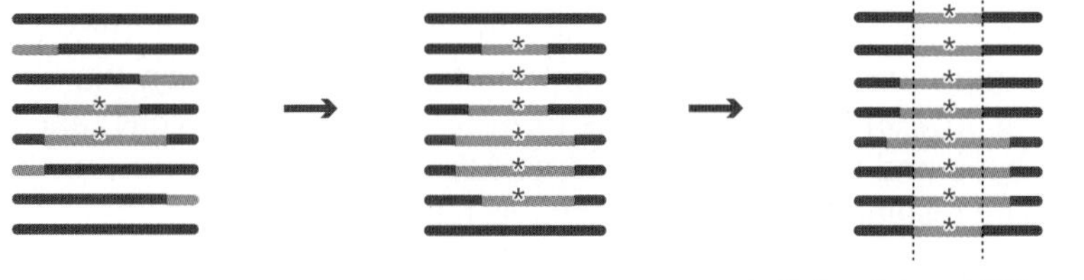

FIGURE 17.18 Selection acts on alleles, but those alleles are contained in DNA molecules. (A) An advantageous mutation appears in a population. Selection acts not just on the new allele that is formed, but on the segment of DNA containing that allele. (B) Crossing-over events that occur in successive generations will gradually reduce the length of the DNA segment that is increasing in frequency. (C) Selection may eventually result in fixation of the advantageous allele. The length of the segment of DNA that is fixed along with the allele depends on how many crossovers have taken place since the mutation occurred.

accompanied by equivalent increases in the frequencies of *a* and *c*, until crossing over occurs in the regions separating gene A from gene B and/or gene B from gene C. If allele b^* becomes fixed before such crossings over occur, then alleles *a* and *c* will also become fixed, even though they themselves confer no increase in fitness (Figure 17.19). This process is called **hitchhiking**—alleles *a* and *c* are looked on as having "hitchhiked" with allele b^* to fixation.

Hitchhiking therefore means that positive selection can cause a disproportionate reduction in the genetic diversity of a population. Not only do we lose variability at the gene under selection, but also at adjacent genes. We refer to this as a **selective sweep**.

Heterozygotes sometimes have a selective advantage

The link between selection and genotype can be quite complex, and there are examples where the most successful genotype is not one of the homozygous forms, but the heterozygote. This is variously called **overdominance**, **heterozygote superiority**, **heterozygote advantage**, or **heterosis**.

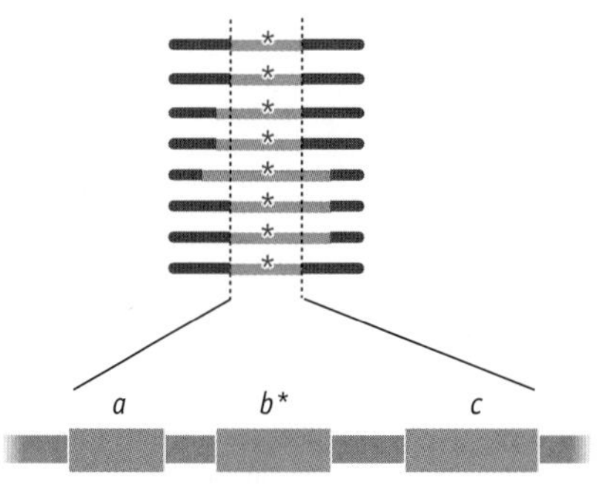

FIGURE 17.19 An example of hitchhiking. Alleles *a* and *c* have become fixed, even though they themselves confer no increase in fitness.

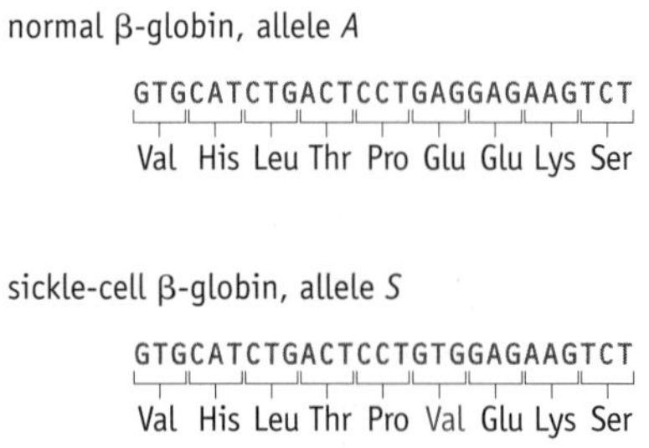

normal β-globin, allele *A*

GTGCATCTGACTCCTGAGGAGAAGTCT

Val His Leu Thr Pro Glu Glu Lys Ser

sickle-cell β-globin, allele *S*

GTGCATCTGACTCCTGTGGAGAAGTCT

Val His Leu Thr Pro Val Glu Lys Ser

FIGURE 17.20 The mutation responsible for sickle-cell anemia. The first nine amino acids of the *A* and *S* variants of human β-globin are shown, along with the corresponding nucleotide sequences. Note that the β-globin gene begins with an ATG initiation codon, but the methionine is removed from the protein by post-translational processing, and so is not shown here.

Human sickle-cell anemia provides an excellent illustration of overdominance. This disease is associated with allele *S* of the β-globin gene, which differs from the normal allele *A* by a single point mutation. This mutation converts the sixth codon in the gene from one that specifies glutamic acid in the normal allele to one that codes for valine in the sickle-cell version (Figure 17.20). The resulting change in the electrical properties of the hemoglobin molecule causes the protein to form fibers when the oxygen tension is low. In the homozygous form *SS*, the red blood cells take on a sickle shape and tend to become blocked in capillaries, where they break down. The resulting sickle-cell anemia is fatal without treatment.

Heterozygotes, with the *AS* genotype, suffer less extensive breakdown of red blood cells and a milder form of anemia. We might therefore anticipate that the normal homozygous genotype *AA* would be the most successful and would have a selection coefficient of 0, with the *SS* genotype displaying a coefficient close to 1, and *AS* at some intermediate value. This is indeed the situation in some parts of the world, but not in those areas where there is a high incidence of malaria. In those regions, the most successful genotype is *AS*. In malarial parts of Africa, for example, the selection coefficients are 0 for *AS*, 0.1 for *AA*, and 0.8 for *SS*. This means that the fitness of *AA* (a homozygote with normal hemoglobin) is nine-tenths that of *AS* (a heterozygote with mild anemia) (Figure 17.21).

The explanation of this unexpected result is that the malaria parasite *Plasmodium falciparum* finds it less easy to multiply within the bloodstream of an *AS* person, possibly because the low red blood cell count simply reduces the capacity of the parasite to reproduce. Individuals with the *AS* genotype are therefore more resistant to malaria than *AA* homozygotes. The increase in fitness conferred by this malarial resistance is greater than the decrease in fitness due to the mild anemia associated with the *AS* genotype. Heterozygotes therefore have a selective advantage compared with either of the homozygous genotypes.

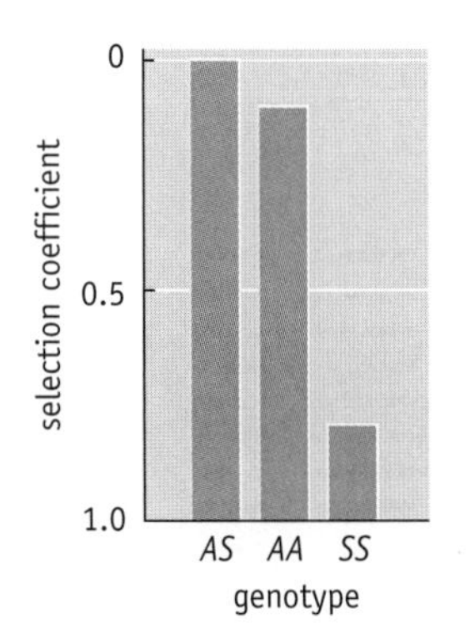

FIGURE 17.21 Selection coefficients for the three combinations of the *A* and *S* alleles of β-globin.

What are the relative impacts of drift and selection on allele frequencies?

Geneticists have long debated the impact of natural selection on populations, in particular the extent to which changes in allele frequencies are influenced by the directional effects of selection as opposed to the random effects of genetic drift. The answer to this question clearly depends on the selection coefficients of the different genotypes specifying an advantageous

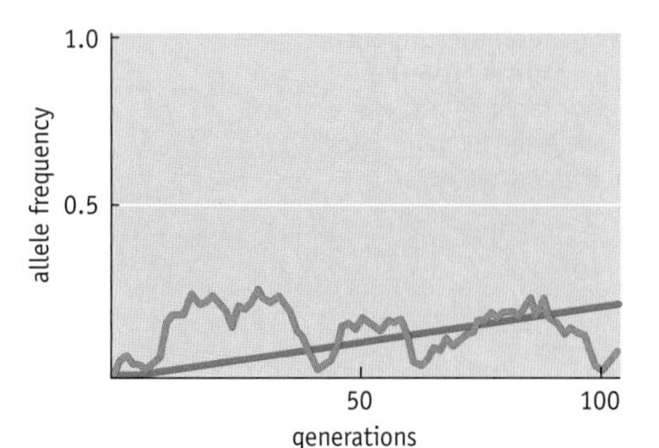

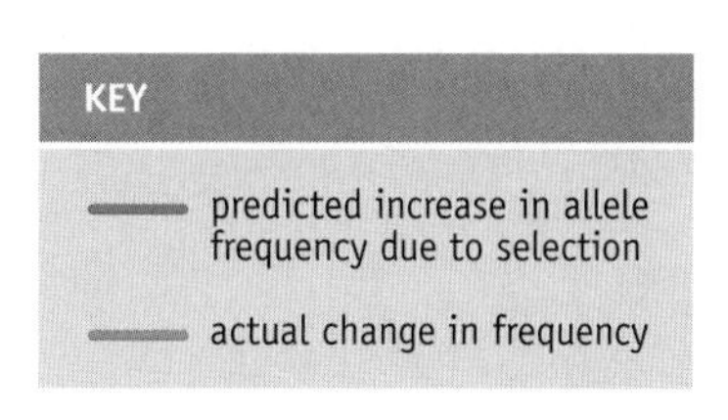

FIGURE 17.22 The effect of weak selection can be masked by genetic drift. In this graph, the gradual increase in allele frequency that is predicted to occur over 100 generations is not seen, because this increase is lesser in extent than the fluctuations in allele frequency occurring as a result of drift.

or disadvantageous trait. If the selection coefficients of the most and least successful genotypes are not hugely different, then the selection pressure will be small and the allele frequencies will change only slowly. If the frequencies of a pair of alleles are changing only slowly as a result of selection, then those changes might be masked by the effects of drift (Figure 17.22).

For a new allele created by mutation, the period immediately after it appears in the population is critical. Initially the allele will have a very low frequency, but if it contributes to a genotype that greatly increases the fitness of the individuals that possess it, then we would expect its frequency to increase rapidly as a result of positive selection. But during the period when its frequency is low a relatively small change resulting from drift could result in elimination of the allele from the population (Figure 17.23). The accidental deaths of just a few individuals before reaching reproductive age could have such an effect. It is believed that most new advantageous alleles that occur in humans are lost in this way.

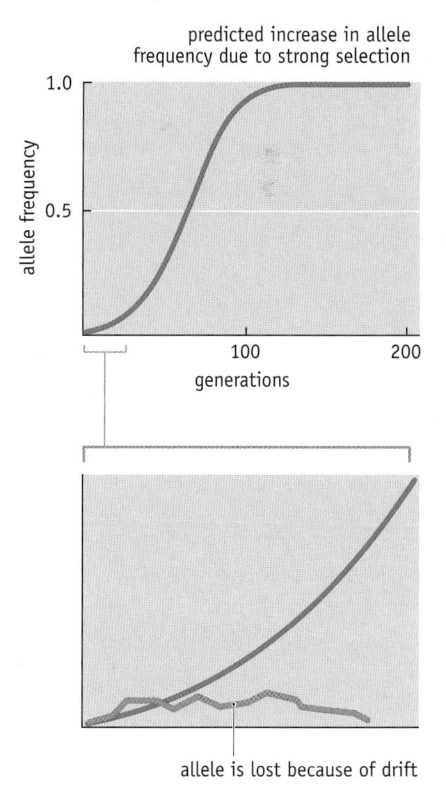

FIGURE 17.23 Even if a new allele is subject to strong selection, it might be lost because of the effects of drift.

Although this area of research is still controversial, many population geneticists favor the **neutral theory**, which postulates that genetic drift is overwhelmingly the most important factor in determining the allele frequencies within a population. Selection has a much more minor impact, with negative selection eliminating the small number of disadvantageous alleles that arise by mutation, but positive selection only rarely resulting in fixation of an advantageous allele.

17.2 THE EFFECTS OF POPULATION BIOLOGY ON ALLELE FREQUENCIES

Our understanding of "population" has gradually developed as we have progressed through this chapter. We began with the infinitely large population on which the Hardy–Weinberg equilibrium is based but quickly realized that all real populations are of finite size and this affects the way in which alleles are inherited within populations. When we considered selection for different variants of the peppered moth we became aware that populations do not occupy static ecological niches and that the environmental factors they are exposed to can change over time. And from our study of sickle-cell anemia we learned that the selective pressures encountered by

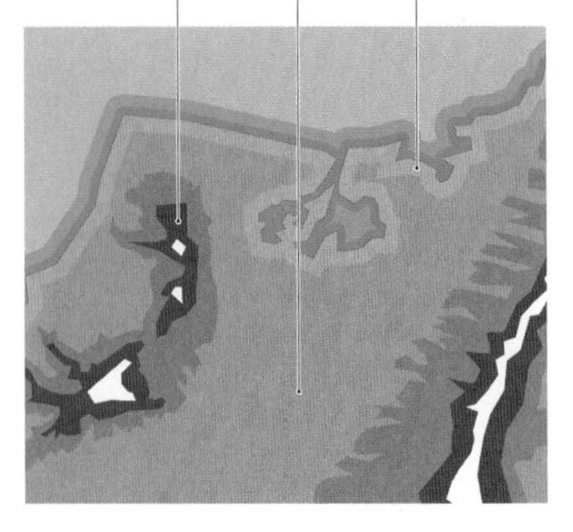

FIGURE 17.24 The genotypes displaying the greatest fitness might vary across the geographical range of a species. Here we consider six genes, A–F, in a population with a varied geographical range. In the mountainous area the greatest fitness is conferred by genotype *AbcDef*, on the inland plain by *ABcDef*, and in the coastal region by *AbCDEf*.

a population might not be the same throughout the entire geographical range occupied by that population, and that different parts of the population might therefore display different allele frequencies. We must now look more closely at some of the complexities that the biological features of populations introduce into population genetics.

Environmental variations can result in gradations in allele frequencies

Many natural populations are large and occupy sizable geographical regions. The environmental conditions throughout the entire range of a large population are unlikely to be uniform, which means that different parts of the population will be subject to different selective pressures. Those genotypes that provide greatest fitness in one part of the population might therefore not be the genotypes that provide greatest fitness in other parts (Figure 17.24). How do these factors affect allele frequencies?

An example is provided by cultivated barley, which is grown in many parts of the world and hence is exposed to many different environmental regimes. In Europe, for example, barley is grown throughout the continent, from the hot, dry areas on the Mediterranean coast all the way to the northern regions where the summers are much cooler and wetter.

In Southern Europe it is advantageous for barley to flower early in the season, in April to May, as this means that seeds are produced before the vegetative parts of the plant begin to wither and die as the temperature rises and the soil becomes dry. In Northern Europe, on the other hand, it is better for the plants to continue growing during the cool, wet summer months, as this means that more nutrients can be taken up and a greater number of seeds produced when the plants eventually flower in August and September (Figure 17.25).

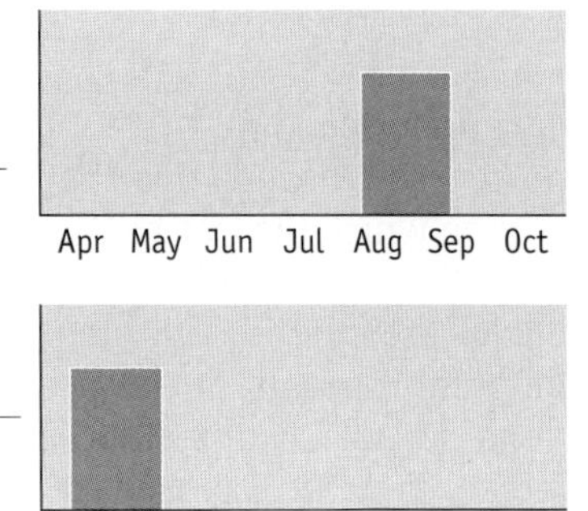

FIGURE 17.25 The most advantageous flowering times for cultivated barley in Southern and Northern Europe.

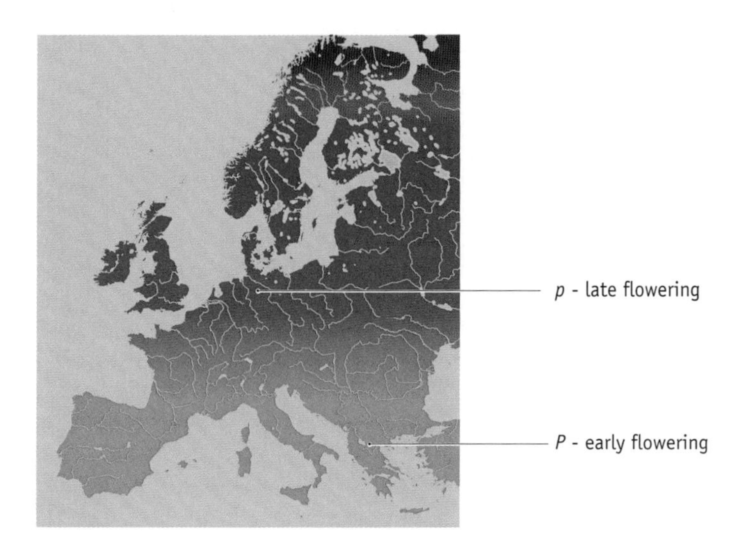

FIGURE 17.26 Frequencies of the two flowering time alleles in barley cultivars growing in different parts of Europe. Allele *p*, associated with late flowering, is more frequent in Northern Europe, and allele *P*, for early flowering, is more frequent in Southern Europe.

The genetic control over flowering time in plants is quite complicated, but plant biologists have discovered that one of the central genes involved in this phenotype is photoperiod-1. This gene enables the plant to respond to daylength, daylength being the trigger that causes the plant to flower. There are two alleles of photoperiod-1, called *P* and *p*. The first of these specifies the daylength response that results in early flowering in April and May. Allele *p*, which differs from *P* by a single point mutation, has an altered daylength response, such that flowering is delayed until later in the year.

When we examine the frequencies of these two alleles in barley cultivars grown in different parts of Europe, we see a gradation from north to south, the frequency of *p* being greatest in the north, and that of *P* being greatest in the south (Figure 17.26). This is called a **cline** and is a feature of many alleles subject to environmental selection in populations that have a large geographical range.

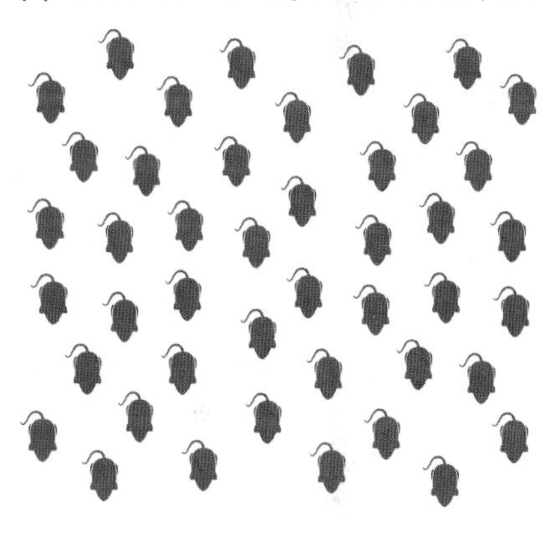

There may be internal barriers to interbreeding within a population

Real populations often contain subdivisions that are brought about and maintained by geographical or other barriers that reduce the amount of interbreeding that can take place between the subpopulations. To take a simple example, imagine a population of small animals inhabiting a broad plain (Figure 17.27A). Initially there are no barriers to interbreeding, so any female in the population could mate with any male, and vice versa. As time passes a river begins to run across the plain, dividing the population into two groups (Figure 17.27B). These small animals cannot swim, and the population therefore becomes divided into two components that do not interbreed. A barrier to interbreeding has sprung up between them.

What will be the effects of this population subdivision on the genetic features of the two subpopulations? The process of genetic drift will still operate within the subpopulations, but its effect on allele frequencies will not be the

FIGURE 17.27 A population subdivision. (A) A population of small animals inhabits a broad plain. Initially, all individuals within the population have access to one another for breeding purposes. (B) If a river begins to flow through the plain, and if the animals are unable to swim, two subpopulations will soon develop on either side of the river that are unable to interbreed with each other.

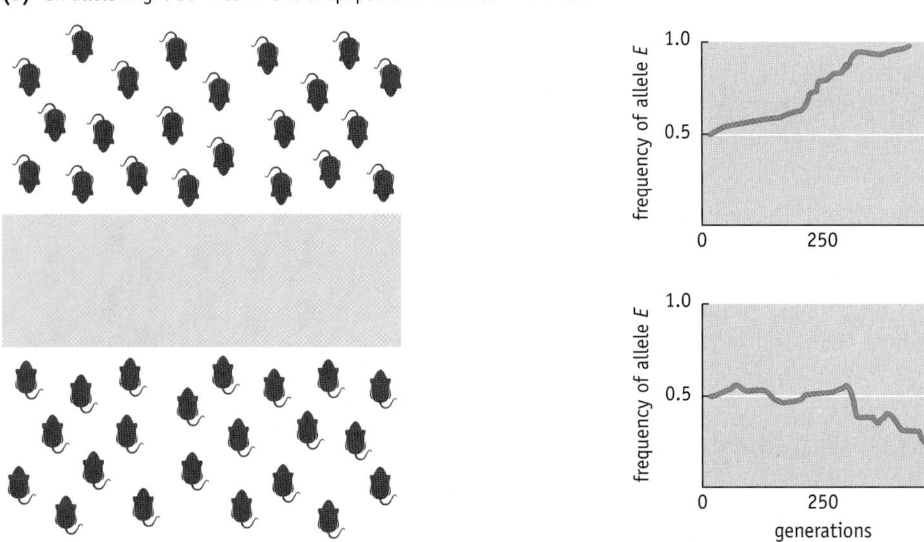

FIGURE 17.28 Drift will have different effects on allele frequencies in two non-interbreeding populations. (A) The changes in frequency for four alleles, *A–D*, are shown. After 100 generations, the allele frequencies are different in the two subpopulations. (B) The frequency of allele *E* is followed in the two subpopulations. In one subpopulation the allele becomes fixed, but in the other it is lost.

same in both. By chance, some alleles that become more frequent in one subpopulation will become less so in the other (Figure 17.28A). There may even be cases where one allele of a pair becomes fixed in one subpopulation and the other allele becomes fixed in the second subpopulation (Figure 17.28B).

If the effective sizes of the two subpopulations are similar, then this differential allele fixation will result in the unusual situation where the two alleles have approximately equal frequencies in the population as a whole but there are no heterozygotes present (Figure 17.29). This pattern would never arise in a single interbreeding population. If there is no barrier to

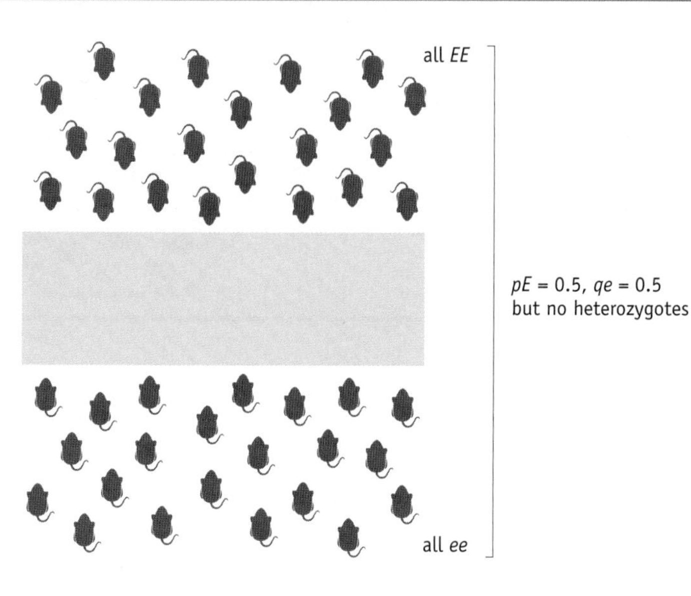

FIGURE 17.29 In this divided population the two alleles *E* and *e* have the same frequency when the population as a whole is considered, but there are no heterozygotes.

interbreeding within a population, then, for any pair of alleles of approximately equal frequency, there will always be many heterozygotes.

The example we have used is an extreme one where the barrier to interbreeding is clearly visible (it is a large river) and it is absolute (the animals cannot swim). In nature, subpopulations might be much less easy to discern, because the barriers between them are less obvious geographical features or because the barriers have a more subtle ecological context. Most natural barriers are not absolute, and there is usually at least a limited amount of interbreeding between members of different subpopulations. How then can we detect the presence of subpopulations?

Even if there is some interbreeding between subpopulations, allele frequencies are still expected to change in different ways because of the random nature of genetic drift. The differences between the predicted and observed numbers of heterozygotes can therefore be used to determine whether population subdivisions exist. This is the basis of the various **fixation indices**, which compare the heterozygosity—the proportion of heterozygotes—within different components of a population. The most commonly used fixation index is the $\boldsymbol{F_{ST}}$ **value**, calculated as:

$$F_{ST} = \frac{H_T - H_S}{H_T}$$

In this equation, H_T is the heterozygosity that is predicted from the allele frequency in the population as a whole, and H_S is the average heterozygosity of the suspected subpopulations. If the F_{ST} value is less than 0.05, then the population as a whole is looked on as having little substructure, but if the value is over 0.25 then there is high genetic differentiation and relatively well-defined subpopulations are present.

Transient reductions in effective population size can result in sharp changes in allele frequencies

Another common feature of real populations is that their sizes might change over time. In particular, a change to the environment or a reduction in the available food supply can result in a rapid and severe drop in the number of individuals in a population. If the cause of the population crash is removed—the environmental conditions return to a more friendly state

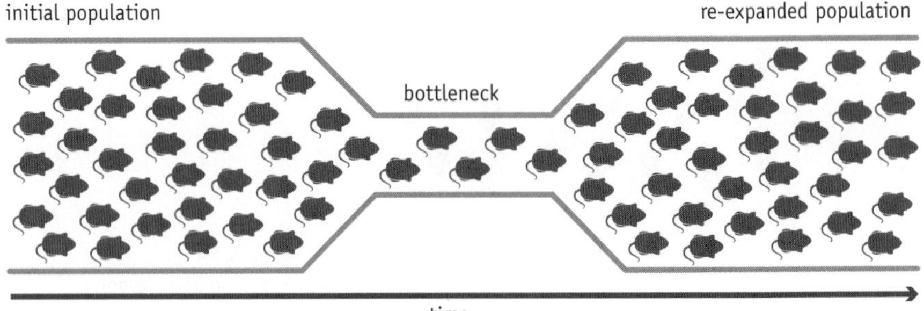

FIGURE 17.30 A population bottleneck.

or the food supply increases—the population can grow again, possibly back to its original size. The period between the decline and recovery is called a **bottleneck** (Figure 17.30).

A number of species are thought to have gone through bottlenecks during the last few thousand years, including the elephant seal and African cheetah (Figure 17.31). The human population is thought to have declined to just 10,000 individuals some 150,000 years ago, during a period of global cooling during the Ice Ages. Additionally, we are aware of many species whose population numbers are currently declining and for which the best future scenario is a severe bottleneck, the alternative being extinction.

A bottleneck has an important influence on the genetic features of a population because of a **sampling effect**. We have already seen that sampling can result in a gradual change in allele frequencies over time, and that this underlies genetic drift. The same principles apply to the sampling that occurs when a population goes through a bottleneck, although the impact can be more extreme (Figure 17.32). The allele frequencies within the bottleneck might be quite different from those displayed by the population before its numbers began to decline, and some alleles, especially rarer ones, might be completely lost. The new expanded population that develops after the bottleneck therefore has different genetic features compared with the population that existed before the bottleneck. The allele frequencies will have shifted and there will probably have been an overall decline

FIGURE 17.31 Two species that have gone through a population bottleneck during the last few thousand years. (A) The northern elephant seal (*Mirounga angustirostris*). (B) The African cheetah (*Acinonyx jubatus*). The elephant seal population is known to have declined to about 30 during the nineteenth century due to hunting, and although the population has now recovered it displays low genetic variability. The African cheetah population also has low genetic variability, but the bottleneck that caused this occurred in the more distant past. (A, courtesy of Robert Schwemmer, National Oceanic and Atmospheric Administration; B, courtesy of Thomas A. Hermann, http://life.nbii.gov)

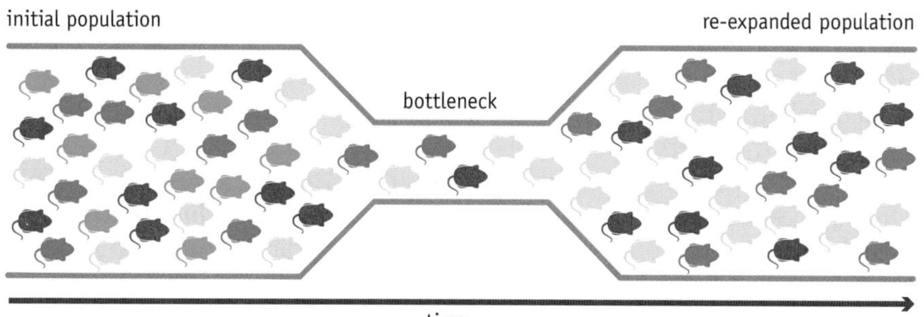

FIGURE 17.32 A bottleneck can cause a change in allele frequencies. Four alleles, each represented here by an animal of a different color, have approximately equal frequencies in the initial population. After the bottleneck, the frequencies are different, due to the sampling effect. Note that the "green" allele is completely eliminated in the bottleneck.

in variability due to allele loss. A bottleneck can therefore cause a rapid change in the genetic features of a population.

Founder effects are similar to bottlenecks but take place in a different context. A founder effect occurs when a few members of a population split away and initiate a new population (Figure 17.33). Typically, this happens when part of the population moves to new territory, previously unoccupied by that species, an event we call **colonization**. Again, there is a sampling effect, because the small number of individuals that act as founders might not possess all the alleles present in the parent population, and might display different frequencies of those that they do possess. Founder events were important in human prehistory. The gradual movement of our species from its original homeland in Africa into Asia, Europe, and the New World is thought to have involved a series of colonizations, each involving a relatively small number of people. The change in allele frequency that occurred during each of these founder events is discernible, to some extent, in modern human populations. This enables the patterns of these past colonizations, and the approximate time when each one took place, to be deduced.

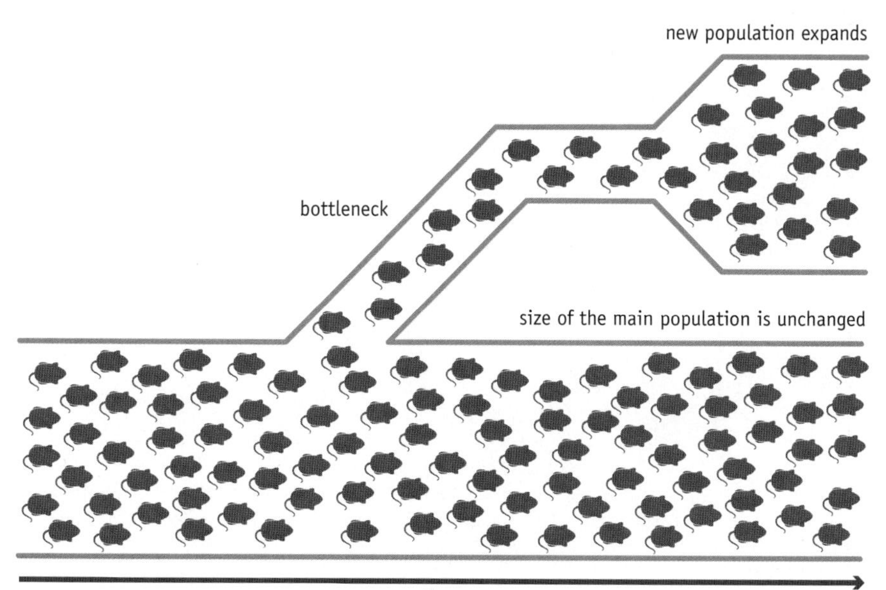

FIGURE 17.33 A founder effect.

A population split can be the precursor of speciation

How do the creation of new alleles, and changes in the frequencies of alleles, relate to the evolution of new species? Our modern concept of species is based on reproductive incompatibility. Two members of the same species, of opposite sexes, are able to reproduce and give rise to viable offspring, whereas members of different species cannot interbreed, or, if they do, the progeny are sterile. To take a simple example, male and female horses produce viable baby horses even if the male and female parents are taken from different parts of the world. This is because all horses are members of the single species *Equus caballus*. In contrast, donkeys form a different species, *Equus asinus* (Figure 17.34). Donkeys and horses can produce offspring—a mule if the horse is the female parent, or a hinny if the horse is the father—but these offspring are sterile. This is an example of **postzygotic isolation**, where the barrier to reproduction occurs after a productive mating. In contrast, with **prezygotic isolation** there is no mating, or mating produces no offspring, not even sterile ones. To understand the genetic basis of **speciation** we must therefore make a link between the microevolutionary events that we have studied in this chapter, involving the factors that influence allele frequencies within populations, and the processes that result in reproductive incompatibility.

A population split is often looked on as the starting point in speciation. A split due to, for example, geographic isolation would not itself constitute speciation, because at this stage members of the two populations, if brought together, could still mate productively. The reproductive isolation does, however, mean that there is no interbreeding between the two populations, which are now free to pursue their own microevolutionary pathways. These might eventually involve differential genetic changes that confer post- or prezygotic isolation.

Understanding how two isolated, microevolving populations can develop reproductive incompatibility in this way has been a challenge for geneticists. Some of the research has focused on the possibility that prezygotic

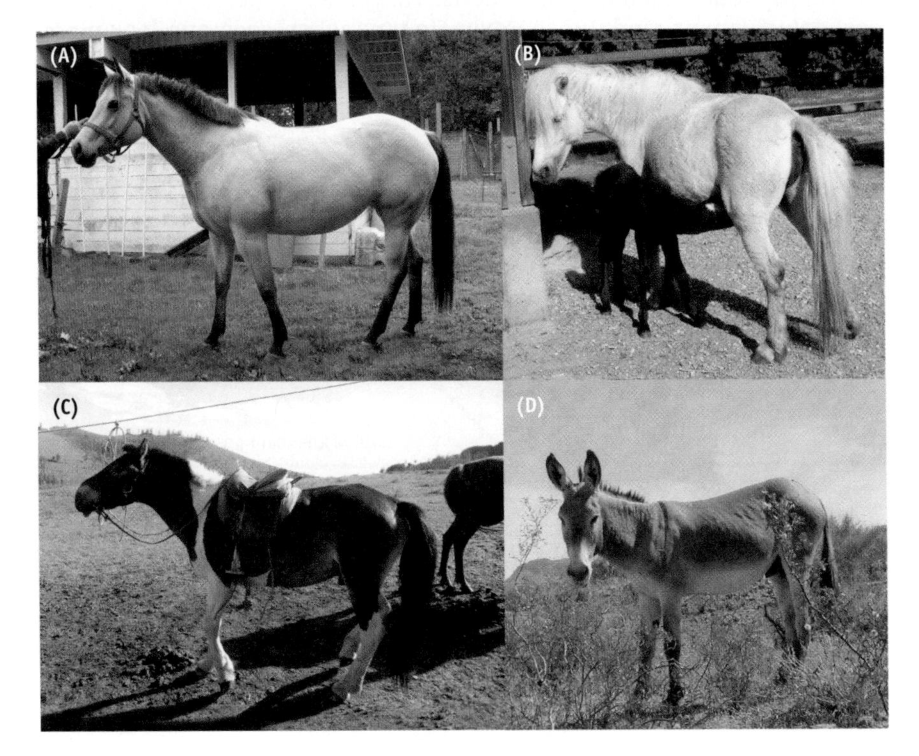

FIGURE 17.34 Three members of the species *Equus caballus*: (A) American quarterhorse, (B) Shetland pony, (C) Mongolian horse; and (D) one member of *Equus asinus*, the donkey. (D, courtesy of BLM/Arizona.)

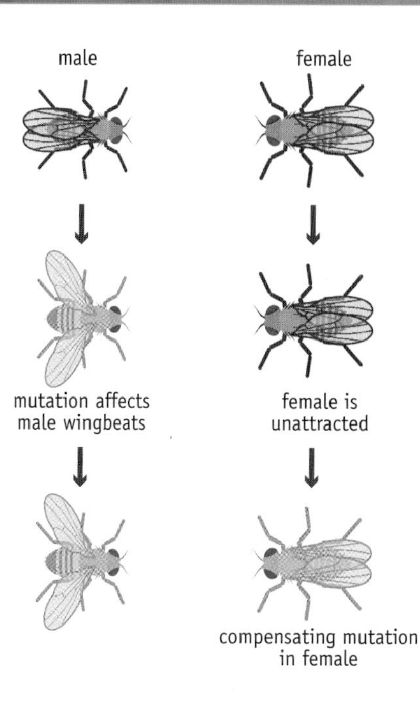

FIGURE 17.35 Prezygotic isolation might occur following a single gene mutation, as indicated here for the *per* gene of *Drosophila*, which controls the pattern of wing movements used by the male to attract a mate. One difficulty with this hypothesis is that a compensating mutation must occur in the genes that specify the female's preference for male wing movements, otherwise the males with abnormal wingbeats will not find partners.

but perhaps not postzygotic isolation could be achieved by a change in a single gene or a small group of genes, at least in less complex animals. An example is provided by the *per* gene of *Drosophila*, which determines the speed and rhythm of the male fly's wingbeats during its courtship behavior. Changes in the male's wing movements reduce its likelihood of attracting a female. If, however, there were a compensating mutation in the genes that specify the female's preference for male wing movements, then males with abnormal wingbeats might find partners again (Figure 17.35). In this scenario, the fruit fly population has split into two subpopulations, separated by prezygotic isolation, possibly without any accompanying geographical isolation.

The difficulty with this model is that the initial change in the behavior of the mutant male flies would have to be compensated, at the same time, by the change in female preference. If the females did not change their preference until a few generations later, then the flies with the new wing movements would simply die out. The same is true of other changes involving possible speciation genes, such as mutations in genes involved in synthesis of the female fly's pheromones—the volatile hydrocarbons that she exudes in order to attract a mate.

Other studies have looked at the possible role that chromosome pairing has on hybrid sterility. Sterility of mules and hinnies is thought to be due to horses and donkeys having different numbers of chromosomes—64 for horses and 62 for donkeys. This difference means that although horse and donkey gametes can fuse to form a hybrid embryo that develops into a mule or a hinny, the hybrid cannot itself produce gametes because the chromosomes cannot pair effectively during meiosis (Figure 17.36). Difficulties in pairing can also arise if the two sets of chromosomes display rearrangements such as inversions and translocations. This type of genetic change might therefore be one of the factors that drive postzygotic isolation and resulting speciation.

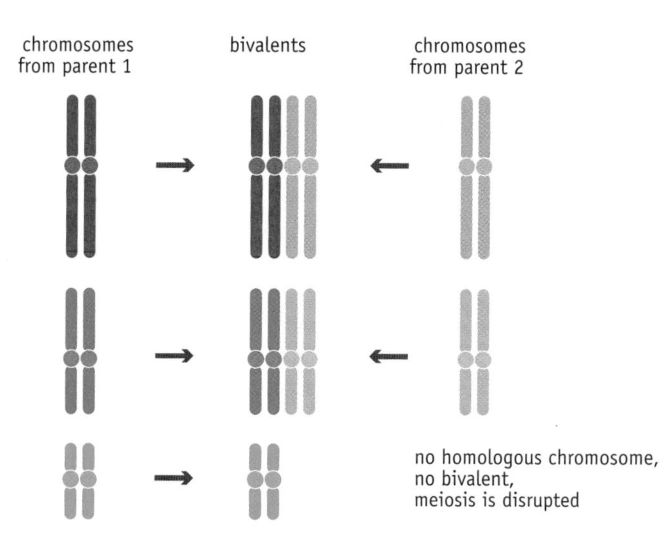

FIGURE 17.36 Hybrid sterility can result from an inability of the paternal and maternal chromosomes to pair effectively during meiosis.

KEY CONCEPTS

- A population is a group of organisms whose members are likely to breed with one another.
- The Hardy–Weinberg equilibrium shows that mating, by itself, has no effect on the allele frequencies present in a population.
- Within a population, the number of individuals that possess a particular allele can change over time. Changes in allele frequencies are referred to as microevolution. They occur as a result of random processes called genetic drift and more directed processes called selection.
- New alleles are continually created by mutation.
- Natural selection favors individuals with high fitness for their environment. Natural selection does not act directly on genes, but the changes that it brings about within a population have important effects on allele frequencies.
- The neutral theory postulates that genetic drift is overwhelmingly the most important factor in determining the allele frequencies within a population. Selection has a much more minor impact, with negative selection eliminating the small number of disadvantageous alleles that arise by mutation, but positive selection only rarely resulting in fixation of an advantageous allele.
- Environmental variations can result in gradations in allele frequencies within a population.
- A transient reduction in effective population size can result in sharp changes in allele frequencies. This might occur if the population goes through a bottleneck or if a few members leave the population and colonize a new geographical region.
- A population split is often looked on as the starting point of speciation. Different microevolutionary pathways within the two subpopulations might eventually result in differential genetic changes that confer post- or prezygotic isolation.

QUESTIONS AND PROBLEMS (Answers can be found at www.garlandscience.com/introgenetics)

Key Terms

Write short definitions of the following terms:

bottleneck
cline
colonization
effective population size
F_{ST} value
fitness
fixation
fixation index
founder effect
genetic drift
Hardy–Weinberg equilibrium
heterosis
heterozygote advantage
heterozygote superiority
hitchhiking
microevolution
natural selection
neutral theory
overdominance
postzygotic isolation
prezygotic isolation
sampling effect
selection
selection coefficient
selective sweep
speciation
stochastic effect

Self-study Questions

17.1 Describe the basis of the misconception that recessive alleles should gradually be lost from a population.

17.2 Explain how the gradual loss of recessive alleles from a population was shown to be a fallacy.

17.3 Why does microevolution occur despite the predictions of the Hardy–Weinberg equilibrium?

17.4 Explain how stochastic effects can lead to genetic drift.

17.5 What is the impact of effective population size on the rate at which alleles become fixed in a population?

17.6 Why is it important to understand the effective population size of an endangered species?

17.7 Explain why most new alleles are rapidly eliminated from a population. Under what circumstances is a new allele likely to increase in frequency?

17.8 Describe how the peppered moth illustrates the principles of natural selection.

17.9 Distinguish between the terms "hitchhiking" and "selective sweep."

17.10 Giving an example, explain why a heterozygote can be the most successful of a set of genotypes.

17.11 According to the neutral theory, what are the relative impacts of genetic drift and natural selection on microevolution?

17.12 Describe how environmental factors can result in a gradation in allele frequencies.

17.13 How might an internal barrier to interbreeding arise in a population? How might such a barrier result in the situation where a pair of alleles have similar frequencies in the population as a whole, but there are no heterozygotes?

17.14 How are F_{ST} values used to study the internal structure of a population?

17.15 Distinguish between a bottleneck and a founder effect.

17.16 Give examples of species thought to have been through a recent population bottleneck.

17.17 What effect does a bottleneck have on the frequency of alleles in a population?

17.18 Describe the differences between prezygotic and postzygotic isolation.

17.19 Give one example of how prezygotic isolation might be achieved by mutation of a single gene.

17.20 How might chromosome rearrangements be involved in speciation?

Discussion Topics

17.21 Is it possible for a real population ever to display the Hardy–Weinberg equilibrium?

17.22 To what extent do studies of microevolution provide an argument against those individuals who contend that evolution is a fallacy?

17.23 Discuss how a knowledge of the extent of genetic diversity displayed by an endangered species can be used to aid the planning of a breeding program aimed at avoiding that species becoming extinct.

17.24 Can you devise an analysis that would enable the length of time that has elapsed since a bottleneck occurred to be deduced from examination of the allele frequencies in a population?

17.25 Evaluate the hypothesis that prezygotic isolation can be achieved by a change in a single gene or a small group of genes.

FURTHER READING

Charlesworth B (2009) Effective population size and patterns of molecular evolution and variation. *Nat. Rev. Genet.* 10, 195–205. *A good description of the importance of effective population size in evolutionary studies.*

Cook LM, Mani GS & Varley ME (1986) Postindustrial melanism in the peppered moth. *Science* 231, 611–613.

Hardy GH (1908) Mendelian proportions in a mixed population. *Science* 28, 49–50. *First description of the Hardy–Weinberg equilibrium.*

Hartl DL & Clark AC (1997) Principles of Population Genetics, 3rd ed. Sunderland, MA: Sinauer.

Holsinger KE & Weir BS (2009) Genetics in geographically structured populations: defining, estimating and interpreting F_{ST}. *Nat. Rev. Genet.* 10, 639–650.

Jones H, Leigh FJ, Bower MA et al. (2008) Population-based resequencing reveals that the flowering time adaptation of cultivated barley originated east of the Fertile Crescent. *Mol. Biol. Evol.* 25, 2211–2219. *The barley photoperiod-1 allele gradation across Europe.*

Kimura M (1985) The Neutral Theory of Molecular Evolution. Cambridge: Cambridge University Press.

Nielsen R (2005) Molecular signatures of natural selection. *Annu. Rev. Genet.* 39, 197–218. *Explains how examination of DNA sequences and genotypes can reveal genes and larger regions of a genome that are under selection.*

Rieseberg LH & Blackman BK (2010) Speciation genes in plants. *Ann. Bot.* 106, 439–455.

Strasser BJ (1999) "Sickle cell anemia: a molecular disease." *Science* 286, 1488–1490. *An account of the discovery of the changes in hemoglobin structure that underlie sickle-cell anemia.*

Tauber E, Roe H, Coosta R, Hennessy JM & Kyriacou CP (2003) Temporal mating isolation driven by a behavioural gene in *Drosophila*. *Curr. Biol.* 13, 140–145. *The possibility of speciation driven by the fruit fly per gene.*

Templeton RA (1980) The theory of speciation via the founder principle. *Genetics* 94, 1011–1038. *Founder effects.*

PART III

GENETICS IN OUR MODERN WORLD

We have completed our study of the two complementary roles of genes as units of biological information and as units of inheritance. Now we will look more broadly at modern genetics and explore some of the areas of research that are responsible for the high profile of the subject in the public perception.

The topics that we will study are not an exhaustive list of all the important areas of current research. It would require several books of this size to cover everything. The choice is based on three factors. The first is to illustrate how genetics works hand in hand with other areas of biology to advance our knowledge of cells and organisms. For this reason, the first chapter in this part of the book is on differentiation and development, an area of research that has made great leaps forward in recent years thanks to the combined endeavours of geneticists, cell biologists, physiologists, and biochemists.

The second factor influencing our choice of subjects is the anthropocentric view that our own species is the most important on the planet. Among all species, humans are probably in a minority of one in holding this view, but nonetheless we will devote three chapters to the genetics of humans. We will study the human genome, the way in which genetics is being used to understand human disease, and the applications of genetics in forensic science and in studies of human history.

These anthropocentric topics include two—genes in medicine and genes in forensics—that also address the third factor that dictates our choice of subjects. This is the need to examine some of the applications to which genetics is being put in our modern world. In addition to medicine and forensics, we will investigate the use of gene cloning as a means of producing important proteins in bacteria, fungi, and animals, and we will also look at the applications of genetic manipulation in agriculture.

In several of these chapters we will encounter areas of modern genetics that are at the forefront of public perception not just because of the excitement of the new discoveries or the importance of the breakthroughs, but also because of concerns that the outcomes of the research might be used unwisely. As students of genetics we cannot ignore these issues. The final chapter of this book is therefore devoted to the ethical issues raised by modern genetics.

Genes in Differentiation and Development CHAPTER 18

In this chapter we will study the genetic basis of differentiation and development. Differentiation and development are inextricably linked, but we must not confuse one with the other. Differentiation is the process by which an individual cell acquires a specialized function. Differentiation therefore requires a change in the pattern of gene expression in a cell. The resulting specialized expression pattern usually remains in place for the remainder of the cell's lifetime and is often inherited by the daughter cells, and the granddaughters, and so on for many cell divisions (Figure 18.1). Development, in contrast, is the pathway that begins with a fertilized egg cell and ends with an adult. A developmental pathway comprises a complex series of genetic, cellular, and physiological events that must occur in the correct order, in the correct cells, and at the appropriate times if the pathway is to reach a successful culmination. These events include the differentiation of many cells into many different specialized types. The human developmental pathway, for example, results in an adult containing 10^{13} cells differentiated into more than 400 specialized types, with the activity of each individual cell being coordinated with that of every other cell. Understanding how genes specify and regulate the differentiation and developmental pathways of multicellular eukaryotes such as humans is one of the biggest challenges in all of genetics.

Developmental biology is a multidisciplinary subject that encompasses areas of genetics, cell biology, physiology, and biochemistry. In this chapter we will focus on the roles of genes in differentiation and development and will attempt to answer two questions. The first of these concerns the genetic basis of differentiation. What genetic changes occur when a cell takes on a specialized role? In particular we want to understand how the specialized pattern of gene expression is retained through many rounds of mitosis. To answer this question we will look at various examples of cellular differentiation, and we will discover that a range of strategies for controlling gene expression are used by different types of specialized cell.

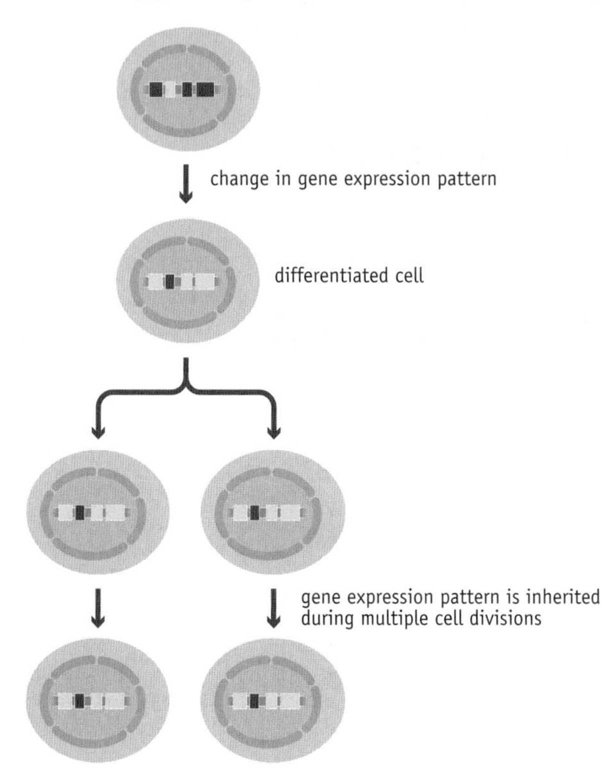

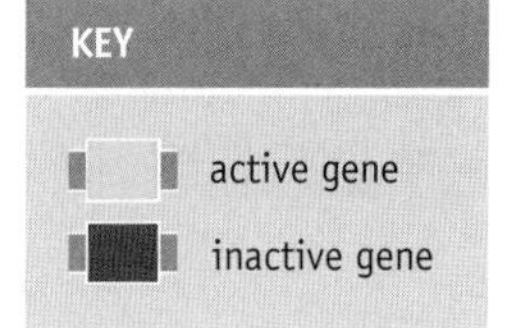

FIGURE 18.1 Differentiation is accompanied by a permanent and heritable change in the gene expression pattern of a cell and its descendants.

The second question asks how changes in gene activity are regulated and coordinated in space and time during development. Spatial and temporal regulation are needed to ensure that development results in the construction of complex body parts at the correct places in the adult organism. We will explore the research that has been carried out on development in model organisms such as the fruit fly, and we will examine how the discoveries made with these relatively simple organisms have provided insights into the more complex pathways in humans.

18.1 CHANGES IN GENE EXPRESSION DURING CELLULAR DIFFERENTIATION

Fewer than 3000 of the 20,500 genes in the human genome are active in all cells. The other genes are differentially expressed, with up to 15,000 active in any particular cell type at any one time (Figure 18.2). These differential patterns of gene expression underlie the specialized biochemistries and physiologies of the 400 cell types that make up the human body plan.

In Chapter 9 we encountered processes by which the expression of individual genes or groups of genes can be modulated or even switched off in response to external and internal stimuli. Most, and possibly all, of these processes result in transient changes in gene activity, changes that occur within the lifetime of a cell and might easily be reversed later in that cell's lifetime (Figure 18.3A). Cellular differentiation requires more permanent changes in gene expression patterns, changes that will last the lifetime of the cell and be maintained even when the stimulus that originally induced them has disappeared. Importantly, the gene expression patterns must be inherited by the descendants of that cell after mitosis (Figure 18.3B). We must therefore broaden our view beyond the types of gene regulation that we have studied so far.

Geneticists are aware of four different ways in which permanent changes in gene activity can be brought about (Figure 18.4). The first is physical rearrangement of the genome, which can result in a change in the structure of the genome inherited by a cell lineage. Second, it is possible for a genetic feedback loop to be established that will maintain a change in gene expression in a cell and its descendants. Third and fourth, there are **chromatin modification**, which involves the chemical alteration of nucleosomes, and **DNA methylation**, both of which are thought to have a general role in maintaining gene expression patterns in differentiated cells.

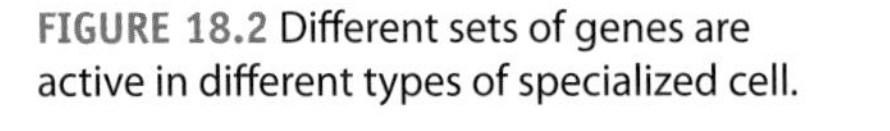

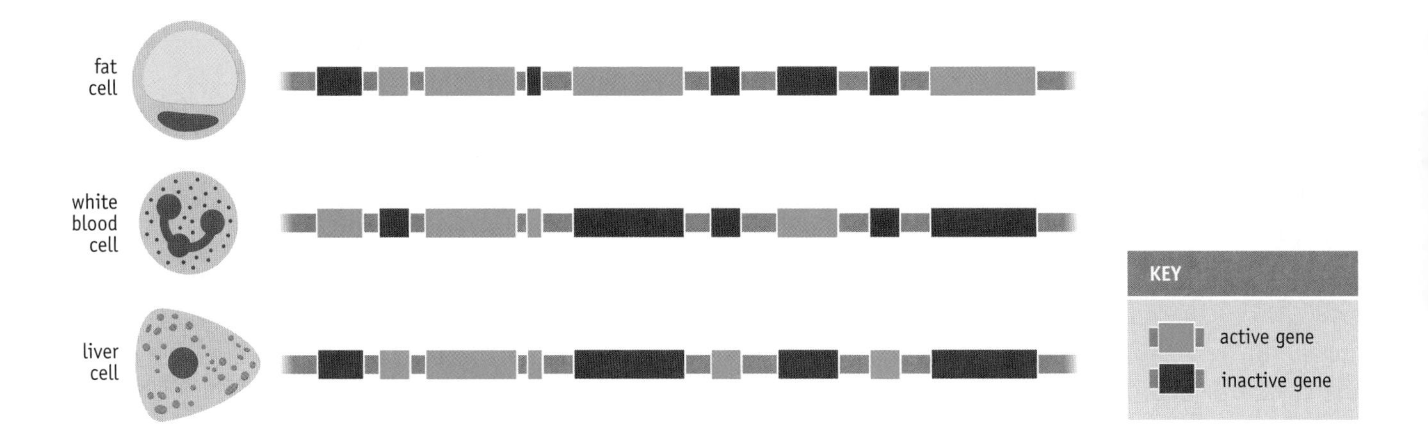

FIGURE 18.2 Different sets of genes are active in different types of specialized cell.

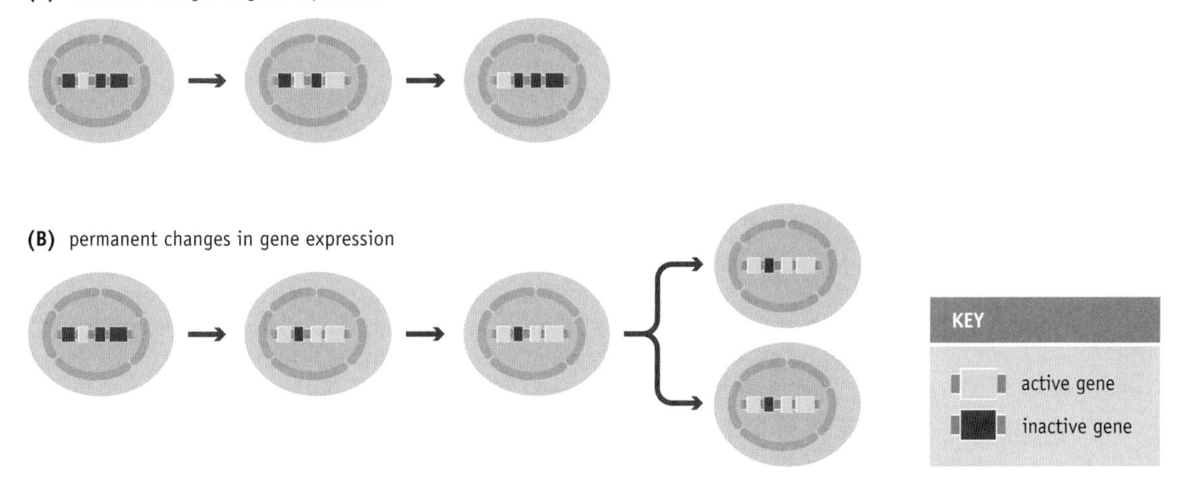

FIGURE 18.3 Transient and permanent changes in gene expression.

Increasingly, the term **epigenesis** is being used to describe these permanent changes in gene expression patterns. Epigenesis is defined as a heritable change to the genome that does not involve mutation.

Immunoglobulin and T-cell diversity in vertebrates results from genome rearrangement

First we will examine an example of the way in which cellular differentiation can be brought about by physical rearrangement of the genome. This example concerns the differentiation of lymphocytes into cells that are specialized for the production of specific components of the mammalian immune system.

Immunoglobulins and T-cell receptors are related proteins that are synthesized by B and T lymphocytes, respectively. Both types of protein become attached to the outer surfaces of their cells, and immunoglobulins are also released into the bloodstream. The proteins help to protect the body against invasion by bacteria, viruses, and other unwanted substances by binding to these **antigens**.

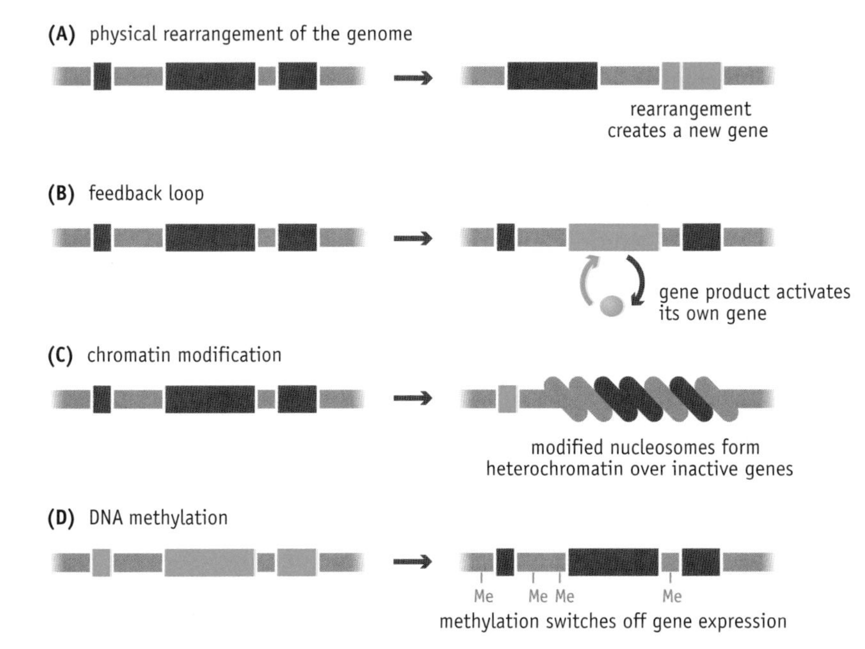

FIGURE 18.4 Four ways in which permanent changes in gene expression can be brought about. (A) Physical rearrangement of the genome can result in a new gene. (B) A genetic feedback loop can maintain a change in gene expression. (C) Chromatin modification involves the chemical alteration of nucleosomes so heterochromatin is formed over inactive genes. (D) DNA methylation switches off expression of adjacent genes.

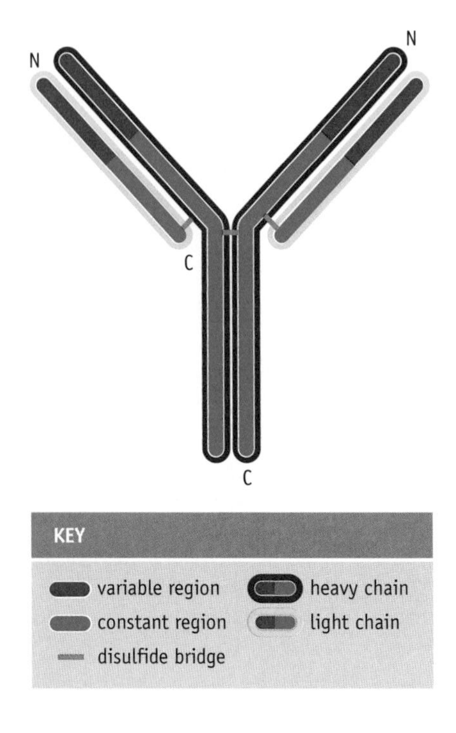

FIGURE 18.5 The structure of a typical immunoglobulin protein. Each heavy chain is 446 amino acids in length and consists of a variable region spanning amino acids 1–108 followed by a constant region. Each light chain is 214 amino acids, again with an N-terminal variable region of 108 amino acids. Additional disulfide bridges form between different parts of individual chains. These and other interactions fold the protein into a more complex three-dimensional structure.

During its lifetime, an organism might be exposed to any number of different antigens, which means that the immune system must be able to synthesize an equally vast range of immunoglobulin and T-cell receptor proteins. Humans are able to make approximately 10^8 different immunoglobulin and T-cell receptor proteins, but there are only 20,500 genes in the human genome, so where do all these proteins come from?

To understand the answer we will look at the structure of a typical immunoglobulin protein. Each immunoglobulin is a tetramer of four polypeptides —two long "heavy" chains and two short "light" chains—linked by disulfide bridges (Figure 18.5). The heavy chains of immunoglobulins specific for different antigens differ mainly in their N-terminal regions, the C-terminal parts being similar, or "constant," in all heavy chains. The same is true for the light chains, except that two families, κ and λ, can be distinguished.

In vertebrate genomes there are no complete genes for the polypeptides that form the immunoglobulin heavy and light chains. Instead, these proteins are specified by gene segments, located on chromosome 14 in humans. There are 11 C_H gene segments, coding for different versions of the constant region, each with a slightly different amino acid sequence. The C_H segments are preceded by 123–129 V_H gene segments, 27 D_H gene segments, and 9 J_H gene segments, coding for different versions of the V (variable), D (diverse), and J (joining) components of the variable part of the heavy chain (Figure 18.6). The entire heavy-chain region stretches over several megabases. A similar arrangement is seen with the light-chain loci on chromosomes 2 (κ locus) and 22 (λ locus), the only difference being that the light chains do not have D segments.

During the early stage of development of a B lymphocyte, the immunoglobulin region of the genome rearranges. Within the heavy-chain region, these rearrangements link one of the D_H gene segments with one of the J_H gene segments, and then link this D–J combination with a V_H gene segment (Figure 18.7). The end result is an exon that contains the complete open reading frame specifying the V, D, and J segments of the immunoglobulin protein. This exon becomes linked to a second exon, containing the C-segment sequence, by splicing during the transcription process. This creates a complete heavy-chain mRNA that can be translated into an immunoglobulin protein that is specific for just that one lymphocyte.

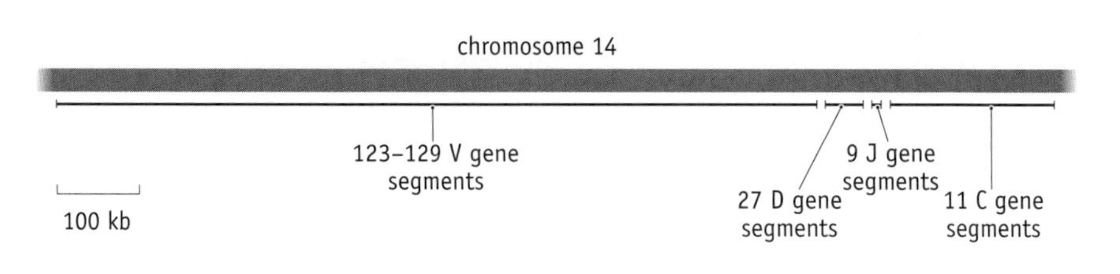

FIGURE 18.6 Organization of the human *IGH* (heavy-chain) region on chromosome 14.

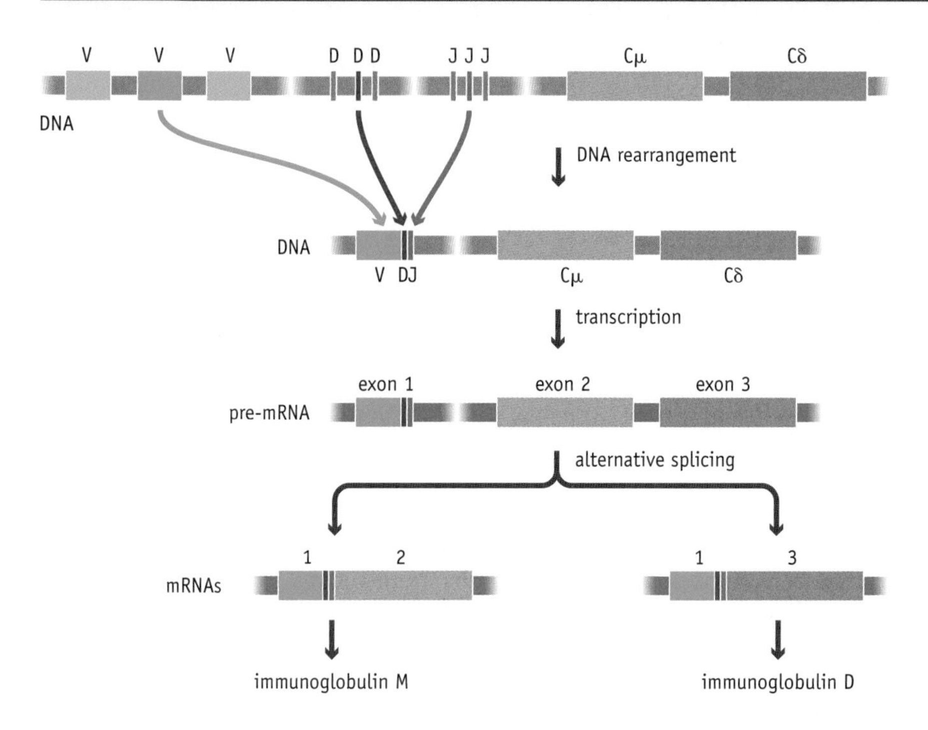

FIGURE 18.7 Synthesis of a specific immunoglobulin protein. DNA rearrangement links V, D, and J segments, which are then linked to a C segment by splicing of the mRNA. In immature B cells the V-D-J exon always becomes linked to the Cμ exon (exon 2) to produce an mRNA specifying a class M immunoglobulin. Later in the development of the B cell, some immunoglobulin D proteins are also produced by alternative splicing that links the V-D-J exon to the Cδ exon. Both types of immunoglobulin become bound to the cell membrane.

A similar series of DNA rearrangements results in the construction of the lymphocyte's light-chain V–J exon at either the κ or λ locus, with splicing once again attaching an exon containing the C segment when the mRNA is synthesized. Each lymphocyte therefore makes its own specific immunoglobulin and passes the information for the synthesis of that immunoglobulin, in the form of its rearranged genome, to all of its progeny.

Diversity of T-cell receptors is based on similar rearrangements that link V, D, J, and C gene segments in different combinations to produce cell-specific genes. Each receptor comprises a pair of β molecules, which are similar to the immunoglobulin heavy chain, and two α molecules, which resemble the immunoglobulin κ light chains. As with immunoglobulins, the T-cell receptors become embedded in the cell membrane and enable each lymphocyte to recognize and respond to its own specific extracellular antigen.

A genetic feedback loop can ensure permanent differentiation of a cell lineage

The second mechanism for bringing about a permanent change in the pattern of gene expression involves the use of a feedback loop. In this system a regulatory protein activates its own transcription so that once its gene has been switched on, it is expressed continuously (Figure 18.8).

One of the best-known examples of feedback regulation concerns the MyoD protein of vertebrates, which is involved in muscle development. A cell becomes committed to becoming a muscle cell when it begins to express the *myoD* gene. This gene codes for a transcription activator that targets several other genes involved in the differentiation of muscle cells. The MyoD protein also binds upstream of *myoD*, ensuring that its own gene is continuously expressed. The result of this positive feedback loop is that the cell continues to synthesize the MyoD protein and remains a muscle cell. The differentiated state is heritable because cell division is accompanied by the transmission of MyoD to the daughter cells, ensuring that these are also muscle cells.

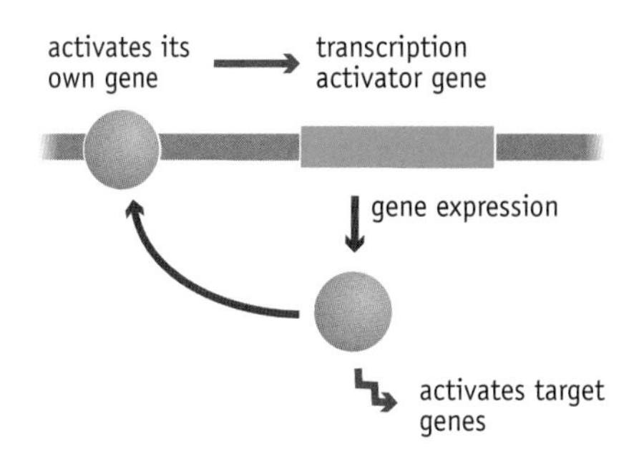

FIGURE 18.8 Feedback regulation of gene expression.

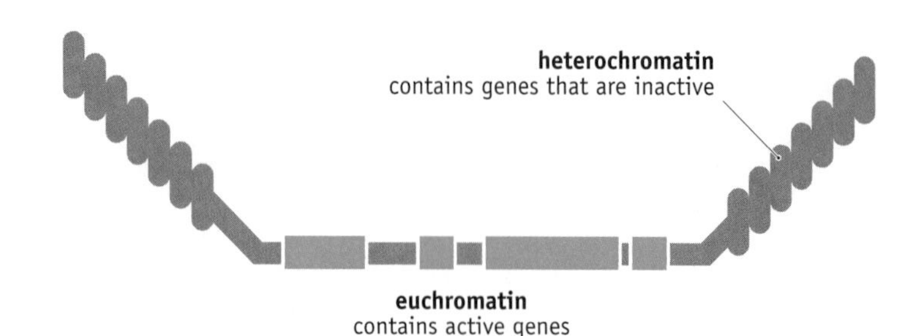

FIGURE 18.9 Genes present in heterochromatin cannot be expressed because the RNA polymerase and other transcription proteins are unable to gain access to them.

A second well-studied example of a genetic feedback loop involves the fruit fly protein called Deformed (Dfd). This protein has an important role during *Drosophila* development, because without it the fly's head fails to develop correctly. To perform its function, Dfd must be continuously expressed in the cells that eventually give rise to the insect's head. This is achieved by a feedback system in which Dfd binds to an enhancer located upstream of the *Dfd* gene. Feedback autoregulation also controls the expression of at least some of the important genes controlling the development of vertebrates.

Chemical modification of nucleosomes leads to heritable changes in chromatin structure

Permanent changes in gene expression can also result from the process that geneticists call chromatin modification. In Chapter 11 we learnt that the DNA in a eukaryotic chromosome is associated with nucleosomes, one nucleosome for every 200 bp or so. Interactions between nucleosomes are responsible for the higher orders of DNA packaging. During interphase, the most compact DNA conformation is heterochromatin, which contains genes that are inactive in a particular cell. The packaging of regions of the genome into heterochromatin therefore influences which genes are expressed and is thought to be one of the primary determinants of cellular differentiation (Figure 18.9). To understand more about differentiation we must therefore explore the factors that cause parts of the genome to become condensed into heterochromatin.

Chemical modification of the histone proteins present in the core octamer of the nucleosome seems to be the main factor that influences the degree of packaging in different regions of the genome. The best-studied of these modifications is acetylation. This is the attachment of an acetyl group to one or more of the lysine amino acids in the N-terminal regions of the histone proteins (Figure 18.10). These N termini form tails that protrude

```
             Ac      Ac
             |       |
H2A   S G R G K Q G G K A R A K A K T R S S R
                       10                  20

             Ac            Ac     Ac      Ac
             |             |      |       |
H2B   P E P S K S A P A P K K G S K K A I T K A
                       10                  20

                     Ac        Ac      Ac        Ac
                     |         |       |         |
H3    A R T K Q T A R K S T G G K A P R K Q L A T K A A R K S A
                       10                  20

             Ac    Ac     Ac      Ac
             |     |      |       |
H4    S G R G K G G K G L G K G G A K R H R K
                       10                  20
```

FIGURE 18.10 The positions at which acetyl (Ac) groups can be attached within the N-terminal regions of the four core histones.

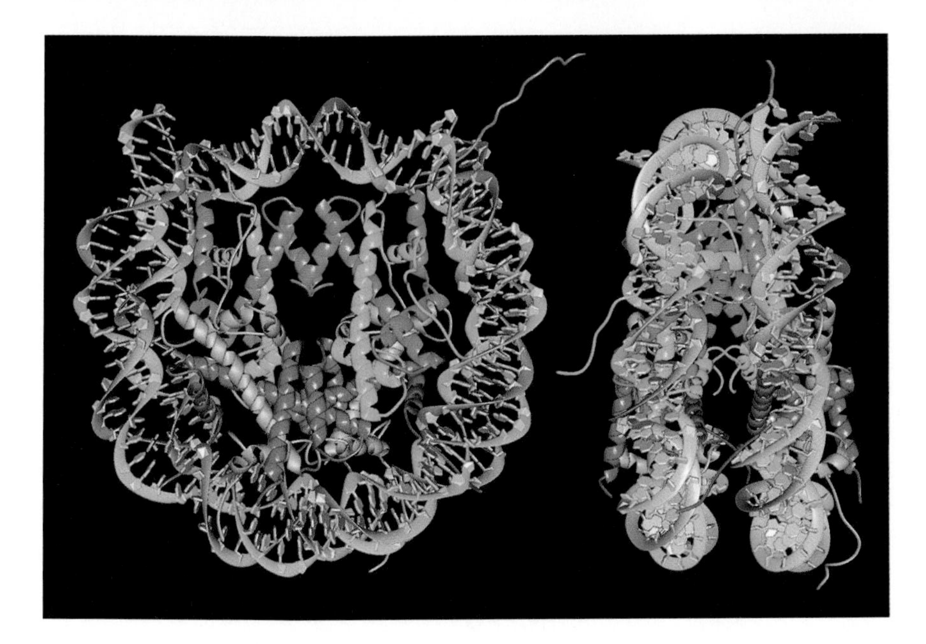

FIGURE 18.11 Two views of the nucleosome core octamer. The view on the left is downward from the top of the barrel-shaped octamer, and the view on the right is from the side. The two strands of the DNA double helix wrapped around the octamer are shown in brown and green. The octamer comprises a central tetramer of two histone H3 (blue) and two histone H4 (bright green) subunits plus a pair of H2A (yellow)–H2B (red) dimers, one above and one below the central tetramer. The N-terminal tails of the histone proteins can be seen protruding from the core octamer. (From K. Luger et al., *Nature* 389: 251–260, 1997. With permission from Macmillan Publishers Ltd.)

from the nucleosome (Figure 18.11). Their acetylation decreases the affinity of the histones for DNA and possibly also decreases the interaction between individual nucleosomes. The histones in heterochromatin are generally unacetylated, whereas those in areas containing active genes are acetylated. This is a clear indication that this type of modification is linked to DNA packaging. Whether or not a histone is acetylated depends on the balance between the activities of two types of enzyme, the **histone acetyltransferases (HATs)**, which add acetyl groups to histones, and the **histone deacetylases (HDACs)**, which remove these groups.

Acetylation is not the only type of histone modification that we know about. Methylation of lysine and arginine residues and phosphorylation of serines in the N-terminal regions of histones are also possible, as is the addition of the small, common ("ubiquitous") protein called **ubiquitin** to lysines in the C-terminal regions. Altogether, 29 sites in the N- and C-terminal regions of the four core histones are known to be subject to covalent modification of one type or another (Figure 18.12).

The modifications at these sites interact with one another to determine the degree of chromatin packaging taken up by a particular stretch of DNA. For example, methylation of lysine 9 (the lysine nine amino acids from the N terminus) of histone H3 forms a binding site for the HP1 protein, which induces chromatin packaging and silences gene expression, but this event is blocked by the presence of two or three methyl groups attached to lysine 4 (Figure 18.13). Methylation of lysine 4 therefore promotes an open chromatin structure and is associated with active genes.

```
      Me  Me      MeAcP      Ac    MeAc       Ac   Me Me P
      |   |         \/ |      |     | |        |     |  | |
H3  A R T K Q T A R K S T G G K A P R K Q L A T K A A R K S A
                    10                  20

    P   Me  Ac    Ac      Ac      Ac      Me
    |   |   |     |       |       |       |
H4  S G R G K G G K G L G K G G A K R H R K
                    10                  20
```

FIGURE 18.12 Modifications of the N-terminal regions of mammalian histones H3 and H4. All of the known modifications occurring in these regions are shown. Abbreviations: Ac, acetylation; Me, methylation; P, phosphorylation.

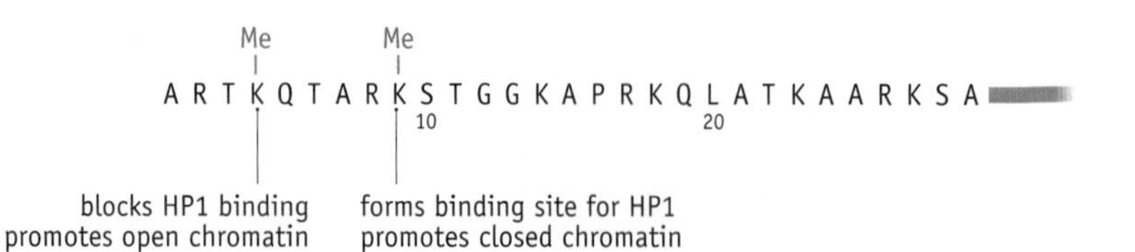

FIGURE 18.13 The differential effects of methylation of lysines 4 and 9 of histone H3.

Our growing awareness of the variety of possible histone modifications, and of the way in which different modifications work together, has led to the suggestion that there is a **histone code**, by which the pattern of chemical modifications specifies which regions of the genome are expressed at a particular time. We know of several examples in which histone modification is directly linked to cellular differentiation (one such example is described below), but what still eludes us is how these patterns of histone modification are inherited during cell division.

The fruit fly proteins called Polycomb and trithorax influence chromatin packaging

The example of differentiation by means of chromatin modification that we will look at involves the Polycomb and trithorax proteins of *Drosophila*. The *Polycomb* multigene family comprises about 30 genes coding for proteins that bind to DNA sequences, called Polycomb response elements, and induce the formation of heterochromatin. Polycomb does this by trimethylating lysines 9 and 27 of histone H3, thereby inducing chromatin packaging.

Each response element is about 10 kb in length. Exactly how it is recognized by Polycomb proteins is unknown, but the outcome is nucleation of heterochromatin around the Polycomb proteins. The heterochromatin then propagates along the DNA for tens of kilobases in both directions (Figure 18.14). The regions that become silenced contain genes that control the development of the individual body parts of the fly—the head, legs, and so on. Those genes that are not needed in any particular body part are permanently switched off by the action of the Polycomb proteins.

Polycomb proteins do not actually determine which genes will be silenced. Expression of these genes is already repressed before the Polycomb proteins bind to their response elements. The role of Polycomb is therefore to *maintain* rather than *initiate* gene silencing. The important point is that the heterochromatin induced by Polycomb is heritable. After division, the two new cells retain the heterochromatin established in the parent cell. This type of regulation of genome activity is therefore permanent not only in a single cell but also in the cell lineage.

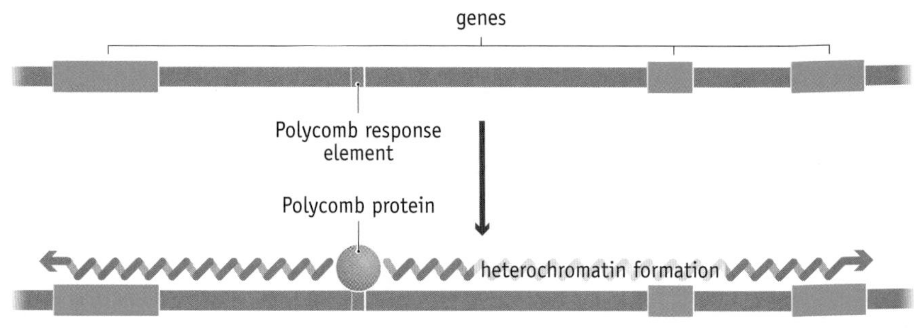

FIGURE 18.14 Polycomb maintains silencing in regions of the *Drosophila* genome by initiating heterochromatin formation.

FIGURE 18.15 Repositioning of nucleosomes in the region upstream of a gene enables the RNA polymerase and other transcription proteins to gain access to the promoter. This is thought to help maintain a gene in an active state.

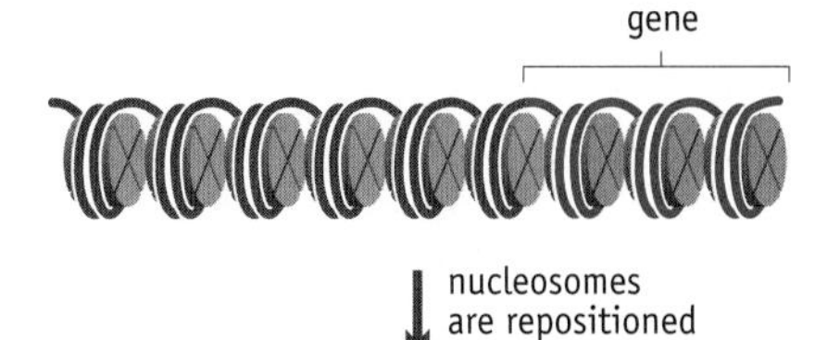

The trithorax proteins have the opposite effect to that of Polycomb. They maintain an open chromatin state in the regions of active genes, the targets including the same genes as those that are silenced in different body parts by Polycomb proteins. The human equivalent of trithorax achieves this by trimethylating lysine 4 of histone H3, promoting an open chromatin structure. In *Drosophila*, trithorax proteins seem to have a different effect, promoting open chromatin by performing **nucleosome remodeling**, which results in the repositioning of nucleosomes within the target region of the genome (Figure 18.15). This does not involve chemical alterations to histone molecules, but instead is induced by an energy-dependent process that weakens the contact between the nucleosome and the DNA with which it is associated.

DNA methylation can silence regions of a genome

Important alterations in gene activity can also be achieved by making chemical changes to the DNA itself. These alterations involve DNA methylation and are associated with the permanent silencing of regions of the genome, possibly entire chromosomes.

In eukaryotes, cytosine bases in chromosomal DNA molecules are sometimes changed to 5-methylcytosine by the addition of methyl groups by enzymes called **DNA methyltransferases** (Figure 18.16). Cytosine methylation is relatively rare in lower eukaryotes but in vertebrates up to 10% of the total number of cytosines in a genome are methylated, and in plants the figure can be as high as 30%. The methylation pattern is not random, instead being limited to the cytosine in some copies of the sequences 5′-CG-3′ and, in plants, 5′-CNG-3′, where N is any of the four nucleotides.A link between DNA methylation and gene expression becomes apparent when the methylation patterns in chromosomal DNAs are examined. These patterns show that active genes are located in unmethylated regions. For example, in humans, 40–50% of all genes are located close to sequences about 1 kb long in which the GC content is greater than the average for the genome as a whole. These are called **CpG islands**. The methylation status of the CpG island reflects the expression pattern of the adjacent gene (Figure 18.17). Housekeeping genes—those that are expressed in all tissues—have unmethylated CpG islands, whereas the CpG islands associated with tissue-specific genes are unmethylated only in those tissues in which the gene is expressed.

With DNA methylation, unlike histone modification, we understand how the methylation pattern is inherited when cells divide. After DNA replication,

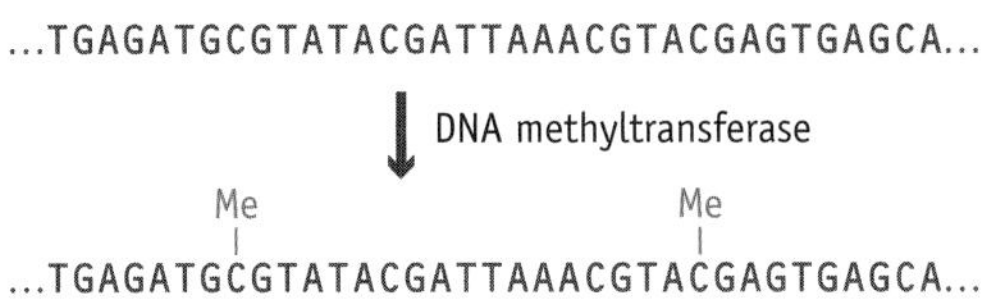

FIGURE 18.16 Methylation of cytosines in the sequence 5′-CG-3′ in eukaryotic DNA.

FIGURE 18.17 A CpG island.

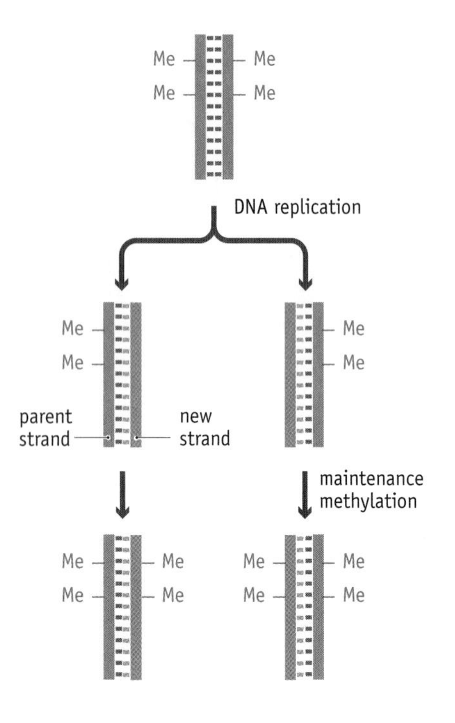

FIGURE 18.18 Maintenance methylation, which ensures that, after DNA replication, the two daughter helices acquire the methylation pattern of the parent molecule.

each daughter double helix comprises one parent strand and one newly synthesized strand. The former is methylated but the latter is not. This situation does not last long because the methyltransferase enzyme called Dnmt1 scans along the new strand, adding methyl groups to cytosines in the CpG sequences that are methylated in the parent strand (Figure 18.18). This **maintenance methylation** ensures that the two daughter double helices retain the methylation pattern of the parent molecule, which means that the pattern is inherited after cell division.

With methylation, the puzzle has been how the modification results in repression of gene activity. A possible clue has been provided by the discovery of proteins called **methyl-CpG-binding proteins** (**MeCPs**), which attach to methylated CpG islands. These work in conjunction with the histone deacetylases, which remove acetyl groups from histones and thereby induce chromatin packaging over the adjacent genes (Figure 18.19).

Genomic imprinting and X inactivation result from DNA methylation

Interesting examples of the role of DNA methylation in gene silencing are provided by **genomic imprinting** and **X inactivation**.

Genomic imprinting is a relatively uncommon but important feature of mammalian genomes in which one of a pair of genes present on homologous chromosomes is silenced by methylation (Figure 18.20). This process also occurs in some insects and some plants. It is always the same member of a pair of genes that is imprinted and hence inactive. For some genes the imprinted copy is the one inherited from the mother, and for other genes it is the paternal copy. Just over 60 genes in humans and mice have been shown to display imprinting, including genes for both protein-coding and noncoding RNA. Imprinted genes are distributed around the genome but tend to occur in clusters. For example, in humans there is a 2.2-Mb segment of chromosome 15 within which there are at least ten imprinted genes, and a smaller, 1-Mb region of chromosome 11 containing eight imprinted genes.

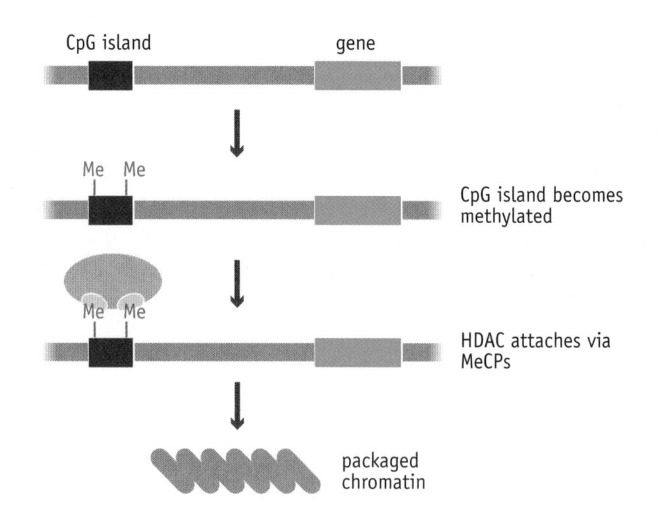

FIGURE 18.19 A model for the link between DNA methylation and genome expression. Methylation of the CpG island upstream of a gene provides recognition signals for the methyl-CpG-binding protein (MeCP) components of a histone deacetylase complex (HDAC). The HDAC modifies the chromatin in the region of the CpG island and hence inactivates the gene.

FIGURE 18.20 Silencing of one of a pair of genes by methylation of its CpG island.

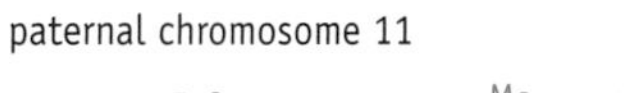

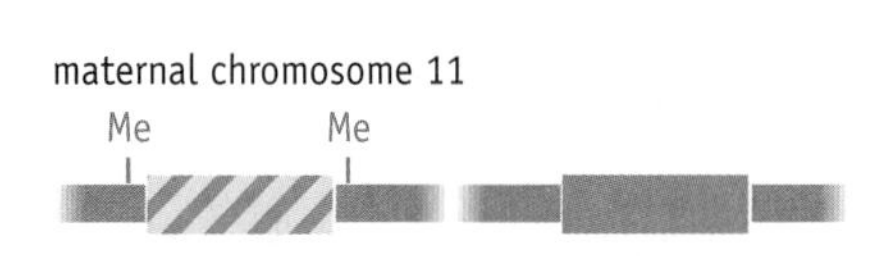

FIGURE 18.21 A pair of imprinted genes on human chromosome 11. *H19* is imprinted on the chromosome inherited from the father, and *Igf2* is imprinted on the maternal chromosome.

An example of an imprinted gene is *Igf2*, which codes for a growth factor, a protein involved in signaling between cells. Only the paternal gene is active (Figure 18.21). On the chromosome inherited from the mother, various segments of DNA in the region of *Igf2* are methylated, preventing expression of this copy of the gene. A second imprinted gene, *H19*, is located about 90 kb away from *Igf2*, but the imprinting is the other way round—the maternal version of *H19* is active and the paternal version is silent.

Imprinting is controlled by **imprint control elements**, DNA sequences that are found within a few kilobases of clusters of imprinted genes. These elements mediate the methylation of the imprinted regions, but how they do this is unknown. The function of imprinting is also a mystery, but it must be important in development because genetically engineered mice that have two copies of the maternal genome fail to develop properly.

X inactivation is a special form of imprinting that leads to inactivation of 80% of the genes on one of the X chromosomes in a female mammalian cell. It occurs because females have two X chromosomes, whereas males have only one. If both of the X chromosomes in females were active, proteins coded by genes on the X chromosome might be synthesized at twice the rate in females compared with males. To avoid this undesirable state of affairs, one of the X chromosomes in females is silenced and is seen in the nucleus as a condensed structure called the **Barr body** (Figure 18.22), which is made up entirely of heterochromatin.

Silencing occurs early in embryo development and is controlled by the X inactivation center (*Xic*), a region present on each X chromosome (Figure 18.23). In a cell undergoing X inactivation, the inactivation center on one of the X chromosomes initiates the formation of heterochromatin, which spreads out from the nucleation point until the entire chromosome is affected, with the exception of a few short segments. The process takes several days to complete. The process involves a gene called *Xist*, located in the inactivation center, which is transcribed into a 25-kb noncoding RNA, copies of which coat the chromosome as the heterochromatin is formed. At the same time, various histone modifications occur and certain DNA sequences become methylated, although methylation seems to occur after the inactive state has been set up. X inactivation is heritable and is present in all cells descended from the initial one within which the inactivation took place.

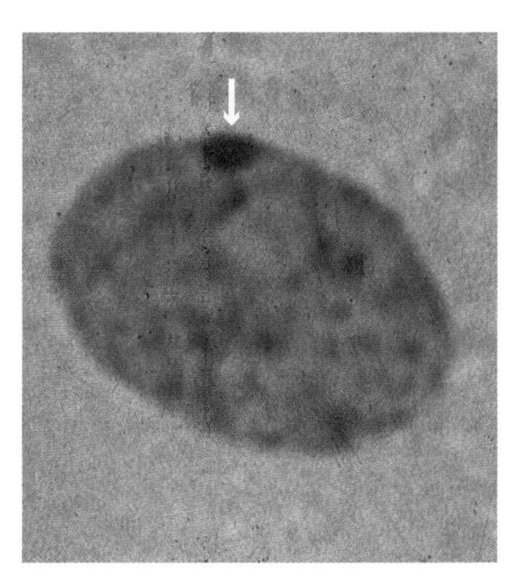

FIGURE 18.22 The Barr body. In this micrograph, the Barr body is seen as a dense object adjacent to the nuclear membrane. (Courtesy of Malcolm Ferguson-Smith, University of Cambridge.)

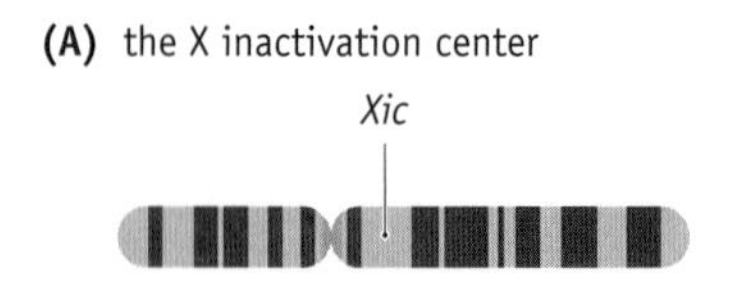

(B) heterochromatin formation spreads from the inactivation center

Xist

heterochromatin formation

FIGURE 18.23 X inactivation. (A) The location of the inactivation center *Xic* on the X chromosome. (B) During X inactivation, heterochromatin spreads from the inactivation center. The gene called *Xist*, located in the inactivation center, is transcribed into RNA which coats the chromosome.

18.2 COORDINATION OF GENE ACTIVITY DURING DEVELOPMENT

Genome rearrangements, feedback loops, chromatin modifications, and DNA methylation, as described above, are the genetic processes thought to be responsible for cellular differentiation. The end products of differentiation—the specialized cell types present in a multicellular organism—must be formed at the correct times during a developmental pathway, and at the appropriate places within the developing organism. To understand the genetic basis of development we must therefore examine how the gene expression patterns within individual cells can be controlled in time and space.

Research into developmental genetics is underpinned by the use of model organisms

Developmental processes, especially in higher eukaryotes such as humans, are extremely complex, but remarkably good progress toward understanding them has been made in recent years. One of the guiding principles of this research has been the assumption that there should be similarities and parallels between developmental processes in different organisms, reflecting their common evolutionary origins. This means that information relevant to human development can be obtained from studies of **model organisms** chosen for the relative simplicity of their developmental pathways.

The use of model organisms is not new. It has underpinned biological research for decades and explains why laboratory animals such as mice and rats have been used so extensively. For many biological questions, such as the causes of cancer, progress can only be made through studying a model organism that is relatively closely related to humans. Developmental genetics is no different in this regard, but a great deal of information directly relevant to humans and other vertebrates has also been obtained from studies of species much lower down the evolutionary scale. Two species in particular are important as models in developmental genetics, the microscopic nematode worm *Caenorhabditis elegans* and the fruit fly *Drosophila melanogaster*.

Research with *C. elegans* (Figure 18.24) was initiated by Sydney Brenner in the 1960s, who chose the organism specifically to act as a model for development. Like all good model organisms, *C. elegans* is easy to grow in the laboratory and has a short generation time, of about 4 days, so there is no delay in obtaining the results of genetic crosses. The worm is transparent at all stages of its life cycle, so internal examination is possible without killing the animal. This is an important point because it has enabled researchers to follow the entire developmental process of the worm at the cellular level.

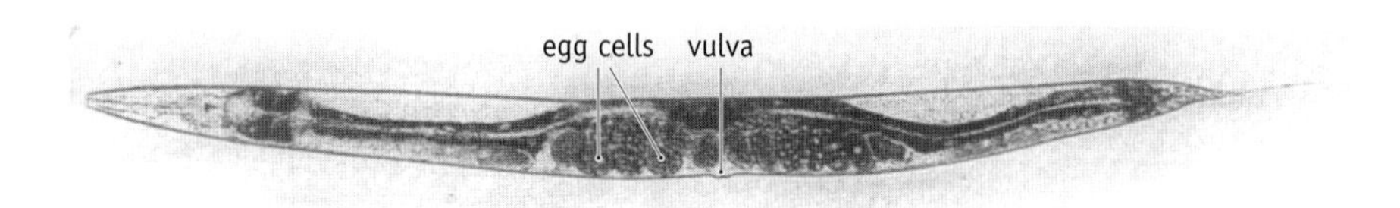

FIGURE 18.24 The nematode worm *Caenorhabditis elegans*. The micrograph shows an adult hermaphrodite worm, about 1 mm in length. The vulva is the small projection located on the underside of the animal, about halfway along. Egg cells can be seen inside the worm's body in the region on each side of the vulva. (From J. Kendrew (ed.) Encyclopedia of Molecular Biology. Oxford: Blackwell, 1994. With permission from Wiley-Blackwell.)

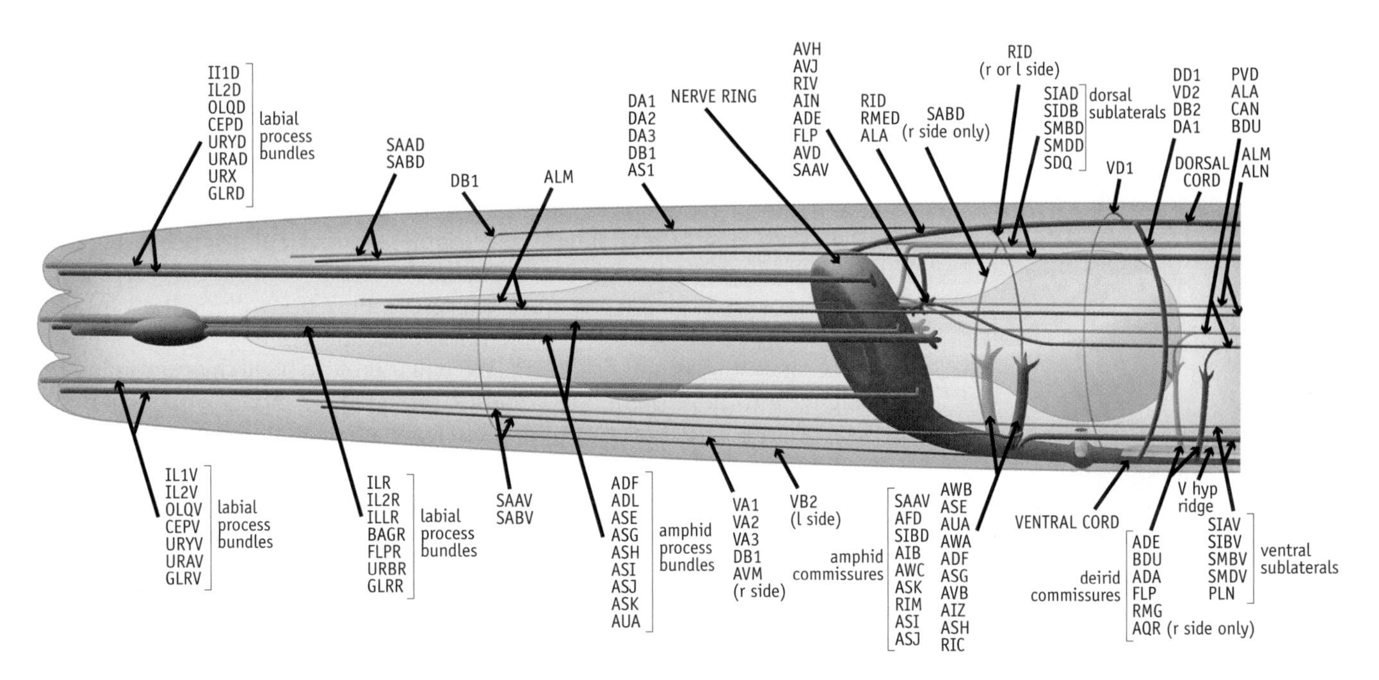

FIGURE 18.25 The nervous system of the anterior region of *C. elegans*. Nerves and other components of the nervous system are shown in red. The yellow structure is part of the alimentary canal. Labels indicate the names of the nerve cells contained in the various structures. (Reprinted with permission from WormAtlas [www.wormatlas.org]. Adapted from J.G. White, E. Southgate, J.N. Thomson and S. Brenner, *Phil. Trans. R. Soc. Lond. B* 314: 1–340, 1986. With permission from Royal Society Publishing.)

Every cell division in the pathway from fertilized egg to adult worm has been charted, and every point at which a cell adopts a specialized role has been identified. In addition, the complete connectivity of the 302 cells that comprise the nervous system of the worm has been mapped (Figure 18.25). Another advantage is that the genome of *C. elegans* is relatively small, just 97 Mb, and the entire sequence is known.

Unlike *C. elegans* the fruit fly was not chosen specifically as a model for developmental studies. It was first used by Thomas Hunt Morgan in 1910 as a convenient organism for studying genetics. For Morgan the advantages were its small size, enabling large numbers to be studied in a single experiment, its minimal nutritional requirements (the flies like bananas), and the presence in natural populations of occasional variants with easily recognized genetic characteristics such as unusual eye colors. Morgan was not aware that other advantages are a small genome (180 Mb) and the fact that gene isolation is aided by the presence in the salivary glands of "giant" chromosomes. These are made up of multiple copies of the same DNA molecule laid side by side, displaying banding patterns that can be correlated with the map of each chromosome to pinpoint the positions of desired genes (Figure 18.26). But Morgan did foresee that *Drosophila* might become an important organism for developmental research, because in 1915 one of his students, Calvin Bridges, isolated a bizarre mutant that he called *Antennapedia*, which has legs where its antennae ought to be (Figure 18.27). This was the first indication that mutation can affect not just

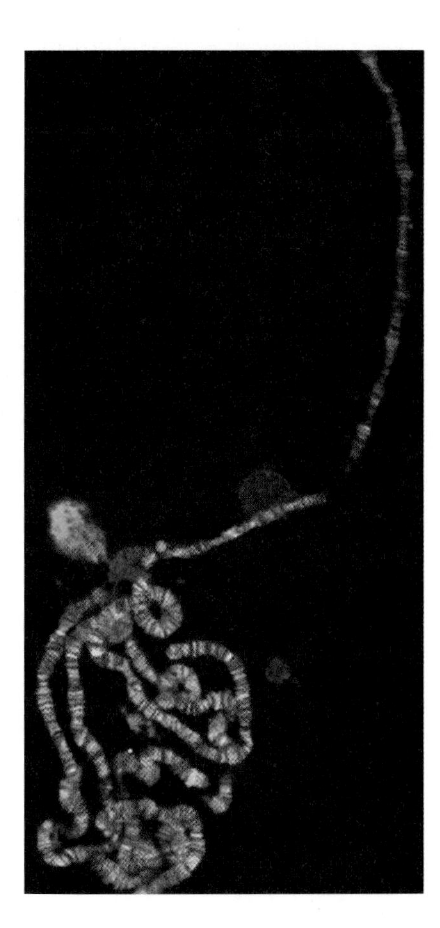

FIGURE 18.26 A giant or polytene chromosome from the salivary glands of a fruit fly larva. The banding pattern can be correlated with a genetic map to identify the positions of individual genes. (Courtesy of LPLT.)

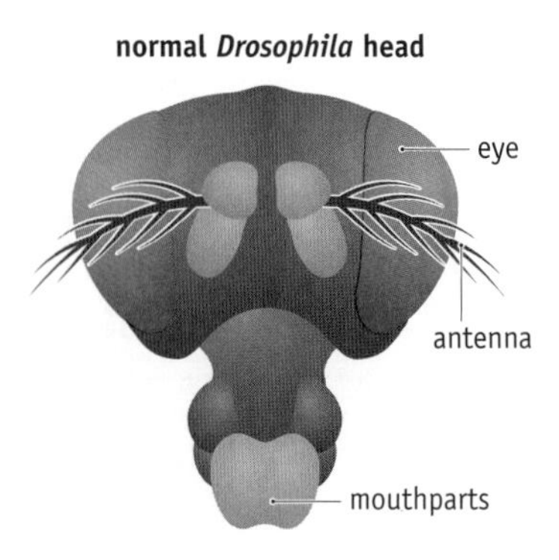

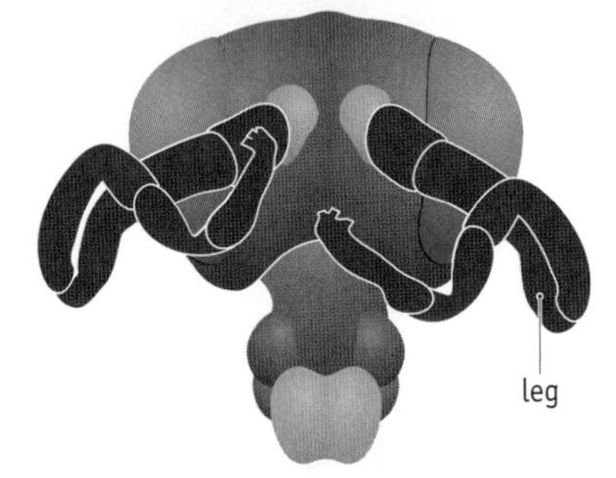

FIGURE 18.27 The *Antennapedia* mutation.

phenotypes such as eye color and wing shape but also the underlying body plan of an organism. We will return to *Antennapedia* later because its study since 1915 has provided some of the foundations for our understanding of the spatial control of gene expression.

The fact that less complex organisms such as worms and insects can be used as models for development in vertebrates indicates that the fundamental processes that underlie development evolved at a very early period, possibly soon after the first multicellular animals appeared. This means that although vertebrates have much more complex body plans than insects, the genetic processes responsible for specification of those body plans are similar in both types of organism. We will therefore begin by examining the insights into developmental genetics provided by model organisms, and then explore how this information has enabled us to understand the equivalent processes in vertebrates.

C. elegans reveals how cell-to-cell signaling can confer positional information

For development to proceed correctly, cells present at particular positions in an embryo must follow their individual differentiation pathways, so that the correct tissue or structure appears at the appropriate place in the adult organism. For this to happen, the cells in the embryo must acquire **positional information**. In simple terms, each cell must know where it is located if it is to know which genes to express (Figure 18.28). Research with *C. elegans* has illustrated that cell-to-cell signaling is critical for the establishment of positional information within a developing embryo.

Most *C. elegans* worms are hermaphrodites, meaning that they have both male and female sex organs. The vulva is part of the female sex apparatus, being the tube through which sperm enter and fertilized eggs are laid. The adult vulva is derived from three vulva progenitor cells, called P5.p, P6.p, and P7.p, which to begin with are located in a row on the undersurface of the developing worm (Figure 18.29). Each of these progenitor cells becomes committed to the differentiation pathway that leads to the production of vulva cells. The central cell, called P6.p, adopts the "primary vulva cell fate" and divides to produce eight new cells. The other two

FIGURE 18.28 A six-day-old human embryo implanting into the wall of its mother's uterus. At this stage of development, the embryo has already acquired the positional information ensuring that the correct structures will appear at the correct positions in the adult. (Courtesy of Yorgos Nikas/Wellcome Images.)

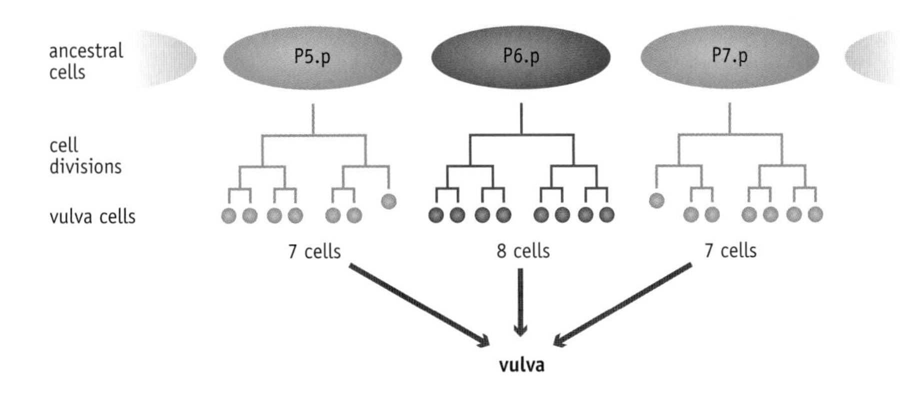

FIGURE 18.29 Cell divisions resulting in production of the vulva cells of *C. elegans*. Three ancestral cells divide in a programmed manner to produce 22 progeny cells, which reorganize their positions relative to one another to construct the vulva.

cells—P5.p and P7.p—take on the "secondary vulva cell fate" and divide into seven cells each. These 22 cells then reorganize their positions to construct the vulva.

Vulva development must occur in the correct position relative to the gonad, the structure containing the egg cells. If the vulva develops in the wrong place, the gonad will not receive sperm and the egg cells will never be fertilized. The positional information needed by the progenitor cells is provided by a cell within the gonad called the anchor cell (Figure 18.30). The importance of the anchor cell has been demonstrated by experiments in which it is artificially destroyed in the embryonic worm. In the absence of the anchor cell, a vulva does not develop. This is because the anchor cell secretes an extracellular signaling compound that induces P5.p, P6.p, and P7.p to differentiate. This signaling compound is a protein called LIN-3, coded by the *lin-3* gene.

Why does P6.p adopt the primary cell fate, whereas P5.p and P7.p take on secondary cell fates? The LIN-3 protein secreted by the anchor cell forms a concentration gradient and therefore has different effects on P6.p, which is closest to the anchor cell, from those on P5.p and P7.p, which are more distant and so are exposed to less of the signaling compound (Figure 18.31). It is also possible that the higher concentration of LIN-3 stimulates P6.p to secrete a second signaling compound that commits P5.p and P7.p to their secondary fates. Finally, there is a third signaling compound, secreted by the hypodermal cell, a multinuclear sheath that surrounds most of the worm's body. This compound has a deactivating effect but cannot overcome the concentrations of LIN-3 to which P5.p, P6.p, and P7.p are exposed. Its function is to prevent the unwanted differentiation of three adjacent cells, P3.p, P4.p, and P8.p, each of which can become committed to vulva development if the repressive signal from the hypodermal cell malfunctions. In combination, these three cell-to-cell signaling processes are therefore able to provide P5.p, P6.p, P7.p, and the adjacent cells with the positional information they need to follow their specific differentiation pathways.

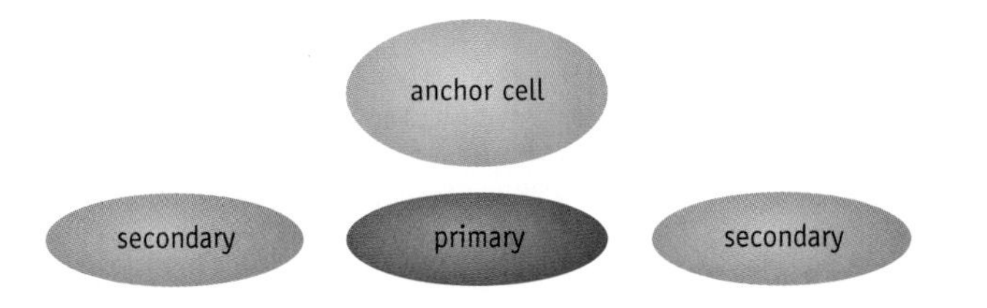

FIGURE 18.30 The relative positions of the anchor cell and the primary and secondary vulva cells.

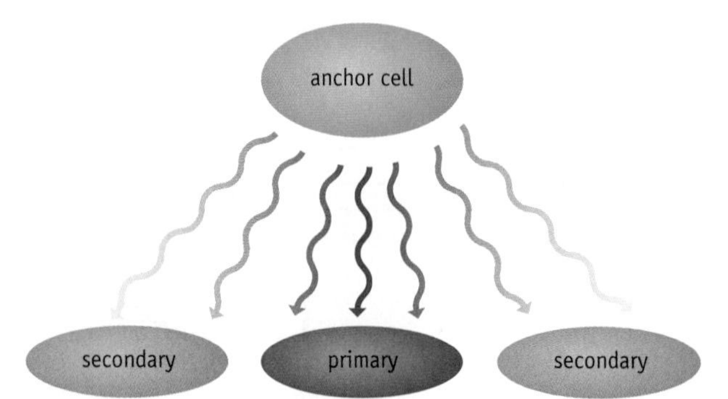

FIGURE 18.31 The postulated role of the anchor cell in determining cell fate during vulva development in *C. elegans*. It is thought that release of the signaling compound LIN-3 by the anchor cell commits P6.p (shown in magenta), the cell closest to the anchor cell, to the primary vulva cell fate. P5.p and P7.p (shown in orange) are farther away from the anchor cell and so are exposed to a lower concentration of LIN-3 and become secondary vulva cells.

Positional information in the fruit fly embryo is initially established by maternal genes

The *C. elegans* vulva provides us with an understanding of the way in which cell-to-cell signaling underlies the establishment of positional information during development. Studies of *Drosophila* have taken this understanding much further.

In one respect *Drosophila* is quite unusual because its embryo is not made up of individual cells, as in most organisms, but instead is a single **syncytium** comprising a mass of cytoplasm and multiple nuclei (Figure 18.32). This structure persists until 13 successive rounds of nuclear division have produced some 1500 nuclei. Only then do individual cells start to appear around the outside of the syncytium, producing the structure called the **blastoderm**. Before the blastoderm stage has been reached, positional information has begun to be established.

Initially the positional information that the embryo needs is the ability to distinguish front (anterior) from back (posterior), as well as similar information relating to up (dorsal) and down (ventral). This information is provided by concentration gradients of proteins that become established in the syncytium. Most of these proteins are not synthesized from genes in the embryo but are translated from mRNAs injected into the embryo by the mother. One of these **maternal effect genes** is called *bicoid*. The *bicoid* gene is transcribed in the maternal nurse cells, which are in contact with the egg cells, and its mRNA is injected into the anterior end of the unfertilized egg. This position is defined by the orientation of the egg cell in the egg chamber. The *bicoid* mRNA remains in the anterior region of the egg cell, attached to the cell's cytoskeleton. Bicoid proteins, translated from the

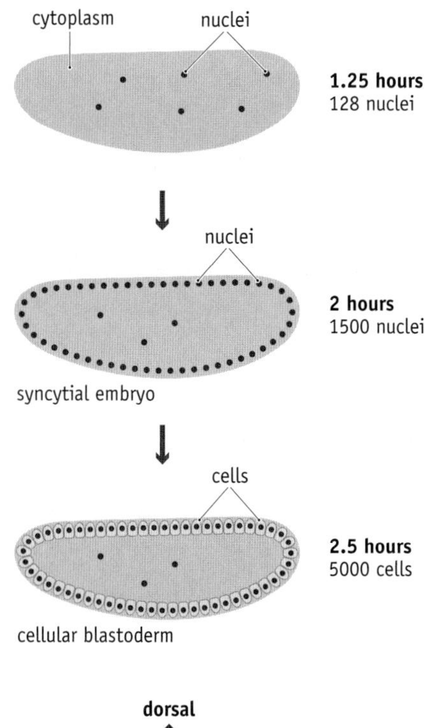

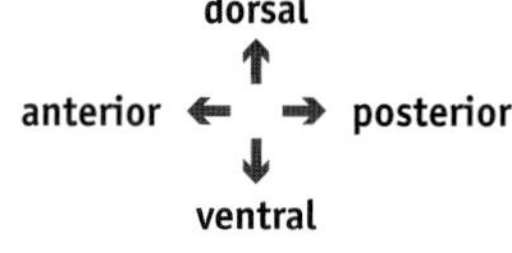

FIGURE 18.32 Early development of the *Drosophila* embryo. To begin with, the embryo is a single syncytium containing a gradually increasing number of nuclei. These nuclei migrate to the periphery of the embryo after about 2 hours, and within another 30 minutes cells begin to be constructed. The embryo is approximately 500 μm in length and 170 μm in diameter.

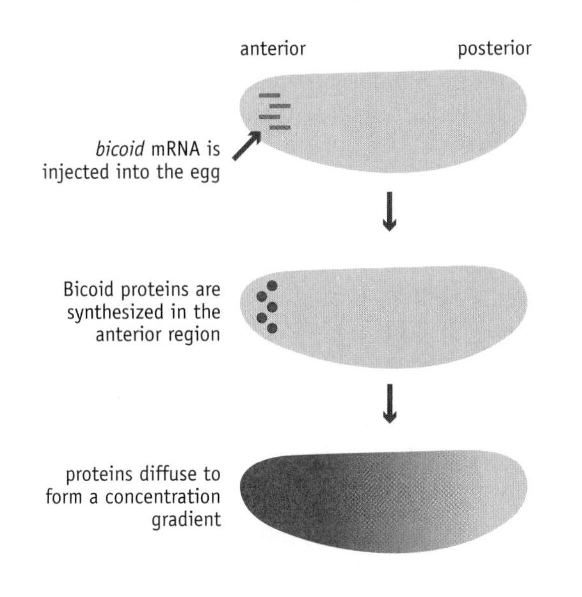

FIGURE 18.33 Establishment of the Bicoid gradient in a *Drosophila* egg cell. The *bicoid* mRNA is injected into the anterior end of the egg. Bicoid proteins, translated from the mRNA, diffuse through the syncytium setting up a concentration gradient.

mRNA, diffuse through the syncytium, setting up a concentration gradient, highest at the anterior end and lowest at the posterior end (Figure 18.33).

Three other maternal effect gene products are also involved in setting up the anterior–posterior gradient. These are the Hunchback, Nanos, and Caudal proteins. All are injected as mRNAs into the anterior region of the unfertilized egg. The *hunchback* and *caudal* mRNAs become distributed evenly through the cytoplasm, but their proteins subsequently form gradients. This is because Bicoid activates the *hunchback* gene in the embryonic nuclei and represses translation of the *caudal* mRNA, increasing the concentration of the Hunchback protein in the anterior region and decreasing that of Caudal (Figure 18.34). The *nanos* mRNA is transported to the posterior part of the egg and attached to the cytoskeleton. Its protein represses the translation of *hunchback* mRNA, contributing further to the anterior–posterior gradient of the Hunchback protein.

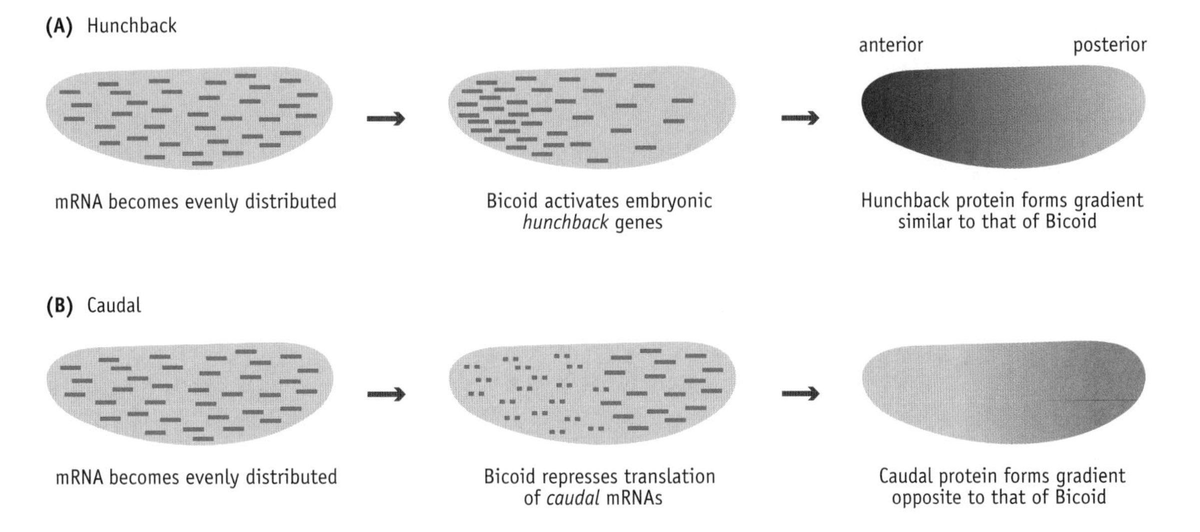

FIGURE 18.34 Establishment of the Hunchback and Caudal gradients. (A) Bicoid activates the *hunchback* gene in the embryonic nuclei, increasing the concentration of the Hunchback protein in the anterior region. (B) Bicoid represses translation of the *caudal* mRNA, decreasing the concentration of the Caudal protein in the anterior region.

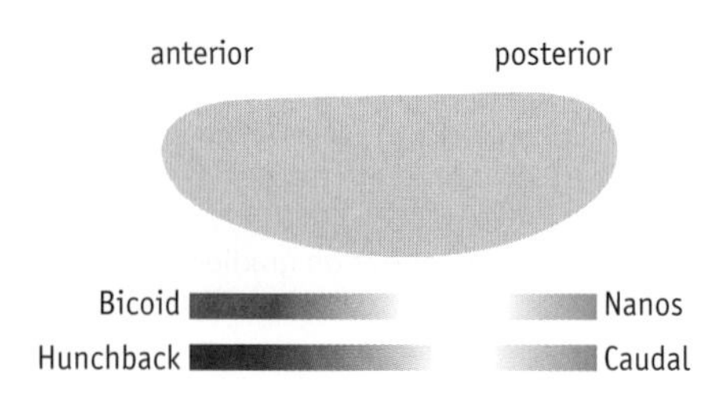

FIGURE 18.35 The protein gradients that provide the initial positional information along the anterior–posterior axis of a *Drosophila* embryo.

The net result is a gradient of Bicoid and Hunchback, greater at the anterior end, and of Nanos and Caudal, greater at the posterior end (Figure 18.35). Similar events result in a dorsal-to-ventral gradient, predominantly of the protein called Dorsal.

A cascade of gene activity converts the positional information into a segmentation pattern

The body plan of the adult fly is built up as a series of segments, each with a different structural role. This is clearest in the thorax, which has three segments, each carrying one pair of legs, and the abdomen, which is made up of eight segments (Figure 18.36). It is also true for the head, even though in the head the segmented structure is less visible. The initial objective of embryo development is therefore the production of a young larva with the correct segmentation pattern.

The gradients established in the embryo by the maternal effect proteins are the first stage in the formation of the segmentation pattern. These gradients provide the interior of the embryo with a basic amount of positional information, with each point in the syncytium now having its own unique protein signature. This positional information is made more precise by expression of the **gap genes**. Three of the anterior–posterior gradient proteins—Bicoid, Hunchback, and Caudal—are transcription activators targeting the gap genes in the nuclei that now line the inside of the embryo (Figure 18.37). The identities of the gap genes expressed in a particular nucleus depend on the relative concentrations of the gradient proteins and hence on the position of the nucleus along the anterior–posterior axis.

Some gap genes are activated directly by Bicoid, Hunchback, and Caudal. Examples are *buttonhead*, *empty spiracles*, and *orthodenticle*, which are activated by Bicoid. Other gap genes, such as *huckebein* and *tailless*, are switched on indirectly, by transcription factors coded by genes activated by the gradient proteins. There are also repressive effects (e.g., Bicoid represses the expression of *knirps*) and the gap genes regulate their own expression in various ways. This complex interplay results in the positional information in the embryo, now carried by the relative concentrations of the gap gene products, becoming more detailed (Figure 18.38).

The next set of genes to be activated, the **pair-rule genes**, establish the basic segmentation pattern. Transcription of these genes responds to the relative concentrations of the gap gene products. The embryo can now be looked on as comprising a series of stripes, each stripe consisting

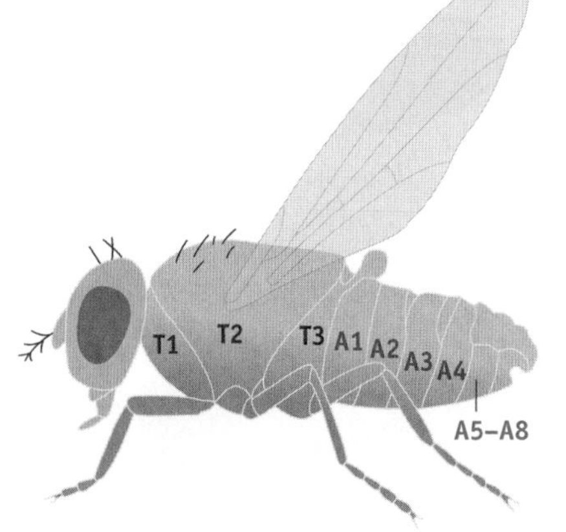

FIGURE 18.36 The segmentation pattern of the adult *Drosophila melanogaster*. T, thorax; A, abdomen. The head is also segmented, but the pattern is not easily discernible from the morphology of the adult fly.

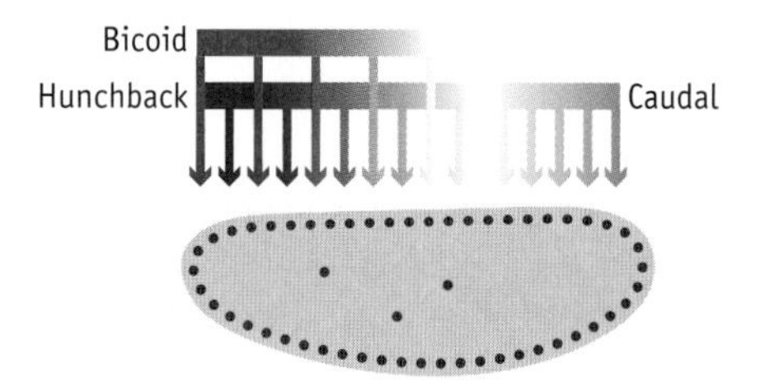

FIGURE 18.37 Bicoid, Hunchback, and Caudal are transcription activators that target gap genes in the nuclei lining the inside of the *Drosophila* embryo. Because each protein has formed a concentration gradient, a differential expression pattern for the gap genes is established along the anterior–posterior axis.

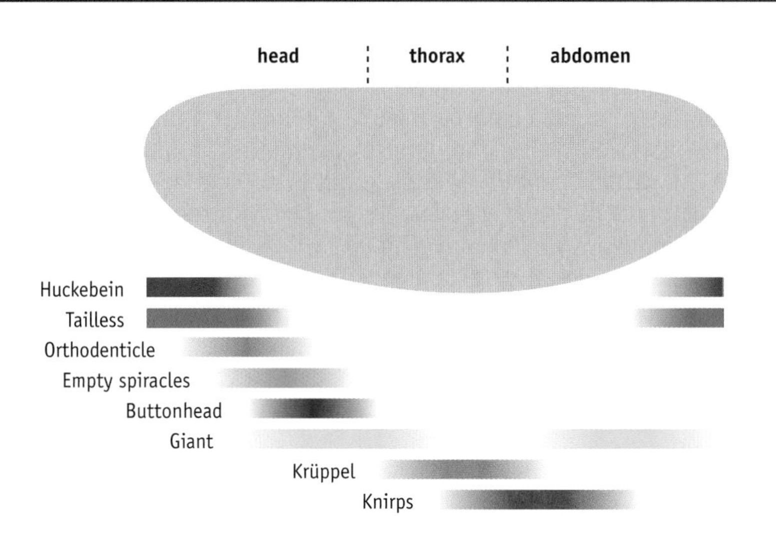

FIGURE 18.38 The concentration gradients for the gap proteins, as established by differential activation of the gap genes by Bicoid, Hunchback, and Caudal.

of a set of cells expressing a particular combination of pair-rule genes (Figure 18.39). In a further round of gene activation, the **segment polarity genes** become switched on, providing greater definition to the stripes by setting the sizes and precise locations of what will eventually be the segments of the larval fly (Figure 18.40). Gradually the positional information of the maternal effect gradients is converted into a sharply defined segmentation pattern.

Segment identity is determined by the homeotic selector genes

The pair-rule and segment polarity genes establish the segmentation pattern of the fruit fly embryo but do not themselves determine the identities of the individual segments. This is the role of the **homeotic selector genes**. These were first discovered because of the extravagant effects of mutations in these genes on the appearance of the adult fly. An example is the *Antennapedia* mutation, which we met earlier, which transforms the head segment that usually produces an antenna into one that makes a leg, so the mutant fly has a pair of legs where its antennae should be (see Figure 18.27). The early geneticists were fascinated by these monstrous **homeotic mutants**, and many were collected during the first few decades of the twentieth century.

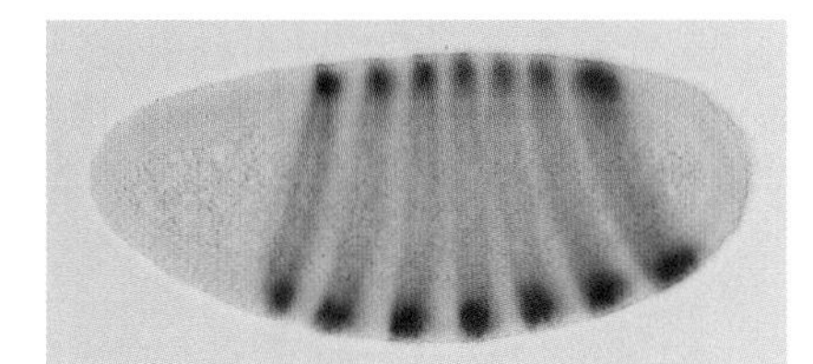

FIGURE 18.39 *Drosophila* embryo showing the cells expressing the pair-rule gene *even-skipped*. The expression pattern has been revealed by staining the embryo with a labeled antibody that binds specifically to the Even-skipped protein. (Courtesy of Michael Levine, University of California, Berkeley and Stephen Small, New York University.)

FIGURE 18.40 *Drosophila* embryo showing the cells expressing the segment polarity gene *engrailed*. The expression pattern has been revealed by staining the embryo with a labeled antibody that binds specifically to the Engrailed protein. (From C. Hama, Z. Ali and T.B. Kornberg, *Genes Dev.* 4: 1079–1093, 1990. With permission from Cold Spring Harbor Laboratory Press.)

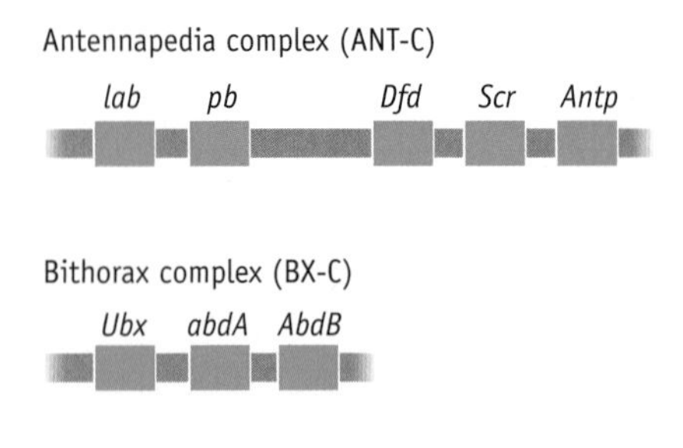

FIGURE 18.41 The Antennapedia and Bithorax gene complexes of *D. melanogaster*. The full gene names are as follows: *lab*, *labial palps*; *pb*, *proboscipedia*; *Dfd*, *Deformed*; *Scr*, *Sex combs reduced*; *Antp*, *Antennapedia*; *Ubx*, *Ultrabithorax*; *abdA*, *abdominal A*; *AbdB*, *Abdominal B*.

Genetic mapping of homeotic mutations has revealed that the selector genes are clustered in two groups on chromosome 3. These clusters are called the Antennapedia complex (ANT-C), which contains genes involved in determination of the head and thorax segments, and the Bithorax complex (BX-C), which contains genes for the abdomen segments (Figure 18.41). The order of genes corresponds to the order of the segments in the fly, the first gene in ANT-C being *labial palps*, which controls the most anterior segment of the fly, and the last gene in BX-C being *Abdominal B*, which specifies the most posterior segment.

The correct selector gene is expressed in each segment because the activation of each is responsive to the positional information provided by the distributions of the gap and pair-rule proteins. The selector gene products are themselves transcription activators that switch on the sets of genes needed to initiate differentiation of the specified segment. Maintenance of the differentiated state is ensured partly by the repressive effect of each homeotic selector gene product on expression of the other selector genes, and partly by the work of Polycomb, which, as we saw above, constructs inactive heterochromatin over the selector genes that are not expressed in a particular cell.

Homeotic selector genes are also involved in vertebrate development

Until the mid-1980s it was not realized that studies of development in the fruit fly were directly relevant to development in vertebrates. This remarkable discovery was made after the *Drosophila* homeotic selector genes were sequenced and it was found that each contains a segment, 180 bp in length, that is very similar in the different genes (Figure 18.42). We now know that this sequence, called the **homeobox**, codes for a DNA-binding structure, the **homeodomain**, that enables these proteins to attach to DNA and hence perform their roles as transcription activators.

The similarities between the homeoboxes of the various *Drosophila* selector genes led researchers in the 1980s to search for other genes containing this sequence. First, the *Drosophila* genome was examined, resulting in isolation of several previously unknown homeobox-containing genes. These have turned out not to be selector genes but other types of gene coding for transcription activators involved in development. Examples include the pair-rule genes *even-skipped* and *fushi tarazu*, and the segment polarity gene *engrailed*.

The presence of additional homeobox genes in *Drosophila* was not particularly surprising. The real excitement came when the genomes of other organisms were searched and it was realized that homeoboxes are present in genes in a wide variety of animals, including humans. Some of the homeobox genes in these other organisms are homeotic selector genes

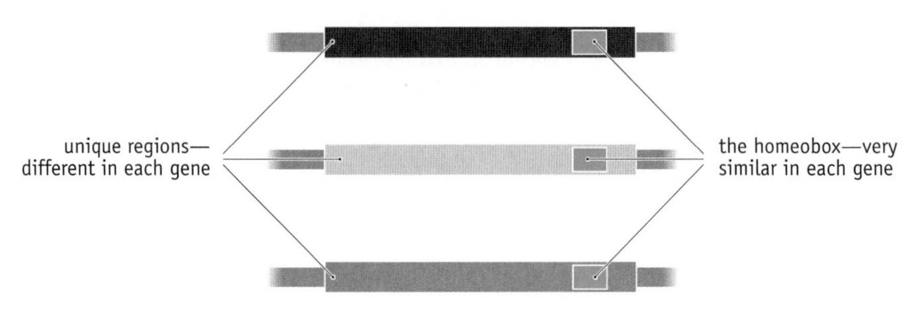

FIGURE 18.42 The homeobox is a nucleotide sequence that is similar in different homeotic selector genes.

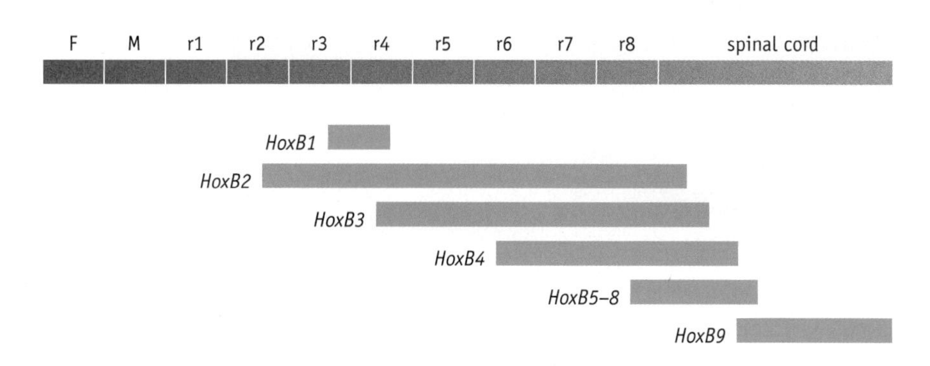

FIGURE 18.43 Specification of the mouse nervous system by selector genes of the HoxB cluster. The nervous system is shown schematically and the positions specified by the individual HoxB genes (*HoxB1* to *HoxB9*) indicated by the green bars. The components of the nervous system are as follows: F, forebrain; M, midbrain; r1–r8, rhombomeres 1–8; followed by the spinal cord. Rhombomeres are segments of the hindbrain seen during development.

organized into clusters similar to ANT-C and BX-C, these genes having equivalent functions to the *Drosophila* versions, specifying construction of the body plan. This is illustrated by the *HoxC8* gene of mouse, mutations in which result in an animal that has an extra pair of ribs, as a result of the conversion of a lumbar vertebra (normally in the lower back) into a thoracic vertebra (from which the ribs emerge). Other Hox mutations in animals lead to limb deformations, such as absence of the lower arm, or extra digits on the hands or feet.

We now look on the ANT-C and BX-C clusters of selector genes in *Drosophila* as two parts of a single complex, the homeotic gene complex or HOM-C. In vertebrates there are four homeotic gene clusters, called HoxA to HoxD. Not all of the vertebrate Hox genes have been ascribed functions, but we believe that the additional versions possessed by vertebrates relate to the added complexity of the vertebrate body plan. As in *Drosophila*, the order of genes in the vertebrate clusters reflects the order of the structures specified by the genes in the adult body. This is clearly seen with the mouse HoxB cluster, which controls the development of the nervous system (Figure 18.43).

When the four vertebrate clusters are aligned with one another and with the *Drosophila* HOM-C cluster, similarities are seen between the genes at equivalent positions (Figure 18.44). This suggests that they have evolutionary relationships. The implication is that in the vertebrate lineage there were two duplications of the original Hox cluster. The first duplication converted the single cluster seen today in fruit flies into a pair of Hox clusters, and the second duplication resulted in the four clusters seen in vertebrates (Figure 18.45). Evidence in support of this idea comes from the discovery

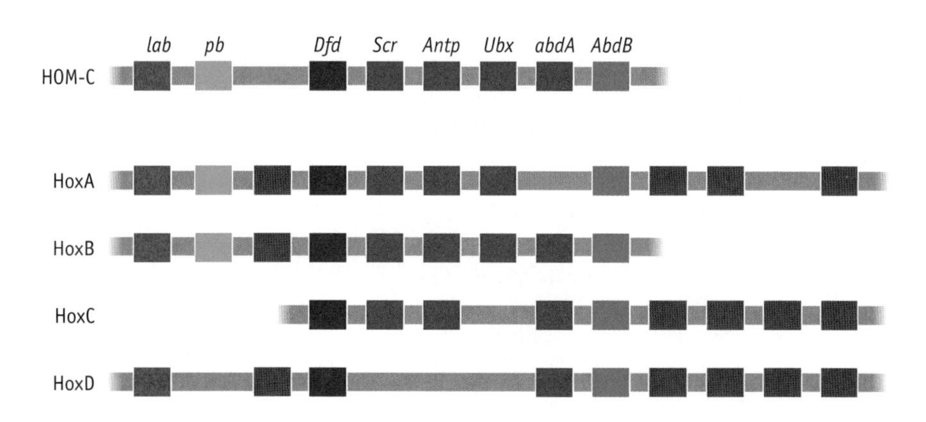

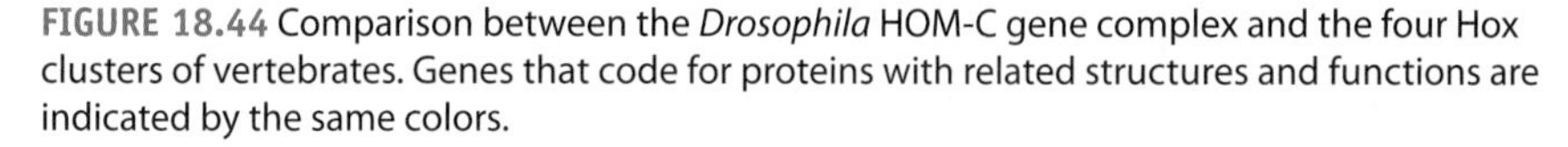

FIGURE 18.44 Comparison between the *Drosophila* HOM-C gene complex and the four Hox clusters of vertebrates. Genes that code for proteins with related structures and functions are indicated by the same colors.

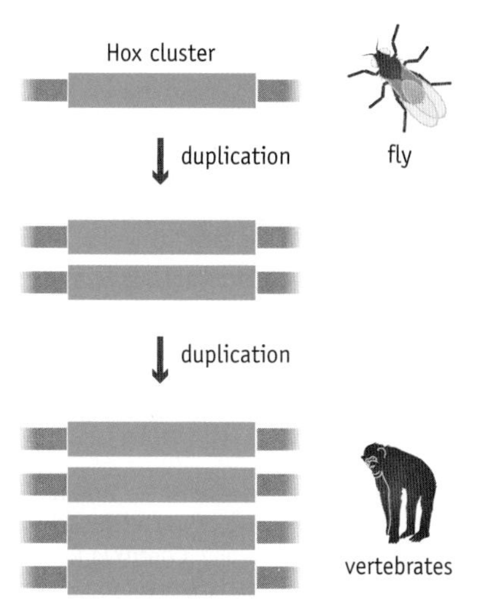

FIGURE 18.45 Duplications in the Hox gene cluster during the evolutionary lineage leading to vertebrates.

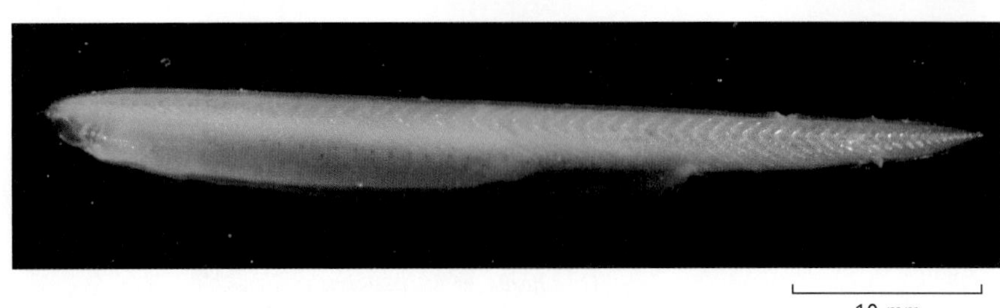

FIGURE 18.46 The amphioxus, a small marine invertebrate with some vertebrate features, for example in the organization of its nervous system. (Courtesy of Sinclair Stammers/Science Photo Library.)

that the amphioxus, an invertebrate with some primitive vertebrate features (Figure 18.46), has two Hox clusters. This is exactly what we might expect for a primitive "protovertebrate."

From our anthropocentric viewpoint, we might smugly assume that four is the maximum number of Hox clusters possessed by any organism on the planet. We would be sadly disillusioned. Ray-finned fishes have seven. Their extra copies of the Hox genes are thought to underlie the vast range of different variations of the basic body plan displayed by these fish (Figure 18.47), which makes them probably the most diverse group of organisms to have evolved.

Homeotic genes also underlie plant development

The power of *Drosophila* as a model system for development extends even beyond vertebrates. Developmental processes in plants are, in many respects, very different from those of fruit flies and other animals, but at the genetic level there are similarities. These are sufficient for knowledge gained from *Drosophila* development to be of value in interpreting developmental processes in plants. In particular, the recognition that a limited number of homeotic selector genes control the *Drosophila* body plan has led to a model for plant development that postulates that the structure of the flower is also determined by a small number of homeotic genes.

FIGURE 18.47 Ray-finned fishes, probably the most diverse group of organisms to have evolved. (A) Giant trevally (*Caranx ignobilis*). (B) Garibaldi damselfish (*Hypsypops rubicundus*). (C) Blackside hawkfish (*Paracirrhites forsteri*). (D) Surge wrasse (*Thalassoma purpureum*). (Courtesy of Claire Fackler, NOAA National Marine Sanctuaries.)

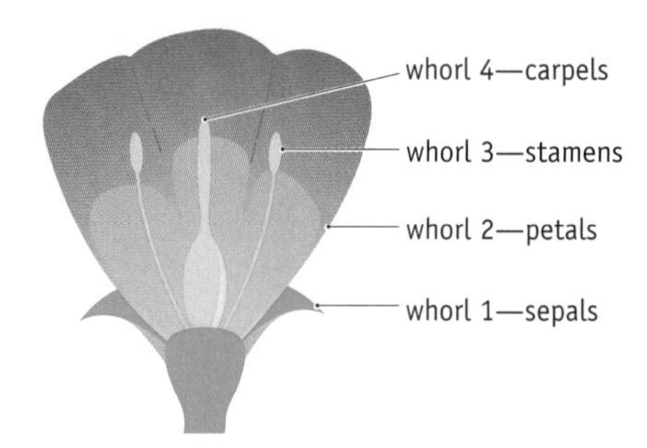

FIGURE 18.48 Flowers are constructed from four concentric whorls.

FIGURE 18.49 *Arabidopsis thaliana*, a model organism for plant genetics. (Courtesy of Dr Jeremy Burgess/Science Photo Library.)

All flowers are constructed along similar lines, made up of four concentric whorls, each comprising a different floral organ (Figure 18.48). The outer whorl, number 1, contains sepals, which are modified leaves that envelop the bud during its early development. The next whorl, number 2, contains the distinctive petals, and within these are whorls 3 (stamens, the male reproductive organs) and 4 (carpels, the female reproductive organs).

Most of the research on plant development has been conducted with *Antirrhinum* (the snapdragon) and *Arabidopsis thaliana*, a small vetch that has been adopted as a model species, partly because it has a genome of only 125 Mb, one of the smallest known among flowering plants (Figure 18.49). Although these plants do not seem to contain homeodomain proteins, they do have genes that, when mutated, lead to homeotic changes in the floral architecture, such as the replacement of sepals by carpels. Analysis of these mutants has led to the **ABC model**, which states that there are three types of homeotic gene—A, B, and C—that control flower development. According to the ABC model, whorl 1 is specified by A-type genes (which include *apetala1* and *apetala2* in *Arabidopsis*), whorl 2 by A genes acting with B genes (such as *apetala3*), whorl 3 by the B genes plus the single C gene (*agamous*), and whorl 4 by the C gene acting on its own (Figure 18.50).

As expected from the work with *Drosophila*, the A, B, and C homeotic gene products are transcription activators. All except the APETALA2 protein contain the same DNA-binding domain, the **MADS box**, which is also found in other proteins involved in plant development. These include SEPALLATA1, 2, and 3, which work with the A, B, and C proteins in defining the detailed structure of the flower. Other similarities to *Drosophila* include the presence in *Arabidopsis* of a gene, called *curly leaf*, whose product acts like Polycomb of *Drosophila*, maintaining the differentiated state of each cell by repressing those homeotic genes that are inactive in a particular whorl.

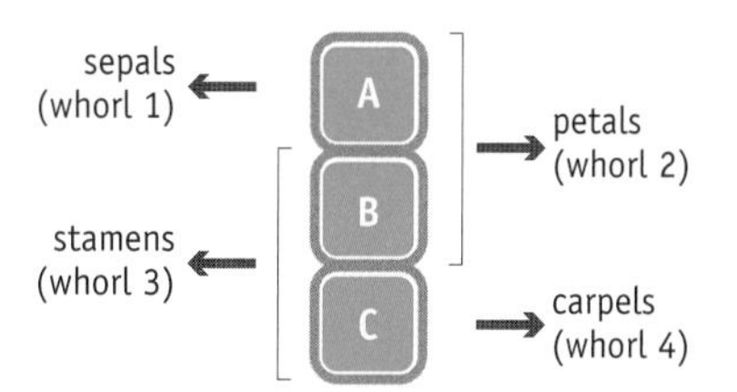

FIGURE 18.50 The ABC model for flower development. According to this model, whorl 1 is specified by A-type genes, whorl 2 by A genes acting with B genes, whorl 3 by the B and C genes, and whorl 4 by the C gene on its own.

KEY CONCEPTS

- Differentiation is the process by which an individual cell acquires a specialized function. Differentiation requires a change in the pattern of gene expression in a cell. The resulting specialized expression pattern usually remains in place for the remainder of the cell's lifetime and is often inherited by the daughter cells, and the granddaughters, and so on for many cell divisions.

- Development is the process that begins with a fertilized egg cell and ends with an adult. A developmental pathway comprises a complex series of genetic, cellular, and physiological events that must occur in the correct order, in the correct cells, and at the appropriate times if the pathway is to reach a successful culmination. These events include the differentiation of many cells into many different specialized types.
- There are four ways in which permanent changes in gene activity can be brought about. Physical rearrangement of the genome can result in a change in the structure of the genome inherited by a cell lineage. A genetic feedback loop can be established to maintain a change in gene expression in a cell and its descendants. Chromatin modification of nucleosomes and DNA methylation result in gene silencing.
- Epigenesis is the term used to describe any heritable change to the genome that does not involve mutation.
- One of the guiding principles of developmental genetics is the assumption that there should be similarities and parallels between developmental processes in different organisms, reflecting their common evolutionary origins. This means that information relevant to human development can be obtained from studies of model organisms chosen for the relative simplicity of their developmental pathways.
- Studies of *Caenorhabditis elegans* have shown how cell-to-cell signaling can confer positional information during embryo development.
- Establishment of the segmentation pattern of a fruit fly embryo involves the sequential expression of different groups of genes. Positional information initially contained in the gradient concentrations of the maternal effect proteins is converted into a sharply defined segmentation pattern.
- The activities of homeotic selector genes define the identities of the fruit fly segments.
- Homeotic selector genes are present in all multicellular animals, including humans.

QUESTIONS AND PROBLEMS (Answers can be found at www.garlandscience.com/introgenetics)

Key Terms

Write short definitions of the following terms:

ABC model
antigen
Barr body
blastoderm
chromatin modification
CpG island
DNA methylation
DNA methyltransferase
epigenesis
gap gene
genomic imprinting
histone acetyltransferase
histone code
histone deacetylase
homeobox
homeodomain
homeotic mutation
homeotic selector gene
imprint control element
MADS box
maintenance methylation
maternal effect gene
methyl-CpG-binding protein
model organism
nucleosome remodeling
pair-rule gene
positional information
segment polarity gene
syncytium
ubiquitin
X inactivation

Self-study Questions

18.1 Distinguish between the terms "differentiation" and "development."

18.2 Describe how immunoglobulin diversity is brought about by physical rearrangement of the human genome.

18.3 Give two examples of genetic feedback loops.

18.4 Distinguish between the activities of histone acetyltransferases and histone deacetylases.

18.5 What types of chemical modification are known to be made to histone proteins? What effects do these modifications have on expression of the genome?

18.6 Explain what is meant by the term "histone code."

18.7 Describe the role of the fruit fly *Polycomb* gene family.

18.8 Using an example, explain how nucleosome remodeling can influence expression of a gene.

18.9 Explain the link between DNA methylation and silencing of the genome.

18.10 Give two examples that illustrate the principles of genomic imprinting.

18.11 What is the purpose of X inactivation, and how is X inactivation brought about?

18.12 Explain why model organisms are important in studies of development.

18.13 Outline the features of *Caenorhabditis elegans* and *Drosophila melanogaster* that make these species good model organisms for the study of development.

18.14 Describe how studies of vulva development in *C. elegans* illustrate how positional information is generated and used during a developmental pathway.

18.15 Explain how positional information in the fruit fly embryo is initially established through the action of the maternal effect genes.

18.16 Distinguish between the roles of the gap genes, pair-rule genes, and segment polarity genes during the development of the fruit fly embryo.

18.17 What is the genetic basis of the *Antennapedia* mutation of the fruit fly?

18.18 Describe the role of homeotic selector genes during development in the fruit fly.

18.19 Explain why the discovery of homeotic selector genes in the fruit fly is relevant to development in vertebrates.

18.20 What is the ABC model, and how does it explain the development of a flower?

Discussion Topics

18.21 Explore and assess the histone code hypothesis.

18.22 In many areas of biology it is difficult to distinguish between cause and effect. Evaluate this issue with regard to nucleosome remodeling and genome expression—does nucleosome remodeling cause changes in genome expression, or is it the effect of these expression changes?

18.23 Maintenance methylation ensures that the pattern of DNA methylation on two daughter DNA molecules is the same as the pattern on the parent molecule. In other words, the methylation pattern, and the information on gene expression that it conveys, is inherited. Other aspects of chromatin structure might also be inherited in a similar way. How do these phenomena affect the principle that inheritance is specified by genes?

18.24 In a normal diploid female, one X chromosome is inactivated and the other remains active. Remarkably, in diploid females with unusual sex chromosome constitutions, the process still results in just a single X chromosome remaining active. For example, in those rare individuals that possess just a single X chromosome no inactivation occurs, and in those individuals with an XXX genotype two of the three X chromosomes are inactivated. What might be the means by which the numbers of X chromosomes in a nucleus are counted so that the appropriate number of X chromosomes can be inactivated?

18.25 What would be the key features of an ideal model organism for the study of development in vertebrates?

18.26 Are *C. elegans* and *D. melanogaster* good model organisms for development in vertebrates?

FURTHER READING

Alt FW, Blackwell TK & Yancopoulos GD (1987) Development of the primary antibody repertoire. *Science* 238, 1079–1087. *Generation of immunoglobulin diversity.*

Amores A, Force A, Yan Y-L et al. (1998) Zebrafish *hox* clusters and vertebrate genome evolution. *Science* 282, 1711–1714. *Hox genes in ray-finned fishes.*

Heard E, Clerc P & Avner P (1997) X-chromosome inactivation in mammals. *Annu. Rev. Genet.* 31, 571–610.

Ideraabdullah FY, Vigneau S & Bartolomei MS (2008) Genomic imprinting mechanisms in mammals. *Mutat. Res.* 647, 77–85.

Ingham PW (1988) The molecular genetics of embryo pattern formation in *Drosophila*. *Nature* 335, 25–34.

Khorasanizedeh S (2004) The nucleosome: from genomic organization to genomic regulation. *Cell* 116, 259–272. *Review of histone modification, nucleosome remodeling, and DNA methylation.*

Kornfeld K (1997) Vulval development in *Caenorhabditis elegans*. *Trends Genet.* 13, 55–61.

Kouzarides T (2007) Chromatin modifications and their function. *Cell* 126, 693–705. *Histone modifications.*

Mahowald AP & Hardy PA (1985) Genetics of *Drosophila* embryogenesis. *Annu. Rev. Genet.* 19, 149–177.

Margueron R, Trojer P & Reinberg D (2005) The key to development: interpreting the histone code? *Curr. Opin. Genet. Dev.* 15, 163–176.

Schwartz YB & Pirrotta V (2007) Polycomb silencing mechanisms and the management of genomic programmes. *Nat. Rev. Genet.* 8, 9–22.

Suzuki MM & Bird A (2008) DNA methylation landscapes: provocative insights from epigenomics. *Nat. Rev. Genet.* 9, 465–476. *The latest ideas on the role of methylation in controlling gene activity.*

Theissen G (2001) Development of floral organ identity: stories from the MADS house. *Curr. Opin. Plant Biol.* 4, 75–85. *Genes involved in the ABC model for flower development.*

Zakany J & Duboule D (2007) The role of *Hox* genes during vertebrate limb development. *Curr. Opin. Genet. Dev.* 17, 359–366.

CHAPTER 19

The Human Genome

Our reasons for devoting a chapter to the human genome are not entirely anthropocentric. It provides an opportunity for us to look at the important features of a eukaryotic genome in more detail than has been possible so far in this book. Complete genome sequences have been obtained for more than 80 eukaryotes, including examples of all the major evolutionary groups, and for some key species sequences have been obtained from different individuals, enabling intraspecific variations to be identified (Table 19.1). Among the eukaryotes, we see extensive differences in gene numbers and in the ways in which the genes are organized within chromosomes. The human genome is a typical example of a mammalian genome, and it also illustrates many of the general features of the genomes of all animals and plants. By focusing on the human genome we can therefore learn more about the biology of our own species while gaining an understanding of how genomes are organized in many other species.

In this chapter we will study four important topics concerning the human genome. First, we will look at the genes that are present and how they are arranged in the genome. We will also make comparisons with the genomes of other organisms to understand the variations that exist among eukaryotes as a whole.

The second subject that we will study is the repetitive DNA content of the human genome. We will learn that most of the genome is made up of these repeated sequences, many of which have no known function. Then we will examine the relationship between the human genome and the genomes of other primates, and we will ask to what extent a knowledge of the sequences of these genomes helps us to understand what it is that makes us human.

The final topic in this chapter is the human mitochondrial genome. In addition to the DNA in the nucleus, humans and almost all eukaryotes also have smaller DNA molecules in their mitochondria, the energy-generating organelles present in the cytoplasm (Figure 19.1). These mitochondrial molecules contain genes, and we must investigate the functions of these genes and ask why they are located in the mitochondria.

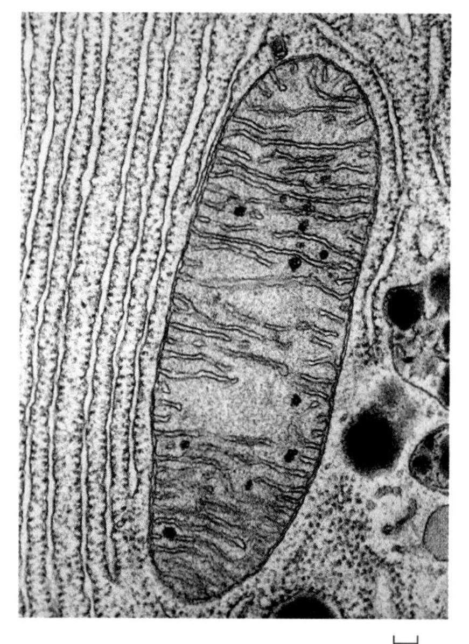

FIGURE 19.1 An electron micrograph of a cross-section of a mitochondrion. (Courtesy of Don Fawcett/Science Photo Library.)

TABLE 19.1 EXAMPLES OF EUKARYOTES WHOSE GENOMES HAVE BEEN COMPLETELY SEQUENCED

Group	Species
Mammals	Human, chimpanzee, cow, dog
Other animals	Zebrafish, frog, sea urchin
Birds	Chicken, zebra finch
Plants	*Arabidopsis thaliana*, rice, maize
Insects	Fruit fly, mosquito, silkworm moth
Fungi	*Saccharomyces cerevisiae*, *Aspergillus nidulans*
Protozoa	*Plasmodium falciparum* (malaria parasite), *Tetrahymena thermophila*, *Entamoeba histolytica*

19.1 THE GENETIC ORGANIZATION OF THE HUMAN NUCLEAR GENOME

The human nuclear genome is made up of about 3200 Mb of DNA, split into 24 chromosomes. Each chromosome contains a single DNA molecule, the shortest being 48 Mb and the longest 250 Mb. About 37.5% of the genome has, or at one time had, a function that is understood. This fraction includes all the genes, pseudogenes, and other types of gene relic, as well as the promoter regions upstream of genes and the introns contained within discontinuous genes (Figure 19.2). The remainder, some 62.5% of the genome, does not seem to have any role. We will begin with the 37.5% of the genome that is understood.

The human genome has fewer genes than expected

The most recent estimates for gene numbers in a range of eukaryotic genomes are given in Table 19.2. These figures indicate that the simplest eukaryotes, typified by the single-celled yeasts, have about 5000–6000 genes. As we move up the scale, it takes about 13,600 genes to make a fruit fly and 20,140 to make a *Caenorhabditis elegans* worm, a surprising result because, intuitively, we might expect a microscopic worm to be less complex than an insect.

We also see that, on the basis of these comparisons, humans are only slightly more complex than *C. elegans*, as our genome has only 20,500 genes. This number is much lower than originally expected, because a "best guess" of 80,000–100,000 was still in vogue up to a few months before the human genome sequence was completed in the year 2000. These early estimates were high because they were based on the supposition that, in most cases, a single gene specifies a single protein. According to this model, the number of genes in the human genome should be similar to the number of different types of protein in human cells, leading to the estimates of 80,000–100,000. The discovery that the number of genes is much lower than this indicates that alternative splicing, the process by which exons are assembled in different combinations so that more than one protein can be produced from a single gene, is more prevalent than was originally appreciated.

The comparisons between species should not therefore be based on the number of genes but on the number of different proteins that can be produced from those genes. This means that humans do indeed show a more satisfying degree of additional complexity than microscopic worms, because the 20,140 genes of *C. elegans* specify only about 22,000 proteins, alternative splicing being much less common in this organism.

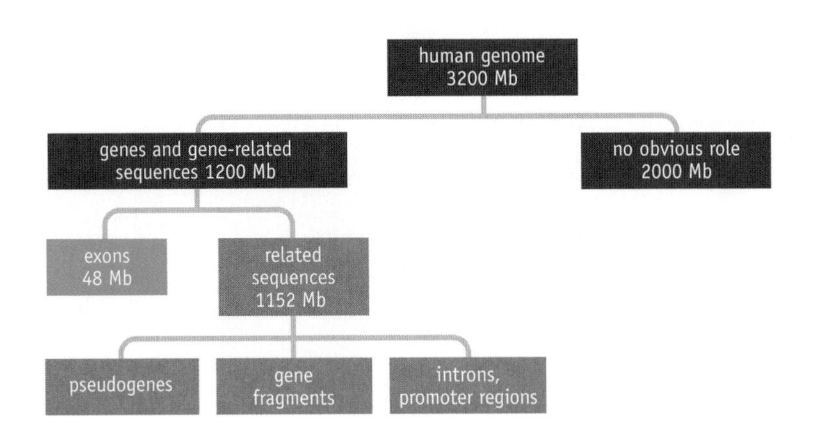

FIGURE 19.2 The composition of the human genome.

TABLE 19.2 GENE NUMBERS FOR VARIOUS EUKARYOTES

Species	Approximate number of genes
Schizosaccharomyces pombe (fission yeast)	4900
Saccharomyces cerevisiae (budding yeast)	6100
Drosophila melanogaster (fruit fly)	13,600
Caenorhabditis elegans (nematode worm)	20,140
Homo sapiens (human)	20,500
Arabidopsis thaliana (plant)	25,500
Zea mays (maize)	32,700

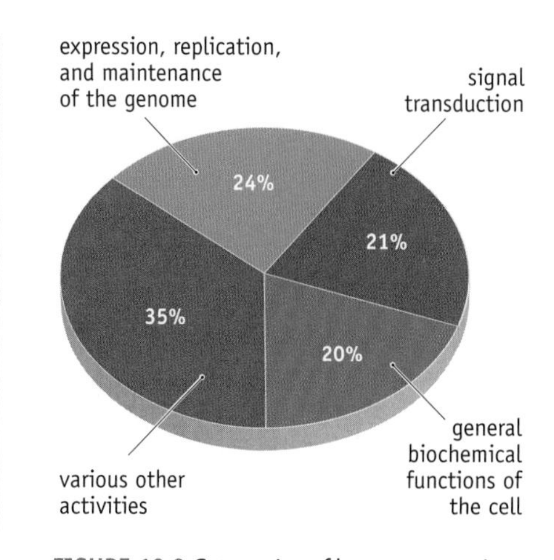

FIGURE 19.3 Categories of known genes in the human genome.

The functions of more than half of the 20,500 human genes are known or can be inferred with a reasonable degree of certainty. Almost a quarter of these genes are involved in the expression, replication, and maintenance of the genome, and another 21% specify components of the signal transduction pathways that regulate genome expression and other cellular activities in response to signals received from outside the cell (Figure 19.3). All of these genes can be looked on as having a function that is involved in one way or another with the activity of the genome. Enzymes responsible for the general biochemical functions of the cell account for another 20% of the known genes, and the remainder are involved in activities such as the transport of compounds into and out of cells, the folding of proteins into their correct three-dimensional structures, the immune response, and the synthesis of structural proteins such as those found in the cytoskeleton and in muscles.

Analysis of the genes in other genomes suggests that all eukaryotes possess the same basic set of genes but that more complex species have a greater number of genes in each category. It currently seems that about one-fifth to one-quarter of the genes in the human genome are unique to vertebrates, and a further quarter are found only in vertebrates and other animals (Figure 19.4). The genes that are unique to vertebrates include several involved in activities such as cell adhesion, electric couplings, and the growth of nerve cells, functions that we look on as conferring the distinctive features of vertebrates compared with other types of eukaryote. But there is a limit to how much useful information we can obtain from this type of comparison—according to the contents of their genomes, humans and chimpanzees are virtually identical.

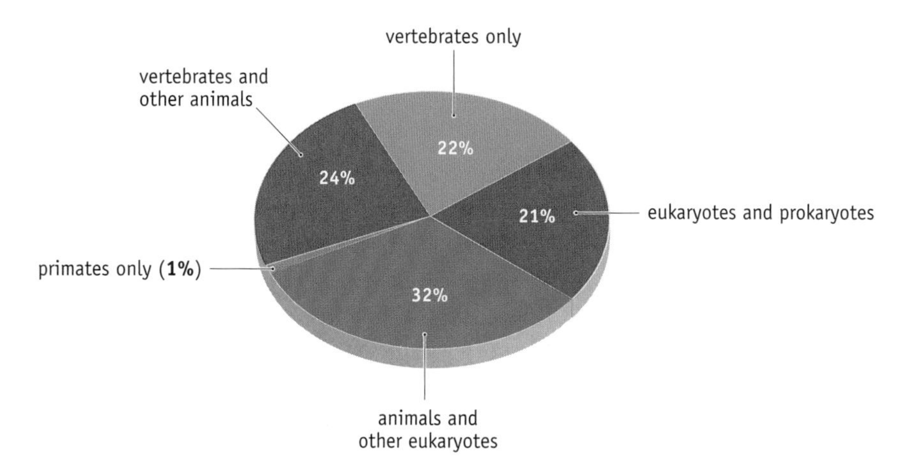

FIGURE 19.4 Relationships between the genes in the human genome and those in the genomes of other types of organism. The pie chart categorizes human genes according to the presence or absence of equivalent genes in other organisms. The chart shows, for example, that 22% of human genes are specific to vertebrates, and that another 24% comprises genes specific to vertebrates and other animals.

There is little logic to the arrangement of genes in the human genome

Human genes cover a spectrum of sizes from less than 100 bp to more than 2000 kb. The tRNA, snRNA, and snoRNA genes are the smallest, in some cases as short as 65–75 bp. The smallest protein-coding genes are slightly longer, for example 406 bp for the histone H4 gene. At the other end of the size spectrum are genes whose bulk is made up not of coding sequence but of introns. The longest human gene is the one coding for dystrophin, a protein that is defective in one type of muscular dystrophy, which contains 78 introns in its 2400-kb length. Even though the dystrophin protein is relatively large (it has a molecular mass of more than 400 kD), its coding sequence (i.e. all of the exons added together) makes up only 0.6% of the length of the gene. A point worth noting is that this single gene is over one-half the length of the entire *E. coli* chromosome.

The genes are not spread evenly throughout the human genome. Instead, the density ranges from 0 to 64 genes per 100 kb, with the less dense regions being those close to the centromeres and to the ends of individual chromosomes. Other than this, there is little obvious order or logic to the locations of individual genes. This point is illustrated by looking at the positions of the members of various multigene families (Figure 19.5).

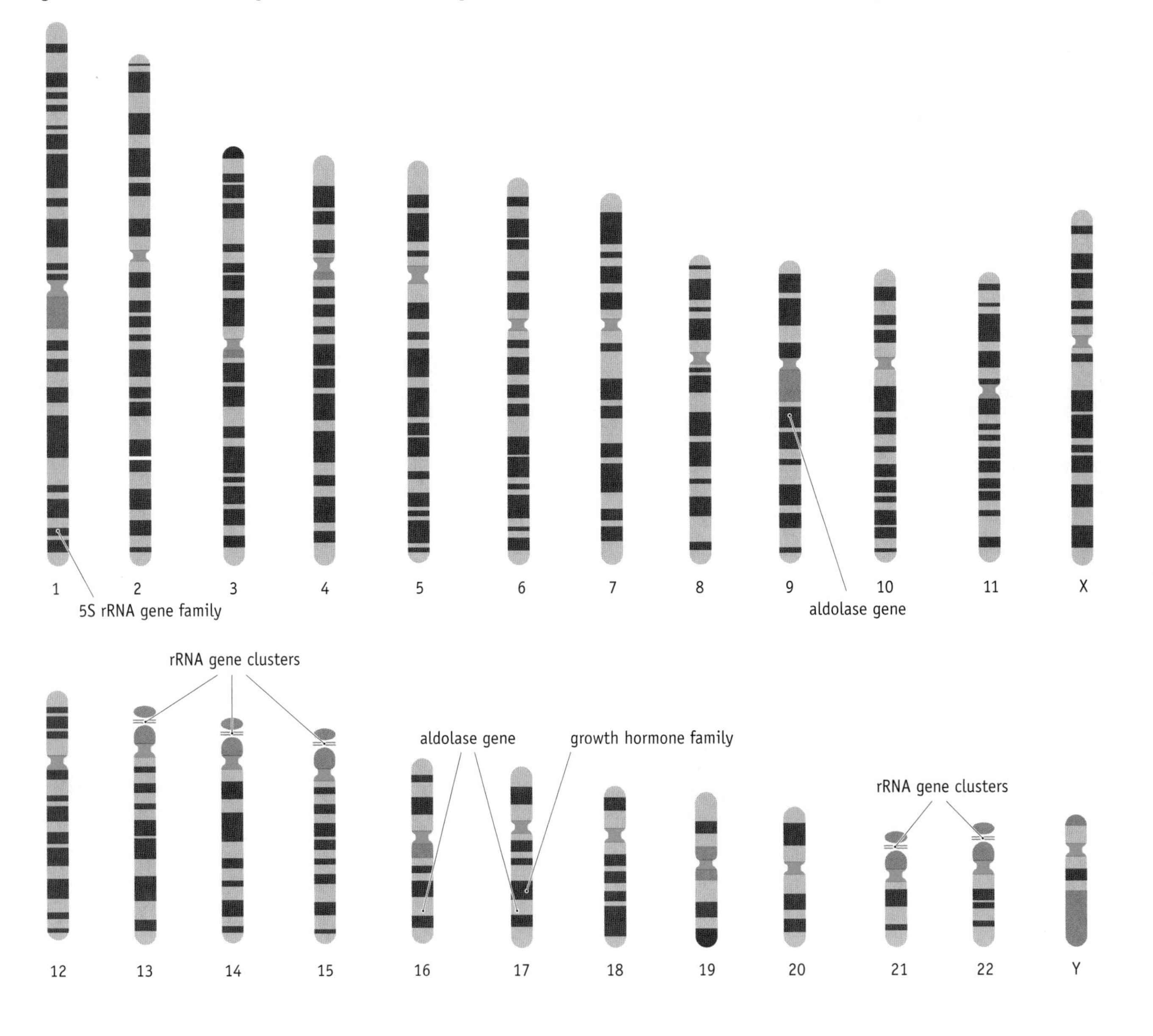

FIGURE 19.5 Locations of some genes and gene families in the human genome.

With some families the individual genes are clustered together at a single position in the genome. Examples are the growth hormone gene family, whose five members are clustered on chromosome 17, and the 5S rRNA gene family, comprising 2000 genes in a tandem array on the long arm of chromosome 1.

With other families the genes are interspersed around the genome. For example, the three members of the aldolase gene family are located on chromosomes 9, 16, and 17. Some large gene families are both clustered and interspersed at the same time. There are about 280 copies of the rRNA transcription unit, grouped into five clusters of 50–70 units each, the clusters being located on the short arms of chromosomes 13, 14, 15, 21, and 22. The implication is that, in most cases, chance events during the past evolution of the genome have been responsible for placing the genes in the positions in which we find them today.

The genes make up only a small part of the human genome

The variations in gene density that occur along the length of a human chromosome mean that it is difficult to identify regions in which the organization of the genes can be looked on as "typical" of the genome as a whole. Bearing this in mind, we will take a closer look at a 50-kb segment of chromosome 12 to understand the detailed organization of the genome (Figure 19.6).

This segment contains four genes. There is no obvious relationship between them, emphasizing the apparently random nature of the distribution of genes in the human genome. The first gene, called *PKP2*, codes for a plakophilin protein, which is involved in the synthesis of desmosomes, structures that act as connection points between adjacent mammalian cells. The second gene is *SYB1*, specifying a protein whose role is to ensure that vesicles fuse with their correct target membranes within the cell. Next there is a gene called *FLJ10143*, whose function has not yet been identified. The last gene in the segment is *CD27*, which codes for a member of the tumor necrosis factor receptor superfamily, a group of proteins that regulate intracellular signaling pathways involved in apoptosis (programmed cell death) and cell differentiation.

We can see from Figure 19.6 how little of the human genome actually contains biological information. Each of the four genes is discontinuous, with the number of introns ranging from two for *SYB1* to eight

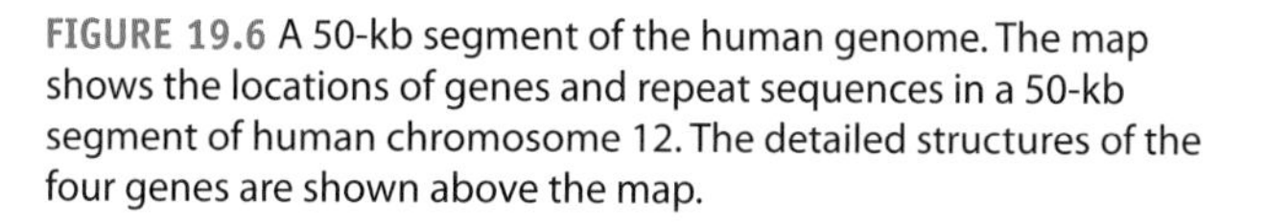

FIGURE 19.6 A 50-kb segment of the human genome. The map shows the locations of genes and repeat sequences in a 50-kb segment of human chromosome 12. The detailed structures of the four genes are shown above the map.

KEY

exon intron repeat sequence

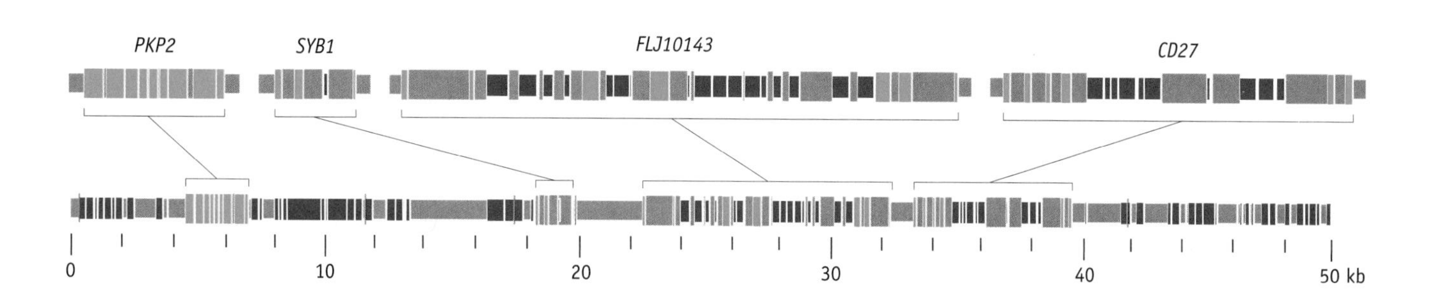

CACACACACACACA

FIGURE 19.7 A microsatellite sequence.

for *PKP2*. When the exons are added together, their total length is just 4745 bp, equivalent to 9.5% of the 50-kb segment. Throughout the genome as a whole, the amount of coding DNA is rather less than this. All the exons in the human genome make up only 48 Mb, just 1.5% of the total (see Figure 19.2).

In addition to the genes there are 88 repeat sequences dispersed at various positions in this segment of chromosome 12. These are examples of sequences that recur at many places in the genome. Most of them are located in the intergenic regions of our segment, but several lie within introns. The distribution of these repeats is typical of the genome as a whole, as we will see later in this chapter when we examine the repetitive DNA in more detail.

There are also seven places in the 50-kb segment where a short motif, such as the dinucleotide 5′-CA-3′, is repeated in tandem (Figure 19.7). These tandem repeat sequences are called **microsatellites**.

Finally, about 30% of our 50-kb segment of the human genome is made up of stretches of nongenic, nonrepetitive, single-copy DNA of no known function or significance.

Genes are more densely packed in the genomes of lower eukaryotes

How extensive are the differences in gene organization among eukaryotes? To explore this question we will compare our 50-kb segment of the human genome with "typical" segments, of the same length, from the genomes of the yeast *Saccharomyces cerevisiae*, the fruit fly *Drosophila melanogaster*, and maize (Figure 19.8).

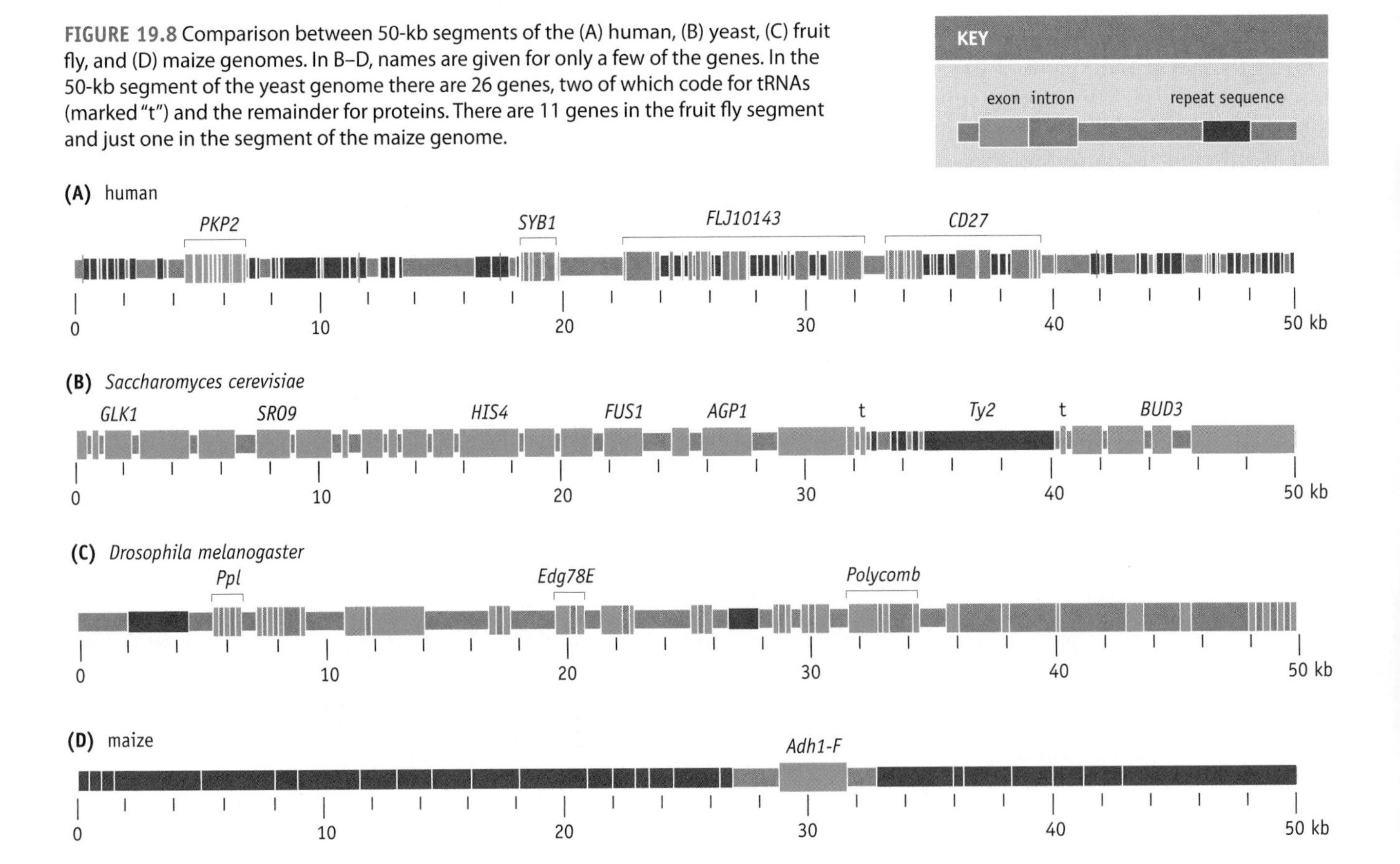

FIGURE 19.8 Comparison between 50-kb segments of the (A) human, (B) yeast, (C) fruit fly, and (D) maize genomes. In B–D, names are given for only a few of the genes. In the 50-kb segment of the yeast genome there are 26 genes, two of which code for tRNAs (marked "t") and the remainder for proteins. There are 11 genes in the fruit fly segment and just one in the segment of the maize genome.

The yeast genome segment comes from chromosome III, the first eukaryotic chromosome to be sequenced. The first striking feature is that it contains more genes than the human segment, 24 genes thought to code for proteins and two that code for tRNAs. None of these genes are discontinuous, reflecting the fact that in the entire yeast genome there are only 300 introns, compared with over 40,000 in the human genome. There are also fewer repeat sequences. This part of chromosome III contains a single copy of an LTR retrotransposon called *Ty2*, and four "delta elements," which are LTR sequences that have become detached from their parent retrotransposon. These five repeat sequences make up 13.5% of the 50-kb segment, but this figure is not typical of the rest of the yeast genome. When all 16 yeast chromosomes are considered, the total amount of sequence taken up by repetitive DNA is only 3.4% of the total.

The picture that emerges is that the genetic organization of the yeast genome is much more economical than that of the human version. The genes themselves are more compact, having fewer introns, and the spaces between the genes are relatively short, with much less space taken up by repetitive DNA and other noncoding sequences.

The hypothesis that less complex eukaryotes have more compact genomes holds true when other species are examined. Figure 19.8 also shows a 50-kb segment of the fruit fly genome. If we agree that a fruit fly is more complex than a yeast but less complex than a human, we would expect the organization of the fruit fly genome to be intermediate between that of yeast and humans, and this is what we find. This 50-kb segment of the fruit fly genome has 11 genes, more than in the human segment but fewer than in the yeast sequence. All of these genes are discontinuous, but seven have just one intron each. The picture is similar when the entire genome sequences of the three organisms are compared (Table 19.3). The gene density in the fruit fly genome is intermediate between that of yeast and humans, and the average fruit fly gene has many more introns than the average yeast gene but still only one-third of the number in the average human gene.

The comparison between the yeast, fruit fly, and human genomes also reveals the differing repetitive DNA contents of these genomes. Repeat sequences make up 3.4% of the yeast genome, about 12% of the fruit fly genome, and 44% of the human genome. It is beginning to become clear that repeat sequences have an intriguing role in dictating the compactness or otherwise of a genome. This is strikingly illustrated by the maize genome, which at 2300 Mb is relatively small for a flowering plant. The maize genome is dominated by repetitive elements. In the segment shown in Figure 19.8 there is just a single gene, coding for an alcohol dehydrogenase enzyme. This is the only gene in this 50-kb region, although there is a second one, of unknown function, roughly 100 kb beyond the right-hand end of the sequence shown here. Instead of genes, the dominant feature of

TABLE 19.3 COMPACTNESS OF THE YEAST, FRUIT FLY, AND HUMAN GENOMES

Feature	Yeast	Fruit fly	Human
Gene density (average number per Mb)	504	76	6.5
Introns per gene (average)	0.04	3	9
Amount of the genome that is taken up by repetitive DNA (%)	3.4	12	44

RESEARCH BRIEFING 19.1

How the human genome was sequenced

The Human Genome Project was the major endeavor in biological research during the 1990s. The project was first discussed in 1984 but did not formally get under way until 1990. The objective was to sequence the 3200 Mb of DNA that makes up the genome, with 2003 set as the target date for completion. Initially this was looked on as over-ambitious, but the goal was reached three years early, in 2000. This was due to the rapid technological advances that the project stimulated. Most important of these was the invention of "sequencing machines," which enabled hundreds of kilobases of sequence to be read in a single day.

Although the Human Genome Project did not begin until 1990, work that formed the foundation to the project had started 10 years earlier, when the first attempts were made to prepare a comprehensive map of the human genome.

The human genome maps showed the positions of DNA markers as well as genes

Until the beginning of the 1980s, a detailed map of the human genome was thought to be unattainable. Although comprehensive gene maps had been constructed for fruit flies and a few other organisms, the problems inherent in the analysis of human pedigrees led many geneticists to doubt that a human map could ever be constructed. Another problem was that relatively few human characteristics can be linked to a single gene. Most characteristics, such as eye and hair color, are specified by groups of genes acting together. Following the inheritance of these characteristics would not give clear recombination frequencies from which a gene map could be constructed.

The impetus for human mapping came from the discovery of **DNA markers**. These are sequences that, like genes, exist as two or more variants. An example is a **restriction fragment length polymorphism** (**RFLP**). An RFLP results from a point mutation that changes the sequence of a recognition site for a restriction endonuclease. This is an enzyme that cuts DNA at a specific sequence. An example is *Bam*HI, which cuts only at the sequence 5′-GGATCC-3′. The two alleles are the version of the RFLP that is cut by the enzyme and the version that is not (Figure 1). Which allele or alleles a person possesses can be determined by testing their DNA with the restriction enzyme. The inheritance of RFLPs can be studied in a human pedigree in the same way as genes. Recombination frequencies between two different RFLPs, or between an RFLP and a gene, can be worked out, and the positions of the RFLPs can be identified on the "gene" map.

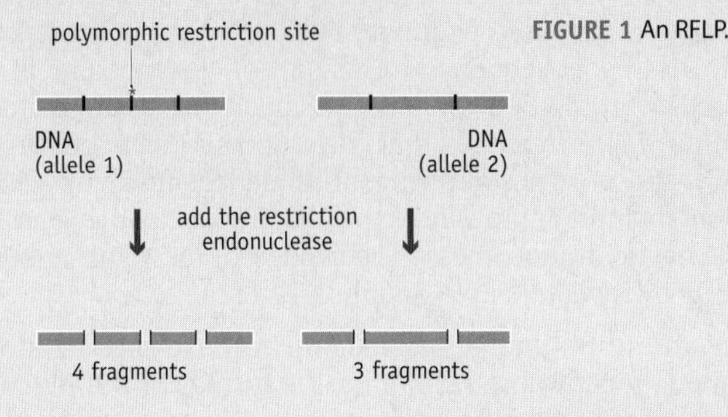

FIGURE 1 An RFLP.

The first human map, showing the positions of 393 RFLPs, was published in 1987. Gradually, more and more detail was added. More markers were mapped by studying pedigrees, and the positions of others were located by **physical mapping**. In physical mapping, a marker is located by direct examination of the DNA. Various techniques are used, including one called **fluorescent *in situ* hybridization** (**FISH**), in which an intact chromosome is examined by probing it with a labeled DNA molecule. The position on the chromosome at which hybridization occurs provides information about the map location of the DNA sequence used as the probe (Figure 2). By 1995, more than 30,000 markers had been mapped.

The map anchors the genome sequence

Why was so much effort put into mapping the human genome? Without the map, it was thought that it would be impossible to assemble the sequence correctly. Remember that each sequencing experiment yields a maximum of 750 bp. Longer sections (contigs) are built up by searching for overlaps between these short sequence reads. Sometimes mistakes might be made in identifying the overlaps, especially for parts of the genome that contain repeat sequences (Figure 3). The map is therefore needed so that assembly of the sequence can be checked. If there is a discrepancy between the location of a marker in the sequence and its position in the map, an error must have occurred. We say the map is used to "anchor" the sequence.

The human genome is so large that it was initially thought that it would be impossible, even with a map, to assemble it correctly. To make the task a little easier, the genome was first split into 300-kb segments, and these were cloned in a special type of vector

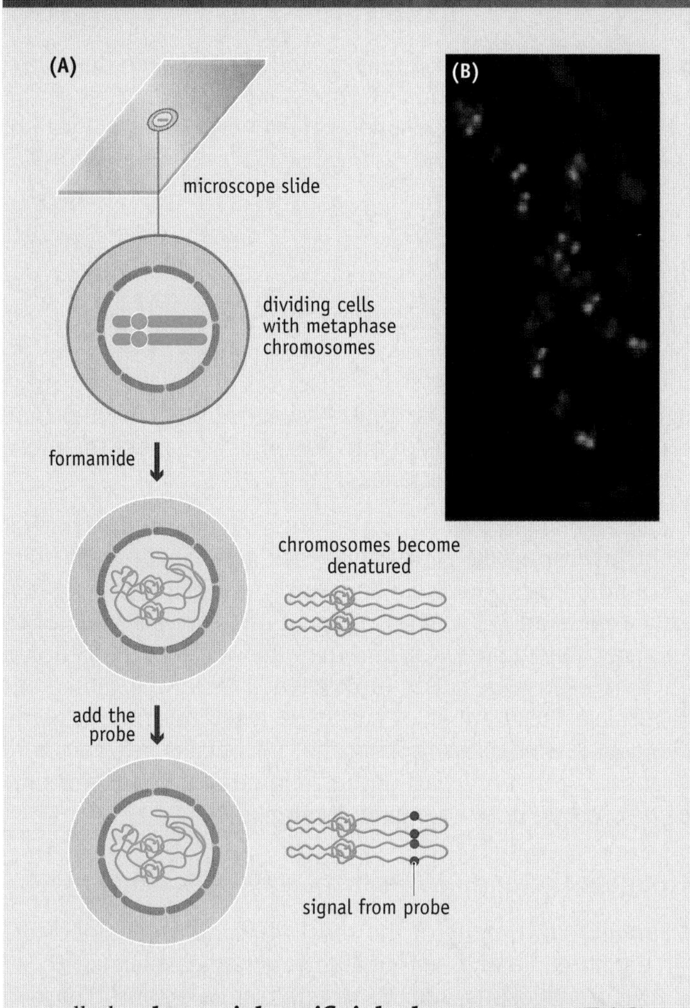

FIGURE 2 FISH. (A) A summary of the method. (B) An example of the result. Eighteen cloned DNA fragments have been labeled with different fluorescent markers and hybridized with a chromosome. The hybridizing positions are seen as doublets because each chromatid contains two DNA copies. (B, courtesy of Octavian Henegariu, Yale University.)

called a **bacterial artificial chromosome** (**BAC**). The position of each BAC clone on the genome map was worked out by identifying which markers were contained in which clones. Each BAC clone was then broken into smaller fragments that were individually sequenced. Overlaps between these fragments enabled contigs comprising all or part of each BAC clone to be assembled. The sequence of each contig was then placed at its correct position on the genome map.

As it turned out, preparation of the BAC library was an unnecessary precaution. At about the time at which the Human Genome Project was moving into the sequence acquisition phase, a second project was begun by a private company with the intention of sequencing the genome by a more random method. The aim of this "shotgun" approach was to obtain 75 million short sequences, each 500 bp or so in length, which would be reassembled to give the genome sequence. This approach would definitely be quicker, and although it might contain more errors it would also be much cheaper.

The draft sequence was completed in 2000

The possibility that the Human Genome Project would not in fact provide the first human genome sequence stimulated the organizers of the Project to bring forward their planned dates for completion. The first sequence of an entire human chromosome (number 22) was published in December 1999 and the sequence of chromosome 21 appeared a few months later. Finally, on 26 June 2000, accompanied by the President of the United States, Francis Collins and Craig Venter, the leaders of the two projects, jointly announced completion of their sequences.

The genome sequences announced in 2000 were drafts, not complete final sequences. The version obtained by the Human Genome Project covered just 90% of the genome, the missing 320 Mb lying predominantly in constitutive heterochromatin. This draft sequence had about 150,000 gaps. The first completely finished chromosome sequences began to appear in 2004, and the entire genome sequence was considered complete a year later. This sequence had a total length of 2850 Mb and lacked just 28 Mb of euchromatin, the latter being present in 308 gaps that at that time had resisted all attempts at closure.

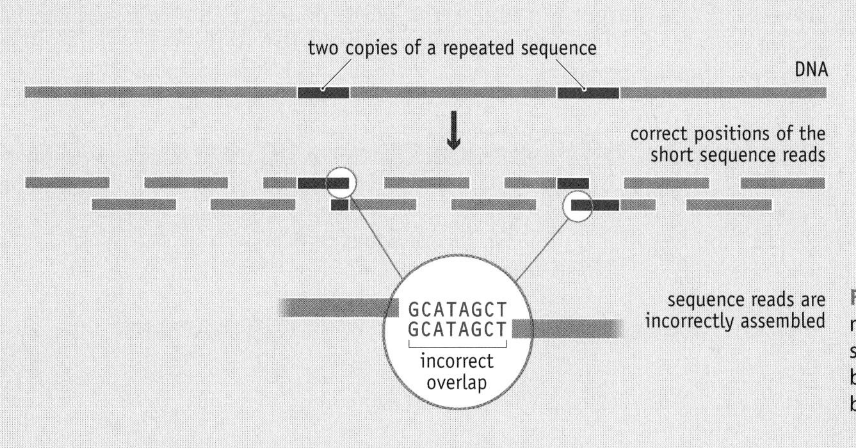

FIGURE 3 The presence of repetitive DNA might lead to the incorrect assembly of sequence reads. In this example, the sequence between the two repeats has accidentally been omitted from the contig.

this genome segment is the repeat sequences, which have been described as forming a sea within which islands of genes are located. Most of the repeat sequences are LTR retrotransposons, which comprise virtually all of the noncoding part of the segment and on their own are estimated to make up about 50% of the maize genome.

19.2 THE REPETITIVE DNA CONTENT OF THE HUMAN NUCLEAR GENOME

The genome segments shown in Figure 19.8 show us that repetitive DNA is an important component of most eukaryotic genomes. We must therefore look more closely at these repeat sequences.

Repetitive DNA can be divided into two categories (**Figure 19.9**). The first is genome-wide or **interspersed repeats**, whose individual repeat units are distributed around the genome in an apparently random fashion. Then there is **tandemly repeated DNA**, whose repeat units are placed next to each other in an array. Repetitive DNA, and the other parts of the genome of no known function, used to be called **junk DNA**. This term is falling out of favor, partly because the number of surprises resulting from genome research over the past few years has meant that geneticists have become less confident in asserting that any part of the genome is unimportant simply because we currently do not know what it might do.

Interspersed repeats include RNA transposons of various types

We are already familiar with many of the most important types of interspersed repeat in the human and other eukaryotic genomes, because these are the cellular RNA transposons, related to viral retroelements, that we studied in Chapter 13. LTR retrotransposons of the *Ty1/copia* and *Ty3/gypsy* families are present in many eukaryotic genomes, but humans and other mammals are slightly unusual in this regard. The bulk of their LTR transposons are decayed endogenous retroviruses rather than the types of LTR retrotransposon that predominate in other eukaryotes. They are all members of the same family of elements, so this is only a small difference.

Not all types of retrotransposon have LTR elements. In mammals the most important types of non-LTR retroelements, or **retroposons**, are the **LINEs** (**long interspersed nuclear elements**) and **SINEs** (**short interspersed nuclear elements**). SINEs have the highest copy number for any type of interspersed repetitive DNA in the human genome, with more than 1.7 million copies comprising 14% of the genome as a whole. LINEs are less frequent, with just over 1 million copies, but because they are longer they make up a larger fraction of the genome, more than 20%. Most of the interspersed repeats in the segment of the human genome shown in Figure 19.6 are LINEs and SINEs.

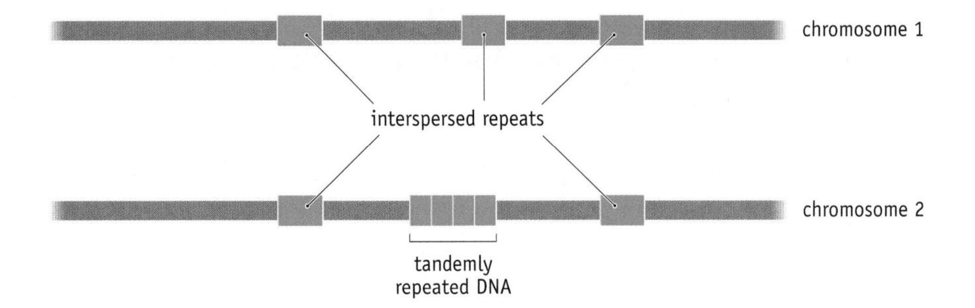

FIGURE 19.9 The two types of repetitive DNA: interspersed repeats and tandemly repeated DNA.

There are three families of LINEs in the human genome, of which one group, LINE-1, is both the most frequent and the only type that is still able to transpose. The LINE-2 and LINE-3 families are made up of inactive relics. A full-length LINE-1 element is 6.1 kb in length and has two genes. One of these codes for a polyprotein similar to the product of the *pol* gene of viral retroelements. This polyprotein includes a reverse transcriptase enzyme (Figure 19.10A). There are no LTRs, but the 3′ end of the LINE is marked by a series of A–T base pairs, giving what is usually referred to as a poly(A) sequence, although of course it is a poly(T) sequence on the other strand of the DNA. Only 1% of the LINE-1 elements in the human genome are full-length versions, with the average size of all the copies being just 900 bp. Most are therefore inactive, and LINE-1 transposition is a rare event, although it has been observed in cultured cells. A relatively recent LINE-1 transposition is responsible for hemophilia in some patients, this transposition having disrupted their factor VIII gene, preventing synthesis of the important factor VIII blood-clotting protein.

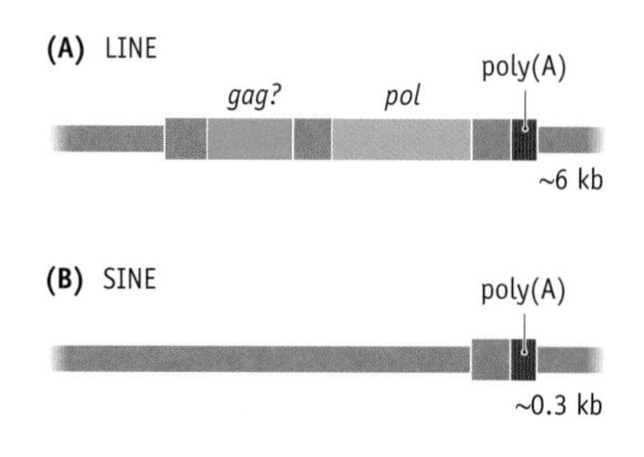

FIGURE 19.10 Non-LTR retroelements in the human genome.

SINEs are much shorter than LINEs, being just 100–400 bp in length and not containing any genes (Figure 19.10B). Instead, SINEs "borrow" enzymes that have been synthesized by LINEs in order to transpose. The commonest SINE in primate genomes is **Alu**, which has a copy number of about 1.2 million in humans. Some Alu elements are actively copied into RNA, providing the opportunity for proliferation of the element.

Alu is derived from the gene for the 7SL RNA, a noncoding RNA involved in the movement of proteins around the cell. The first Alu element may have arisen by the accidental reverse transcription of a 7SL RNA molecule and integration of the DNA copy into the genome. Other SINEs are derived from tRNA genes, which, like the gene for the 7SL RNA, are transcribed by RNA polymerase III in eukaryotic cells, suggesting that some feature of the transcripts synthesized by this polymerase makes these molecules prone to occasional conversion into retroposons.

Tandemly repeated DNA forms satellite bands in density gradients

Tandemly repeated DNA is also called **satellite DNA** because DNA fragments containing tandemly repeated sequences form "satellite" bands when genomic DNA is centrifuged in a density gradient (see Research Briefing 10.1).

The buoyant density of a DNA molecule, which determines the position it takes up in a density gradient, depends on its GC content, with human DNA (GC content 40.3%) having a buoyant density of 1.701 g cm^{-3}. Human DNA therefore forms a band at the 1.701 g cm^{-3} position in a density gradient. This is not, however, the only band that appears. With human DNA there are also three "satellite" bands, at 1.687, 1.693, and 1.697 g cm^{-3} (Figure 19.11). These additional bands contain repetitive DNA. They form because

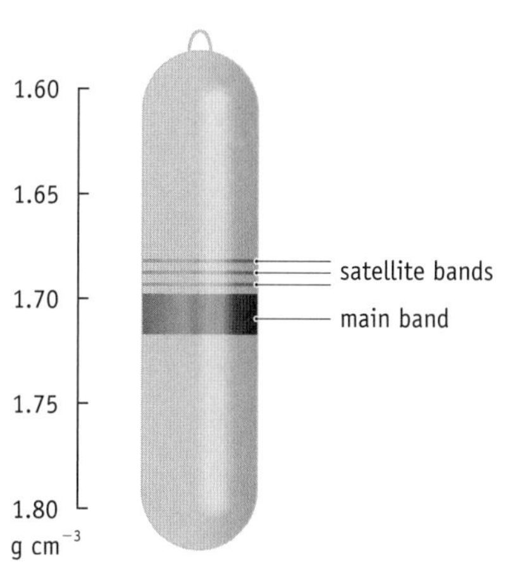

FIGURE 19.11 Satellite DNA from the human genome. Human DNA has an average GC content of 40.3% and average buoyant density of 1.701 g cm^{-3}. Fragments made up mainly of single-copy DNA have a GC content close to this average and are contained in the main band in the density gradient. The satellite bands at 1.687, 1.693, and 1.697 g cm^{-3} consist of fragments containing repetitive DNA. The GC contents of these fragments depend on their repeat motif sequences and are different from the genome average, meaning that these fragments have different buoyant densities to single-copy DNA and migrate to different positions in the density gradient.

the long chromosomal molecules become cleaved into fragments, 50 kb or so in length, when the cell is broken open. Fragments containing predominantly single-copy DNA have GC contents close to the "standard" human value of 40.3%, and so move to the 1.701 g cm^{-3} position in the density gradient. However, fragments containing large amounts of repetitive DNA behave differently. For instance, a fragment made up entirely of ATTAC repeats has a GC content of just 20%, lower than the standard value, and so has a buoyant density somewhat less than 1.701 g cm^{-3}. These repetitive DNA fragments therefore migrate to a satellite position, above the main band in a density gradient. A single genome can contain several different types of satellite DNA, each with a different repeat unit, these units being anything from <5 to >200 bp in length. The three satellite bands in human DNA include at least four different repeat types.

We have already encountered one type of human satellite DNA, the alphoid DNA repeats found in the centromere regions of chromosomes. Although some satellite DNA is scattered around the genome, most is located in the centromeres, where it may have a structural role, possibly as binding sites for one or more of the special centromere proteins.

Minisatellites and microsatellites are special types of tandemly repeated DNA

Although not appearing as satellite bands in density gradients, two other types of tandemly repeated DNA are also classed as "satellite" DNA. These are **minisatellites** and **microsatellites**. Minisatellites form clusters up to 20 kb in length, with each repeat unit up to 25 bp. Microsatellite clusters are shorter, usually <150 bp, and the repeat unit is usually 13 bp or less.

Minisatellite DNA is a second type of repetitive DNA with which we are already familiar because of its association with structural features of chromosomes. Telomere DNA, which in humans comprises hundreds of copies of the motif 5′-TTAGGG-3′ (see Figure 11.18), is an example of a minisatellite. In addition to telomere minisatellites, some eukaryotic genomes contain various other clusters of minisatellite DNA, many, although not all, near the ends of chromosomes. The functions of these other minisatellite sequences have not been identified.

Microsatellites are also examples of tandemly repeated DNA. In a microsatellite the repeat unit is short—up to 13 bp in length. Dinucleotide repeats are the commonest, with roughly 140,000 microsatellites of this type in the genome as a whole. About half of these are repeats of the motif 5′-CA-3′. Microsatellites with single-nucleotide repeats (e.g. AAAAA) are the next most common, with about 120,000 in total.

As with interspersed repeats, it is not clear whether microsatellites have a function. They are, however, very useful to geneticists. Many microsatellites are variable, meaning that the number of repeat units in the array is different in different members of a species. This is because "slippage" sometimes occurs when a microsatellite is copied during DNA replication, leading to the insertion or, less frequently, the deletion of one or more of the repeat units (see Figure 16.9). No two humans alive today, except for genetically identical twins, have exactly the same combination of microsatellite length variants. This means that if enough microsatellites are examined a unique **genetic profile** can be established for every person (Figure 19.12). Genetic profiling is well known as a tool in forensic science, but the identification of criminals is a fairly trivial application of microsatellite variability. More sophisticated methodology makes use of the fact that a person's genetic profile is inherited partly from the mother and partly from

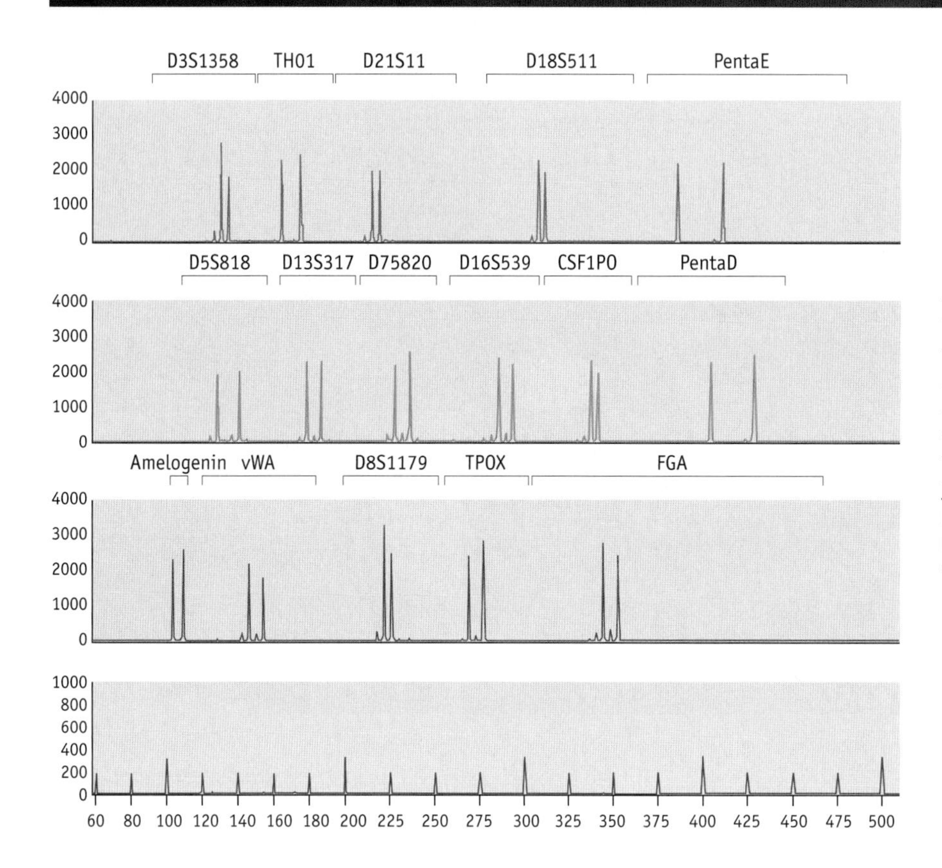

FIGURE 19.12 A genetic profile. The positions of the peaks in the upper three graphs indicate the length variants of 15 microsatellites in the genome of the person whose DNA has been tested. The microsatellites are called D3S1358, TH01, etc., as shown in the labels above the peaks. The two peaks labeled "Amelogenin" are not derived from microsatellites but from the amelogenin gene on the X and Y chromosomes. A single peak indicates that only the X chromosome is present and that the person is female. Two peaks, as in this case, indicate a male. The lower graph shows a series of size markers, used to calibrate the peaks in the profile. The *y*-axis is arbitrary units used for measuring peak height. We will examine how a genetic profile is obtained in Chapter 21. (Courtesy of Promega Corporation.)

the father. This means that microsatellites can be used to establish family relationships and population affinities, not only for humans but also for other animals, and for plants.

19.3 HOW DIFFERENT IS THE HUMAN GENOME FROM THAT OF OTHER ANIMALS?

Both philosophers and biologists have debated for centuries what it is that makes us human. Why are we, at least to our own eyes, a special type of species in possession of important biological attributes that make us very different from all other animals? For a geneticist, metaphysical explanations such as the possession of a soul are unsatisfactory, and the answer has to be sought in the genome. To address this question, we will start by examining our evolutionary position in the tree of life, to identify our closest relatives in the animal kingdom. We will then ask whether the human genome provides any clues about what makes us different from these animals.

Chimpanzees are our closest relatives in the animal kingdom

Darwin was the first biologist to speculate on the evolutionary relationships between humans and other animals. His view—that humans are closely related to the chimpanzee, gorilla, and orangutan—was controversial when it was first proposed, and it fell out of favor in the following decades, even among Darwin's supporters. Indeed, biologists at that time were among the most ardent advocates of a nonscientific interpretation of our place in the world. Realization that humans are indeed just another type of animal was driven largely by studies of the fossil record.

FIGURE 19.13 Representatives of the four genera of the Hominidae. (A) Two humans (*Homo sapiens*). (B) Chimpanzee (*Pan troglodytes*). (C) Gorilla (*Gorilla gorilla*). (D) Orangutan (*Pongo pygmaeus*). (A, courtesy of BLM/Oregon; C, courtesy of Richard Ruggiero, US Fish and Wildlife Service.)

In evolutionary terms, humans are looked on as **hominids**, part of the taxonomic family called the Hominidae. The living members of this family are humans, chimpanzees, gorillas, and orangutans (Figure 19.13). The exact relationships within the Hominidae have been worked out by a combination of paleontology and **molecular phylogenetics**. The latter uses the molecular clock to make comparisons between DNA sequences from different species to work out their evolutionary relationships. The tree resulting from these studies (Figure 19.14) reveals that chimpanzees are our closest relative. The split between the lineages leading to humans on the one hand, and chimpanzees on the other, is thought to have occurred about 4.6–5.0 million years ago, though some evolutionary biologists favor a slightly older date.

Since the split, the human lineage has embraced several genera and a number of species, not all of which were on the direct line of descent to *Homo sapiens* (Figure 19.15). As we follow this evolutionary pathway in the paleontological and archaeological records we see the first indications of our special human attributes gradually appearing. The ability to walk upright, as opposed to the knuckle-walking locomotion of chimpanzees, was first possessed by *Ardipithecus ramidus*, which lived in east Africa 4.4 million years ago. The first stone tools were manufactured about 2.5 million years ago by *Homo habilis*, the earliest member of our own genus. Humans just like us—members of *Homo sapiens*—first appeared in Africa about 195,000 years ago and gradually spread across the globe. We do not know when speech evolved, but humans started sculpting stone for artistic

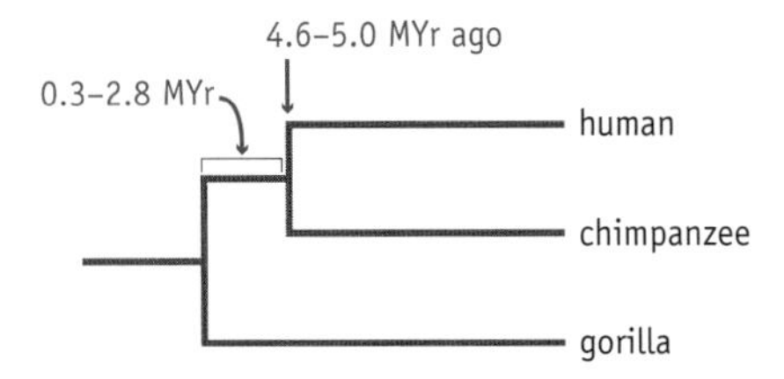

FIGURE 19.14 Phylogenetic tree showing the evolutionary relationships between humans, chimpanzees, and gorillas. The split between the lineages leading to humans and chimpanzees occurred 4.6–5.0 million years ago. The split between the human–chimp and gorilla lineages occurred 0.3–2.8 million years earlier. MYr, million years.

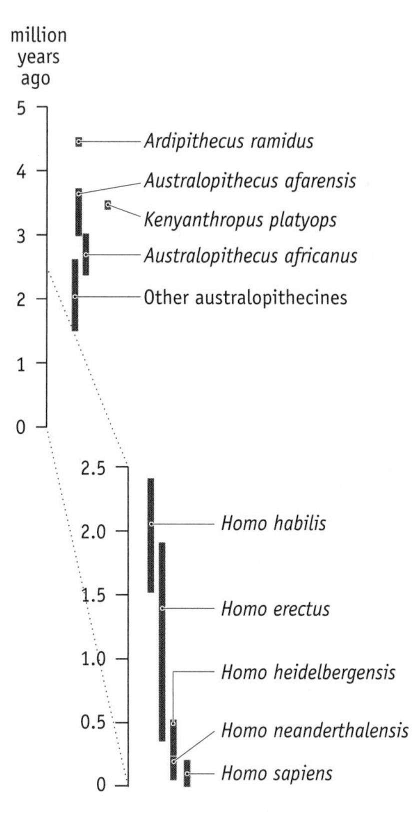

FIGURE 19.15 The periods spanned by the fossil remains thought to form the evolutionary lineage leading to our species, *Homo sapiens*.

rather than utilitarian purposes about 50,000 years ago, and at the same time might have been making music. What changes in the human genome accompanied our intellectual, cultural, and social evolution?

Comparisons between the human and chimpanzee genomes fail to reveal what makes us human

As far as our genomes are concerned, the difference between humans and chimpanzees is 1.73%, this being the extent of the nucleotide sequence dissimilarity between humans and chimpanzees. Indeed, when the human and chimpanzee genomes are compared it is much easier to find similarities than differences. The degree of sequence identity within the coding DNA is greater than 98.5%, with 29% of the genes in the human genome coding for proteins whose amino acid sequences are identical to the sequences of their counterparts in chimpanzees. Even in the noncoding regions of the genome the nucleotide identity is rarely less than 97%.

Gene order is almost the same in the two genomes, and the chromosomes have very similar appearances. At this level, the most marked difference is that human chromosome 2 is two separate chromosomes in chimpanzees (**Figure 19.16**), so chimpanzees have 24 pairs of chromosomes whereas humans have just 23 pairs. The alphoid DNA sequences present at human centromeres are quite different from the equivalent sequences in chimpanzee chromosomes, and Alu elements are more prevalent in the human genome. These features probably tell us more about the evolution of repetitive DNA than about the differences between humans and chimpanzees.

Comparisons of the human and chimpanzee genomes have also failed to reveal changes to individual genes that might somehow be key to the special attributes of humans. These analyses have revealed significant differences in genes associated with amino acid breakdown, in line with the greater proportion of meat eaten by humans than by chimpanzees, and in genes providing protection against human diseases such as tuberculosis and malaria. But no genes with clear roles in brain or neuronal development have been uncovered by this type of analysis. The only substantial difference in gene structure is that humans lack a 92-bp segment of the gene for the *N*-glycolyl-neuraminic acid hydroxylase. This means that we cannot synthesize the hydroxylated form of *N*-glycolyl-neuraminic acid, which is present on the surfaces of some chimpanzee cells. This may have an effect on the ability of certain pathogens to enter human cells, and could possibly influence some types of cell–cell interaction, but the difference is not thought to be particularly significant. Two other genes that are functional in chimpanzees seem to have been inactivated by point mutations in the human genome. One of these codes for a T-cell receptor and the other for a hair keratin protein, but neither of these changes is likely to have

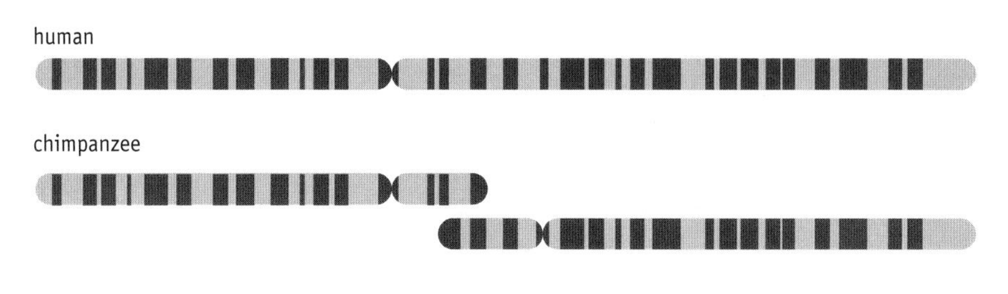

FIGURE 19.16 Human chromosome 2 is a fusion of two chromosomes that are separate in chimpanzees.

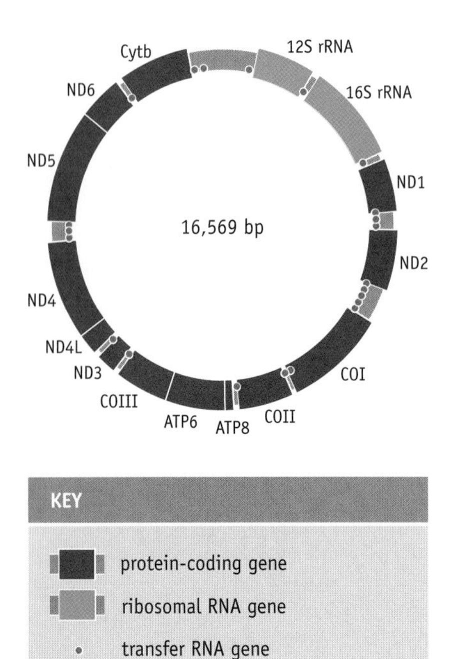

FIGURE 19.17 The human mitochondrial genome. ATP6, ATP8, genes for ATPase subunits 6 and 8; COI, COII, COIII, genes for cytochrome *c* oxidase subunits I, II, and III; Cytb, gene for apocytochrome *b*; ND1–ND6, genes for NADH hydrogenase subunits 1–6.

had any substantial impact. In any case, it seems impossible that the special features of humans could have arisen as a result of the loss of a gene function.

Instead we need to find genes whose activities have changed rather than been lost. In this regard, considerable interest has centered on the gene for the FOXP2 transcription factor. Defects in this protein result in the human disability called dysarthria, characterized by a difficulty in articulating speech. This gene might therefore underlie the human ability for language. There are indeed two amino acid differences between the FOXP2 proteins in humans and chimpanzees, but a direct link between these amino acid differences and human linguistic ability is elusive.

It is now becoming clear that many, if not all, of the key differences between humans and chimpanzees are likely to lie not with the genes themselves but with the way in which the genes are expressed. Attention is therefore moving from the genomes to the transcriptomes and proteomes. These studies are beginning to suggest that the pattern of gene expression in the brain has undergone significant change in the human lineage since the divergence between humans and chimpanzees. Of course, this is precisely what we might expect, as it is clearly our brains that distinguish us from chimpanzees and other animals. The key question, which has not yet been answered, is whether the identities of the genes that are up-regulated or down-regulated in the human brain are informative in any way.

19.4 THE MITOCHONDRIAL GENOME

The possibility that some genes might be located outside the nucleus—**extrachromosomal genes** as they were initially called—was first raised in the 1950s as a means of explaining the unusual inheritance patterns of certain genes in lower eukaryotes. Electron microscopy and biochemical studies at about the same time provided hints that DNA molecules might be present in mitochondria and chloroplasts. Eventually, in the early 1960s, these various lines of evidence were brought together and the existence of mitochondrial and chloroplast genomes, independent of and distinct from the eukaryotic nuclear genome, was accepted.

The human mitochondrial genome is packed full of genes

Each human mitochondrion contains about 10 identical copies of its genome, which means that there are about 8000 copies per cell. The genome is circular, 16,569 bp in length, and contains genes for 13 proteins, 2 rRNAs, and 22 tRNAs (Figure 19.17).

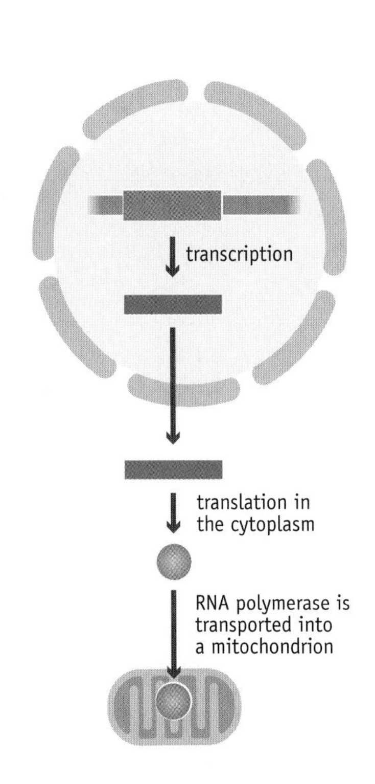

FIGURE 19.18 The mitochondrial RNA polymerase is coded by a nuclear gene. The mRNA is translated in the cytoplasm and the protein transported into the mitochondrion.

The mitochondrial DNA is transcribed by an RNA polymerase that is coded by a nuclear gene. This RNA polymerase is synthesized in the cytoplasm and then moves into the mitochondrion (Figure 19.18). The initial transcripts are long molecules that are cut into segments to give the mature mRNAs, rRNAs, and tRNAs. The rRNAs assemble with one another to make ribosomes that remain in the mitochondrion and, with the aid of the tRNAs, translate the mRNAs into protein. There are only 22 tRNAs, fewer than the minimum usually needed to read the genetic code. This means that the standard codon–anticodon pairing rules have to be supplemented by "superwobble," which allows some tRNAs to recognize all four codons of a single family (Figure 19.19). The genetic code that is used is slightly different from the one that operates in the rest of the cell (see Table 7.2).

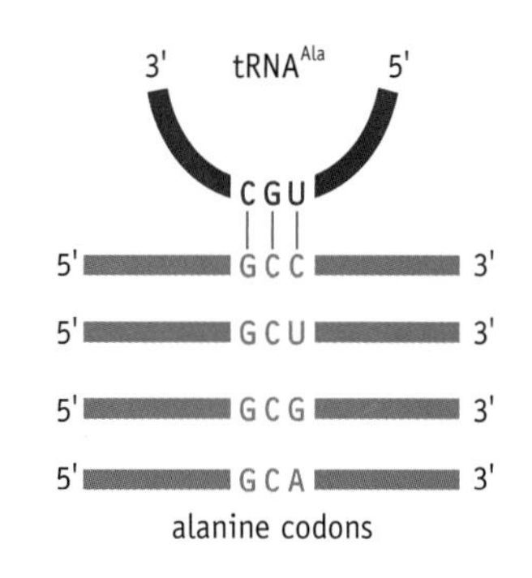

FIGURE 19.19 Superwobble enables a single mitochondrial tRNA to recognize all four codons in a four-codon family.

The various genes account for 15,368 bp of the mitochondrial DNA, with most of the remainder being taken up by the replication origin. Very little of the genome, only 87 bp or so, is genetically unimportant. In fact the human mitochondrial genome seems to have become as compact as possible. The rRNA genes are among the smallest known, coding for a large subunit rRNA with a sedimentation coefficient of only 16S and a small subunit RNA of 12S. Compare these figures with the typical values given in Figure 6.4. There are no intergenic regions, so the mRNAs just comprise the coding sequence of the gene. Several mRNAs are so truncated that the termination codons are incomplete. Five mRNAs end with just a U or UA, and the rest of the termination codon is provided by polyadenylation after transcription (Figure 19.20). The small number of tRNAs also seems to result from a need to keep the genome as small as possible. Why the genome has to be so small is not understood.

Mitochondrial genomes as a whole show great variability

Almost all eukaryotes have mitochondrial genomes. Initially, it was thought that virtually all of these were circular DNA molecules, like the human mitochondrial genome. We still believe this to be true for most organisms, but we now recognize that there is a great deal of variability. In many eukaryotes the circular molecules coexist with linear versions, and in some microbial eukaryotes (e.g. *Paramecium*, *Chlamydomonas*, and several yeasts) the mitochondrial genome is always linear.

Mitochondrial genomes vary in size (Table 19.4), the size unrelated to the complexity of the organism. Most multicellular animals have small mitochondrial genomes with a compact genetic organization, the genes being close together with little space between them. The human mitochondrial genome is typical of this type. In contrast, most lower eukaryotes such as

TABLE 19.4 SIZES OF MITOCHONDRIAL GENOMES

Species	Type of organism	Genome size (kb)
Plasmodium falciparum	Protozoan (malaria parasite)	6
Chlamydomonas reinhardtii	Green alga	16
Homo sapiens	Vertebrate (human)	17
Drosophila melanogaster	Insect (fruit fly)	19
Aspergillus nidulans	Fungus	33
Reclinomonas americana	Protozoan	69
Saccharomyces cerevisiae	Yeast	75
Brassica oleracea	Flowering plant (cabbage)	160
Arabidopsis thaliana	Flowering plant (vetch)	367
Zea mays	Flowering plant (maize)	570
Cucumis melo	Flowering plant (melon)	2500

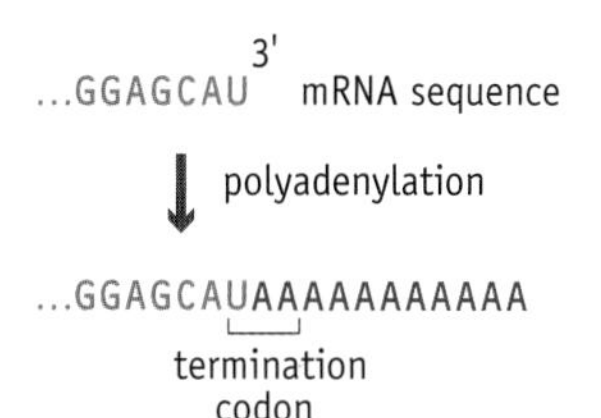

FIGURE 19.20 Some human mitochondrial mRNAs lack a complete termination codon. The codon sequence is completed by polyadenylation.

TABLE 19.5 GENE CONTENTS OF MITOCHONDRIAL GENOMES

Feature	*Plasmodium falciparum*	*Chlamydomonas reinhardtii*	*Homo sapiens*	*Saccharomyces cerevisiae*	*Arabidopsis thaliana*	*Reclinomonas americana*
Total number of genes	**5**	**12**	**37**	**35**	**52**	**92**
Types of gene						
Protein-coding genes	3	7	13	8	27	62
Respiratory complex	3	7	13	7	17	24
Ribosomal proteins	0	0	0	1	7	27
Transport proteins	0	0	0	0	3	6
RNA polymerase	0	0	0	0	0	4
Translation factor	0	0	0	0	0	1
Noncoding RNA genes	2	5	24	27	25	30
Ribosomal RNA genes	2	2	2	2	3	3
Transfer RNA genes	0	3	22	24	22	26
Other RNA genes	0	0	0	1	0	1

S. cerevisiae, as well as flowering plants, have larger and less compact mitochondrial genomes, with several of the genes containing introns.

Mitochondrial genomes show great variability in gene contents, ranging from 5 for the malaria parasite *Plasmodium falciparum* to 92 for the protozoan *Reclinomonas americana* (Table 19.5). All mitochondrial genomes contain genes for the rRNAs and at least some of the protein components of the respiratory chain, the latter being the main biochemical feature of the mitochondrion. The more gene-rich genomes also code for tRNAs, ribosomal proteins, RNA polymerase, and proteins involved in the transport of other proteins into the mitochondrion from the surrounding cytoplasm.

A general feature of mitochondrial genomes emerges from Table 19.5. These genomes specify some of the proteins found in mitochondria, but not all of them. The other proteins are coded by nuclear genes, synthesized in the cytoplasm, and transported into the mitochondria. If the cell has mechanisms for transporting proteins into mitochondria, then why not have all the mitochondrial proteins specified by the nuclear genome? We do not yet have a convincing answer to this question, although it has been suggested that at least some of the proteins coded by the mitochondrial genome are extremely hydrophobic and cannot be transported through the membranes that surround the mitochondrion, and so simply cannot be moved into the organelle from the cytoplasm. The only way in which the cell can get them into the mitochondrion is to make them there in the first place.

Mitochondria and chloroplasts were once free-living prokaryotes

The discovery of genomes in mitochondria and chloroplasts led to much speculation about their origins. Today most biologists accept that the **endosymbiont theory** is correct, at least in outline, even though it was considered quite unorthodox when first proposed in the 1960s.

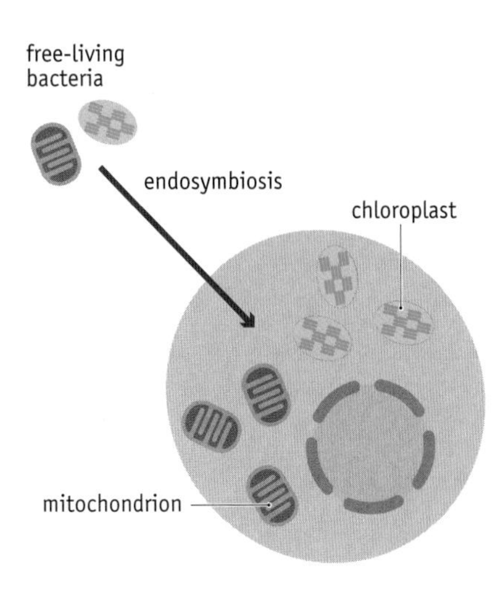

FIGURE 19.21 The endosymbiont theory. According to this theory, mitochondria and chloroplasts are the relics of free-living bacteria that formed a symbiotic association with the precursor of the eukaryotic cell.

The endosymbiont theory is based on the observation that the gene expression processes occurring in mitochondria and chloroplasts are similar in many respects to the equivalent processes in bacteria. In addition, when nucleotide sequences are compared, the genes in these organelles are found to be more similar to equivalent genes from bacteria than they are to eukaryotic nuclear genes. The endosymbiont theory therefore holds that mitochondria and chloroplasts are the relics of free-living bacteria that formed a symbiotic association with the precursor of the eukaryotic cell, way back at the very earliest stages of evolution (Figure 19.21).

Support for the endosymbiont theory has come from the discovery of organisms that seem to exhibit stages of endosymbiosis that are less advanced than those seen with mitochondria and chloroplasts. For example, an early stage in endosymbiosis is shown by the protozoan *Cyanophora paradoxa*, whose photosynthetic structures, called **cyanelles**, are different from chloroplasts and instead resemble ingested cyanobacteria (Figure 19.22A). Similarly, *Rickettsia* bacteria, which live inside eukaryotic cells, might be modern versions of the bacteria that gave rise to mitochondria (Figure 19.22B).

If mitochondria and chloroplasts were once free-living bacteria, then since the endosymbiosis was set up there must have been a transfer of genes from the organelle into the nucleus. We do not understand how this occurred, or indeed whether there was a mass transfer of many genes at once or just a gradual trickle from one site to the other. But we do know that DNA transfer from organelle to nucleus, and indeed between organelles, still occurs. This was discovered in the early 1980s, when the first partial sequences of chloroplast genomes were obtained. It was found that in some plants the chloroplast genome contains segments of DNA, often including entire genes, that are copies of parts of the mitochondrial genome. The implication is that this so-called **promiscuous DNA** has been transferred from one organelle to the other.

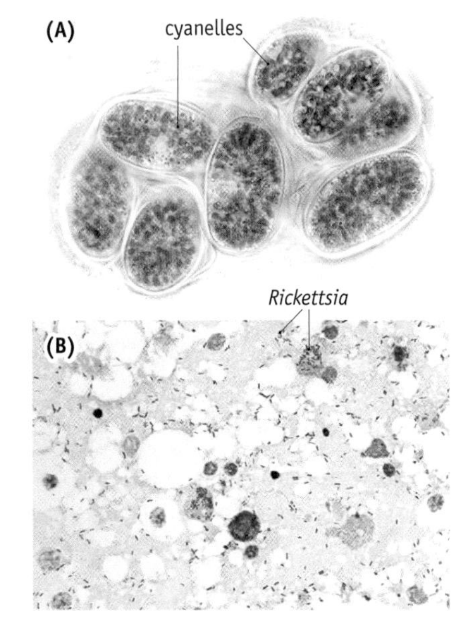

FIGURE 19.22 Possible living clues to the origins of chloroplasts and mitochondria. (A) Cyanelles in cells of the protozoan *Cyanophora*. (B) *Rickettsia* bacteria in a yolk sac smear, stained so that the bacteria show up as tiny pink rods. (A, courtesy of Michael Abbey/Science Photo Library; B, courtesy of CDC/Billie Ruth Bird.)

We now know that transfer of DNA between genomes has occurred in the evolutionary histories of many species. This point is illustrated by the plant *Arabidopsis thaliana*. Its mitochondrial genome contains various segments of nuclear DNA as well as 16 fragments of the chloroplast genome, including six tRNA genes that have retained their activity after transfer to the mitochondrion. The nuclear genome of *Arabidopsis* includes several short segments of the chloroplast and mitochondrial genomes and also a 270-kb piece of mitochondrial DNA located within the centromere region of chromosome 2. The transfer of mitochondrial DNA to vertebrate nuclear genomes has also been documented.

KEY CONCEPTS

- The human genome has about 20,500 genes but codes for 80,000–100,000 proteins. Because of alternative splicing, many genes code for more than one protein.
- All eukaryotes possess the same basic set of genes, but the more complex species have a greater number of genes for specialized functions.

- The genes are not spread evenly throughout the human genome. Regions close to the centromeres and to the ends of individual chromosomes have a lower gene density. Other than this, there is little obvious order or logic to the locations of individual genes.
- As a general rule, the more complex types of eukaryote have less compact genomes, with a greater number of introns, longer intergenic regions, and a greater proportion of repetitive DNA.
- Repetitive DNA can be divided into two categories. The first is genome-wide or interspersed repeats, whose individual repeat units are distributed around the genome in an apparently random fashion. The second category is tandemly repeated DNA, whose repeat units are placed next to each other in an array.
- Comparisons between the human and chimpanzee genomes have failed to reveal important differences in the sequences of individual genes. The special attributes of humans are more likely to be due to differences in gene expression patterns.
- Almost all eukaryotes have mitochondrial genomes, and photosynthetic species also have genomes in their chloroplasts.
- The endosymbiont theory for the origins of organelle genomes holds that mitochondria and chloroplasts are the relics of free-living bacteria that formed a symbiotic association with the precursor of the eukaryotic cell.

QUESTIONS AND PROBLEMS (Answers can be found at www.garlandscience.com/introgenetics)

Key Terms

Write short definitions of the following terms:

Alu
bacterial artificial chromosome
cyanelles
DNA marker
endosymbiont theory
extrachromosomal gene
fluorescent *in situ* hybridization
genetic profile
hominid
interspersed repeat
junk DNA
LINE
microsatellite
minisatellite
molecular phylogenetics
physical mapping
promiscuous DNA
restriction fragment length polymorphism
retroposons
satellite DNA
SINE
tandemly repeated DNA

Self-study Questions

19.1 Why is the human genome able to specify many more types of protein than are present in *Caenorhabditis elegans*, even though the two species have similar gene numbers?

19.2 Give examples that illustrate the different lengths of the genes present in the human genome.

19.3 Describe the differences between the organizations of the human growth hormone and aldolase gene families.

19.4 How are genes for the rRNA molecules arranged in the human genome?

19.5 Describe the key differences in gene organization in humans, yeast, and the fruit fly.

19.6 What is unusual about the repetitive DNA content of the maize genome?

19.7 Distinguish between interspersed and tandemly repeated DNA.

19.8 Describe the difference between a LINE and SINE.

19.9 What are the key features of the LINE-1 family of sequences in the human genome?

19.10 Explain how Alu sequences are thought to have arisen.

19.11 Describe how the buoyant density of a DNA molecule is measured.

19.12 Give examples of satellite and minisatellite DNA sequences in the human genome.

19.13 What are the special features of microsatellite DNA?

19.14 Outline the evolutionary relationships between humans and chimpanzees.

19.15 Describe the similarities and differences between the human and chimpanzee genomes, and explain why comparisons between these two genomes have failed to reveal what it is that makes us human.

19.16 Describe the important features of the human mitochondrial genome.

19.17 Outline how the human mitochondrial genome is similar to or different from the genomes of other organisms.

19.18 According to the endosymbiont theory, how did organelle genomes originate?

19.19 Explain how the features of *Cyanophora* and *Rickettsia* lend support to the endosymbiont theory.

19.20 Distinguish between the roles of genetic and physical methods in mapping the human genome.

19.21 Describe how the human genome was sequenced, distinguishing between the approaches of the official Human Genome Project and the project led by Craig Venter.

Discussion Topics

19.22 Devise a research project whose objective is to identify all parts of the human genome that are transcribed into RNA.

19.23 Discuss the theory that repetitive DNA is simply "junk."

19.24 Explain how the study of microsatellites might be used to establish whether a male and female adult and three children are all members of a single family.

19.25 How might the comparisons between the human and chimpanzee genomes be supplemented with other types of study in order to understand what makes us human?

19.26 How do we know that human chromosome 2 arose by fusion of two chromosomes that are separate in chimpanzees, rather than the two chimpanzee chromosomes arising from a single human chromosome that split into two?

19.27 What reasons might there be for having some proteins coded by genes in the mitochondrion and others coded in the nucleus?

FURTHER READING

Anderson S, Bankier AT, Barrell BG et al. (1981) Sequence and organization of the human mitochondrial genome. *Nature* 290, 457–465.

Clamp M, Fry B, Kamal M et al. (2007) Distinguishing protein-coding and noncoding genes in the human genome. *Proc. Natl. Acad. Sci. USA* 104, 19428–19433. *Shows that the human genome contains approximately 20,500 genes.*

Csink AK & Henikoff S (1998) Something from nothing: the evolution and utility of satellite repeats. *Trends Genet.* 14, 200–204.

IHGSC (International Human Genome Sequencing Consortium) (2001) Initial sequencing and analysis of the human genome. *Nature* 409, 860–921. *The draft sequence obtained by the "official" Human Genome Project.*

IHGSC (International Human Genome Sequencing Consortium) (2004) Finishing the euchromatic sequence of the human genome. *Nature* 431, 931–945. *The strategy used to obtain the final sequence.*

Lang BV, Gray MW & Burger G (1999) Mitochondrial genome evolution and the origin of eukaryotes. *Annu. Rev. Genet.* 33, 351–397.

Li WH & Saunders MA (2005) The chimpanzee and us. *Nature* 437, 50–51. *Describes the key differences between the human and chimpanzee genomes.*

Ostertag EM & Kazazian HH (2005) LINEs in mind. *Nature* 435, 890–891. *Brief review of recent research into LINEs.*

Tennyson CN, Klamut HJ & Worton RG (1995) The human dystrophin gene requires 16 hours to be transcribed and is cotranscriptionally spliced. *Nature* 9, 184–190.

Venter JC, Adams MD, Myers EW et al. (2001) The sequence of the human genome. *Science* 291, 1304–1351. *The draft sequence obtained by the shotgun approach.*

A great deal of biological research is directly or indirectly aimed at medical questions. This is just as true with genetics as it is in any other area of biology. Throughout this book we have touched on the relevance of different aspects of genetics to understanding human health and disease. In this chapter we will pull several of these threads together by considering some of the links between genetics and medicine. Our focus will be on the genetic basis of **inherited disease** and of cancer, and the genetic approaches, such as **gene therapy**, that are being explored as possible ways of curing these illnesses.

20.1 THE GENETIC BASIS OF INHERITED DISEASE

An inherited disease is one that is caused by a defect in the genome and that, like other genetic features, can be passed from parents to offspring. Sometimes these are called **genetic diseases** to emphasize that the underlying defect lies within a gene or other part of the genome. The term must, however, be used with caution because there are some genetic diseases, cancer being an example, that are not inherited.

There are more than 6000 inherited diseases that result from defects in individual genes. These are called **monogenic** disorders. About one in every 200 births are of a child with one or other of these disorders (Table 20.1). The commonest include the lung disease cystic fibrosis, Huntington's disease, which is a neurological disorder characterized by uncoordinated body movements, and Marfan syndrome, which affects connective tissue and leads to various problems including heart disease. Other diseases are much rarer, with just a few cases being reported worldwide, possibly affecting just a few families. Inherited diseases also affect other animals, in particular those subject to inbreeding such as some pedigree dogs.

TABLE 20.1 SOME OF THE COMMONEST MONOGENIC DISORDERS

Disease	Symptoms	Frequency (UK births per year)
Inherited breast cancer	Cancer	1 in 300 females
Cystic fibrosis	Lung disease	1 in 2000
Huntington's disease	Neurodegeneration	1 in 2000
Duchenne muscular dystrophy	Progressive muscle weakness	1 in 3000 males
Hemophilia A	Blood disorder	1 in 4000 males
Sickle-cell anemia	Blood disorder	1 in 10,000
Phenylketonuria	Mental retardation	1 in 12,000
β-Thalassemia	Blood disorder	1 in 20,000
Retinoblastoma	Cancer of the eye	1 in 20,000
Hemophilia B	Blood disorder	1 in 25,000 males
Tay-Sachs disease	Blindness, loss of motor control	1 in 200,000

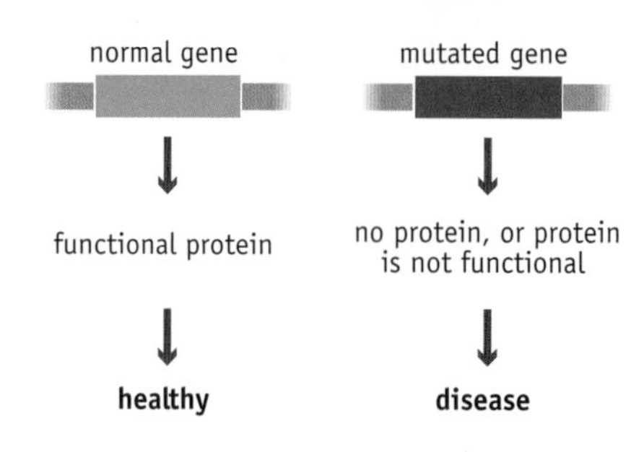

FIGURE 20.1 The genetic basis of inherited disease.

Many inherited diseases are due to loss-of-function mutations

In Chapter 16 we learnt that mutations in a multicellular organism can result in either a loss of function or a gain of function, with the former being much more common. This is as true with inherited diseases as with any other phenotype, and the vast majority of human monogenic disorders are due to a loss-of-function mutation that inactivates a particular gene. The protein coded by that gene is absent, or modified in some way that prevents it from functioning correctly, and the resulting defect is manifested as the inherited disease (Figure 20.1).

A loss of function leading to an inherited disease can arise in a number of different ways (Figure 20.2). The loss of function could be caused by a point mutation, such as a nonsense mutation that creates an internal termination codon within a gene. It could equally well be caused by a small deletion or insertion within the gene, or by a large deletion that removes the entire gene. The mutation might also be in a regulatory sequence upstream of the gene so that no transcription initiation occurs, or within one of the internal sequences that control splicing.

A single disorder is often caused by more than one mutation, with different individuals carrying different mutations but still having the same disease. Cystic fibrosis is a good example. This disease, which affects approximately 8000 people in the UK and 30,000 in the USA, can be caused by any one of 1400 different mutations. Each of these mutations occurred in a sex cell of a person who lived at some period in the past, and has now been passed down to living descendants of that person. The commonest of these mutations is the one called ΔF508 (Table 20.2). The notation indicates that the mutation results in deletion (Δ) of a phenylalanine (F) from position 508 of the protein coded by the cystic fibrosis gene (Figure 20.3A). This protein is the cystic fibrosis transmembrane regulator, which is involved in the transport of chloride ions into and out of the cell. The ΔF508 mutation prevents the protein from attaching to the cell membrane, and this loss of function leads to inflammation and the accumulation of mucus in the lungs, the primary symptoms of cystic fibrosis.

The next most common cystic fibrosis mutation, G542X, acts in a completely different way. This is a point mutation that converts the codon for a glycine (G) normally present at position 542 in the protein into a termination codon (X) (Figure 20.3B). The presence of this internal termination

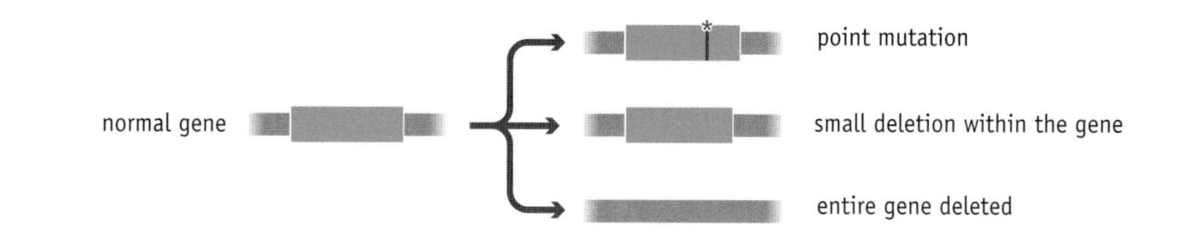

FIGURE 20.2 Three ways in which a loss of function leading to an inherited disease can arise.

TABLE 20.2 THE COMMONEST CYSTIC FIBROSIS MUTATIONS

Mutation	Frequency in cystic fibrosis patients	Effect on protein structure	Effect on protein function
ΔF508	68%	Deletion of phenylalanine	Does not attach to cell membrane
G542X	2.5%	Not synthesized	No protein
G551D	1.5%	Replacement of glycine with aspartic acid	Low rate of chloride transport

codon in the mRNA transcribed from the mutant *CFTR* gene switches on the surveillance system that degrades mutated mRNAs. This means that the CFTR protein is not synthesized, again resulting in the loss of function.

The third most common mutation, G551D, has a different effect again. Replacing glycine 551 with aspartic acid does not prevent the protein from being synthesized or from folding correctly, but changes its kinetic properties so it now transports chloride ions at only 4% of the rate of the unmutated protein (Figure 20.3C). Again, this results in a loss of function. Although the protein is still able to transport chloride ions, it does so at such a slow rate that the physiological function of the protein is lost.

Inherited diseases resulting from a gain of function are much less common

Gain of function is much less common among inherited diseases, simply because there are relatively few types of underlying mutation that can cause a gain of function. One possibility is overexpression of a gene, so that its protein product accumulates in the cell in greater than normal quantities. Overexpression could be due to a mutation in one of the regulatory sequences that control transcription initiation, but more frequently it arises from gene duplication. This is the cause of Charcot–Marie–Tooth (CMT) disease, one of the commonest inherited neurological disorders, affecting the peripheral nerves and muscle strength so that actions such as walking are affected. The increase from two to three copies of the myelin protein

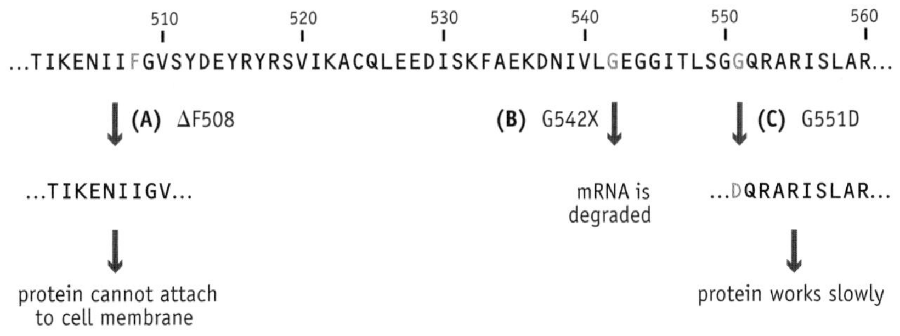

510 520 530 540 550 560

...TIKENIIFGVSYDEYRYRSVIKACQLEEDISKFAEKDNIVLGEGGITLSGGQRARISLAR...

(A) ΔF508
(B) G542X
(C) G551D

...TIKENIIGV...
mRNA is degraded
...DQRARISLAR...

protein cannot attach to cell membrane
protein works slowly

FIGURE 20.3 The three commonest cystic fibrosis mutations. The amino acid sequence of the CFTR protein from positions 501 to 560 is shown. (A) The ΔF508 mutation results in deletion of the phenylalanine at position 508, and gives rise to a protein that cannot attach to the cell membrane. (B) The G542X mutation replaces the codon for G542 with a termination codon. The mRNA transcribed from this mutant gene is degraded before any protein is made. (C) The G551D mutation replaces the glycine at position 551 with an aspartic acid, giving rise to a protein with only 4% of the activity of the unmutated version.

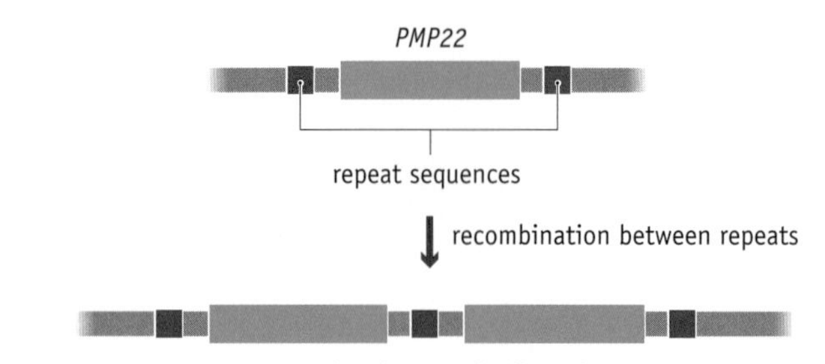

FIGURE 20.4 Duplication of the *PMP22* gene.

gene *PMP22* is sufficient, in some unknown way, to cause the symptoms of this disease. Duplication of *PMP22* arises by recombination between repeat sequences on each side of the 1.5-Mb region of chromosome 17 that contains this gene (Figure 20.4). The recombination can also lead to the deletion of *PMP22* on one of the chromosome copies, reducing the gene number from two to one. This leads to a *loss* of function, manifested as a completely different disease called tomaculous neuropathy, again affecting the peripheral nerves but in this case causing numbness and unusual sensitivity to pressure.

A second way of acquiring a gain of function is by the mutation of a protein that acts as a cell surface receptor and that relays a message into the cell when it binds an external signaling compound, such as a hormone (Figure 20.5). The signal passed on by the surface receptor might have various effects inside the cell, including changes in gene expression patterns. A normal cell will respond to the external stimulus in an appropriate way, but some mutations can affect a cell surface receptor so that it remains active even when its external signaling compound is absent. The cell therefore acts as though it were continuously being affected by the signaling compound, resulting in the gain in function. Early onset of male puberty can be caused in this way by mutation of the luteinizing hormone receptor, and the very rare Jansen's disease, with only 20 known cases in the entire world, is due to this kind of mutation in the parathyroid hormone receptor, leading to growth defects.

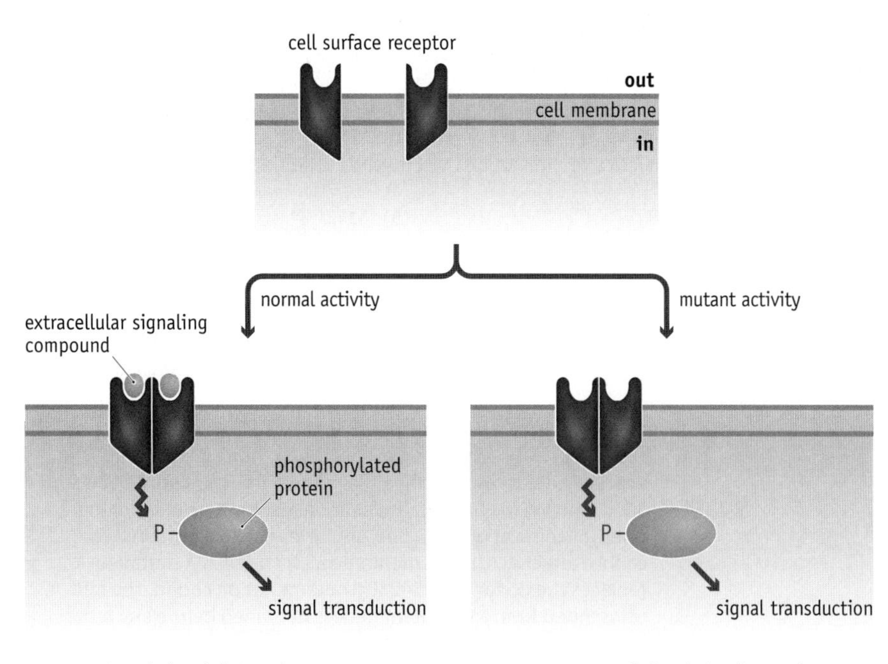

FIGURE 20.5 A mutation in the gene for a receptor protein will lead to a gain of function if the mutated receptor sends its signal into the cell in the absence of the extracellular signaling compound.

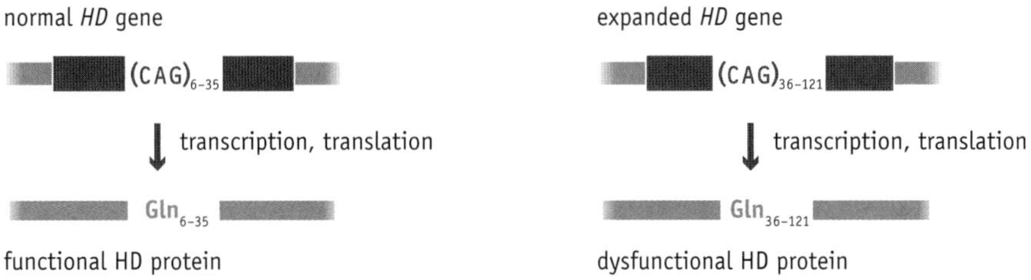

normal *HD* gene

(CAG)$_{6-35}$

transcription, translation

Gln$_{6-35}$

functional HD protein

expanded *HD* gene

(CAG)$_{36-121}$

transcription, translation

Gln$_{36-121}$

dysfunctional HD protein

FIGURE 20.6 The genetic basis of Huntington's disease.

Although rare among inherited diseases, gain-of-function mutations are relatively common causes of cancer. We will therefore return to them later in the chapter.

Trinucleotide repeat expansions can cause inherited disease

One particular type of DNA mutation that often leads to inherited disease deserves a special mention. This is trinucleotide repeat expansion, in which a relatively short series of trinucleotide repeats becomes elongated to two or more times its normal length. These expansions are associated with several neurodegenerative diseases, including Huntington's disease (HD), one of the commonest inherited disorders (see Table 20.1). The normal *HD* gene contains the sequence 5′-CAG-3′ repeated between 6 and 35 times in tandem, coding for a series of glutamines in the protein product (**Figure 20.6**). In Huntington's disease this repeat expands to a copy number of 36–121, increasing the length of the polyglutamine tract and resulting in a loss of function of the HD protein. The exact biochemical role of this protein has not yet been discovered but it is thought to act within the nerve cells of the brain.

Several other inherited diseases are also caused by expansions of polyglutamine codons (**Table 20.3**). There are also two examples of diseases resulting from expansion of trinucleotides that lie outside the coding region of a gene. In Friedreich's ataxia, a 5′-GAA-3′ expansion within the

TABLE 20.3 EXAMPLES OF HUMAN TRINUCLEOTIDE REPEAT EXPANSIONS

Gene	Repeat sequence		Associated disease
	Normal	Mutated	
Polyglutamine expansions (within the coding regions of genes)			
HD	$(CAG)_{6-35}$	$(CAG)_{36-121}$	Huntington's disease
AR	$(CAG)_{9-36}$	$(CAG)_{38-62}$	Spinal and bulbar muscular atrophy
DRPLA	$(CAG)_{6-35}$	$(CAG)_{49-88}$	Dentatorubral–pallidoluysian atrophy
SCA1	$(CAG)_{6-44}$	$(CAG)_{39-82}$	Spinocerebellar ataxia type 1
SCA3	$(CAG)_{12-40}$	$(CAG)_{55-84}$	Machado–Joseph disease
Other expansions (outside the coding regions of genes)			
X25	$(GAA)_{7-34}$	$(GAA)_{34-\text{over } 200}$	Friedreich's ataxia
DMPK	$(CTG)_{5-37}$	$(CTG)_{50-3000}$	Myotonic dystrophy
EPM1	$(CCCCGCCCCGCG)_{2-3}$	$(CCCCGCCCCGCG)_{\text{over } 12}$	Progressive myoclonus epilepsy

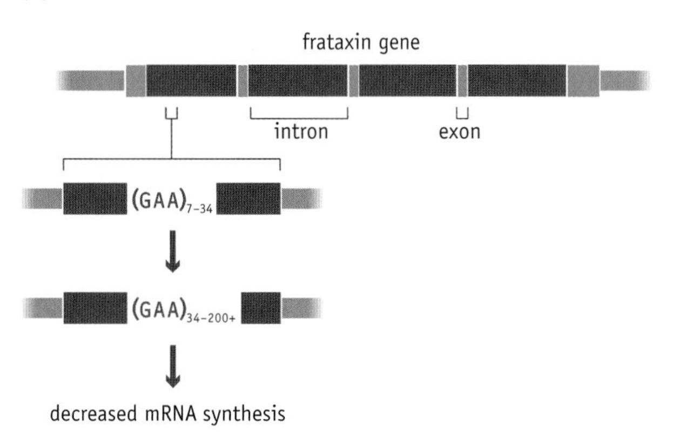

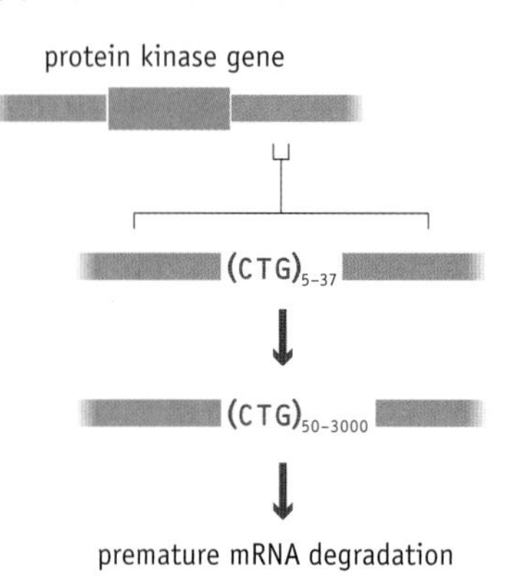

FIGURE 20.7 The genetic basis of (A) Friedreich's ataxia and (B) myotonic dystrophy.

first intron of the frataxin gene decreases synthesis of the mature mRNA (Figure 20.7). This results in a loss of function that leads to a decrease in the number of nerve cells in the spinal cord. Similarly, a 5′-CTG-3′ expansion in the trailer region of the myotonic dystrophy protein kinase gene leads to premature degradation of the mRNA, resulting in muscle wasting.

A few disease-causing mutations involve expansions of longer sequences. An example is progressive myoclonus epilepsy. This is caused by the copy number of the sequence 5′-CCCCGCCCCGCG-3′ being increased from 2–3 to more than 12. This sequence lies in the promoter region of the cystatin B gene, the product of which is involved in regulating the activities of proteases within human cells.

Trinucleotide expansions are generated by errors in DNA replication, but exactly how this happens is not understood. Once an expansion reaches a certain length it seems to become susceptible to further expansion in subsequent rounds of DNA replication, so that the disease becomes increasingly severe in succeeding generations of an affected family. Studies of similar trinucleotide expansions in yeast have shown that these are more prevalent when the *RAD27* gene is inactivated. This is an interesting observation, because *RAD27* is the yeast version of the mammalian gene for FEN1, the protein involved in the processing of Okazaki fragments. The implication is that a trinucleotide repeat expansion is caused by an aberration in lagging-strand synthesis during DNA replication.

There are dominant and recessive inherited diseases

The mutated version of a gene that leads to an inherited disease is an allele, and like all other alleles it will have a dominant–recessive relationship with the other alleles of the gene. Those inherited diseases that are dominant are displayed by individuals who either are homozygous for the disease allele or are heterozygotes with one disease allele and one normal allele (Figure 20.8A). At least one of the parents of an affected child must also have the disease, which in turn tells us that the disease, however unfortunate its symptoms, is not so severe as to prevent at least some afflicted individuals from reaching reproductive age and having children. It is therefore not surprising that many dominant inherited diseases either have incomplete penetrance, so not all individuals with a disease genotype actually develop the symptoms, or have a delayed onset and so do not affect the person until later in their life. Examples are Huntington's disease and Marfan syndrome.

(A) dominant inherited disease

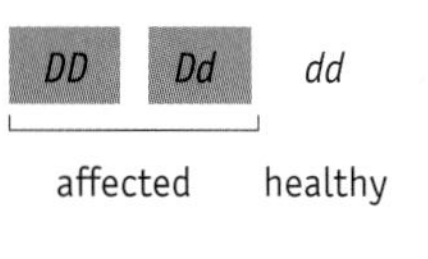

affected healthy

(B) recessive inherited disease

DD Dd dd

healthy affected

FIGURE 20.8 Genotypes resulting in disease for (A) a dominant and (B) a recessive autosomal inherited disease.

If the disease allele is recessive, as is the case for cystic fibrosis, then only homozygotes will have the disease (Figure 20.8B). With recessive diseases it is therefore quite possible for an affected child to have two nonaffected parents. Both parents would be heterozygotes, **carriers** who have one normal and one disease allele. Recessive diseases are often more severe in their effects than dominant ones and it is quite possible that most affected people die before being able to have their own children. The disease persists in the population because its allele survives in the pool of unaffected carriers.

In the examples shown in Figure 20.8, the disease gene lies on one of the autosomes. The inheritance pattern will be different if the gene is on one of the sex chromosomes. If the gene is on the X chromosome then it will always be expressed by males who possess the defective allele. This is because males have just one copy of the X chromosome, making it immaterial whether the disease allele is dominant or recessive (Figure 20.9). If the disease is dominant then all females with a disease allele will also be affected, but if it is recessive then the disease will be expressed only by homozygous females. X-linked dominant diseases are in fact quite rare. Most X-linked diseases are recessive, the best known being hemophilia A. This disease gained notoriety in the late nineteenth and early twentieth centuries because it affected some of the descendants of Queen Victoria, leading to various succession crises among the ruling families of Europe

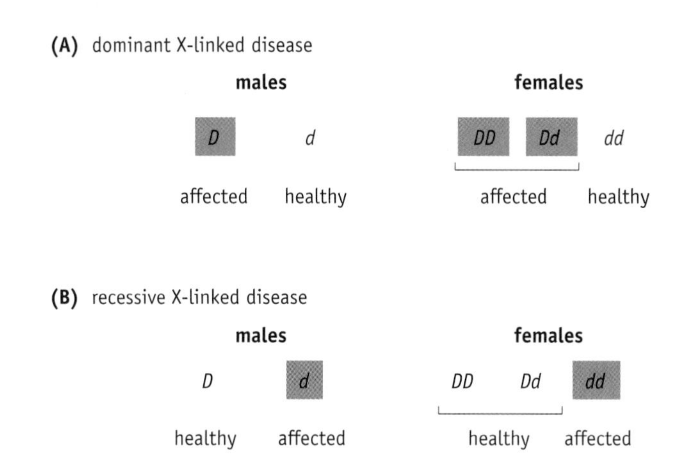

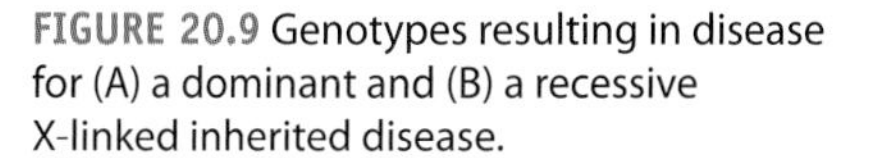
FIGURE 20.9 Genotypes resulting in disease for (A) a dominant and (B) a recessive X-linked inherited disease.

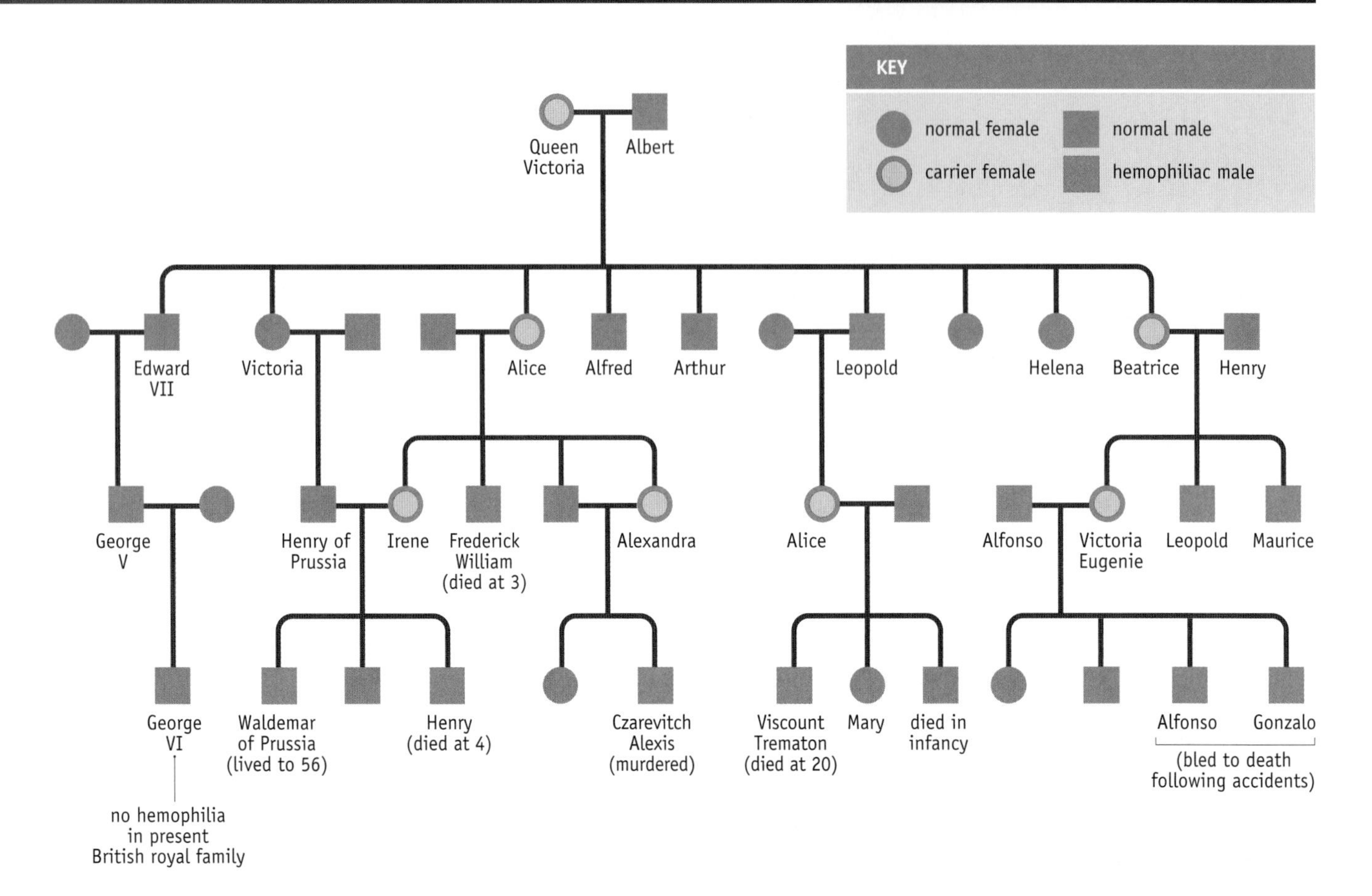

FIGURE 20.10 Partial pedigree of the descendants of Queen Victoria, indicating those females who were carriers for hemophilia, and those males who had the disease.

as male heirs with the disease died young (Figure 20.10). Y-linked diseases are uncommon because the Y chromosome has very few genes—fewer than 100—so the scope for mutation is limited. Most Y-linked diseases affect fertility and all, of course, are restricted to males.

The dominant–recessive relationship can tell us something about the basis of a disease

Understanding the basis of inherited disease is clearly important, so how can dominance and recessiveness help us in this regard? A simple principle is that a disease that results from a gain-of-function mutation is likely to be dominant, and one due to a loss-of-function mutation is likely to be recessive. The rationale behind this statement is that in a heterozygote a new activity resulting from a gain-of-function mutation is likely to mask the effect of the normal allele, and so the gain of function is dominant. In contrast, a loss of function can often be compensated for by the presence of the single unmutated allele, so a loss of function is recessive.

Exceptions to this principle are likely to be interesting for one reason or another. For example, many of the diseases that result from trinucleotide expansions within the coding region of a gene give rise to a loss of function, but the expanded alleles are dominant and the normal alleles are recessive. The reason for this is not known, but it might relate to the tendency of the abnormal protein products coded by the expanded genes to form insoluble aggregates within nerve cells (Figure 20.11). Dominance would be explained if these aggregates have any role in manifestation of the disease, because the presence of the normal allele would not prevent or mask the pathological effect of the aggregate.

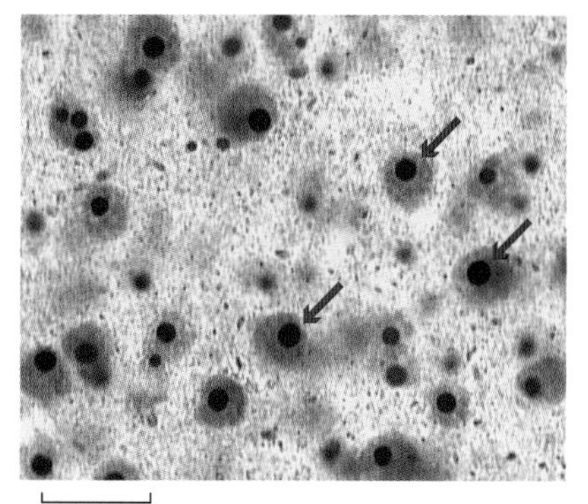

FIGURE 20.11 Insoluble aggregates of the huntingtin protein seen as brown nuclear inclusions in neurons from the brain of a mouse with the murine equivalent of Huntington's disease. (Courtesy of Dr Vanita Chopra, Massachusetts General Hospital.)

A few other loss-of-function mutations are neither dominant nor recessive, but instead display haploinsufficiency. This is the situation in which the phenotype displayed by the heterozygote is affected by the roughly 50% reduction in protein activity caused by the presence of the mutated allele (Figure 20.12A). This is not a common situation with inherited diseases because the single normal allele possessed by a heterozygote is usually able to direct the synthesis of enough protein to maintain the healthy state. A gene could display haploinsufficiency simply because large amounts of its protein are needed in the cell and a single allele cannot satisfy this need, but diseases of this type are uncommon. Supravalvular aortic stenosis, an inherited narrowing of the aorta brought about by mutation of the gene for the connective tissue protein elastin, is thought to be an example.

Most other examples of haploinsufficiency are more subtle and involve a gene whose protein product must be present in the cell in a carefully controlled amount, possibly because it is a signaling protein whose oversynthesis would disrupt its signaling pathway (Figure 20.12B). As the protein is normally synthesized at just the correct level, a 50% reduction would cause a significant loss of activity. This is probably the explanation of the haploinsufficiency displayed in Alagille syndrome, which causes heart and liver problems. The syndrome results from mutation of the gene for the CD339 protein, leading to disruption of a cell-to-cell signaling system that has an important role in embryo development.

Inherited diseases can also be caused by large deletions and chromosome abnormalities

Not all inherited diseases result from the mutation of a single gene. Many have more complex origins that involve, for example, the deletion of regions of the genome containing several genes, or the duplication of a chromosome.

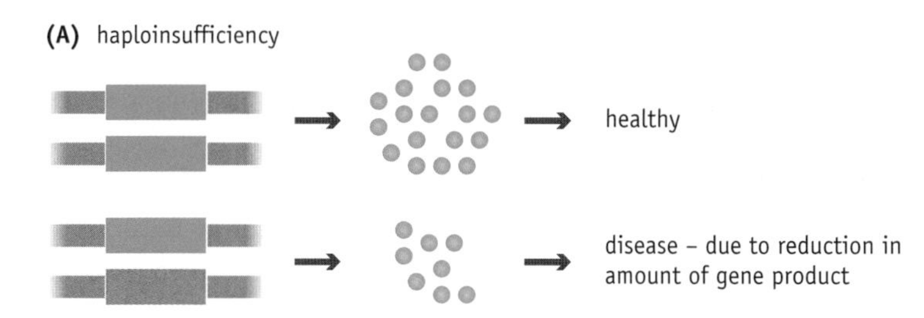

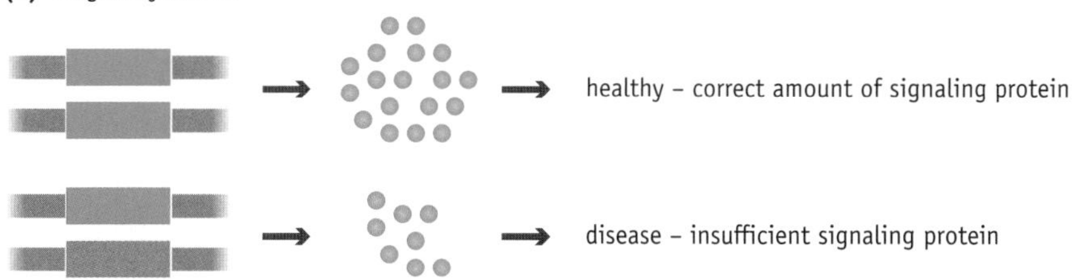

FIGURE 20.12 Haploinsufficiency and its role in Alagille syndrome.

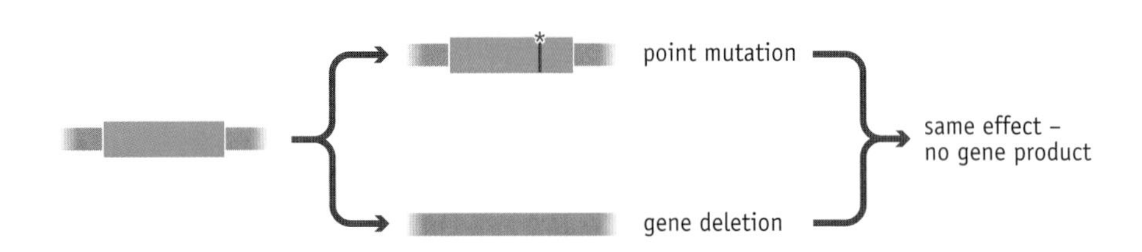

FIGURE 20.13 A point mutation and a gene deletion can give rise to the same inherited disease.

On average, there are 6.5 human genes per megabase of the genome, so deletions of more than 1 Mb are likely to result in the loss of at least a few genes. Often this will be so disastrous that the fertilized egg will not develop and the deletion will never appear in the population. On occasions, however, deletions of a few megabases can be tolerated, although associated with some kind of genetic disease. Many of these involve the deletion of a single important gene, possibly along with a few adjacent genes whose loss can be tolerated. Often, such a deletion will give rise to symptoms indistinguishable from those resulting from a point mutation within the disease gene (Figure 20.13). Several of the diseases mentioned so far in this chapter are known to arise both by simple mutation and by gene deletion. This is the case with Alagille syndrome, one of the examples of haploinsufficiency, which results from a deletion of the CD339 gene in 93% of cases and a point mutation in the same gene in the other 7%.

Occasionally a deletion removes a segment of the genome containing two or more important genes that are individually associated with inherited disease. The unfortunate person will then display a range of symptoms characteristic of different disorders. There has been one recorded case where the deletion of a large segment of the X chromosome led to a combination of retinitis pigmentosa, chronic granulomatous disease, and Duchenne muscular dystrophy, the genes for these disorders occupying adjacent positions on the chromosome (Figure 20.14). Such multiple conditions are exceedingly rare.

More common are genetic diseases arising from the duplication of a chromosome, resulting in a cell that contains three copies of one chromosome and two copies of all the others. This condition is called **trisomy** and is probably lethal for most chromosomes, because for several it is never seen. Trisomy of chromosome 21 or chromosome 18 is, however, relatively common, resulting in, respectively, Down syndrome (about 1 in 800 births) and Edwards syndrome (1 in 6000 births). Probably, the resulting increase in copy number for the genes on the duplicated chromosome leads to an imbalance of their gene products and disruption of the cellular biochemistry, giving rise to the developmental abnormalities that characterize trisomy diseases.

Trisomies are genetic but not inherited diseases. They result from aberrations during the meiosis that produces the affected reproductive cell and occur spontaneously in the population as a whole (Figure 20.15). The same is true of the disorders resulting from abnormal numbers of the sex chromosomes. These include Klinefelter's syndrome, which affects males with an extra X chromosome (XXY), and the self-explanatory XYY syndrome.

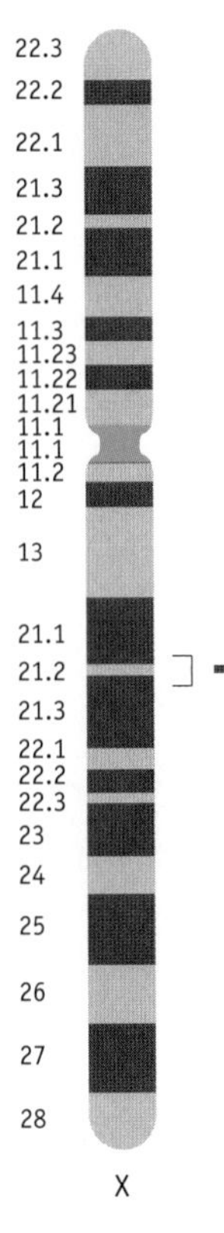

FIGURE 20.14 Location of the single deletion that can give rise to a combination of retinitis pigmentosa, chronic granulomatous disease, and Duchenne muscular dystrophy.

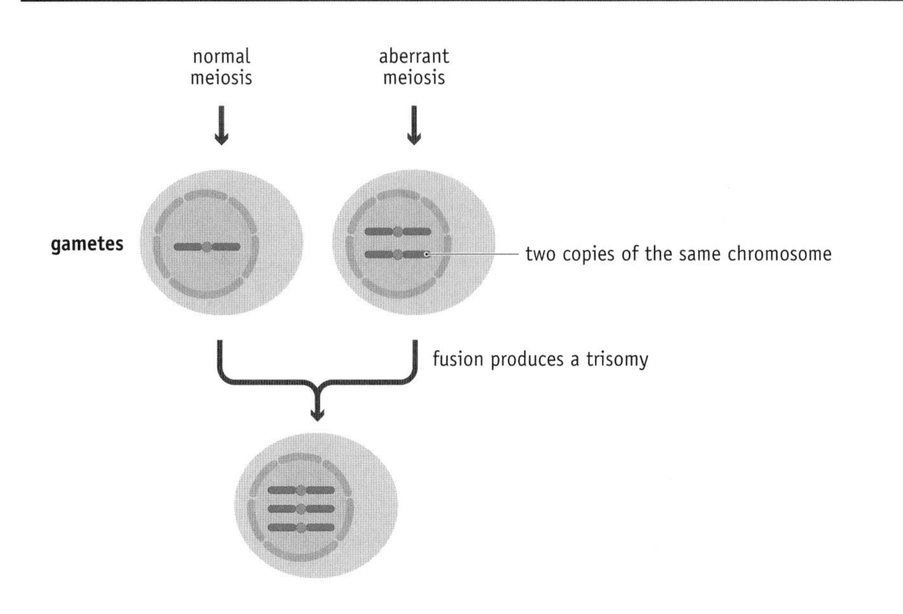

FIGURE 20.15 Production of trisomy by the fusion of one normal and one abnormal gamete.

20.2 THE GENETIC BASIS OF CANCER

Cancer is a group of diseases characterized by uncontrolled division of a somatic cell. Cancer is a *genetic* disease in the sense that the underlying defects are gene mutations and the inheritance of certain mutations gives a predisposition to cancer. But cancer is not an *inherited* disease, because the cancerous state is not directly passed on from parent to offspring.

A full appreciation of cancer requires an understanding not only of its genetic basis but also aspects of cell biology, biochemistry, physiology, and developmental biology, as well as a consideration of the myriad different types of cancer that are known to occur. Of necessity, therefore, our discussion of cancer will be limited in extent. We will consider just two central questions, concerning the nature of the genetic changes that can give rise to cancer and the multistep model for the development of a cancer.

Cancers begin with the activation of proto-oncogenes

One of the most important breakthroughs in understanding cancer came in the 1960s when the acute transforming viruses were discovered. As we learnt in Chapter 13, these are retroviruses that carry a copy of a human gene and which when they infect a cell, can transform that cell into the cancerous state. It was realized that this type of transformation is brought about by the uncontrolled expression of the gene carried by the retrovirus, which overrides the strictly controlled expression pattern of the cellular version of the gene (Figure 20.16). The transformation process is called oncogenesis, and the gene carried by the virus is called an **oncogene**.

A second breakthrough was made in the early 1980s when the normal cellular function of a viral oncogene was identified for the first time. It was

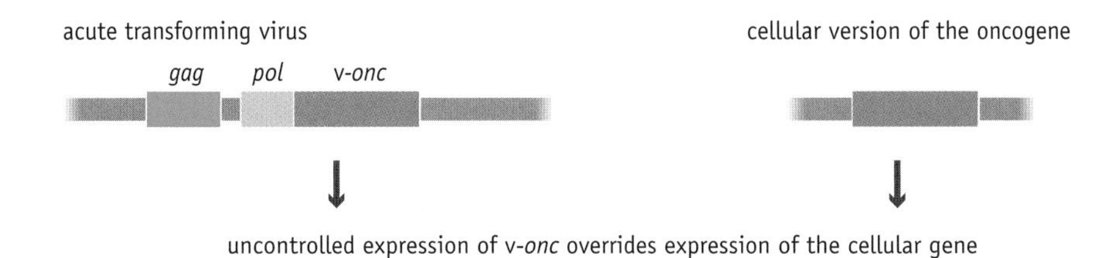

FIGURE 20.16 Uncontrolled expression of a viral oncogene (v-*onc*) can override the controlled expression of the equivalent cellular gene.

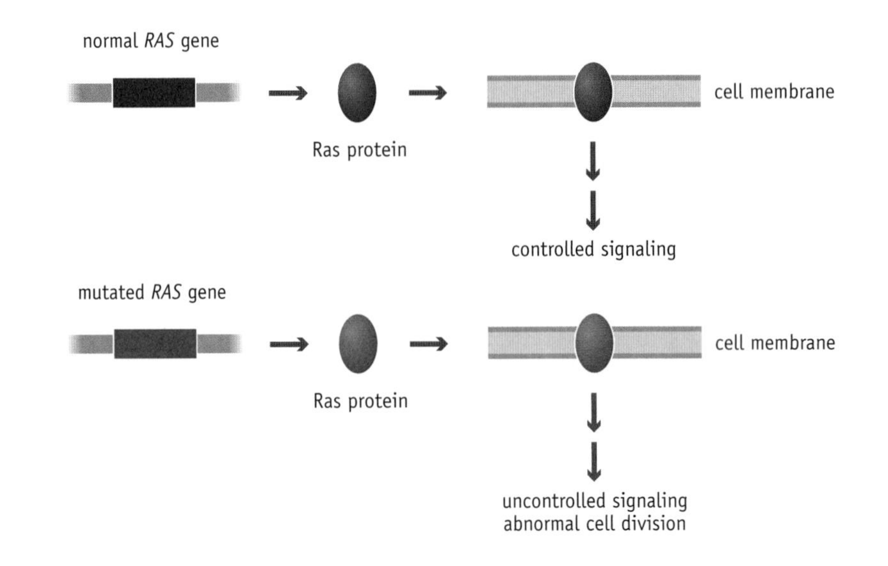

FIGURE 20.17 The *RAS* proto-oncogene. The normal Ras protein transmits signals into the cell in a controlled manner. "Activation" of the *RAS* proto-oncogene, for example by point mutation, gives rise to an abnormal Ras protein whose signaling is no longer controlled.

discovered that the cellular version of the v-*sis* oncogene is the gene for the platelet-derived growth factor B protein, which has a central role in the control of cell growth and division. Other viral oncogenes were subsequently found to be versions of cellular genes involved in activities such as intracellular signaling, regulation of transcription, and control of the cell cycle.

The terminology we now use is to refer to the normal cellular versions of these genes as **proto-oncogenes**. In its normal state, a proto-oncogene is not harmful to the cell—quite the reverse—but activation of a proto-oncogene converts it into the oncogene capable of initiating a cancer. One way of activating a proto-oncogene is, as we have seen, to place it in a retrovirus genome where its normal expression pattern is lost and the gene becomes overexpressed.

Activation can, however, occur without the intervention of the retrovirus. Point mutations are common activating factors for proto-oncogenes such as the *RAS* genes, which code for cell surface receptors that transmit external signals into the cell. Typically, the point mutation results in the receptor sending signals into the cell when it should not, one possible consequence being that the cell divides even though there has been no external stimulus for it to do so (Figure 20.17).

Activation can also occur by duplication and further increases in the copy number of the proto-oncogene, possibly resulting from a DNA replication error, so that greater than normal amounts of protein product are synthesized. Many breast cancers derive from amplification of the *ERBB2* gene, which codes for a cell surface receptor. In the activated state, multiple copies of *ERBB2* are found side by side on chromosome 17 (Figure 20.18).

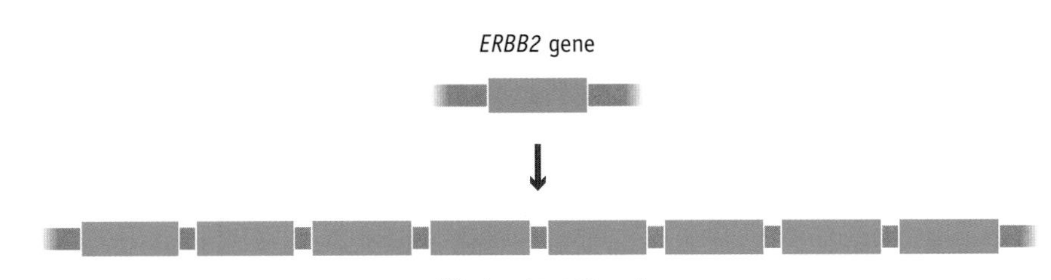

FIGURE 20.18 Amplification of the *ERBB2* gene, one of the causes of breast cancer.

Chromosome translocation, which results in one segment of DNA being exchanged between chromosomes, can also activate a proto-oncogene. For reasons that are unknown, some translocations are relatively common, including one between human chromosomes 9 and 22, resulting in the abnormal product called the **Philadelphia chromosome**. The breakpoint in chromosome 9 lies within the *ABL* gene, which codes for a tyrosine kinase enzyme that participates in intracellular signaling by attaching phosphate groups to target proteins. In the Philadelphia chromosome, the *ABL* segment is fused with part of a second gene normally resident on chromosome 22 (Figure 20.19). This gene is called the "breakpoint cluster region" because its precise function is unknown, although it also seems to be involved in protein phosphorylation. The product of the hybrid gene created by the translocation resembles the *ABL* tyrosine kinase but has altered properties that result in cell transformation.

Note that the above activation processes all lead to a gain of function. They are therefore dominant and need affect only one of a pair of alleles for cell transformation to occur.

Tumor suppressor genes inhibit cell transformation

The **tumor suppressors** are a second type of gene that, when defective, can contribute to the development of a cancer. This class of genes was first discovered as a result of studies of retinoblastoma, an unpleasant cancer of the eye that affects children and can lead to blindness. Sporadic retinoblastoma occurs occasionally but rarely throughout the population, in contrast to familial retinoblastoma, which is much more prevalent but is only found in certain families. The members of these families clearly have an inherited predisposition to this type of cancer.

The "two-hit" model was devised to explain the difference between sporadic and familial retinoblastoma. According to this model, the retinoblast cells in the eye, being rapidly dividing, have a predisposition to becoming cancerous. In unaffected people, the cancer is prevented from arising

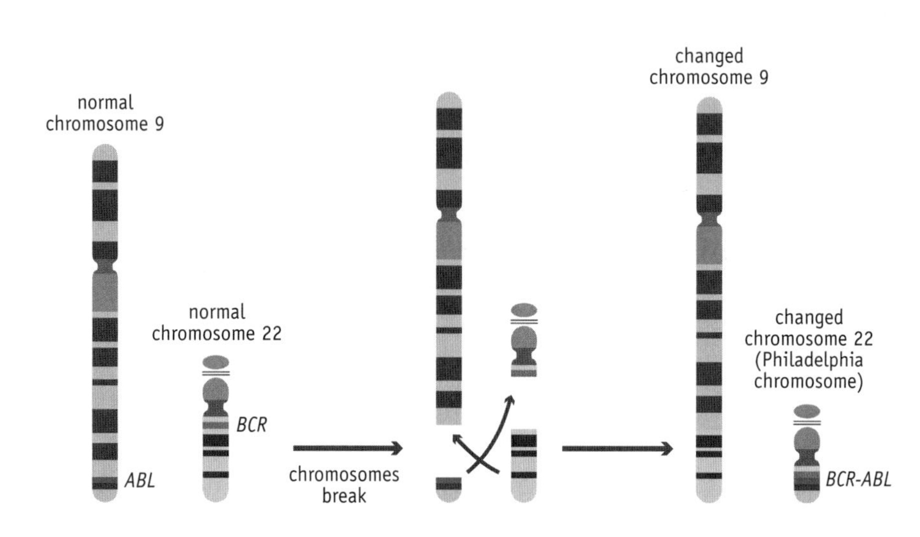

FIGURE 20.19 Details of the reciprocal translocation that results in production of the Philadelphia chromosome. The breakpoints lie within the *ABL* gene on chromosome 9 and the breakpoint cluster region (*BCR*) of chromosome 22.

RESEARCH BRIEFING 20.1

Identifying a gene for an inherited disease

For any inherited disease, identification of the causative gene is a very important objective. Once the gene has been isolated, the biochemical basis of the disease can be studied, possibly leading to ways of treating the disease by conventional means or by gene therapy. Identification of the mutations that result in the disease is equally important because it enables a screening program to be devised. The defective gene can then be identified in individuals who have not yet developed the disease or who are carriers. So how do we go about isolating the gene for an inherited disease?

Linkage analysis can reveal the region of the genome in which the gene is located

Usually the first step is to identify the map position of the gene in the genome. This can be achieved by examining the inheritance of the gene within affected families. The analysis seeks to establish linkage between the disease gene and other markers whose positions in the human genome are already known. The markers can be other genes, or they can be DNA markers such as RFLPs.

The accuracy with which a disease gene can be mapped depends on several factors including, importantly, the number of affected families whose pedigrees are available for study. Usually it is possible to identify a region of less than 5 Mb within which the disease gene is located. On average, there are about 6.5 genes per megabase of the human genome, so the region that has been identified might include quite a few genes. The next step is therefore to examine the relevant part of the genome sequence to find out exactly what it contains.

Examination of the genome sequence is not as arduous as it might sound. **Annotation** of the genome, to locate the positions of all the genes, was performed as part of the Human Genome Project. The entire genome was searched with computer programs designed to look for **open reading frames** (**ORFs**), sequences that consist of a series of codons, starting with an initiation codon and ending with a termination codon, and long enough to code for a protein. The programs are very accurate with bacterial genomes, but they are less effective with eukaryotes because of the presence of introns. If a gene contains one or more introns, it does not appear as a continuous ORF in the genome sequence. This means that the programs must also search for the special sequence features of exon–intron boundaries, but unfortunately the distinctiveness of these sequences is not so great as to make their location a trivial task. Sequence inspection is therefore combined with other approaches, such as RNA transcript mapping, in which RNA transcripts are located within the genome sequence. The sequences of these transcripts reveal the positions of exons.

The annotated genome can be examined online at various **genome browser** websites. These include the UCSC Genome Browser developed by the University of California Santa Cruz, and Ensembl, which is run jointly by the European Bioinformatics Institute and the UK Sanger Centre. A typical example of the information provided by a genome browser is shown in Figure 1.

Various criteria are used to test candidate genes to identify the correct one

The browser reveals the locations of the genes in the part of the genome we are interested in, but this information on its own will probably not tell us which gene is the one responsible for the disease. Even if the functions of the genes are known, the identity of the correct one might not be obvious. Any genes that cannot be discounted are therefore candidate genes that must be looked at more closely.

In Research Briefing 3.1, we learnt how expression profiling was used to confirm the identity of the cystic fibrosis gene. Expression profiling is very useful in this context, but the results can sometimes be misleading. It is quite possible for a disease that manifests itself in one or a limited number of tissues to be caused by a gene that is expressed more widely in the body. More specific tests are therefore needed.

One approach is to sequence the candidate gene in normal and affected people and see whether the genes from the latter have inactivating mutations. Unfortunately, identifying a critical mutation is not always easy. If the defect is a large deletion or a nonsense mutation then it can be recognized, but some diseases are caused by point mutations that affect just a single amino acid, the importance of which might not be immediately obvious. The G551D mutation in the cystic fibrosis gene is an example. In these situations it is necessary to compare genes from affected individuals in unrelated families, because the presence of the same mutation in different families provides strong evidence that the correct gene has been identified.

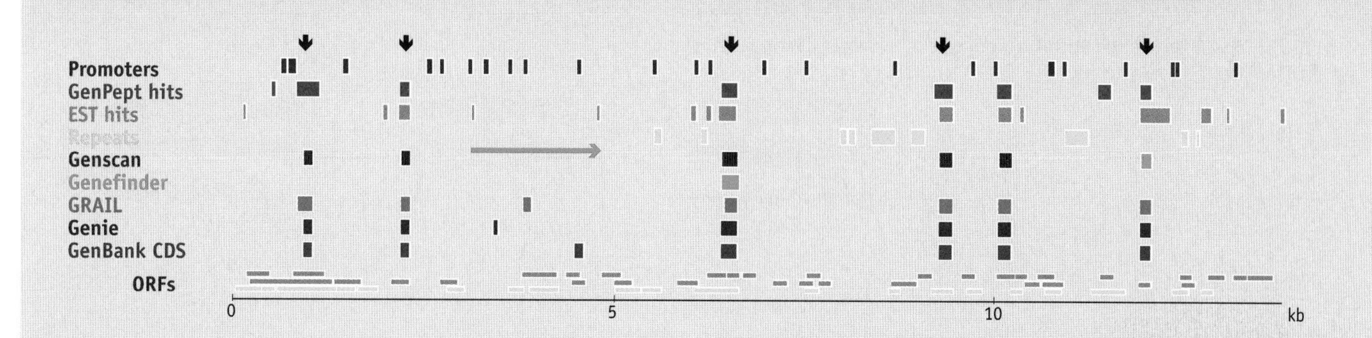

FIGURE 1 Part of the display from a typical genome browser. The example shown is the Genotator browser annotation for a 15-kb segment of the human genome. From the top, the analyses are: locations of possible promoters; sequences able to code for human proteins in the GenPept database; sequences corresponding to human ESTs (expressed sequence tags), which are derived from RNA transcripts; location of repeat sequences; exon predictions by the Genscan, Genefinder, GRAIL, and Genie programs; sequences corresponding to human genes in the GenBank database; and ORFs in each of the three reading frames. The green arrow indicates the direction in which the genome sequence is being read. The annotation reveals the positions of five possible exons, as indicated by the arrows at the top. (Courtesy of Nomi Harris, Lawrence Berkeley National Laboratory.)

An alternative way of testing a candidate gene is to study the effect that mutation has on the equivalent gene in mice. The mouse is frequently used as a model organism for humans because the mouse genome contains many of the same genes. If the disease arises from a loss of function, inactivating the equivalent gene in the mouse genome might give rise to an animal with symptoms similar to the human disease. A piece of DNA containing the mutated version of the mouse gene is therefore injected into a mouse **embryonic stem (ES) cell** (Figure 2). Once inside the nucleus, homologous recombination inserts the DNA fragment into the mouse genome, so the endogenous copy of the gene becomes replaced with the inactivated version. Unlike most mouse cells, ES cells are **totipotent**, meaning that they are not committed to a single developmental pathway and can give rise to all types of differentiated cell. The engineered ES cell is therefore transferred into a mouse embryo, which is implanted in a foster mother. The embryo continues to develop and eventually gives rise to a **chimera**, a mouse whose cells are a mixture of mutant ones, derived from the engineered ES cells, and non-mutant ones, derived from all the other cells in the embryo. This chimeric mouse is allowed to mate with a normal one. Some of the chimeric mouse's gametes will carry the mutated gene. After fusion with a gamete from the normal partner, these will produce heterozygous offspring, which have one copy of the mutated gene in every cell. A further cross between a pair of heterozygotes will give some new offspring who are homozygous for the mutation. These are **knockout mice**, and with luck their phenotypes will provide the desired information on the function of the candidate gene.

This strategy is not foolproof because some genes that give rise to inherited disease in humans can be inactivated in mice without any discernible effect on the health of the animal. Knockout mice are, however, extremely valuable not only for the assessment of candidate genes but also for more detailed studies of inherited disease, because frequently the mouse responds in a similar way to humans. Much important progress in devising treatments for human disorders has been made through the use of mouse models.

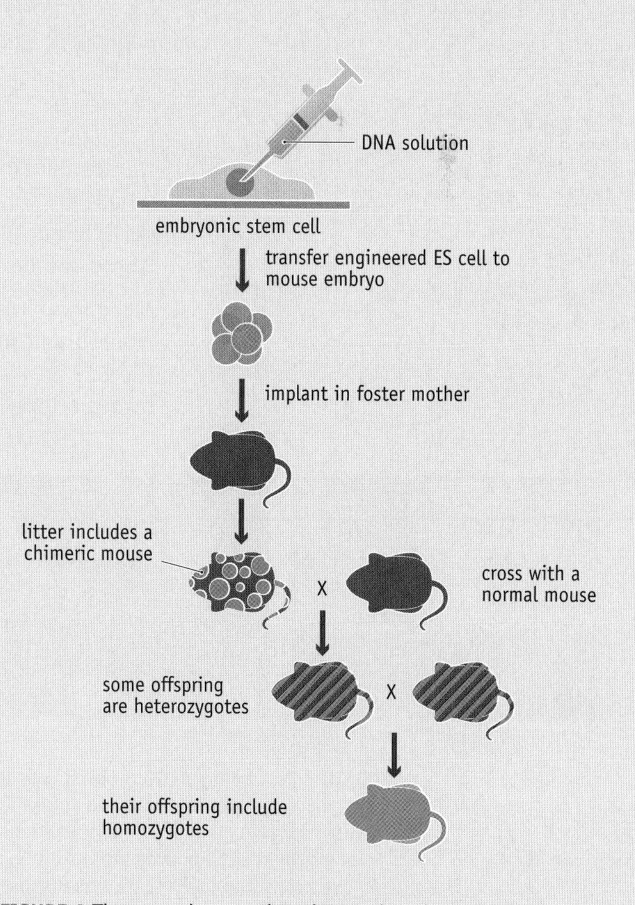

FIGURE 2 The procedure used to obtain a knockout mouse.

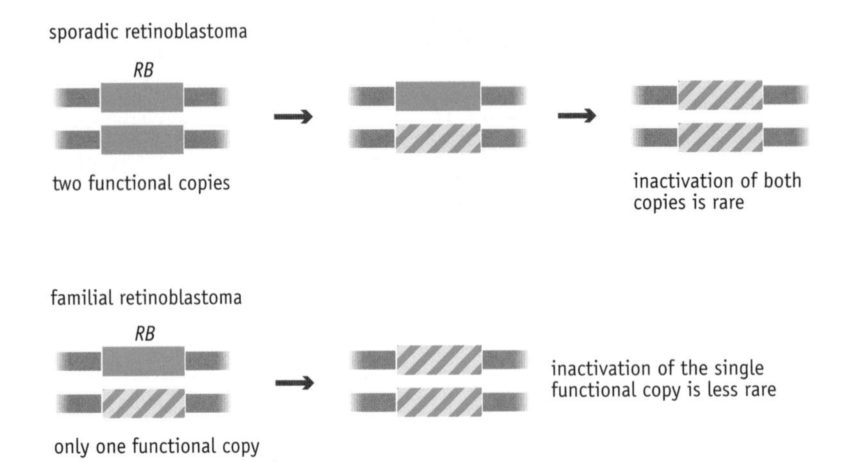

FIGURE 20.20 The genetic basis of sporadic and familial retinoblastoma.

by the retinoblastoma gene *RB* (**Figure 20.20**). Because it can suppress the cancer, *RB* is called a tumor suppressor gene. Most people have two functioning copies of *RB*. It is very unlikely that inactivating mutations will occur in both genes of a single cell, so these homozygotes virtually never develop the disease. The sporadic version of retinoblastoma is therefore rare. Heterozygotes, in contrast, are more vulnerable to retinoblastoma. They have just one active copy of *RB* per cell, and so a single inactivating mutation in any retinal cell could lead to the disease. Familial retinoblastoma therefore occurs in families whose members include heterozygotes who have inherited a defective allele from one of their parents and may pass that allele on to their offspring.

Most tumor suppressor genes are involved in regulation of the cell cycle, especially in the detection of and prevention of aberrant cell division. If a problem of some kind is detected, division is halted, and if the problem cannot be corrected then the cell might be induced to undergo apoptosis (**Figure 20.21**). One role of tumor suppressors is therefore to counter the transforming effects of an oncogene by causing the death of a cell in which an oncogene has been activated. Mutation of a tumor suppressor gene leads to at least partial loss of this surveillance process, enabling a rogue cell to escape and initiate tumor formation.

More than 150 tumor suppressor genes are known. We now know that the retinoblastoma protein prevents abnormal cells from dividing by inhibiting the E2F transcription factors that normally initiate DNA replication. Other tumor suppressors include the protein called p53 (the name indicates that

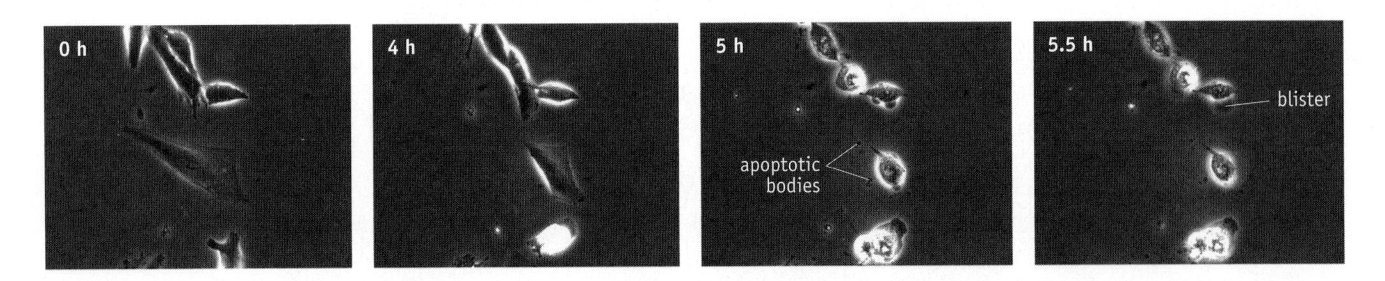

FIGURE 20.21 Human trophoblast cells undergoing apoptosis induced by treatment with spermine. After 5 h the cells have become rounded and appear bright in these phase contrast micrographs. The membranes start to bleb and blister but remain intact. Cell fragments called apoptotic bodies may bud off and are seen as tiny dark spots around some dying cells. Apoptosis is looked on as a "clean" form of cell death, compared with necrosis, which gives rise to broken cells that often cannot be disposed of by macrophages. (Courtesy of Guy Whitley, St George's University of London.)

it is 53 kD in size), which monitors the genome for DNA damage, initiating the repair of small-scale defects and inducing apoptosis if the damage is too great to repair.

Inactivation of tumor suppressor genes often occurs by point mutation or the deletion of some part of the coding sequence. Additionally, tumor suppressor genes can become inactive not as a result of a standard mutation but because of DNA methylation. In some transformed cells the CpG island upstream of a tumor suppressor gene is heavily methylated, and occasionally this is the only reason why the gene is nonfunctional—the coding region remains intact and lacks point mutations. We do not yet understand whether this methylation occurs by chance or whether these particular CpG islands have an abnormal predisposition toward methylation.

The development of cancer is a multistep process

If every mutation in a proto-oncogene or pair of mutations in a tumor suppressor gene were to give rise to a mature cancer, the frequency of deaths in the population would be much higher than it is. Most cancers develop via a multistep process that involves several mutations and other genome modifications that occur over a lengthy period. Often, detection of the developing cancer at an early stage in the process enables more effective treatment, and better survival rates, than is possible if detection is delayed until the cancer is mature.

The multistep development of a tumor has been most clearly described for the colon cancer called familial adenomatous polyposis (**Figure 20.22**). The morphological changes occurring during this process involve the initial formation of an adenoma, a small bulge in the epithelial tissue in the wall of the colon. This is followed by growth of the adenoma into a polyp a centimeter or so in diameter, giving rise finally to a tumor called a carcinoma, which contains metastatic cells, ones that can spread to other places in the body and initiate new tumors. The genetic events accompanying progression through these stages are not the same in all cases, but usually they begin with inactivation of the tumor suppressor gene called *APC*. As described above for retinoblastoma, there are sporadic and familial versions of colon cancer, the latter being more frequent because it occurs in a person who has inherited a defective *APC* allele from one parent and so needs just one additional mutation to inactivate the *APC* gene completely.

Subsequent events are variable but often there is a decrease in the overall degree of DNA methylation within the adenoma, followed by activation of the *KRAS* proto-oncogene, which codes for an intracellular signaling protein. Progression to this stage gives rise to a structure intermediate between the adenoma and polyp. Inactivation of the tumor suppressor gene *SMAD4*, coding for another signaling protein, gives rise to the polyp, and the carcinoma is produced if *p53* is inactivated. Metastasis might occur now or after other genetic events. The multistep process reflects the number of controls that exist within the cell to prevent aberrant division, controls that must be broken down one by one for a metastatic tumor to develop.

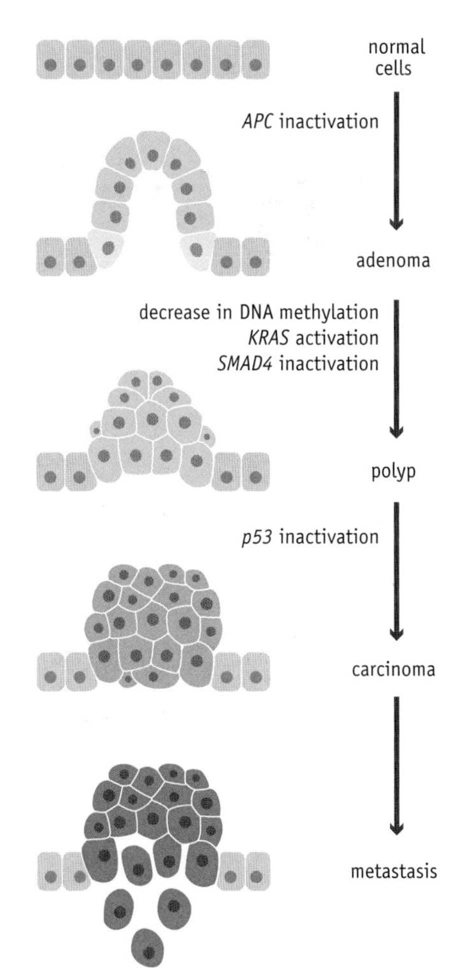

FIGURE 20.22 The multistep pathway leading to tumor formation in colorectal cancer. The pathway begins with inactivation of the tumor suppressor gene *APC* and formation of an adenoma. The adenoma develops into a polyp and then a carcinoma. These changes are associated with genetic events such as a decrease in DNA methylation, activation of the *KRAS* proto-oncogene, inactivation of the tumor suppressor gene *SMAD4*, and inactivation of *p53*. After the carcinoma has formed, metastasis might occur.

20.3 GENE THERAPY

A knowledge of the genetic basis of an inherited disease is important for two reasons. First, it provides the information needed to devise a screening program so that the defective gene can be identified in individuals who are carriers or who have not yet developed the disease. Carriers can receive counseling regarding the chances that their children will inherit the disease. Early identification in individuals who have not yet developed the disease allows appropriate precautions to be taken to reduce the risk of the disease becoming expressed.

Genetics also can lead to strategies for treating an inherited disease. Once the gene responsible for the disease has been identified it might be possible to work out its normal function in healthy individuals, thereby indicating the biochemical basis of the disease state. This might enable drugs to be designed that will reverse or alleviate the biochemical defect that causes the disease.

Treatment is, however, different from cure. The only way of curing an inherited disease is to replace the mutated gene with a nondefective version. This is the objective of gene therapy.

There are two approaches to gene therapy

There are two basic approaches to gene therapy. These are **germ-line therapy** and **somatic cell therapy**. In germ-line therapy a fertilized egg is provided with a copy of the correct version of the defective gene and reimplanted into the mother. The new gene will be present in all cells of the individual that develops from that egg. Germ-line therapy is usually performed by **microinjection** of the gene into a somatic cell followed by transfer of the nucleus of that cell into an oocyte whose own nucleus has been removed (Figure 20.23). The nuclear transfer is needed because successful microinjection directly into a fertilized egg is very difficult to

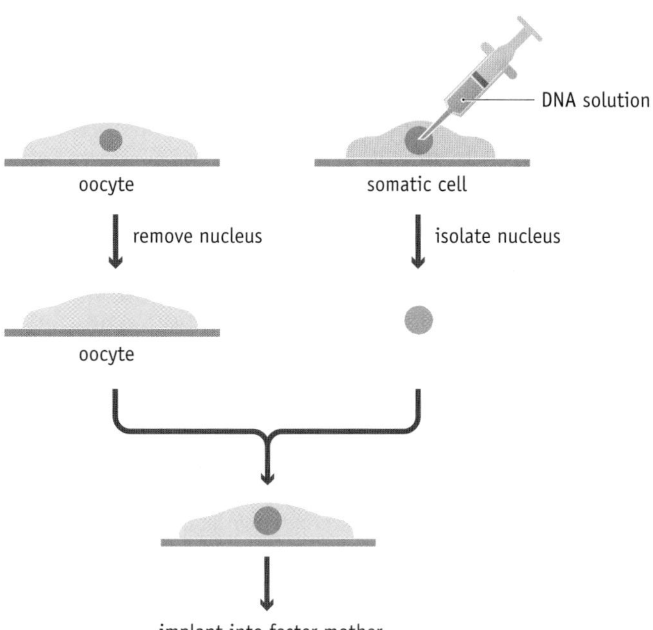

FIGURE 20.23 The nuclear transfer process for the introduction of new DNA into an oocyte.

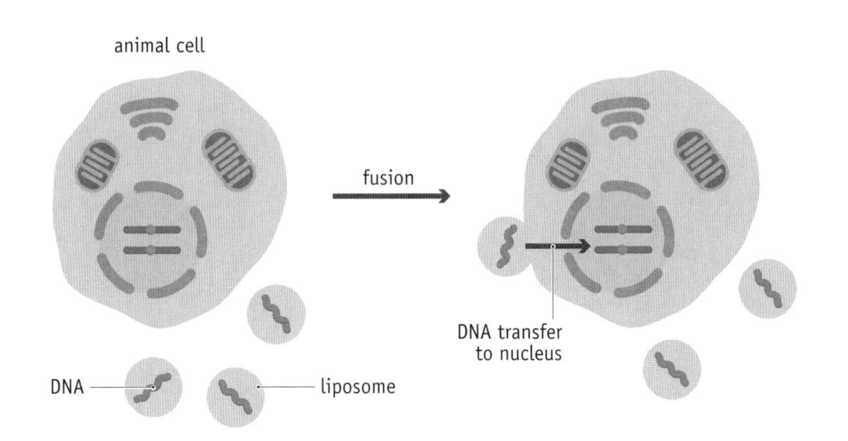

FIGURE 20.24 The use of liposomes to introduce new DNA into an animal cell.

achieve. Theoretically, germ-line therapy could be used to treat any inherited disease, but it is considered ethically unacceptable to manipulate the human germ line in this way, and at present this technology is used only to produce genetically engineered animals.

The transfer of new genes into somatic cells is less controversial because it only results in changes to the treated individual, and the new genes are not passed on to offspring. One way of performing the transfer is to enclose DNA containing the new gene in membrane-bound vesicles called **liposomes**. These can then be fused with the membrane of the target cell (Figure 20.24). Usually, DNA that is transported into a cell in this way is not stable, although the genes it contains might be expressed for a few days or weeks. The method has been tried, with limited success, as a way of treating lung diseases such as cystic fibrosis, with the DNA-containing liposomes being introduced into the respiratory tract via an inhaler.

To achieve a more permanent type of gene therapy it is necessary for the transferred gene to survive in the host cell for much longer than a few days. Ideally the gene should also be passed to daughter cells when the treated cell divides. Viruses are therefore often used as **vectors** for the transfer of genes into animal cells. The rationale is that a gene inserted into a virus genome will be expressed in the target cell for as long as that cell, and its descendants, remain infected with the virus. The virus used as the vector is modified so it can no longer cause its own disease but is still able to replicate. **Adenoviruses** (Figure 20.25) have been used for this purpose, because they can take up semi-permanent residence within the nucleus of an infected cell and are passed to daughter cells when the infected cell divides. Even better is a virus that transfers its genome into one of the host chromosomes, as this will lead to long-term expression of a therapeutic gene. Retroviruses can do this, but they insert at random positions, which is a disadvantage if the outcome of the transfer must be checked rigorously, as is the case with gene therapy. Attention is now focused on **adeno-associated viruses**, because these always insert their DNA at the same position in the human genome.

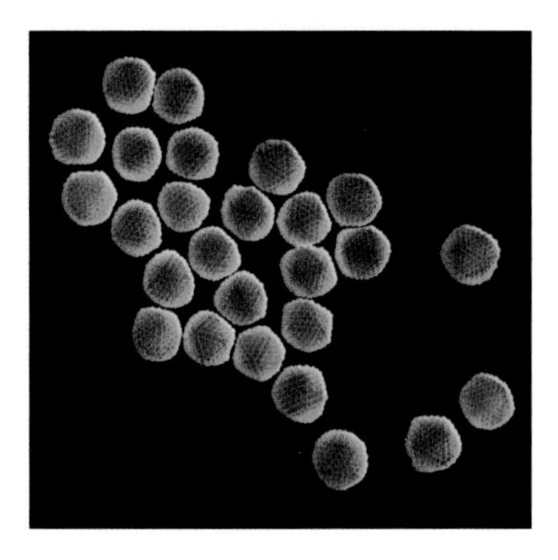

FIGURE 20.25 Adenoviruses, sometimes used as a vector for gene therapy. (Courtesy of Dr Hans Gelderblom, Visuals Unlimited/Science Photo Library.)

Somatic cell therapy using a virus vector has been used in the treatment of inherited blood diseases, such as hemophilia and thalassemia. The new gene is inserted into stem cells from the bone marrow, because these cells give rise to all the specialized cell types in the blood (Figure 20.26). The replication of the treated stem cells therefore results in the presence of the new gene throughout the blood system.

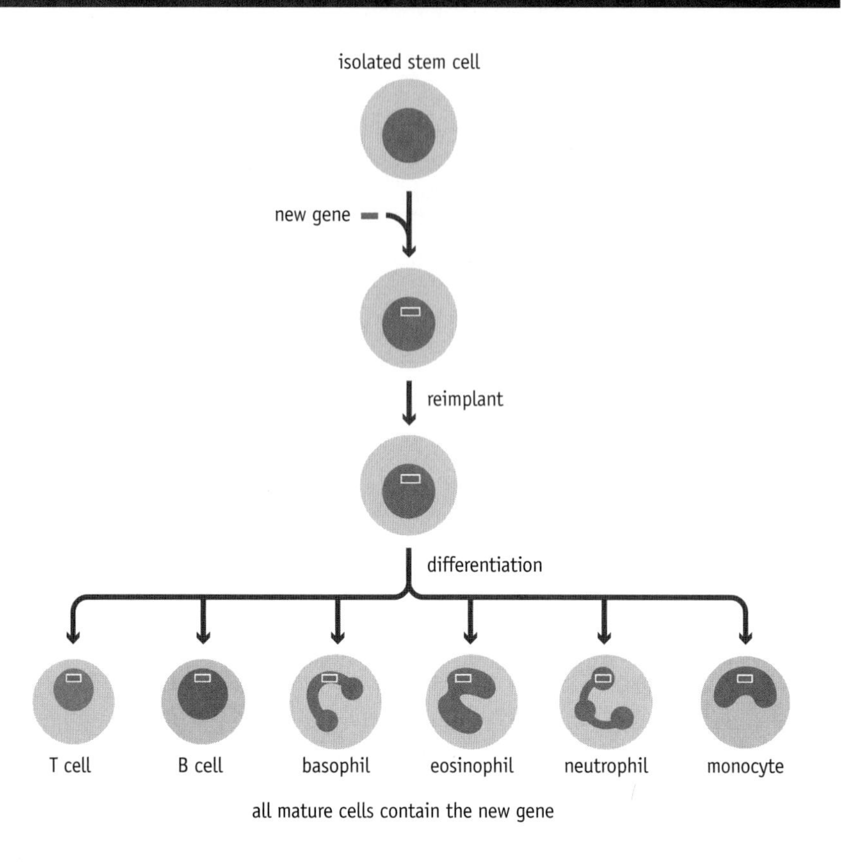

FIGURE 20.26 The insertion of a new gene into the stem cells in the bone marrow gives rise to mature blood cells carrying this new gene.

Those examples that we have considered so far concern recessive diseases, which can be treated simply by providing the correct gene, without any requirement that the defective copies be removed. For dominant diseases, this approach will not work because the defective allele will still specify the disease (Figure 20.27). A gene therapy for a dominant disease must therefore include not only the addition of the correct gene but also the removal of the defective version. This is a much more difficult prospect, and broadly applicable procedures have not yet been developed.

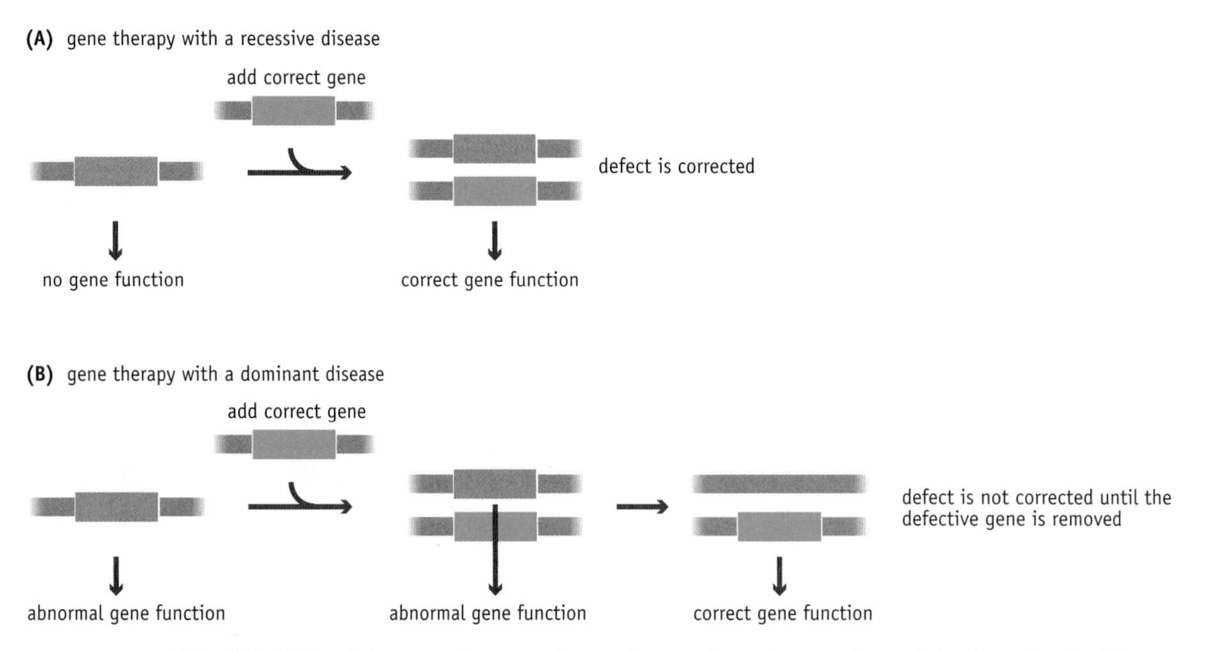

FIGURE 20.27 The different requirements of gene therapy directed at recessive and dominant inherited diseases. (A) A recessive disease can be treated by addition of the correct gene. (B) To treat a dominant disease the defective gene must also be removed.

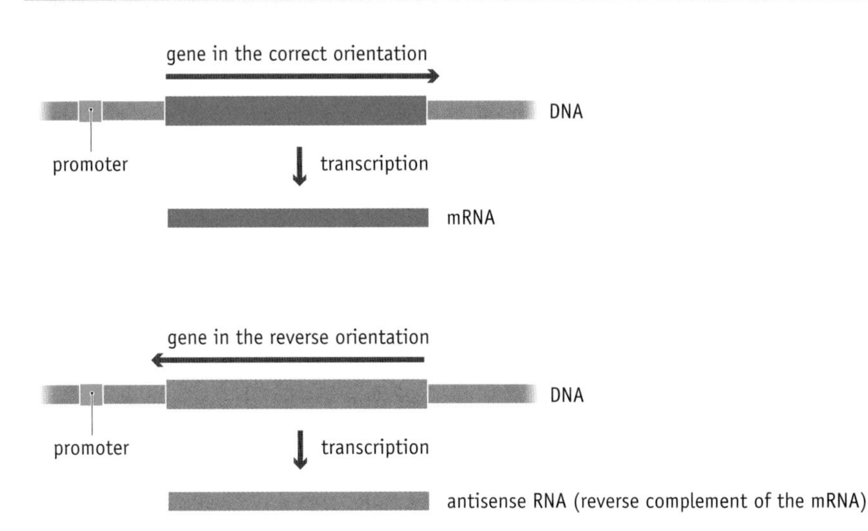

FIGURE 20.28 Synthesis of an antisense RNA.

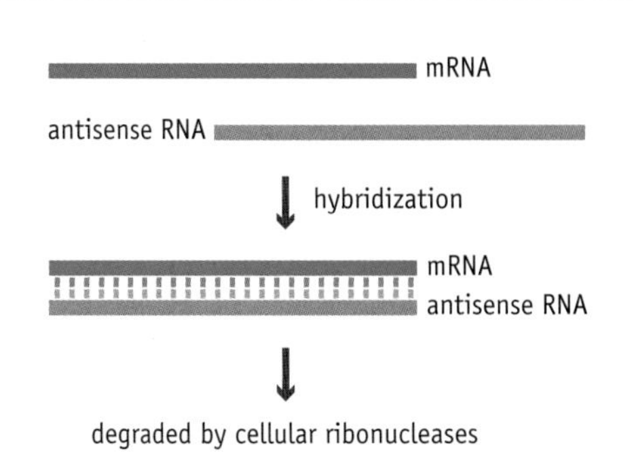

FIGURE 20.29 The probable mechanism by which an antisense RNA prevents the expression of its target gene.

Gene therapy can also be used to treat cancer

The clinical uses of gene therapy are not limited to the treatment of inherited diseases. There have also been attempts to use this technique as a treatment for cancer.

Inactivation of a tumor suppressor gene could be reversed by introduction of the nondefective version of the gene by one of the methods described above for inherited disease. Inactivation of an oncogene would, however, require a more subtle approach, because the objective would be to prevent expression of the oncogene, not to replace it with a nondefective copy. One possible way of doing this would be to introduce into a tumor a gene specifying an **antisense** version of the mRNA transcribed from the oncogene (Figure 20.28). An antisense RNA is the reverse complement of a normal RNA and can prevent synthesis of the protein coded by the gene it is directed against. It probably does this by hybridizing to the mRNA producing a double-stranded RNA molecule that is rapidly degraded by cellular ribonucleases (Figure 20.29). The target is therefore inactivated.

An alternative would be to introduce a gene that selectively kills cancer cells or promotes their destruction by drugs administered in a conventional fashion. This is called **suicide gene therapy** and is looked on as an effective general approach to cancer treatment, because it does not require a detailed understanding of the genetic basis of the particular disease being treated. Many genes that code for toxic proteins are known, and there are also examples of enzymes that convert nontoxic precursors of drugs into the toxic form (Figure 20.30). Introduction of the gene for one of these toxic

FIGURE 20.30 Suicide gene therapy. (A) The mode of action of the suicide gene. (B) When the drug precursor is applied, the tumor cells are killed.

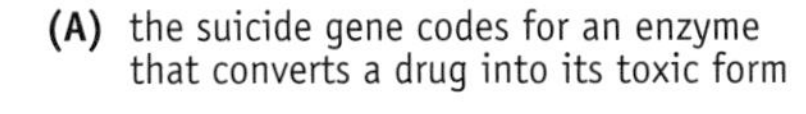

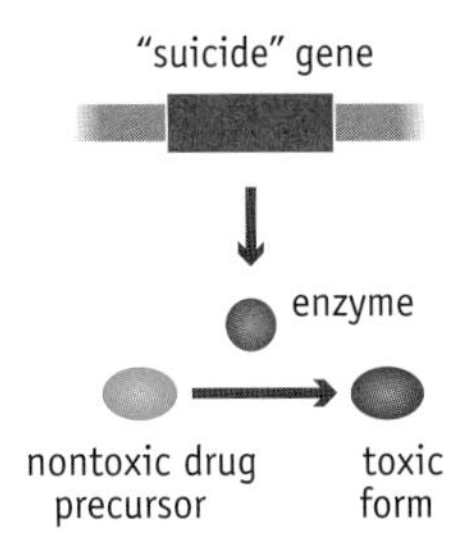

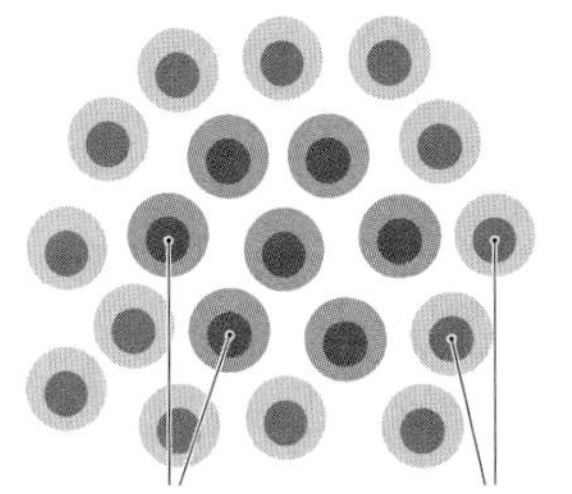

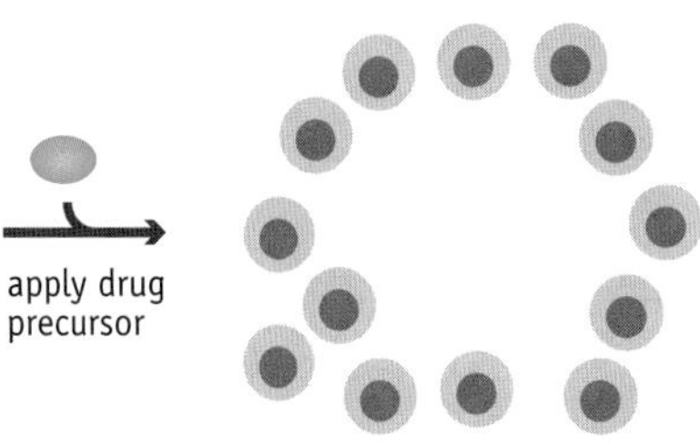

proteins or enzymes into a tumor should result in the death of the cancer cells, either immediately or after drug administration. It is obviously important that the introduced gene is targeted accurately at the cancer cells, so that healthy cells are not killed. This requires a very precise delivery system, such as direct inoculation into the tumor, or some other means of ensuring that the gene is expressed only in the cancer cells.

KEY CONCEPTS

- An inherited disease is one that is caused by a defect in the genome and that, like other genetic features, can be passed from parents to offspring.
- Most human inherited diseases are caused by a loss-of-function mutation that inactivates a particular gene. The protein coded by that gene is absent or is modified in some way that prevents it from functioning correctly, and the resulting defect is manifested as the inherited disease.
- There are dominant and recessive types of inherited disease. Understanding whether a disease is dominant or recessive can help in understanding the basis of the disease.
- Cancer is a group of diseases characterized by uncontrolled division of a somatic cell. Cancer is a genetic disease, because the underlying defects are gene mutations and the inheritance of certain mutations gives a predisposition to cancer. But cancer is not an inherited disease, because the cancerous state is not directly passed on from parent to offspring.
- Cancer begins with the activation of a proto-oncogene, a normal gene that can give rise to uncontrolled cell division when mutated.
- Tumor suppressor genes are involved in the regulation of the cell cycle, especially in the detection of and prevention of aberrant cell division. Mutation of a tumor suppressor gene leads to an at least partial loss of this surveillance process, increasing the likelihood that a cell will become cancerous.
- In germ-line gene therapy a fertilized egg is provided with a copy of the correct version of the defective gene and reimplanted into the mother. The new gene will be present in all cells of the individual that develops from that egg. Theoretically, germ-line therapy could be used to treat any inherited disease, but it is considered ethically unacceptable to manipulate the human germ line in this way, and at present this technology is used only to produce genetically engineered animals.
- Somatic gene therapy is less controversial because it only results in changes to the treated individual. The new genes are not passed on to offspring.

QUESTIONS AND PROBLEMS (Answers can be found at www.garlandscience.com/introgenetics)

Key Terms

Write short definitions of the following terms:

adeno-associated virus
adenovirus
annotation
antisense RNA
carrier
chimera
embryonic stem cell
gene therapy
genetic disease
genome browser
germ-line therapy
inherited disease
knockout mouse
liposome
microinjection

monogenic disorder
oncogene
open reading frame
Philadelphia chromosome
proto-oncogene
somatic cell therapy
suicide gene therapy
totipotent
trisomy
tumor suppressor

Self-study Questions

20.1 Describe the ways in which a loss-of-function mutation could lead to an inherited disease.

20.2 Using cystic fibrosis as an example, explain how a single genetic disease can be caused by more than one mutation.

20.3 Why are inherited diseases that result from a gain-of-function mutation relatively uncommon?

20.4 Describe the special features of trinucleotide repeat expansion diseases.

20.5 Give examples of (a) a dominant inherited disease, and (b) a recessive inherited disease.

20.6 How will a disease whose causative gene is on the X chromosome be inherited? What will be the inheritance pattern if the gene is on the Y chromosome?

20.7 Describe how a knowledge of whether a disease is dominant or recessive can be used to make inferences about the molecular basis to the disorder. Why are exceptions to a simple dominant–recessive relationship particularly informative?

20.8 Give examples of inherited diseases caused by large deletions.

20.9 Why is trisomy not a type of inherited disease?

20.10 Explain how activation of a proto-oncogene can lead to cancer.

20.11 What is the Philadelphia chromosome and why is its possession linked with cancer?

20.12 Describe the role of tumor suppressor genes in preventing cancer.

20.13 Outline the "two-hit" model for the onset of retinoblastoma.

20.14 Describe the multistep process that leads to colon cancer.

20.15 Distinguish between germ-line therapy and somatic cell therapy.

20.16 Give examples of the way in which new DNA is introduced into an organism to perform gene therapy.

20.17 Describe how gene therapy might be used to treat cancer.

20.18 Explain what is meant by the term "suicide gene therapy."

20.19 Describe how a genome browser can be used as a complement to linkage analysis in a project designed to identify a gene for an inherited disease.

20.20 What is a knockout mouse and why are these animals useful in studies of human genetic disease?

Discussion Topics

20.21 Why are some inherited diseases more common than others?

20.22 Explore the current knowledge concerning trinucleotide repeat expansion diseases, including hypotheses that attempt to explain why triplet expansion in these genes leads to disease.

20.23 If two parents are both carriers for an inherited disease, what are the chances that their next child will suffer from the disease? What are the chances that the next child but one will have the disease?

20.24 A pharmaceutical company has invested a great deal of time and money to isolate the gene for a genetic disease. The company is studying the gene and its protein product and is working to develop drugs to treat the disease. Does the company have the right (in your opinion) to patent the gene? Justify your answer.

20.25 Is gene therapy a realistic approach to the treatment of inherited disease?

20.26 Discuss the ethical issues that are raised by the development of gene therapy techniques. In your discussion, make sure that you distinguish between somatic cell therapy and germ-line therapy.

FURTHER READING

Bobadilla JL, Macek M, Fine JP & Farrell PM (2002) Cystic fibrosis: a worldwide analysis of *CFTR* mutations—correlation with incidence data and application to screening. *Hum. Mutat.* 19, 575–606.

Hassold TJ & Jacobs PA (1984) Trisomy in man. *Annu. Rev. Genet.* 18, 69–97.

Hubbard T, Barker D, Birney E et al. (2002) The Ensembl genome database project. *Nucl. Acids Res.* 30, 38–41. *A genome browser.*

Koretzky GA (2007) The legacy of the Philadelphia chromosome. *J. Clin. Invest.* 117, 2030–2032. *Describes the discovery of the Philadelphia chromosome and what its study has told us about cancer.*

Lupski JR, de Oca-Luna RM, Slaugenhaupt S et al. (1991) A duplication associated with Charcot-Marie-Tooth disease type 1A. *Cell* 66, 219–232.

Mirkin AM (2007) Expandable DNA repeats and human disease. *Nature* 447, 932–940. *Trinucleotide repeat expansion diseases.*

Schepelmann S, Ogilvie LM, Hedley D et al. (2007) Suicide gene therapy of human colon carcinoma xenografts using an armed oncolytic adenovirus expressing carboxypeptidase G2. *Cancer Res.* 67, 4949–4955. *An example of suicide gene therapy.*

Smith KR (2003) Gene therapy: theoretical and bioethical concepts. *Arch. Med. Res.* 34, 247–268.

van Deutekom JCT & van Ommen GJB (2003) Advances in Duchenne muscular dystrophy gene therapy. *Nat. Rev. Genet.* 4, 774–783. *An example of the use of gene therapy.*

Vogelstein B & Kinzler KW (1993) The multistep nature of cancer. *Trends Genet.* 9, 138–141.

Yeo CJ (1999) Tumor suppressor genes: a short review. *Surgery* 125, 363–366.

DNA in Forensics and Studies of Human History CHAPTER 21

In the public perception, probably the most important application of genetics is its use in forensic science. Hardly a week goes by without a report in the media of a murder or rape that has been solved thanks to DNA. The increasing sophistication of the technology means that not only are recent crimes being solved, but so also are many "cold cases" dating back tens of years. As well as the reports of real events on the front pages of newspapers, the television guides are also full of dramas in which glamorous geneticists use DNA to fight the forces of evil.

We will begin this chapter by exploring the science behind the real-life crime stories and the fictional TV dramas. In doing so we will discover that there is an unbroken progression from the use of DNA in criminal cases to its application in archaeology. This is because the methods used to identify individuals from the DNA in hairs, bloodstains, and semen can also be used with **ancient DNA**—DNA extracted from fossil or archaeological remains. We will include some of these archaeological applications in our study of forensic genetics.

It is also possible to use DNA from living people to study archaeological questions. The variations that exist in the human genome reveal evolutionary patterns that can be related to the origins of our species and the routes followed by our ancestors as they migrated out of Africa and colonized the rest of the world, migrations that began almost 100,000 years ago. These evolutionary studies are referred to as **archaeogenetics**. In the second part of this chapter we will look at the information emerging from this new discipline.

21.1 DNA AND FORENSIC SCIENCE

In the popular media, the DNA techniques that are used in forensic science are often called "genetic fingerprinting," but the more accurate term for the procedure used today is **genetic profiling**. The key to genetic profiling is the immense variability that exists within the human genome, which makes it highly unlikely that any two people have exactly the same genome sequence. The only exceptions are identical twins who have developed from the same fertilized egg. We must therefore begin our study of genetic profiling by examining the variability that is present in the human genome.

Every human genome is unique

Although every copy of the human genome has the same genes in the same order, and the same stretches of intergenic DNA between those genes, the genomes possessed by different people are by no means identical. This is because the human genome, as well as those of other organisms, contains many **polymorphisms**, positions where the nucleotide sequence is not the same in every member of the population. Many of these variable positions are single nucleotide polymorphisms (SNPs), where either of two different nucleotides might be present (Figure 21.1). There are believed to be at least 15 million SNPs in the human genome, and it is likely that many others will be discovered in the next few years as the genomes of more and more people are sequenced.

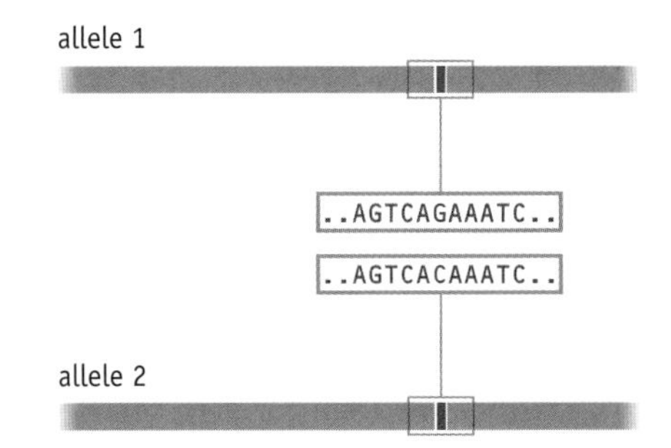

FIGURE 21.1 Two alleles of a single nucleotide polymorphism.

Additional variability between genomes is provided by the short tandem repeats called microsatellites (Figure 21.2). An error when a microsatellite is being replicated occasionally results in an increase or decrease in

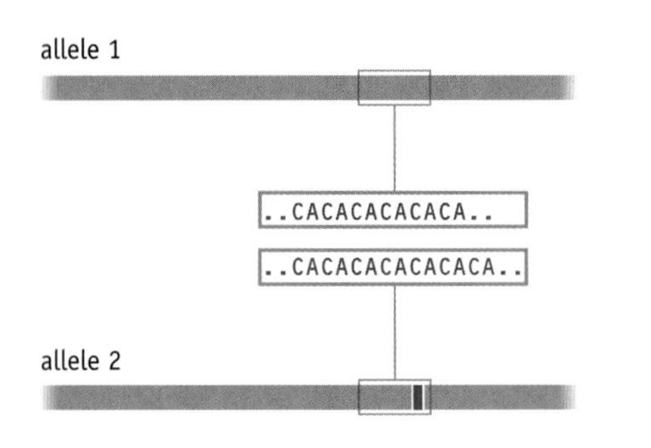

FIGURE 21.2 Two alleles of a microsatellite.

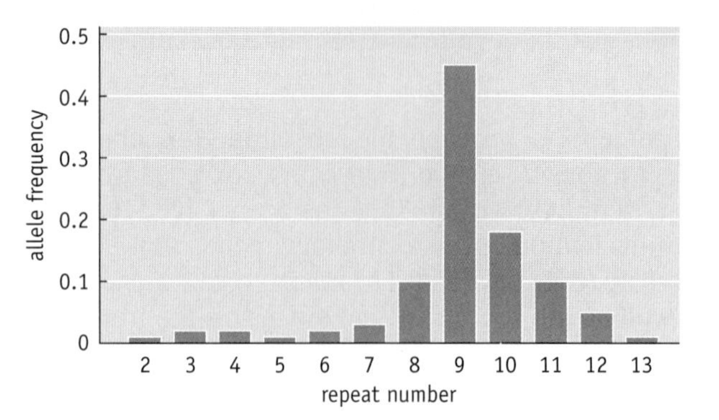

FIGURE 21.3 The allele frequencies for a typical human microsatellite. The histogram shows the frequencies for 12 alleles of a microsatellite, with repeat numbers from 2 to 13. For many microsatellites, the frequencies form a Gaussian distribution, as shown here.

the number of repeat units, creating a new allele that might survive in the population. This means that most microsatellites are multiallelic, with 10 or more alleles present in the population as a whole (Figure 21.3). Because there are more than 500,000 microsatellites in the human genome, the number of possible allele combinations is huge. There is only a remote chance that two people will have the same combination, unless those people are identical twins. Everybody else will have their own, unique combination.

A genetic profile is simply a list of the alleles present at a set of microsatellites. This set does not need to be particularly large to establish a profile that is likely to be unique to a particular individual. In most countries, genetic profiles are obtained by typing the alleles present at just 13 microsatellites (Figure 21.4). These are called the CODIS (Combined DNA Index System) set. They have sufficient variability to give only a 1 in 10^{15} chance that two individuals, other than identical twins, have the same profile. Because the world population is about 6×10^9, the likelihood that any two people alive today have the same profile is very small.

A match between a profile obtained from a crime scene and that of a suspect is therefore looked on as highly suggestive. In most countries, a match is not considered enough to secure a conviction on its own, but if the prosecution can provide another form of evidence linking the suspect with the crime, the DNA evidence is often taken as conclusive. Equally importantly, the absence of a match can exclude suspects from an enquiry, and in particular has resulted in the release from prison of people who were wrongly convicted at a trial held before the days of DNA testing.

A genetic profile is obtained by multiplex PCR

How do we identify the alleles present at a microsatellite sequence? The answer is to use the polymerase chain reaction (PCR—see Research Briefing 10.2). The primers for the PCR are designed so that they anneal on each side of the microsatellite. This means that the length of the molecule that is produced during the PCR will indicate how many repeats are present and will therefore identify the allele (Figure 21.5). One of the primers used in the PCR is labeled with a fluorescent marker so that the length of the PCR product can be measured by capillary gel electrophoresis. One band, indicating a single allele, will be detected if the person whose DNA is being tested is homozygous for the microsatellite. Two bands will be seen if the individual is a heterozygote.

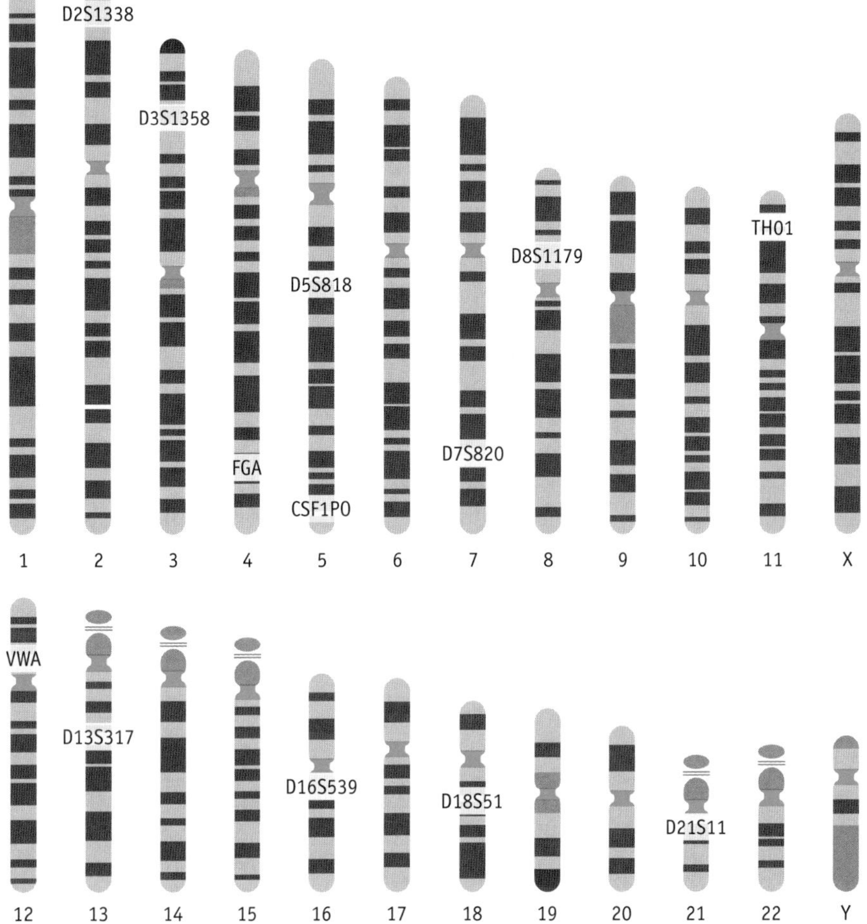

FIGURE 21.4 The CODIS set of microsatellites, indicating their locations in the human genome.

Performing 13 PCRs in parallel is time-consuming, an important consideration in a forensic laboratory that might have to process hundreds of samples a week. To save time, the profile is usually constructed by performing all the PCRs together in a single reaction. This **multiplex PCR** will result in 13–26 products, one or two from each microsatellite. These can easily be separated in the capillary gel, but a method is needed of identifying which molecule comes from which microsatellite. One possibility would be to use a different fluorescent label for each individual PCR in the multiplex, but this makes the detection process overly complicated. Instead, just four different labels are used, the results of PCRs with the same label being distinguished by ensuring that the product sizes do not overlap (Figure 21.6).

The methods for performing multiplex PCR and generating genetic profiles have been available since the early 1990s. The big advances that have occurred in forensic genetics in recent years have been due to improvements

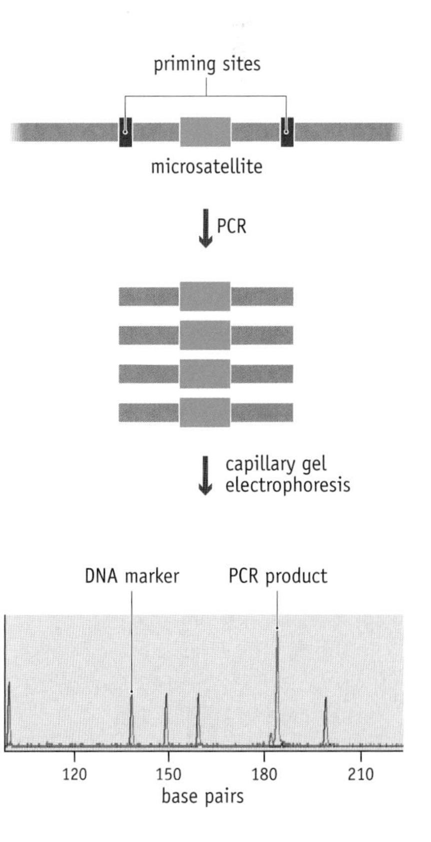

FIGURE 21.5 Identifying the alleles present at a microsatellite by PCR followed by capillary gel electrophoresis. The length of the PCR product indicates the number of repeats present in the allele that has been detected. In this example there is just one product, indicating that the individual being tested is homozygous for this particular microsatellite.

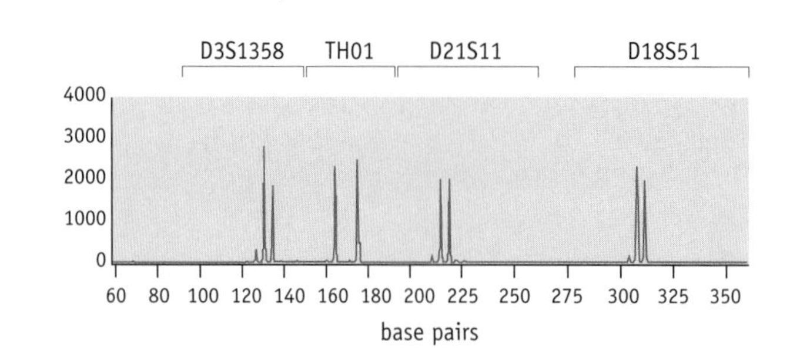

FIGURE 21.6 Part of a genetic profile, showing the PCR products for four microsatellites from the CODIS set. The labels above the profile show that these are D3S1358, TH01, D21S11, and D18S51. All four sets of PCR products are labeled with the same fluorescent marker, but they can be distinguished after capillary gel electrophoresis because their sizes are different. All four microsatellites are heterozygous. The *y*-axis is arbitrary units used to measure the peak heights. (Courtesy of Promega Corporation.)

in the techniques used for purifying DNA, along with modifications to the PCR systems to make these more sensitive. Until the mid-2000s, a profile could only be constructed from samples such as hairs and semen, which contain relatively large amounts of DNA. Now it is possible to use less substantial remains, including small bloodstains and saliva taken from clothing. The new methods not only increase the chances of finding suitable samples at a crime scene, they also allow DNA testing to be conducted with material that has been kept from older unsolved crimes. Genetic profiling can therefore be applied to these "cold cases" in the hope that new evidence that might lead to a conviction can be obtained.

DNA testing can also identify the members of a single family

Genetic profiling is not only used in criminal investigations. The same techniques can also identify family relationships. This is because each person inherits one allele of a microsatellite from their mother, and one from their father. Siblings—children with the same parents—will therefore have allele combinations that are similar to one another and to those of their parents (Figure 21.7A). The most common application of this type of genetic profiling is in paternity testing, when there is uncertainty about the identity of the father of a child (Figure 21.7B).

The 13 microsatellites of the CODIS set are all located on autosomes (see Figure 21.4). Sometimes it is an advantage to follow just the male or female line in a family. The male line can be followed by typing one or more microsatellites located on the Y chromosome, because the Y chromosome is passed from a father to his sons. This means that male siblings possess the same alleles for a set of Y microsatellites, these being the alleles that their father has (Figure 21.8A).

How can we follow the female line in a family? We could type microsatellites located on the X chromosome but this would provide only part of the answer. Male children always inherit an X chromosome from their mother, but one of the X chromosomes inherited by a female child comes from the father (Figure 21.8B).

The female line is more precisely traced by studying mitochondrial DNA. Unlike the nuclear DNA, the mitochondrial genome is inherited in a **uniparental** manner, with only the mother contributing her genome to the

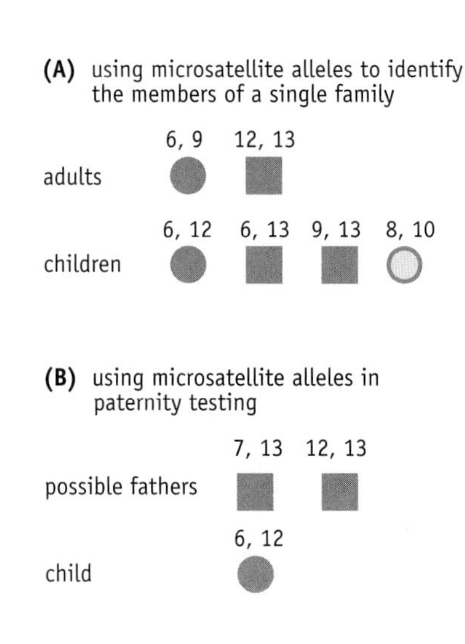

FIGURE 21.7 The use of genetic profiling (A) in the identification of the members of a single family and (B) in paternity testing. Squares represent males and circles represent females. A single microsatellite has been typed in each individual, and the alleles are indicated by the numbers. In (A), three of the children could be siblings and the two adults could be their parents. The fourth child, shown as the open circle, cannot be a member of this family. In (B), only the second of the two males could be the father of the child.

FIGURE 21.8 Following the male or female lineages within a family. Squares represent males and circles represent females. (A) The male line can be followed by typing microsatellites on the Y chromosome. In this example, a single Y chromosome microsatellite has been typed. Two of the children have the same allele as the male adult and could be his sons. The third child, shown as an open square, cannot be his son. (B) The maternal line cannot be precisely followed by typing microsatellites on the X chromosomes. The combinations shown here are compatible with the hypothesis that the two adults are the parents of three of the children, but the inheritance pattern is not strictly maternal because the female child inherits an X chromosome from her father. The fourth child cannot be the son of either parent. Although he has the same X chromosome allele as the male adult, he cannot have inherited it from that person. (C) The female line can be followed by typing the mitochondrial genome. Three children have the same mitochondrial DNA haplotype as the female adult and so could be her daughter and sons. The fourth child cannot be a member of this family group. Although his mitochondrial haplotype is the same as that of the male adult, he cannot have inherited it from that person.

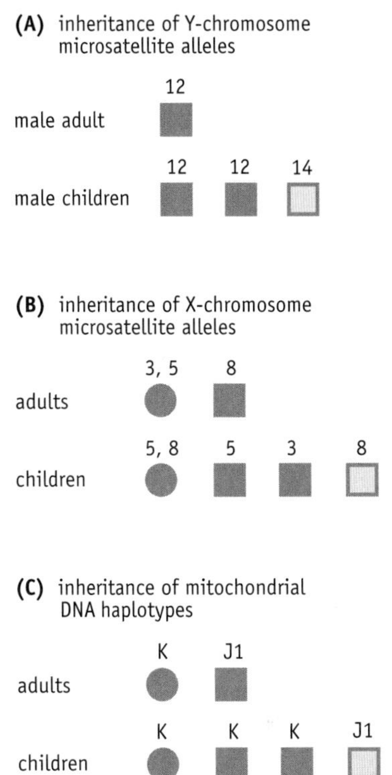

offspring. The paternal mitochondrial genome is present in the male sperm but does not penetrate the egg cell during fertilization, or if it does it is degraded soon afterward. Each of a set of siblings therefore has the same version of the mitochondrial genome, this being the version possessed by their mother (Figure 21.8C). There are no microsatellites in the human mitochondrial DNA, but there are many SNPs, especially in the region either side of the replication origin, the only part of the genome that lacks genes (Figure 21.9). The sequence of this **hypervariable region** defines the mitochondrial DNA haplotype, and it is this haplotype that is shared by a set of siblings and their mother.

Genetic profiling was used to identify the Romanov skeletons

As an illustration of the way in which genetic profiling can be used to identify a family group we will examine the work that was performed in the 1990s with the skeletons of the Romanovs. The Romanovs and their descendants ruled Russia from the early seventeenth century until the time of the Russian Revolution, when Tsar Nicholas II was deposed and he and his wife, the Tsarina Alexandra, and their five children were imprisoned (Figure 21.10). On 17 July 1918, all seven, along with their doctor and three servants, were murdered and their bodies disposed of in a shallow roadside grave near Yekaterinburg in the Urals. In 1991, after the fall of communism, the remains were recovered with the intention that they should be given a more fitting burial.

Although it was suspected that the bones were indeed those of the Romanovs, the possibility that they belonged to some other unfortunate group of people could not be discounted. Nine skeletons were present in the grave. Six of these were adults and three children. Examination of the bones showed that four of the adults were male and two were female, and that the three children were all female. This meant that if they were the Romanov group, their son Alexei and one of their daughters were, for some reason, absent. The skeletons showed signs of violence, consistent with reports of their treatment during and after death. Some of the remains were clearly aristocratic because their teeth were filled with porcelain, silver, and gold, dentistry well beyond the means of the average Russian of the early twentieth century. Could DNA tests confirm the identities of the remains?

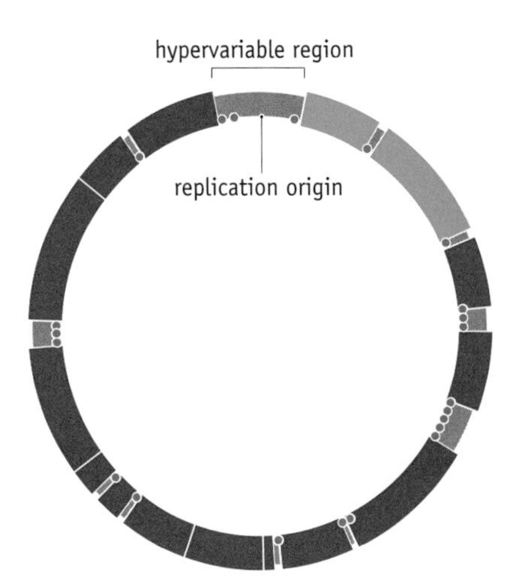

FIGURE 21.9 The human mitochondrial genome, showing the positions of the replication origin and the hypervariable region.

FIGURE 21.10 The Romanovs. The Tsar and Tsarina are seated with Alexei at the front. From left to right the daughters are Olga, Tatiana, Maria, and Anastasia. (Courtesy of Buyenlarge/Getty Images.)

Five microsatellites were typed in each of the skeletons to test the hypothesis that the three children were siblings and that two of the adults were their parents, as would be the case if indeed these were the Romanovs (Table 21.1). The results immediately showed that the three children could be sisters, because they had identical genotypes for two microsatellites, the ones called VWA/31 and FES/FPS, and shared alleles for each of the three others.

Which of the adults could be the parents? If we examine the TH01 results we see that female adult 2 cannot be the mother of the children because she possesses only allele 6, which none of the children has. Female adult 1, however, has allele 8, which all three children share. Examination of

TABLE 21.1 MICROSATELLITE GENOTYPES OBTAINED FROM THE SKELETONS THOUGHT TO INCLUDE THE ROMANOVS

	Microsatellite				
Skeleton	VWA/31	TH01	F13A1	FES/FPS	ACTBP2
Male adult 1	14, 20	9, 10	6, 16	10, 11	Not done
Male adult 2	17, 17	6, 10	5, 7	10, 11	11, 30
Male adult 3	15, 16	7, 10	7, 7	12, 12	11, 32
Male adult 4	15, 17	6, 9	5, 7	8, 10	Not done
Female adult 1	15, 16	8, 8	3, 5	12, 13	32, 36
Female adult 2	16, 17	6, 6	6, 7	11, 12	Not done
Child 1	15, 16	8, 10	5, 7	12, 13	11, 32
Child 2	15, 16	7, 8	3, 7	12, 13	11, 36
Child 3	15, 16	8, 10	3, 7	12, 13	32, 36

The table shows the repeat numbers for the two alleles for each microsatellite in each individual. For example, "14, 20" indicates that male adult 1 was heterozygous for VWA/31, one allele having 14 repeats and the other having 20 repeats. The entry "17, 17" for male adult 2 indicates that this person was homozygous for VWA/31, both alleles having 17 repeats.

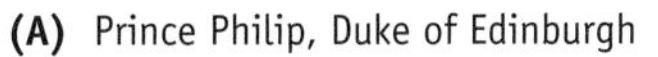

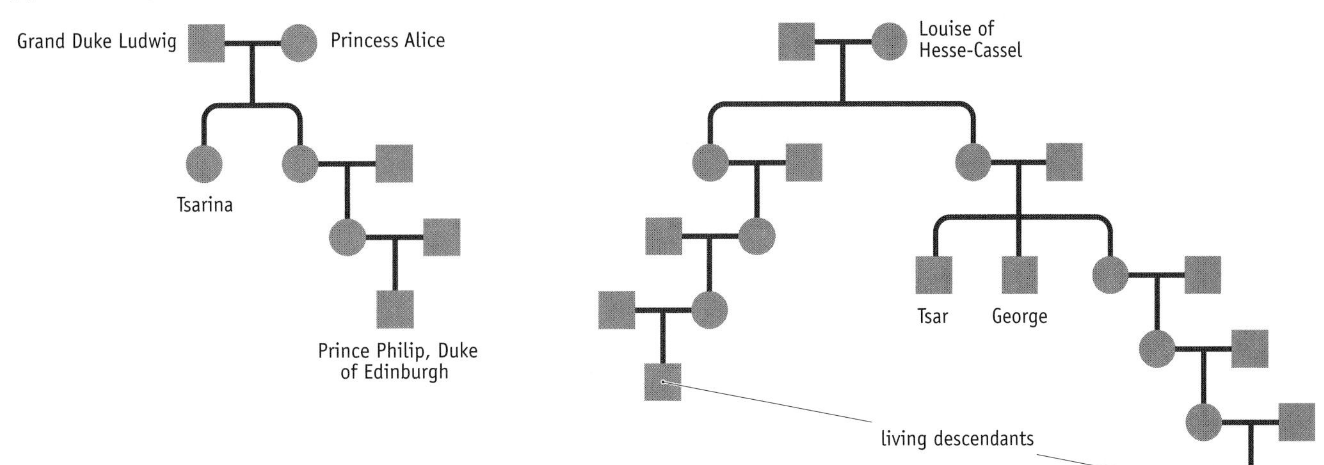

FIGURE 21.11 Genealogies showing those living relatives of the Romanovs whose DNA was tested. (A) Prince Philip, the Duke of Edinburgh, whose maternal great-grandmother was Princess Alice, the Tsarina Alexandra's mother. (B) Two living matrilineal descendants of the Tsar's grandmother, Louise of Hesse-Cassel.

the other microsatellites confirms that female adult 1 could indeed be the mother of each of the children. Turning to the male skeletons, the TH01 results exclude male adult 4 as a possible father, and the VWA results exclude male adults 1 and 2. This leaves only male adult 3. Working across the table, we see that for each microsatellite, he could provide an allele that would combine with one from female adult 1 to give the genotypes possessed by child 1. The same is true for the other two children. Male adult 3 could therefore be their father.

Microsatellite analysis therefore established a possible kinship pattern among the remains. Confirmation that these were indeed the Romanovs was obtained by comparing their mitochondrial genomes with those of their living relatives. For the Tsarina and the three children this was straightforward, because the haplotype of all four matched that of Prince Philip, the Duke of Edinburgh, whose maternal great-grandmother was Princess Alice, the Tsarina Alexandra's mother (Figure 21.11). With the adult male thought to be the Tsar the analysis was more complicated, because two sequences were obtained from his mitochondrial DNA, suggesting that he displayed **heteroplasmy**, an uncommon situation in which two different mitochondrial haplotypes coexist within the same cells. Two living matrilineal descendants of the Tsar's grandmother, Louise of Hesse-Cassel, agreed to be tested, and each was found to have one of these haplotypes. In addition, DNA from the remains of the Tsar's brother, Grand Duke George Alexandrovich, who died in 1899, showed that he also displayed heteroplasmy. On balance, the evidence suggested that the Tsar's remains had been correctly identified. The results were considered sufficiently convincing for the skeletons identified as the Tsar, Tsarina, and their three daughters to be given a state funeral in St. Petersburg in 1998.

The final mystery concerned the whereabouts of the other two children. During the middle decades of the twentieth century several women claimed to be a Romanov princess, because even before the bones were recovered there had been rumors that one of the girls, Anastasia, had escaped the clutches of the Bolsheviks and fled to the West. One of the most famous of these claimants was Anna Anderson, whose case was first widely publicized in the 1960s. Anna Anderson died in 1984 but she left an archived tissue sample whose mitochondrial haplotype does not match that of the Tsarina. There have also been various people claiming to be descended from Tsarevich Alexei. But these stories are romances, because the partly cremated bodies of two other children, found near Yekaterinburg in 2007, have now been examined, and their mitochondrial haplotypes suggest that they are the missing Romanov children.

DNA testing can also be used to identify family groups among archaeological remains

It is possible to go much further back in time than the early twentieth century in our use of genetic profiling. It has been known for some years that DNA molecules can be preserved for centuries and possibly millennia after the death of the organism in which they are contained, being recoverable as short degraded fragments preserved in bones and other biological remains. There is never very much DNA in an archaeological specimen, possibly only a few picograms in a gram of bone, but that need not concern us because we can always use PCR to amplify these tiny amounts into larger quantities from which we can obtain DNA sequences. Extreme care must be taken to avoid contamination with modern DNA from the hands of researchers who handle the bones, but when the appropriate precautions are taken it is sometimes possible to extract authentic ancient DNA from specimens up to about 50,000 years in age. This means that many archaeological questions can be tackled by DNA analysis.

Typing microsatellites in ancient DNA is difficult because there is so little DNA and it is often partly degraded, with some nucleotide positions unidentifiable because of chemical changes that obscure the original DNA sequence. Mitochondrial DNA is easier to study than the nuclear genome because it has a high copy number—approximately 8000 copies in each cell—which increases the chances that some undamaged molecules remain in a specimen. Information on family relationships therefore has to be built up carefully, using a combination of approaches.

An example of what can be achieved is provided by research on skeletons from a 4600-year-old cemetery at Eulau, in central Germany. Four graves were excavated, containing a total of 13 skeletons, a mixture of adult males, adult females, and children. The arrangement of the skeletons in the graves suggested that each was a family group comprising one or both parents and their children. For two of the graves this hypothesis was supported by ancient DNA results. In grave 99, for example, the female adult had mitochondrial haplotype K, and the male adult possessed the combination of Y microsatellite alleles that are designated haplotype R1a. The two children, both males, also possessed haplotypes K and R1a (Figure 21.12). The children therefore had the combination of alleles that would be expected if the two adults were their parents.

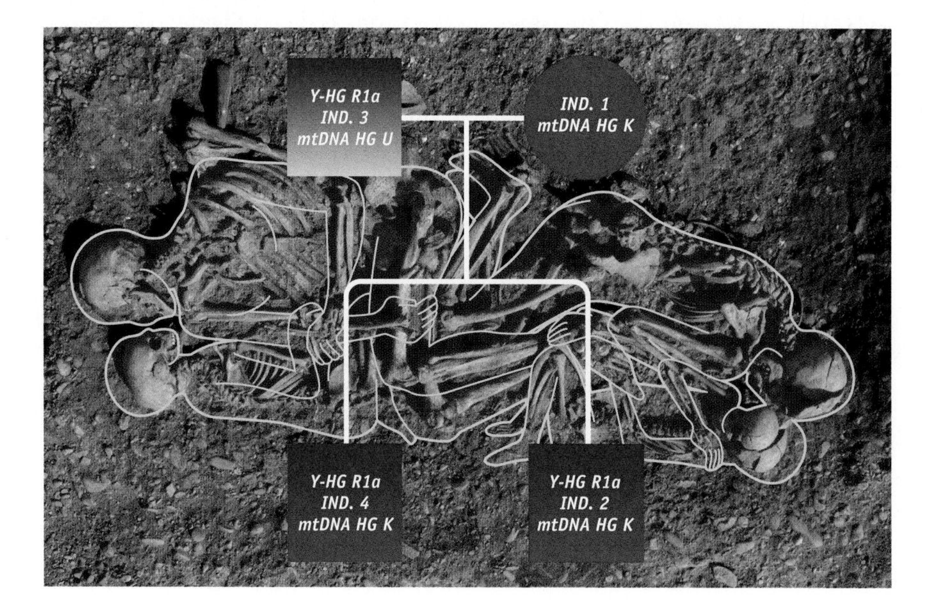

FIGURE 21.12 The genetic relationships between the four skeletons in grave 99 at Eulau. The outline drawing shows the two adult skeletons at the top, with the female on the right and the male on the left. The two children, both boys, are below. In the family tree, squares represent males and the circle represents the female adult. The mitochondrial and Y-chromosome haplotypes that were identified for each individual are shown. (Courtesy of Andrea Hörentrup, Landesamt für Denkmalpflege und Archäologie Sachsen-Anhalt, State Office for Heritage Management and Archaeology Saxony-Anhalt.)

The skeletons at Eulau date from the period when people in northern Europe were abandoning their traditional lifestyle of hunting wild animals and gathering wild plants, and were becoming farmers. Hunting and gathering requires a great deal of mobility because the animals must be followed as they migrate from place to place. Farming, in contrast, enables a more settled lifestyle, with a community staying in one place for years on end. The DNA results show that people in these early farming communities were organizing themselves into family groups similar to ones that we are familiar with today.

An interesting sideline was revealed by analyzing the strontium isotope composition of the teeth of some of the skeletons. The strontium in teeth reflects the geological isotope composition of the area where a person is living when their teeth form. The results with the Eulau skeletons suggested that although the adult males had been born and raised in the area surrounding Eulau, the females—their wives in our modern parlance—came from the Harz mountains about 40 km away. Anthropologists refer to this as an exogamous and patrilocal marriage system. Exogamy is where marriage is between members of two different social groups, and it is often practiced as a way of forging alliances and reducing conflict. Patrilocal marriage is where the female partner moves and the new couple lives close to the man's family. In modern societies patrilocal marriage typically occurs when there is patrilineal inheritance, a married son staying close to his parents so he can take over the homestead when his father dies.

21.2 USING GENETICS TO TRACE THE ORIGINS OF OUR SPECIES

Now we will look at some of the ways in which genetics is being used to peer further back in time in order to understand the origins of our species.

We know that humans originated in Africa because it is here that all of the oldest prehuman fossils have been found. The paleontological evidence reveals that humans first moved outside Africa about 1.8 million years ago, but these were not members of our own species, they were an earlier type called *Homo erectus* (Figure 21.13). The first examples of *Homo sapiens*, or "anatomically modern humans", do not appear in the fossil record until 195,000 years ago. Where did we come from?

There have been two alternative hypotheses for the origins of modern humans

Before genetics was applied to the question, two different hypotheses for the origins of modern humans had been proposed, based on comparisons using fossil skulls and bones. The first of these hypotheses held that the

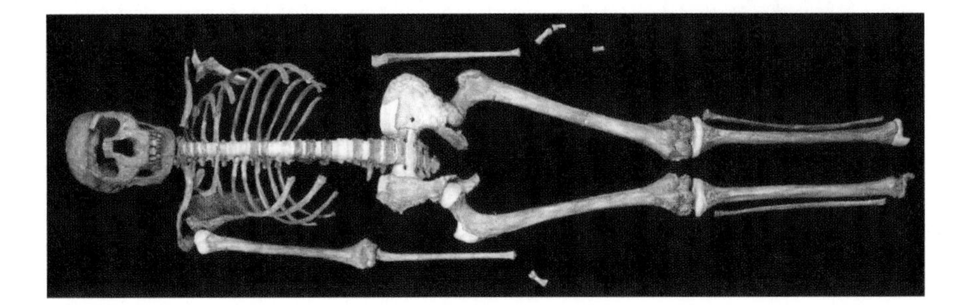

FIGURE 21.13 Skeleton of an adolescent male *Homo erectus*. This specimen, also called the "Nariokotome boy," is about 1.6 million years in age and was found near Lake Turkana, Kenya. (Courtesy of American Museum of Natural History Library.)

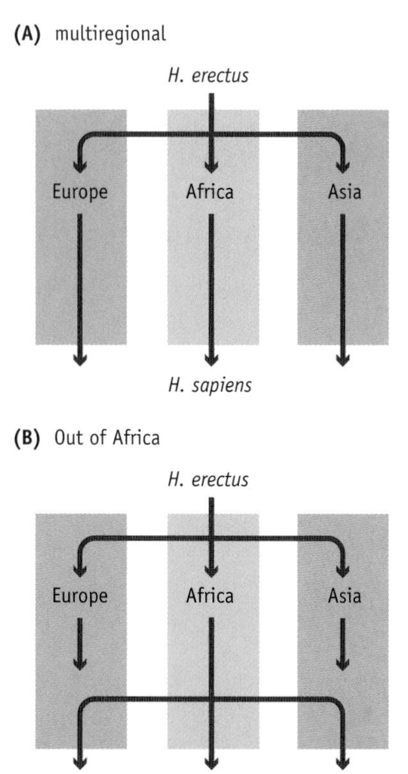

FIGURE 21.14 The multiregional (A) and Out of Africa (B) hypotheses for the origins of modern humans.

H. erectus populations that became located in different parts of the Old World gave rise to the modern human populations of those areas by a process called **multiregional evolution** (Figure 21.14A). There may have been a certain amount of interbreeding between humans from different geographic regions but, to a large extent, these various populations remained separate throughout their evolutionary history. One outcome of the multiregional hypothesis is that the geographical variations that we see today among people from different parts of the world, and which largely reflect adaptation to different climates and other environmental conditions, are relatively ancient, having evolved over a period of 1 million years or more.

The second hypothesis, first proposed in the 1980s, has a very different view of the origins of modern humans. Rather than evolving in parallel throughout the world, as suggested by the multiregional hypothesis, the **Out of Africa** hypothesis states that *H. sapiens* originated solely in Africa. Members of our species then began to move into the rest of the Old World less than 100,000 years ago, displacing the descendants of *H. erectus* that they encountered (Figure 21.14B). According to the Out of Africa model, the differences between modern human populations are relatively recent variations that evolved mostly during the past 50,000 years.

The multiregional and Out of Africa hypotheses are amenable to testing by genetic analysis because they make different predictions about our ancestry. These predictions concern the time when our most recent common ancestor (MRCA) lived. This is the person from whom all the people alive today are descended. The MRCA would not have been the only male or female alive at the time—indeed, the population of which they formed a part could have been quite large. But he or she was the only person from that period whose descendants are alive today. All of the lineages derived from everyone else alive at that time have now died out (Figure 21.15).

According to the multiregional hypothesis, our MRCA must have lived almost 2 million years ago, because after that time our ancestors were distributed across the Old World and no single person could have given rise

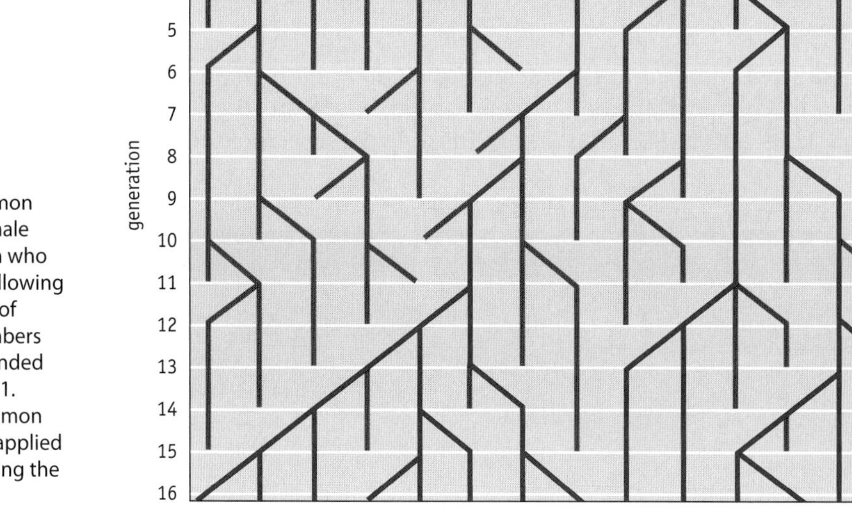

FIGURE 21.15 The most recent common ancestor. The drawing shows the female lineages descending from 15 women who make up generation 1. During the following 15 generations the lineages from 14 of these women die out. All of the members of generation 16 are therefore descended from a single member of generation 1. This woman is their most recent common ancestor. The same principle can be applied to any group of living people, including the entire population of the world.

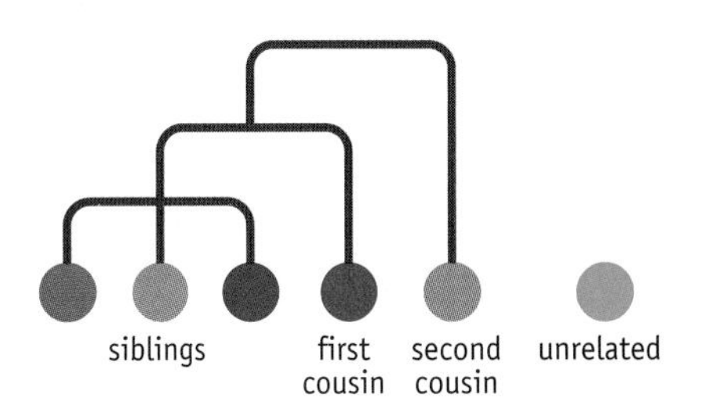

FIGURE 21.16 The degree of similarity between the genomes of different individuals reflects the closeness of the genetic relationship between them. The three siblings have closely related genomes. Their first cousin has a slightly different genome, and that of their second cousin is slightly different again. The unrelated person has a significantly more different genome.

to all of the present-day human population. In contrast, if the Out of Africa hypothesis is correct, our MRCA could have lived much more recently, possibly just 100,000–200,000 years ago, before modern humans left Africa.

The differing predictions of the multiregional and Out of Africa hypotheses have been tested by genetic studies, and we will examine the results in a moment. First we must understand the approaches and methods used in this type of work.

Comparisons between DNA sequences can be used to infer evolutionary relationships

The evolutionary history of a species can be studied by making comparisons between DNA sequences from living members of that species. This is called **molecular phylogenetics**. The underlying assumption is that the degree of similarity between the genomes of two individuals reflects the closeness of the genetic relationship between them. We expect brothers and sisters, for example, to have more closely similar genomes than cousins, who in turn have more similar genomes than individuals with no close family relationship (Figure 21.16).

When we make these comparisons we would, ideally, compare the entire genome sequences of different people, but at present this is not feasible. It might be possible in the future when more and more individual genomes have been sequenced, but for the time being we have to use much shorter sequences, containing SNPs or other polymorphisms, that we hope are representative of the genome as a whole.

Various methods can be used to depict the evolutionary relationships between DNA sequences. The most popular is to construct a **phylogenetic tree** (Figure 21.17). The branching pattern of the tree indicates the relationships between the sequences that are being compared. To construct a tree, the DNA sequences are first aligned so that the differences between them are revealed (Figure 21.18). The alignment is then converted into numerical data that can be analyzed mathematically to produce the tree. The simplest method involves conversion of the sequence data into a **distance matrix**. This is a table showing the evolutionary distances between all pairs of sequences, these distances being calculated from the number of nucleotide differences between each pair.

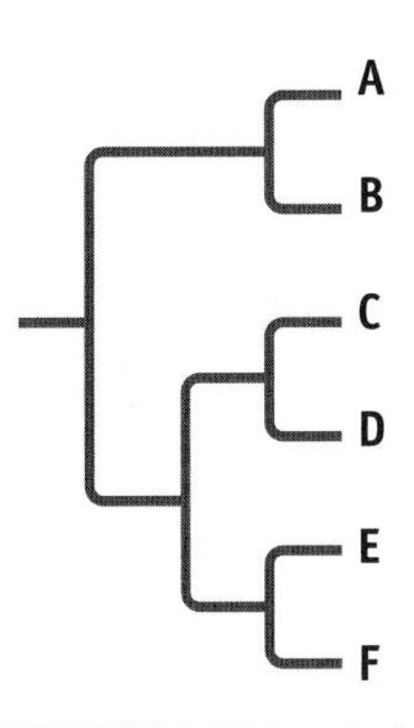

FIGURE 21.17 A typical phylogenetic tree, showing the evolutionary relationships between six DNA sequences labeled A–F.

There are several computer programs for converting distance matrices into trees, two of the most popular methods being **neighbor joining** and **maximum parsimony**. These differ in the degree of rigor that they apply to the analysis, and whether or not they simply display the relative similarities between the sequences being studied, or whether they genuinely attempt to infer an evolutionary history. The problem is that a completely rigorous analysis requires a great deal of data handling, so that even with data sets containing just 10 sequences an immense amount of computing power is required. Most tree-building programs therefore employ various shortcuts to make the analysis feasible, but with the danger that the resulting tree will contain errors.

sequence alignment

1	A G G C C A A G C C A T A G C T G T C C
2	A G G C A A A G A C A T A C C T G A C C
3	A G G C C A A G A C A T A G C T G T C C
4	A G G C A A A G A C A T A C C T G T C C

distance matrix

	1	2	3	4
1	–	0.20	0.05	0.15
2		–	0.15	0.05
3			–	0.10
4				–

FIGURE 21.18 Obtaining the data for constructing a phylogenetic tree. The alignment reveals four polymorphisms among this set of sequences. In the distance matrix, these polymorphisms are converted into numerical data. For example, sequences 1 and 2 differ at four positions out of the 20 that are shown, so the evolutionary distance is 0.20 (4/20 = 1/5 = 0.20). Sequences 1 and 3 differ at just one position, an evolutionary distance of 0.05 (1/20 = 0.05).

The molecular clock enables the time of divergence of ancestral sequences to be estimated

The topology of a phylogenetic tree reveals the pattern of evolutionary relationships between the DNA sequences that are being compared. Can we also work out when the ancestral sequences diverged to give the modern sequences (Figure 21.19)? This type of analysis will be necessary if we are going to use a phylogenetic tree to estimate the time of the MRCA of the human population.

To assign dates to branch points in a phylogenetic tree we must make use of a **molecular clock**. The molecular clock hypothesis, first proposed in the early 1960s, assumes that **substitutions** occur at a constant rate. Substitutions are those mutations that escape the repair processes and result in a permanent change in a DNA sequence. If substitutions occur at a constant rate, the degree of difference between two sequences can be used to assign a date to the branch point that gave rise to those two sequences.

For us to be able to do this, the molecular clock must be calibrated so that we know how many substitutions to expect in a particular time period. Calibration is usually achieved by reference to the fossil record. For example, fossils suggest that the most recent common ancestor of humans and orangutans lived 13 million years ago. To calibrate the human molecular clock we therefore compare human and orangutan DNA sequences to determine the number of substitutions that have occurred. We then divide this figure by 13, followed by 2, to obtain a substitution rate per million years (Figure 21.20).

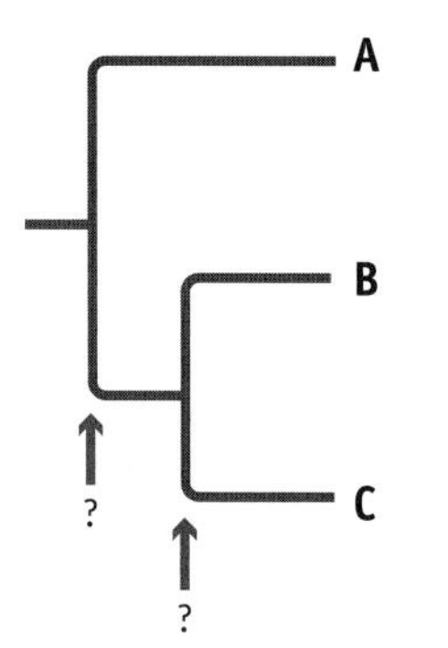

FIGURE 21.19 A phylogenetic tree showing the relationships between three DNA sequences A–C. Can we also work out when the ancestral sequences diverged to give the modern sequences? To do so we would have to assign dates to the branch points in a phylogenetic tree.

At one time it was thought that there might be a universal molecular clock that applied to all genes in all organisms. Now we realize that molecular clocks are different in different species and are variable even within a single organism. The differences between species might be the result of generation times, because a species with a short generation time is likely to accumulate DNA replication errors more quickly than a species with a longer generation time. This probably explains the observation that rodents have a faster molecular clock than primates. Within a single genome the variations reflect the effects of natural selection. Nonsynonymous substitutions (which change the amino acid sequence of the resulting protein) occur more slowly than synonymous ones. This is because a mutation that results in a change in the amino acid sequence of a protein might be deleterious to the organism, so the accumulation of nonsynonymous mutations in the population is reduced by the processes of selection. This means that when gene sequences in two species are compared, there are usually fewer nonsynonymous than synonymous substitutions.

In mammals such as humans, the molecular clock for mitochondrial DNA is faster than that for DNA in the nucleus. This is probably because mitochondria lack many of the DNA repair systems that operate in the nucleus, enabling a greater proportion of the mutations that occur to be propagated

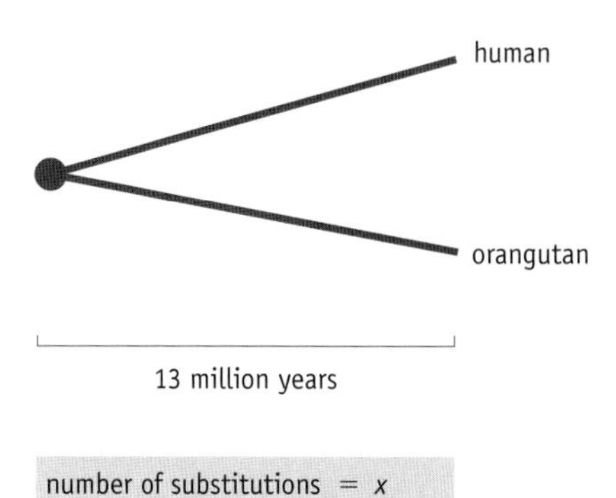

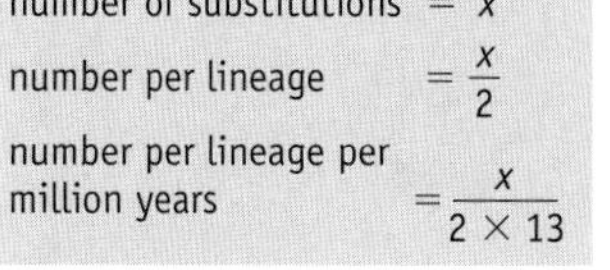

FIGURE 21.20 Calibrating the human molecular clock. The fossil record suggests that the most recent common ancestor of humans and orangutans lived 13 million years ago. We therefore compare human and orangutan DNA sequences to determine the number of substitutions (x) that have occurred. We then divide this figure by 13, followed by 2, to obtain a substitution rate per million years.

in the population. Mitochondrial DNA therefore evolves relatively rapidly, with the hypervariable regions being the fastest of all. Because of this rapid evolutionary rate, mitochondrial DNA has often been used in intraspecific studies—ones in which all the sequences come from the members of a single species. Even quite closely related populations, whose nuclear allele frequencies are very similar, will display variations in mitochondrial haplotype frequencies.

DNA analysis supports the Out of Africa hypothesis

Now we can examine how DNA analysis has been used to test the predictions of the multiregional and Out of Africa hypotheses for the origins of modern humans. This problem was first addressed in 1987, which was before PCR had become established as a means of obtaining sequences from specified regions of a DNA molecule. Instead it was necessary to make use of the less direct method of obtaining sequence data called **restriction fragment length polymorphism (RFLP) analysis**. RFLPs result from substitutions that change the sequence of a recognition site for a restriction endonuclease, so that the DNA is no longer cut at that position when treated with the enzyme (Figure 21.21). The substitution results in two fragments remaining joined together. It is revealed by examining the banding pattern obtained when the restriction digest is analyzed by agarose gel electrophoresis.

RFLP data were obtained from the mitochondrial genomes of 147 humans, from all parts of the world. The phylogenetic tree produced from these data (Figure 21.22) has a root representing a woman (remember, mitochondrial DNA is inherited only through the female line) whose mitochondrial genome is ancestral to all the 147 modern mitochondrial DNAs that were tested. This woman has been called **mitochondrial Eve**. Of course, she was not equivalent to the Biblical character and was by no means the only woman alive at the time. She is simply our female MRCA, the person who carried the ancestral mitochondrial DNA that gave rise to all the mitochondrial DNAs in existence today.

Mitochondrial Eve lived in Africa. This can be deduced because the segment of the tree lying below the ancestral sequence (as drawn in Figure 21.22) is composed solely of African mitochondrial DNAs. This topology suggests

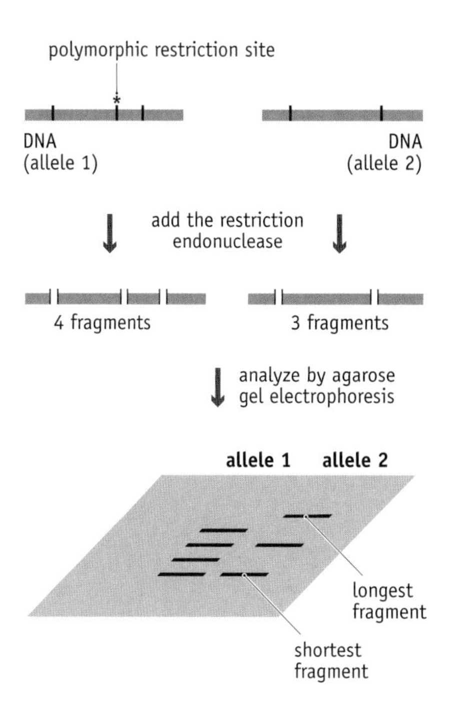

FIGURE 21.21 A restriction fragment length polymorphism (RFLP). An RFLP results from a substitution that changes the sequence of a recognition site for a restriction endonuclease. This is a type of endonuclease that only cuts DNA at a specific sequence, such as the enzyme called *Eco*RI, which cuts only at the sequence 5′-GAATTC-3′. In this drawing, the polymorphic restriction site is only present in allele 1. The two alleles can therefore be distinguished by the number of fragments that are given after treatment with the restriction endonuclease. The restriction digest is analyzed by electrophoresis in a slab of agarose gel. Agarose gel electrophoresis is similar to polyacrylamide gel electrophoresis, but is used to separate longer molecules, ones up to 20 kb.

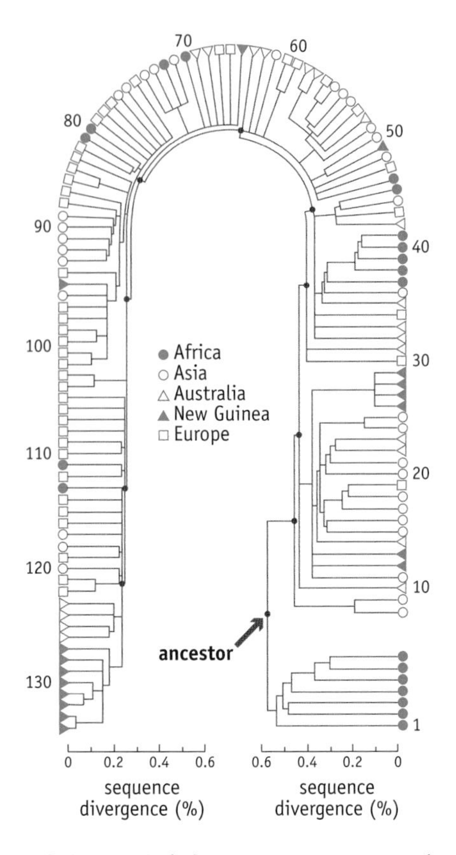

FIGURE 21.22 Phylogenetic tree constructed from mitochondrial RFLP data obtained from 147 modern humans. The scale bars at the bottom indicate sequence divergence, from which, using the mitochondrial molecular clock, it is possible to assign a date to the time when the ancestor was alive. (From R.L. Cann, M. Stoneking and A.C. Wilson, *Nature* 325: 31–36, 1987. With permission from Macmillan Publishers Ltd.)

that the ancestor was also located in Africa. She lived there between 140,000 and 290,000 years ago. This is the date obtained when the mitochondrial molecular clock is applied to the tree. The figure is compatible with the prediction made by the Out of Africa hypothesis, but incompatible with the multiregional model.

Inevitably, such an important result was subject to close scrutiny, and this led to the uncovering of one important problem. When the RFLP data were examined by other molecular phylogeneticists it became clear that the original computer analysis had been flawed, and that several quite different trees could be reconstructed from the data. These other trees gave greatly varying dates for the MRCA, some old enough to agree with the multiregional hypothesis. The problem arose from the immense amount of computer power needed to construct a phylogenetic tree. As mentioned above, constructing such a tree causes difficulties even today, and the problem was much more acute with the limited hardware available back in the 1980s. To reduce the computational requirement, the programs then used employed various shortcuts, not all of which are now looked on as sound.

So, is the discovery of mitochondrial Eve correct or not? Since 1987 there have been numerous additional projects examining mitochondrial DNA in modern humans. Many of these have used PCR so that DNA sequences, rather than RFLPs, can be obtained. All of these later studies confirm the relatively recent date for mitochondrial Eve. To take one example, when the complete mitochondrial genome sequences of 53 people, again from all over the world, were compared, a date of 120,000–220,000 years ago for the MRCA was obtained. An interesting complement has been provided by studies of the Y chromosome. This work has revealed that the paternal MRCA—Y-chromosome Adam—also lived in Africa between 40,000 and 140,000 years ago. DNA therefore provides strong support for one of the predictions of the Out of Africa hypothesis by showing that our maternal and paternal MRCAs were still living in Africa 1.5 million years or so after the spread of *Homo erectus* into the rest of the Old World.

Neanderthals are not the ancestors of modern humans

We have seen how the dates of our mitochondrial and Y-chromosome ancestors support the Out of Africa hypothesis. A second way of testing this hypothesis would be to establish the evolutionary relationship between Neanderthals and modern humans. Neanderthals (**Figure 21.23**) are an extinct type of human who lived in Europe and parts of Asia between 200,000 and 30,000 years ago. According to the multiregional hypothesis, Neanderthals are the direct ancestors of the *H. sapiens* populations living in those regions today. According to the Out of Africa hypothesis, they are not.

For us to be able to use genetics to test whether Neanderthals are ancestral to modern humans, it will be necessary to obtain DNA sequences from Neanderthal fossils. We have already seen how ancient DNA is being used to address archaeological questions such as the relationships between groups of people who were buried together. Ancient DNA is also being used to obtain sequences of the mitochondrial and nuclear genomes of Neanderthals and other extinct human species.

The first breakthrough came in 1997 when ancient DNA was extracted from the Neanderthal type specimen, believed to be between 30,000 and 100,000 years old. A series of nine overlapping PCRs enabled a 377-bp sequence to be built up from the hypervariable region of the Neanderthal mitochondrial genome. A phylogenetic tree was then constructed to compare the sequence obtained from the Neanderthal bone with the sequences of the

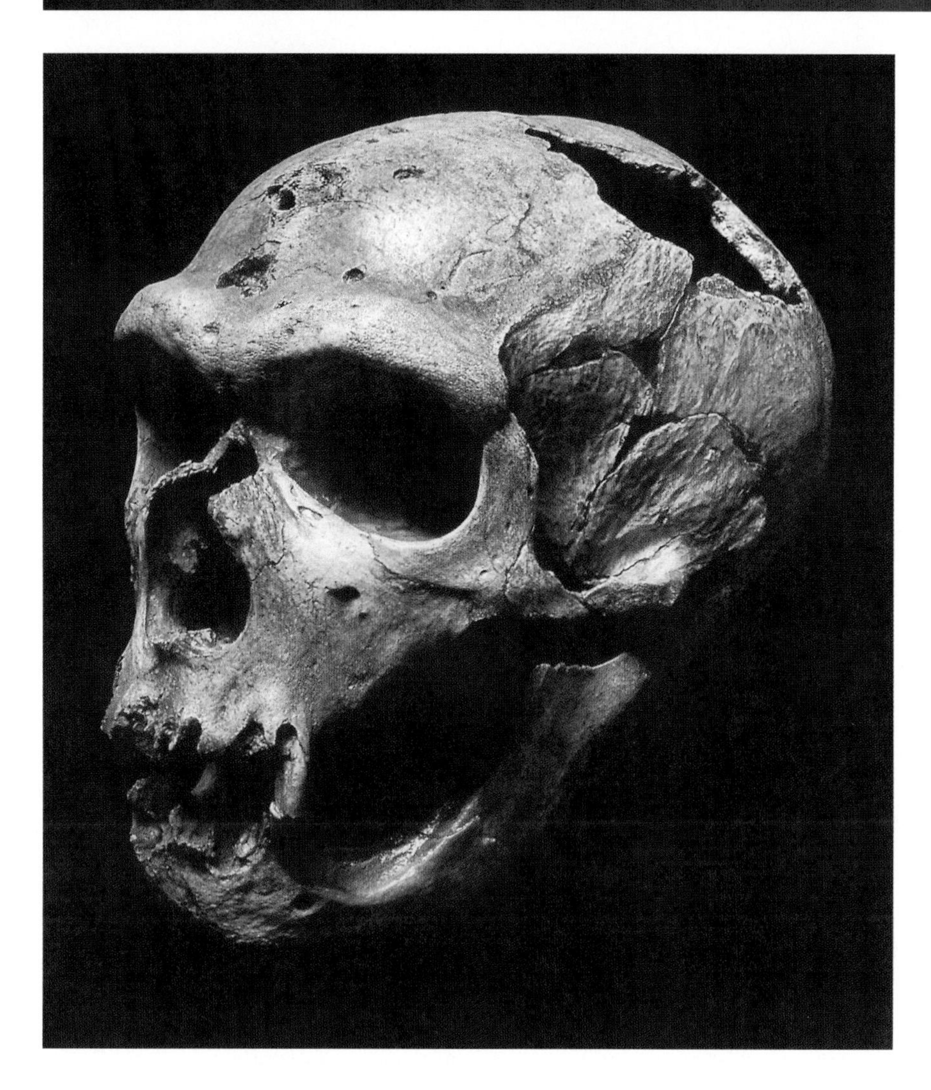

FIGURE 21.23 Skull of a 40–50-year-old male Neanderthal. The specimen is about 50,000 years old and was found at La Chapelle-aux-Saints, France. (Courtesy of John Reader/Science Photo Library.)

hypervariable regions from modern humans. The Neanderthal sequence was positioned on a branch of its own, connected to the root of the tree but not linked directly to any of the modern human sequences (**Figure 21.24A**). This result was confirmed in 2008 when a complete mitochondrial DNA sequence was obtained from three small pieces of Neanderthal bone, dating to 38,000 years ago, found in a cave at Vindija in Croatia. The phylogenetic tree constructed from this sequence again places Neanderthals on a branch of their own, separate from modern humans (**Figure 21.24B**). The Neanderthal mitochondrial genome therefore falls outside the modern human range, which would be unexpected if it were ancestral to modern mitochondrial DNAs as predicted by the multiregional hypothesis. Instead, the sequence comparisons suggest that the Neanderthals and modern humans form two separate branches of the *Homo* evolutionary tree, a scenario entirely in keeping with the Out of Africa hypothesis.

21.3 THE SPREAD OF MODERN HUMANS OUT OF AFRICA

As well as establishing that our common ancestor lived in Africa relatively recently, genetics is also helping to trace and date the migrations by which modern humans colonized the rest of the planet. The earliest fossil human remains found outside of Africa were discovered in the Qafzeh and Skhul

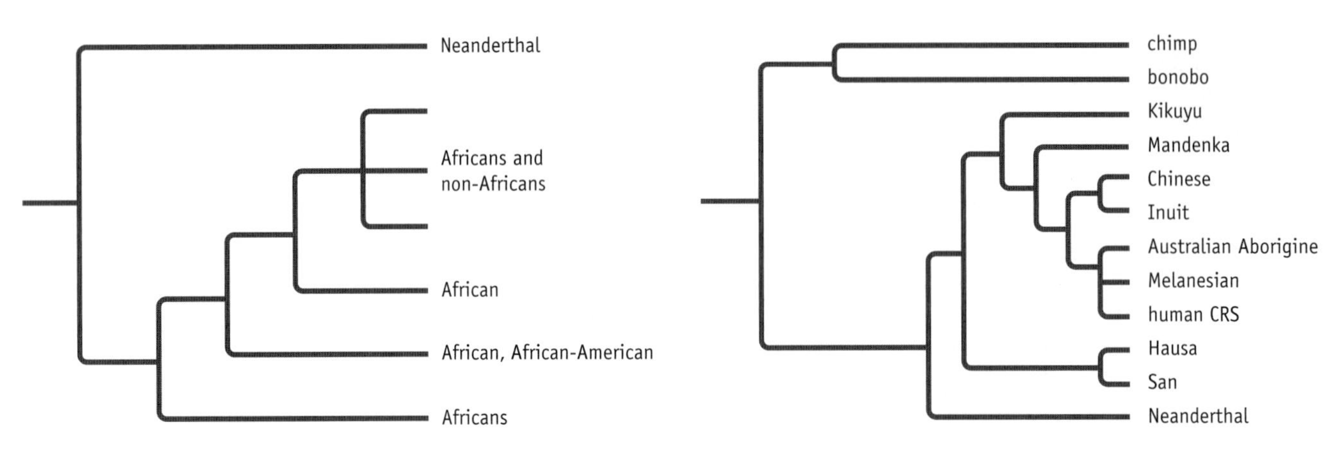

FIGURE 21.24 Phylogenetic trees constructed from (A) the Neanderthal hypervariable region and (B) the complete Neanderthal mitochondrial genome. In (A) the human mitochondrial sequences were obtained from various African and non-African groups. In (B) the human sequences came from populations associated with different parts of the world. "Human CRS" refers to the "Cambridge Reference Sequence," which was the first human mitochondrial sequence to be obtained, and was from a person of European origin.

caves near Nazareth in modern Israel. These are dated to 90,000–100,000 years ago. Are they the remains of some of the first members of *H. sapiens* to migrate out of Africa? Before we attempt to answer this question we must, once again, first consider some of the technical issues involved in using DNA to study the human past.

Phylogenetic trees and the molecular clock are not always the best tools for studying the human past

Phylogenetic tree construction was originally used as a way of studying the relationships between different species, and in some respects it is not entirely appropriate for studying the relationships between individual members of a single species. One assumption of tree building is that all of the sequences are located at ends of the branches. Often this assumption is invalid when members of the same species are compared, because now it is possible that ancestral sequences, represented by branch points within a tree, still exist in the population. If this is the case, a tree will not give a true depiction of the relationships (Figure 21.25).

One solution to this problem is to use a network rather than a tree to illustrate the evolutionary relationships between the sequences being examined (Figure 21.26). A network enables ancestral sequences that still exist to be identified and clearly illustrates their relationship with the sequences descended from them. This is the approach that is now taken in studies of human mitochondrial DNA.

We now recognize that the various mitochondrial DNA sequences that exist in the human population can be divided into a hundred or so **haplogroups**, each comprising a mixture of sequences that share certain substitutions that define that haplogroup. These defining substitutions enable the evolutionary relationships between haplogroups to be depicted as a network. The haplogroup called J1, for example, is distinguished from J because it has a substitution at position 261 in the mitochondrial DNA sequence. Haplogroup J1a is different again because of additional substitutions at

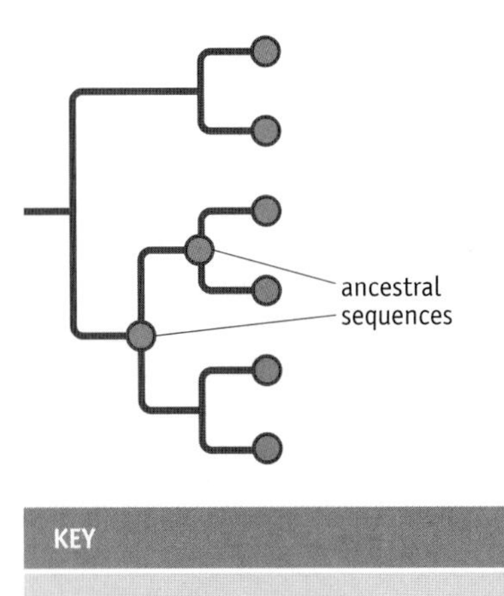

FIGURE 21.25 A phylogenetic tree will not give a true depiction of the relationships between a set of sequences if one or more ancestral sequences are still present in the population. In this example, two ancestral sequences still exist, but the tree places them at internal branch points, implying that they are extinct.

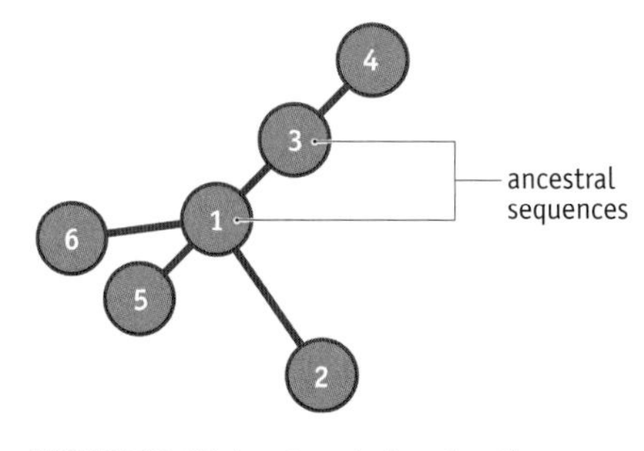

FIGURE 21.26 A network showing the evolutionary relationships between six sequences, two of which are ancestral ones.

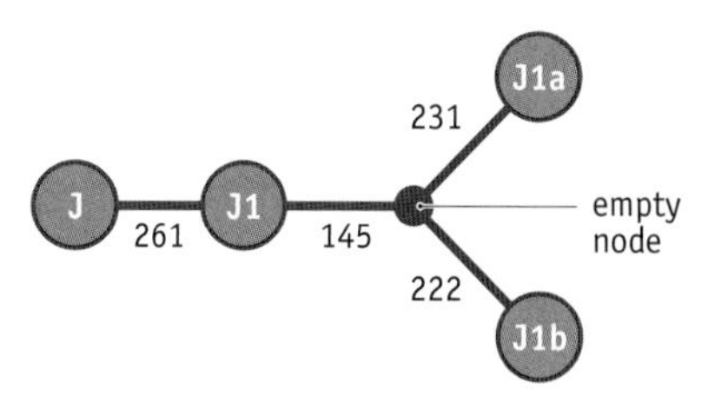

FIGURE 21.27 The relationships between mitochondrial haplogroups J, J1, J1a, and J1b. The numbers indicate the positions in the mitochondrial hypervariable region of the nucleotide substitutions that distinguish between pairs of haplogroups. The empty node represents a haplogroup that is unknown in the modern human population. It might exist in a very few individuals whose DNA has not yet been sequenced, or it might have existed in the past but become extinct.

positions 145 and 231. Haplogroup J1b has the same substitution at position 145, and a different one at 222 (Figure 21.27). All of these substitutions lie within the hypervariable region. Substitutions elsewhere in the genome are less common but are sometimes used to distinguish between certain pairs of haplogroups.

The members of a single haplogroup are not identical. They share the substitutions that define their haplogroup, but the individual haplotypes within a haplogroup differ from one another as a result of sporadic substitutions that have occurred since the group itself originated (Figure 21.28). We are now familiar with the concept of the molecular clock and its use of substitution rate as a means of assigning a date to a branch point in a phylogenetic tree. In a similar, though not identical, manner we can use the overall degree of diversity among a set of haplotypes to estimate the **coalescence time** for their haplogroup. This is an estimate of the time that has elapsed since the haplogroup first came into existence (Figure 21.29A). The reasoning is that the greater the diversity among the haplotypes, the larger the number of substitutions that has occurred, and the more ancient the coalescence time. An extension of this concept is **founder analysis**, which uses haplogroup diversities to estimate the time of a population split, such as occurs when a group of people leave their parent population and migrate to another location (Figure 21.29B).

Armed with these new ways of studying mitochondrial DNA sequences, we can start to follow the migrations of modern humans out of Africa.

One model holds that modern humans moved rapidly along the south coast of Asia

The direction taken by modern humans when they first left Africa is still a subject of intense debate. It now seems likely that the modern humans who lived near Qafzeh and Skhul 90,000–100,000 years ago were not members

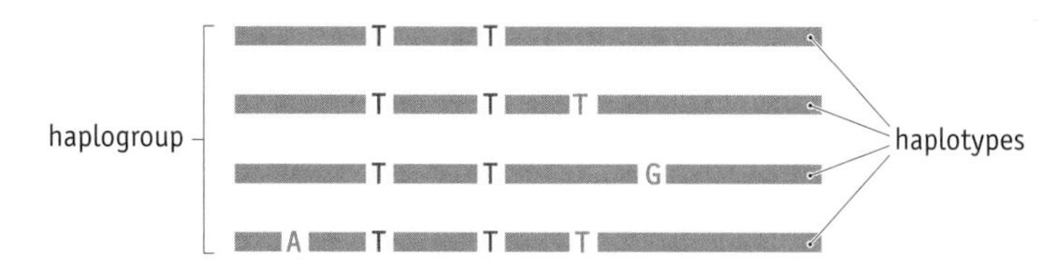

FIGURE 21.28 The relationship between a haplogroup and the haplotypes that it contains. The distinguishing substitutions of this haplogroup are the two T bases shown in black. The individual haplotypes within this haplogroup all have these two T bases, but some also have their own variations, shown in orange.

(A) the basis of coalescence analysis

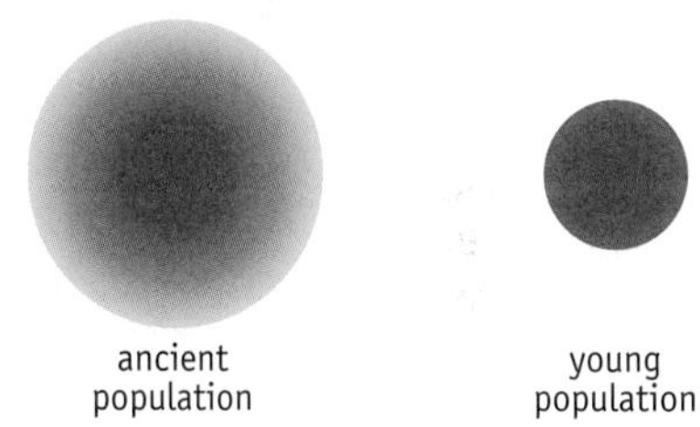

(B) the basis of founder analysis

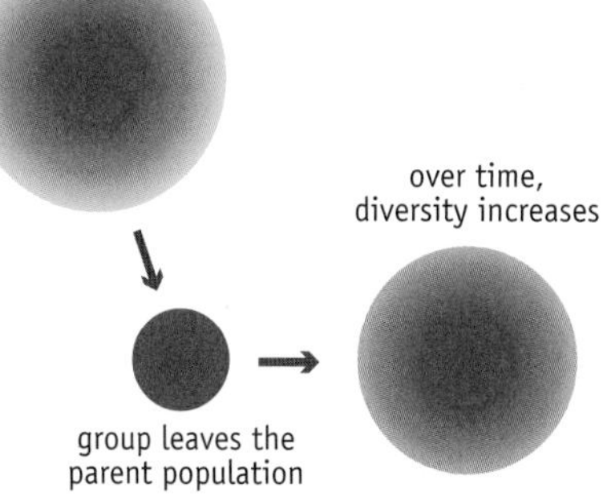

FIGURE 21.29 The basis of (A) coalescence analysis and (B) founder analysis. Panel (A) shows that an ancient population has more genetic diversity than a young one. In (B) the new group initially has less diversity than its parent population, but the diversity of this group increases over time.

RESEARCH BRIEFING 21.1

Sequencing the Neanderthal genome

The discovery of ancient DNA in Neanderthal fossils has opened up a new field called **paleogenomics**, the study of extinct genomes. One of the goals of paleogenomics is to sequence the Neanderthal genome. This is a realistic possibility because of new developments in sequencing technology. These include the invention of high-throughput methods that enable millions of short sequences, totaling several megabases, to be obtained in a single experiment.

Next-generation methods are ideal for genome sequencing

The new technology is referred to as **next-generation sequencing**. There are several types of next-generation method, but each works in a similar way (Figure 1). The DNA is broken into fragments no more than 500 bp in length, and short oligonucleotides, called **adaptors**, are ligated to each end. These adaptors enable the fragments to be attached to metallic beads. This is because one of the adaptors has a small protein called **biotin** attached to its 5′ end. Biotin has a strong affinity for a second protein, **streptavidin**, with which the beads are coated. DNA fragments therefore become attached to the beads via biotin–streptavidin linkages. The ratio of DNA fragments to beads is set so that, on average, a single fragment becomes attached to each bead. The beads are then placed in water droplets in an oil emulsion, again with one bead per droplet, and the attached DNA fragment is amplified by PCR with primers that anneal to the adaptor sequences. This means that every fragment can be amplified by the same pair of primers, each one in its individual water droplet, and many thousands of PCRs can be performed in parallel.

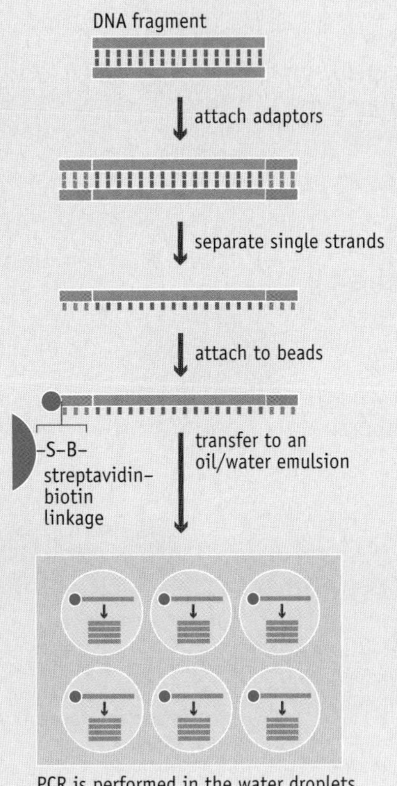

FIGURE 1 Next-generation sequencing.

The PCR products are then sequenced by the **pyrosequencing** method. This is different from the method based on dideoxynucleotides that we studied in Chapter 2. In pyrosequencing, a DNA molecule is copied by a DNA polymerase (Figure 2). As the new strand is being made, the order in which nucleotides are incorporated is detected, so the sequence can be read as the reaction proceeds. The addition of a nucleotide to the end of the growing strand is detectable because it is accompanied by the release of a molecule of pyrophosphate, which can be converted by the enzyme sulfurylase into a flash of chemiluminescence. Each nucleotide is added separately, one after the other, with a nucleotidase enzyme also present so that if a nucleotide is not incorporated into the polynucleotide it is degraded before the next one is added. This makes it possible to follow the order in which the nucleotides are added to the growing strand.

The Neanderthal genome is being sequenced from 38,000-year-old bones

The Neanderthal sequence is being generated from ancient DNA extracted from small pieces of bone from the Vindija cave in Croatia. These fragments are 38,000 years old and come from three females. Although Neanderthal fossils are known from all over Eurasia, these particular specimens were chosen because they are free from contamination with modern human DNA. This is important because there is never much ancient DNA in a specimen. Any contaminating DNA from skin cells, deposited on a specimen during handling, will be amplified more efficiently and so will dominate the reaction, possibly to the exclusion of the ancient DNA. This could lead to mistakes in assembling the genome sequence. Previous attempts to sequence the Neanderthal genome suffered from this problem, with some of the published results containing segments of Neanderthal and modern human sequences all mixed together.

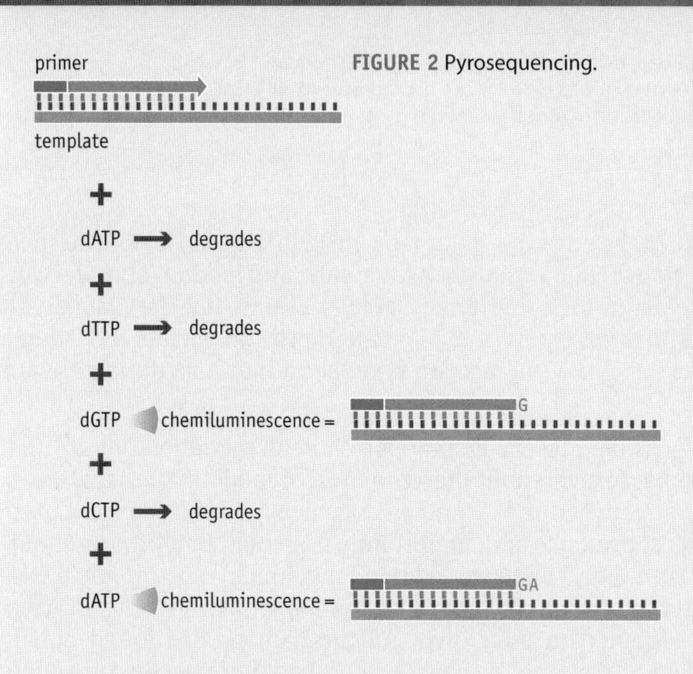

FIGURE 2 Pyrosequencing.

A draft version of the Neanderthal genome was completed in 2010. This draft was assembled from 4100 Mb of Neanderthal DNA. The genome is, like ours, 3200 Mb in length, so some parts have been sequenced more than once. This multiple coverage is important because ancient DNA sequences are not as accurate as those obtained with modern DNA, because some of the nucleotides have degraded. It will be necessary to sequence each part of the Neanderthal genome at least five times to exclude all such errors.

Neanderthals and modern humans might have interbred

Despite its draft status, the Neanderthal genome is already providing new information about the relationships between Neanderthals and modern humans. In particular, the intriguing possibility that Neanderthals and modern humans interbred is being addressed. Modern humans arrived in Europe about 45,000 years ago, and Neanderthals did not become extinct until about 30,000 years ago, so they must have lived in the same regions for several thousand years.

Tentative evidence for interbreeding has been obtained from both archaeological and genetic studies. In 1998, the skeleton of a 4-year-old child was found at Abrigo do Lagar Velho in Portugal. This skeleton seems to have both Neanderthal and modern human features. Its status as a hybrid might, however, be doubtful because the burial has been dated to 24,500 years ago, about 5000 years after the last definite evidence for Neanderthals in Europe.

Genetic evidence for interbreeding comes from an examination of modern human genes that have unusual evolutionary histories. One example is the microcephalin gene, which is thought to influence brain size. Today, there are several versions of the microcephalin gene, which we refer to as haplogroups rather than alleles because each is made up of a family of related sequences. Haplogroup D is predominant, being present in 70% of modern humans. The degree of diversity shown by the haplogroup suggests that it has been in existence for 1.7 million years but did not begin to increase in frequency until 37,000 years ago. Some geneticists believe that this is because haplogroup D was originally a Neanderthal variant that was transferred to us by interbreeding 37,000 years ago (Figure 3). Once in the modern human population it experienced a positive selective pressure that did not act on Neanderthals, resulting in its present frequency.

What does the Neanderthal genome reveal? The initial analyses suggest that some interbreeding did indeed occur. The evidence comes from comparisons between the Neanderthal genome and that of modern humans. These show that the similarity between the genomes of Neanderthals and present-day Eurasians is slightly greater than that between Neanderthals and present-day Africans. This suggests that there was interbreeding between Neanderthals and prehistoric Eurasians. If there had been no interbreeding, Eurasians and Africans should be indistinguishable when compared with Neanderthals.

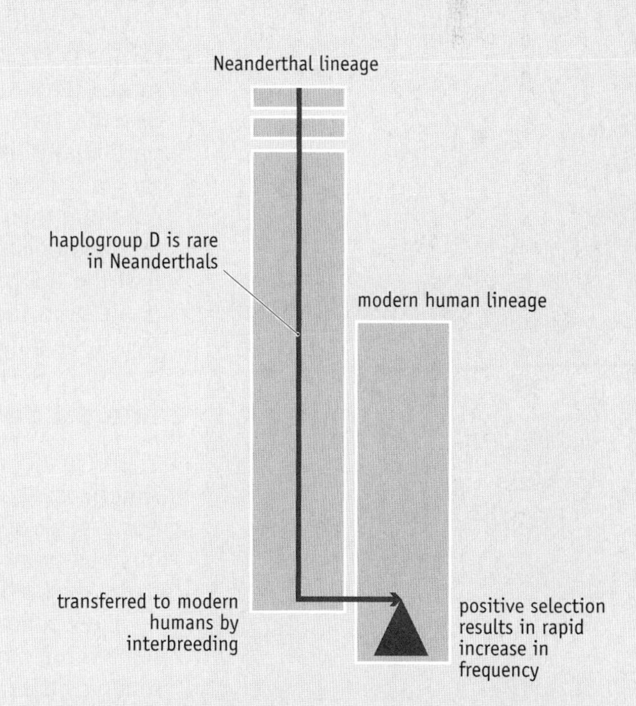

FIGURE 3 Haplogroup D of the microcephalin gene might show evidence for interbreeding between Neanderthals and modern humans.

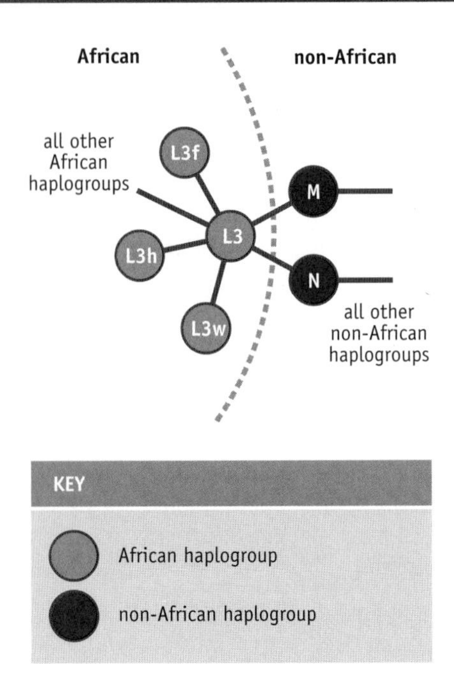

FIGURE 21.30 The key position of haplogroup L3 in the human mitochondrial DNA network. L3 is an African haplogroup, to which all of the haplogroups found outside of Africa are linked.

of the first wave of the advance of humans through the Old World. The caves were reoccupied by Neanderthals at a later date, and modern humans were not seen there again for some time. Sadly, it seems that the Qafzeh and Skhul people either were part of a failed attempt at colonization or merely represent an extension of the African population into Asia during a warm period.

It now seems possible that the first modern human migration out of Africa left from Ethiopia, further south than the area of modern Suez that forms the physical link between Africa and Asia. This hypothesis is based partly on archaeological evidence and, importantly from our point of view, is supported by the DNA data. When we examine the human mitochondrial DNA network, we see that all the haplogroups that are common today in Africa are clustered together on their own. All the haplogroups found outside Africa are linked to just one of these African haplogroups, the one called L3 (Figure 21.30). The immediate non-African descendants of haplogroup L3 are M and N, which are found mainly in Asia, suggesting that the initial migration was into Asia. The modern Africans whose L3 haplotypes have the greatest similarity to M and N live in or close to Ethiopia, so the migration probably originated in the area we now call Ethiopia, possibly by boat across the entrance of the Red Sea to the southern coast of Arabia (Figure 21.31). The sea levels might have been lower at that time, making the crossing less arduous than it would be today. Founder analysis suggests that haplogroups M and N both originated 50,000–70,000 years ago, suggesting that the migration into Asia began around that time.

What did these first migrants do once they had reached southern Asia? It seems that they moved fairly rapidly, in relative terms, along the coast toward Australia, which archaeologists believe might have been occupied by 50,000–60,000 years ago. Again there is DNA evidence to support this model. The Andaman Islanders, who live in the Bay of Bengal to the east of India, have M haplotypes that coalesce at about 60,000 years ago. This particular population therefore seems to have originated around then, not long after the initial migration from Africa and in good time for humans to continue their coastal journey to reach Australia by the time suggested by archaeologists. Combining various pieces of evidence it has been estimated that the migration moved at a rate of 0.7–4.0 km per year and involved a population including some 500–2000 females—remember, mitochondrial DNA tells only about our female past.

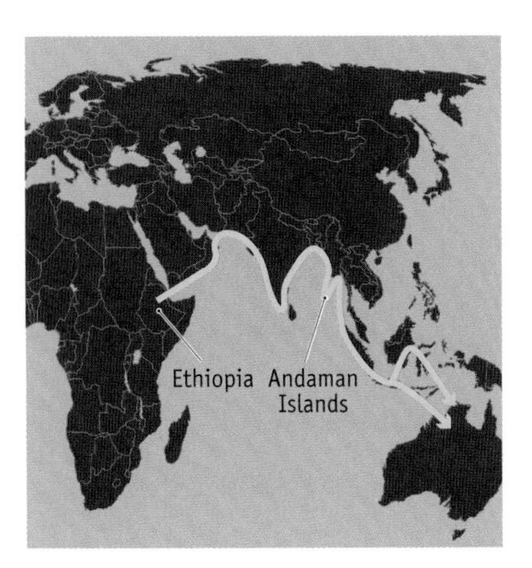

FIGURE 21.31 The possible route by which modern humans first migrated out of Africa.

Into the New World

There is no evidence for the presence of *H. erectus* anywhere in the Americas. Possibly this reflects the difficulty in reaching the continent without a sea voyage, which *H. erectus* could not or would not attempt. The obvious point of entry is across the Bering Strait, which separates Siberia from Alaska (Figure 21.32). The Bering Strait is quite shallow and if the sea level dropped by 50 m it would be possible to walk across from one continent to the other. It is believed that this Beringian land bridge was the route taken by the first humans to venture into the New World.

The sea was 50 m or more below its current level for most of the last Ice Age, between about 60,000 and 11,000 years ago, but for a large part of this time the route would have been impassable because of the build-up of ice, not on the land bridge itself but in the areas that are now Alaska and northwest Canada. In addition, the glacier-free parts of northern America would have been arctic during much of this period, providing few game animals for the migrants to hunt and very little wood with which they could make fires. Only for a brief period around 14,000–12,000 years ago was the Beringian land bridge open, at a time when the climate was warming and the glaciers were receding. During these 2000 years there was an ice-free corridor leading from Beringia to central North America (Figure 21.33). The implication is that the first modern humans might have reached the Americas about 13,000 years ago.

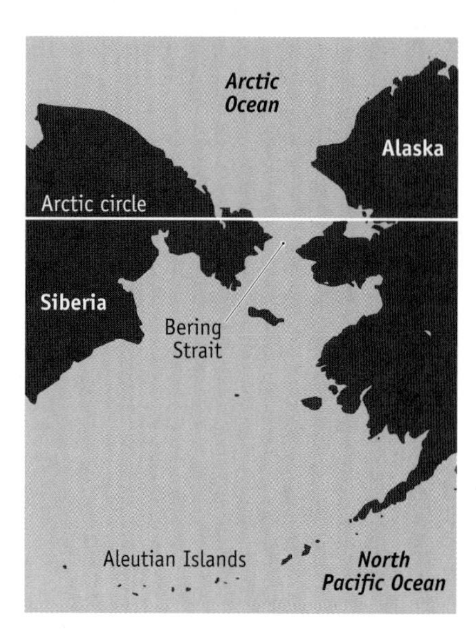

FIGURE 21.32 The Bering Strait between modern Siberia and Alaska.

The supposition that humans entered the New World via Beringia is supported by DNA analysis, because the vast majority of Native Americans have one of four mitochondrial DNA haplogroups, namely A, B, C, and D. These haplogroups are widely scattered in the mitochondrial DNA network, but all four are common in east Asia, suggesting that east Asia is the source of the populations that colonized the Americas. This is exactly what we would expect if these people entered the New World across the Bering Strait.

Was there a single migration that included a high proportion of each of the four haplogroups, or were there two or more separate migrations? This question has not yet been answered satisfactorily. The most recent common ancestors of the American members of the A, B, C, and D haplogroups each lived between 14,000 and 24,000 years ago, which some researchers think is evidence that the four haplogroups entered the New World together. The problem is that the implied date of their arrival is about 20,000 years ago, much earlier than we expected from our consideration of the period when the route through the ice was open.

This debate is still ongoing, and centers largely on the archaeological interpretation of sites in the Americas that might have been occupied by humans earlier than 12,000 years ago. The conventional view has been that the first paleoindians were the Clovis people, who made large spearheads called Clovis points, named after the site in New Mexico from which they were first extensively recovered (Figure 21.34). The first Clovis points date to about 13,000 years ago, so if the Clovis people were the first Americans, humans did not cross the Bering Strait until about that time. A growing body of archaeologists are starting to question the "Clovis-first" hypothesis, arguing that there is evidence of human occupation at sites that date to earlier than 13,000 years ago. The problem is that skeletons are rarely found, so the evidence that humans were present is indirect, including such things as the possible discovery of the remains of a hearth where a fire was tended and food was cooked.

The questions regarding the early occupation of the New World have been addressed by a DNA study. Coprolites—fossilized excrement—dating to more than 14,000 years ago were discovered in a cave in Oregon. Judging by the size, shape, and color of the coprolites they could be human in origin, or they might be from wild dogs. Excrement contains DNA from the animal or person that deposited the material, and coprolites have previously been shown to be a good source of ancient DNA, so could DNA be extracted

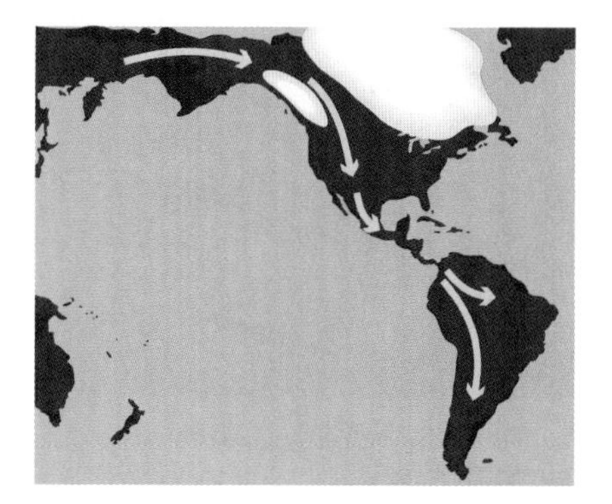

FIGURE 21.33 A possible route for the migration of humans into the New World around 14,000–12,000 years ago. The white areas are glaciers. The map shows the land bridge linking Siberia and Alaska, and the ice-free corridor leading to the southern part of the American continent.

FIGURE 21.34 Clovis points. (Courtesy of The Gault School of Archaeological Research.)

from these specimens to determine whether they are human or canine in origin? The answer is yes—DNA was present, and it was clearly human mitochondrial DNA. This discovery pushes the date for the first migration into the Americas back to 15,000 years or so. Gradually the combination of archaeology and genetics is resolving these long-standing questions about the peopling of the New World.

The origins of modern Europeans

The archaeological record shows that modern humans were present throughout most of Europe by 45,000 years ago. The controversial issue in European prehistory is whether these populations were displaced about 35,000 years later by other humans migrating into Europe from southwest Asia.

The question centers on the process by which agriculture spread into Europe. The transition from hunting and gathering to farming occurred in the Fertile Crescent of southwest Asia about 10,000 years ago, when early Neolithic villagers began to cultivate crops such as wheat and barley (Figure 21.35). After becoming established in the Fertile Crescent, farming spread into other parts of Asia and into Europe and northern Africa. By searching for evidence of agriculture at archaeological sites, for example by looking for the remains of cultivated plants or for implements used in farming, it has been possible to trace the expansion of farming along two routes through Europe. One of these routes leads around the coast to Italy and Spain and the second through the Danube and Rhine valleys to northern Europe (Figure 21.36).

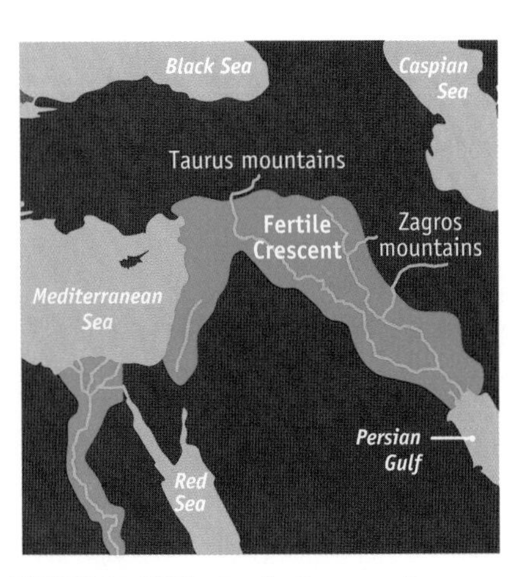

FIGURE 21.35 The Fertile Crescent of southwest Asia.

How did farming spread? The simplest explanation is that farmers migrated from one part of Europe to another, taking with them their implements, animals, and crops, and displacing the indigenous pre-agricultural communities that were present in Europe at that time. This **wave of advance** model was initially favored by geneticists because of the results of a large-scale study of the allele frequencies for 95 nuclear genes in populations from across Europe. Such a large and complex data set cannot be analyzed in any meaningful way by conventional phylogenetics but instead has to be examined by more advanced statistical methods, ones based more on population biology. One such procedure is **principal component analysis**, which attempts to identify patterns in a data set that correspond to the uneven geographic distribution of alleles, an uneven distribution possibly being indicative of a past population movement. The most striking pattern within the European data set, accounting for about 28% of the total genetic variation, is indeed a gradation of allele frequencies across

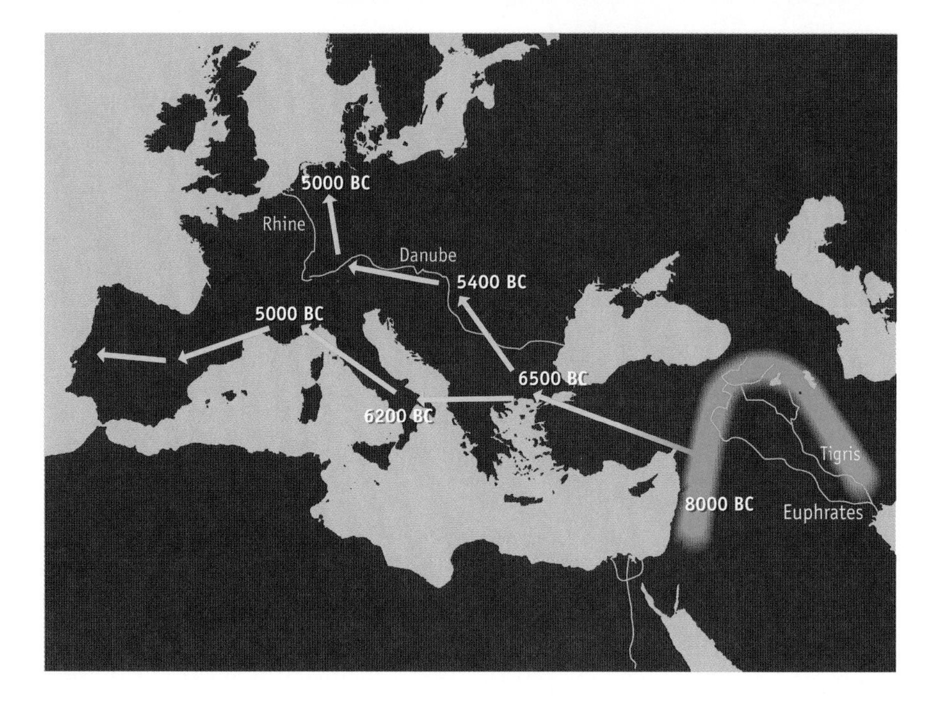

FIGURE 21.36 The routes along which agriculture is believed to have spread from southwest Asia to Europe.

the continent (Figure 21.37). This pattern implies that a migration of people occurred either from southwest Asia to northeast Europe or in the opposite direction. Because the former coincides with the expansion of farming, as revealed by the archaeological record, the results have been looked on as providing strong support for the wave of advance model.

The analysis looks convincing but has an important limitation. The data provide no indication of when the inferred migration took place. The link with the spread of agriculture is therefore based solely on the pattern of the allele gradation, not on any complementary evidence relating to the period when this gradation was set up.

A second study of European human populations, one that does include a time dimension, has reached a rather different conclusion. This study looked at mitochondrial DNA haplogroups in 821 individuals from various populations across Europe. It failed to confirm the gradation of allele frequencies detected in the nuclear DNA data set, and instead suggested that European populations have remained relatively static over the past 20,000 years. A refinement of this work led to the discovery that 11 mitochondrial DNA haplogroups predominate in the modern European population. Founder analyses suggest that these haplogroups entered Europe at different times (Figure 21.38). The most ancient haplogroup, called U, first appeared in Europe about 50,000 years ago, coinciding with the period when, according to the archaeological record, the first modern humans moved into the continent as the ice sheets withdrew to the north at the end of the last major glaciation. The youngest haplogroups, J and T1, which at 9000 years in age could correspond to the origins of agriculture, are possessed by just 8.3% of the modern European population, suggesting that the spread of farming into Europe was not the huge wave of advance indicated by the principal component study. Instead, it is now thought that farming was brought into Europe by a smaller group of "pioneers" who interbred with the existing pre-farming communities rather than displacing them.

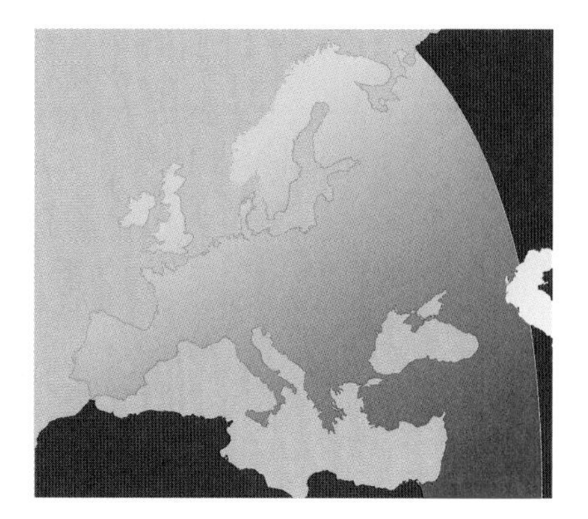

FIGURE 21.37 A genetic gradation across Europe, as revealed by principal component analysis of 95 nuclear genes.

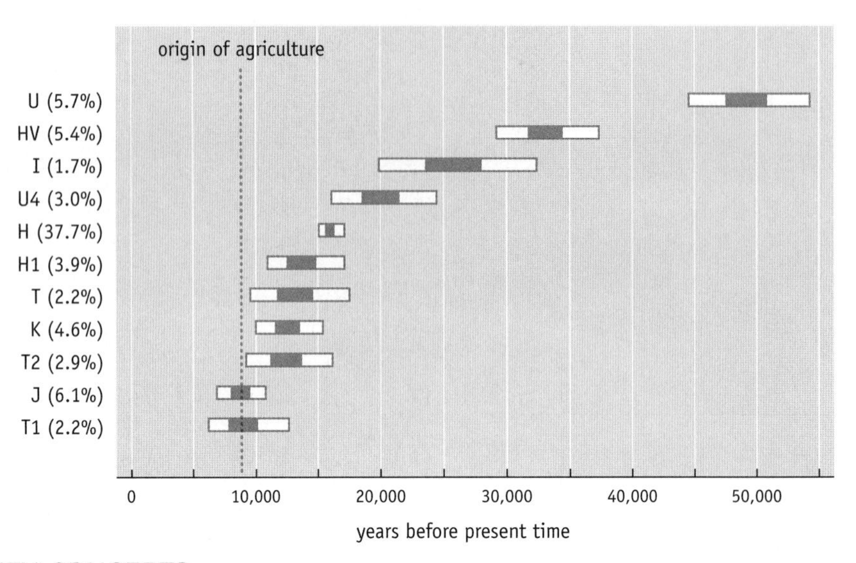

FIGURE 21.38 The eleven major European mitochondrial haplogroups. The calculated time of origin for each haplogroup is shown, the closed and open parts of each bar indicating different degrees of confidence. The percentages refer to the frequency of each haplogroup in the modern European population.

KEY CONCEPTS

- Genetic profiling is possible because of the immense variability that exists within the human genome, which makes it highly unlikely that any two people, other than identical twins, have exactly the same genome sequence.
- A genetic profile is constructed by typing the alleles present at 13 microsatellites. These have sufficient variability to give only a 1 in 10^{15} chance that two individuals have the same allele combinations. As the world population is about 6×10^9, the likelihood that any two people alive today have the same genetic profile is very small.
- Genetic profiling is also used to identify family relationships. The most common application of this type of genetic profiling is in paternity testing, when there is uncertainty regarding the identity of the father of a child.
- Genetic profiling can also be used to identify family groups among archaeological remains.
- The evolutionary history of a species can be studied by making comparisons between DNA sequences from living members of that species. Often this is done by constructing a phylogenetic tree displaying the relationships between a group of DNA sequences.
- The molecular clock is a measure of the rate at which substitutions accumulate in a DNA sequence. It can be used to assign dates to the branch points in a phylogenetic tree.
- The construction of phylogenetic trees, and the dating of branch points within those trees, have shown that all the people alive today are descended from a woman who lived in Africa between 220,000 and 120,000 years ago.
- Sequence analysis of ancient DNA from Neanderthal fossils has shown that Neanderthals are not the direct ancestors of modern Europeans.
- The degree of diversity for a set of mitochondrial DNA haplotypes can be used to deduce the age of the haplogroup to which those haplotypes belong.
- Mitochondrial haplogroup analysis has been used to trace the migrations of modern humans out of Africa and into Asia, the New World, and Europe.

QUESTIONS AND PROBLEMS (Answers can be found at www.garlandscience.com/introgenetics)

Key Terms

Write short definitions of the following terms:

adaptor
ancient DNA
archaeogenetics
biotin
coalescence time
distance matrix
founder analysis
genetic profiling
haplogroup
heteroplasmy
hypervariable region
maximum parsimony
mitochondrial Eve
molecular clock
molecular phylogenetics
multiplex PCR
multiregional evolution
neighbor joining
next-generation sequencing
Out of Africa hypothesis
paleogenomics
phylogenetic tree
polymorphism
principal component analysis
pyrosequencing
restriction fragment length polymorphism analysis
streptavidin
substitution
uniparental inheritance
wave of advance model

Self-study Questions

21.1 Why is every human genome considered to be unique? Are there any exceptions to this rule?

21.2 Explain why the microsatellites of the CODIS set are sufficient to establish a unique genetic profile for every person alive today.

21.3 Describe how a genetic profile is obtained.

21.4 Outline how genetic profiling can be used to establish that individuals are members of a single family.

21.5 Describe how the male and female lines in a family can be followed by DNA testing.

21.6 Explain how DNA testing was used to identify the skeletons of the Romanovs.

21.7 Give an example of how DNA testing has been used to study family relationships in an archaeological context.

21.8 Distinguish between the multiregional and Out of Africa hypotheses for the origins of modern humans.

21.9 Describe how comparisons between DNA sequences can be used to infer evolutionary relationships.

21.10 How is the molecular clock calibrated? Why is the molecular clock not the same for all genes in all organisms?

21.11 Explain how molecular phylogenetics has been used to provide support for the Out of Africa hypothesis.

21.12 Describe how ancient DNA has established that Neanderthals are not the ancestors of modern humans.

21.13 Outline the methods that are being used to sequence the Neanderthal genome.

21.14 What is the evidence for interbreeding between Neanderthals and modern humans?

21.15 Describe how the study of mitochondrial DNA can be used to infer the past migrations of modern humans.

21.16 Give a genetic justification for the model that modern humans left Africa about 50,000–70,000 years ago and then spread rapidly along the south coast of Asia.

21.17 What has genetics told us about the time and routes taken by the first humans to enter the New World?

21.18 Describe how the Clovis-first hypothesis has been questioned by the study of ancient DNA.

21.19 Explain how principal component analysis of human allele frequencies led to the wave of advance model for the spread of agriculture into Europe.

21.20 Describe how mitochondrial DNA studies have contributed to our understanding of the process by which agriculture spread into Europe.

Discussion Topics

21.21 Discuss the possible reasons why the allele frequencies for a microsatellite often form a Gaussian distribution.

21.22 The forensic use of DNA requires that the genetic profile of the criminal be obtained so that it can be matched with DNA from the crime scene. This process would be easier if the profiles of all members of the public were held in a national database. Discuss the ethical implications raised by the concept of a national DNA database.

21.23 Should a match between the DNA at a crime scene and that of a suspect ever be accepted as the sole type of evidence on which to base a conviction?

21.24 How reliable are molecular clocks?

21.25 Phylogenetic studies of mitochondrial DNA assume that this genome is inherited through the maternal line and that there is no recombination between maternal and paternal genomes. Assess the validity of this assumption and describe how the hypotheses regarding the origins and migrations of modern humans would be affected if recombination between maternal and paternal genomes was shown to occur.

21.26 Mitochondrial DNA is inherited only through the female line. What are the implications of this fact for the use of mitochondrial DNA to study past human migrations? What other types of study might be performed to address any weaknesses in the mitochondrial approach?

FURTHER READING

Bromham L & Penny D (2003) The molecular clock. *Nat. Rev. Genet.* 4, 216–224.

Butler J (2006) Genetics and genomics of core STR loci used in human identity testing. *J. Forensic Sci.* 51, 253–265. *Describes the CODIS set of microsatellites used in forensic science.*

Cann RL, Stoneking M & Wilson AC (1987) Mitochondrial DNA and human evolution. *Nature* 325, 31–36. *The first discovery of mitochondrial Eve.*

Coble MD, Loreille OM, Wadhams MJ et al. (2009) Mystery solved: the identification of the two missing Romanov children using DNA analysis. *PLoS ONE* 4, e4838.

Evans PD, Mekel-Bobrov N, Vallender EJ, Hudson RR & Lahn BT (2006) Evidence that the adaptive allele of the brain size gene *microcephalin* introgressed into *Homo sapiens* from an archaic *Homo* lineage. *Proc. Natl. Acad. Sci. USA* 103, 18178–18183.

Gilbert MTP, Jenkins DL, Götherström A et al. (2008) DNA from pre-Clovis human coprolites in Oregon, North America. *Science* 320, 786–789.

Gill P, Ivanov PL, Kimpton C et al. (1994) Identification of the remains of the Romanov family by DNA analysis. *Nat. Genet.* 6, 130–135.

Green RE, Krause J, Briggs AW et al. (2010) A draft sequence of the Neandertal genome. *Science* 328, 710–722.

Green RE, Malaspinas A-S, Krause J et al. (2008) A complete Neandertal mitochondrial genome sequence determined by high-throughput sequencing. *Cell* 134, 416–426.

Haak W, Brandt G, de Jong HN et al. (2008) Ancient DNA, strontium isotopes, and osteological analyses shed light on social and kinship organization of the Later Stone Age. *Proc. Natl. Acad. Sci. USA* 105, 18226–18231. *Studies of the archaeological remains at Eulau.*

Jobling MA & Gill P (2004) Encoded evidence: DNA in forensic analysis. *Nat. Rev. Genet.* 5, 739–751.

Mellars P (2006) Going East: new genetic and archaeological perspectives on the modern human colonization of Eurasia. *Science* 313, 796–800.

Soares P, Achilli A, Semino O et al. (2010) The archaeogenetics of Europe. *Curr. Biol.* 20, R174–R183. *Includes a section on genetic studies of the spread of agriculture.*

Some of the most important and controversial applications of genetics in our modern world lie in the areas of industry and agriculture. The industrial use of biology is called **biotechnology**. The common perception is that biotechnology is a new subject, but the industry has much more ancient roots. Fermentation processes that make use of living yeast cells to produce ale and mead have been exploited by humans for at least 4000 years. Since the start of the twentieth century, microorganisms have been used extensively for the industrial production of important organic compounds and enzymes such as glycerol and invertase. The discovery of penicillin by Alexander Fleming in 1929 led to a further expansion of the biotechnology industry driven by the large-scale production of antibiotics by fungi and bacteria.

The biotechnology industry was transformed in the 1970s by the first use of gene cloning to expand the repertoire of products that can be obtained from microorganisms. Cloning enables a gene for an important animal protein to be taken from its normal host and introduced into a bacterium. The transfer can be designed so that the animal gene remains active in the bacterium, which now synthesizes the protein that the gene specifies (Figure 22.1). In this way, large quantities of proteins with important pharmaceutical applications can be obtained relatively cheaply. We will examine this technology during the first part of this chapter.

More recently, gene cloning has been used with animals or plants as the host organisms. In some cases the objective is once again the production of an important protein, and in others it is to change the biological characteristics of the organism in some way that is beneficial for its use by humans. Examples include the development of **genetically modified (GM) plants** that are resistant to herbicides. We will study these aspects of biotechnology in the second part of this chapter.

22.1 PRODUCTION OF RECOMBINANT PROTEIN BY MICROORGANISMS

The term **recombinant protein** is used to describe a protein that is produced from a cloned gene. Microorganisms are looked on as ideal hosts for recombinant protein production because they can be grown at high density, enabling large amounts of the protein to be obtained. One possibility is simply to grow the engineered microbe in a large culture vessel

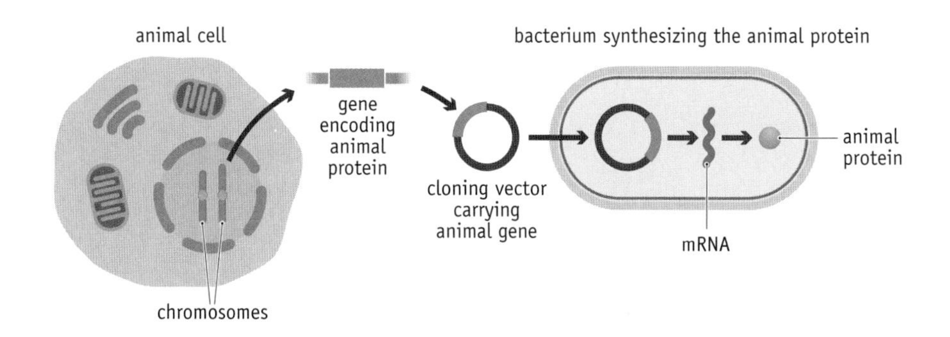

FIGURE 22.1 An outline of the way in which DNA cloning is used to synthesize an animal protein in a bacterium.

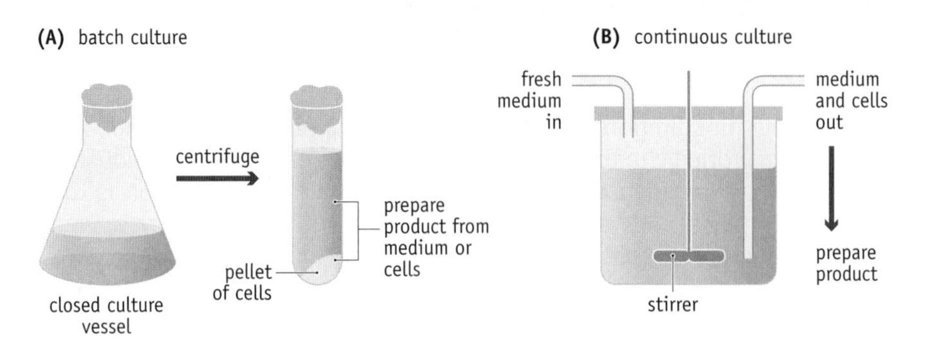

FIGURE 22.2 Two different methods for the large-scale culture of a microorganism. (A) In batch culture, the microorganism is grown in a large culture vessel from which the product is purified. (B) In continuous culture, a constant supply of product is obtained.

from which the product is purified after the cells have been removed (Figure 22.2A). This **batch culture** method gives good yields, but those yields are limited by the size of the culture vessel. Alternatively, the microbes can be grown in **continuous culture**, with a **fermenter**, from which samples are continuously drawn off (Figure 22.2B). This provides a nonstop supply of the product.

High yields of product are therefore possible, but how is a microorganism engineered so that it synthesizes a recombinant protein? We will look first at the procedures that underlie this type of genetic engineering, and then at how these procedures are used to make recombinant proteins such as insulin and factor VIII, as well as vaccines that give protection against hepatitis B and other viruses.

Special cloning vectors are needed for the synthesis of recombinant protein

The first microorganism to be used for the synthesis of recombinant protein was, not surprisingly, *Escherichia coli*. This is the microbe whose physiology and metabolism is best understood, and hence it was the natural choice for this new venture.

The animal gene coding for the recombinant protein is transferred into *E. coli* by DNA cloning (Figure 22.3). First the animal gene is inserted into a cloning vector, a DNA molecule that is able to replicate inside an *E. coli* cell. Usually the vector is a modified version of a naturally occurring plasmid. The recombinant plasmids are then mixed with competent *E. coli* cells. Some of the cells become transformed with a recombinant plasmid, which replicates inside the bacterium and is passed to the daughter bacteria when the cell divides. If a few transformed cells are used to start a batch or continuous culture, then each of the billions of bacteria present in the culture when it reaches maximum density will contain a copy of the animal gene.

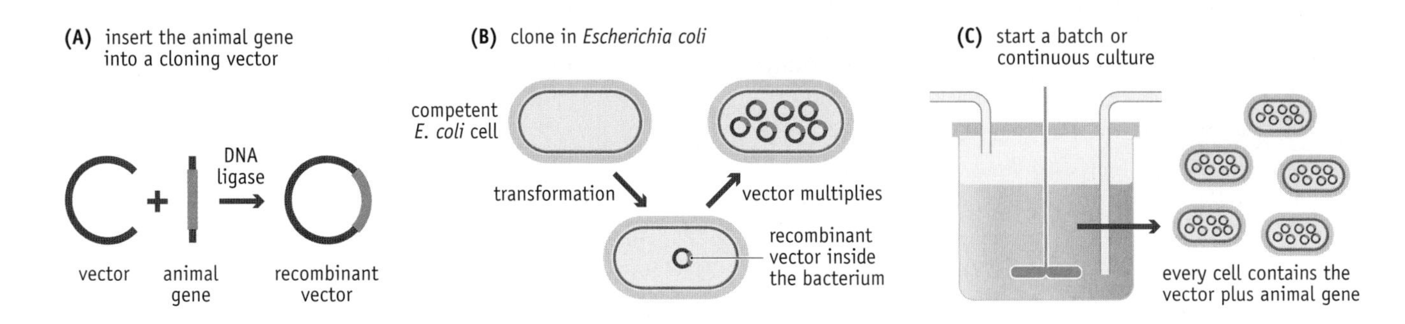

FIGURE 22.3 Cloning an animal gene in *E. coli* to obtain recombinant protein. (A) The animal gene is inserted into a cloning vector. (B) The recombinant vector is taken up by *E. coli* cells and replicates inside the bacteria. (C) A few transformed cells are used to start a batch or continuous culture. All of the bacteria in the culture will contain a copy of the animal gene.

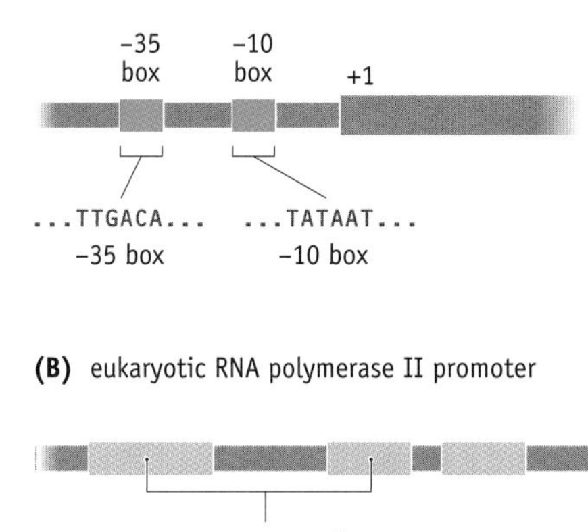

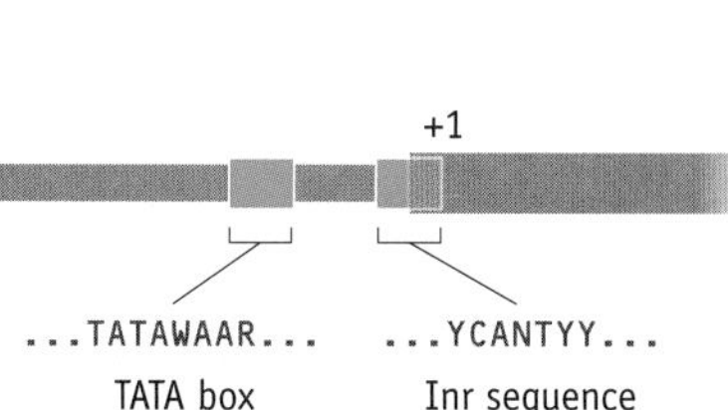

FIGURE 22.4 Comparison between the structures of the promoters of (A) *E. coli* and (B) a eukaryotic protein-coding gene transcribed by RNA polymerase II. Consensus sequences are shown, where "N" is any nucleotide, "R" is a purine, "W" is A or T, and "Y" is a pyrimidine.

The question is, will the animal gene be expressed so that recombinant protein is produced? An animal gene will not itself possess the correct signals for expression in *E. coli*. To be transcribed and translated, the gene must be preceded by an *E. coli* promoter and a ribosome binding sequence, and followed by a sequence able to act as a transcription termination site. Eukaryotic genes possess all of these signals but the sequences are different from the *E. coli* versions. This is illustrated by comparing the consensus sequences of the promoters from protein-coding genes in *E. coli* and eukaryotes. There are similarities, but it is unlikely that an *E. coli* RNA polymerase would recognize and attach to a eukaryotic promoter (Figure 22.4). Most animal genes are therefore inactive in *E. coli*.

To solve this problem, a cloning vector that will be used for recombinant protein production must contain a set of *E. coli* expression signals. These are positioned in such a way that when the animal gene is inserted into the vector it is placed downstream of a promoter and ribosome binding site and upstream of a termination sequence (Figure 22.5). Plasmids that provide these signals are called **expression vectors**.

Of these three signals, the promoter is the one that is most important. There are two reasons for this. First, the promoter determines how rapidly a gene is expressed. The strongest *E. coli* promoters direct 1000 times as many productive initiations of transcription as the weakest ones. For recombinant protein synthesis, in which high yields are important, the promoter must be as strong as possible.

The second reason why the promoter is important is because it provides a means of controlling gene expression. This can be an advantage in the production of recombinant protein, especially if the protein is harmful to the bacteria, in which case its synthesis must be carefully regulated to prevent the protein from accumulating to lethal levels. Even if the recombinant protein has no harmful effects, it is still useful to be able to regulate the expression of the cloned gene. This is because a continuously high level of transcription might interfere with plasmid replication and the partitioning of plasmids during division of the host bacterium. The recombinant plasmid might therefore be lost gradually from the culture.

The ideal promoter is therefore one that directs a high rate of transcription initiation and can be controlled by the addition of a regulatory chemical to the culture medium. There are several possible choices, but the most popular is the lactose promoter. This is the promoter that, in normal *E. coli* cells, controls transcription of the lactose operon. It is a strong promoter

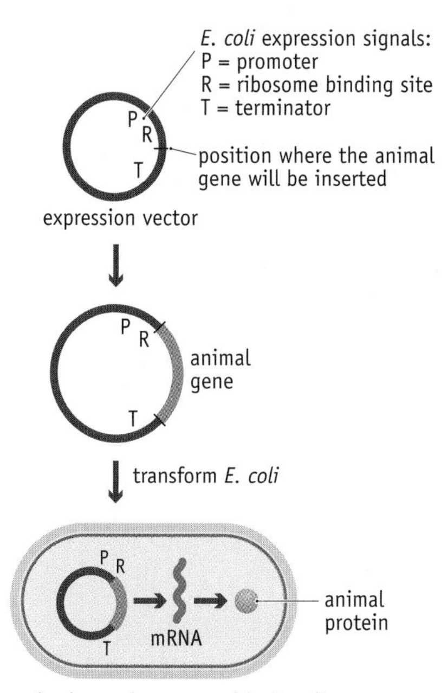

FIGURE 22.5 An expression vector and the way in which it is used.

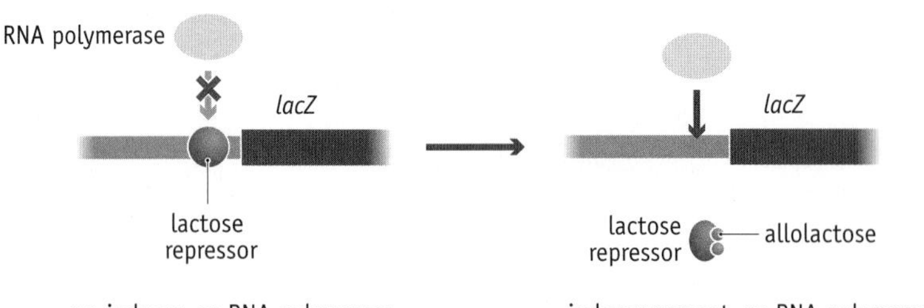
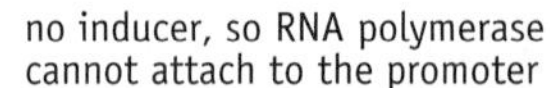

FIGURE 22.6 Induction of the lactose promoter. In the absence of an inducer, the repressor prevents the RNA polymerase from accessing the promoter.

and so directs a large amount of protein synthesis, and it is regulated by the lactose repressor. In the absence of an inducer, the repressor prevents the RNA polymerase from accessing the promoter (Figure 22.6). In the natural system, the inducer is allolactose, which binds to the repressor, changing its conformation so it detaches from the DNA. The polymerase is now able to gain access to the promoter and gene expression is switched on.

These regulatory events still occur even after the segment of DNA containing the promoter has been removed from the lactose operon and placed in a cloning vector. An animal gene that is inserted just downstream of the lactose promoter will be expressed at a high level, with expression switched on by adding an inducer to the culture medium (Figure 22.7). We could use allolactose as the inducer, but this compound is not particularly stable. It would be necessary to continually add fresh allolactose to prevent the animal gene from being switched off. Instead an artificial inducer, such as isopropylthiogalactoside (IPTG), is used. IPTG is an analog of allolactose (Figure 22.8) and can therefore bind to the repressor. It is much more stable than allolactose, so it does not need continual replenishment to maintain expression of the animal gene.

Insulin was one of the first human proteins to be synthesized in *E. coli*

The initial goal of "recombinant" biotechnology was to use *E. coli* to produce large quantities of important therapeutic proteins. Insulin, needed for the treatment of diabetes mellitus, provides a good example. Insulin is synthesized by the β-cells of the islets of Langerhans in the pancreas, and controls the level of glucose in the blood. An insulin deficiency leads to diabetes, which can be fatal if untreated. Fortunately, many forms of diabetes can be controlled by insulin injections, supplementing the limited amount of hormone synthesized by the patient's pancreas.

The insulin used in this treatment was originally obtained from the pancreas of pigs and cows slaughtered for meat production. Pig and cow insulins have very similar amino acid sequences to the human version (one and three amino acid differences, respectively) and both are able to control blood sugar levels after injection into humans. Although this is an effective way of treating diabetes, purification of insulin from pancreas tissue is not straightforward, and carryover of contaminants occasionally leads to allergic reactions. Recombinant insulin would avoid this problem because purification from a bacterial culture is a much simpler process and is less prone to contamination.

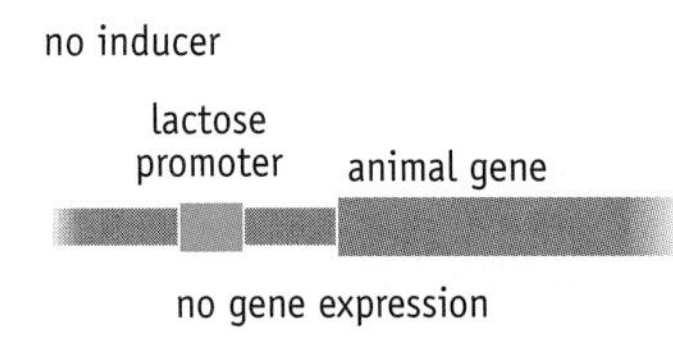
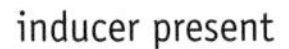
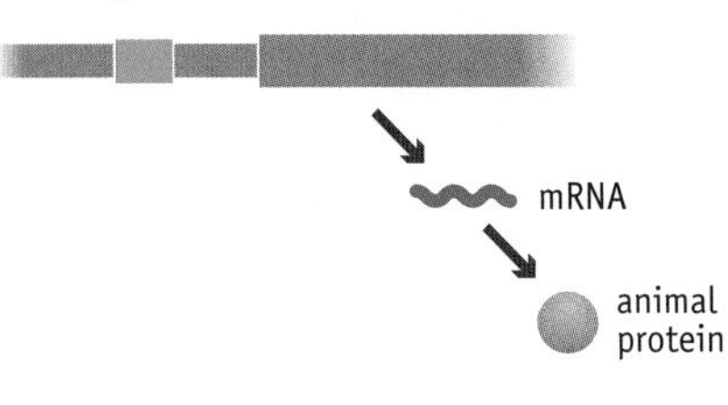

FIGURE 22.7 The use of the lactose promoter in the synthesis of animal protein in *E. coli*.

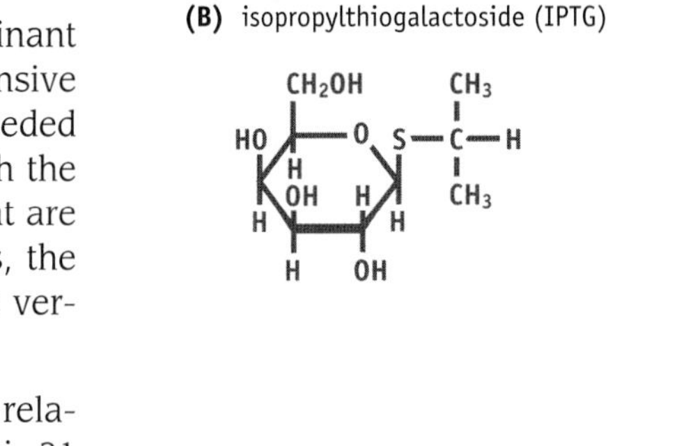

FIGURE 22.8 Structures of (A) allolactose, the natural inducer of the lactose promoter, and (B) isopropylthiogalactoside (IPTG), which is often used to induce the promoter when recombinant protein is being synthesized. Allolactose is less stable than IPTG because *E. coli* can break it down into its constituent monosaccharides.

Insulin has two features that facilitate its production by recombinant methods. The first is that the human protein does not undergo extensive post-translational modification. Bacteria lack many of the enzymes needed to perform these modifications, and in particular are unable to attach the sugar side chains that are possessed by those eukaryotic proteins that are modified by glycosylation. Because insulin lacks these modifications, the recombinant version synthesized by *E. coli* should be identical to the version made by the human pancreas.

The second important feature is the size of the molecule. Insulin is a relatively small protein, made up of two polypeptides, the A chain, which is 21 amino acids in length, and the B chain of 30 amino acids (Figure 22.9). In the pancreas these chains are synthesized as a precursor called preproinsulin, which contains the A and B segments linked by a third chain (the C chain) and preceded by a signal peptide. The signal peptide is removed after synthesis and the C chain is excised, leaving the A and B polypeptides linked to each other by two disulfide bonds.

Several strategies have been used to obtain recombinant insulin. In one of the first projects, in the 1970s, two artificial genes were synthesized, one specifying the A chain and the other the B chain. Two recombinant plasmids were constructed, one carrying the artificial gene for the A chain and one the gene for the B chain. In each case the artificial gene was inserted into an expression vector so that its reading frame became contiguous with the first few codons of the gene for the *E. coli* β-galactosidase (Figure 22.10). This is the first gene in the lactose operon and is immediately downstream of the lactose promoter. The insulin genes were therefore placed under the control of the lactose promoter, and the A and B chains were synthesized as **fusion proteins**, each consisting of the first few amino acids of

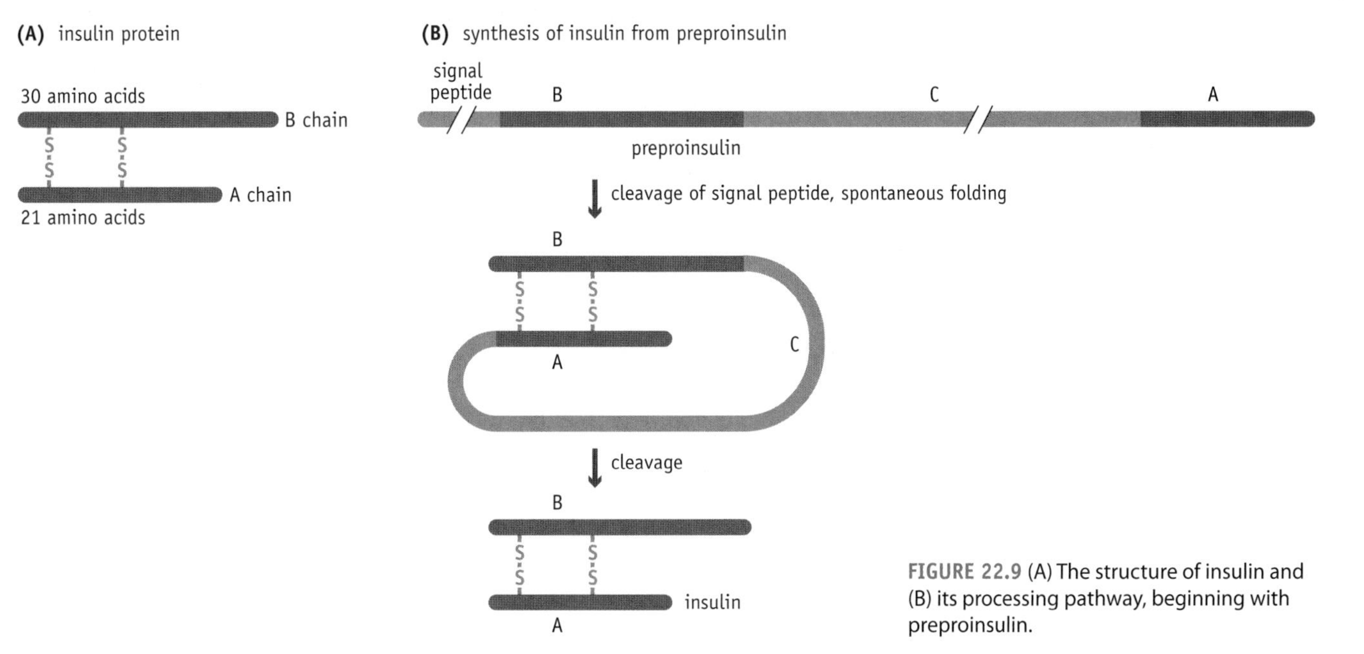

FIGURE 22.9 (A) The structure of insulin and (B) its processing pathway, beginning with preproinsulin.

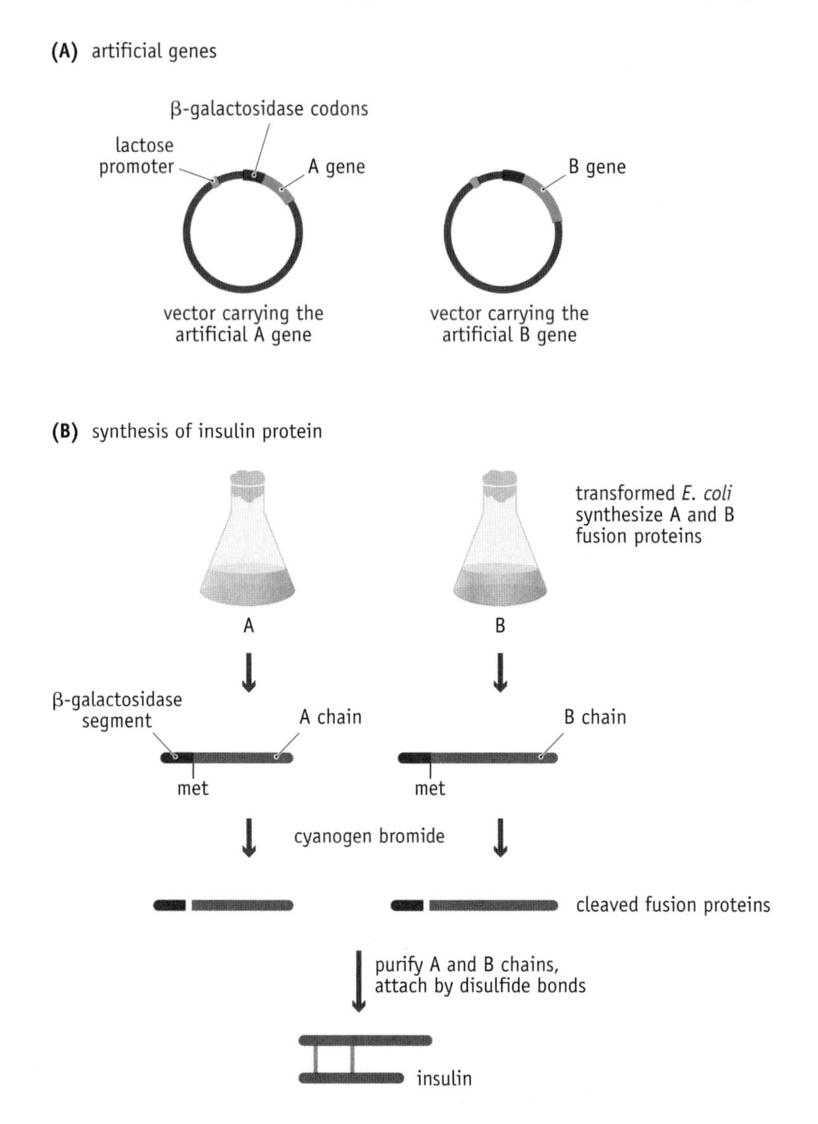

FIGURE 22.10 Synthesis of insulin as a recombinant protein in *E. coli*. (A) Two recombinant vectors were constructed, one carrying an artificial gene for the A chain and the other carrying a gene for the B chain. Each of these artificial genes was fused with the first few codons of the *E. coli* β-galactosidase gene. (B) *E. coli* was transformed with the recombinant vectors and the fusion proteins purified. The proteins were cleaved with cyanogen bromide, the β-galactosidase fragments discarded, and the A and B chains attached to each other by formation of the disulfide bonds.

β-galactosidase followed by the A or B polypeptides. Each fusion gene was designed so that the β-galactosidase and insulin segments of the protein were separated by a methionine amino acid. Methionine is attacked by cyanogen bromide, so after purification from the culture each fusion protein was treated with this chemical to cut apart the insulin polypeptide and the β-galactosidase fragment. The β-galactosidase fragment was discarded, and the purified A and B chains were attached to each other by treating with an oxidizing agent, resulting in the formation of the interchain disulfide bonds.

E. coli is not an ideal host for recombinant protein production

Despite the development of sophisticated expression vectors, recombinant protein synthesis in *E. coli* is a difficult and not always successful procedure. There are two types of problem, the first relating to the differences between animal and *E. coli* genes, and the second concerning the biochemical and physiological limitations of bacteria.

We have already seen how it is necessary to place the animal gene under the control of an *E. coli* promoter, ribosome binding site, and transcription termination sequence. This is not a major problem and is solved by the

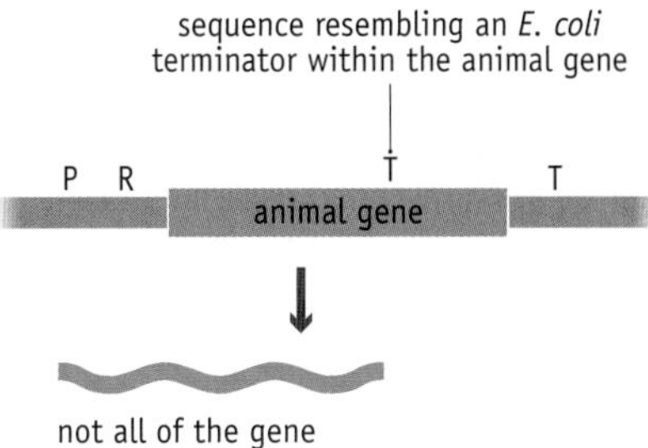

FIGURE 22.11 The problem caused by a termination signal within a cloned animal gene. P, promoter; R, ribosome binding site; T, termination signal.

design of the expression vector. But there are three other ways in which the sequence of an animal gene might hinder its expression in *E. coli*. First, the animal gene might contain introns. This is a problem because *E. coli* genes do not contain introns and therefore the bacterium does not possess the necessary machinery for removing introns from mRNA molecules. The second problem is that the animal gene might contain sequences that act as transcription termination signals in *E. coli* (Figure 22.11). These sequences would have no function when the gene was inside its animal cell, but after transfer to the bacterium might result in the premature termination of transcription and a loss of gene expression.

A final possibility is that the codon bias of the gene may not be ideal for translation in *E. coli*. Virtually all organisms use the same genetic code, but each species has a bias toward preferred codons. For example, leucine is specified by six codons (TTA, TTG, CTT, CTC, CTA, and CTG), but in human genes leucine is most frequently coded by CTG and is only rarely specified by TTA or CTA. Similarly, of the four valine codons, human genes use GTG four times more frequently than GTA. The biological reason for codon bias is not understood, but all organisms have a bias, which is different in different species. If an animal gene contains a high proportion of codons that are relatively rare in *E. coli*, recombinant protein synthesis might only occur at a low rate. The reason why codon bias should affect the rate of protein synthesis is not understood, but it probably relates to the precise sequences of the anticodons carried by the cell's tRNAs.

None of these problems are insoluble. If the animal gene contains introns, then an intron-free version can be obtained by converting its mRNA into **complementary DNA (cDNA)**. This is achieved by first copying the RNA strand into DNA, with the RNA-dependent DNA polymerase called reverse transcriptase. The new DNA strand is then copied with a DNA-dependent DNA polymerase to give double-stranded DNA that can be inserted into the vector (Figure 22.12). Because the mRNA version of a gene lacks introns, the double-stranded cDNA will also be intron-free. If the gene still contains termination sequences, or if codon bias is a problem, then **site-directed mutagenesis** techniques can be used to alter the nucleotide sequence (Figure 22.13). Alternatively, if the gene is less than 1 kb in length then it might be possible to synthesize an entirely artificial version in the test tube, designed to ensure that introns and termination sequences are absent, and that only the preferred *E. coli* codons are used.

The limitations that are placed on recombinant protein synthesis by the biochemical and physiological properties of bacteria are more difficult to

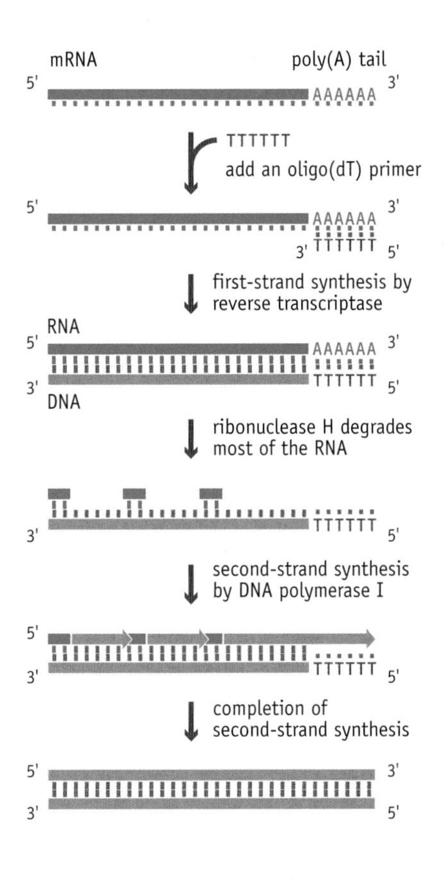

FIGURE 22.12 One method for preparing a double-stranded cDNA copy of an mRNA. The poly(A) tail at the 3′ end of the mRNA is used as the priming site for the first stage of cDNA synthesis, performed by reverse transcriptase. The oligo(dT) primer is a short DNA oligonucleotide made up entirely of T bases. When first-strand synthesis has been completed, the preparation is treated with ribonuclease H, which degrades part of the RNA molecule. The short RNA segments that remain prime the synthesis of the second DNA strand, catalyzed by DNA polymerase I.

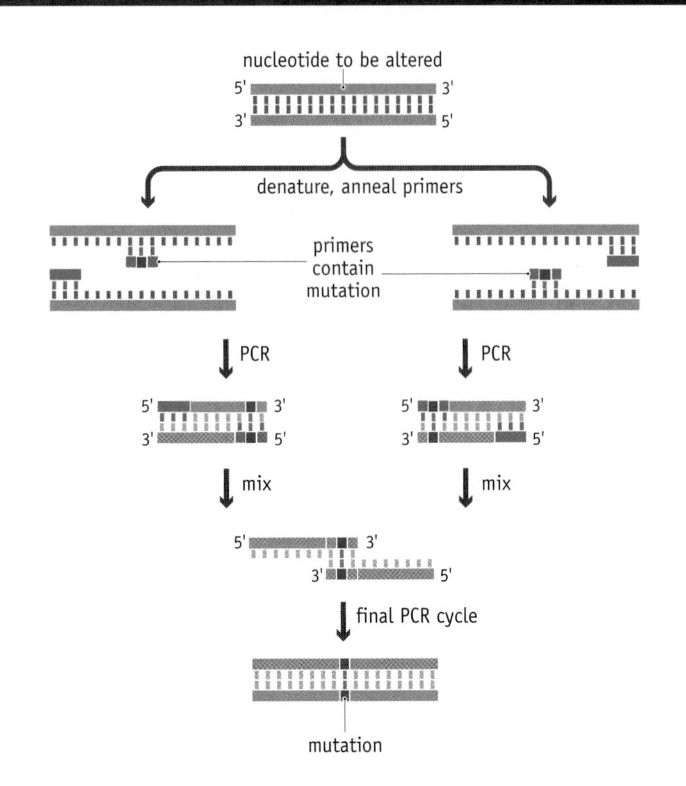

FIGURE 22.13 One method for site-directed mutagenesis—the alteration of one or more nucleotides in a DNA sequence. Two PCRs are performed. Each reaction uses one normal primer, which forms a fully base-paired hybrid with the template DNA, and a "mutagenic primer," which contains a single base-pair mismatch corresponding to the mutation. There are two mutagenic primers, which base-pair to identical positions in the DNA, but on opposite strands. After PCR, the two products are mixed together and a final PCR cycle is performed to construct the full-length, mutated DNA molecule.

solve. The protein-folding processes occurring in bacterial cells are less sophisticated than those in eukaryotes. An animal protein synthesized in *E. coli* might therefore not be folded correctly, in which case it usually becomes insoluble and forms an **inclusion body** within the bacterium (Figure 22.14). It is not difficult to recover the protein from an inclusion body, but subsequently converting it into its correctly folded form in the test tube might be impossible. An incorrectly folded protein is, of course, inactive.

The inability of *E. coli* to perform some of the complex chemical modifications that occur with animal proteins is an equally severe problem. In particular, glycosylation is extremely uncommon in bacteria, and recombinant proteins synthesized in *E. coli* are never glycosylated correctly. Glycosylation is not always essential for the function of the animal protein, but incorrect or deficient glycosylation might significantly reduce the stability of the protein when injected into a patient, and it might also induce an allergic reaction. So far the absence of glycosylation in *E. coli* has proved insurmountable, limiting this host to the synthesis of those few important animal proteins that do not need to be processed in this way.

Recombinant protein can also be synthesized in eukaryotic cells

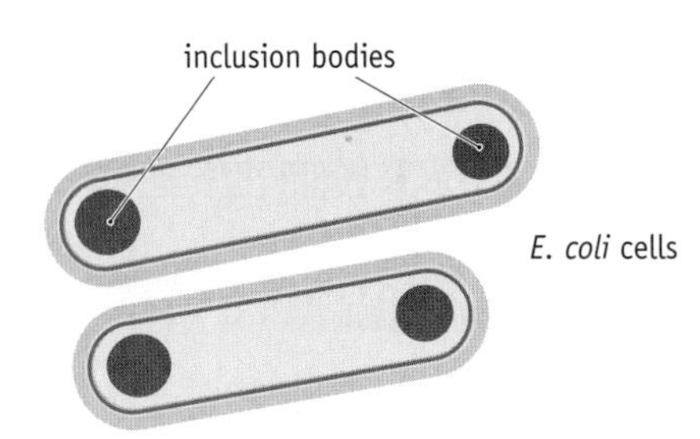

FIGURE 22.14 Inclusion bodies of recombinant protein in *E. coli*.

The problems associated with obtaining high yields of active recombinant proteins from animal genes cloned in *E. coli* have led to the development of expression systems for eukaryotic cells. Microbial eukaryotes, such as yeast and filamentous fungi, can be grown almost as easily as bacteria in continuous culture, and because they are more closely related to animals they might be able to deal with recombinant protein synthesis more efficiently than *E. coli*.

Expression vectors are still required because the promoters and other expression signals for animal genes do not in general work efficiently in these lower eukaryotes. *Saccharomyces cerevisiae* is currently the most popular microbial eukaryote for recombinant protein synthesis, partly because it is accepted as a safe organism for the production of proteins for use in medicines or in foods, and partly because of the wealth of knowledge built up over the years regarding its biochemistry and genetics. Many yeast expression vectors carry the *GAL* promoter, from the gene coding for galactose epimerase, an enzyme involved in the metabolism of galactose. The *GAL* promoter is induced by galactose, providing a convenient system for regulating the expression of a cloned animal gene (**Figure 22.15**).

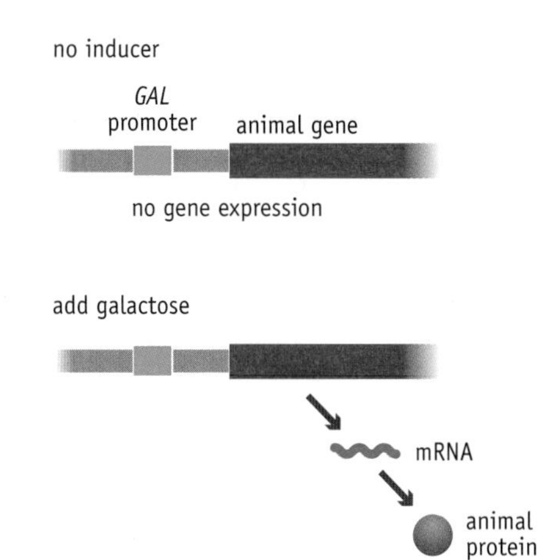

FIGURE 22.15 Induction of the *S. cerevisiae GAL* promoter.

Although yields of recombinant protein are relatively high, *S. cerevisiae* does not glycosylate animal proteins correctly, often hyperglycosylating—adding too many sugar units (**Figure 22.16**). *S. cerevisiae* also lacks an efficient system for secreting proteins into the growth medium, which means that recombinant proteins remain in the cell and are less easy to purify. A second species of yeast, *Pichia pastoris*, suffers less from these problems and has been used in some important recombinant protein projects. *P. pastoris* can synthesize large amounts of recombinant protein, up to 30% of the total cell protein. It does not glycosylate proteins in exactly the same way as animal cells, but the differences are relatively trivial and do not usually result in an allergic reaction when injected into the bloodstream. The alcohol oxidase promoter, which is induced by methanol, is often used to drive expression of the cloned gene.

Although *S. cerevisiae* and *P. pastoris* are frequently used as hosts for recombinant protein synthesis, they do not allow the efficient production of all animal proteins. For proteins with complex and essential glycosylation structures, an animal cell might be the only type of host within which the active protein can be synthesized. Culture methods for animal cells have gradually been developed over the past 50 years, to the stage where today it is possible to grow at least some types of cell in continuous systems at relatively high cell densities. Yields of recombinant protein are much lower than is possible with microorganisms, but this can be tolerated if it is the only way of obtaining the active protein. Mammalian cell lines derived from humans or hamsters are often used, which means that there are few problems with gene expression and protein synthesis, and the proteins that are made are indistinguishable from the nonrecombinant versions. The expression vector is needed only to maximize yields and enable protein synthesis to be regulated. The promoter is often from a virus such as simian virus 40, cytomegalovirus, or Rous sarcoma virus. Although this is the most reliable approach to the synthesis of active proteins, it is also the most expensive, especially because the possible presence of viruses in the cell

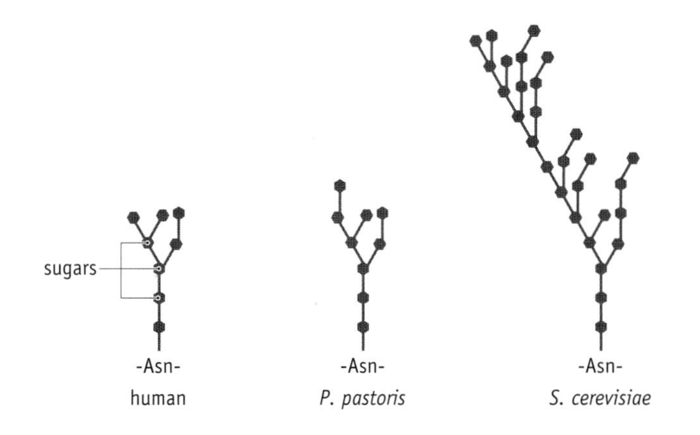

FIGURE 22.16 Comparison of typical *N*-linked glycosylation structures synthesized by human cells and by *P. pastoris* and *S. cerevisiae*.

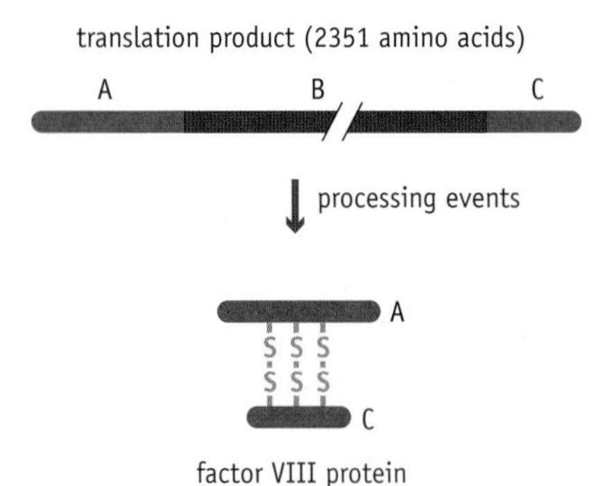

FIGURE 22.17 Processing of the human factor VIII protein. The active protein has 17 disulfide bonds and is extensively glycosylated.

lines means that rigorous quality control procedures must be employed to ensure that the purified protein is safe.

Factor VIII is an example of a protein that is difficult to produce by recombinant means

Synthesis of factor VIII as a recombinant protein has been one of the major goals of modern biotechnology. Factor VIII is the protein that is defective in the commonest form of hemophilia, leading to a breakdown in the blood clotting pathway and the well-known symptoms associated with the disease. Hemophilia can be treated by regular injections of purified factor VIII, but until recently the only source of the protein was human blood provided by donors. Purification of factor VIII directly from the blood is a complex procedure that is beset with difficulties, in particular the need to remove viruses that might be present in the donated blood. Sadly, both hepatitis and AIDS have been passed on to hemophiliacs via factor VIII injections. Recombinant factor VIII, free from contamination problems, would be a significant achievement for biotechnology.

Factor VIII has proved difficult to obtain by recombinant methods. This is mainly because of the complexity of the pathway that converts the initial translation product of the factor VIII gene into the active protein. The translation product is a polypeptide of 2351 amino acids, which is cut into three segments (Figure 22.17). The central segment is discarded and the two outer ones are joined by 17 disulfide bonds. While these events are taking place, the protein becomes extensively glycosylated. As might be expected for a protein with such a complex processing pathway, it has proved impossible to synthesize an active version of recombinant factor VIII in *E. coli* or yeast. Attention has therefore focused on mammalian cells.

Initially a cDNA version of the full-length factor VIII gene was cloned in hamster cells. Yields were disappointingly low, probably because the post-translational modifications, although carried out correctly in hamster cells, did not convert all of the initial translation product into an active form. To try to improve yields, two fragments from the cDNA were cloned, the aim being to synthesize the A and C subunits of the protein separately. Each cDNA fragment was inserted into an expression vector, downstream of a promoter and upstream of a polyadenylation signal (Figure 22.18). The two recombinant plasmids were then introduced into a hamster cell line. Within the hamster cells, the A and C protein segments were synthesized separately and then joined together by the normal processing pathway. The yields were more than 10 times greater than those from cells containing the complete cDNA, and the resulting factor VIII protein was functionally indistinguishable from the nonrecombinant version.

To produce sufficient amounts of recombinant factor VIII for commercial needs, the recombinant hamster cells are grown in continuous culture. The cells secrete the recombinant factor VIII into the culture medium, from which it is purified. The cell debris is removed, and then various chromatography and filtration procedures are performed to remove impurities and to stabilize the protein so it can be stored before use by patients.

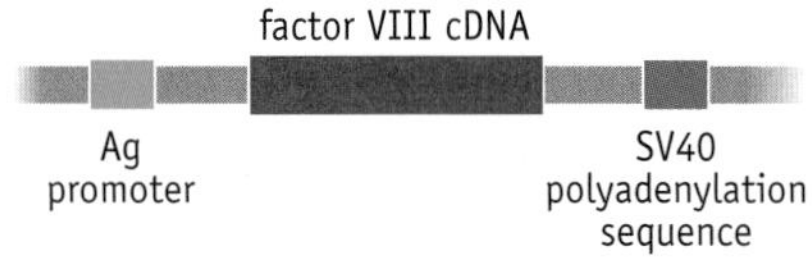

factor VIII cDNA

Ag promoter

SV40 polyadenylation sequence

FIGURE 22.18 The construct used to obtain the expression of a factor VIII cDNA in hamster cells. The Ag promoter is a hybrid of the promoter sequences of the chicken β-actin and rabbit β-globin genes. The polyadenylation sequence is from simian virus 40 (SV40).

Vaccines can also be produced by recombinant technology

Recombinant techniques are not confined to the synthesis of proteins used to treat diseases—they have also been used to synthesize vaccines. After inoculation into the bloodstream, a vaccine stimulates the immune system to synthesize antibodies that protect the body against subsequent infection with the virus, bacterium, or other pathogenic organism against which the vaccine is targeted. Most vaccines contain an inactivated form of the infectious organism, such as heat-killed virus particles. The large amounts of virus that are needed to prepare these vaccines are usually obtained from infected tissue cultures. Some viruses, such as the one that causes hepatitis B, cannot be grown in tissue culture, so obtaining enough material to make a vaccine is difficult. A second problem with these conventional vaccines is the possibility that a few viruses survive the heat treatment. If a live virus is present in a vaccine then there is the risk that inoculation will cause the disease rather than providing protection. This has happened with vaccines for the cattle disease foot-and-mouth.

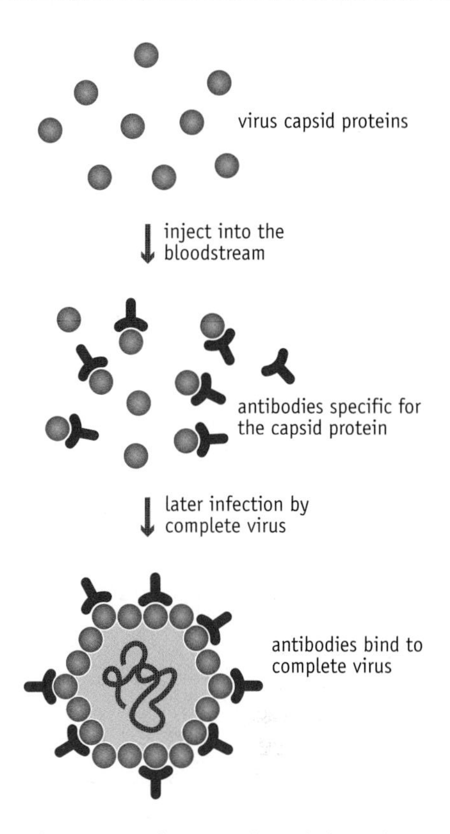

FIGURE 22.19 The rationale underlying the use of virus capsid proteins as a vaccine.

Vaccines can be effective even if they do not contain intact virus particles. Isolated components of a virus, in particular purified preparations of the capsid proteins, will also stimulate the immune system and give protection against a later infection (Figure 22.19). This provides the basis of recombinant vaccine production. The gene coding for a virus capsid protein is transferred to a microbial or animal cell, and the recombinant protein is then used as the vaccine. A vaccine prepared in this way is entirely free from intact virus particles and can be obtained in large quantities.

This approach has been most successful with hepatitis B virus. Hepatitis B is endemic in many tropical parts of the world and causes liver disease. If the infection persists, the patient might develop cancer of the liver, which has a very low survival rate. More than half a million people die every year from primary liver cancer, and 80% of these contracted the cancer as a result of hepatitis B infection.

Fortunately, most people recover from hepatitis B infection with no long-term effects. These individuals are immune to future infection because their blood contains antibodies against the capsid protein called the hepatitis B surface antigen (HBsAg) (Figure 22.20). Recombinant HBsAg has been synthesized in *S. cerevisiae* and hamster cells, and in both cases it was obtained in reasonably high quantities. When injected into test animals the recombinant proteins provided protection against hepatitis B. Both the yeast and hamster-cell vaccines have been approved for use in humans, and the yeast version is being produced commercially by two companies.

A second strategy for recombinant vaccine production makes use of engineered vaccinia viruses. Vaccinia is the cowpox virus, which is harmless to humans but stimulates immunity to smallpox when inoculated as live virus particles. Recombinant vaccinia viruses have been constructed by inserting new genes, such as the gene for HBsAg, into the vaccinia genome (Figure 22.21). The idea is that after injection into the bloodstream, replication of the recombinant virus will result not only in new vaccinia particles but also in significant quantities of HBsAg. Although this approach has not yet been approved for use with humans, it has been shown to work with animals. A

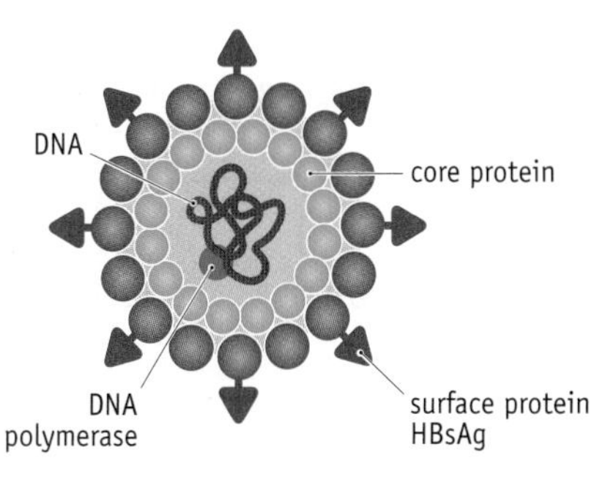

FIGURE 22.20 The structure of the hepatitis B virus, showing the locations of the HBsAg proteins on the surface of the virus.

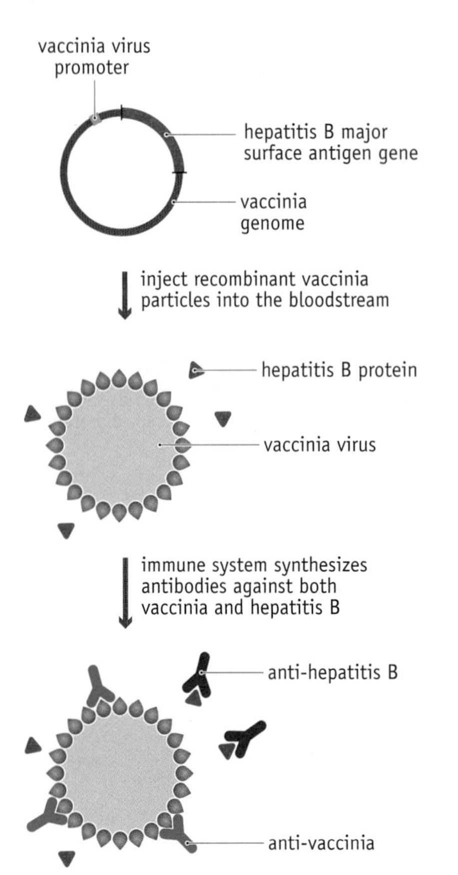

FIGURE 22.21 The rationale underlying the use of a recombinant vaccinia virus that synthesizes HBsAg proteins.

vaccinia virus expressing a coat protein from the rabies virus is being used as a vaccine to control rabies in Europe and North America.

22.2 GENETIC ENGINEERING OF PLANTS AND ANIMALS

So far in this chapter, we have only considered examples of DNA cloning in which a new gene is introduced into a microorganism or into eukaryotic cells that are being grown in culture. It is also possible to introduce new genes into animals and plants. This procedure is called **genetic engineering**.

The objective of genetic engineering is to change the phenotype of an animal or plant in a way that is considered beneficial for the applied use of the organism. Genetic engineering is being extensively used to breed improved varieties of domesticated plants and farm animals as one possible means of helping to address the food shortages facing the world in decades to come. If the engineering involves the introduction of a new gene, the product is referred to as a **transgenic** plant or animal. In the case of plants, the popular term "genetically modified" or "GM" is widely used.

Several projects are being carried out around the world, many by biotechnology companies, aimed at exploiting the potential of genetic engineering in crop and animal improvement. In this section we will investigate a representative selection of these projects.

Crops can be engineered to become resistant to herbicides

In commercial terms the most important genetically engineered organisms are crop plants that have been modified so that they are resistant to the herbicide called glyphosate. This herbicide is widely used because it is nontoxic to animals and insects and has a short residence time in soil, breaking down after a few days into harmless products. It is therefore less damaging to the environment than some other types of herbicide that have been used in the past. The problem for the farmer is that glyphosate is a broad-spectrum herbicide. It kills all plants, not just weeds, and so has to be applied to fields very carefully to avoid harming the crop plants. If crop plants could be engineered so that they are resistant to the effects of glyphosate, then a field could be treated with herbicide simply by spraying the entire area, not worrying about avoiding contact with the crop. This less rigorous application process would save time and money.

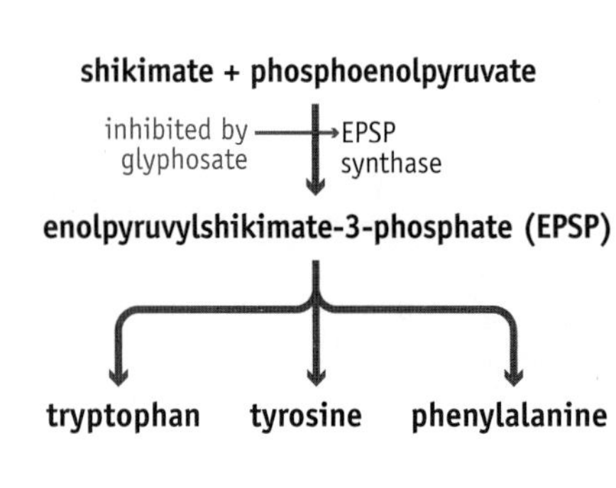

FIGURE 22.22 Inhibition of aromatic amino acid synthesis by the herbicide glyphosate.

Glyphosate kills plants by inhibiting the enzyme called enolpyruvyl-shikimate-3-phosphate synthase (EPSP synthase). Enolpyruvylshikimate-3-phosphate is a substrate for the synthesis of the aromatic amino acids tryptophan, tyrosine, and phenylalanine (Figure 22.22). Inhibition of EPSP synthase therefore leads to a deficiency in these amino acids and the subsequent death of the plant. The first crops to be engineered for glyphosate resistance were not, strictly speaking, transgenic because they did not contain any genes from other species. The aim of the genetic modification was simply to increase the amount of EPSP synthase that was made by the crop plant, in the expectation that plants overproducing the enzyme would be able to withstand higher levels of glyphosate than the nonengineered

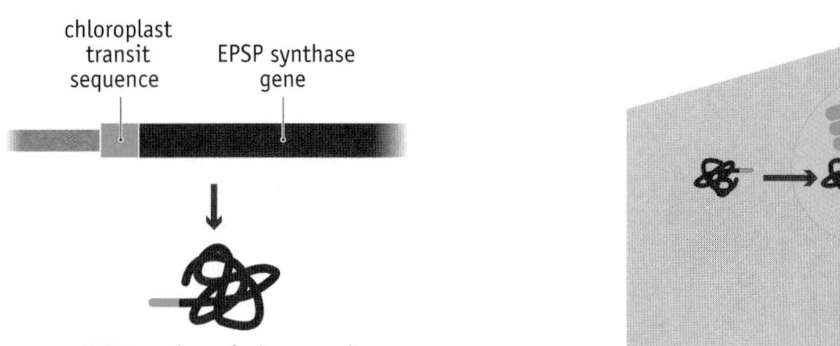

FIGURE 22.23 The transgenic strategy used to create glyphosate-resistant soybean plants. (A) The EPSP synthase gene was fused with a leader segment coding for a chloroplast transit sequence. (B) After synthesis, the transit sequence directs the fusion protein into the chloroplast. Note that the transit sequence is detached during the transport process.

weeds. As it turned out, this approach was unsuccessful. Engineered plants that made up to 80 times the normal amount of EPSP synthase were obtained, but the resulting increase in glyphosate tolerance was not sufficient to protect these plants from herbicide application in the field.

A transgenic strategy was therefore adopted, making use of organisms whose EPSP synthase enzymes are naturally resistant to glyphosate. Transfer of the EPSP synthase gene from such an organism into a crop plant should confer resistance to the herbicide. The gene that was chosen was not from a plant but from a soil bacterium called *Agrobacterium* strain CP4. EPSP synthase is located in the plant chloroplasts, so the *Agrobacterium* gene was inserted into a cloning vector in such a way that it became fused with a set of codons specifying a chloroplast transit sequence. This is an amino acid sequence that, after translation, directs a protein across the chloroplast membrane and into the organelle (Figure 22.23). The recombinant plasmid was introduced into soybean cells by **biolistics**. In this method, small gold microprojectiles are coated with the plasmid and literally fired into the cells with a gun (Figure 22.24). If done carefully, this does not cause terminal damage to the cells. Plant cells are totipotent, so a single transformed cell can give rise to a mature plant, all of whose cells will contain the new gene (Figure 22.25). These plants were three times more resistant to glyphosate than nonengineered soybean. Similar experiments have more recently been performed with maize.

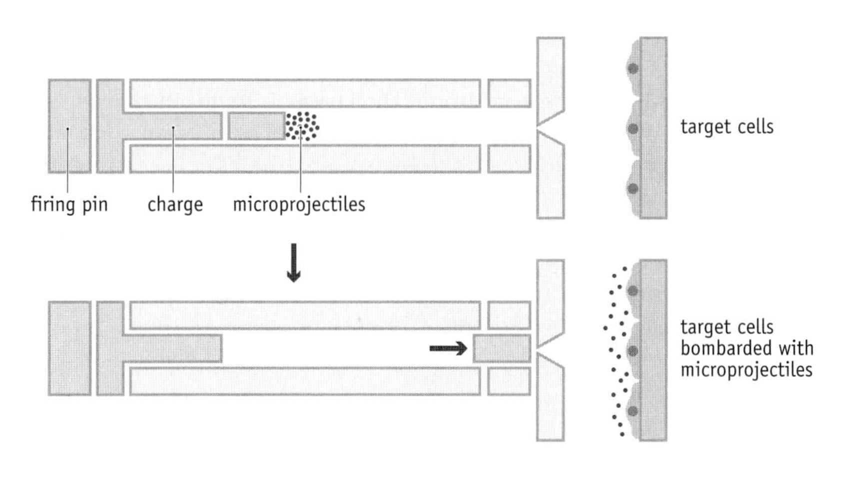

FIGURE 22.24 Biolistics, the process by which DNA is fired into plant cells on gold microprojectiles.

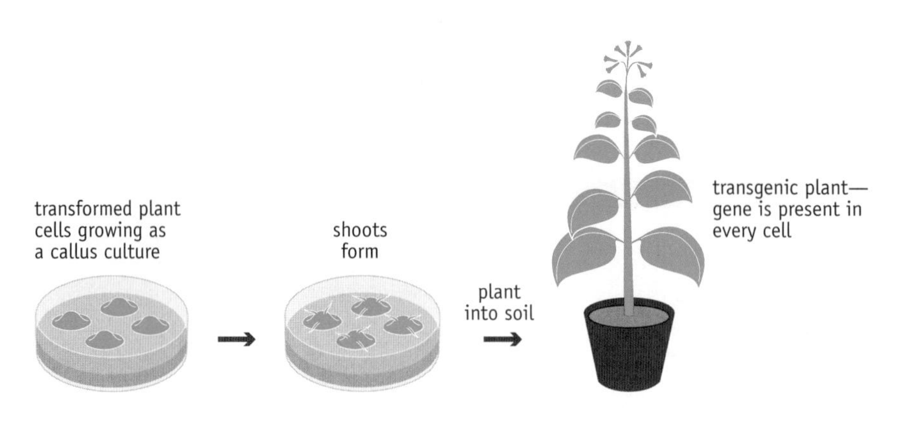

FIGURE 22.25 Regeneration of a transgenic plant from transformed plant cells. The transformed cells are initially plated onto solid agar and grown into small callus cultures. The addition of plant growth compounds, such as 6-benzylaminopurine and naphthalene acetic acid, to the agar stimulates the calluses to produce shoots that can be propagated to give the mature plant.

Various GM crops have been produced by gene addition

Herbicide resistance is just one of the ways in which the addition of new genes into plant genomes has been used to breed crops with advantageous properties. One of the first examples of this **gene addition** approach to genetic engineering involved the introduction into plants of genes for insect resistance. The conventional way of protecting crops from the potentially devastating effects of insect attack is to spray regularly with insecticides. Most conventional insecticides, such as pyrethroids and organophosphates, are nonspecific and kill many different types of insect, not just the ones eating the crop. Several of these insecticides also have potentially harmful side effects for other species, including humans. To make matters worse, the treatment is sometimes ineffective because insects that live within the plant, or on the undersurfaces of leaves, may avoid the insecticide altogether.

The search for more environmentally friendly insecticides has focused largely on the δ-endotoxins of the soil bacterium *Bacillus thuringiensis*. The δ-endotoxins are proteins that the bacteria make to protect themselves from being eaten by insects. The proteins are some 80,000 times more poisonous than chemical insecticides, and are synthesized initially as a harmless precursor that is converted into the toxic form only after ingestion by the insect. They are selective, with different strains of *B. thuringiensis* synthesizing proteins effective against the larvae of different groups of insects. All of these properties make the δ-endotoxins very attractive as environmentally friendly insecticides. It is no surprise that the first patent for their use was taken out as far back as 1904.

In 1904 nobody had any idea that it might be possible to use gene cloning to engineer plants to make their own δ-endotoxins. Instead, up to the 1980s, they were used as conventional insecticides, and were not particularly popular because their instability meant that they had to be sprayed repeatedly onto crops. Transfer of δ-endotoxin genes into plants was first achieved in 1993, resulting in GM maize plants with significant levels of resistance to the European corn borer, the larva of the moth *Ostrinia nubilalis* (Figure 22.26). This caterpillar tunnels into the plant from eggs laid on the undersurfaces of the leaves, thereby evading the effects of insecticides applied by spraying. The average length of the tunnels made by larvae in the GM plants was just 6.3 cm, compared with 40.7 cm for the unmodified controls. In addition to maize, "Bt" (derived from the name *B. thuringiensis*) versions of potato and cotton have also been produced, displaying resistance to a range of insects and hence improving yields of these important crops.

FIGURE 22.26 The European corn borer, the larva of the moth *Ostrinia nubilalis*. (Courtesy of Keith Weller/USDA.)

Gene addition has also been used to improve the nutritional content of crop plants. Rice, for example, has been engineered so that the grains produce

geranylgeranyl diphosphate

↓ phytoene synthase—from maize

phytoene

↓ carotene desaturase—from *E. uredovora*

ζ-carotene

↓ carotene desaturase—from *E. uredovora*

lycopene

↓ lycopene cyclase—already present in rice

β-carotene

FIGURE 22.27 The pathway leading from geranylgeranyl diphosphate to β-carotene. The sources of the enzymes used to synthesize β-carotene in transgenic rice are indicated.

β-carotene, which humans use as a precursor for vitamin A synthesis. The objective is to produce a GM variety of rice for those parts of the world where there is a high incidence of vitamin A deficiency. Almost half a million children every year become blind because they do not have enough vitamin A in their diet. Engineering rice grains to produce β-carotene is a more complicated project than the ones we have studied so far, because two genes must be transferred—one from maize and one from the bacterium *Erwinia uredovora*. The enzymes specified by these two genes work together to convert geranylgeranyl diphosphate into lycopene. This compound can then be converted into β-carotene by an enzyme that is already present in rice (Figure 22.27). The higher β-carotene content gives the grains a yellow color, and the GM variety is called "golden rice" (Figure 22.28).

Other examples of GM plants produced by gene addition are listed in Table 22.1. These projects include a second approach to insect resistance, using genes coding for proteinase inhibitors. These are small proteins that interfere with enzymes in the insect gut, preventing the insect from feeding on the GM plants. Proteinase inhibitors are produced naturally by legumes such as cowpeas and common beans, and genes from these species have successfully been transferred to other crops that do not normally make the proteins in significant amounts. Proteinase inhibitors are particularly effective against beetle larvae that feed on stored seeds. In a different sphere of commercial activity, ornamental plants with novel flower colors are being produced by transferring genes involved in pigment synthesis from one species to another.

FIGURE 22.28 Comparison of golden rice (right) with white rice (left). (Courtesy of Golden Rice Humanitarian Board. http://www.goldenrice.org)

TABLE 22.1 EXAMPLES OF PLANT GENETIC ENGINEERING PROJECTS INVOLVING GENE ADDITION

Added gene	Source organism	Characteristic conferred on the engineered plants
δ-Endotoxin	*Bacillus thuringiensis*	Insect resistance
Acyl carrier protein thioesterase	California bay laurel tree	Modified fat and oil content
Chitinase	Rice	Fungal resistance
Δ12-Desaturase	Soybean	Modified fat and oil content
Dihydroflavanol reductase	Various flowering plants	Modified flower color
Enolpyruvylshikimate-3-phosphate synthase	*Agrobacterium* spp.	Herbicide resistance
Flavonoid hydroxylase	Various flowering plants	Modified flower color
Glyphosate *N*-acetyltransferase	*Bacillus licheniformis*	Herbicide resistance
Methionine-rich protein	Brazil nuts	Improved sulfur content
Monellin	Sweet prayers plant	Sweetness
Ornithine carbamyltransferase	*Pseudomonas syringae*	Bacterial resistance
Phosphatidylinositol-specific phospholipase C	Maize	Drought tolerance
Phytoene synthase + carotene desaturase	Maize, *Erwinia uredovora*	Improved nutritional content
Proteinase inhibitors	Various legumes	Insect resistance
Thaumatin	Sweet prayers plant	Sweetness
Virus coat proteins	Various viruses	Virus resistance

Plant genes can also be silenced by genetic modification

Plants can also be genetically engineered so that targeted genes are partly or completely suppressed. This is sometimes called **gene subtraction**, but the term is inaccurate because the gene is inactivated, not removed. There are several possible ways of inactivating a single, chosen gene in a living plant, the most successful being the use of antisense RNA technology, which we met in Chapter 20 as a potential means of silencing disease genes by gene therapy.

Antisense RNA technology has been used to inactivate ethylene synthesis in plants that are likely to spoil as a result of overripening during storage. Ethylene is a regulatory compound that switches on the genes involved in the later stages of fruit ripening in plants such as tomato. Fruit ripening can therefore be delayed by inactivating one or more of the genes involved in ethylene synthesis. Fruits on these plants initially develop as normal but are unable to complete the ripening process, and so can be transported to the marketplace with less danger of the crop spoiling. Before the fruits are sold to the consumer, artificial ripening is induced by spraying them with ethylene.

Ethylene is synthesized from a precursor called 1-aminocyclopropane-1-carboxylic acid (ACC), which in turn is made from *S*-adenosylmethionine (SAM) (Figure 22.29). In tomato, the strategy for delaying fruit ripening involved inactivation of the gene for ACC synthase, the enzyme that converts SAM into ACC. This gene was inserted into a cloning vector in the reverse orientation, so that the cloned sequence directed the synthesis of an antisense version of the ACC synthase mRNA (Figure 22.30). The modified plants made only 2% of the amount of ethylene produced by non-engineered ones, and their fruit were unable to complete the ripening process until treated with ethylene.

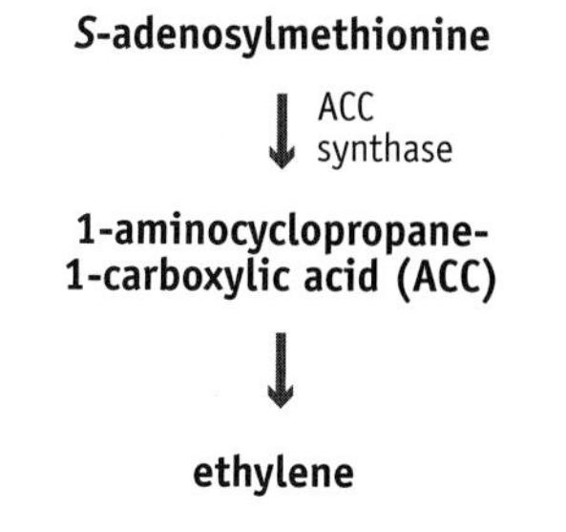

FIGURE 22.29 The ethylene biosynthesis pathway.

The applications of gene subtraction in plant genetic engineering are less broad than those of gene addition. This is simply because there are relatively few genes whose inactivation would be advantageous to the farmer or consumer. Most gene subtraction projects have been aimed at reducing crop spoilage, but a few ways of using this approach to improve nutritional value have also been devised (Table 22.2).

Engineered plants can be used for the synthesis of recombinant proteins, including vaccines

Now we return to the subject with which we started this chapter, the use of cloning to transfer genes into new hosts to obtain recombinant protein. We did not discuss the possibility that plants might be used for this purpose. Plants would be a good choice because they have similar protein-processing activities to those in animals, so most animal proteins produced in plants would be expected to undergo the correct post-translational modifications and to be fully functional.

This approach to recombinant protein production has been used with a variety of crops, such as maize, tobacco, rice, and sugarcane. By inserting the new gene downstream of a promoter that is active only in developing seeds, large amounts of recombinant protein can be obtained in a form that is easily harvested and processed. Recombinant proteins have also been synthesized in the leaves of tobacco and alfalfa and in the tubers of potatoes.

TABLE 22.2 EXAMPLES OF PLANT GENETIC ENGINEERING PROJECTS INVOLVING GENE INACTIVATION

Inactivated gene	Modified characteristic
1-Aminocyclopropane-1-carboxylic acid synthase	Modified fruit ripening in tomato
1-D-*myo*-Inositol 3-phosphate synthase	Reduction of indigestible phosphorus content of rice grains
Chalcone synthase	Modification of flower color in various decorative plants
Δ12-Desaturase	High oleic acid content in soybean
Polygalacturonase	Delay of fruit spoilage in tomato
Polyphenol oxidase	Prevention of discoloration in fruits and vegetables
Starch synthase	Reduction of starch content in vegetables

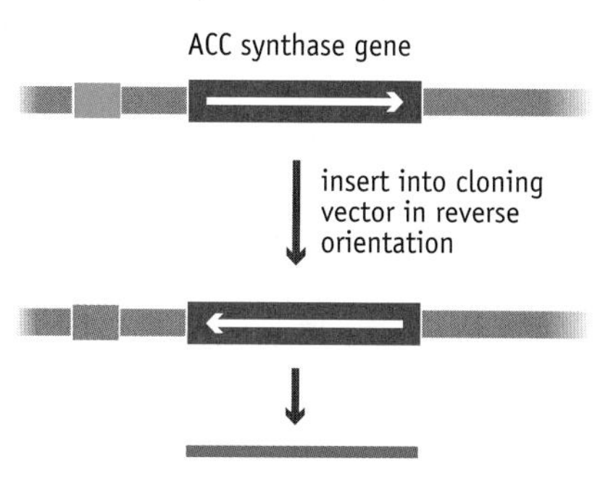

FIGURE 22.30 Synthesis of an antisense version of the ACC synthase mRNA.

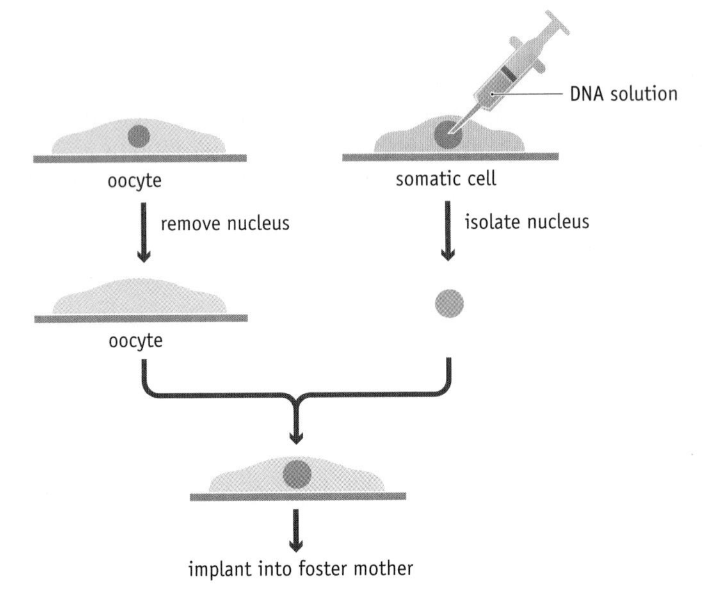

FIGURE 22.31 The nuclear transfer process that is used to produce a transformed oocyte from which a transgenic animal is obtained.

Whichever production system is used, plants offer a low-technology means of mass-producing recombinant proteins. One promising possibility is that plants could be used to synthesize vaccines, providing the basis for a cheap and efficient vaccination program. If the recombinant vaccine is effective after oral administration, then immunity could be acquired simply by eating part or all of the transgenic plant. The feasibility of this approach has been demonstrated by trials with vaccines such as HBsAg and the capsid proteins of measles virus and respiratory syncytial virus. With each of these, immunity has been conferred by feeding the transgenic plant to test animals. The main difficulty is in ensuring that the amount of recombinant protein synthesized by the plant is sufficient to stimulate complete immunity against the target disease. This requires vaccine yields of 8–10% of the soluble protein content of the part of the plant that is eaten, but in practice yields achieved so far are much less than this, usually not more than 0.5%.

Recombinant proteins can be produced in animals by pharming

It is also possible to obtain transgenic animals by genetic engineering, using the technology that we encountered when we considered the means by which germ-line gene therapy might be carried out. With mice, it is possible to produce a transgenic animal by direct microinjection of the gene to be cloned into a fertilized egg cell, but this is not possible with many other mammals. With these, the more sophisticated nuclear transfer procedure is required, in which the gene is first injected into the nucleus of a somatic cell (Figure 22.31).

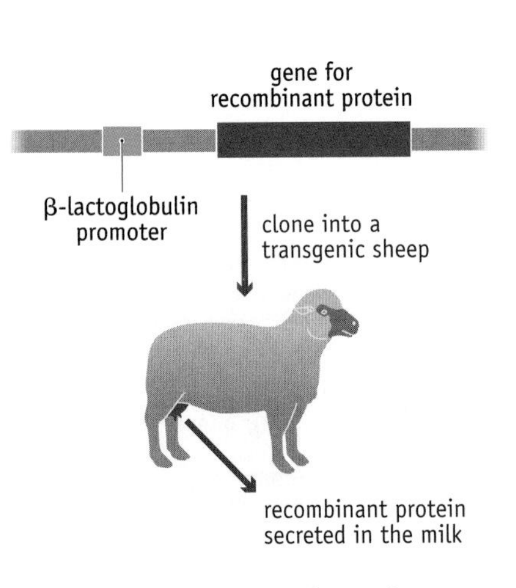

FIGURE 22.32 Production of recombinant protein in the milk of a sheep. In this example, the gene for the recombinant protein is attached to the β-lactoglobulin promoter, which is active only in mammary tissue.

Recombinant proteins have been produced in the blood of transgenic animals, and in the eggs of transgenic chickens, but the most successful approach has been to obtain the protein from the milk of farm animals such as sheep or pigs. This is possible if the gene is attached to a promoter that is active in mammary tissue (Figure 22.32). Sheep and pigs are, of course, mammals, and most human proteins that have been produced in this way undergo the correct post-translational modifications and are fully active. The process is called **pharming**. The first recombinant protein obtained by pharming and approved for medical use was antithrombin III, from goat's milk. This is an anticoagulant sometimes used to prevent blood clotting during heart surgery. Other blood proteins, including factor VIII, have also been synthesized in the milk of transgenic animals.

KEY CONCEPTS

- DNA cloning enables a gene for an important animal protein to be taken from its normal host and introduced into a bacterium. The transfer can be designed so that the animal gene remains active. The term "recombinant protein" is used to describe a protein that is produced in this way from a cloned gene.
- Microorganisms are looked on as ideal hosts for recombinant protein production because they can be grown at high density, enabling large amounts of the protein to be obtained.
- An expression vector contains a bacterial promoter, ribosome binding site, and termination sequence, and is used to ensure that a cloned animal gene is active after transfer to a bacterium.
- Simple proteins such as insulin can be synthesized in *E. coli*, but more complex ones might not be processed correctly.
- Yeast and fungi are used as hosts for the production of recombinant proteins because they are able to carry out post-translational modifications such as glycosylation, but some complex proteins, such as human factor VIII, can only be produced in animal cells.
- Virus coat proteins synthesized as recombinant proteins have been used as vaccines.
- The objective of genetic engineering is to change the phenotype of an animal or plant in a way that is considered beneficial for the applied use of the organism.
- In commercial terms the most important genetically engineered organisms are crop plants that have been modified so that they are resistant to the herbicide called glyphosate.
- Plant genes can be inactivated by genetic engineering techniques. This approach has been used to delay fruit ripening by inactivating genes involved in ethylene synthesis.
- Recombinant proteins, including vaccines, have been produced in transgenic plants, and others have been obtained from the milk of transgenic animals.

QUESTIONS AND PROBLEMS (Answers can be found at www.garlandscience.com/introgenetics)

Key Terms

Write short definitions of the following terms:

batch culture
biolistics
biotechnology
complementary DNA
continuous culture
expression vector
fermenter
fusion protein
gene addition
gene subtraction
genetic engineering
genetically modified plant
inclusion body
pharming
recombinant protein
site-directed mutagenesis
transgenic

Self-study Questions

22.1 Distinguish between the batch and continuous methods for culture of a microorganism.

22.2 Describe the key features of an expression vector and explain why this type of vector is needed when an animal gene is cloned in *E. coli*.

22.3 What are the desirable features of the promoter carried by an expression vector? What specific features of the lactose promoter make it a good choice for a vector used for expressing animal genes in *E. coli*?

22.4 Outline the procedure that was used to obtain recombinant insulin from *E. coli*.

22.5 Why is codon bias sometimes a problem when an animal gene is being expressed in *E. coli*?

22.6 Describe the possible solutions to the problems caused by introns, internal termination sequences, and codon bias during expression of an animal gene in *E. coli*.

22.7 What is an inclusion body and why is the formation of inclusion bodies sometimes a problem when *E. coli* is used as a host for recombinant protein synthesis?

22.8 Explain why glycosylation can be a problem when an animal protein is synthesized in *E. coli*. To what extent is this problem solved by using *S. cerevisiae* or *P. pastoris* as the host?

22.9 Describe the key features of the promoters used for recombinant protein production in *S. cerevisiae* and *P. pastoris*.

22.10 Outline the procedure that was used to obtain human factor VIII from hamster cells. Why was the production of factor VIII a challenge for recombinant technology?

22.11 Describe how recombinant protein synthesis is used to prepare a vaccine.

22.12 Describe how a recombinant vaccinia virus is constructed and give an example of the successful use of this technology.

22.13 Outline the advantages of a glyphosate-resistant crop compared with a nonresistant crop, and describe how glyphosate-resistant soybean was produced by genetic engineering.

22.14 Describe how plants that synthesize δ-endotoxins have been obtained by genetic engineering.

22.15 List the various characteristics that have been conferred on genetically modified plants by gene addition.

22.16 Giving examples, describe how antisense RNA technology has been used to inactivate plant genes.

22.17 Why are there relatively few applications of gene subtraction in plant genetic engineering?

22.18 Give examples of the production of recombinant protein in plants.

22.19 Outline how a transgenic animal is obtained.

22.20 Explain what is meant by the term "pharming," and give examples of proteins that have been obtained by pharming.

Discussion Topics

22.21 Most cloning vectors carry one or more genes for antibiotic resistance. Suggest a way in which the presence of such genes could be used to distinguish bacteria that have taken up a vector molecule from those that have not been transformed.

22.22 With most cloning vectors, insertion of a new piece of DNA disrupts a gene carried by the vector. Suggest a way in which this insertional inactivation could be used to distinguish bacteria that have taken up a recombinant vector molecule from those that have taken up a vector that contains no inserted DNA.

22.23 Speculate on the advantages of obtaining a recombinant protein as a fusion with an endogenous bacterial protein such as β-galactosidase.

22.24 What strategies might be used to prevent or reduce the development among insect populations of resistance to the δ-endotoxin produced by a genetically modified crop plant?

22.25 Discuss the concerns that have been raised about the possibility that GM crops might harm the environment.

22.26 Discuss the ethical issues raised by the development of pharming.

FURTHER READING

Brocher B, Kieny MP, Costy F et al. (1991) Large-scale eradication of rabies using recombinant vaccinia-rabies vaccine. *Nature* 354, 520–522.

Feitelson JS, Payne J & Kim L (1992) *Bacillus thuringiensis*: insects and beyond. *Biotechnology* 10, 271–275. *Details of* δ*-endotoxins and their potential as conventional insecticides and in genetic engineering.*

Goeddel DV, Kleid DG, Bolivar F et al. (1979) Expression in *Escherichia coli* of chemically synthesized genes for human insulin. *Proc. Natl. Acad. Sci. USA* 76, 106–110.

Houdebine L-M (2009) Production of pharmaceutical proteins by transgenic animals. *Comp. Immunol. Microbiol. Infect. Dis.* 32, 107–121.

Kind A & Schnieke A (2008) Animal pharming, two decades on. *Transgen. Res.* 17, 1025–1033. *Reviews the progress and controversies with animal pharming.*

Liu MA (1998) Vaccine developments. *Nat. Med.* 4, 515–519. *Describes the development of recombinant vaccines.*

Matas AJ, Gapper NE, Chung M-Y, Giovannoni JJ & Rose JKC (2009) Biology and genetic engineering of fruit maturation for enhanced quality and shelf-life. *Curr. Opin. Biotechnol.* 20, 197–203.

Padgette SR, Kolacz KH, Delannay X et al. (1995) Development, identification, and characterization of a glyphosate-tolerant soybean line. *Crop Sci.* 35, 1451–1461. *Transfer of the* Agrobacterium *EPSP synthase gene into soybean.*

Robinson M, Lilley R, Little S et al. (1984) Codon usage can affect efficiency of translation of genes in *Escherichia coli*. *Nucleic Acids Res.* 12, 6663–6671.

Sørensen HP & Mortensen KK (2004) Advanced genetic strategies for recombinant protein expression in *Escherichia coli*. *J. Biotechnol.* 115, 113–128. *A general review of the use of* E. coli *for recombinant protein production.*

Stoger E, Ma JK-C, Fischer R & Christou P (2005) Sowing the seeds of success: pharmaceutical proteins from plants. *Curr. Opin. Biotechnol.* 16, 167–173.

Tiwari S, Verma PC, Sing PK & Tuli R (2009) Plants as bioreactors for the production of vaccine antigens. *Biotechnol. Adv.* 27, 449–467.

Yonemura H, Sugawara K, Nakashima K et al. (1993) Efficient production of recombinant human factor VIII by co-expression of the heavy and light chains. *Protein Eng.* 6, 669–674.

The Ethical Issues Raised by Modern Genetics

CHAPTER 23

It is no longer possible for a student of genetics to consider only the academic aspects of the subject. The ethical issues raised by the applications of genetics in our modern world must also be examined. To many people, some of the recent advances in genetics are alarming, with the potential to do great harm if the discoveries are not used wisely. For some, any form of science is a threat, and opposition to genetics is merely part of a broader philosophy that science is bad. However, for most people, concerns about genetics are based on well-reasoned arguments. These views are important. They must be, because they are held by many geneticists.

We must therefore examine the ethical issues raised by modern genetics. We have already begun to do this in the *Questions and Problems* for the previous three chapters, where you were asked to think about the broader implications of the topics that you had just studied. Discovering these issues for yourself is important, because everybody, whether or not they are studying genetics, must make up their own mind about the possible dangers. Simply following the lead of others is not an option, especially for a student of genetics who is aware of the underlying science and so is able to form an opinion based on an expert knowledge of the subject. Bear in mind that you now have a much better understanding of genetics than that possessed by most politicians and many leaders of public thought.

In this chapter we will look first at some of the areas of genetics that cause concern among the general public. As far as possible, we will attempt to do this in a nonjudgmental manner. In the second part of the chapter we will explore the more general framework within which the ethical aspects of genetics should be debated.

23.1 AREAS OF CONCERN WITHIN MODERN GENETICS

Four areas of modern genetics are of particular concern to the general public (Table 23.1). These are the development of GM crops, the use of personal DNA data, the possible misuse of gene therapy, and pharming. The

TABLE 23.1 SOME OF THE AREAS OF CONCERN IN MODERN GENETICS

Topic	Specific issues
GM crops	Impact on human health and on the environment
	Use of genetic use restriction technology to prevent GM (and other) crops being grown for a second season from seeds collected from the previous harvest
Personal DNA data	Retention of genetic profiles of individuals who have not committed any crime
	Use of genetic data by insurance companies, and the more general discrimination against individuals with "bad" alleles of particular genes
	Personal freedom to know, or not know, one's genetic susceptibility to delayed-onset diseases
Gene therapy	Use of germ-line gene therapy to create "designer babies"
Pharming	Creation of animals specifically for use as producers of recombinant protein
	Suffering associated with the manipulations used to produce transgenic animals

FIGURE 23.1 Three overripe tomatoes (left) compared to three FlavrSavr tomatoes (right). The FlavrSavr tomato was the first type of GM crop to be approved for sale to the public. (Courtesy of Martyn F. Chillmaid/Science Photo Library.)

last two topics are related in that the public concerns center on the underlying technology, which is the same for both germ-line gene therapy and the production of pharmed animals. We will examine the issues relating to each of these areas of genetics in turn.

There have been concerns that GM crops might be harmful to human health

The first issue that we will consider is the perceived hazard to human health that might result from the consumption of GM crops. The first GM crop to be approved for sale to the public was a type of tomato that had been engineered so that its ripening was delayed, enabling the fruits to be transported from grower to consumer with less spoilage. These "FlavrSavr" tomatoes appeared in supermarkets in 1994 (Figure 23.1). They raised public awareness about genetics because, up to that point, it was not widely realized that plants and animals could be modified by genetic engineering. Not surprisingly, when faced with GM tomatoes, many people asked the question "are these safe to eat?"

FlavrSavr tomatoes had been engineered by the introduction of an antisense version of the gene for the polygalacturonase enzyme. This enzyme slowly breaks down the pectin molecules present in the cell walls in the fruit pericarp (Figure 23.2), resulting in the gradual softening that makes the fruit palatable but which, if allowed to proceed too far leads to spoilage. By inhibiting the synthesis of polygalacturonase, fruit ripening is delayed. The rationale is exactly the same as that behind the use of antisense technology to inhibit ethylene production in developing fruits, the only difference being that another part of the fruit ripening pathway is targeted.

It seems inconceivable that introduction of the polygalacturonase antisense RNA into these tomatoes would cause a health hazard. But this was not the point. Of much greater concern was the fact that these tomatoes also contained a gene for the enzyme called aminoglycoside 3-phosphotransferase or APH(3′). This enzyme confers resistance to aminoglycoside antibiotics

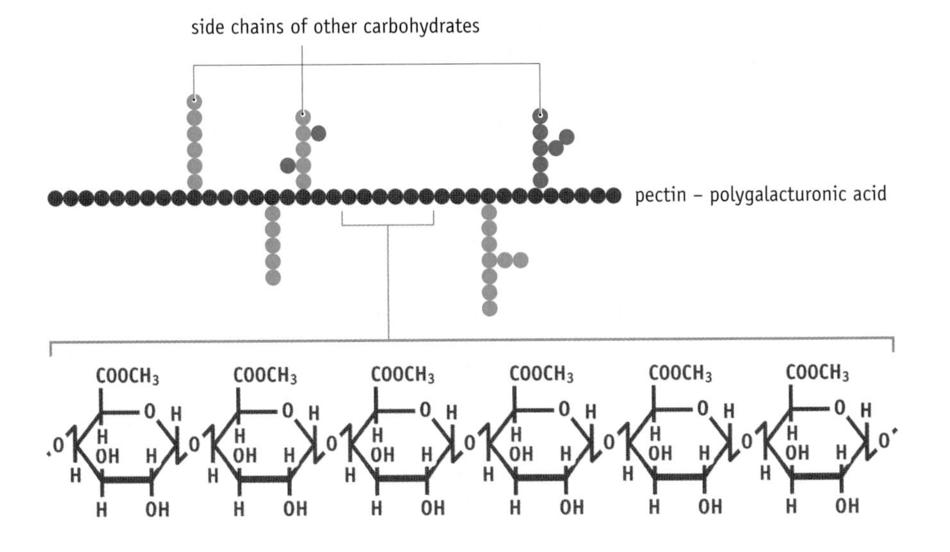

FIGURE 23.2 The structure of pectin. This component of the plant cell wall is a carbohydrate polymer made up of modified galacturonic acid units, with side chains of other carbohydrates. Polygalacturonase breaks the pectin chains in the cell wall of the tomato pericarp, leading to softening of the fruit.

such as kanamycin. Why was its gene present in the engineered tomatoes? We must return to the procedure used to construct a transgenic plant if we are to answer this question. This procedure begins with a recombinant plasmid that contains the gene that we wish to introduce into the plant (Figure 23.3). Various methods can be used to transport this plasmid into a plant cell, one method being the biolistics procedure that fires the plasmid through the cell wall on a gold microprojectile. However this step is performed, only a few cells will take up a copy of the recombinant plasmid. Many cells are unaffected by the process. To obtain our engineered plant we must grow a callus culture from a recombinant cell, and then a plant from that callus. To ensure that only the recombinant cells—the ones that contain a recombinant plasmid—give rise to calluses, the cells are initially plated onto agar medium that contains kanamycin. The presence of the APH(3′) gene in the recombinant plasmid means that the recombinant cells are resistant to kanamycin and are able to grow on this medium. The nonrecombinant cells, lacking the plasmid and therefore having no APH(3′) gene, cannot withstand the antibiotic and die. The APH(3′) gene is called a **selectable marker**. It enables the recombinant cells to be selected from the mixture of cells obtained after the DNA introduction step.

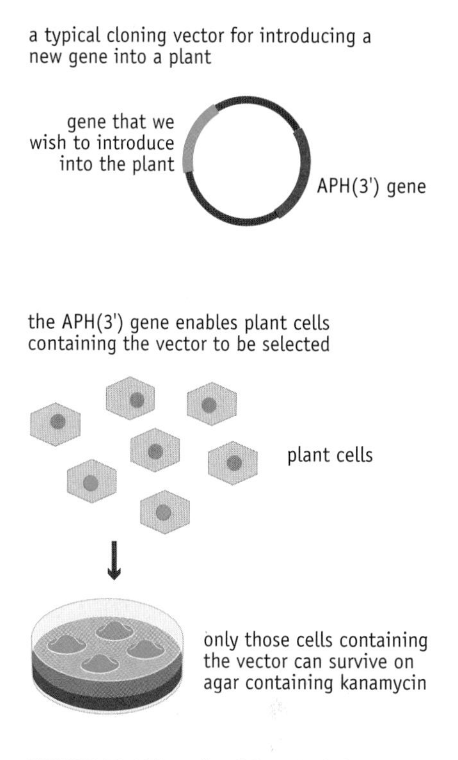

FIGURE 23.3 The role of the APH(3′) gene as a selectable marker during a gene cloning experiment with a plant cell.

After the selection process, the APH(3′) gene remains in the tomato cells and will be present in every cell of the engineered plant, along with the gene for the polygalacturonase antisense RNA. Besides kanamycin, the APH(3′) gene confers resistance to amikacin, tobramycin, and other aminoglycoside antibiotics that are used to treat bacterial infections. The concern, therefore, is that the APH(3′) gene might possibly be transferred, after ingestion, from the tomato fruit to the bacteria present in the human gut. Further movement of the gene between different bacteria might lead to antibiotic resistance spreading to the types of harmful bacteria that these antibiotics are currently used against.

When these fears were first raised, there were debates about the actual likelihood that a gene could be transferred from a foodstuff into gut bacteria. Most of the latter live in the large and small intestines, whereas the DNA in the tomato fruits would be degraded by the acid conditions in the stomach. But the spread of antibiotic resistance, from whatever source, is a real concern. Geneticists have therefore devised ways of deleting the APH(3′) gene after the recombinant cells have been selected. One of these strategies makes use of an enzyme called Cre, which comes from bacteriophage P1. The Cre enzyme catalyzes a type of recombination event that removes a segment of DNA that is flanked by 34-bp recognition sequences (Figure 23.4).

To make use of the Cre enzyme, two recombinant plasmids are introduced into a single plant cell. One plasmid carries the gene being added to the plant along with its APH(3′) selectable marker, the latter being flanked by the Cre target sequences. The second plasmid carries the Cre gene. Once inside the plant cell, the activity of the Cre enzyme results in excision of the APH(3′) gene from the plant DNA (Figure 23.5). There is still time, however, to test the cells for kanamycin resistance so that recombinant ones can be selected.

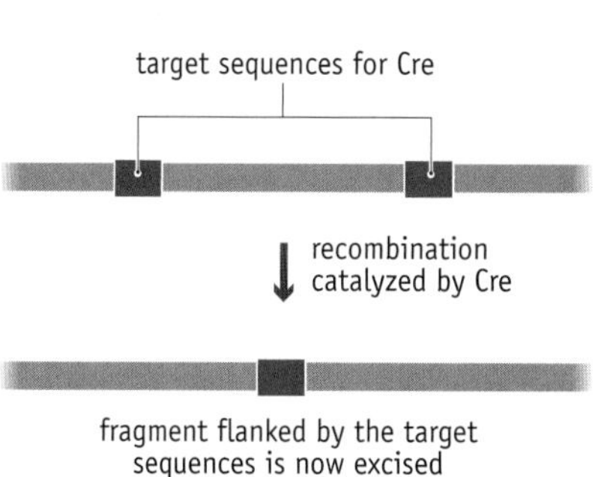

FIGURE 23.4 The mode of action of the Cre protein.

What if the Cre gene is itself hazardous in some way? This is not an issue, because the two recombinant plasmids would probably become inserted into different chromosomes. Random segregation during sexual reproduction will therefore result in a new generation of plants that contain one inserted plasmid but not the other. It is therefore possible to obtain a plant that does not contain either the Cre or APH(3′) gene but does contain the important gene that we wish to add to the plant's genome.

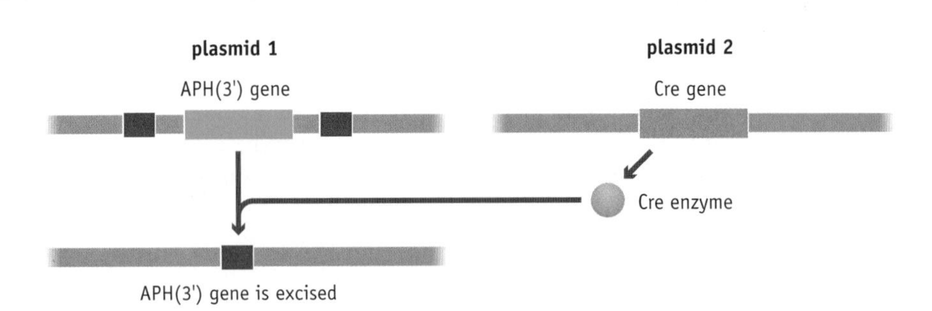

FIGURE 23.5 Excision of the APH(3′) gene by the Cre enzyme in a recombinant plant cell.

There are also concerns that GM crops might be harmful to the environment

The second area of concern about GM plants is that they might be harmful to the environment in some way. The development of maize, soybean, and oilseed rape engineered for increased herbicide resistance, and cotton and potato varieties that make δ-endotoxin insecticides, has resulted in large increases in the amounts of land given over to growth of GM plants. What impact might these plants have on the ecology of the environment in which they are growing?

Two aspects of the environmental impact of GM crops must be considered. The first is whether the introduced gene can move from the crop into the natural vegetation, possibly leading to versions of wild plants that are resistant to insects and herbicides. For this to happen, pollen from the GM crop must be able to fertilize the wild species. This type of cross-hybridization is rare, but it can occur between a crop and closely related wild plants. Examples are oilseed rape (*Brassica napus*) and the weed species called charlock (*Sinapis arvensis*) (**Figure 23.6**). Transfer of herbicide resistance from rape to charlock has been observed in experimental cultivations.

Movement of a gene from a GM crop to a wild plant must be distinguished from a second, and much commoner, way in which so-called "superweeds" can arise. This is through the natural process of evolution. Experience has shown that, when cultivating a herbicide-resistant GM crop, some farmers are tempted to use larger amounts of herbicide than in the past. This ensures that as many weeds as possible are killed, without any risk of harming the crop. The problem is that increased application of herbicide places a selective pressure on the environment that encourages the propagation of those rare, naturally resistant weed varieties that already exist or that evolve through the generation of new alleles by mutation events. Once these weeds become common in a field, the advantage of growing the engineered crop is lost.

FIGURE 23.6 (A) Oilseed rape and (B) its wild relative, charlock. (A, courtesy of James King-Holmes/Science Photo Library; B, courtesy of Bjorn Rorslett/Science Photo Library.)

This is an issue specific to herbicide resistance, and although it affects farming practices it is not obvious that these weeds will have any longer-term impact on the natural environment. Outside an area in which the herbicide is being used, resistance confers no selective advantage and will probably be lost from the plant population.

The evolution of superweeds in response to the overuse of herbicides leads us to the second potential environmental impact of GM crops. It is possible that cultivation of these plants might be accompanied by changes in farming practices that will have indirect impacts on the local environment and affect the abundance and diversity of farmland wildlife. This is, of course, an issue for any change in the way in which land is managed, but because of the controversies much more work has been done on the effects of growing GM crops than on other farming practices. One such study was commissioned by the UK Government in 1999. An independent investigation was conducted into how genetically modified herbicide-tolerant (GMHT) crops, whose growth in the UK was not at that time permitted, might affect the environment. The study involved 273 field trials throughout England, Wales, and Scotland, and included glyphosate-resistant sugar beet as well as maize and spring rape engineered for resistance to a second herbicide, glufosinate-ammonium. The summary of the results, as described in the official report, illustrates the difficulties in understanding the impact that GM crops will have on the environment:

"The team found that there were differences in the abundance of wildlife between GMHT crop fields and conventional crop fields. Growing conventional beet and spring rape was better for many groups of wildlife than growing GMHT beet and spring rape. There were more insects, such as butterflies and bees, in and around the conventional crops because there were more weeds to provide food and cover. There were also more weed seeds in conventional beet and spring rape crops than in their GM counterparts. Such seeds are important in the diets of some animals, particularly some birds. In contrast, growing GMHT maize was better for many groups of wildlife than conventional maize. There were more weeds in and around the GMHT crops, more butterflies and bees around at certain times of the year, and more weed seeds. The researchers stress that the differences they found do not arise just because the crops have been genetically modified. They arise because these GM crops give farmers new options for weed control. That is, they use different herbicides and apply them differently. The results of this study suggest that growing such GM crops could have implications for wider farmland biodiversity. However, other issues will affect the medium- and long-term impacts, such as the areas and distribution of land involved, how the land is cultivated and how crop rotations are managed. These make it hard for researchers to predict the medium- and large-scale effects of GM cropping with any certainty. In addition, other management decisions taken by farmers growing conventional crops will continue to impact on wildlife."

GM technology can be used to ensure that farmers have to buy new seed every year

Now we turn our attention to a different type of ethical issue that is raised by our ability to introduce new genes into crop plants. In addition to the development of plants with properties such as insect or herbicide resistance, genetic engineering has also been used to introduce genes that are of benefit not to the farmer or the consumer but to the company that produces the seed.

An example of this type of engineering is the **genetic use restriction technology** (**GURT**). This is a process that enables a company that develops and markets GM crops to protect its financial investment by ensuring that farmers must buy new seed every year. In some parts of the world, in particular poorer regions, it is quite common for a farmer to save some seed from the crop and to sow this second-generation seed the following year. This is, of course, cheaper than buying new seed from the plant breeding company every year.

The intellectual property rights of plant breeding companies have been debated for many years. Some argue that a new variety of crop should be looked on as an "invention," and should therefore be covered by patent laws that would give the company control over the use of its invention. Currently, plant breeders' rights are slightly different, and in many countries "saved seed" is specifically excluded from those rights. This applies to any type of new crop variety, including ones produced by conventional breeding. The controversy arises because the exclusion of saved seed from plant breeders' rights does not necessarily mean that a company is obliged to market crops that produce viable seeds. If the seeds are nonviable, the farmer cannot grow them the following year and so must keep on buying new stocks from the breeding company. GURT is a suite of methods that can be used to engineer crops so that their seeds are nonviable.

There are different versions of GURT. One of these is called the **terminator technology** and makes use of the gene for the **ribosome inactivating protein** (**RIP**). This protein blocks the translation stage of gene expression by cutting one of the ribosomal RNA molecules into two pieces (**Figure 23.7A**). Any cell in which the ribosome inactivating protein is active will die. Genetic engineering has been used to place the RIP gene under the control of a promoter that is active only during embryo development. These GM plants grow normally and will produce seeds, but the seeds will not be able to germinate because they do not contain embryos—the part that gives rise to the new plant.

Why are the first-generation seeds, the ones sold to farmers, not sterile? The RIP gene is nonfunctional in the plants that produce these seeds because it is disrupted by a segment of DNA (**Figure 23.7B**). This DNA is flanked by the 34-bp recognition sequences for the Cre enzyme. In these plants the gene for the Cre enzyme is attached to a promoter that is switched on by tetracycline. The plants grow normally and produce seeds with viable embryos. Once these seeds have been obtained, the supplier activates the Cre enzyme by placing the seeds in a tetracycline solution. The blocking DNA is deleted from the RIP gene. The seeds are still able to germinate, because the RIP gene is not expressed until its promoter becomes active during the next round of embryogenesis, when the next generation of seeds are being produced by the adult plants. These are the seeds that the farmer would normally save.

The terminator technology was initially developed as a means of preventing GM crops from cross-hybridizing with wild plants. The RIP protein will kill the embryos in seeds resulting from cross-pollination between a GM plant and a wild relative, so the new genes carried by the GM plant will never be transferred to the natural environment. But today it is looked on more as a means of preventing farmers from saving seeds, and although there are moves to introduce the technology, it is currently banned in most countries.

Several issues surround the use of personal DNA data

The second area of modern genetics whose controversial aspects we will examine is the use of personal DNA data. By "personal DNA data" we mean

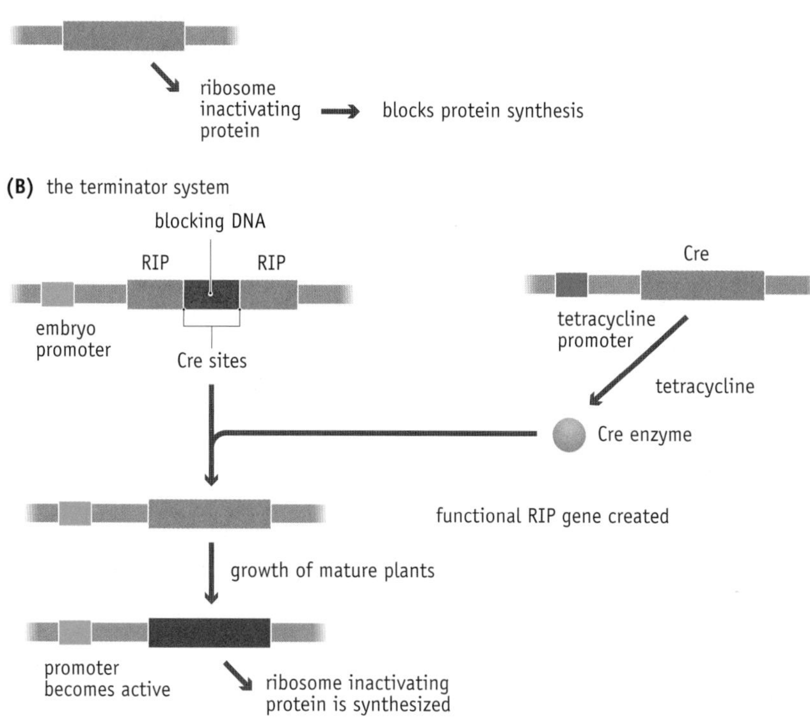

FIGURE 23.7 The terminator technology. (A) The RIP gene codes for the ribosome inactivating protein, which prevents protein synthesis by cutting one of the ribosomal RNA molecules into two pieces. (B) The RIP gene is modified so it contains a piece of blocking DNA flanked by Cre recognition sites, and is preceded by an embryo-specific promoter. The plants are also transformed with a Cre gene under control of a promoter that is switched on by tetracycline. When tetracycline is added, Cre is synthesized and the blocking DNA is removed from the RIP gene. Later, when the plants begin to produce seed, the embryo promoter is switched on and the RIP protein is made.

any information about an individual's genome. These data include both genetic profiles and DNA sequences.

In most countries, the authorities are allowed to retain genetic profiles that have been obtained as part of a criminal investigation. National legislations differ on issues such as the circumstances under which a DNA sample must be given, the length of time that a profile can be kept on record, and whether profiles of individuals who are eventually cleared of having committed any offence can be kept. The argument for having as large a database as possible is that it increases the chances of a new crime being solved quickly. If a genetic profile from the scene of a crime matches a profile in the database, that person can be interviewed without delay. If there is no match, the investigation will clearly take longer while suspects are tested, and quite possibly a conviction will never be made. There have been several reports of criminals only being apprehended after a period of years, often because they have been forced to give a DNA sample after committing a relatively minor offence such as a traffic violation. Their sample when tested matches the profile from an unsolved crime. In some cases, it has emerged that the individual concerned committed other crimes in the interim, some of a serious nature. Prevention of a rapist from reoffending on multiple occasions is a powerful argument in favor of large databases of genetic profiles. The arguments against these databases center on interpretations of personal privacy, in particular for individuals who have never been convicted of a crime.

A genetic profile, being simply a list of short tandem repeat alleles, contains no information on an individual's biological characteristics. More detailed types of DNA data include information on genes and so can be used to identify at least a few of these characteristics. This kind of DNA typing has been used for several years to test unborn children for the presence of mutations associated with inherited disease, so that the parents can, if they wish, make an informed choice about continuing a pregnancy. The methods for performing these screens are being improved year by year, so that it is now feasible to check for the presence of many mutations in an individual sample of DNA. From the resulting information, susceptibility to various types of genetic disease can be predicted. Should this information be available to insurance companies that wish to judge the risk they are taking by offering a life insurance policy? Life insurance can be refused if a regular health check discovers ailments that reduce life expectancy, such as heart disease, and smokers may be asked to pay higher premiums in the expectation that they will die from lung cancer at a relatively young age. Is there a difference between this type of health information and the information on health that is obtained from DNA sequences?

Should a person be forced to know their genetic disposition to disease? With late-onset diseases such as Alzheimer's and Parkinson's, which at present are not preventable, is there any justification for informing a person in their twenties that they have a high risk of developing the disease when they reach their sixties? This would inevitably be the case if such information were available to an insurance company that, because of the perceived risk, decided not to offer a life policy or requested higher premiums. These are among the most difficult and pressing of the ethical issues surrounding modern genetics, especially as "personal genetics testing" services are now widely available. The proliferation of these services illustrates the complexity of the ethical issues. In addition to a right not to know one's disposition to genetic disease, it can be argued that everyone has a right to know these risks so they can plan their lifestyle accordingly. In the past, when such tests could only be performed by trained professionals, the results would be presented to the patient with a careful explanation of their implications. Many people, for example, are unsure what a phrase such as "70% lifetime risk" actually means. A personal test conducted voluntarily but without access to professional counseling might be as ethically unsound as a test forced on an individual by an insurance company.

Gene therapy and pharming raise questions about the use of technology

Gene therapy and pharming are two additional areas of modern genetics that raise ethical concerns. The underlying issues are very similar, and we can therefore deal with them together.

When considering gene therapy, it is important to distinguish between the somatic and germ-line versions. Somatic cell therapy is less controversial because in this technique the genetic manipulation is not inherited. It is difficult to sustain an argument against the application via a respiratory inhaler of correct versions of the cystic fibrosis gene as a means of managing this disease. The new gene will be active only in the lungs of the patient, and for only a short period of time before another administration is needed. Other than the fact that the active ingredient is DNA, this kind of treatment seems to be no different from the use of any other type of drug.

Germ-line therapy is more controversial, because with this procedure the change that is made to the genome is heritable and so will be passed to all the descendants of the person who has been treated. The concerns do not center so much on the use of germ-line therapy to eradicate terrible genetic diseases from the human population. The problem is that the techniques used for germ-line correction of inherited diseases are exactly the same techniques that could be used for germ-line manipulation of other inherited characteristics. This leads to fears that one day it will be possible to create "designer babies" whose features match the preferences of the parents, or which are produced for less benign purposes. Our understanding of the link

FIGURE 23.8 A genetically modified salmon that has been engineered to produce a greater body mass than an unmodified salmon. (Courtesy of AquaBounty Technologies.)

between DNA sequences and the complex traits that define most human characteristics would have to advance quite dramatically before germ-line gene therapy could be used in this way, but the future misuse of a technology is an issue that cannot be ignored.

What distinction should be made between humans and other animals? The techniques that might one day be used to produce designer babies are already being applied to farmed animals and fish to make "improvements" such as greater meat yield (Figure 23.8). They also underlie the creation of transgenic animals for pharming. Here, concerns have been raised that the procedures used might cause suffering to the animals. Animals produced by somatic nuclear transfer often have such major developmental defects that they are either stillborn or die soon after birth. Some that are healthy do not display the desired phenotype and are destroyed. This type of genetics is therefore accompanied by a high "wastage."

Those animals that survive to adulthood sometimes seem to undergo premature aging. This is thought to have been the case with "Dolly the sheep," which, although not transgenic, was the first animal to be produced by nuclear transfer (Figure 23.9). Dolly was a Finn Dorset, a breed that usually has a life expectancy of up to 12 years. Dolly had to be put down when she was six because she was found to be suffering from terminal lung disease, which is normally found only in much older sheep. The previous year, when she was just five years old, she had developed arthritis. Her premature aging might have been related to the age of the somatic cell that was used as the source of the nucleus that gave rise to Dolly. This cell came from a six-year-old sheep, so in effect Dolly was already six years old when she was born. This is not such a fanciful notion when we recall the link between telomere shortening and aging.

The technology for the creation of engineered animals by somatic nuclear transfer has moved on dramatically since 1997, when Dolly was born, but the welfare issues regarding transgenic animals have not been resolved. This application of modern genetics remains at the forefront of public awareness.

FIGURE 23.9 Dolly the sheep, the first animal to be produced by somatic nuclear transfer. (Courtesy of Ph. Plailly/Eurelios/Science Photo Library.)

23.2 REACHING A CONSENSUS ON ETHICAL ISSUES RELATING TO MODERN GENETICS

How should the ethical issues that are raised by the applications of modern genetics be resolved? There is no easy answer to this question. Indeed, the problem is so complex that **bioethics** has emerged as a distinct subdiscipline of biology. Entire textbooks and university departments are now devoted to an academic discussion of ethical issues and how they should be dealt with. This raises the danger that "procedures" might be devised for resolving ethical concerns. Policy makers tend to like "procedures" because they allow a decision to be justified not on the basis of whether it is correct or not, but because it was reached by the agreed procedure.

We will not attempt here to devise a procedure for dealing with ethical issues. We will limit ourselves to a consideration of the fundamental requirements that must be met by any debate on the ethical problems raised by the applications of modern genetics.

Ethical debates must be based on science fact, not science fiction

The principal requirement when the ethical implications of genetics are debated is that the discussion is based in the real world and does not stray into the realms of science fiction. It is, of course, important to be forward-looking and to recognize issues that might arise as a result of the future development of the subject. With humans, for example, germ-line gene therapy is not yet possible because the cells resulting from somatic nuclear transfer do not divide to produce **blastocysts**. A blastocyst is the structure, comprising 100 or so cells, that attaches to the wall of the uterus during a normal pregnancy (Figure 23.10). The failure to form blastocysts means that, at present, it is technically impossible to use somatic nuclear transfer as a means of modifying the genetic make-up of an unborn child. Geneticists are uncertain whether this problem can ever be solved, but many agree that a method for human germ-line therapy will eventually become a reality. The ethical issues that germ-line therapy raises should therefore be debated now, so that when the procedure becomes possible there is already an understanding of the rules and regulations under which it should be used.

The need to be forward-looking does not imply that we should consider the ethical implications of areas of science that are not yet plausible and have little prospect of becoming so in the near future, if at all. It might be amusing to debate, for example, whether there should be any exceptions to the *Star Trek* prime directive that humans in interstellar spaceships must not interfere with alien cultures that have not yet developed spaceflight. Whole websites are devoted to the subject. But that is not a debate on scientific ethics, it is a debate on science fiction.

FIGURE 23.10 A 5-day-old human blastocyst. The technical difficulties in getting a human oocyte created by somatic nuclear transfer to develop into a blastocyst currently limit the prospects of human germ-line therapy.

The *Star Trek* analogy might seem obvious and trivial, but a search of the Internet reveals many other web pages devoted to debates on scientific issues that an expert will immediately recognize as being fictional, but which the debaters passionately believe to be realistic. A good way to find these sites is to search for emotive terms such as "Frankenstein food." It is easy to be critical of the lack of scientific understanding of people who believe that any kind of GM crop has the potential to cause brain damage. One might consider such views to be extremist, but it is important to be aware that the other end of the spectrum is the equally extreme view that a particular application of genetics, such as GM crops, is harmless and benign simply because that is the way we would like things to be. A focus on science fact is not an invitation to minimize or ignore potential issues.

The line of reasoning that we have followed leads us to the key conclusion that debates on the ethical issues surrounding any area of science must be driven by facts. Scientists therefore have a critical role in these debates. The issues surrounding GM crops, personal DNA data, germ-line gene therapy, and pharming must be resolved by the community as a whole, but geneticists must take the lead in these discussions. It is ethically unsound for geneticists to live their life in a research lab and not engage in the wider debates that their subject generates.

Few geneticists, probably none at all, are mad or evil

If geneticists must lead the debate on the ethical issues raised by their research, then the public must trust geneticists. The public perception of scientists is an interesting subject in its own right. Geneticists currently benefit from the assumption, driven by popular television dramas, that although many of us are geeks, most of us are also rather glamorous once we take off our glasses. The truth is that geneticists are just ordinary people and are as concerned about the ethical outcomes of their work as are any other members of the public.

Geneticists, in fact, have a good record in self-regulation and in open discussion of the possible dangers of their subject. This has not always been the case, but it is true for the period beginning with the discovery of the double-helix structure of DNA. We have already considered the potential problems that might arise if a gene for antibiotic resistance were transferred from a GM foodstuff into bacteria living in the human gut. This is not solely an issue for GM crops, because antibiotic resistance genes are used as selectable markers in virtually all cloning experiments, including those used to introduce genes into bacteria such as *Escherichia coli*. Because *E. coli* is a component of the gut microflora, the transfer of an antibiotic resistance gene, or any other type of gene, from an engineered *E. coli* to bacteria in the gut is a much greater possibility than the transfer from a foodstuff. Steps must therefore be taken to prevent it from happening. This problem was understood by the geneticists who designed the first cloning vectors and who developed the techniques for their use. So concerned were these geneticists that they voluntarily halted their work until a scientific conference could be held to discuss the hazards of gene cloning and how these ought to be avoided. That conference, held in Asilomar, California, in February 1975, established a set of guidelines for ensuring that cloned genes could not be transferred from *E. coli* to other bacteria. These guidelines enabled the new technology to continue to develop in a safe way while the hazards were explored more comprehensively.

The potential problems resulting from modern genetics were also a theme of the Human Genome Project. When this project was first conceived, in the 1980s, it was agreed that funds totaling 3–5% of the total annual budget would be set aside every year for studying the ethical, legal, and social issues (**ELSI**) that might arise as a result of knowing the complete sequence of the human genome (Table 23.2). This funding continues today, under the auspices of the National Human Genome Research Institute, part of the US National Institutes of Health.

The implication is that geneticists can be trusted to lead discussions on the ethical issues raised by their work. But there must be a caution. The Asilomar and ELSI initiatives are positives in the recent history of genetics, but there have also been negatives. Genetics has become big business, not just because of the commercial applications but also because of the prestige that academic researchers acquire from large research grants and publications in high-impact journals. The recent history of science is littered with cases in which the temptations of prestige and fame have become too great for a researcher, who has made up results, or suppressed

TABLE 23.2 THE ETHICAL, LEGAL, AND SOCIAL GOALS OF THE HUMAN GENOME PROJECT AS SET OUT IN 1988

1.	Examine the issues surrounding the completion of the human DNA sequence and the study of human genetic variation
2.	Examine issues raised by the integration of genetic technologies and information into health care and public health activities
3.	Examine issues raised by the integration of knowledge about genomics and gene–environment interactions into nonclinical settings
4.	Explore ways in which new genetic knowledge may interact with a variety of philosophical, theological, and ethical perspectives
5.	Explore how socioeconomic factors and concepts of race and ethnicity influence the use, understanding, and interpretation of genetic information, the utilization of genetic services, and the development of policy

others, to achieve the objective that they are searching for. And there are cases in which a scientist has become so passionately convinced that their own favorite hypothesis is correct that they fail to recognize when other research proves it to be wrong. Geneticists should therefore lead the debate on the ethical implications of genetics, but they should not be the sole arbiters of that debate. In the end, these issues must be resolved by the public at large, with all members of the community contributing, and all views given a balanced hearing based on their merits.

KEY CONCEPTS

- Students of genetics must be aware of, and form opinions on, the ethical issues arising from the applications of genetics in the modern world.
- Geneticists have developed methods for ensuring that the antibiotic resistance genes carried by the cloning vectors used in the production of GM foodstuffs cannot be transferred to bacteria in the human gut.
- The environmental impacts of GM crops have been studied extensively, but they are complex.
- The terminator technology has been proposed as a method for enabling a plant breeding company to protect its financial investment in the development of a new variety of crop produced by GM or conventional methods.
- The ease with which personal DNA data can be acquired raises issues of privacy, the use of data by insurance companies, and the right to know—or not know—one's susceptibility to genetic disease.
- Germ-line gene therapy and pharming raise issues concerning the manipulation of human characteristics for cosmetic purposes, and the welfare of transgenic animals.
- Any debate on the ethical implications of genetics must be based on facts, not fiction.
- Geneticists should lead debates on the ethical implications of genetics but should not be the sole arbiters of those debates.

FURTHER READING

Berg P (2008) Meetings that changed the world: Asilomar 1975: DNA modification secured. *Nature* 455, 290–291.

Greely HT (1998) Legal, ethical, and social issues in human genome research. *Annu. Rev. Anthropol.* 27, 473–502.

Hills MJ, Hall L, Arnison PG & Good AG (2007) Genetic use restriction technologies (GURT): strategies to impede transgene movement. *Trends Plant Sci.* 12, 177–183.

Miki B & McHugh S (2003) Selectable marker genes in transgenic plants: applications, alternatives and biosafety. *J. Biotechnol.* 107, 193–232.

Prainsack B & Reardon J (2008) Misdirected precaution: personal-genome tests are blurring the boundary between experts and lay people. *Nature* 456, 34–35.

Shelton AM, Zhao JZ & Roush RT (2002) Economic, ecological, food safety, and social consequences of the deployment of Bt transgenic plants. *Annu. Rev. Entomol.* 47, 845–81. *Various issues relating to δ-endotoxin plants.*

Shiels PG, Kind AJ, Campbell HS et al. (1999) Analysis of telomere length in Dolly, a sheep derived by nuclear transfer. *Cloning* 1, 119–125.

Simoncelli T (2006) Dangerous excursions: the case against extending forensic DNA databases to innocent persons. *J. Law Med. Ethics* 34, 390–397.

Various authors (2003) Theme issue: the Farm Scale Evaluations of spring-sown genetically modified crops. *Phil. Trans. R. Soc. Lond. B* 358, 1775–1889 (2003). *A series of papers on the environmental impact of GM crops.*

α helix One of the commonest secondary structural conformations taken up by segments of polypeptides.

β-*N*-glycosidic bond The linkage between the base and sugar of a nucleotide.

β sheet One of the commonest secondary structural conformations taken up by segments of polypeptides.

β turn A sequence of four amino acids, the second usually glycine, which causes a polypeptide to change direction.

γ complex A component of DNA polymerase III comprising subunit γ in association with δ, δ′, χ, and ψ.

2-aminopurine A base analog that can cause mutations by replacing adenine in a DNA molecule.

3′ -OH terminus The end of a polynucleotide that terminates with a hydroxyl group attached to the 3′-carbon of the sugar.

3′ → 5′ exonuclease An enzyme that sequentially removes nucleotides in the 3′ → 5′ direction from the end of a nucleic acid molecule.

30 nm chromatin fiber A relatively unpacked form of chromatin consisting of a possibly helical array of nucleosomes in a fiber approximately 30 nm in diameter.

5-bromouracil (5-bU) A base analog that can cause mutations by replacing thymine in a DNA molecule.

5′ -P terminus The end of a polynucleotide that terminates with a mono-, di-, or triphosphate attached to the 5′-carbon of the sugar.

5′ → 3′ exonuclease An enzyme that sequentially removes nucleotides in the 5′ → 3′ direction from the end of a nucleic acid molecule.

(6–4) lesion A dimer between two adjacent pyrimidine bases in a polynucleotide, formed by ultraviolet irradiation.

(6–4) photoproduct photolyase An enzyme involved in photoreactivation repair.

A-form of DNA A structural configuration of the double helix, present but not common in cellular DNA.

ABC model A model for flower development based on three groups of homeotic genes that work together to specify the structures of the flower.

acceptor arm Part of the structure of a tRNA molecule.

acceptor or A site The splice site at the 3′ end of an intron.

acridine dye A chemical compound that causes a frameshift mutation by intercalating between adjacent base pairs of the double helix.

acrocentric A chromosome whose centromere is positioned toward one end.

acute transforming virus A retrovirus that has captured an oncogene and which is therefore a very efficient inducer of cell transformation.

acylation The attachment of a lipid sidechain to a polypeptide.

Ada enzyme An *Escherichia coli* enzyme that is involved in the direct repair of alkylation mutations.

adapter A synthetic, double-stranded oligonucleotide used to attach sticky ends to a blunt-ended molecule.

adenine A purine base found in DNA and RNA.

adeno-associated virus (AAV) A virus that is unrelated to adenovirus but which is often found in the same infected tissues, because AAV makes use of some of the proteins synthesized by adenovirus to complete its replication cycle.

adenovirus An animal virus, derivatives of which have been used to clone genes in mammalian cells.

adenylate cyclase The enzyme that converts ATP to cyclic AMP.

alkylating agent A mutagen that acts by adding alkyl groups to nucleotide bases.

allele One of two or more alternative forms of a gene.

alphoid DNA The tandemly repeated nucleotide sequences located in the centromeric regions of human chromosomes.

alternative polyadenylation The use of two or more different sites for polyadenylation of an mRNA.

alternative promoter One of two or more different promoters acting on the same gene.

alternative splicing The production of two or more mRNAs from a single pre-mRNA by joining together different combinations of exons.

Alu A type of SINE found in the genomes of humans and related mammals.

amino acid One of the monomeric units of a protein molecule.

amino terminus The end of a polypeptide that has a free amino group.

aminoacyl or A site The site in the ribosome occupied by the aminoacyl-tRNA during translation.

aminoacylation Attachment of an amino acid to the acceptor arm of a tRNA.

aminoacyl-tRNA synthetase An enzyme that catalyzes the aminoacylation of one or more tRNAs.

anaphase The period during the division of a cell when the pairs of replicated chromosomes separate.

ancient DNA DNA preserved in ancient biological material.

annotation The process by which the genes, control sequences and other interesting features are identified in a genome sequence.

anticodon The triplet of nucleotides, at positions 34–36 in a tRNA molecule, that base-pairs with a codon in an mRNA molecule.

anticodon arm Part of the structure of a tRNA molecule.

antigen A substance that elicits an immune response.

antiparallel Refers to the arrangement of polynucleotides in the double helix, these running in opposite directions.

antisense RNA An RNA molecule that is the reverse complement of a naturally occurring mRNA, and which can be used to prevent translation of that mRNA in a transformed cell.

antitermination A bacterial mechanism for regulating the termination of transcription.

antiterminator protein A protein that attaches to bacterial DNA and mediates antitermination.

AP endonuclease An enzyme involved in base excision repair.

AP (apurinic/apyrimidinic) site A position in a DNA molecule where the base component of the nucleotide is missing.

apoptosis Programmed cell death.

archaea One of the two main groups of prokaryotes, mostly found in extreme environments.

archaeogenetics The use of DNA analysis to study the human past.

attenuation A process used by some bacteria to regulate the expression of an amino acid biosynthetic operon in accordance with the levels of the amino acid in the cell.

autoradiography The detection of radioactively labeled molecules by exposure of an X-ray-sensitive photographic film.

autosome A chromosome that is not a sex chromosome.

B-DNA The commonest structural conformation of the DNA double helix in living cells.

back mutation A mutation that reverses the effect of a previous mutation by restoring the original nucleotide sequence.

backtracking The reversal of an RNA polymerase a short distance along its DNA template strand.

bacteria One of the two main groups of prokaryotes.

bacterial artificial chromosome (BAC) A high-capacity cloning vector based on the F plasmid of *Escherichia coli*.

bacterial chromosome One of the DNA molecules (possibly the only one) in a bacterial cell that carries essential genes and is located in the nucleoid.

bacteriophage A virus that infects a bacterium.

Barr body The highly condensed chromatin structure taken on by an inactivated X chromosome.

basal promoter element Sequence motifs that are present in many eukaryotic promoters and set the basal level of transcription initiation.

basal rate of transcription The number of productive initiations of transcription occurring per unit time at a particular promoter.

base analog A compound whose structural similarity to one of the bases in DNA enables it to act as a mutagen.

base excision repair A DNA repair process that involves the excision and replacement of an abnormal base.

base pair The hydrogen-bonded structure formed by two complementary nucleotides. The shortest unit of length for a double-stranded DNA molecule.

base pairing The attachment of one polynucleotide to another, or one part of a polynucleotide to another part of the same polynucleotide, by base pairs.

baseless site A position in a DNA molecule where the base component of the nucleotide is missing.

batch culture Growth of bacteria in a fixed volume of liquid medium in a closed vessel, with no additions or removals made during the period of incubation.

beads-on-a-string An unpacked form of chromatin consisting of nucleosome beads on a string of DNA.

bioethics Study of the ethical issues raised by the applications of biology.

biolistics A means of introducing DNA into cells that involves bombardment with high-velocity microprojectiles coated with DNA.

biological information The information that is contained in the genome of an organism and which directs the development and maintenance of that organism.

biotechnology The use of living organisms, often (but not always) microbes, in industrial processes.

biotin A molecule that can be incorporated into DNA and used as a nonradioactive label.

bivalent The structure formed when a pair of homologous chromosomes lines up during meiosis.

blastocyst The embryonic structure that attaches to the wall of the uterus during a normal pregnancy in a mammal.

blastoderm The structure that forms when individual cells start to appear around the outside of the syncytium of a developing fruit fly embryo.

bottleneck The period between the decline and recovery in the size of a population.

branch migration A step during homologous recombination, involving exchange of polynucleotides between a pair of recombining double-stranded DNA molecules.

buoyant density The density possessed by a molecule or particle when suspended in an aqueous salt or sugar solution.

C terminus The end of a polypeptide that has a free carboxyl group.

CAAT box A basal promoter element.

candidate gene A gene, identified by any of various means, that might be responsible for a genetic disease.

cap binding complex The complex that makes the initial attachment to the cap structure at the beginning of the scanning phase of eukaryotic translation.

cap structure The chemical modification at the 5′ end of most eukaryotic mRNA molecules.

capillary gel electrophoresis Polyacrylamide gel electrophoresis carried out in a thin capillary tube, providing high resolution.

capsid The protein coat that surrounds the DNA or RNA genome of a virus.

carboxy terminus The end of a polypeptide that has a free carboxyl group.

carcinogen An environmental agent that causes cancer.

carrier An individual who has one normal and one defective allele for a gene responsible for an inherited disease, but does not have the disease because the defective allele is recessive.

cascade system A system in which the completion of one event triggers the initiation of a second event. In genetics, usually a system for controlling the order in which genes are expressed.

catabolite activator protein A regulatory protein that binds to various sites in a bacterial genome and activates transcription initiation at downstream promoters.

cell cycle The series of events occurring in a cell between one division and the next.

cell transformation The alteration in morphological and biochemical properties that occurs when an animal cell is infected by an oncogenic virus.

cell-free protein synthesizing system A cell extract containing all the components needed for protein synthesis and able to translate added mRNA molecules.

cell-surface receptor A protein located in the cell membrane that responds to an external signal by causing a biochemical change within the cell.

centimorgan (cM) The unit used to describe the distance between two genes on a chromosome. 1 cM is the distance that corresponds to a 1% probability of recombination in a single meiosis.

centromere The constricted region of a chromosome that is the position at which the pair of chromatids are held together.

centrosome The structures on opposite sides of the mitotic spindle from which microtubules radiate during cell division.

chain termination method A DNA sequencing method that involves enzymatic synthesis of polynucleotide chains that terminate at specific nucleotide positions.

chaperonin A multisubunit protein that forms a structure that aids the folding of other proteins.

charging Aminoacylation of a tRNA.

chi form An intermediate structure seen during recombination between DNA molecules.

chi site A repeated nucleotide sequence in the *Escherichia coli* genome that is involved in the initiation of homologous recombination.

chiasmata The crossover points visible when a pair of recombining chromosomes are observed by microscopy.

chimera An organism composed of two or more genetically different cell types.

chromatid The arm of a chromosome.

chromatin The complex of DNA and histone proteins found in chromosomes.

chromatin modification Regulation of genome expression by chemical alteration of nucleosomes.

chromatosome A subcomponent of chromatin made up of a nucleosome core octamer with associated DNA and a linker histone.

chromid A bacterial DNA molecule that has the characteristic features of a plasmid but which carries essential genes.

chromosome One of the DNA–protein structures that contains part of the nuclear genome of a eukaryote. Less accurately, the DNA molecule(s) containing a bacterial genome.

clastogen An environmental agent that causes breaks in DNA molecules.

cleavage and polyadenylation specificity factor (CPSF) A protein that has an ancillary role during polyadenylation of eukaryotic mRNAs.

cline A geographical gradation in the allele frequencies within a population.

clone A group of cells that contain the same recombinant DNA molecule.

clone library A collection of clones, possibly representing an entire genome, from which individual clones of interest are obtained.

closed promoter complex The structure formed during the initial step in assembly of the transcription initiation complex. The closed promoter complex consists of the RNA polymerase and/or accessory proteins attached to the promoter, before the DNA has been opened up by breakage of base pairs.

cloverleaf A two-dimensional representation of the structure of a tRNA molecule.

co-repressor A molecule that must bind to a bacterial repressor before the repressor can attach to its operator site.

coalescence time An estimate of the time that has elapsed since a haplogroup first came into existence, based on the degree of divergence of the haplotypes in that haplogroup.

codominance The relationship between a pair of alleles that both contribute to the phenotype of a heterozygote.

codon A triplet of nucleotides coding for a single amino acid.

codon–anticodon recognition The interaction between a codon on an mRNA molecule and the corresponding anticodon on a tRNA.

cohesin The protein that holds sister chromatids together during the period between genome replication and nuclear division.

Col plasmid A plasmid that carries one or more genes coding for colicins.

colicin A toxic bacterial protein often coded for by a gene carried by a plasmid.

colonization Movement of a population to new territory previously unoccupied by that species.

compatibility The ability of two different types of plasmid to coexist in the same cell.

competent Refers to a culture of bacteria that have been treated, for example by soaking in calcium chloride, so that their ability to take up DNA molecules is enhanced.

complementary Refers to two nucleotides or nucleotide sequences that are able to base-pair with one another.

complementary DNA (cDNA) A double-stranded DNA copy of an mRNA molecule.

complementary gene action The situation in which a particular combination of alleles at two separate genes is needed in order to produce a phenotype.

complex multigene family A type of multigene family in which the genes have similar nucleotide sequences but are sufficiently different to code for proteins with distinctive properties.

concatamer A DNA molecule made up of linear genomes or other DNA units linked head to tail.

conjugation Transfer of DNA between two bacteria that come into physical contact with one another.

consensus sequence A nucleotide sequence that represents an 'average' of a number of related but nonidentical sequences.

conservative replication A hypothetical mode of DNA replication in which one daughter double helix is made up of the two parental polynucleotides and the other is made up of two newly synthesized polynucleotides.

constitutive heterochromatin Chromatin that is permanently in a compact organization.

context-dependent codon reassignment Refers to the situation whereby the DNA sequence surrounding a codon changes the meaning of that codon.

contig A contiguous set of overlapping DNA sequences.

continuous culture The culture of microorganisms in liquid medium under controlled conditions, with additions to and removals from the medium over a lengthy period.

copy number The number of molecules of a plasmid contained in a single cell.

core enzyme The version of *Escherichia coli* RNA polymerase, subunit composition $\alpha_2\beta\beta'$, that performs RNA synthesis but is unable to locate promoters efficiently.

core octamer The central component of a nucleosome, made up of two subunits each of histones H2A, H2B, H3, and H4, around which DNA is wound.

core promoter The position within a eukaryotic promoter where the initiation complex is assembled.

CpG island A GC-rich DNA region located upstream of approximately 56% of the genes in the human genome.

cross-fertilization Fertilization of a female gamete with a male gamete derived from a different individual.

crossing over The exchange of DNA between chromosomes during meiosis.

cryptic splice site A site whose sequence resembles an authentic splice site and which might be selected instead of the authentic site during aberrant splicing.

cryptogene One of several genes in the trypanosome mitochondrial genome that specify abbreviated RNAs that must undergo pan-editing to become functional.

C-terminal domain (CTD) A component of the largest subunit of RNA polymerase II, important in activation of the polymerase.

cyanelle Photosynthetic structures, resembling ingested cyanobacteria, inside the cells of *Cyanophora* protozoa.

cyclic AMP (cAMP) A modified version of AMP in which an intramolecular phosphodiester bond links the 5′ and 3′ carbons.

cyclin A regulatory protein whose abundance varies during the cell cycle and which regulates biochemical events in a cell-cycle-specific manner.

cyclobutyl dimer A dimer between two adjacent pyrimidine bases in a polynucleotide, formed by ultraviolet irradiation.

Cys_2His_2 finger A type of zinc-finger DNA-binding domain.

cytochemistry The use of compound-specific stains, combined with microscopy, to determine the biochemical content of cellular structures.

cytokinesis The period during the division of a cell when the two daughter cells form around the divided nuclei.

cytosine One of the pyrimidine bases found in DNA and RNA.

D arm Part of the structure of a tRNA molecule.

D-loop The structure formed when a DNA double helix is invaded by a single-stranded DNA or RNA molecule, which forms a region of base pairing with one of the polynucleotides of the helix.

dark repair A type of nucleotide excision repair process that corrects cyclobutyl dimers.

deadenylation-dependent decapping A process for degradation of eukaryotic mRNAs that is initiated by removal of the poly(A) tail.

deaminating agent A mutagen that acts by removing amino groups from nucleotide bases.

defective virus A viral genome that has lost essential nucleotide sequences and so cannot replicate on its own. A defective virus may give rise to progeny by making use of proteins synthesized from a nondefective virus genome present in the same cell.

degeneracy Refers to the fact that the genetic code has more than one codon for most amino acids.

degradative plasmid A plasmid whose genes allow the host bacterium to metabolize unusual molecules such as toluene and salicylic acid.

delayed-onset mutation A mutation whose effect is not apparent until a relatively late stage in the life of the mutant organism.

deletion mutation A mutation resulting from deletion of one or more nucleotides from a DNA sequence.

denaturation Breakdown by chemical or physical means of the noncovalent interactions, such as hydrogen bonding, that maintain the secondary and higher levels of structure of proteins and nucleic acids.

density gradient centrifugation A technique in which a cell fraction is centrifuged through a dense solution, in the form of a gradient, so that individual components are separated.

deoxyribonuclease An enzyme that cleaves phosphodiester bonds in a DNA molecule.

deoxyribonucleic acid One of the two forms of nucleic acid in living cells; the genetic material for all cellular life forms and many viruses.

diauxie The phenomenon whereby a bacterium, when provided with a mixture of sugars, uses up one sugar before beginning to metabolize the second sugar.

Dicer The ribonuclease that has a central role in RNA interference.

dideoxynucleotide A modified nucleotide that lacks the 3′ hydroxyl group and so terminates strand synthesis when incorporated into a polynucleotide.

differential gene expression Refers to one or more genes that are switched on only when the biological information that they contain is needed.

diploid A nucleus that has two copies of each chromosome.

direct repair system A DNA repair system that acts directly on a damaged nucleotide.

discontinuous gene A gene that is split into exons and introns.

dispersive replication A hypothetical mode of DNA replication in which both polynucleotides of each daughter double helix are made up partly of parental DNA and partly of newly synthesized DNA.

displacement replication A mode of replication that involves continuous copying of one strand of the helix, the second strand being displaced and subsequently copied after the synthesis of the first daughter strand has been completed.

distance matrix A table showing the evolutionary distances between all pairs of nucleotide sequences in a dataset.

disulfide bridge A covalent bond linking cysteine amino acids on different polypeptides or at different positions on the same polypeptide.

DNA Deoxyribonucleic acid, one of the two forms of nucleic acid in living cells; the genetic material for all cellular life forms and many viruses.

DNA adenine methylase (Dam) An enzyme involved in methylation of *Escherichia coli* DNA.

DNA cloning Insertion of a fragment of DNA into a cloning vector, and subsequent propagation of the recombinant DNA molecule in a host organism.

DNA cytosine methylase (Dcm) An enzyme involved in methylation of *Escherichia coli* DNA.

DNA glycosylase An enzyme that cleaves the β-*N*-glycosidic bond between a base and the sugar component of a nucleotide as part of the base excision and mismatch repair processes.

DNA gyrase A type II topoisomerase of *Escherichia coli*.

DNA ligase An enzyme that synthesizes phosphodiester bonds as part of DNA replication, repair, and recombination processes.

DNA marker A DNA sequence that exists as two or more readily distinguished versions and which can therefore be used to mark a map position on a genetic, physical, or integrated genome map.

DNA methylation Refers to the chemical modification of DNA by the attachment of methyl groups.

DNA methyltransferase An enzyme that attaches methyl groups to a DNA molecule.

DNA photolyase A bacterial enzyme involved in photoreactivation repair.

DNA polymerase An enzyme that synthesizes DNA.

DNA polymerase α The enzyme that primes DNA replication in eukaryotes.

DNA polymerase δ The main eukaryotic DNA-replicating enzyme.

DNA polymerase γ The enzyme responsible for replication of the mitochondrial genome.

DNA polymerase I The bacterial enzyme that completes the synthesis of Okazaki fragments during DNA replication.

DNA polymerase II A bacterial DNA polymerase involved in DNA repair.

DNA polymerase III The main DNA replicating enzyme of bacteria.

DNA repair The biochemical processes that correct mutations arising from replication errors and the effects of mutagenic agents.

DNA sequence The order of nucleotides in a DNA molecule.

DNA sequencing The technique for determining the order of nucleotides in a DNA molecule.

DNA topoisomerase An enzyme that introduces or removes turns from the double helix by the breakage and reunion of one or both polynucleotides.

DNA-binding motif The part of a DNA-binding protein that makes contact with the double helix.

DNA-dependent DNA polymerase An enzyme that makes a DNA copy of a DNA template.

DNA-dependent RNA polymerase An enzyme that makes an RNA copy of a DNA template.

dominant Refers to the allele that is expressed in a heterozygote.

donor site The splice site at the 5′ end of an intron.

double helix The base-paired double-stranded structure that is the natural form of DNA in the cell.

double heterozygote A nucleus that is heterozygous for two genes.

double homozygote A nucleus that is homozygous for two genes.

early genes Bacteriophage genes that are expressed during the early stages of infection of a bacterial cell; the products of early genes are usually involved in replication of the phage genome.

effective population size The size of that part of the population that participates in reproduction.

elongation factor A protein that has an ancillary role in the elongation step of transcription or translation.

ELSI Ethical, legal, and social issues, in particular those that might arise as a result of knowing the complete sequence of the human genome.

embryonic stem (ES) cell A totipotent cell from the embryo of a mouse or other organism.

endogenous retrovirus (ERV) An active or inactive retroviral genome integrated into a host chromosome.

endonuclease An enzyme that breaks phosphodiester bonds within a nucleic acid molecule.

endosymbiont theory A theory that states that the mitochondria and chloroplasts of eukaryotic cells are derived from symbiotic prokaryotes.

enhancer A regulatory sequence that increases the rate of transcription of a gene or genes located some distance away in either direction.

enzyme A protein that catalyzes one or more biochemical reactions.

epigenesis A heritable change to the genome that does not involve mutation.

episome A plasmid capable of integration into the host cell's chromosome.

epistasis The situation in which the alleles present at one gene mask or cancel those at a second gene.

ethidium bromide A type of intercalating agent that causes mutations by inserting between adjacent base pairs in a double-stranded DNA molecule.

ethylmethane sulfonate (EMS) A mutagen that acts by adding alkyl groups to nucleotide bases.

euchromatin Regions of a eukaryotic chromosome that are relatively uncondensed, thought to contain active genes.

eukaryote An organism whose cells contain membrane-bound nuclei.

excision repair A DNA repair process that corrects various types of DNA damage by excising and resynthesizing a region of polynucleotide.

exit or E site A position within a bacterial ribosome to which a tRNA moves immediately after deacylation.

exon A coding region within a discontinuous gene.

exon skipping Aberrant splicing in which one or more exons are omitted from the spliced RNA.

exonic splicing enhancer (ESE) A nucleotide sequence that has a positive regulatory role during splicing of GU–AG introns.

exonic splicing silencer (ESS) A nucleotide sequence that has a negative regulatory role during splicing of GU–AG introns.

exonuclease An enzyme that sequentially removes nucleotides from the ends of a nucleic acid molecule.

expression proteomics The methodology used to identify the proteins in a proteome.

expression vector A special type of cloning vector designed for the synthesis of recombinant protein.

extrachromosomal gene A gene in a mitochondrial or chloroplast genome.

F^+ cell A bacterium that carries an F plasmid.

F^- cell A bacterium that does not carry an F plasmid.

F' cell A bacterium that carries a modified F plasmid, one carrying a small piece of DNA derived from the host bacterial DNA molecule.

F plasmid A fertility plasmid that directs conjugal transfer of DNA between bacteria.

facultative heterochromatin Chromatin that has a compact organization in some but not all cells, thought to contain genes that are inactive in some cells or at some periods of the cell cycle.

FEN1 The 'flap endonuclease' involved in replication of the lagging strand in eukaryotes.

fermenter A vessel used for the large-scale culture of microorganisms.

filamentous A bacteriophage or virus capsid in which the protomers are arranged in a helix, producing a rod-shaped structure.

fitness The ability of an organism or allele to survive and reproduce.

fixation Refers to the situation that occurs when a single allele reaches a frequency of 100% in a population.

fixation index A statistic indicating the degree of difference between the allele frequencies present in two populations.

fluorescent *in situ* hybridization (FISH) A technique for locating markers on chromosomes by observing the hybridization positions of fluorescent labels.

foldback RNA The precursor RNA molecules that are cleaved to produce microRNAs.

folding pathway The series of events, involving partly folded intermediates, that results in an unfolded protein attaining its correct three-dimensional structure.

founder analysis An analysis that uses haplogroup diversities to estimate the time of a population split.

founder effect The situation that occurs when a few members of a population split away and initiate a new population.

frameshift mutation A mutation resulting from the insertion or deletion of a group of nucleotides that is not a multiple of three and which therefore changes the frame in which translation occurs.

F_{ST} value A statistic indicating the degree of difference between the allele frequencies present in two populations.

fusion protein A protein that consists of a fusion of two polypeptides, or parts of polypeptides, normally coded for by separate genes.

G1 phase The first gap period of the cell cycle.

G2 phase The second gap period of the cell cycle.

gain-of-function mutation A mutation that results in an organism acquiring a new function.

gamete A reproductive cell, usually haploid, that fuses with a second gamete to produce a new cell during sexual reproduction.

gap genes Developmental genes that have a role in establishing positional information within the fruit fly embryo.

Gaussian distribution The distribution pattern commonly called a bell-shaped curve.

GC box A type of basal promoter element.

gene A DNA segment containing biological information and hence coding for an RNA and/or polypeptide molecule.

gene addition A genetic engineering strategy that involves the introduction of a new gene or group of genes into an organism.

gene amplification The production of multiple copies of a DNA segment so as to increase the rate of expression of a gene carried by the segment.

gene expression The series of events by which the biological information carried by a gene is released and made available to the cell.

gene subtraction A genetic engineering strategy that involves the inactivation of one or more of an organism's genes.

gene therapy A clinical procedure in which a gene or other DNA sequence is used to treat a disease.

genetic code The rules that determine which triplet of nucleotides codes for which amino acid during protein synthesis.

genetic disease A disease caused by a defect in the genome.

genetic drift The changes in allele frequencies that occur in a population over time due to random processes.

genetic engineering The use of experimental techniques to produce DNA molecules containing new genes or new combinations of genes.

genetic material The chemical material of which genes are made, now known to be DNA in most organisms, and RNA in a few viruses.

genetic profile The banding pattern revealed after electrophoresis of the products of PCRs directed at a range of microsatellite loci.

genetic profiling The process by which a genetic profile is obtained.

genetic use restriction technology (GURT) A process that enables a company that develops and markets GM crops to protect its financial investment by ensuring that farmers must buy new seed every year.

genetically modified (GM) plant A plant whose phenotype has been altered by genetic engineering.

genetics The branch of biology devoted to the study of genes.

genome The entire genetic complement of a living organism.

genome browser A website that allows the user to examine a genome for the locations of features such as genes, transcripts, and repeated sequences.

genomic imprinting Inactivation by methylation of a gene on one of a pair of homologous chromosomes.

genotype A description of the genetic composition of an organism.

germ-line therapy A type of gene therapy in which a fertilized egg is provided with a copy of the correct version of the defective gene and reimplanted into the mother.

glycosylation The attachment of sugar units to a polypeptide.

guanine One of the purine nucleotides found in DNA and RNA.

guanine methyltransferase The enzyme that attaches a methyl group to the 5′ end of a eukaryotic mRNA during the capping reaction.

guanylyl transferase The enzyme that attaches a GTP to the 5′ end of a eukaryotic mRNA at the start of the capping reaction.

guide RNA A short RNA that specifies the positions at which one or more nucleotides are inserted into an abbreviated RNA by pan-editing.

half-life A measure of the rate of degradation of a molecule or the rate of decay of an atom.

haplogroup A set of related haplotype sequences.

haploid A nucleus that has a single copy of each chromosome.

haploinsufficiency The situation in which the inactivation of a gene on one of a pair of homologous chromosomes results in a change in the phenotype of the mutant organism.

haplotype A sequence variant of a gene, used as an alternative to "allele" when there are several variants that confer the same phenotype.

Hardy–Weinberg equilibrium The equilibrium between allele frequencies that occurs in a population of infinite size for genes that are not subject to selection.

head-and-tail A bacteriophage capsid made up of an icosahedral head, containing the nucleic acid, and a filamentous tail that facilitates entry of the nucleic acid into the host cell.

heat shock module A regulatory sequence upstream of genes involved in the protection of a cell from heat damage.

helicase An enzyme that breaks base pairs in a double-stranded DNA molecule.

helix–turn–helix motif A common structural motif for attachment of a protein to a DNA molecule.

heredity The passage of characteristics from parents to offspring.

heterochromatin Chromatin that is relatively condensed and is thought to contain DNA that is not being transcribed.

heteroduplex An intermediate in recombination in which two double-stranded DNA molecules are linked together by a pair of Holliday structures.

heterogenous nuclear RNA (hnRNA) The nuclear RNA fraction that comprises unprocessed transcripts synthesized by RNA polymerase II.

heteroplasmy The situation in which an individual possesses two haplotypes of the mitochondrial genome.

heterosis The situation in which the heterozygote is the most successful genotype in a population.

heterozygote A diploid cell or organism that contains two different alleles for a particular gene.

heterozygote advantage or superiority The situation in which the heterozygote is the most successful genotype in a population.

Hfr cell A bacterium whose DNA molecule contains an integrated copy of the F plasmid.

histone One of the basic proteins found in nucleosomes.

histone acetyltransferase (HAT) An enzyme that attaches acetyl groups to core histones.

histone code The hypothesis that the pattern of chemical modification on histone proteins influences various cellular activities.

histone deacetylase (HDAC) An enzyme that removes acetyl groups from core histones.

hitchhiking The process by which alleles that are not subject to selection become fixed or lost because they are adjacent to an allele that is under selection.

Holliday structure An intermediate structure formed during recombination between two DNA molecules.

holoenzyme The version of the *Escherichia coli* RNA polymerase, subunit composition $\alpha_2\beta\beta'\sigma$, that is able to recognize promoter sequences.

homeobox A conserved sequence element, coding for a DNA-binding motif called the homeodomain, found in several genes believed to be involved in the development of eukaryotic organisms.

homeodomain A DNA-binding motif found in many proteins involved in the developmental regulation of gene expression.

homeotic mutation A mutation that results in the transformation of one body part into another.

homeotic selector gene A gene that establishes the identity of a body part such as a segment of the fruit fly embryo.

hominid A member of the taxonomic family called the Hominidae, comprising humans, chimpanzees, gorillas, and orangutans.

homologous chromosomes Two or more identical chromosomes present in a single nucleus.

homozygote A diploid cell or organism that contains two identical alleles for a particular gene.

horizontal gene transfer Transfer of a gene from one species to another.

hormone response element A nucleotide sequence upstream of a gene that mediates the regulatory effect of a steroid hormone.

housekeeping protein A protein that is continually expressed in all or at least most cells of a multicellular organism.

Hsp70 chaperone A family of proteins that bind to hydrophobic regions in other proteins so as to aid their folding.

hybridization probing A technique that uses a labeled nucleic acid molecule as a probe to identify complementary or homologous molecules to which it base pairs.

hydrogen bond A weak electrostatic attraction between an electronegative atom such as oxygen or nitrogen and a hydrogen atom attached to a second electronegative atom.

hypervariable region One of the two very variable regions of the noncoding part of the mitochondrial genome.

icosahedral A bacteriophage or virus capsid in which the protomers are arranged into a three-dimensional geometric structure that surrounds the nucleic acid.

immunoelectron microscopy An electron microscopy technique that uses antibody labeling to identify the positions of specific proteins on the surface of a structure such as a ribosome.

imprint control element A DNA sequence, found within a few kilobases of clusters of imprinted genes, which mediates the methylation of the imprinted regions.

inclusion body A crystalline or paracrystalline deposit within a cell, often containing substantial quantities of insoluble protein.

incompatibility group Comprises several different types of plasmid, often related to each other, that are unable to coexist in the same cell.

incomplete dominance Refers to a pair of alleles, neither of which displays dominance, the phenotype of a heterozygote being intermediate between the phenotypes of the two homozygotes.

incomplete penetrance A mutation that provides a predisposition to a phenotype, which appears only in response to a later trigger.

inducer A molecule that induces the expression of a gene or operon by binding to a repressor protein and preventing the repressor from attaching to the operator.

inducible operon An operon that is switched on by an inducer molecule that prevents the repressor from attaching to its DNA-binding site.

induction (1) Of a gene: the switching on of the expression of a gene or group of genes in response to a chemical or other stimulus. (2) Of λ phage: the excision of the integrated form of λ and accompanying switch to the lytic mode of infection, in response to a chemical or other stimulus.

inheritance The transmission of biological information from parent to offspring.

inherited disease A disease caused by an inherited defect in the genome.

initiation codon The codon, usually but not exclusively 5′-AUG-3′, found at the start of the coding region of a gene.

initiation complex The complex of proteins that initiates transcription. Also the complex that initiates translation.

initiation factor A protein that has an ancillary role during initiation of translation.

initiation region A region of eukaryotic chromosomal DNA within which replication initiates at positions that are not clearly defined.

initiator (Inr) sequence A component of the RNA polymerase II core promoter.

insertion mutation A mutation that arises by insertion of one or more nucleotides into a DNA sequence.

integron A set of genes and other DNA sequences that enable plasmids to capture genes from bacteriophages and other plasmids.

intercalating agent A compound that can enter the space between adjacent base pairs of a double-stranded DNA molecule, often causing mutations.

intergenic DNA The regions of a genome that do not contain genes.

internal ribosome entry site (IRES) A nucleotide sequence that enables the ribosome to assemble at an internal position in some eukaryotic mRNAs.

interphase The period between cell divisions.

interrupted mating The artificial cessation of conjugation between bacteria, often brought about by vigorous shaking, used as part of the procedure for conjugation mapping of bacterial genes.

interspersed repeat A sequence that recurs at many dispersed positions within a genome.

intrinsic terminator A position in bacterial DNA where termination of transcription occurs without the involvement of Rho.

intron A noncoding region within a discontinuous gene.

inversion A mutation that involves the excision of a portion of a DNA molecule followed by its reinsertion at the same position but in the reverse orientation.

inverted repeat Two identical nucleotide sequences repeated in opposite orientations in a DNA molecule.

iron-response element A regulatory sequence upstream of a gene involved in iron uptake or storage.

isoaccepting tRNAs Two or more tRNAs that are aminoacylated with the same amino acid.

isoelectric focusing Separation of proteins in a gel containing chemicals that establish a pH gradient when the electrical charge is applied.

isoelectric point The position in a pH gradient where the net charge of a protein is zero.

isotope-coded affinity tag (ICAT) Markers, containing normal hydrogen and deuterium atoms, used to label individual proteomes.

junk DNA One interpretation of the intergenic DNA content of a genome.

karyogram The entire chromosome complement of a cell, with each chromosome described in terms of its appearance at metaphase.

kilobase pair (kb) 1000 base pairs.

kinetochore The part of the centromere to which spindle microtubules attach.

knockout mouse A mouse that has been engineered so that it carries an inactivated gene.

Kozak consensus The nucleotide sequence surrounding the initiation codon of a eukaryotic mRNA.

labeling The incorporation of a marker nucleotide into a nucleic acid molecule. The marker is often, but not always, a radioactive or fluorescent label.

lagging strand The strand of the double helix that is copied in a discontinuous fashion during DNA replication.

late genes Bacteriophage genes that are expressed during the later stages of the infection cycle. Late genes usually code for proteins needed for synthesis of new phage particles.

latent period The period between injection of a phage genome into a bacterial cell and the time at which cell lysis occurs.

leading strand The strand of the double helix that is copied in a continuous fashion during DNA replication.

lethal allele An allele that contains a lethal mutation and so causes the death of an organism at a very early stage in its development.

lethal mutation A mutation that results in the death of the cell or organism.

leucine zipper A dimerization domain commonly found in DNA-binding proteins.

LINE (long interspersed nuclear element) A type of interspersed repeat, often with transposable activity.

linkage The physical association between two genes that are on the same chromosome.

linkage analysis The procedure used to assign map positions to genes by genetic crosses.

linkage group A group of genes that display linkage. With eukaryotes a single linkage group usually corresponds to a single chromosome.

linker DNA The DNA that links nucleosomes: the 'string' in the 'beads-on-a-string' model for chromatin structure.

linker histone A histone, such as H1, that is located outside the nucleosome core octamer.

liposome A lipid vesicle sometimes used to introduce DNA into an animal or plant cell.

lod score A statistical measure of linkage as revealed by pedigree analysis.

long patch repair A nucleotide excision repair process of *Escherichia coli* that results in the excision and resynthesis of up to 2 kb of DNA.

long terminal repeat (LTR) Repeat sequences present at the ends of certain types of transposable element.

loss-of-function mutation A mutation that reduces or abolishes a protein's activity.

M phase The stage of the cell cycle when mitosis or meiosis occurs.

MADS box A DNA-binding domain found in several transcription factors involved in plant development.

maintenance methylation Addition of methyl groups to positions on newly synthesized DNA strands that correspond to the positions of methylation on the parent strand.

major groove The larger of the two grooves that spiral around the surface of the B-form of DNA.

MAP kinase system A signal transduction pathway.

map unit A unit used to describe the distance between two genes on a chromosome, now superseded by centimorgan.

mass spectrometry An analytical technique in which ions are separated according to their charge–mass ratios.

maternal-effect gene A fruit fly gene that is expressed in the parent and whose mRNA is subsequently injected into the egg, after which it influences development of the embryo.

matrix-assisted laser desorption ionization time-of-flight (MALDI-TOF) A type of mass spectrometry used in proteomics.

matrix-associated region (MAR) An AT-rich segment of a eukaryotic genome that acts as an attachment point to the nuclear matrix.

maximum parsimony method A method for construction of phylogenetic trees.

mediator A protein complex that forms a contact between various activators and the C-terminal domain of the largest subunit of RNA polymerase II.

megabase pair (Mb) 1000 kb; 1,000,000 bp.

meiosis The series of events, involving two nuclear divisions, by which diploid nuclei are converted to haploid gametes.

messenger RNA (mRNA) The transcript of a protein-coding gene.

metabolome The complete collection of metabolites present in a cell under a particular set of conditions.

metacentric A chromosome whose centromere is positioned in the middle.

metagenome The collection of genomes present in an environmental sample that contains many different organisms.

metaphase The period during the division of a cell when the chromosomes line up in the middle of the nuclear region.

metaphase chromosome A chromosome at the metaphase stage of cell division, when the chromatin takes on its most condensed structure and features such as the banding pattern can be visualized.

metastasis The process that results in cells from one cancer spreading to other places in the body and initiating new tumors.

methyl-CpG-binding protein (MeCP) A protein that binds to methylated CpG islands and may influence the acetylation of nearby histones.

MGMT (O^6-methylguanine-DNA methyltransferase) An enzyme involved in the direct repair of alkylation mutations.

microRNA (miRNA) A class of short RNAs involved in regulation of gene expression in eukaryotes, and which act by a pathway similar to RNA interference.

microbiome The components of a microbial community, such as the one living on the surface of and inside an animal such as a human being.

microevolution The changes in allele frequencies that occur in a population over time.

microinjection A method of introducing new DNA into a cell by injecting it directly into the nucleus.

microsatellite A type of repeat sequence comprising tandem copies of, usually, di-, tri-, or tetranucleotide repeat units. Also called a simple tandem repeat (STR).

minisatellite A type of repetitive DNA comprising tandem copies of repeats that are a few tens of nucleotides in length. Also called a variable number of tandem repeats (VNTR).

minor groove The smaller of the two grooves that spiral around the surface of the B-form of DNA.

mismatch A position in a double-stranded DNA molecule where base pairing does not occur because the nucleotides are not complementary; in particular, a non-base-paired position resulting from an error in replication.

mismatch repair A DNA repair process that corrects mismatched nucleotide pairs by replacing the incorrect nucleotide in the daughter polynucleotide.

mitochondrial Eve The woman who lived in Africa between 140,000 and 290,000 years ago and who carried the ancestral mitochondrial DNA that gave rise to all the mitochondrial DNAs in existence today.

mitochondrial genome The genome present in the mitochondria of a eukaryotic cell.

mitosis The series of events that results in nuclear division.

mitotic spindle The microtubular arrangement that, during cell division, occupies the region of the cell previously taken up by the nucleus.

model building An experimental approach in which possible structures of biological molecules are assessed by building scale models of them.

model organism An organism that is relatively easy to study and hence can be used to obtain information that is relevant to the biology of a second organism that is more difficult to study.

molecular approach The approach to the study of genetics that takes as its starting point the function of DNA rather than the inheritance of genes.

molecular chaperone A protein that helps other proteins to fold.

molecular clock A device based on the inferred mutation rate that enables times to be assigned to gene duplication events and to the branch points in a phylogenetic tree.

molecular phylogenetics A set of techniques that enable the evolutionary relationships between DNA sequences to be inferred by making comparisons between those sequences.

monogenic An inherited disease that is caused by a defect in a single gene.

monomer One of the structural units that are joined together to form a polymer.

motor protein A protein involved in cell motility.

mRNA surveillance A RNA degradation process in eukaryotes.

multigene family A group of genes, clustered or dispersed, with related nucleotide sequences.

multiplex PCR PCR with two or more primer pairs in a single reaction.

multipoint cross A genetic cross in which the inheritance of three or more markers is followed.

multiregional evolution A hypothesis that modern humans in the Old World are descended from *Homo erectus* populations that left Africa over 1 million years ago.

mutagen A chemical or physical agent that can cause a mutation in a DNA molecule.

mutasome A protein complex that is constructed during the SOS response of *Escherichia coli*.

mutation An alteration in the nucleotide sequence of a DNA molecule.

N-degron An N-terminal amino acid sequence that influences the degradation of a protein in which it is found.

N-linked glycosylation The attachment of sugar units to an asparagine residue in a polypeptide.

N terminus The end of a polypeptide that has a free amino group.

natural selection The preservation of favorable alleles and the rejection of injurious ones.

neighbor-joining method A method for the construction of phylogenetic trees.

neutral theory The theory that genetic drift is the most important factor in determining the allele frequencies within a population.

next generation sequencing Recently developed methods for DNA sequencing that enable millions of short sequences to be obtained in a single experiment.

noncoding RNA An RNA molecule that does not code for a protein.

nonhomologous end-joining (NHEJ) A process for repairing a double-strand break in a DNA molecule.

nonpolar A hydrophobic (water-hating) chemical group.

nonsense mutation An alteration in a nucleotide sequence that changes a triplet coding for an amino acid into a termination codon.

nonsynonymous mutation A mutation that converts a codon for one amino acid into a codon for a different amino acid.

normal distribution The distribution pattern commonly called a bell-shaped curve.

nuclear matrix A proteinaceous scaffold-like network that permeates the cell.

nuclease An enzyme that degrades a nucleic acid molecule.

nuclease protection experiment A technique that uses nuclease digestion to determine the positions of proteins on DNA or RNA molecules.

nucleic acid The term first used to describe the acidic chemical compound isolated from the nuclei of eukaryotic cells. Now used specifically to describe a polymeric molecule comprising nucleotide monomers, such as DNA and RNA.

nucleic acid hybridization Formation of a double-stranded hybrid by base pairing between complementary polynucleotides.

nucleoid The DNA-containing region of a prokaryotic cell.

nucleolus The region of the eukaryotic nucleus in which rRNA transcription occurs.

nucleoplasm A general name for the complex mixture of molecules that comprises the ground substance of the nucleus of a living cell.

nucleoside A purine or pyrimidine base attached to a five-carbon sugar.

nucleosome The complex of histones and DNA that is the basic structural unit in chromatin.

nucleosome remodeling A change in the positioning of a nucleosome, associated with a change in access to the DNA to which the nucleosome is attached.

nucleotide A purine or pyrimidine base attached to a five-carbon sugar, to which a mono-, di-, or triphosphate is also attached. The monomeric unit of DNA and RNA.

nucleotide excision repair A repair process that corrects various types of DNA damage by excising and resynthesizing a region of a polynucleotide.

O-linked glycosylation The attachment of sugar units to a serine or threonine in a polypeptide.

octamer sequence A basal promoter element.

Okazaki fragment One of the short segments of RNA-primed DNA synthesized during replication of the lagging strand of the double helix.

oligonucleotide A short synthetic single-stranded DNA molecule.

'ome Any of a number of different types of biological dataset that enables all of the constituents of a system to be studied collectively; for example, the transcriptome is all the messenger RNAs in a cell, and the proteome is all the proteins.

oncogen An environmental agent that causes tumor formation.

oncogene A gene that when carried by a retrovirus is able to cause cell transformation.

one-step growth curve A single infection cycle for a lytic bacteriophage.

open promoter complex A structure formed during assembly of the transcription initiation complex consisting of the RNA polymerase and/or accessory proteins attached to the promoter, after the DNA has been opened up by breakage of base pairs.

open reading frame (ORF) A series of codons starting with an initiation codon and ending with a termination codon. The part of a protein-coding gene that is translated into protein.

operator The nucleotide sequence to which a repressor protein binds to prevent the transcription of a gene or operon.

operon A set of adjacent genes in a bacterial genome, transcribed from a single promoter and subject to the same regulatory regime.

origin of replication A site on a DNA molecule where replication initiates.

origin recognition complex (ORC) A set of proteins that bind to the origin recognition sequence.

origin recognition sequence A component of a eukaryotic origin of replication.

Out of Africa A hypothesis that modern humans evolved in Africa, moving to the rest of the Old World between 100,000 and 50,000 years ago, displacing the descendants of *Homo erectus* that they encountered.

overdominance The situation in which the heterozygote is the most successful genotype in a population.

overlapping genes Two genes whose coding regions overlap.

peptidyl or P site The site in the ribosome occupied by the tRNA attached to the growing polypeptide during translation.

pair-rule genes Developmental genes that establish the basic segmentation pattern of the fruit fly embryo.

paleogenomics The study of the genomes of extinct species.

pan-editing The extensive insertion of nucleotides into an abbreviated RNA, resulting in a functional molecule.

paper chromatography A chromatography method in which a mixture of compounds is placed at one end of a paper strip, and an organic solvent is then allowed to soak along the strip to separate the compounds.

pararetrovirus A viral retroelement whose capsid contains the DNA version of the viral genome.

parental genotype The genotype possessed by one or both of the parents in a genetic cross.

pentose A sugar containing five carbon atoms.

peptide bond The chemical link between adjacent amino acids in a polypeptide.

peptide mass fingerprinting Identification of a protein by examination of the mass spectrometric properties of peptides generated by treatment with a sequence-specific protease.

peptidyl transferase The enzyme activity that synthesizes peptide bonds during translation.

PEST sequences Amino acid sequences that influence the degradation of proteins in which they are found.

pharming Genetic modification of a farm animal so that the animal synthesizes a recombinant pharmaceutical protein, often in its milk.

phenotype The observable characteristics displayed by a cell or organism.

Philadelphia chromosome An abnormal chromosome resulting from a translocation between human chromosomes 9 and 22, a common cause of chronic myeloid leukemia.

phosphodiester bond The chemical link between adjacent nucleotides in a polynucleotide.

phosphodiesterase A type of enzyme that can break phosphodiester bonds.

photoproduct A modified nucleotide resulting from the treatment of DNA with ultraviolet radiation.

photoreactivation A DNA repair process in which cyclobutyl dimers and (6–4) photoproducts are corrected by a light-activated enzyme.

phylogenetic tree A tree depicting the evolutionary relationships between a set of DNA sequences, species or other taxa.

physical mapping Location of a marker by direct examination of a DNA molecule.

pilus A structure involved in bringing a pair of bacteria together during conjugation, and possibly the tube through which DNA is transferred.

plasmid A usually circular piece of DNA often found in bacteria and some other types of cell.

plectonemic Refers to a helix whose strands can only be separated by unwinding.

point mutation A mutation that results from a single nucleotide change in a DNA molecule.

polar A hydrophilic (water-loving) chemical group.

poly(A) polymerase The enzyme that attaches a poly(A) tail to the 3′ end of a eukaryotic mRNA.

poly(A) tail A series of A nucleotides attached to the 3′ end of a eukaryotic mRNA.

polyacrylamide A polymer made up of acrylamide monomers used as matrix for gel electrophoresis.

polyadenylation site The position within a eukaryotic mRNA to which the poly(A) tail is attached.

polymer A compound made up of a long chain of identical or similar units.

polymerase chain reaction (PCR) A technique that results in exponential amplification of a selected region of a DNA molecule.

polymorphic Refers to a locus that is represented by several different alleles or haplotypes in the population as a whole.

polynucleotide A single-stranded DNA or RNA molecule.

polynucleotide kinase An enzyme that adds phosphate groups to the 5′ ends of DNA molecules.

polypeptide A polymer of amino acids.

polyprotein A translation product consisting of a series of linked proteins that are processed by proteolytic cleavage to release the mature proteins.

polypyrimidine tract A pyrimidine-rich region near the 3′ end of a GU–AG intron.

polysome An mRNA molecule that is being translated by more than one ribosome at the same time.

positional information Information that enables a cell to follow the differentiation pathway that is appropriate for its particular location in a developing organism.

post-replication complex (post-RC) A complex of proteins, derived from a pre-RC, that forms at a eukaryotic origin of replication during the replication process and ensures that the origin is used just once in each cell cycle.

post-translational chemical modification Chemical modification of a protein that occurs after that protein has been synthesized by translation of an mRNA.

postzygotic isolation The situation in which members of two species can interbreed but the offspring are sterile.

pre-initiation complex The structure comprising the small subunit of the ribosome, the initiator tRNA plus ancillary factors that forms the initial association with the mRNA during protein synthesis in eukaryotes. Also the structure that forms at the core promoter of a gene transcribed by RNA polymerase II.

pre-replication complex (pre-RC) A protein complex that is constructed at a eukaryotic origin of replication and enables the initiation of replication to occur.

pre-rRNA The primary transcript of a gene or group of genes specifying rRNA molecules.

prepriming complex A protein complex, initially comprising six copies of DnaB and six copies of DnaC, which initiates the construction of a replication fork on a DNA molecule.

prezygotic isolation The situation in which interbreeding between members of two species produces no offspring.

Pribnow box A component of the bacterial promoter.

primary structure The sequence of amino acids in a polypeptide.

primase The RNA polymerase enzyme that synthesizes RNA primers during bacterial DNA replication.

primer A short oligonucleotide that is attached to a single-stranded DNA molecule to provide a start point for strand synthesis.

primosome A protein complex involved in DNA replication.

principal component analysis A procedure that attempts to identify patterns in a large dataset of variable character states.

prion An unusual infectious agent that consists purely of protein.

processing Events that change the chemical or physical structure of an RNA or protein molecule.

proflavin A chemical mutagen, one of the acridine dyes, that is frequently used to induce mutations for experimental purposes.

prokaryote An organism whose cells lack a distinct nucleus.

proliferating cell nuclear antigen (PCNA) An accessory protein involved in DNA replication in eukaryotes.

prometaphase The period during the division of a cell when the chromosomes begin to attach to microtubules.

promiscuous DNA DNA that has been transferred from one organelle genome to another.

promoter The nucleotide sequence, upstream of a gene, to which RNA polymerase binds to initiate transcription.

proofreading The $3' \rightarrow 5'$ exonuclease activity possessed by some DNA polymerases, enabling the enzyme to replace a misincorporated nucleotide.

prophage The integrated form of the genome of a lysogenic bacteriophage.

prophase The period during the division of a cell when the chromosomes condense.

protease An enzyme that degrades protein.

proteasome A multisubunit protein structure that is involved in the degradation of other proteins.

protein electrophoresis Separation of proteins in an electrophoresis gel.

protein profiling The methodology used to identify the proteins in a proteome.

proteome The collection of proteins in a living cell.

proteomics A variety of techniques used to study proteomes.

proto-oncogene The normal cellular version of an oncogene.

protomer One of the individual polypeptide subunits that combine to make the protein coat of a virus.

pseudogene An inactivated and hence nonfunctional copy of a gene.

punctuation codon A codon that specifies either the start or the end of a gene.

Punnett square A tabular analysis for predicting the genotypes of the progeny resulting from a genetic cross.

pure breeding Refers to a population of homozygous plants that, when self-fertilized, give rise to progeny whose phenotypes are identical to those of the parents.

purine One of the two types of nitrogenous base found in nucleotides.

pyrimidine One of the two types of nitrogenous base found in nucleotides.

pyrosequencing A novel DNA sequencing method in which the addition of a nucleotide to the end of a growing polynucleotide is detected directly by conversion of the released pyrophosphate into a flash of chemiluminescence.

quantitative trait A phenotype, such as height, that has a continuous distribution pattern within a population, and which is typically determined by the combined effects of several genes.

quaternary structure The structure resulting from the association of two or more polypeptides.

R group The variable group in the structure of an amino acid.

reading frame A series of triplet codons in a DNA sequence.

readthrough mutation A mutation that changes a termination codon into a codon specifying an amino acid, and hence results in readthrough of the termination codon.

RecA An *Escherichia coli* protein involved in homologous recombination.

RecBCD enzyme An enzyme complex involved in homologous recombination in *Escherichia coli*.

recessive The allele that is not expressed in a heterozygote.

reciprocal strand exchange The exchange of DNA between two double-stranded molecules, occurring as a result of recombination, such that the end of one molecule is exchanged for the end of the other molecule.

recognition helix An α helix in a DNA-binding protein, one that is responsible for recognition of the target nucleotide sequence.

recombinant A progeny member that possesses neither of the combinations of alleles displayed by the parents.

recombinant DNA molecule A DNA molecule created in the test tube by ligating pieces of DNA that are not normally joined together.

recombinant protein A protein synthesized in a recombinant cell as the result of the expression of a cloned gene.

recombination The outcome of crossing over between pairs of homologous chromosomes, and the physical process involving breakage and reunion of DNA molecules that underlies crossing over.

recombination frequency The proportion of recombinant progeny arising from a genetic cross.

recombination hotspot A region of a chromosome where crossovers occur at a higher frequency than the average for the chromosome as a whole.

regulatory gene A gene that codes for a protein, such as a repressor, involved in regulation of the expression of other genes.

regulatory protein A protein that controls one or more cellular activities.

relaxed plasmid A plasmid whose replication is not linked to replication of the host genome, and which therefore can exist as multiple copies within the cell.

release factor A protein that has an ancillary role during the termination of translation.

renaturation The return of a denatured molecule to its natural state.

replication factory A large structure attached to the nuclear matrix; the site of DNA replication.

replication fork The region of a double-stranded DNA molecule that is being opened up to enable DNA replication to occur.

replication licensing factors (RLFs) A set of proteins that regulate DNA replication, in particular by ensuring that only one round of DNA replication occurs per cell cycle.

replication protein A (RPA) The main single-strand binding protein involved in the replication of eukaryotic DNA.

replication slippage An error in replication that leads to an increase or decrease in the number of repeat units in a tandem repeat such as a microsatellite.

replisome A complex of proteins involved in DNA replication.

repressible operon An operon that is switched off by the repressor working in conjunction with a co-repressor molecule.

repressor A regulatory protein involved in the control of expression of a gene or operon in a bacterium.

resistance or R plasmid A plasmid that carries genes conferring on the host bacterium resistance to one or more antibacterial agents.

resolution Separation of a pair of recombining double-stranded DNA molecules.

restriction fragment length polymorphism (RFLP) A restriction fragment whose length is variable because of the presence of a polymorphic restriction site.

restriction fragment length polymorphism (RFLP) analysis Analysis of the positions of RFLPs in a genome, often used as a means of obtaining information on the evolutionary relationships between organisms.

retroposon A retroelement that does not have LTRs.

retrotransposition Transposition via an RNA intermediate.

retrovirus A viral retroelement whose capsid contains the RNA version of the genome.

reverse transcriptase A polymerase that synthesizes DNA on an RNA template.

Rho A protein involved in the termination of transcription of some bacterial genes.

Rho dependent A position in bacterial DNA where termination of transcription occurs with the involvement of Rho.

ribonuclease An enzyme that degrades RNA.

ribosomal protein One of the protein components of a ribosome.

ribosomal RNA (rRNA) The RNA molecules that are components of ribosomes.

ribosome One of the protein–RNA assemblies on which translation occurs.

ribosome binding site The nucleotide sequence that acts as the attachment site for the small subunit of the ribosome during the initiation of translation in bacteria.

ribosome inactivating protein (RIP) A protein that blocks the translation stage of gene expression by cutting one of the ribosomal RNA molecules into two pieces.

ribosome recycling factor (RRF) A protein responsible for disassembly of the ribosome at the end of protein synthesis in bacteria.

ribozyme An RNA molecule that has catalytic activity.

RNA Ribonucleic acid, one of the two forms of nucleic acid in living cells; the genetic material for some viruses.

RNA editing A process by which nucleotides not encoded by a gene are introduced at specific positions in an RNA molecule after transcription.

RNA-induced silencing complex (RISC) A complex of proteins that cleaves and hence silences an mRNA as part of the RNA interference pathway.

RNA interference (RNAi) An RNA degradation process in eukaryotes.

RNA polymerase I The eukaryotic RNA polymerase that transcribes ribosomal RNA genes.

RNA polymerase II The eukaryotic RNA polymerase that transcribes protein-coding and snRNA genes.

RNA polymerase III The eukaryotic RNA polymerase that transcribes tRNA and other short genes.

RNA polymerase An enzyme that synthesizes RNA on a DNA or RNA template.

RNA silencing An RNA degradation process in eukaryotes.

RNA splicing The process by which introns are removed from RNA molecules.

RNA transposon A transposable element that transposes via an RNA intermediate.

RNA-dependent DNA polymerase An enzyme that makes a DNA copy of an RNA template; a reverse transcriptase.

RNA-dependent RNA polymerase An enzyme that makes an RNA copy of an RNA template.

rolling circle replication A replication process that involves continual synthesis of a polynucleotide that is 'rolled off' a circular template molecule.

S phase The stage of the cell cycle at which DNA synthesis occurs.

sampling effect The process that can result in the allele frequencies of a small number of sampled individuals differing from those of the population from which they are taken.

satellite DNA Repetitive DNA that forms a satellite band in a density gradient.

satellite RNA An infective RNA molecule some 320–400 nucleotides in length that does not encode its own capsid proteins, instead moving from cell to cell within the capsid of a helper virus.

scaffold attachment region (SAR) An AT-rich segment of a eukaryotic genome that acts as an attachment point to the nuclear matrix.

scanning A system used during the initiation of eukaryotic translation, in which the pre-initiation complex attaches to the 5′-terminal cap structure of the mRNA and then scans along the molecule until it reaches an initiation codon.

second-site reversion A second mutation that reverses the effect of a previous mutation in the same gene but without restoring the original nucleotide sequence.

secondary structure The conformations, such as α helix and β sheet, taken on by a polypeptide.

sedimentation coefficient The value used to express the velocity at which a molecule or structure sediments when centrifuged in a dense solution.

segment polarity genes Developmental genes that provide greater definition to the segmentation pattern of the fruit fly embryo established by the action of the pair-rule genes.

segmented genome A virus genome that is split into two or more DNA or RNA molecules.

segregation The separation of homologous chromosomes, or members of allele pairs, into different gametes during meiosis.

selectable marker A gene carried by a vector and conferring a recognizable characteristic on a cell containing the vector or a recombinant DNA molecule derived from the vector.

selection A process that brings about the preservation of favorable alleles and the rejection of injurious ones.

selection coefficient A measure of the relative fitness of a genotype.

selective sweep A disproportionate decrease in the genetic diversity of a population caused by the loss of variability of genes adjacent to one that is under selection.

self-fertilization Fertilization of a female gamete with a male gamete derived from the same individual.

semiconservative replication The mode of DNA replication in which each daughter double helix is made up of one polynucleotide from the parent and one newly synthesized polynucleotide.

senescence The period in a cell lineage when the cells are alive but no longer able to divide.

sex cell A reproductive cell; a cell that divides by meiosis.

sexual reproduction Reproduction that involves the fusion of male and female gametes.

Shine–Dalgarno sequence The ribosome binding site upstream of an *Escherichia coli* gene.

short interfering RNA (siRNA) An intermediate in the RNA interference pathway.

short patch repair A nucleotide excision repair process of *Escherichia coli* that results in the excision and resynthesis of about 12 nucleotides of DNA.

signal peptide A short sequence at the N terminus of some proteins that directs the protein across a membrane.

signal transduction Control of cellular activity, including gene expression, via a cell-surface receptor that responds to an external signal.

silent mutation A change in a DNA sequence that has

no effect on the expression or functioning of any gene or gene product.

simple multigene family A multigene family in which all of the genes are the same.

SINE (short interspersed nuclear element) A type of interspersed repeat, typified by the Alu sequences found in the human genome.

single nucleotide polymorphism (SNP) A point mutation that is carried by some individuals of a population.

single-strand binding protein (SSB) One of the proteins that attach to single-stranded DNA in the region of the replication fork, preventing base pairs from forming between the two parent strands before they have been copied.

site-directed mutagenesis Techniques used to produce a specified mutation at a predetermined position in a DNA molecule.

small cytoplasmic RNA (scRNA) A type of short eukaryotic RNA molecule with various roles in the cell.

small interfering RNA (siRNA) A type of short eukaryotic RNA molecule involved in the control of gene expression.

small nuclear ribonucleoprotein (snRNP) Structures involved in RNA splicing and other RNA processing events, comprising one or two snRNA molecules complexed with proteins.

small nuclear RNA (snRNA) A type of short eukaryotic RNA molecule involved in RNA splicing and other RNA processing events.

small nucleolar RNA (snoRNA) A type of short eukaryotic RNA molecule involved in the chemical modification of rRNA.

somatic cell A nonreproductive cell; a cell that divides by mitosis.

somatic cell therapy A type of gene therapy in which the correct version of a gene is introduced into a somatic cell.

sonication A procedure that uses ultrasound to cause random breaks in DNA molecules.

SOS response A series of biochemical changes that occur in *Escherichia coli* in response to damage to the genome and other stimuli.

speciation The process that gives rise to new species.

spliced leader RNA (SL RNA) A transcript that donates a leader segment to several RNAs by trans-splicing.

spliceosome The protein–RNA complex involved in RNA splicing.

splicing The removal of introns from the primary transcript of a discontinuous gene.

SR protein A protein that has a role in splice-site selection during the RNA splicing.

STAT (signal transducer and activator of transcription) A type of protein that responds to the binding of an extracellular signalling compound to a cell surface receptor by activating a transcription factor.

stem cell A progenitor cell that divides continually throughout the lifetime of an organism.

steroid hormone A type of extracellular signalling compound.

steroid receptor A protein that binds a steroid hormone after the latter has entered the cell, and then moves to the nucleus where it activates expression of target genes.

stochastic effects The deviations from statistical perfection that occur with real events in the real world.

streptavidin A protein with a strong affinity for biotin, used to purify biotin-labeled molecules from a mixture.

stringent Refers to a plasmid with a low copy number of perhaps just one or two per cell.

strong promoter A promoter that directs a relatively large number of productive initiations per unit time.

structural gene A gene that codes for an RNA molecule or protein other than a regulatory protein.

structural protein A protein that has a structural role, for example by giving rigidity to the framework of an organism.

submetacentric A chromosome whose centromere is positioned a little off-center.

substitution A mutation that escapes the repair processes and results in a permanent change in a DNA sequence

suicide gene therapy A type of gene therapy, aimed at cancerous cells, involving the introduction of a gene that selectively kills the cell or promotes its destruction by drugs administered in a conventional fashion.

supercoiling A conformational state in which a double helix is overwound or underwound so that superhelical coiling occurs.

suppression Refers to a mutation in one gene that reverses the effect of a mutation in a second gene.

syncytium A cell-like structure comprising a mass of cytoplasm and many nuclei.

synonymous mutation A mutation that changes a codon into a second codon that specifies the same amino acid.

TΨC arm Part of the structure of a tRNA molecule.

TAF- and initiator-dependent cofactor (TIC) A type of protein involved in the initiation of transcription by RNA polymerase II.

tandem array A set of identical or very similar genes that are arranged one after the other in a group.

tandemly repeated DNA Direct repeats that are adjacent to each other.

TATA box A component of the RNA polymerase II core promoter.

TATA-binding protein (TBP) A component of the transcription factor TFIID, that makes contact with the TATA box of the RNA polymerase II promoter.

tautomers Structural isomers that are in dynamic equilibrium.

TBP-associated factor (TAF) One of several components of the transcription factor TFIID, with ancillary roles in the recognition of the TATA box.

telocentric A chromosome whose centromere is positioned very close to one end.

telomerase The enzyme that maintains the ends of eukaryotic chromosomes by synthesizing telomeric repeat sequences.

telomere The end of a eukaryotic chromosome.

telophase The period during the division of a cell when the new nuclear membranes are formed.

temperate bacteriophage A bacteriophage that is able to follow a lysogenic mode of infection.

template-dependent DNA synthesis Synthesis of a DNA molecule on a DNA or RNA template.

teratogen An environmental agent that causes developmental abnormalities.

termination codon One of the three codons that mark the position where translation of an mRNA should stop.

terminator sequence One of several sequences on a bacterial genome involved in the termination of genome replication.

terminator technology A recombinant DNA process that results in synthesis of the ribosome inactivating protein in plant embryos, used to prevent GM crops from producing seeds.

tertiary structure The structure resulting from folding the secondary structural units of a polypeptide.

test cross A genetic cross between a double heterozygote and a double homozygote that carries recessive alleles of the two genes being studied.

thermostable A protein that is able to function at temperatures higher than those that can be withstood by normal proteins.

thymine One of the pyrimidine bases found in DNA.

topoisomerase An enzyme that introduces or removes turns from the double helix by the breakage and reunion of one or both polynucleotides.

totipotent Refers to a cell that is not committed to a single developmental pathway and can hence give rise to all types of differentiated cell.

trans-splicing Splicing between exons that are contained within different RNA molecules.

transcription The synthesis of an RNA copy of a gene.

transcription bubble The non-base-paired region of the double helix, maintained by RNA polymerase, within which transcription occurs.

transcription factor IID (TFIID) The protein complex, including the TATA-binding protein, that recognizes the core promoter of a gene transcribed by RNA polymerase II.

transcription initiation complex The complex of proteins, including the enzyme that performs transcription, that is assembled on the DNA adjacent to a eukaryotic gene that is going to be expressed.

transcriptome The entire mRNA content of a cell.

transduction Transfer of bacterial genes from one cell to another by packaging in a phage particle.

transfer RNA (tRNA) A small RNA molecule that acts as an adapter during translation and is responsible for decoding the genetic code.

transformation The acquisition by a cell of new genes by the uptake of naked DNA.

transforming principle The compound, now known to be DNA, responsible for the transformation of an avirulent *Streptococcus pneumoniae* bacterium into a virulent form.

transgenic organism An organism that carries a cloned gene.

transition A point mutation that replaces a purine with another purine, or a pyrimidine with another pyrimidine.

translation The synthesis of a polypeptide whose amino acid sequence is determined by the nucleotide sequence of an mRNA in accordance with the rules of the genetic code.

translocation The movement of a ribosome along an mRNA molecule during translation.

transversion A point mutation that involves the replacement of a purine by a pyrimidine, or vice versa.

trinucleotide repeat expansion disease A disease that results from the expansion of an array of trinucleotide repeats in or near a gene.

triplet binding assay A technique for determining the coding specificity of a triplet of nucleotides.

triplex A DNA structure comprising three polynucleotides.

trisomy The presence of three copies of a homologous chromosome in a nucleus that is otherwise diploid.

tRNA nucleotidyltransferase The enzyme responsible for the post-transcriptional attachment of the triplet 5′-CCA-3′ to the 3′ end of a tRNA molecule.

***trp* RNA-binding attenuation protein (TRAP)** A protein involved in the attenuation regulation of some operons in bacteria such as *Bacillus subtilis*.

tumor suppressor gene A gene coding for a tumor suppressor protein.

tumor suppressor protein A protein, often one involved in control of the cell cycle, whose inactivation can lead to tumor formation.

turnover Complete or partial degradation of a set of RNA or protein molecules.

Tus The protein that binds to a bacterial terminator sequence and mediates the termination of genome replication.

two-dimensional gel electrophoresis A method for the separation of proteins, used especially in studies of the proteome.

type 0 cap The basic cap structure, consisting of 7-methylguanosine attached to the 5′ end of an mRNA.

type 1 cap A cap structure comprising the basic 5′-terminal cap plus an additional methylation of the ribose of the second nucleotide.

type 2 cap A cap structure comprising the basic 5′-terminal cap plus methylation of the riboses of the second and third nucleotides.

type I topoisomerase A topoisomerase that makes a single-strand break in a double-stranded DNA molecule.

type II topoisomerase A topoisomerase that makes a double-strand break in a double-stranded DNA molecule.

ubiquitin A 76-amino-acid protein that, when attached to a second protein, acts as a tag directing that protein for degradation.

uniparental The inheritance pattern displayed by mitochondrial DNA, in which only the mother contributes her genome to the offspring.

unit factor Mendel's term for a gene.

upstream Toward the 5′ end of a polynucleotide.

upstream control element A component of an RNA polymerase I promoter.

UvrABC endonuclease A multienzyme complex involved in the short-patch repair process of *Escherichia coli*.

variable loop Part of the structure of a tRNA molecule.

vector A DNA molecule, capable of replication in a host organism, into which a gene is inserted to construct a recombinant DNA molecule.

viral retroelement A virus whose genome replication process involves reverse transcription.

viroid An infectious RNA molecule 240–375 nucleotides in length that contains no genes and never becomes encapsidated, spreading from cell to cell as naked DNA.

virulence plasmid A plasmid whose genes confer pathogenicity on the host bacterium.

virulent bacteriophage A bacteriophage that follows the lytic mode of infection.

virus An infective particle, composed of protein and nucleic acid, that must parasitize a host cell to replicate.

virusoid An infectious RNA molecule some 320–400 nucleotides in length that does not encode its own capsid proteins, instead moving from cell to cell within the capsid of a helper virus.

wave of advance A hypothesis that the spread of agriculture into Europe was accompanied by a large-scale movement of human populations.

wild-type A gene, cell, or organism that displays the typical phenotype and/or genotype for the species and is therefore adopted as a standard.

wobble hypothesis The process by which a single tRNA can decode more than one codon.

X inactivation Inactivation by methylation of most of the genes on one copy of the X chromosome in a female nucleus.

X-ray crystallography A technique for determining the three-dimensional structure of a large molecule.

X-ray diffraction pattern The pattern obtained after the diffraction of X-rays through a crystal.

Z-DNA A conformation of DNA in which the two polynucleotides are wound into a left-handed helix.

zinc finger A common structural motif for the attachment of a protein to a DNA molecule.

INDEX

D

H

I

J

K

L

M

N

O

P

Q

R

S

T

U

V

W

X

Y

Z